Biofuel and Biorefinery Technologies

Volume 13

This book series provides detailed information on recent developments in biofuels & bioenergy and related research. The individual volumes highlight all relevant biofuel production technologies and integrated biorefinery methods, describing the merits and shortcomings of each, including cost-efficiency. All volumes are written and edited by international experts, academics and researchers in the respective research areas.

Biofuel and Biorefinery Technologies will appeal to researchers and post-graduates in the fields of biofuels & bioenergy technology and applications, offering not only an overview of these specific fields of research, but also a wealth of detailed information.The ideas and technologies presented in this book series contribute to the UN Sustainable Development Goal 7: *Affordable and Clean Energy.*

Carlos Ricardo Soccol •
Satinder Kaur Brar •
Kugenthiren Permaul •
Kannan Pakshirajan • Júlio Cesar de Carvalho
Editors

Biohydrogen - Advances and Processes

Springer

Editors
Carlos Ricardo Soccol
Department of Bioprocess Engineering and Biotechnology
Federal University of Paraná
Curitiba, Brazil

Kugenthiren Permaul
Department of Biotechnology and Food Science
Durban University of Technology
Durban, South Africa

Júlio Cesar de Carvalho
Department of Bioprocess Engineering and Biotechnology
Federal University of Paraná
Curitiba, Brazil

Satinder Kaur Brar
Department of Civil Engineering
Lassonde School of Engineering, York University
Toronto, ON, Canada

Kannan Pakshirajan
Department of Biosciences and Bioengineering
Indian Institute of Technology Guwahati
Guwahati, India

ISSN 2363-7609 ISSN 2363-7617 (electronic)
Biofuel and Biorefinery Technologies
ISBN 978-3-031-49817-6 ISBN 978-3-031-49818-3 (eBook)
https://doi.org/10.1007/978-3-031-49818-3

This Springer imprint is published by the registered company Springer Nature Switzerland AG
The registered company address is: Gewerbestrasse 11, 6330 Cham, Switzerland

Contents

Chapter 1
Introduction to Biohydrogen: Advancements, Challenges, and Perspectives of the Cleanest Fuel

Júlio Cesar de Carvalho, Adriane Bianchi Pedroni Medeiros, Walter José Martinez Burgos, Marcela Moreira Albuquerque, and Carlos Ricardo Soccol

Abstract Hydrogen, the simplest fuel, is also the one with the highest energy density, and the only one whose combustion produces only water as a byproduct. Because it is non-polluting and can be produced sustainably, it is gaining importance and is the object of new developments backed by government policies worldwide. Although most of the so-called green hydrogen is not bio-based, the same fuel, aptly called *bio*hydrogen, can be produced using bioprocesses. Waste biomass, an unavoidable product of human existence even in the most sustainable societies, can be used to produce biohydrogen via dark fermentation with no additional energy input. Other approaches, such as biophotolysis or microbial electrolysis, require an energy input—ideally renewable—to reclaim even more chemical energy from organic matter. One kilogram of carbohydrate-based organic matter can produce up to 1.5 kWh worth of hydrogen via dark fermentation while producing a stream of diluted organic acids that can be reclaimed for sustainable processing. Biohydrogen is not only a fuel; it can be used in industry as a reducing agent and a raw material for ammonia production. This chapter describes the basics of biohydrogen production technology and the trends and challenges to its widespread production soon.

Keywords Dark fermentation · Photofermentation · MFCs · Energy density · Microbiota · Hydrogen colors · Residues

J. C. de Carvalho (✉) · A. B. P. Medeiros · W. J. Martinez Burgos · M. M. Albuquerque · C. R. Soccol (✉)
Centro Politécnico, Department of Bioprocess Engineering and Biotechnology, Federal University of Paraná, Curitiba, Paraná, Brazil
e-mail: jccarvalho@ufpr.br; soccol@ufpr.br

C. R. Soccol et al. (eds.), *Biohydrogen - Advances and Processes*, Biofuel and Biorefinery Technologies 13, https://doi.org/10.1007/978-3-031-49818-3_1

1.1 Introduction

1.1.1 The Need for Renewable Energy and BioH$_2$

Increasing the sustainability of the world energy matrix may be the most critical challenge humankind faces today. Energy is central to human activities, from water treatment and agriculture to manufacturing and transportation. And yet, much of the energy used today is fossil-based, resulting in the emission of greenhouse gases. Affordable and clean energy is paramount for sustainable development, and it was included as one of the UN SDGs (United Nations Sustainable Development Goals) in 2015 (Falcone et al. 2021).

The world uses about 170 PWh of energy annually (IEA 2022), with about 80% supplied by fossil fuels and only 11% from renewable sources such as solar, wind, hydro, and bioenergy. However, the Sun provides a staggering 174 PW of energy to the Earth's upper atmosphere (Trenberth et al. 2009), and on average, 185 W/m^2 of radiant energy at the Earth's surface (Forster et al. 2021)—which translates into 826,500 PWh of energy yearly. If only 0.1% of the Earth's land mass were used to capture solar energy, that would still be enough to power all human needs in the following decades, at 217 PWh by 2050 (projection extrapolated from data of the International Energy Agency (IEA 2022). Considering that a significant fraction of the energy from the Sun powers wind and water cycles, that energy can be harnessed by hydro and wind generation. Hence, a net zero future is possible. The reader may notice that hydrogen is not considered an energy source in this scenario—because, strictly speaking, *it is not.* Hydrogen is more appropriately an energy vector, a fuel whose ultimate energy comes from solar energy through several routes.

Fuel remains essential even in a future that depends on renewable electricity and batteries for portable energy. The higher energy density of H_2 compared to batteries provides advantages such as higher autonomy and independence from electrical grids, which is vital in remote areas or during transitional periods while infrastructure is being developed. Figure 1.1 summarizes hydrogen properties, including its energy density compared to other fuels and batteries:

As shown in Fig. 1.1, hydrogen has a low specific gravity relative to air (0.0696), and even in liquefied form, its actual density is very low, 71 kg/m^3. Therefore, the specific energy density of hydrogen, 142 MJ/kg, is much higher than that of other fuels. This compensates for its lower volumetric energy density compared to that of liquid fuels such as gasoline or ethanol. Besides, hydrogen burns cleanly, leaving behind only water—making it a perfect fuel to avoid pollution in cities. Compressed hydrogen has a volumetric energy content comparable to that of batteries and can be similarly used in electric vehicles if they use fuel cells to generate electricity. With all this potential, hydrogen clearly has a place in the future economy. That is reflected in the number of projects for biohydrogen production commissioned worldwide until 2030: 652 new planned or in construction as registered by the IEA (IEA—International Energy Agency 2021). That is more than double the amount

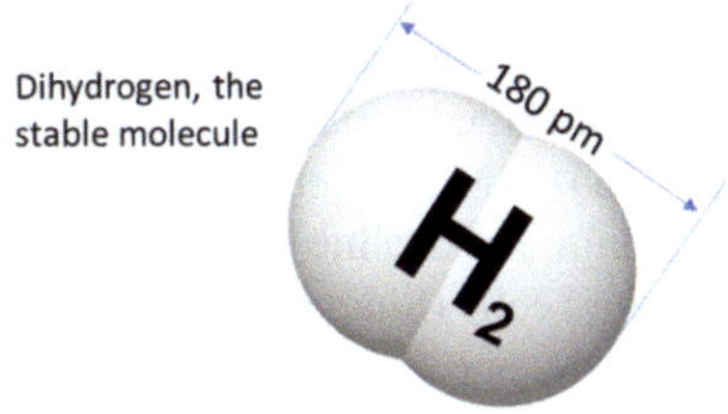

Molecular weight	2016
Specific Gravity, air = 1	0.070
Specific Volume	12.1 m³/kg
Density of liquid at 1 atm	71.0 kg/m³
Absolute viscosity	0.009 cP
Sound velocity in gas	1315 m/s
Specific Heat - cp	14310 J/kg.K
Specific Heat Ratio - cp/cv	1405
Gas constant - R	4126 J/kg°C
Thermal Conductivity	0.182 W/m°C)
Boiling Point (bp) at 1 atm	20.4 K
Latent Heat of Evaporation at bp	447 kJ/kg
Freezing / Melting Point at 1 atm	-259.1 °C
Latent Heat of Fusion	58 kJ/kg
Critical Temperature	-240.0 °C
Critical Pressure	1300 kPa
Critical Volume	0.033 m³/kg
Heat of combustion	144000 kJ/kg

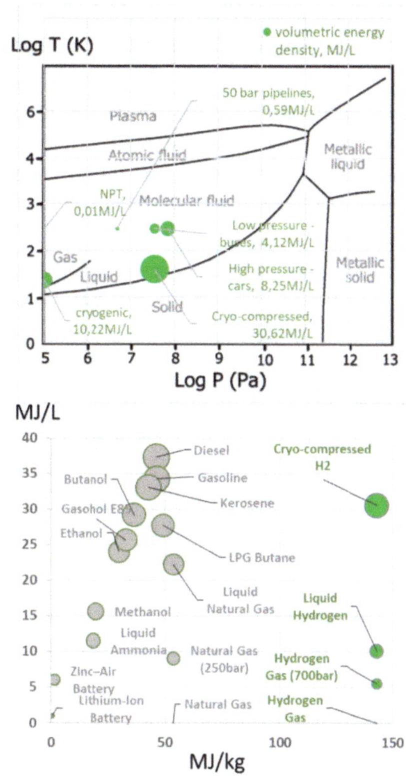

Fig. 1.1 Hydrogen (H_2) properties and the energy density of selected fuels. Chemically, H_2 has the same properties regardless of the source. Compiled from (Dial 2008; NIST, 2010; O'Neil 2013). Hydrogen phase diagram licensed under Creative Commons Attribution-Share Alike 3.0 Unported license, available at https://creativecommons.org/licenses/by-sa/3.0/deed.en. Original available at https://commons.wikimedia.org/wiki/File:Phase_diagram_of_hydrogen.png

registered in the prior decades, and most projects are feasibility studies, proof of concept, or operational units. The technologies vary widely, from renewable hydrogen from electrolysis powered by wind or solar energy to processes using biomass or fossil fuels.

1.1.1.1 Hydrogen as a Fuel

Hydrogen is considered by many the "fuel of the future." But why hydrogen? The main reason is that burning hydrogen releases much energy (242 kJ/mol or 121 kJ/g) (Grangeiro et al. 2023; Rosa et al. 2020) and produces water as a byproduct:

$$H_{2(g)} + ½O_{2(g)} \rightarrow H_2O_{(g)}, \quad \Delta H^0{}_c = -242\text{kJ/mol} \tag{1.1}$$

It is hard to imagine something farther from a pollutant than water. In this reaction, the product could, in principle, be discarded without much concern. The burning of hydrogen (H_2) can be done the same way as other fuels, such as LPG or natural gas, although hydrogen flame reaches 2254K (Molina et al. 2023), a bit higher than what is achieved with natural gas, ethanol, or gasoline combustion with air as the oxidizer. Hydrogen can also be used in fuel cells (Singla et al. 2021), in a reversal of electrolysis where the reaction between H_2 and O_2 generates water and an electric current.

In the context of fuels and chemicals, hydrogen refers to the diatomic molecule H_2, a gas at ambient conditions. As an element, hydrogen is *the* most abundant in the Universe, although, on Earth, it accounts for less than 0.88% by weight (0.14% on Earth's crust and 11% in oceans) (Haynes 2014). Still, it represents the third most abundant element in atom count, with 15.4%. Hydrogen can be obtained from water: one kilogram of water can provide up to 111g of gaseous hydrogen (H_2), which, when burned, releases the same amount of energy as 0.4 liters of gasoline or 0.63 l of anhydrous alcohol. Energy production is not the primary role of hydrogen in industry today: most of the 94 million tons of the gas used in 2021 were for the refining and chemical sectors and produced using fossil fuels such as coal (water gas shift reaction) and natural gas (steam reforming), and thus responsible for significant annual CO_2 emissions (IEA 2022). The gas is essential for the production of ammonia, the basis for the nitrogen fertilizer industry, and is growing in importance in siderurgy.

The hydrogen found in nature is virtually all oxidized, bonded to carbon, oxygen, or other elements, and therefore useless as a fuel. A hydrogen economy requires hydrogen in the reduced form, H_2. Most of the so-called *green hydrogen* is obtained from water, "investing" energy, e.g., hydroelectric, wind, or solar. Taking into account energy losses in the process, the following reaction, electrolysis, consumes more than 16 MJ (megajoules) per kilogram of water:

$$H_2O_{(l)} \rightarrow H_{2(g)} + ½O_{2(g)} \quad \Delta H^0 = -285.6\text{kJ/mol or } 15.9\text{MJ/kg} \tag{1.2}$$

Most organic molecules contain hydrogen bonded to carbon. It is possible to recover that hydrogen in reduced form, provided that some other atom is oxidized, in an oxidation-reduction reaction. Consider the production of hydrogen from a polysaccharide, restricted at the Thauer limit of 4 mols H_2/mol hexose (Kumaraswamy et al. 2019; Thauer et al. 1977):

$$\begin{aligned} &C_5H_{10}O_{5(s)} + 3H_2O \rightarrow 4H_{2(g)} + 2CO_{2(g)} + 2CH_3COOH_{(aq)} \\ &\Delta H^0 = +79.2; \Delta G^0 = -206.3; \text{in kJ/mol} \end{aligned} \tag{1.3}$$

In this reaction, hydrogen is reduced from an oxidation state (OS) of +1 in the carbohydrate to 0 in H_2, while C is oxidized from 0 in glucose to +4 in CO_2. In terms of energy conservation, considering the standard enthalpies of combustion, only 40% of the energy in the carbohydrate is conserved in H_2. However, it becomes available in a practical form—a gaseous fuel.

The fact that gaseous hydrogen spontaneously separates from a complex fermentation mixture, requiring a simple downstream processing, is one of the significant advantages of conversion of residual biomass via dark fermentation. The gas produced actually also contains about 50% in volume of CO_2 and low amounts of other gases such as H_2, N_2, and water(Full et al. 2021; Marks et al. 2020; Torii et al. 2023). To be made into a fuel, CO_2 should be separated to increase the H_2 concentration and energy density, and H_2S must be reduced to avoid corrosion or catalyst poisoning downstream. These technologies are well known and can be adapted from the treatment of flue gases used in other industries.

1.1.1.2 The Many Colors of Hydrogen

All hydrogen burns cleanly, but not all of it is *green*. There are many routes to produce the gas, from splitting water by electrolysis to producing it chemically or through fermentation. These routes have been assigned color codes—a now widespread practice (Ajanovic et al. 2022; Incer-Valverde et al. 2023), but with an origin hard to pinpoint; possibly it started with green hydrogen, and other colors were added using sources of matter and energy as a logical basis—e.g., black or brown hydrogen derived from coals, and yellow derived more directly from light through electricity. Table 1.1 presents the most common interpretation of these colors:

Each of the technologies elicited in Table 1.1 has its subcategories. Although many of these technologies have low carbon emissions, some rely on sensitive technology (nuclear) and depend on high capital costs (fracking) or installations with a comparatively high environmental impact (solar panels). Biohydrogen does not yet have a consensual color of its own, and some propose that it could be called "greener" or "dark green" hydrogen or, conversely, a "light green" hydrogen (Squadrito et al. 2023). Other references suggest that biohydrogen from bioenergy could be called orange (Thomann et al. 2022).

1.1.1.3 Why Biohydrogen Is Green Hydrogen

Regardless of the "formal" color code, biohydrogen can be viewed as a type of green hydrogen: it is produced from renewable sources, as other sustainable, energy-driven, green hydrogens are. Biohydrogen reclaims chemical energy already present in organic substrates, preferably food, municipal, or agroindustrial residues. This organic matter and energy were produced initially via photosynthesis by primary producers in the biosphere, such as plants, algae, and cyanobacteria. Thus, the carbon dioxide emitted in biohydrogen production is neutral, not contributing to pollution.

Biohydrogen energy *is solar energy*, temporarily fixed in biomass and recovered by biological processes: dark fermentation, biophotolysis, and biocatalyzed electrolysis. The first method is the most common, although the other two have their

Table 1.1 Hydrogen colors and sources. Despite the code colors, the table is in growing order of greenness regarding CO_2 emissions during operation, not considering the life cycle of the installations

Hydrogen color	H source	Energy source	Mechanism	CO_2 Emissions
Black or brown	Water	Coal	Steam reforming	Yes
Gray	Water and fossil fuel natural gas	Fossil fuel natural gas	Steam reforming	Yes
Turquoise[a]	Methane, may be biomethane	Variable	High-temperature methane pyrolysis	low, forms solid carbon
Blue	Natural gas and water	Natural gas	Steam reforming	Yes, but sequestrated
Red	Water	Nuclear	Thermal dissociation	No
Purple	Water	Nuclear	Chemo-thermal electrolysis	No
Pink	Water	Nuclear	Electrolysis	No
White	Naturally occurring	-	Fracking deposits at high depths	No
Green	Water	Renewable electricity (hydro, wind)	Electrolysis	No
Yellow	Water	Renewable electricity (specifically, solar)	Electrolysis	No
Greener or dark green	Biomass and water	Biomass	Fermentation	No

[a]Turquoise hydrogen emissions depend on the methane source, the permanent separation of C, and the energy source for pyrolysis

advantages. All are being actively researched and present peculiarities and challenges that will be overviewed in the following sections.

1.2 Biological Processes for Hydrogen Production

Hydrogen can be produced from fossil fuels, water, and biomass through thermochemical, electrochemical, and biological processes (Cheng et al. 2022). Approximately 500 billion m^3 of hydrogen are produced annually (Rajesh et al. 2021), of which 95% are derived from fossil fuels, using steam reforming of methane and naphtha, as well as coal gasification (Chandrasekhar et al. 2020; Chi and Yu 2018), while the other 5% are derived from the electrolysis of water.

However, these methods are not environmentally friendly due to the emission of greenhouse gases when produced with fossil fuels (Manish and Banerjee 2008). Furthermore, processes such as water electrolysis require substantial amounts of

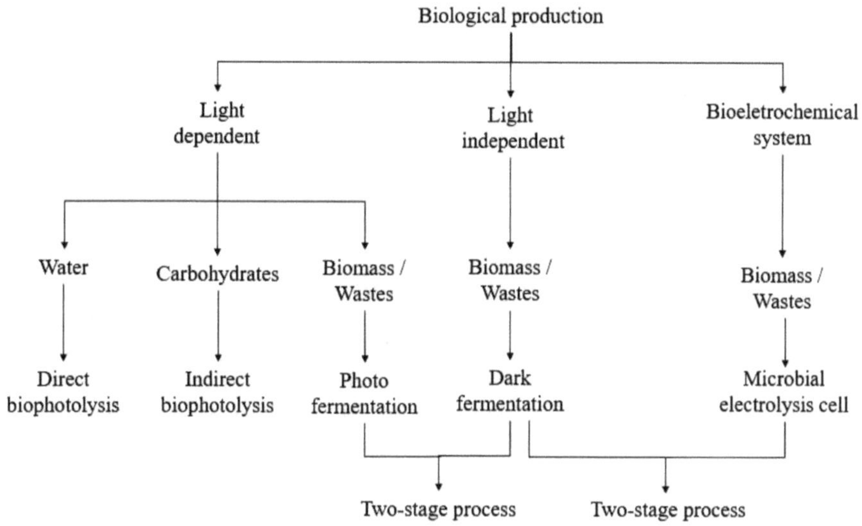

Fig. 1.2 Biological processes of hydrogen production

energy, with energy costs accounting for approximately 80% of the total process cost (Armor 1999).

Thus, the biological production of hydrogen presents itself as a more sustainable alternative and an inexhaustible source for obtaining hydrogen compared to the previously mentioned processes, with its production conducted under ambient temperature and pressure, using a series of residues as substrate, culminating in its treatment and lower energy consumption (Aziz et al. 2021; Nguyen et al. 2020).

The biological processes of hydrogen production can be classified into biophotolysis (direct and indirect), photofermentation, dark fermentation, integrated hybrid system (dark fermentation followed by photofermentation), and bioelectrochemical systems (Sivaramakrishnan et al. 2021) (Fig. 1.2). Each route has its specific production parameters, advantages, and disadvantages (Sivaramakrishnan et al. 2021).

1.2.1 Biophotolysis

Biophotolysis is a light-dependent route for biohydrogen production carried out by autotrophic microorganisms such as microalgae and cyanobacteria (Da Silva et al. 2017), and it can be classified as either direct or indirect biophotolysis (Nabgan et al. 2021).

1.2.2 Direct Biophotolysis

Direct biophotolysis is performed by microalgae such as *Chlamydomonas reinhardtii*, *Chlorella fusca*, *Scenedesmus obliquus*, *Chlorococcum littorale*, and *Platymonas subcordiformis* (Nagarajan et al. 2017) under anaerobic conditions (Nabgan et al. 2021) through a series of reactions that culminate in the division of the water molecule and generation of hydrogen and oxygen.

This process begins with the photolysis of water by photosystem II (PSII) in chloroplasts during photosynthesis, causing the release of oxygen, protons, and electrons, which are then directed to photosystem I (PSI) through different protein complexes such as plastoquinone and cytochrome b6f in the thylakoid membrane (Javed et al. 2022). Electrons in PSI pass through a transport chain to reach ferredoxin (Nagarajan et al. 2017), which subsequently transfers these electrons to the hydrogenase enzyme and combines them with protons in the chloroplast environment, producing molecular hydrogen (Allakhverdiev et al. 2010; Huesemann et al. 2010). Equation (1.4) represents the production of hydrogen through direct biophotolysis:

$$2H_2O + \text{light energy} \rightarrow 2H_2 + O_2 \tag{1.4}$$

This process is not associated with the emission of greenhouse gases. One of its most significant advantages is the fact that it does not require a source of organic carbon in the medium since the process only uses sunlight, CO_2, and water, which are used to generate H_2 by the enzyme hydrogenase (Javed et al. 2022; Putatunda et al. 2023).

However, biohydrogen production through direct biophotolysis still faces numerous challenges, such as the inhibition of the enzyme hydrogenase in the presence of oxygen, which can occur during the photosynthesis step (Akkerman et al. 2002). Thus, during direct biophotolysis, hydrogen production is discontinuous, occurring only for a short period after lighting until the presence of oxygen causes the inhibition of the process (Javed et al. 2022), resulting in limited hydrogen production.

1.2.3 Indirect Biophotolysis

Indirect biophotolysis is most commonly performed by cyanobacteria such as those of the genera *Oscillatoria*, *Calothrix*, *Anabaena*, *Nostoc*, *Synechocystis*, *Synechococcus*, and *Gleobacter* (Nabgan et al. 2021), differing from direct biophotolysis mainly because it uses carbohydrates as a substrate for the production of biohydrogen (Pewnual et al. 2022).

In this way, indirect biophotolysis is divided into two phases. During the first phase, photosynthesis occurs under light energy and, therefore, the production of carbohydrates and their storage. Then, under dark conditions, the second phase occurs, like what happens during direct biophotolysis; the carbohydrates produced during photosynthesis are degraded through a glycolytic pathway generating NADPH that will release electrons to the pool of plastoquinones (PQ) bypassing photosystem II (PSII) (Aziz et al. 2021; Nagarajan et al. 2017; Pewnual et al. 2022). The use of electrons by the electron transport chain occurs during the presence of oxygen. Once oxygen is depleted, cells are in anaerobic conditions that cause electrons to be transferred to photosystem I (PSI), ferredoxin (Fd), and finally to the enzyme hydrogenase for hydrogen to be produced (Sivaramakrishnan et al. 2021).

The production of biohydrogen through indirect biophotolysis can be observed through Eqs. (1.5) and (1.6):

$$12H_2O + 6CO_2 + \text{light energy} \rightarrow C_6H_{12}O_6 + 6O_2 \tag{1.5}$$

$$C_6H_{12}O_6 + 12H_2O + \text{light energy} \rightarrow 12H_2 + 6CO_2 \tag{1.6}$$

As with direct biophotolysis, the efficiency of biohydrogen production by indirect photolysis is limited, making it unsuitable for large-scale biohydrogen production (Sivaramakrishnan et al. 2021).

In both direct and indirect biophotolysis processes, the high costs for microalgae cultivation, requirement for specially designed photobioreactors, high capital costs, and low production efficiency make photoautotrophic processes for biohydrogen production still unfeasible (Akkerman et al. 2002; Nagarajan et al. 2021).

1.2.4 Photofermentation

Photofermentation is characterized as a metabolic route to produce biohydrogen, which is also anaerobic and dependent on light, carried out by non-sulfur nitrogen-fixing purple bacteria (PNSB) (Ghosh et al. 2017) such as those of the genera *Rhodobium*, *Rhodobacter*, *Rhodospirillum*, *Rhodopseudomonas*, and others (Aziz et al. 2021; Kanmani 2022).

These bacteria preferentially carry out fermentation of organic substrates such as short-chain fatty acids (SCFA's) or volatile fatty acids (VFA's) such as acetate, butyrate, formate, lactate, malate, and others, as well as carbohydrates (Ghosh et al. 2017; Kanmani 2022). However, the literature reports the preference of non-sulfur purple bacteria for using substrates such as volatile fatty acids to produce hydrogen to the detriment of pure sugars such as glucose (Tao et al. 2008).

In the presence of N_2 and using ATP as an energy source, non-sulfur purple bacteria catalyze the reduction of N_2 into NH_3 through the enzyme nitrogenase, producing one molecule of hydrogen per molecule of fixed nitrogen, and the reaction can be seen in Eq. (1.7) (Tiang et al. 2020). On the other hand, under conditions of N_2 deficiency, nitrogenase catalyzes an alternative function of H_2 production that can be observed in Eq. (1.8). Under these conditions, non-sulfur purple bacteria ferment the substrate that serves as an electron donor using ATP as an energy source, causing the transfer of electrons to the enzyme nitrogenase so that hydrogen is produced (Putatunda et al. 2023).

$$N_2 + 8H^+ + 8e^- + 16ATP \rightarrow 2NH_3 + H_2 + 16ADP + 16Pi \quad (1.7)$$

$$8H^+ + 8e^- + 16ATP \rightarrow 4H_2 + 16ADP + 16Pi \quad (1.8)$$

Photofermentation has been gaining prominence due to the wide range of substrates that can produce biohydrogen and the high efficiency of the process, which can be conducted on a wide scale and at ambient temperature and pressure. Furthermore, as will be discussed later, photofermentation has been used in conjunction with dark fermentation, increasing the efficiency of the process (Zhang et al. 2020).

1.2.5 Dark Fermentation

Unlike the processes described above, dark fermentation occurs regardless of the presence of light and under anaerobic conditions (Putatunda et al. 2023).

The anaerobic digestion process involves the steps of hydrolysis, acidogenesis, acetogenesis, and methanogenesis (Anukam et al. 2019). During the hydrolysis step, microorganisms such as *Bacteroides*, *Clostridia*, and other facultative organisms break down a wide variety of complex substrates which may come from agro-industrial waste (Martinez-Burgos et al. 2022), waste from food processing (Chaiprapat and Laklam 2011; Maaroff et al. 2019; Martinez-Burgos et al. 2021a, b; Rosa et al. 2020), urban wastewater (Ferreira et al. 2018), waste from livestock (Hernández et al. 2014), and others, releasing monomers such as monosaccharides, amino acids, and other simpler compounds (Aziz et al. 2021).

Subsequently, these substrates can be used by microorganisms of different types, such as *Clostridium*, *Ethanoligens*, *Escherichia*, *Citrobacter*, and others (Lee et al. 2022), converting them into hydrogen and other compounds such as volatile fatty acids (VFA'S) (Aziz et al. 2021).

In general, in dark fermentation, the substrate is converted into pyruvate through the glycolytic pathway, simultaneously producing ATP and NADH (Yaashikaa et al. 2022). Subsequently, pyruvate is converted into Acetyl-CoA, generating reduced ferredoxin, which is a direct electron donor to hydrogenase, the main hydrogen catalytic enzyme (Khanna and Das 2013), resulting in the conversion of the substrate into hydrogen, volatile fatty acids such as butyric acid and acetic acid, as well as alcohols and CO_2 (Aziz et al. 2021) as demonstrated by Eq. (1.9).

$$C_6H_{12}O_6 + 2H_2O \rightarrow 2CH_3COOH + 4H_2 + 2CO_2 \quad (1.9)$$

The theoretical maximum yield of dark fermentation is about 4 mol of H_2/mol of glucose (Manish and Banerjee 2008). Although the yield is modest, dark fermentation has some advantages, such as (1) continuous hydrogen production, (2) use of residual substrates as a source of carbon and, therefore, culminating in its treatment, (3) the need for less energy input, (4) ease of scaling up, and (5) need for simpler reactor designs (Putatunda et al. 2023). To increase the efficiency of the process, dark fermentation can be integrated with photofermentation and microbial electrolysis cells (Khongkliang et al. 2017).

1.2.6 Fermentation in Two Stages (Dark Fermentation and Photofermentation)

As seen previously, the production of biohydrogen by dark fermentation leads to the production of many fermentation byproducts, such as volatile fatty acids that are rich in energy and remain unused and decrease the pH of the medium, limiting the production of biohydrogen (Rezaeitavabe et al. 2020). In this way, to achieve higher yields, the dark fermentation step can be coupled with a second step, such as photofermentation (Kanmani 2022), since the preferred substrate for biohydrogen production by non-sulfur purple bacteria are these volatile fatty acids produced during the dark fermentation (Ghosh et al. 2017). The coupling can be performed using two sequential bioreactors, estimating that the yields of biohydrogen can reach about 12 mols of hydrogen/mol of glucose, significantly improving the efficiency of the process (Aziz et al. 2021).

1.2.7 Microbial Electrolysis Cells (MECs)

Microbial electrolysis cells (MEC) are bioelectrochemical biohydrogen production systems that combine microbial metabolic processes with electrochemical systems (Bora et al. 2022). MECs comprise one or two chambers, an anode, and a cathode, that may or may not be separated by a membrane (Samsudeen et al. 2020; Rani et al. 2022).

Under anaerobic conditions, exoelectrogenic bacteria such as those of the genera *Shewanella*, *Acetobacterium*, *Geobacter*, *Sphingomonas*, *Ochrobactrum*, *Rhodopseudomonas*, *Rhodoferax* (Putatunda et al. 2023), and others, form biofilms on the anode and carry out the oxidation of organic matter-releasing electrons that are transferred to the cathode through an external circuit, and under the external application of a low voltage leads to the evolution of hydrogen by the recombination of protons and electrons (Kadier et al. 2022). The biohydrogen production reaction through microbial electrolysis cells (MECs) can be observed through Eq. (1.10):

$$CH_3COO^- + 4H_2O \rightarrow 2HCO_3^- + H^+ + 4H_2 \tag{1.10}$$

One of the most attractive points of biohydrogen production through MEC is the possibility of using a series of residues as substrate, such as residues from industrial processing (Rani et al. 2022), lignocellulosic biomass (Zhang et al. 2019), urban wastewater (Katuri et al. 2019), and others, resulting concomitantly in its treatment and production of biohydrogen (Son et al. 2021). In addition, compared to water electrolysis processes for hydrogen production, MECs require less energy and produce a gas with high purity (Bora et al. 2022).

Still, just like photofermentation, microbial electrolysis cells can also be used in the production of biohydrogen in two stages, with biohydrogen production occurring in the first stage from dark fermentation, and the effluent resulting from this step can be used in microbial electrolysis cells to oxidize the remaining substrate (Khongkliang et al. 2017). This two-step production can result in a hydrogen yield of about 12 mol of hydrogen/mol of glucose (Rajesh et al. 2021).

1.3 How to Produce BioH_2 in Practice Through Dark Fermentation

Dark fermentation is one of the most promising biological processes to produce biohydrogen since it uses effluent or waste substrates. In addition, it uses simple process configurations, such as traditional fermentation bioreactors, which can operate in batch mode, feed batch, or in continuous mode (Jarunglumlert et al. 2018; Rosa et al. 2020). This anaerobic process uses organic compounds such as carbohydrates, lipids, and proteins to produce hydrogen by the action of microorganisms in a setup similar to classical biogas production. This makes dark fermentation one of the easiest and fastest processes for biohydrogen production compared to other techniques. However, the microbiota and process conditions differ from that of biogas production and require the careful development of three steps in small scale, before any attempt at implementation of the field (Fig. 1.3).

In the first step, the inoculum is produced, which can be a pure culture or a microbial consortium. In the case of consortiums, these can be isolated from different sources, such as sludge, sediments, animal manure, or even the same effluents. However, these must be treated to eliminate hydrogen-consuming bacteria, such as methanogenic and sulfate-reducing bacteria (Martinez-Burgos et al. 2019). There are physical, chemical, and physicochemical methods to eliminate hydrogen-consuming bacteria. One of the most straightforward techniques is to subject the inoculum to temperatures close to 95 °C for 15–60 min. In these processes, only bacteria capable of sporulating and producing hydrogen survive. Among the microorganisms used to produce biohydrogen are the genera *Clostridium*, *Acidovorax*, *Kluyvera*, *Thermoanaerobacterium*, *Caloramator*, etc.

Then, in the second stage, the fermentation medium is made, which must contain sources of macronutrients (carbon, nitrogen, phosphorus, among others), micronutrients, and, on some occasions, to improve the yield, the medium is supplemented with amino acids, vitamins, or nanoparticles that are enzymatic cofactors of hydrogenases (Do Nascimento Junior et al. 2021), in this step, the pH is also adjusted to be

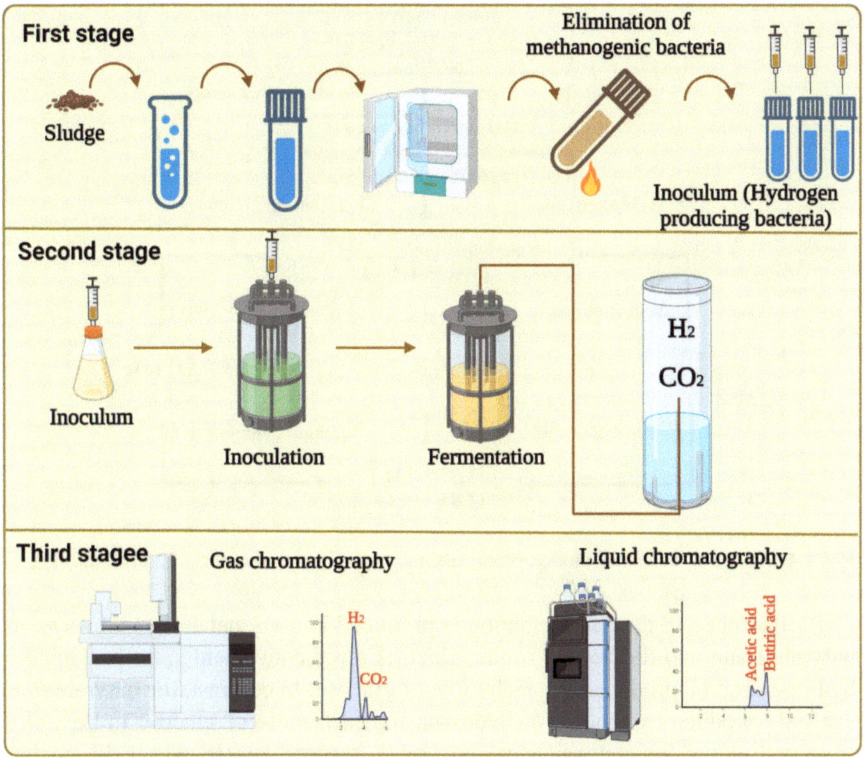

Fig. 1.3 Steps for hydrogen production via dark fermentation

between 6 and 7 and the medium is purged (CO_2, nitrogen, or argon) to guarantee anaerobic conditions in the process; finally the medium is pasteurized or sterilized to eliminate the autochthonous microbial flora from the effluents.

Then, inoculation is carried out with the consequent fermentation of the medium and the production of biogas rich in hydrogen and CO_2; generally, the inoculum rate varies between 5 and 20% of the medium volume. The process can be carried out at different temperature conditions. The production of biogas occurs because of microbial growth; according to Martinez-Burgos et al. (2019), hydrogen in dark fermentation is linked to maintaining the redox balance during cellular energy production. Biohydrogen is an intermediate metabolite of anaerobic digestion (Fig. 1.4). The process takes place through several biochemical reactions in which enzymatic complexes participate, among which hydrogenases stand out, which are the enzymes responsible for the synthesis of hydrogen (Agarwal and Kaur-Ghumaan 2020; Martinez-Burgos et al. 2021a, b).

Anaerobic digestion takes four steps: hydrolysis, acidogenesis, acetogenesis, and methanogenesis (De Carvalho et al. 2023; Karp et al. 2022). In the case of dark fermentation, only the first three steps are performed since it does not have the necessary microorganisms to carry out the methanogenesis process.

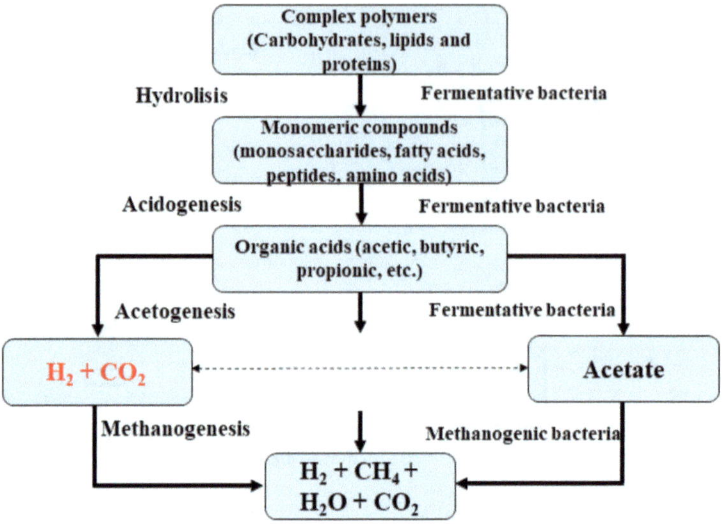

Fig. 1.4 Phases of anaerobic biodigestion. Adapted from Karp et al. (2022); Li et al. (2019)

Biohydrogen in dark fermentation is produced by two metabolic pathways, the butyrate route and the acetate route, which is more convenient since one mole of hydrogen can be produced for each mole of glucose. In contrast, the butyrate route can only produce two moles of hydrogen for each mole of glucose in Eqs. (1.9) (repeated below for convenience) and (1.11), respectively (Saady 2013; Sydney et al. 2020).

Finally, in step iii, evaluation of the composition of the biogas and the fermented broth through gas and liquid chromatography, respectively.

$$C_6H_{12}O_6 + 2H_2O \rightarrow 2CH_3COOH + 4H_2 + 2CO_2 \tag{1.9}$$

$$C_6H_{12}O_6 \rightarrow 2CH_3CH_2CH_2COOH + 2H_2 + 2CO_2 \tag{1.11}$$

Different types of bioreactors have been evaluated for hydrogen production via dark fermentation. Depending on the operating principle, the bioreactors used in dark fermentation can be classified as suspended or immobilized. Among the first are: continuous stirred tank reactors (CSTR), anaerobic sequential batch reactors (ASBR), and anaerobic membrane reactors (AnMBR), and among the immobilized systems are: packed-bed or fixed-bed reactors (PBR), upflow anaerobic sludge blanket reactors (UASBR) (Carolin et al. 2021).

Among the main advantages of dark fermentation is the simplicity of the process, the fact that it requires small amounts of energy, and the reuse of waste with the consequent removal of pollutants. However, the main bottleneck is the yield for which further research is needed, focused on the thermodynamic parameters that influence biohydrogen production.

1.4 Trends and Perspectives in $BioH_2$ Production

Although conceptually simple, biohydrogen production has several challenges that must be addressed through research and pilot processes. These aspects are highlighted below and discussed in detail in the other chapters of this book.

Microbial Systems and Additives Biohydrogen production depends on a stable and efficient microbiota (Gunasekaran et al. 2019;Dzulkarnain et al. 2022) whose composition can change if anaerobiosis is not maintained, increasing the proportion of methanogens and changing the gas composition. Understanding the structure of microbial communities and how to develop reliable and efficient inocula to start or amend reactors is essential and can be done using microbiological and metabolomic tools (Martinez-Burgos et al. 2020). Genetic modification of selected strains may be a way to enhance the production of $BioH_2$ beyond the Thauer limit of 4 mol H_2/mol glucose (Kumaraswamy et al. 2019), dramatically increasing energy recovery from biomass. However, this can be a trick to implement in large-scale systems.

Biowaste Pretreatment and Utilization There is intensive research in biomass conversion into biohydrogen, from food and municipal solid wastes to agroindustrial wastes such as sugarcane vinasse and palm oil mill effluent (Lopez-Hidalgo et al. 2018; Martinez-Burgos et al. 2021a, b; Rosa et al. 2020). This has a good reason: residues must have a destination, and biohydrogen production requires substrates for conversion. However, using residues requires specific pretreatments and microbiota simply because residue composition is variable. Even if some pretreatments, such as size reduction, homogenization, or hydrolysis, are conceptually similar for different substrates, the process conditions must be fine-tuned (Reginatto and Antônio 2015; Saravanan et al. 2022). Thus, there is space for research on the production of $BioH_2$ from many residues, and a logical approach for choosing a substrate for process development is to evaluate its availability and whether hydrogen production can be connected to existing processes, increasing circularity. Sugarcane vinasse, produced in large volumes, is the object of active research nowadays (De Carvalho et al. 2023; Rego et al. 2022). Co-digestion appears in some research articles (Martinez-Burgos et al. 2021a, b; Moreno-Andrade et al. 2023) but requires geographically compatible resources to be feasible.

Fermentation Conditions Besides substrate pretreatment and tweaking with the microbiota, another logical step in process development is optimizing physicochemical conditions. Aspects such as pH and temperature are systematically investigated in any fermentative process, which also occurs in the case of $BioH_2$ production. Optimization of the medium composition is a worthy step for a new residue and is more or less case-specific, leading to improved processes as in Hamed et al. (2020). However, two research themes stand out in biohydrogen production conditions. They can potentially affect any process: the effect of pressure and its manipulation during fermentation and the use of nanoparticle additives. The effect of pressure is essential because pressure affects hydrogen solubility and, therefore, its concentration, reaction rates, and equilibrium (RamKumar et al. 2021); pressure also affects bubble formation and coalescence, which is an important aspect in the phase separation of H_2 (Jiao et al. 2023). The effect of nanoparticles is proving relevant in enhancing electron transfer and other less-known phenomena that may affect metabolic pathways for hydrogen production. Additives such as iron-based nanoparticles and modified nanobiochars seem promising (Ren et al. 2022; Wang et al. 2022).

Biohydrogen Scale-Up and Purification Although many projects are commissioned for renewable hydrogen production worldwide, most rely on electrolysis, and few focus on biological hydrogen production (IEA—International Energy Agency 2021). This is bound to change because abundant municipal solid wastes and lignocellulosic waste biomass are available at 2 and 200 billion tons per year, respectively (Kaur et al. 2021; Bhatia et al. 2022). However, before large-scale production can be considered, it is necessary to develop the process at intermediate scales. The scaling up into a pilot scale can show some of the problems that can be expected, from process stability with variable feed to microbiological succession and its effect on $BioH_2$ evolution. Problems not addressed in small, controlled laboratory settings, such as thermal insulation, gas, purification, and storage, must be faced in these intermediate scales (Sahrin et al. 2022; Usman et al. 2019). The downstream processing of $BioH_2$ also requires larger-scale testing, with purification through adsorption or absorption being promising (Muin et al. 2020). Translational research from biogas production plants can prove promising for large-scale $BioH_2$ production ventures.

Techno-economic and Life Cycle Analysis Based on microbial transformation, most $BioH_2$ processes require low capital costs: large but simple digesters, low or no-cost raw materials, and biomass energy. However, how does its throughput compare to other green energy projects? Techno-economic analyses can illustrate how feasible the process is economically (Jovan and Dolanc 2020; Keeley et al. 2022). Realistically, if the processes are not economically viable, they are unlikely to be implemented. Another critical question is the environmental impacts of $BioH_2$ production and how it compares with other green energy ventures. Even if waste biomass is used, are fossil fuels involved in its transportation for processing? Is there a risk of emitting methane—a potent greenhouse gas—in decomposing secondary residues? These are but a few questions that can be addressed in a life-cycle analysis (Letti et al. 2016). Both analysis tools depend on material and energy balances. These models must use information from pilot plants. They can include biohydrogen uses beyond energy with integration into biorefineries—e.g., a reducing agent for producing fuel hydrocarbons from fatty acids.

Biogas, Biomethane, and Biohythane Although $BioH_2$ is a drop-in product in a hydrogen economy, the production of biogas and its purification into biomethane can be a further process development. While the preferred glycolytic pathways in microorganisms allow the recovery of about 40% of the substrate energy as $BioH_2$, the recovery can be improved by producing methane from the leftover volatile fatty acids (Corneli et al. 2016; Thauer et al. 1977). This biogas can be used directly for energy generation in situ or purified and injected into natural gas pipelines—greening the fossil fuel. Methane can also be thermochemically converted into turquoise hydrogen (Sharma et al. 2023; Ajanovic et al. 2022), intensifying the biofuel production at the expense of some energy for conversion and eventually sequestering carbon for good as a solid. Such a bioprocess is interesting because it goes beyond carbon neutrality—it has *negative* emissions. The mixture of biomethane, with a share of 10–30% of H_2, is called *biohythane* and is gaining importance because of its transport and fuel properties (Bolzonella et al. 2018; Hans and Kumar 2019), such as high energy density.

Perspectives Several other research themes are gaining importance, from using surfactants and novel solvents in substrate pretreatment to immobilizing hydrolytic enzymes or microflora for fermentation. All these aspects apply to dark fermentation and photofermentation, but dark fermentation is leading research efforts. What is certain is that biohydrogen production is a promising technology that is bound to join other green hydrogen technologies as a sustainable way to reclaim chemical energy from biomass that inevitably results from human activity. However, there is a lack of pilot scale studies, and considering how far the knowledge about microbiology and biochemistry of biohydrogen production already progressed, it is time to invest in the replication of these technologies while scaling up the processes.

Acknowledgments This work was supported by the multilateral project BRICS STI FP 2021—BIOGEN (293), financed in Brazil by the National Council for Scientific and Technological Development, CNPq, grant no. 440138/2022-1, also with the support of the Coordination of Improvement of Higher Education Personnel—CAPES, project Print-BioADD.

References

Agarwal T, Kaur-Ghumaan S (2020) Mono- and dinuclear mimics of the [FeFe] hydrogenase enzyme featuring bis(monothiolato) and 1,3,5-triaza-7-phosphaadamantane ligands. Inorg Chim Acta 504:119442. https://doi.org/10.1016/j.ica.2020.119442

Ajanovic A, Sayer M, Haas R (2022) The economics and the environmental benignity of different colors of hydrogen. Int J Hydrog Energy 47(57):24136–24154

Akkerman I, Janssen M, Rocha J, Wijffels RH (2002) Photobiological hydrogen production: photochemical efficiency and bioreactor design. Int J Hydrog Energy 27(11–12):1195–1208. https://doi.org/10.1016/S0360-3199(02)00071-X

Allakhverdiev SI, Thavasi V, Kreslavski VD, Zharmukhamedov SK, Klimov VV, Ramakrishna S, Los DA, Mimuro M, Nishihara H, Carpentier R (2010) Photosynthetic hydrogen production. J Photochem Photobiol C Photochem Rev 11(2–3):101–113. https://doi.org/10.1016/j.jphotochemrev.2010.07.002

Anukam A, Mohammadi A, Naqvi M, Granström K (2019) A review of the chemistry of anaerobic digestion: Methods of accelerating and optimizing process efficiency. Processes 7(8):1–19. https://doi.org/10.3390/PR7080504. MDPI AG

Armor JN (1999). The multiple roles for catalysis in the production of H2. Applied Catalysis A: General, 176(2), 159–176

Aziz M, Darmawan A, Juangsa FB (2021) Hydrogen production from biomasses and wastes: A technological review. Int J Hydrog Energy 46(68):33756–33781. Elsevier. https://doi.org/10.1016/j.ijhydene.2021.07.189

Bhatia L, Sarangi PK, Singh AK, Prakash A, Shadangi KP (2022) Lignocellulosic waste biomass for biohydrogen production: future challenges and bio-economic perspectives. Biofuels Bioprod Biorefin 16(3):838–858

Bolzonella D, Battista F, Cavinato C, Gottardo M, Micolucci F, Lyberatos G, Pavan P (2018) Recent developments in biohythane production from household food wastes: a review. Bioresour Technol 257:311–319

Bora A, Mohanrasu K, Angelin Swetha T, Ananthi V, Sindhu R, Chi NTL, Pugazhendhi A, Arun A, Mathimani T (2022) Microbial electrolysis cell (MEC): Reactor configurations, recent advances and strategies in biohydrogen production. Fuel 328. https://doi.org/10.1016/j.fuel.2022.125269

Carolin CF, Kumar PS, Vo DVN, Joshiba GJ (2021) A review on critical assessment of advanced bioreactor options for sustainable hydrogen production. Int J Hydrog Energy 46(10):7113–7136. https://doi.org/10.1016/j.ijhydene.2020.11.244

Chaiprapat S, Laklam T (2011) Enhancing digestion efficiency of POME in anaerobic sequencing batch reactor with ozonation pretreatment and cycle time reduction. Bioresour Technol 102(5):4061–4068. https://doi.org/10.1016/J.BIORTECH.2010.12.033

Chandrasekhar K, Kumar S, Lee BD, Kim SH (2020) Waste based hydrogen production for circular bioeconomy: Current status and future directions. Bioresour Technol 302. https://doi.org/10.1016/j.biortech.2020.122920. Elsevier

Cheng D, Ngo HH, Guo W, Chang SW, Nguyen DD, Deng L, Chen Z, Ye Y, Bui XT, Hoang NB (2022) Advanced strategies for enhancing dark fermentative biohydrogen production from biowaste towards sustainable environment. Bioresour Technol 351. https://doi.org/10.1016/j.biortech.2022.127045. Elsevier

Chi J, Yu H (2018) Water electrolysis based on renewable energy for hydrogen production. Cuihua Xuebao/Chin J Catal 39(3):390–394. https://doi.org/10.1016/S1872-2067(17)62949-8. Science Press

Corneli E, Dragoni F, Adessi A, De Philippis R, Bonari E, Ragaglini G (2016) Energy conversion of biomass crops and agroindustrial residues by combined biohydrogen/biomethane system and anaerobic digestion. Bioresour Technol 211:509–518

Da Silva T, Mozer TS, da Costa Rubim Messeder dos Santos D, da Silva César A (2017) Hydrogen: Trends, production and characterization of the main process worldwide. Int J Hydrog Energy 42(4):2018–2033. https://doi.org/10.1016/J.IJHYDENE.2016.08.219

De Carvalho JC, de Souza Vandenberghe LP, Sydney EB, Karp SG, Magalhães AI, Martinez-Burgos WJ, Medeiros ABP, Thomaz-Soccol V, Vieira S, Letti LAJ, Rodrigues C, Woiciechowski AL, Soccol CR (2023) Biomethane production from sugarcane vinasse in a circular economy: developments and innovations. Fermentation 9(4):1–27, 349. https://doi.org/10.3390/fermentation9040349

Dial S (2008) A plot of selected energy densities (excluding oxidizers). MATLAB. https://commons.wikimedia.org/wiki/File:Energy_density.svg

Do Nascimento Junior JR, Zevallos Torres LA, Medeiros ABP, Woiciechowski AL, Martinez-Burgos WJ, Soccol CR (2021) Enhancement of biohydrogen production in industrial wastewaters with vinasse pond consortium using lignin-mediated iron nanoparticles. Int J Hydrog Energy 46(54):27431–27443. https://doi.org/10.1016/j.ijhydene.2021.06.009

Dzulkarnain ELN, Audu JO, Wan Dagang WRZ, Abdul-Wahab MF (2022) Microbiomes of biohydrogen production from dark fermentation of industrial wastes: Current trends, advanced tools and future outlook. Bioresour Bioprocess 9(1):16

Falcone PM, Hiete M, Sapio A (2021) Hydrogen economy and sustainable development goals: Review and policy insights. Curr Opin Green Sustain Chem 31:100506. https://doi.org/10.1016/J.COGSC.2021.100506

Ferreira A, Marques P, Ribeiro B, Assemany P, de Mendonça HV, Barata A, Oliveira AC, Reis A, Pinheiro HM, Gouveia L (2018) Combining biotechnology with circular bioeconomy: From poultry, swine, cattle, brewery, dairy and urban wastewaters to biohydrogen. Environ Res 164:32–38. https://doi.org/10.1016/J.ENVRES.2018.02.007

Forster P, Storelvmo T, Armour K, Collins W, Dufresne J-L, Frame D, Lunt D, Mauritsen T, Palmer M, Watanabe M (2021) The earth's energy budget, climate feedbacks, and climate sensitivity. In: Masson-Delmotte V, Zhai P, Pirani A, Connors SL, Péan C, Berger S, Caud N, Chen Y, Goldfarb L, Gomis MI, Huang M, Leitzell K, Lonnoy E, Matthews JBR, Maycock TK, Waterfield T, Yelekçi O, Yu R, Zhou B (eds) Climate change 2021: the physical science basis. Contribution of Working Group I to the Sixth Assessment Report of the Intergovernmental Panel on Climate Change. Cambridge University Press, Cambridge

Full J, Merseburg S, Miehe R, Sauer A (2021) A new perspective for climate change mitigation—introducing carbon-negative hydrogen production from biomass with carbon capture and storage (Hybeccs). Sustainability 13(7):4026

Ghosh S, Dairkee UK, Chowdhury R, Bhattacharya P (2017) Hydrogen from food processing wastes via photofermentation using Purple Non-sulfur Bacteria (PNSB)—a review. Energy Convers Manag 141:299–314. https://doi.org/10.1016/j.enconman.2016.09.001. Elsevier

Grangeiro LC, de Mello BS, Rodrigues BCG, Rodrigues CV, Marin DFC, de Carvalho Junior RP, Pires LO, Maintinguer SI, Sarti A, Dussán KJ (2023) Dark fermentation and principal routes to produce hydrogen. In: Inamuddin, Altalhi T, Adnan SM, Amin MA (eds) Materials for hydrogen production, conversion, and storage, 181–223. Scrivener, Beverly, MA

Gunasekaran M, Merrylin J, Usman TMM, Kumar G, Kim S-H, Banu JR (2019) Biohydrogen production from industrial wastewater. In: Biofuels: Alternative feedstocks and conversion processes for the production of liquid and gaseous biofuels. Elsevier, Oxford, pp 733–760

Hamed SM, Kapoore RV, Raut MP, Vaidyanathan S, Wright PC (2020) Influence of nutrient status on the biohydrogen and lipid productivity in Parachlorella kessleri: a biorefinery approach. Appl Microbiol Biotechnol 104:10293–10305

Hans M, Kumar S (2019) Biohythane production in two-stage anaerobic digestion system. Int J Hydrog Energy 44(32):17363–17380

Haynes WM (2014) CRC handbook of chemistry and physics. CRC, Boca Raton, FL

Hernández MA, Rodríguez Susa M, Andres Y (2014) Use of coffee mucilage as a new substrate for hydrogen production in anaerobic co-digestion with swine manure. Bioresour Technol 168:112–118. https://doi.org/10.1016/J.BIORTECH.2014.02.101

Huesemann MH, Hausmann TS, Carter BM, Gerschler JJ, Benemann JR (2010) Hydrogen generation through indirect biophotolysis in batch cultures of the nonheterocystous nitrogen-fixing cyanobacterium plectonema boryanum. Appl Biochem Biotechnol 162(1):208–220. https://doi.org/10.1007/s12010-009-8741-6

IEA—International Energy Agency (2021) Hydrogen projects database. https://www.iea.org/reports/hydrogen-projects-database.

IEA—International Energy Agency (2022) Global hydrogen review 2022 (IEA, ed). IEA, Paris, France, p 284

Incer-Valverde J, Korayem A, Tsatsaronis G, Morosuk T (2023) "Colors" of hydrogen: Definitions and carbon intensity. Energy Convers Manag 291:117294

Jarunglumlert T, Prommuak C, Putmai N, Pavasant P (2018) Scaling-up bio-hydrogen production from food waste: Feasibilities and challenges. Int J Hydrog Energy 43(2):634–648. https://doi.org/10.1016/j.ijhydene.2017.10.013

Javed MA, Zafar AM, Aly Hassan A, Zaidi AA, Farooq M, El Badawy A, Lundquist T, Mohamed MMA, Al-Zuhair S (2022) The role of oxygen regulation and algal growth parameters in hydrogen production via biophotolysis. J Environ Chem Eng 10(1):107003. https://doi.org/10.1016/J.JECE.2021.107003

Jiao Y, Wang Z, Liu L, Liu X, Pan X, Li G, Zhang Q, Lu C, He C (2023) Numerical simulation of the interaction and coalescence of inline hydrogen bubbles in biohydrogen production by photofermentation with corncob. Int J Hydrog Energy 48(52):19879–19896

Jovan DJ, Dolanc G (2020) Can green hydrogen production be economically viable under current market conditions. Energies 13(24):6599

Kadier A, Singh R, Song D, Ghanbari F, Zaidi NS, Prihartini Aryanti PT, Jadhav DA, Islam MA, Kalil MS, Nabgan W, Hamid AA, Hasan HA, Ma PC (2022) A novel pico-hydro power (PHP)-Microbial electrolysis cell (MEC) coupled system for sustainable hydrogen production during palm oil mill effluent (POME) wastewater treatment. Int J Hydrog Energy. https://doi.org/10.1016/j.ijhydene.2022.09.023

Kanmani S (2022) Enhancement of biological hydrogen production from organic wastes with the application of nanomaterials. Int J Energy Res 46(13):18929–18946. https://doi.org/10.1002/er.8616. Wiley

Karp SG, Burgos WJM, Vandenberghe LPS, Diestra KV, Torres LAZ, Woiciechowski AL, Letti LAJ, Pereira GVM, Thomaz-Soccol V, Rodrigues C, De Carvalho JC, Soccol CR (2022) Sugarcane: a promising source of green carbon in the circular bioeconomy. Sugar Tech 24(4):1230–1245. https://doi.org/10.1007/s12355-022-01161-z

Katuri KP, Ali M, Saikaly PE (2019) The role of microbial electrolysis cell in urban wastewater treatment: integration options, challenges, and prospects. Curr Opin Biotechnol 57:101–110. https://doi.org/10.1016/j.copbio.2019.03.007. Elsevier

Kaur A, Bharti R, Sharma R (2021) Municipal solid waste as a source of energy. Mater Today Proc. https://doi.org/10.1016/j.matpr.2021.04.286

Keeley AR, Takeda S, Managi S (2022) A systematic review of the techno-economic assessment of various hydrogen production methods of power generation. Front Sustain 3:943145

Khanna N, Das D (2013) Biohydrogen production by dark fermentation. Wiley Interdiscip Rev Energy Environ 2(4):401–421. https://doi.org/10.1002/wene.15

Khongkliang P, Kongjan P, Utarapichat B, Reungsang A, O-Thong, S. (2017) Continuous hydrogen production from cassava starch processing wastewater by two-stage thermophilic dark fermentation and microbial electrolysis. Int J Hydrog Energy 42(45):27584–27592. https://doi.org/10.1016/j.ijhydene.2017.06.145

Kumaraswamy KG, Krishnan A, Ananyev G, Zhang S, Bryant DA, Dismukes GC (2019) Crossing the Thauer limit: rewiring cyanobacterial metabolism to maximize fermentative H 2 production. Energy Environ Sci 12(3):1035–1045

Lee HS, Xin W, Katakojwala R, Venkata Mohan S, Tabish NMD (2022) Microbial electrolysis cells for the production of biohydrogen in dark fermentation—a review. Bioresour Technol 363:127934. https://doi.org/10.1016/J.BIORTECH.2022.127934

Letti LAJ, De Carvalho JC, Da Costa SJ, Martins TS, da Silva Ramos N, Dotto MC, Woiciechowski AL, Soccol CR (2016) Life-cycle assessment of biofuels. Green Energy Technol. https://doi.org/10.1007/978-3-319-30205-8_20

Li Y, Chen Y, Wu J (2019) Enhancement of methane production in anaerobic digestion process: A review. Appl Energy 240:120–137. https://doi.org/10.1016/j.apenergy.2019.01.243

Lopez-Hidalgo AM, Alvarado-Cuevas ZD, De Leon-Rodriguez A (2018) Biohydrogen production from mixtures of agro-industrial wastes: chemometric analysis, optimization and scaling up. Energy 159:32–41

Maaroff RM, Md Jahim J, Azahar AM, Abdul PM, Masdar MS, Nordin D, Abd Nasir MA (2019) Biohydrogen production from palm oil mill effluent (POME) by two stage anaerobic sequencing batch reactor (ASBR) system for better utilization of carbon sources in POME. Int J Hydrog Energy 44:3395–3406. https://doi.org/10.1016/j.ijhydene.2018.06.013

Manish S, Banerjee R (2008) Comparison of biohydrogen production processes. Int J Hydrog Energy 33(1):279–286. https://doi.org/10.1016/j.ijhydene.2007.07.026

Marks S, Dach J, Garcia-Morales JL, Fernandez-Morales FJ (2020) Bio-energy generation from synthetic winery wastewaters. Appl Sci 10(23):8360

Martinez-Burgos WJ, Sydney EB, Brar SK, Tanobe VODA, Pedroni Medeiros AB, Cesar de Carvalho J, Soccol CR (2019) The effect of hydrolysis and sterilization in biohydrogen production from cassava processing wastewater medium using anaerobic bacterial consortia. Int J Hydrog Energy 44(47):25551–25564. https://doi.org/10.1016/j.ijhydene.2019.08.085

Martinez-Burgos WJ, Sydney EB, de Paula DR, Medeiros ABP, de Carvalho JC, Soccol VT, de Souza Vandenberghe LP, Woiciechowski AL, Soccol CR (2020) Biohydrogen production in cassava processing wastewater using microbial consortia: process optimization and kinetic analysis of the microbial community. Bioresour Technol 309:123331

Martinez-Burgos WJ, Junios JR, Medeiros ABP, Herrmann LW, Sydney EB, Soccol CR (2021a) Biohydrogen production from agro-industrial wastes using Clostridium beijerinckii and isolated bacteria as inoculum. Bioenergy Res 15:987–997. https://doi.org/10.1007/s12155-021-10358-1

Martinez-Burgos WJ, Sydney EB, de Paula DR, Medeiros ABP, de Carvalho JC, Molina D, Soccol CR (2021b) Hydrogen production by dark fermentation using a new low-cost culture medium composed of corn steep liquor and cassava processing water: Process optimization and scale-up. Bioresour Technol 320. https://doi.org/10.1016/j.biortech.2020.124370

Martinez-Burgos WJ, do Nascimento Junior JR, Medeiros ABP, Herrmann LW, Sydney EB, Soccol CR (2022) Biohydrogen production from agro-industrial wastes using Clostridium beijerinckii and isolated bacteria as inoculum. Bioenergy Res 15(2):987–997. https://doi.org/10.1007/s12155-021-10358-1

Molina S, Novella R, Gomez-Soriano J, Olcina-Girona M (2023) Study on hydrogen substitution in a compressed natural gas spark-ignition passenger car engine. Energy Convers Manag 291:117259. https://doi.org/10.1016/j.enconman.2023.117259

Moreno-Andrade I, Berrocal-Bravo MJ, Valdez-Vazquez I (2023) Biohydrogen production from food waste and waste activated sludge in codigestion: Influence of organic loading rate and changes in microbial community. J Chem Technol Biotechnol 98(1):230–237

Muin NAA, Isah AN, Asli UA, Sadikin AN, Norazahar N, Kamaruddin MJ, Hassim MH, Shin HW, Azman NR (2020) A short review on various purification techniques suitable for biohydrogen-mixed gases. J Energy Saf Technol (JEST) 3(2)

Nabgan W, Tuan Abdullah TA, Nabgan B, Jalil AA, Nordin AH, Ul-Hamid A, Hassan NS, Hussain I, Coelho A, Amin A, Ikram M (2021) Catalytic biohydrogen production from organic waste materials: A literature review and bibliometric analysis. Int J Hydrog Energy 46(60):30903–30925. https://doi.org/10.1016/J.IJHYDENE.2021.04.100

Nagarajan D, Lee DJ, Kondo A, Chang JS (2017) Recent insights into biohydrogen production by microalgae—from biophotolysis to dark fermentation. Bioresour Technol 227:373–387. https://doi.org/10.1016/j.biortech.2016.12.104. Elsevier

Nagarajan D, Di Dong C, Chen CY, Lee DJ, Chang JS (2021) Biohydrogen production from microalgae—Major bottlenecks and future research perspectives. Biotechnol J 16(5). https://doi.org/10.1002/biot.202000124. Wiley

Nguyen PKT, Das G, Kim J, Yoon HH (2020) Hydrogen production from macroalgae by simultaneous dark fermentation and microbial electrolysis cell. Bioresour Technol 315. https://doi.org/10.1016/j.biortech.2020.123795

NIST (2010) Standard reference database 69: NIST chemistry webbook. National Institute of Standards and Technology, Gaithersburg, MD

O'Neil MJ (2013) The Merck index: an encyclopedia of chemicals, drugs, and biologicals. RSC, Cambridge, UK

Pewnual T, Jampapetch N, Saladtook S, Raksajit W, Klinsalee R, Maneeruttanarungroj C (2022) Response of green alga Tetraspora sp. CU2551 under potassium deprivation: a new promising strategy for hydrogen production. J Appl Phycol 34(2):811–819. https://doi.org/10.1007/s10811-021-02672-0

Putatunda C, Behl M, Solanki P, Sharma S, Bhatia SK, Walia A, Bhatia RK (2023) Current challenges and future technology in photofermentation-driven biohydrogen production by utilizing algae and bacteria. Int J Hydrog Energy 48(55):21088–21109. https://doi.org/10.1016/j.ijhydene.2022.10.042

Rajesh J, Ginni G, Kavitha S, Yukesh Kannah R, Adish Kumar S, Bhatia SK, Kumar G (2021) Integrated biorefinery routes of biohydrogen: Possible utilization of acidogenic fermentative effluent. Bioresour Technol 319. https://doi.org/10.1016/j.biortech.2020.124241

RamKumar N, Anupama PD, Nayak T, Subudhi S (2021) Scale up of biohydrogen production by a pure strain; Clostridium butyricum TM-9A at regulated pH under decreased partial pressure. Renew Energy 170:1178–1185
Rani G, Banu JR, Kumar G, Yogalakshmi KN (2022) Statistical optimization of operating parameters of microbial electrolysis cell treating dairy industry wastewater using quadratic model to enhance energy generation. Int J Hydrog Energy 47(88):37401–37414. https://doi.org/10.1016/j.ijhydene.2022.03.120
Reginatto V, Antônio RV (2015) Fermentative hydrogen production from agroindustrial lignocellulosic substrates. Braz J Microbiol 46:323–335
Rego GC, Ferreira TB, Ramos LR, de Menezes CA, Soares LA, Sakamoto IK, Varesche MBA, Silva EL (2022) Bioconversion of pretreated sugarcane vinasse into hydrogen: new perspectives to solve one of the greatest issues of the sugarcane biorefinery. Biomass Convers Biorefinery 12(12):5527–5541
Ren Y, Si B, Liu Z, Jiang W, Zhang Y (2022) Promoting dark fermentation for biohydrogen production: potential roles of iron-based additives. Int J Hydrog Energy 47(3):1499–1515
Rezaeitavabe F, Saadat S, Talebbeydokhti N, Sartaj M, Tabatabaei M (2020) Enhancing biohydrogen production from food waste in single-stage hybrid dark-photo fermentation by addition of two waste materials (exhausted resin and biochar). Biomass Bioenergy 143. https://doi.org/10.1016/j.biombioe.2020.105846
Rosa D, Medeiros ABP, Martinez-Burgos WJ, do Nascimento JR, de Carvalho JC, Sydney EB, Soccol CR (2020) Biological hydrogen production from palm oil mill effluent (POME) by anaerobic consortia and Clostridium beijerinckii. J Biotechnol 323:17–23. https://doi.org/10.1016/j.jbiotec.2020.06.015
Saady NMC (2013) Homoacetogenesis during hydrogen production by mixed cultures dark fermentation: Unresolved challenge. Int J Hydrog Energy 38(30):13172–13191. https://doi.org/10.1016/j.ijhydene.2013.07.122
Sahrin NT, Khoo KS, Lim JW, Shamsuddin R, Ardo FM, Rawindran H, Hassan M, Kiatkittipong W, Abdelfattah EA, Da Oh W (2022) Current perspectives, future challenges and key technologies of biohydrogen production for building a carbon-neutral future: A review. Bioresour Technol 364:128088
Samsudeen N, Spurgeon J, Matheswaran M, Satyavolu J (2020) Simultaneous biohydrogen production with distillery wastewater treatment using modified microbial electrolysis cell. Int J Hydrog Energy 45(36):18266–18274. https://doi.org/10.1016/j.ijhydene.2019.06.134
Saravanan A, Kumar PS, Aron NSM, Jeevanantham S, Karishma S, Yaashikaa PR, Chew KW, Show PL (2022) A review on bioconversion processes for hydrogen production from agro-industrial residues. Int J Hydrog Energy 47(88):37302–37320
Sharma P, Jain A, Bora BJ, Balakrishnan D, Show PL, Ramaraj R, Ağbulut Ü, Khoo KS (2023) Application of modern approaches to the synthesis of biohydrogen from organic waste. Int J Hydrog Energy. https://doi.org/10.1016/j.ijhydene.2023.03.029
Singla MK, Nijhawan P, Oberoi AS (2021) Hydrogen fuel and fuel cell technology for cleaner future: a review. Environ Sci Pollut Res 28:15607–15626
Sivaramakrishnan R, Shanmugam S, Sekar M, Mathimani T, Incharoensakdi A, Kim SH, Parthiban A, Edwin Geo V, Brindhadevi K, Pugazhendhi A (2021) Insights on biological hydrogen production routes and potential microorganisms for high hydrogen yield. Fuel 291. https://doi.org/10.1016/j.fuel.2021.120136
Son S, Koo B, Chai H, Tran HVH, Pandit S, Jung SP (2021) Comparison of hydrogen production and system performance in a microbial electrolysis cell containing cathodes made of non-platinum catalysts and binders. J Water Process Eng 40:101844. https://doi.org/10.1016/J.JWPE.2020.101844
Squadrito G, Maggio G, Nicita A (2023) The green hydrogen revolution. Renew Energy 216(6):119041
Sydney EB, Duarte ER, Martinez Burgos WJ, de Carvalho JC, Larroche C, Soccol CR (2020) Development of short chain fatty acid-based artificial neuron network tools applied to

biohydrogen production. Int J Hydrog Energy 45(8):5175–5181. https://doi.org/10.1016/j.ijhydene.2019.10.128

Tao Y, He Y, Wu Y, Liu F, Li X, Zong W, Zhou Z (2008) Characteristics of a new photosynthetic bacterial strain for hydrogen production and its application in wastewater treatment. Int J Hydrog Energy 33(3):963–973. https://doi.org/10.1016/j.ijhydene.2007.11.021

Thauer RK, Jungermann K, Decker K (1977) Energy conservation in chemotrophic anaerobic bacteria. Bacteriol Rev 41(1):100–180

Thomann J, Marscheider-Weidemann F, Die AS, Giz LL, Hank C, Weise F, Edenhofer L, Thiel Z (2022) Background paper on sustainable green hydrogen and synthesis products. Notes

Tiang MF, Fitri Hanipa MA, Abdul PM, Jahim JMD, Mahmod SS, Takriff MS, Lay CH, Reungsang A, Wu SY (2020) Recent advanced biotechnological strategies to enhance photo-fermentative biohydrogen production by purple non-sulphur bacteria: An overview. Int J Hydrog Energy 45(24):13211–13230. https://doi.org/10.1016/J.IJHYDENE.2020.03.033

Torii K, Hamazaki M, Kumon S, Sato K, Kato S, Dowaki K (2023) A system analysis of a biohydrogen production system using granulated mine residue as a H2S adsorbent. Energies 16(6):2625

Trenberth KE, Fasullo JT, Kiehl J (2009) Earth's global energy budget. Bull Am Meteorol Soc 90(3):311–324

Usman TMM, Banu JR, Gunasekaran M, Kumar G (2019) Biohydrogen production from industrial wastewater: an overview. Bioresour Technol Rep 7:100287

Wang Y, Xiao G, Wang S, Su H (2022) Application of nanomaterials in dark or light-assisted fermentation for enhanced biohydrogen production: a mini-review. Bioresour Technol Rep 21:101295

Yaashikaa PR, Keerthana Devi M, Senthil Kumar P (2022) Biohydrogen production: An outlook on methods, constraints, economic analysis and future prospect. Int J Hydrog Energy 47(98):41488–41506. https://doi.org/10.1016/j.ijhydene.2022.07.082

Zhang L, Wang YZ, Zhao T, Xu T (2019) Hydrogen production from simultaneous saccharification and fermentation of lignocellulosic materials in a dual-chamber microbial electrolysis cell. Int J Hydrog Energy 44(57):30024–30030. https://doi.org/10.1016/j.ijhydene.2019.09.191

Zhang T, Jiang D, Zhang H, Jing Y, Tahir N, Zhang Y, Zhang Q (2020) Comparative study on bio-hydrogen production from corn stover: Photo-fermentation, dark-fermentation and dark-photo co-fermentation. Int J Hydrog Energy 45(6):3807–3814. https://doi.org/10.1016/J.IJHYDENE.2019.04.170

Chapter 2
Molecular Hydrogen (H_2) Metabolism in Microbes: A Special Focus on Biohydrogen Production

Karan Kumar, Avinash Anand, and Vijayanand S. Moholkar

Abstract In recent years, biohydrogen production (produced from renewable sources either biologically or photo-biologically) has attracted substantial interest from researchers worldwide because of its potential application as an alternative and clean fuel. Microbes not only produce molecular hydrogen (H_2) but can also utilize it as an electron acceptor with the help of hydrogenases. The capability of microorganisms to use or produce H_2 is deeply encrypted in their genes, which are transcribed as mRNA and finally translated to protein (some of which act as enzymes). Enzymes play a pivotal role during cellular metabolism that decides the fate of any metabolite. One needs to dig down at the genetic and proteomic levels to understand these complex cellular metabolisms and regulate the production of a metabolite (or product) of interest. In this chapter, we have discussed the microbial metabolism of molecular H_2 with special emphasis on its production. Herein, we have first focused on genes for H_2 metabolism and their origins across various species. Then, metabolic pathways involved in hydrogen metabolism have been discussed with an applied view—cofactors and coenzymes. Further, building blocks and targets for synthetic biology for the overproduction of H_2 have been considered.

Keywords Biohydrogen · Hydrogenases · Metabolism · Synthetic biology · Microorganism · Microbial production · H_2

K. Kumar
School of Energy Science and Engineering, Indian Institute of Technology Guwahati, Guwahati, Assam, India

A. Anand
Department of Chemical Engineering, Indian Institute of Technology Guwahati, Guwahati, Assam, India

V. S. Moholkar (✉)
School of Energy Science and Engineering, Indian Institute of Technology Guwahati, Guwahati, Assam, India

Department of Chemical Engineering, Indian Institute of Technology Guwahati, Guwahati, Assam, India
e-mail: vmoholkar@iitg.ac.in

C. R. Soccol et al. (eds.), *Biohydrogen - Advances and Processes*, Biofuel and Biorefinery Technologies 13, https://doi.org/10.1007/978-3-031-49818-3_2

2.1 Introduction

Rapid economic growth coupled with expeditious global urbanization has increased the concerns over the supply of fossil fuels and their adverse environmental impacts worldwide (Malani et al. 2019; Kumar et al. 2021; Singh et al. 2022). Relying heavily on fossil fuels has triggered environmental challenges, including global warming, air pollution, and acid rain (Kumar et al. 2023). This dependence has negatively impacted the nation's economic stability, leading to elevated inflation rates and energy crises (Kumar et al. 2022a, b). Therefore, it is high time for a developing country like India to find a clean and sustainable alternative renewable energy source (Das 2001; Banerjee 2006; Pandey et al. 2013; Das et al. 2014; Kumar et al. 2023a, b, c, 2022a). While discussing the alternate sustainable and clean energy source, one must consider the potential of hydrogen gas as fuel (Christl et al. 1992; Sarma et al. 2016, 2017; Younas et al. 2022).

Hydrogen is the first and most abundant element that has existed since the creation of the universe after, about 3,80,000 years of the Big Bang (Dincer 2012). Hydrogen makes up about three-fourth (73%) of the mass of the universe (Vignais and Billoud 2007; Pandey et al. 2013). Most hydrogen is found in molecular states (H_2) on Earth's surface, primarily as water and organic matter, constituting approximately 0.14% of all compounds on the planet's surface (Vignais and Billoud 2007; Arizzi et al. 2021). It has numerous applications as (1) an essential and indispensable raw material in chemical, coal, and petroleum industries; (2) a shielding gas used in the welding process; (3) a potential energy carrier with an energy density of 142 kJ/g; and (4) ideal fuel because it is highly flammable and has the capability of burning in the air at various concentrations ranging from 4 to 75% by volume with the only product as water. Table 2.1 summarizes the metadata of molecular hydrogen in various databases (e.g., KEGG and PubChem).

$$H_2 + \frac{1}{2}O_2 \rightarrow H_2O \quad \Delta G = -286\,\text{kJ}/\text{mol}$$

Various chemical or biological processes produce molecular hydrogen (H_2) by utilizing fossil fuels, biomass, renewable energy, and nuclear energy sources (Dincer 2012). Figure 2.1 represents a graphical outline of various routes and sources for H_2 production. This consists of electrolysis, thermochemical water splitting, gasification, cracking, reforming, microbial electrolysis cell, dark fermentation, photofermentation, biophotolysis, etc. Among these techniques, H_2 produced from renewable energy sources and biomass sources are much more promising than other techniques because they solve crucial challenges in contemporary times, such as waste management, environmental pollution, zero carbon emission, and low-cost methods (Das et al. 2014).

Broadly, H_2 production methods can be classified into four categories, viz. (1) steam reforming, (2) Electrolysis, (3) Thermochemical or thermal chemical, and (4) Biological. Considering the associated environmental issues of the use of fossil

Table 2.1 Metadata of molecular hydrogen in various databases (e.g., KEGG and PubChem)

Property	Entry ID	Description
Name	C00282 (KEGG)	Hydrogen (H_2)
Molecular weight and exact mass		2.0159 and 2.0157 Da
Involved pathways	map02020	Two-component system
	map04973	Adsorption and digestion of carbohydrates
	map04974	Adsorption and digestion of proteins
Chemical safety	783 (PubChem CID)	Flammable
Canonical smiles		[HH]
Physical descriptions		Colorless, odorless, easily ignited, burns pale blue (almost invisible flame)
Boiling point		–423 °F at 760 mmHg
Solubility in water		1.62 mg/L at 21 °C
Density		0.071 at –423.4 °F
Relative vapor density		0.07 (air = 1)
Heat of combustion		–285.8 °C
Isotopes		^{1}H (protium, 98.9844%), ^{2}H (deuterium, 0.0156%), and ^{3}H (tritium, traces only)

fuels and the high energy requirements of the first three methods of H_2 production, the biological route for H_2 production (or biohydrogen production) using renewable biomass has proven to be a green and sustainable approach (Aydin et al. 2021). Bio-H_2 production can be classified mainly into four categories: photofermentation, biophotolysis, microbial electrolysis, and dark fermentation (Younas et al. 2022). Most

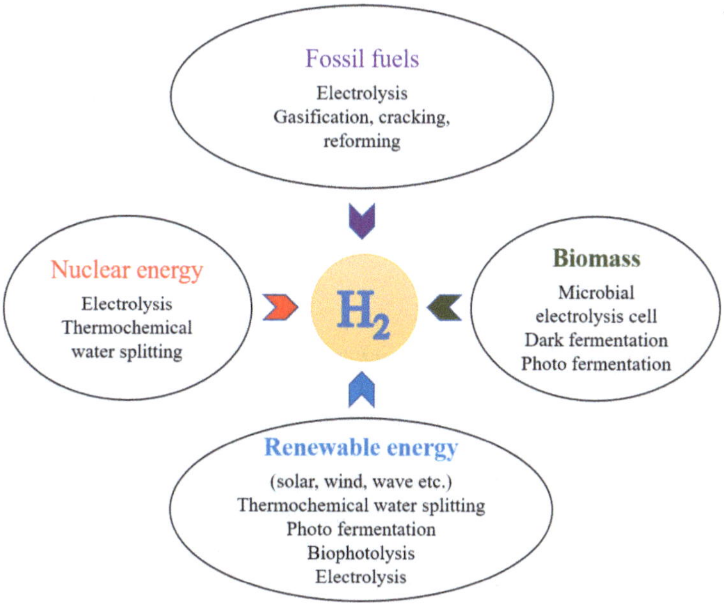

Fig. 2.1 Various routes and sources for molecular hydrogen (H_2) production

of the demand for H_2 in chemical industries is for producing methanol and ammonia (Dincer 2012). Despite this, the potential application of H_2 as a transportation fuel can always be addressed, as it will significantly help improve air quality, manifest social equality, and yield economic stability (Adamson and Pearson 2000). Numerous pathways, feedstocks, and technologies have been illustrated by scientists around the globe for bio-H_2 production. Several studies have demonstrated fermentative biohydrogen production (FBP) at the laboratory scale as well as pilot scales, which is expected to have the potential for FBP at the industrial scale (Dincer 2012; Aydin et al. 2021; Arizzi et al. 2021; Younas et al. 2022). The main bottlenecks for FBP are a moderate rate of production and a low yield on the substrate (Chen et al. 2021). However, for industrial-scale FBP, significant product, and volumetric production rate improvement surpass the current achievements (Bakonyi et al. 2014).

In recent decades, FBP has garnered significant attention and recognition within the academic and research communities due to its capacity to generate H_2 from diverse renewable sources, including wastewater, through an energy-efficient continuous process (Barca et al. 2015; Dzulkarnain et al. 2022). Microorganisms are not only the hosts for FBP, but some can utilize H_2 as an electron donor with the help of hydrogenases (*hyd*) (Appel and Schulz 1998; Schuchmann et al. 2018). The capability of microorganisms to use or produce H_2 is deeply encrypted in their DNA (or genes), which are transcribed as mRNA and finally translated to functional proteins (or enzymes). Enzymes play a pivotal role during cellular metabolism that decides the fate of any metabolite (Brink et al. 1988; Leskovac 2003; Singh et al. 2016). Numerous important metabolites are produced extracellularly during metabolism, which includes acids (e.g., lactate, acetate, and butyrate) and alcohols (e.g., ethanol and butanol) in the course of fermentation, resulting in an intricate metabolic network (Balderas-Hernandez et al. 2020; Ahmed et al. 2021). Only a little has been understood to date, excluding the recognition of the intricacy of metabolic networks (Cai et al. 2011; Ahmed et al. 2021).

Molecular H_2 metabolism is mainly coupled with anaerobic digestion, a complex process involving various microbial groups, including acetogens and methanogens (Ahmed et al. 2021). Acetogens are bacteria that produce acetate through the breakdown of organic matter, while methanogens are microbes that produce methane from acetate, hydrogen, or carbon dioxide. Certain acetogenic microorganisms, like *C. acetobutylicum* and *C. pasteurianum*, generate acetate, organic acids, and hydrogen as by-products during the growth phase of their life cycle (Sarma et al. 2019; Kumar et al. 2023a, c). In contrast, other acetogens such as *C. ljungdahlii* and *C. carboxydivorans* utilize H_2 gas and CO_2 as substrates to synthesize acetate via the Wood-Ljungdahl (WL) pathway (Kumar et al. 2023a, c; Greening et al. 2014). Acetogens are essential in producing biofuels, such as bio-H_2, as they can consume H_2 gas and generate acetate, which can then be converted to H_2 by other microorganisms in a mixed culture. Conversely, methanogens from the archaea family can generate methane as a by-product during their metabolic processes (Adamson and Pearson 2000; Yan and Dong 2018; Du et al. 2020). These methanogens can use H_2 gas and CO_2, or other organic compounds, as substrates to produce methane through methanogenesis. Methanogens are important in FBP as they can compete with hydrogen-consuming microorganisms and decrease hydrogen production yield. However, methanogens can also be used in bio-H_2 production as a means of removing H_2 from the system and preventing the build-up of inhibitory concentrations of hydrogen (Fu et al. 2009; Salimi et al. 2010; Barca et al. 2016; Du et al. 2020).

In this chapter, we have summarized the microbial metabolism of molecular H_2 with particular emphasis on FBP. Herein, we have discussed the microorganisms capable of H_2 metabolism, then focused on genes for H_2 metabolism, their origins, and conservation across various species. Furthermore, the critical operational factors that influence FBP metabolism are discussed. Next, metabolic pathways involved in hydrogen metabolism have been discussed with an applied view—cofactors and coenzymes. Further, building blocks and targets for synthetic biology for the overproduction of H_2 have been considered. In-depth research on engineering the microbial metabolic pathways to enrich the FBP is strongly recommended. We anticipate that this chapter may provide fundamental principles and technological updates to scientists and bioengineers working in fermentative hydrogen or biohydrogen production.

2.2 Microbes Capable of Molecular Hydrogen (H_2) Metabolism

During evolution, many microorganisms diverged genetically with new metabolic capabilities to utilize various energy resources (Vignais and Billoud 2007; Benoit et al. 2020). These changes resulted in sustainable, self-regulatory, and stable systems that convert complex organic or inorganic matter to diverse value-added products, usually including H_2 gas (Vignais and Billoud 2007; Wolf et al. 2016; Angikath et al. 2021). Several studies have suggested that early Earth was hydrogen-rich; if it is true, then thinking the cause of contemporary conditions (favorable environmental conditions to sustain life on Earth) as *hyd* (the enzymes responsible for utilizing molecular hydrogen) is not irrational (Adamson and Pearson 2000; Oh et al. 2008; Argun et al. 2009; Schuchmann et al. 2018). This *hyd* would have been invented by the earliest life (bacteria) on earth (Ooshima et al. 1998; Guo et al. 2008; Dzulkarnain et al. 2022). Presently, the microorganisms are not only able to utilize H_2 as an energy source with the aid of *hyd* but also are capable of producing H_2 and setting up oxygen and carbon-free ecosystems (as a product of photosynthesis) (Vyas 1995; François et al. 2021). One such ecosystem has been witnessed in an active oceanic hydrothermal field (Central Indian Ridge). The accumulated H_2 and CO_2 generated in this hydrothermal field (Takai et al. 2006), which is the geological and abiotic environment, can support a hydrogen-driven ecosystem termed as HYPERthermophilic Subsurface LIthoautotrophic Microbial Ecosystem (HyperSLIME). Few studies on microbial species capable of carbon monoxide oxidization have revealed that numerous bacterial and archaeal species can grow autotrophically in CO-rich environments by utilizing CO and producing molecular H_2 as an end-product (Das 2009; Benoit et al. 2020; Hu 2021; Hughes et al. 2021).

In a microbial ecosystem, the microorganisms possessing hydrolytic characteristics take help from their symbiotic partner, as they are poor hydrolyzers (Banerjee et al. 1989; Nelson and Stegen 2015) for solubilizing the insoluble constituents of organic compounds (commonly known as anaerobic digestion). Next, microbes capable of producing H_2 utilize these soluble intermediates as substrates (Cai et al. 2011; Aydin et al. 2021). Widespread microbes in various domains of life, including archaea, strict and facultative anaerobe, lower eukaryotes, and cyanobacteria such as protists and green algae, can produce H_2 in their life cycle (Cornish et al. 2015; Fakhimi et al. 2020). Habitational and nutritional adaptability of microbes assists them to survive severe circumstances, e.g., complete darkness, high temperature, and absence of oxygen (Ahmed et al. 2021; Hu 2021; Dzulkarnain et al. 2022).

Microbes from the archaeal domain are extremophiles that can sustain harsh environments such as salt lakes and hot springs (Pandey et al. 2013; Das et al. 2014; Gong et al. 2021). They are organotrophs, lithotrophs, or phototrophs located in a versatile habitat. However, bacterial species have their peculiar hydrogen life cycle (François et al. 2021; Li et al. 2021; Arias et al. 2021). The syntrophic association of H_2 consumers and producers plays a crucial and indispensable role in maintaining a low partial pressure of H_2 (pH_2) (Vignais and Billoud 2007). The majority of H_2-producing organisms in the bacterial realm exhibit heterotrophic characteristics and generate hydrogen as a by-product of their fermentative metabolism. Certain organisms in this category can operate independently of photon-derived energy and resist O_2 deprivation, making them "dark fermentative bacteria." In contrast, light energy is essential for H_2 production in strictly anaerobic conditions designated as photo-fermentative bacteria for a few of them. Cyanobacteria found in oceanic and fresh water are photoautotrophic algae. The hydrogen is produced by biophotolysis in the oxygen-deficit environment in cyanobacterial species. A few anaerobic protozoa, such as *Trichomonas*, are also identified for H_2 production. Figure 2.2 represents the evolutionary relationship of different life forms on earth, which consists of three principal groups of organisms, resulting in the tree of Life. Figure 2.3 illustrates the different classifications of microorganisms capable of metabolizing molecular hydrogen (Table 2.2).

2.3 Biochemistry and Physiology of H_2 Metabolism

The prebiotic Earth's atmosphere was rich in hydrogen. Consequently, the hypothesis proposing the presence of early biotic species capable of metabolizing hydrogen seems plausible. This theory was first validated in the late nineteenth century (Tian et al. 2005). In the year 1931, the enzyme responsible for H_2 metabolism was discovered as identified as *hyd* by two scientists, Stephenson and Stickland (Harris and Reddy 1977). The key H_2 metabolizing enzyme, *hyd*, catalyzes the most

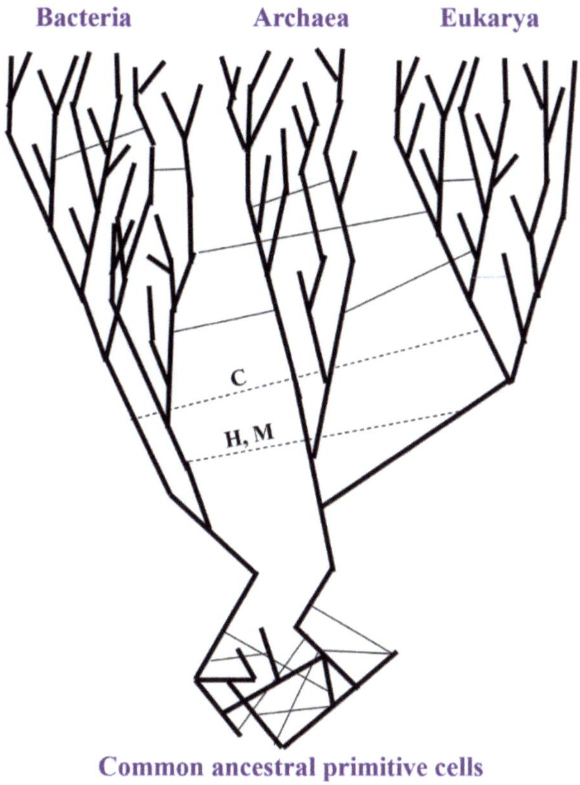

Fig. 2.2 Schematic depiction of the evolutionary relationship between different life forms on earth. This tree consists of three principal groups of organisms, resulting in the tree of Life. Dashed lines represent the endosymbiotic events leading to the emergence of chloroplast (C) in plants and mitochondria (M) and hydrogenosomes (H) in eukaryotic cells (source: Vignais and Billoud 2007, reproduced with minor modification with permission)

fundamental reversible reaction: hydrogen reductive formation from protons and electrons (Eq. 2.1).

$$H_2 \leftrightharpoons \left[H^+ + H^-\right] \leftrightharpoons 2H^+ + 2e^- \tag{2.1}$$

Regardless of the simplicity of the reaction given in Eq. (2.1), the *hyd* has a highly complex structure and catalytic mechanism (Schuchmann and Müller 2012). The predominant reaction direction is contingent upon the redox characteristics of the substrates engaging with *hyd*. When an electron acceptor is present, *hyd* facilitates the uptake of H_2, while the presence of an electron donor transforms *hyd* into an H_2 producer (Bowman et al. 2014). The reaction that uptake H_2 generates protons and electrons, essential for generating ATP and reducing power. Simultaneously, the processes generating H_2 encompass the reduction of protons (H^+) through the utilization of low oxidation state electrons produced during fermentation (Amin et al.

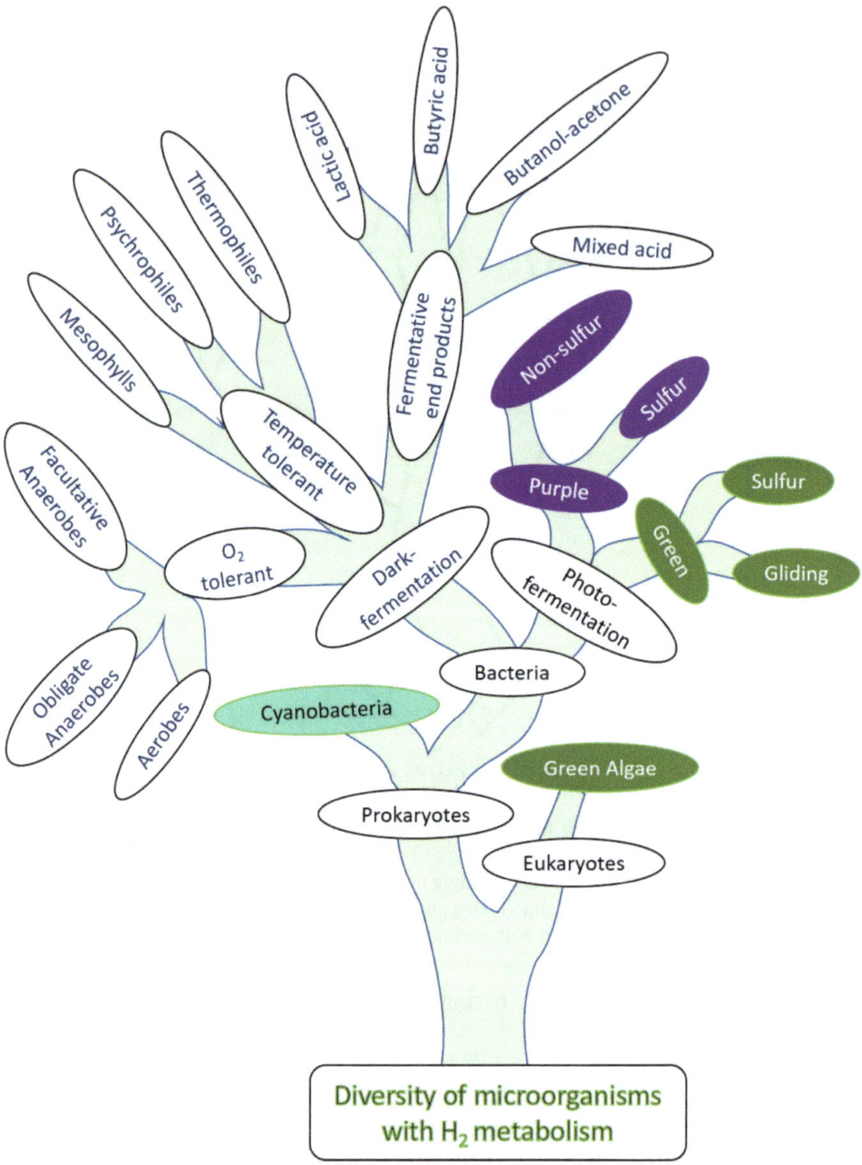

Fig. 2.3 Schematic representation for diversity and classification of microorganisms capable of fermentative biohydrogen production

2021). Under cellular conditions, the H_2 oxidation is kinetically challenging but thermodynamically favorable. The generation of an ion gradient during the reaction is crucial for effective enzyme catalysis (Volbeda et al. 2012). Microbes tackle this problem with the help of two different classes of metal *hyd*, the [NiFe]-*hyd* and [FeFe]-*hyd*. In most microbes, these metal *hyd* is responsible for the H_2

Table 2.2 List of microorganisms with respective hydrogenase types present in them

Microorganisms	Enzymes	Regulatory genes	Potential functions	References
Microbes having H_2-consuming capabilities				
Clostridium botulinum	*[NiFe]-hyd Group (Gr.) 1a*	*hysAB*	H_2-dependent sulfite respiration	da Silva et al. (2008)
Bilophila wadsworthia				
Helicobacter pylori	*[NiFe]-hyd Gr. 1b*	*hynABC*	Respiration processes reliant on H_2 as well as fumarate under aerobic conditions	Olson and Maier (2002); de Vries et al. (1984); Maier et al. (1996); Weerakoon et al. (2009)
Campylobacter jejuni		*hynABC*		
Escherichia coli	*[NiFe]-hyd Gr. 1c*	*hybOABC*	H_2-dependent fumarate respiration; may be proton motive	Maier et al. (2013); Lukey et al. (2010); Beaton et al. (2018)
	[NiFe]-hyd Gr. 1d	*hyaABC*	H_2-dependent aerobic and fumarate respiration; oxygen tolerant	Zbell and Maier (2009); Bowman et al. 2014; Volbeda et al. (2012)
Salmonella typhimurium	*[NiFe]-hyd Gr. 1c*	*hybOABC*	Fumarate respiration driven by H_2 dependency, potentially a proton motive force (PMF)	Beaton et al. (2018); Lukey et al. (2010)
	[NiFe]-hyd Gr. 1d	*hyaABC*	Respiration processes are fueled by H_2 and fumarate in an aerobic environment, showcasing tolerance to oxygen	Zbell and Maier (2009); Bowman et al. (2014); Volbeda et al. (2012)
Corynebacterium diphtheriae	*[NiFe]-hyd Gr. 1f*	*hyoSLE*	Uncertain; potentially involved in H_2-dependent aerobic respiration	Myers and King (2016)
Mycobacterium gordonae	*[NiFe]-hyd Gr. 1h*	*hhySLE*	H_2-dependent aerobic respiration; oxygen tolerant, high affinity	Meredith et al. (2014); Greening et al. (2014)
	[NiFe]-hyd Gr. 2a	*hucLSAEB*		Park and Decicco (1974); Greening et al. (2014)
Rhodococcus equi	*[NiFe]-hyd Gr. 1h*	*hhySLE*		Meredith et al. (2014)
Eggerthella lenta	*[NiFe]-hyd Gr. 1i*	*vhzGAC*	Uncertain; potentially contributing to H_2-dependent anaerobic respiration	Søndergaard et al. (2016)
Arcobacter butzleri	*[NiFe]-hyd Gr. 2d*	*huaSL*		Brugna-Guiral et al. (2003)
Microbes having H_2-production capabilities				

(continued)

Table 2.2 (continued)

Microorganisms	Enzymes	Regulatory genes	Potential functions	References
Bilophila wadsworthia	*[NiFe]-hyd Gr. 4c*	*cooMKLXUH*	CO-coupled H_2 production; maybe a proton motive	Fox et al. (1996)
Eggerthella lenta	*[NiFe]-hyd Gr. 4e*	*echABCDEF*	Ferredoxin-coupled H_2 production; reversible and proton motive	Harris et al. (2018)
Trichomonas vaginalis	*[NiFe]-hyd Gr. A1*	*hydA*		Morra et al. (2016)
Clostridium perfringens	*[FeFe]-hyd Gr. A1*	*hydA*	H_2 production coupled with ferredoxin; specific instances exhibit activity in the reverse direction	Morra et al. (2016)
Treponema denticola	*[FeFe]-hyd Gr. A2*	*hydAgltB*	Not yet understood; possibly functionally connected to glutamate synthase	Calusinska et al. (2010)
Clostridioides difficile	*[FeFe]-hyd Gr. A3*	*hydABC*	A reversible process that splits electrons from H_2 to both NAD and ferredoxin	Schuchmann and Müller (2012); Köpke et al. (2013)
	[FeFe]-hyd Gr. A4	*hytAE1E2*	H_2 production dependent on formate; characterized by physiological reversibility	Schuchmann and Müller (2013)
Brachyspira pilosicoli	*[FeFe]-hyd Gr. A3*	*hydABC*	A reversible process that bifurcates electrons from H_2 to both NAD and ferredoxin	Köpke et al. (2013); Schuchmann and Müller (2012)
Entamoeba histolytica	*[FeFe]-hyd Gr. B*	*hydM*	Ferredoxin-coupled H_2 production	Nixon et al. (2003); Harris and Reddy (1977)
Bacteroides fragilis	*[FeFe]-hyd Gr. B*	*hydM*		Harris and Reddy (1977); Nixon et al. (2003)
Mycobacterium gordonae	*[NiFe]-hyd Gr. 3b*	*hyhBGSL*	NADPH-coupled H_2 production maybe physiologically reversible	Berney et al. (2014a, b)
Rhodococcus equi	*[NiFe]-hyd Gr. 3b*	*hyhBGSL*		
Escherichia coli	*[NiFe]-hyd Gr. 4a*	*hycBCDEFG*	Formate-dependent H_2 production; maybe a proton motive	McDowall et al. (2014); Lamichhane-Khadka et al. (2015)
Salmonella typhimurium	*[NiFe]-hyd Gr. 4a*	*hycBCDEFG*		

consumption and generation during their life cycle (Meredith et al. 2014; Doğan-Güner and Koku 2022). The catalytic core of these hydrogenases comprises two metals that play a pivotal role in coordinating and facilitating the heterolytic cleavage of H_2. This cleavage involves generating a $H^+$and hydride (H^-) anion, where the metal center's interaction with a base lead to increased acidity. The released H^+ is captured by a base p within the catalytic core and then transferred through a sequence of amino acids within the hydrogenase's structure to the surrounding aqueous environment. Nevertheless, the destiny of a hydride anion contrasts with that of a H^+. The H^- species carries two electrons $2e^-$, which are transferred through a [FeS] cluster to downstream acceptors in the electron transport chain (e.g., cytochromes). Simultaneously, the resulting H^+ ion is discharged from the enzyme's catalytic site. Several *hyd* during this reversible catalytic mechanism reduces the H^+ to molecular H_2. The ultimate direction of the *hyd*-catalyzed reaction is governed by the intrinsic redox potential of the interacting molecules with *hyd*.

To study the biochemical and structural features of *hyd*, three techniques have been utilized, viz., crystallography, biochemical, and molecular or genetic. During the later half of the twentieth century, multiple types of *hyd* were isolated, purified, and characterized. With the emergence of recombinant DNA technology in the late twentieth century, *hyd* classification based on its genome sequence prevailed, which streamlined numerous studies on identifying features of *hyd* enzymes. Understanding genome sequences in the late twentieth century paved the avenue for a structural engineering approach for *hyd* and associated genes for cell maturation. During the decade of 1990s, were able to develop the first crystal structure of [NiFe]-*hyd* isolated from *Desulfovibrio vulgaris* (*D. vulgaris*). Out of the prevalently found *hyd* enzymes, viz. [NiFe] and [FeFe], the double iron *hyd*, are known to be present exclusively in a few green algae, facultative, and obligate anaerobic bacteria. Though currently [FeFe] type *hyd* has not been identified in Archaea, they are present in lower eukaryotes such as green algae and protozoa. In eukaryotes, they have been found in their subcellular organelles, e.g., chloroplasts of green algae and hydrogenosomes of protozoa. Initially, the classification of *hyd* was made based on the presence of specific electron donors and acceptors in their core catalytic sites, viz. coenzyme F_{420} (EC1.12.99.1), NAD (*hyd* of EC 1.12.1.12), ferredoxins (EC 1.18.99.1), or cytochromes (EC 1.12.2.1). As discussed above, the major *hyd* found in nature are iron-sulfur enzymes consisting of either two Fe atoms (in case of [FeFe]-*hyd* class) or a Ni atom and a Fe atom (in case of [NiFe]-*hyd* class). Scientists have also discovered a third type of *hyd* in a few bacteria capable of gas fermentation, e.g., methanogens. The principle of this *hyd* is similar to H_2-producing bacteria commonly known as methylene-tetra-hydromethanopterin dehydrogenase (Hmd, EC 1.12.99.4). The Hmd class does not contain any metal apart from the Fe-S cluster. Therefore, it is also called iron-sulfur-cluster-free *hyd* or [Fe]-*hyd* (previously known as "metal-free dehydrogenase").

2.3.1 Enzymes Essential for H_2 Metabolism

Presently, the sequences of altogether more than 650 *hyd* are available in the protein database (Bernstein et al. 1977; Berman et al. 2000). Extensive diversity exists among *hyd* enzymes, encompassing factors such as size, electron donors and acceptors, and quaternary structure. Despite this diversity, they can be categorized into three distinct classes based on phylogeny: [FeFe]-, [NiFe]-, and [Fe]-*hyds*. Each class, as depicted in Fig. 2.4, possesses a unique functional core that remains conserved within its respective group. This classification is supported by studies conducted by Vignais et al. (2001), Shima and Thauer (2007), and Meyer (2007). Within the enzyme, the functional core is composed of subunits or domains that provide the space for the catalytic site. This minimal configuration is essential for maintaining both the structural integrity and the operational functionality of the enzyme (Vignais and Billoud 2007). Metal content and sequence similarity are thus a reliable classification criterion. Since the presence of [Fe]-*hyd* or Hmd (Shima and Thauer 2007) are only restricted to some methane metabolizing microbes (aka., methanogens), we have not adequately discussed their phylogenetic relationships, and therefore, only

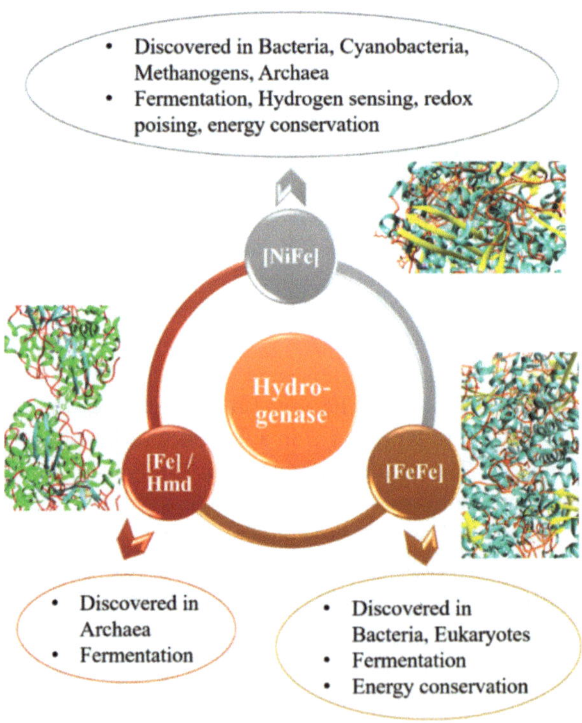

Fig. 2.4 Schematic representation of classification, source, and role of hydrogenase in the commonly found in microbes

the *hyd* from [FeFe]- and [NiFe]-family is discussed in this chapter. Table 2.3 summarizes the biological reactions for H_2 metabolism that exist in nature.

2.3.1.1 [Fe]-*hyd* or Hmd

The initial instance of the Hmd enzyme was uncovered in *Methanothermobacter marburgensis*, a species extensively examined within this category of hydrogenases (Adamson and Pearson 2000; McDowall et al. 2014). Hmd functions as a catalyst, facilitating the reversible conversion of methenyltetrahydromethanopterin (methenyl-H4MPT$^+$) into methylene-H4MPT and H^+ through the utilization of H_2. This enzyme plays a crucial intermediary role in the process of converting CO_2 to methane using H_2 (Xie et al. 2008; Amin et al. 2011; Galadima and Muraza 2015). Under specific growth conditions, the activity of these enzymes is contingent upon the presence of nickel. In the absence of Ni atom, the synthesis of F_{420}-reducing [NiFe]-hyd (Frh) ceases. The [Fe]-hyd comprises two identical subunits of 38 kDa each, encoded by a monocistronic gene. This [Fe]-hyd lacks an iron-sulfur cluster and contains two iron atoms per homodimer (Shima and Thauer 2007; Berney et al. 2014b; Bowman et al. 2014). The presence of this enzyme has been marked in a dozen species from the methanogenic microbial family. Initially, it was wrongly thought that Hmd functions as a pure organic catalyst; but it has now come to know that its catalytic activity heavily depends on a cofactor containing iron in its core (Shima and Thauer 2007; Schuchmann and Müller 2012). The structural details of the *Hmd* apoenzyme have recently been recorded and are available in protein databases. In brief, the [Fe]-*hyd* substantially differs from the other two hyd types, namely [FeFe]- and [NiFe]-hyd. This distinction extends not only to their structural characteristics but also to their catalytic mechanisms (Shima and Thauer 2007; Meyer 2007; Berney et al. 2014b; McDowall et al. 2014). The iron present in [Fe]-*hyd* does not exhibit redox activity necessary for enzyme activity. Consequently, their catalytic behavior differs from the [NiFe]- and [FeFe]-class *hyd* enzymes. As a result, *Hmd* enzymes do not facilitate the reversible H_2 production reaction:

$$2H^+ + 2e^- \leftrightarrow H_2$$

The three-dimensional structures of Hmd or only [Fe]-*hydrogenase* from *D. vulgaris* (PDB ID: 1HFE) are given in Fig. 2.5. The binding mode and interactions of catalytic amino acid residues with ligands, cofactors, and ions bounds with the crystal structure of 1HFE are shown in Fig. 2.5b, c.

2.3.1.2 [NiFe]-*hyd*

The *hyd* from this class is most abundant and extensively studied among all types of *hyd* from the bacterial domain (Brugna-Guiral et al. 2003). The central [NiFe]-*hyd* structure comprises an αβ heterodimer, where the α-subunit (larger, 60 kDa) houses

Table 2.3 Biological reactions for molecular hydrogen metabolism that exist in nature (as per the KEGG database)

Reaction name	Reaction ID	Enzyme ID	Chemical equation
Hydrogen: ferredoxin oxidoreductase	R00019	EC 1.12.7.2	2 Reduced ferredoxin + 2 H+ ↔ H_2 + 2 Oxidized ferredoxin
Reduced ferredoxin: dinitrogen oxidoreductase	R00067	EC 1.18.6.1	2 e^- + 2 H^+ ↔ H_2
Hydrogen: NAD^+ oxidoreductase	R00700	EC 1.12.1.2; EC 1.12.1.5	H_2 + NAD^+ ↔ NADH + H^+
Hydrogen: quinone oxidoreductase	R02965	EC 1.12.5.1	Menaquinone + H_2 ↔ Menaquinol
Hydrogen: coenzyme F_{420} oxidoreductase	R03025	EC 1.12.98.1	Coenzyme F420 + H_2 ↔ Reduced coenzyme F_{420}
Hydrogen: ferricytochrome-c3 oxidoreductase	R04015	EC 1.12.2.1	2 Ferricytochrome c3 + H_2 ↔ 2 Ferrocytochrome c3 + 2 H^+
Hydrogen: N5, N10-methenyltetraydro-methanopterin oxidoreductase	R04455	EC 1.12.98.2	5,10-Methenyltetrahydromethanopterin + H_2 ↔ 5,10-Methylenetetrahydromethanopterin + H^+
Reduced ferredoxin: dinitrogen oxidoreductase (ATP-hydrolyzing)	R05185	EC 1.18.6.1	16 ATP + N_2 + 8 Reduced ferredoxin + 8 H^+ + 16 H_2O ↔ 16 Orthophosphate + 16 ADP + 8 Oxidized ferredoxin + 2 NH_3 + H_2
Reduced flavodoxin: dinitrogen oxidoreductase (ATP-hydrolyzing)	R05186	EC 1.19.6.1	16 ATP + N_2 + 4 Reduced flavodoxin + 16 H_2O ↔ 16 Orthophosphate + 16 ADP + 4 Oxidized flavodoxin + 2 NH_3 + H_2
Hydrogen: $NADP^+$ oxidoreductase	R07181	EC 1.12.1.3; EC 1.12.1.5	H_2 + NADP+ ↔ NADPH + H^+
Hydrogen: (acceptor) oxidoreductase	R07182	EC 1.12.99.6	H_2 + Acceptor ↔ Reduced acceptor
Hydrogen: 2-(2,3-dihydropentaprenyloxy) phenazine oxidoreductase	R09095	EC 1.12.98.3	Methanophenazine + H_2 ↔ Dihydromethanophenazine
Ferredoxin oxidoreductase	R09508	EC 1.12.1.4	2 H_2 + NAD^+ + 2 Oxidized ferredoxin ↔ 5 H^+ + NADH + 2 Reduced ferredoxin
H_2: polysulfide oxidoreductase	R10390	EC 1.12.98.4	H_2 + Polysulfide(n) ↔ H_2S + Polysulfide(n-1)
Ferredoxin: H_2 oxidoreductase	R11943	EC 1.8.98.5	Coenzyme B + Coenzyme M + 2 Reduced ferredoxin + 2 H^+ ↔ Coenzyme M 7-mercapto-heptanoylthreonine-phosphate heterodisulfide + 2 Oxidized ferredoxin + 2 H_2

(continued)

Table 2.3 (continued)

Reaction name	Reaction ID	Enzyme ID	Chemical equation
Ferredoxin: dinitrogen oxidoreductase	R12084	EC 1.18.6.2	12 Reduced ferredoxin + 12 H^+ + N_2 + 40 ATP + 40 $H_2O \leftrightarrow$ 12 Oxidized ferredoxin + 3 H_2 + 2 NH_3 + 40 ADP + 40 Orthophosphate
CO_2 + Hydrogen ↔ formate	R12754	EC1.17.98.-	$CO_2 + H_2 \leftrightarrow$ Formate

the bimetallic active site and the β-subunit (smaller, 30 kDa) contains the iron-sulfur clusters (depicted in Fig. 2.6a). An intriguing observation is the reduced size of the α- and β-subunits within multimeric hydrogenases. Significant progress has been made in comprehending the overall structure and the characteristics of the binuclear NiFe active site, thanks to the accessibility of crystal structures derived from hyd enzymes found in *Desulfovibrio* (Volbeda et al. 2012; Arias et al. 2021). The structural analysis of [NiFe]-*hyd* obtained from *D. desulfuricans* ATCC 27774 (PDB ID: 1E3D) has provided insights into the interaction between the two subunits. These subunits have a substantial interface that forms a compact globular heterodimer (as depicted in Fig. 2.6a). The catalytic core, represented by the bimetallic nickel-iron center, is deeply embedded within the larger subunit. This core is bound to the enzyme through the coordination of four cysteine amino acid residues. The FTIR analysis provided evidence of three non-protein ligands positioned around the iron atom, including interactions with SO_4, CO, and CN^- groups. Furthermore, the combination of FTIR and EPR analyses unveiled the presence of two extra CN^- ligands. One of these CN^- ligands is attached to nickel, indicating a structural arrangement

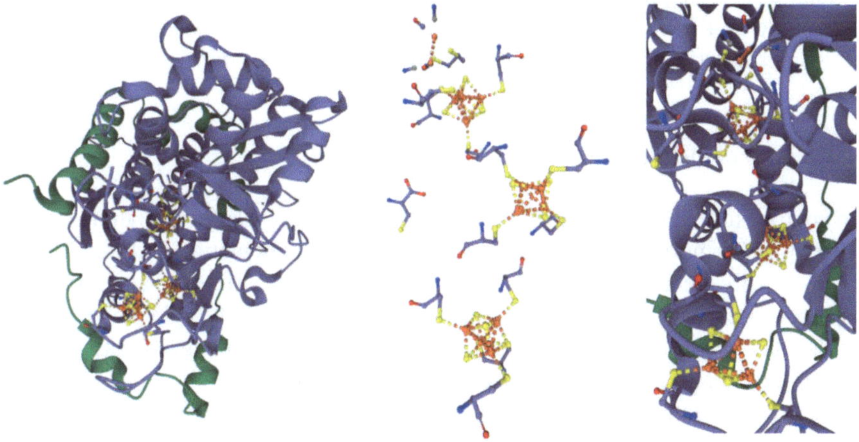

(a) (b) (c)

Fig. 2.5 Three-dimensional (3D) structures of [Fe]-hydrogenase from *Desulfovibrio vulgaris* (PDB ID: 1HFE). (**a**) Complete 3D structure of 1HFE visualized in Chimera, (**b**) interaction of active site amino acid residue with all types of ligands present in 1HFE without showing polymer chain, and (**c**) interaction of active site amino acid residue with ligand present in 1HFE with polymer chain

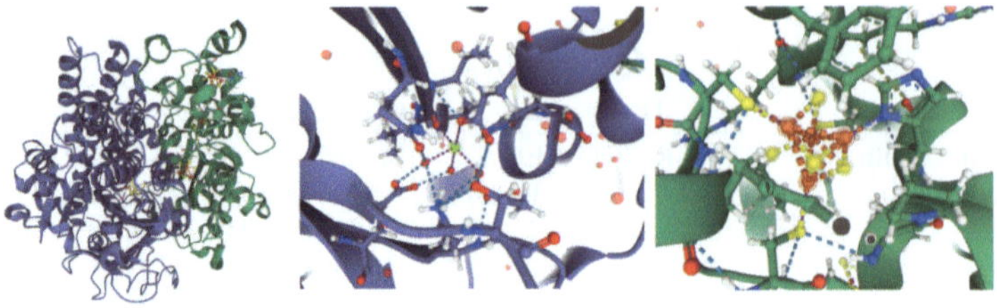

(a) (b) (c)

Fig. 2.6 The 3D models of [NiFe]-*hyd* derived from *D. desulfuricans* ATCC 27774 (PDB ID-1E3D). (**a**) Complete 3D structure of 1E3D visualized in Chimera, (**b**) interaction of active site amino acid residue with Mg^{2+} ion present in 1E3D, and (**c**) interaction of active site amino acid residue with iron-sulfur ([4Fe-3S]) cluster present in 1E3D

of $Ni(CN)Fe(CN)_3(CO)$ within the cytoplasmic NAD-reducing hydrogenase of *Ralstonia eutropha*. It is important to note that *Ralstonia eutropha* was previously recognized as *Alcaligenes eutrophus* and has now been renamed *Cupriavidus necator*.

The β-subunit is characterized by a linear configuration that accommodates as many as three cubane iron-sulfur clusters of the [4Fe-4S] type. These clusters play a crucial role in facilitating the transfer of electrons between the H_2-activating center and the *hyd*'s electron acceptor or donor in the physiological context. The crystal structure of [NiFe]-hydrogenase from *D. desulfuricans* ATCC 27774 (PDB ID: 1E3D) has provided insights indicating that the [4Fe-4S] cluster is positioned in close proximity to the bimetallic center. This bimetallic center has been subject to changes, involving the addition of three oxygen atoms and loss of one sulfur atom, leading to a [4Fe-3S-3O] configuration. Figure 2.6 provides visual representations of the 3D structures of the [NiFe]-*hyd* sourced from *D. desulfuricans* ATCC 27774.

2.3.1.3 [FeFe]-Hydrogenases

In contrast to [NiFe]-*hyd*, which is composed of not less than two subunits, several [FeFe]-*hyd* comprise only a catalytic subunit and are monomeric structures (Berney et al. 2014b; Schuchmann et al. 2018). While dimeric, trimeric, and tetrameric hydrogenase (hyd) structures are recognized, the smallest [FeFe]-hyd, with a molecular weight of 45–48 kDa, has been identified in green algae (Fakhimi et al. 2020). *Hyd* of this category is also found in anaerobic microbes, such as lower eukaryotes, sulfate-reducing bacteria, and *clostridia*. The [FeFe]-*hyd* are exclusive *hyd* types found in eukaryotes, and only in organelles, that is, in hydrogenosomes or chloroplasts (Meyer 2007; Schuchmann et al. 2018). In contrast to [NiFe] bimetallic *hyd* enzymes, the catalytic subunits of [FeFe]-*hyd* differ significantly in size. Despite the presence of the active site (H-cluster) that are conserved domains, these enzymes often consist of additional catalytic domains, which accommodate iron-sulfur clusters. In those, the H-clusters comprise a binuclear [FeFe] center, which is bound to a [4Fe-4S] cluster through a bridging cysteine of the [FeFe]-*hyd*. Non-protein components, such as CN^{-} and CO, are bound to the Fe atoms of the center. The iron atoms also engage with two bridging sulfur ligands from a small molecule comprising five atoms, potentially a di(thiomethyl)amine molecule, HN(CH2-S^{-})$_2$. The Fe^{2+} atom is positioned distally to the [Fe_4-S_4] cluster c, and this cluster features an unoccupied coordination site that becomes occupied by a competitive inhibitor, CO. Therefore, it is postulated that the configuration of the enzyme inhibited by CO corresponds to the site where the binding of H_2 or H^{-} occurs during the enzymatic turnover process. In the structural analyses of both the [FeFe]-*hyd* from *D. desulfuricans* ATCC 775764 and the *hyd* I from *C. pasteurianum*, a single hydrophobic channel was notably identified. Extending from the molecular surface to the active site, this channel directs toward the Fe^{2+} atom. Similar to the [NiFe]-hyd, a plausible H^{+} pathway has been hypothesized for the [FeFe]-*hyd*. The 3D structures of [FeFe]-*hyd* from *C. pasteurianum* ATCC 6013 (PDB ID: 4XDD) and *Hyd*-1 from *Escherichia coli* (PDB ID: 5LMM) are given in Fig. 2.7. This image also provides the detailed information of ligands and cofactors bound to the active sites of 4XDD and 5LMM.

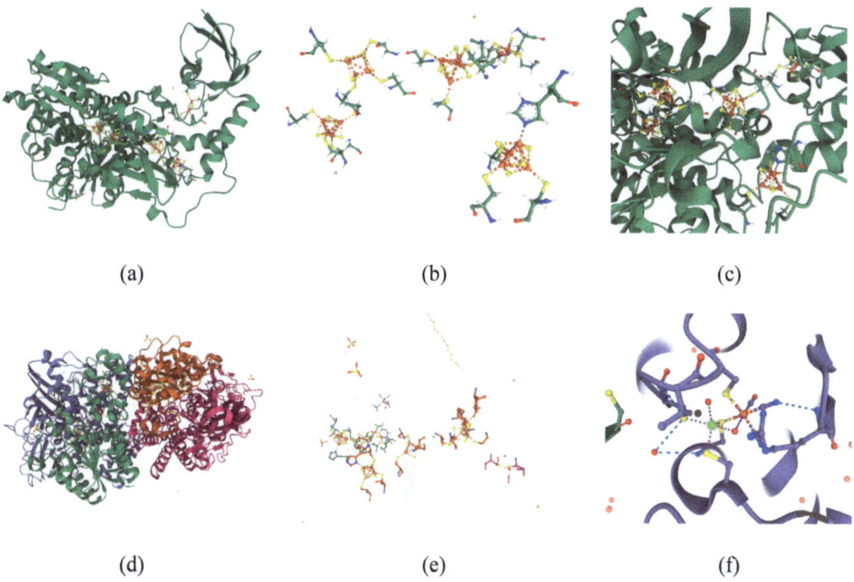

Fig. 2.7 The 3D models of [FeFe]-*hyd* from *C. pasteurianum* ATCC 6013 (PDB ID: 4XDD) and Hydrogenase-1 from *Escherichia coli* (PDB ID: 5LMM). (**a**) Complete 3D structure of 4XDD visualized in Chimera, (**b**) interaction of active site amino acid residue all types of ligands present in 4XDD without showing polymer chain, (**c**) interaction of active site amino acid residue with iron-sulfur ([4Fe-3S]) cluster present in 4XDD, (**d**) Complete 3D structure of 5LMM visualized in Chimera, (**e**) interaction of active site amino acid residue with all types of ligands present in 5LMM without showing polymer chain, and (**f**) interaction of active site amino acid residue with Ni^{+} ion present in 5LMM

2.4 Metabolic Pathways and Molecular Mechanisms of H_2 Metabolism

As discussed above, major microbial families capable of metabolizing H_2 are prokaryotes that belong to the Archaeal and Bacterial domain of life. A list of metabolic pathways in microorganisms for H_2 metabolism is provided in Table 2.4. They comprise a group of photosynthetic prokaryotes, fermentative organisms, autotrophs, heterotrophs, a(na)erobes, etc. A few lower eukaryotic cells, such as *Trichomonas vaginalis* (parasitic protozoa), *Neocallimastix frontalis* (an anaerobic fungus), *Chlamydomonas*, and *Scenedesmus* (unicellular green algae), are also able to produce H_2 with the help of [FeFe]-*hyd*(s). These microbes do not contain mitochondria or chloroplasts; their genomes consist of hydrogenosomes. The bacterial domain comprises the majority of prokaryotes, encompassing medically relevant and environmentally significant microbes. As discussed above, despite the simplicity of the hydrogen formation reaction, *hyd* is extremely complicated concerning its structural features and mechanism of action. Furthermore, the sophisticated action mechanism of enzymes is compulsory to utilize the e^- and H^+ derived from the oxidation of H_2 to generate an ion gradient (aka. chemiosmosis). As a notable illustration of convergent evolution, the molecular *hyd* found in microbes display mechanistic and structural similarities despite their diverse phylogenetic ancestries. Several authors have documented a detailed recent review on this topic (Vignais et al. 2001; Meyer 2007; Vignais and Billoud 2007; Calusinska et al. 2010; Berney et al. 2014b).

The bimetallic centers of both *hyd* ([NiFe] and [FeFe]) coordinate H_2 and catalyze its heterolytic cleavage into a H^+ and a H^-. At the start, a base near the catalytic center captures the H^+ ion, which is subsequently conveyed to the surrounding aqueous environment through a sequence of amino acids. Conversely, the two electrons conveyed by the H^- traverse iron-sulfur clusters to reach downstream receptors like respiratory cytochromes, while the resulting proton is also liberated from the enzyme. This process is bidirectional, where numerous hydrogenases predominantly facilitate the reduction of H^+ to form H_2. The ultimate direction of the *hyd*-catalyzed reaction is determined by their intrinsic redox chemistry (known as catalytic bias) and their association with external electron donors or acceptors. The [NiFe]-*hyd* are enzymes that are widely present in various bacterial species.

Found widely across both aerobic and anaerobic pathogens, these enzymes have significant roles in the pathogenicity of *H. pylori*, *S. typhimurium*, and *C. jejuni*. Due to comprehensive biochemical, structural, spectroscopic, and electrochemical analyses, there exists a relatively solid understanding of these enzymes. The primary subunit of these enzymes attaches to a catalytic center containing nickel and iron, wherein coordination is achieved through the binding of one CO molecule, two CN^- groups, and four cysteine amino acids. The presence of diatomic groups (e.g., $CN^-$and CO) alters the redox chemistry of the two metals at this specific site, enabling the cleavage of H_2. Each [NiFe]-*hyd* incorporates a compact subunit capable of accommodating a maximum of three [Fe-S] clusters. The flow of electrons

Table 2.4 List of metabolic pathways in microorganisms for H_2 metabolism (as per KEGG database)

Pathway name	Chemical equation	Enzymes involved	Microbial genus family	Type of metabolism
Ferredoxin	$Fd_{rd} \rightarrow Fd_{ox} + H_2$	[FeFe]-*hyd* group A1 and B	*Firmicutes, Bacteroidetes, Fusobacteria, Eukaryotes*	Production
Bifurcation	$Fd_{rd} + NADH \rightarrow Fd_{ox} + NAD^+ + H_2$	[FeFe]-*hyd* group A3 and A4	*Firmicutes, Bacteroidetes, Fusobacteria*	
Formate	$Formate \rightarrow CO_2 + H_2$	[NiFe]-*hyd* group 4a and 4f	*Gammaproteobacteria and Epsilonproteobacteria*	
Nicotinamide	$NAD(P)H \rightarrow NAD(P)^+ + H_2$	[NiFe]-*hyd* group 3b and 3c	*Actinobacteria*	
Acetogenesis	$CO_2 + H_2 \rightarrow Acetate$	[FeFe]-*hyd* group A3 and A4	*Firmicutes*	Consumption
Methanogenesis	$CO_2 + H_2 \rightarrow Methane$	[NiFe]-*hyd* group 3b, 3c, 4h, and 4i and [Fe]-*hyd*	*Euryarchaeota*	
Aerobic respiration	$O_2 + 2H_2 \rightarrow 2H_2O$	[NiFe]-*hyd* group 1b, 1d, 1f, 1h, and 2a	*Gammaproteobacteria, Epsilonproteobacteria and Actinobacteria*	
Fumarate respiration	$Fumarate + H_2 \rightarrow Succinate$	[NiFe]-*hyd* group 1b, 1c, and 1d	*Gammaproteobacteria and Epsilonproteobacteria*	
Sulfate respiration	$SO_4^{2-} + H_2 \rightarrow H_2S$	[NiFe]-*hyd* group 1a and 1b	*Deltaproteobacteria*	

between the active site and the physiological electron donor or acceptor hinges on the orientation of the enzymes, and this transfer occurs through these clusters. As detailed further, an extensive array of variations within this category of hydrogenases exist, encompassing factors like catalytic traits, subunit configuration, and redox collaborators. This diversity equips them to fulfill various functions in bacterial respiration, fermentation, and other processes. While the active sites of [NiFe]-*hyd* are generally prone to oxygen inhibition, various classes of these enzymes have been discovered to operate under ambient conditions, and some of them are even capable of supporting aerobic respiration. Numerous mechanisms have been identified that enable these enzymes to prevent or eliminate O_2 from their active sites, as extensively documented in separate studies (Benoit et al. 2020; Hughes et al. 2021). In contrast, [FeFe]-hydrogenases are primarily linked to obligate anaerobes. They are widespread among numerous bacterial pathogens that engage in fermentation (such as *C. perfringens*) and have also been found in a range of protist species, such as *T. vaginalis*.

Structural investigations reveal that this *hyd* class comprises a core domain binding to the catalytic cofactor (H-cluster) and a variable C-terminal domain linking to 2–4 [Fe-S] clusters. The binuclear iron catalytic center is connected by an azadithiolate bridge, one CO, and two CN^- ligands, and a [4Fe-4S] cluster through bridging cysteine residues. This crucial site plays a vital role in enabling the heterolytic cleavage of H_2, where the H^- is tightly bound to the metal center and the H^+ is accepted by the azadithiolate nitrogen. In general, [FeFe]-*hyd*(s) are specialized for catalyzing H_2 production and commonly employ reduced ferredoxin (Fd) as their electron donor. These enzymes are especially advantageous for obligate fermentative microorganisms like *Clostridia*. Furthermore, certain microorganisms possess trimeric [FeFe]-*hyd*(s) that exhibit the ability to reversibly bifurcate electrons from H_2 to Fd. Extensive investigations have been conducted on numerous uptake [FeFe]-*hyd*(s), showcasing their remarkable efficiency within the cytosol of sulfate-reducing bacteria. In comparison to [NiFe]-*hyd*(s), [FeFe]-*hyd*(s) tend to be operational exclusively in the absence of oxygen, and they are particularly vulnerable to permanent impairment upon exposure to O_2. The sites responsible for H_2 binding in both types of hydrogenases are intricate formations of metal ions meticulously coordinated by unique ligands. The production of these enzymes is notably specialized and carried out through dedicated metabolic pathways. The development of [NiFe]-hydrogenases encompasses a multipart process that demands the coordinated contribution of no less than seven elements (HupD, HypD, HypC, HypA, HypF, HypE, and HypB). A recent investigation has significantly unraveled the maturation path of [FeFe]-hydrogenases, revealing the involvement of three pivotal enzymes (HydE, HydG, and HydF). The intricate intricacies of the maturation procedures of these enzymes extend beyond the coverage of this chapter. Other enzymes are also indispensable for the functional performance of hydrogenases, including those accountable for Ni and Fe transportation, the generation of [Fe-S] clusters, and the relocation of *hyd* into distinct cellular compartments.

Integration and Regulation of H_2 Metabolism Microbes utilize *hyd* in various physiological circumstances. As recapitulated in Table 2.2 and Fig. 2.8, the [Ni Fe]- and [Fe Fe]-*hyd* vary considerably into distinct groups concerning their phylogenetic relations and functionality. They vary in key traits, such as catalytic mechanism, oxygen tolerance, genetic organization, cellular localization, and redox partners. The majority of pathogens that incorporate H_2 into their growth cycle and facilitate hydrogenotrophic growth (characterized by the capacity to utilize molecular hydrogen as an energy source during growth) accomplish this by harnessing the H_2 oxidation potential to lower the redox level of ultimate respiratory electron acceptors. For instance, in the context of aerobic respiration, this involves reducing O_2, while in the case of anaerobic respiration, it involves fumarate reduction (Weerakoon et al. 2009; Berney et al. 2014a). This phenomenon heavily depends on generating a proton gradient that results in proton motive force (PMF) around the plasma or cell membrane, thereby invigorating the synthesis of ATP through the F_1F_o-ATPase (i.e., oxidative phosphorylation). The PMF is generated as a result of proton translocation because of two principal chemical events that occur across the plasma membrane (1) the difference in H^+ concentration (aka. pH gradient or Δ pH), and (2) the difference in electrical potential (aka., membrane potential, Δ). Most hydrogenotrophic microbes utilize group 1b, 1c, and 1d [NiFe]-*hyd* to catalyze H_2 oxidation. These [NiFe]-*hyd* (specifically group 1b, 1c, and 1d) are bound on the membrane and oriented toward periplasm *hyd*. Due to H_2 oxidation events, the periplasmic protons directly participate in PMF generation.

The metabolic activity of a few microbes also depends extensively on H_2 oxidation for carbon fixation and invigorating energy-demanding processes, e.g., swarming motility, protein secretion, and substrate uptake. Similarly, hydrogenase enzymes also function critically in dispersing reducing agents in the shape of H_2 throughout FBP. In the course of this process, the energy obtained from the partial oxidation of carbohydrates into both organic and gaseous end products serves as the source for generating ATP through substrate-level phosphorylation. Only through this way does an obligate fermentative microbe, such as *Trichomonads* and *Clostridia*, sustain its growth. The intracellular [FeFe]-hydrogenases, encompassing the A1, A3, and B subtypes, present within these microorganisms play a role in the reoxidation of Fd and nicotinamides, which become reduced during the process of carbon compound breakdown. In certain protists like *T. vaginalis*, this process is confined to hydrogenosomes. The objective of this process is believed to sustain redox equilibrium, regulate the pH of the cytoplasm, and potentially generate a proton motive force (PMF). In general, the metabolism of H_2 is under tight regulation. Some obligate fermentative pathogens are believed to consistently produce H_2 throughout their lifecycle, thus constantly synthesizing their hydrogenases. However, for the majority of other bacteria, H_2 metabolism is a facultative trait that is triggered in response to internal and external cues. The genes responsible for regulating H_2 metabolism tend to be organized in clusters, promoting coordinated expression. An illustrative instance can be seen in the hyn and hyp operons of *C. jejuni*. Within this pathogen, three structural subunits and a maturation

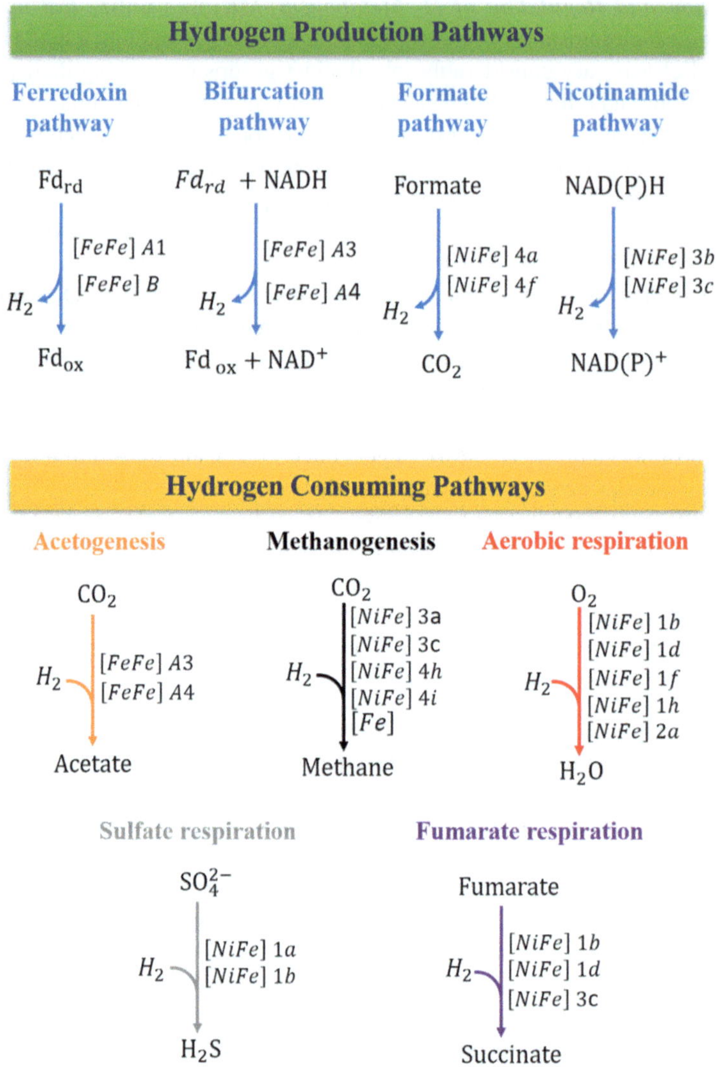

Fig. 2.8 Microbial hydrogen metabolism—pathways of production and consumption

endopeptidase of its group 1b [NiFe]-hydrogenase are encoded in hyn operon (hynABCD). A second operon, hyp (hypFBCDEA), contains the genetic information for the six essential proteins involved in synthesizing the [NiFe] cofactor. These two operons are triggered during infection, with the maturation-related genes being expressed at levels lower than the structural genes. Various activators and repressors can attach to the promoters of hydrogenase operons in reaction to both internal and external cues. Indeed, many signals and regulatory factors have been demonstrated to govern the expression of hydrogenase operons in diverse bacterial species. These factors encompass aspects such as redox status, metal presence, energy availability, oxygen concentration, as well as developmental phase (e.g., Spo0A). The intricacies of this regulation, both at the cellular and molecular levels, are extensively discussed in other sources (Appel and Schulz 1998; Schuchmann et al. 2018; El-Dalatony et al. 2020). Some environmental organisms additionally modulate the expression of hydrogenase genes by hydrogen partial pressures (pH_2), although this aspect holds less significance for most pathogens, primarily because hydrogen is generally abundant within host tissues.

2.5 Molecular Biology and Metabolic Engineering of Enzymes Involved in FBP

H_2-producers have exclusive H_2 metabolic systems compared to non-H_2-producers (Calusinska et al. 2010; Cai et al. 2011). *hyd*, the rate-limiting step for H_2 release during energy metabolism, performs a vital candidate in H_2 metabolism. The levels of *hyd* gene expression and the active phase of hydrogenase directly exert an influence on the energy metabolism of hydrogen-producing organisms, thereby directly affecting the rate of hydrogen release and its overall yield. The elevation of *hyd* expression is advantageous for augmenting the hydrogen production rate (Pugazhendhi et al. 2019). In 1994, Menon et al. conducted a study involving the cloning, sequencing, and mutation analysis of the *hyb* operon responsible for encoding *E. coli hyd* 2 (*hyd2*). Their investigation unveiled that the genes governing the two structural subunits of *E. coli* hyd 2 (*hyd2*) were organized in a cohesive operon (*hyb*), encompassing a total of seven open reading frames (ORFs). All seven of these ORFs were indispensable for reinstating the active *hyd2* to wild-type levels in AP3. Gorwa et al. (1996) analyzed the transcription and molecular traits of the potential *hyd* gene found in *C. acetobutylicum* (*Cac*) ATCC 824. The researchers reported that the 1.74 kbp hydA segment encoded a 64 kDa enzyme in *Cac*, which displayed remarkable similarity to the [Fe]-hydrogenase genes observed in *Desulfovibrio*. Miyake et al. (2001) undertook the genetic transformation of the [FeFe]-hydrogenase gene from *C. pasteurianum* (*Cpa*) ATCC 6013 into *Synechococcus elongatus* PCC7942. Their findings indicated that hydrogen production surged to three times that of the wild type under the influence of a potent promoter. Moreover, the recombinant exhibited an up to 1.7-fold enhancement in hydrogen gas productivity as compared to the wild type. In 2004, Mishra et al. achieved the successful isolation of a novel [Fe]-hydrogenase gene from the bacterium *Enterobacter cloacae* IIT-BT 08, renowned for its high-rate hydrogen production capabilities. They then proceeded to conduct an overexpression of this [Fe]-hydrogenase gene. In a related investigation, Morimoto et al. (2005) delved into the consequences of overexpressing *hyd* in *C. paraputrificum*, resulting in a notable enhancement of H_2 production. The *hydA* gene, stemming from the [Fe]-hydrogenase of *C. paraputrificum* M-21, was introduced into *E. coli* to achieve this outcome. The process involved utilizing a conserved DNA sequence sourced from *Clostridial hyd* genes, which was amplified through PCR and employed as a probe. Subsequently, the gene was integrated into a shuttle vector, specifically pJIR751, and subsequently expressed within the framework of *C. paraputrificum*. González-Pajuelo et al. (2005) delved into the domain of metabolic engineering by focusing on the potential of *Cac* for the industrial-scale synthesis of 1,3-propanediol from glycerol. The researchers integrated the 1,3-propanediol pathway from *C. butyricum* into numerous engineered strains of *Cac* using a plasmid. This endeavor led to substantial alterations in product formation, with particular attention given to the pSPD5 recombinant strain. Notably, chemostat cultures of the pSPD5 recombinant strain, cultivated in a glucose-rich medium, yielded a composition dominated by acids (including acetate, butyrate, and lactate) alongside a considerable amount of hydrogen production.

2.6 Synthetic Biology Targets for H_2 Overproduction

Currently, the genome of several microbial species has been completely sequenced, making the research of microbial genetic engineering extensively promising (Sarma et al. 2019). A limited number of investigations into microbial FBP have focused on the theoretical modulation and control of cellular metabolic pathways, aiming to heighten hydrogen production efficiency at a microbially molecular scale (El-Dalatony et al. 2020). To realize industrial-scale microbial FBP and mitigate the depletion of humanity's finite resources, upcoming research endeavors should be expanded and concentrated in the subsequent domains.

2.6.1 The Innovation of Methodologies for Cultivating Hydrogen-Producing Bacteria

Apart from *E. coli*, numerous microbial species offer potential as host organisms for genetic or metabolic engineering purposes. The expression and modification of Gram-positive bacteria, including *Bacillus* and lactic acid bacteria, have been effectively demonstrated (Hu 2021). Recent decades have witnessed the successful development of novel methodologies and analytical tools. These sophisticated techniques encompass real-time quantitative PCR, gene chips, two-dimensional gel electrophoresis for protein analysis, microarrays, multidimensional liquid chromatography, and surface plasmon resonance. These advancements have assumed a pivotal role in the realm of microbial genetic engineering and molecular biology (Benoit et al. 2020; Li et al. 2021; Hughes et al. 2021).

Nevertheless, there remains room for enhancements in the transformation and expression systems of microorganisms engaged in fermentative butyrate production

(Berney et al. 2014a; Kothari 2017; Chen et al. 2021). Research pertaining to the improvement of fermentative butyrate production through metabolic and genetic engineering is still in its nascent stages. Thus, devising inventive techniques for the cultivation of hydrogen-producing bacteria stands as a significant undertaking.

2.6.2 Utilization of Novel Genes and Strains

Extensive research has been dedicated to the utilization of molecular engineering techniques in model organisms well-suited for genetic engineering purposes, including *E. coli*, yeast, *B. subtilis*, filamentous fungi, and *Actinomycete*. Nevertheless, the pool of microbial species capable of hydrogen production remains limited, encompassing only a select few like *Fusobacterium*, *Clostridium*, *Zymomonas*, and *Bacteroides* (Kothari 2017; François et al. 2021). The foundation of metabolic engineering for these microorganisms has been relatively fragile, thus constraining the hydrogen output of engineered microbes (Mathews and Wang 2009; Young 2014; Balderas-Hernandez et al. 2020). The exploration of enzymes and microorganisms with extremophilic attributes has gained rapid momentum across the globe. Researchers have successfully sourced diverse microbial strains with enzyme potential from extreme oceanic microorganisms, thereby paving the way for heightened hydrogen yields.

2.6.3 Escalating the Usage of Microbial Molecular Engineering for H_2 Production

Microbial FBP has developed comparatively with advanced fermentation machinery and flow techniques. However, H_2-production efficiency still needs to improve due to low yields. Hence, redirecting research toward recombinant microbes generated through contemporary biotechnological methods such as protein engineering, genetic engineering, and metabolic engineering is crucial to enhancing the H_2 production efficiency by microorganisms. Because O_2 restraints the *hyd* expression, rational mutation of transcriptional regulatory proteins of *hyd* can be undertaken. Additionally, Photo System II of light-harvesting complex (LHC) can be enclosed in a compartment having inefficient oxidases (You et al. 2015). The electron bifurcation networks guide electrons toward a "Photosystem I-Fd-*hyd* compartment," with oxygen anticipated to supply electrons for reducing H^+ and O_2. In decoupling, a certain electron expense is necessary to detoxify O_2 to establish a steady FBP. H_2 production via the photoheterotrophic technique depends heavily on nitrogenase activity, the continuous ATP supply through photophosphorylation, and the flow of electrons and carbon. These factors are firmly regulated and strongly interconnected under most physiological conditions.

The majority of investigations have remained confined to specific segments of the comprehensive metabolic networks (such as PHB synthesis, H_2 production stemming from pyruvate, or photophosphorylation) or have focused solely on the individual biochemical constituents (notably, uptake hydrogenases and/or nitrogenase), without considering the intricate and interdependent interactions spanning the entire network. To elevate the efficiency of photoheterotrophic H_2 production, a comprehensive strategy of metabolic engineering is essential, encompassing the broader context of microbial metabolism on a global scale. These approaches include the prevailing strategies of integrated metabolic engineering discussed in this chapter. This effort also encompasses the integration of broader substrate utilization pathways for non-native carbon sources (particularly intricate sugars and lignocellulosic biomass). Additionally, it involves the advancement of more comprehensive in-silico metabolic models that account for regulatory mechanisms and photosystems. Furthermore, the engineering of nitrogenase enzymes to enhance interactions with ferredoxin and/or catalytic turnover rates constitutes another facet. In dark fermentative butyrate production, a key research focus revolves around augmenting the yield of H_2 production from sugars. Previous investigations have indicated that, in the absence of NAD(P)H-dependent H_2 production pathways, the H_2 yield remains confined to a maximum of 4 mol H_2/mol glucose (Liu and Moon 2009; Liu et al. 2017).

Several theoretical examinations have suggested that attaining a yield of 9 mol H_2/mol glucose is a plausible objective achievable primarily through H_2 production facilitated by NAD(P)H-linked hydrogenases (Fontaine et al. 2002; Liu et al. 2017; Ahmed et al. 2021). The H_2 production process via NAD(P)H-dependent hydrogenases is subject to diverse factors, encompassing the H_2 partial pressure, the concentration of H^+ ions, the ratio of NAD(P)H to NAD(P), and temperature. Nonetheless, the research on these parameters remains limited, often characterized by qualitative and disjointed analyses. To harness the potential of the NAD(P)H-dependent pathway effectively, a comprehensive understanding of the factors impacting this pathway and their intricate interactions with enzymes demands quantitative investigation. Notably, in the pursuit of surmounting the thermodynamic constraints associated with the reduction of H_2 pressure, NAD(P)H-dependent H_2 production reactions and elevation of H^+ concentration emerge as pivotal considerations. A significant challenge lies in expanding substrate utilization pathways to facilitate enhanced H_2 production. The viability of the dark FBP process on an economic front is widely acknowledged to hinge on the availability of cost-effective substrates, notably encompassing organic wastes and lignocellulosic biomass. The ongoing exploration of heightened rates and yields in H_2 production from these resilient carbon sources is still nascent research, presenting a critical facet of metabolic engineering. While significant hurdles loom on the path to commercializing FBP processes, the promise of H_2 as a "clean energy" source is poised to continue motivating researchers dedicated to metabolic engineering for H_2-producing organisms.

2.7 Overview and Prospects

Being a carbon-free alternate fuel, H_2 comes with several benefits; therefore, it is in the spotlight of Research and Development initiatives of various academic institutions and industries (Singh and Wahid 2015). In February 2022, the Ministry of Power, New & Renewable Energy, Government of India (GoI) announced the "green H_2 and green NH_3 policy" to boost the domestic green H_2 production by setting the target to achieve 5 million tonnes production by 2030 (Singh 2022; Ahluwalia 2022; ETEnergyWorld 2022). This policy was announced to make India an export hub for clean energy. Apart from this, GoI has also notified several attractive subsidy schemes and rewards for those working in the production and storage of the green hydrogen sector.

In biohydrogen generation, FBP can offer a sustainable H_2 stream essential for developing a potential hydrogen economy (Li et al. 2021; Hughes et al. 2021; Dzulkarnain et al. 2022). Various routes for microbial H_2 production have been identified, and many potential methods are under extensive investigation. Currently, the domain is adequately advanced to determine and remedy the factors limiting speeding up the development of FBP to power a potential biohydrogen economy. Given that knowledge required to uncover the molecular mechanisms and synthesis of the various *hyd* are now satisfactorily comprehended, a solid base for rationally designing the upgraded catalysts has been set up. However, many bottlenecks still need to be improved for the effective implementation of practical biophotolysis, including the sensitivity of *hyd* to oxygen and low conversion efficiencies of the biophotolysis process.

Similarly, the photofermentation process writhes from the low efficacy of light transformation. Potential advancements can be achieved through numerous strategies, which include the *hyd* substitution in place of nitrogenase, which requires extra energy. Comprehending the fundamental factors that impose limitations is essential in the pursuit of elevating the overall process efficiency. A rigorous investigation of H_2 production is ongoing. Several state-of-the-art techniques have been explored to answer a few of the relentless bottlenecks. Such methods include metabolic engineering that supplies metabolic energy to overcome challenges of thermodynamic and kinetic barriers, heterologous protein expression, including *hyd*, rewiring metabolic networks for higher substrate consumption, thermodynamic barriers, and escalated electron flux to reduce a proton gradient. These techniques are being advanced for greater H_2 yields (Pandey et al. 2013).

An interdisciplinary approach is imperative for establishing a sustainable H_2 technology (Seol et al. 2008; Hu 2021; Varjani et al. 2021). The productivity of any biological system depends on the operational parameters, which need to be engineered and optimized to influence FBP substantially. The performance of FBP has significantly enhanced with the accomplishment of this technology over the past two decades, which strengthens hopes that commercial production and application of FBP could be realized shortly.

Acknowledgments Mr. Karan Kumar is grateful to the Prime Minister's Research Fellowship provided by the Ministry of Education, Government of India.

Compliance with Ethical StandardsWe confirm that we have no conflict of interest to declare, and this work is original and has not been published elsewhere, or is it currently under consideration for publication elsewhere.

References

Adamson K-A, Pearson P (2000) Hydrogen and methanol: a comparison of safety, economics, efficiencies and emissions. J Power Sour 86:548–555. https://doi.org/10.1016/S0378-7753(99)00404-8

Ahluwalia S (2022) India's green hydrogen policy: tentative beginnings. ORF. https://www.orfonline.org/expert-speak/indias-green-hydrogen-policy/. Accessed 7 June 2022

Ahmed SF, Rafa N, Mofijur M et al (2021) Biohydrogen production from biomass sources: metabolic pathways and economic analysis. Front Energy Res 9

Amin AM, Croiset E, Epling W (2011) Review of methane catalytic cracking for hydrogen production. Int J Hydrog Energy 36:2904–2935. https://doi.org/10.1016/j.ijhydene.2010.11.035

Amin MM, Taheri E, Fatehizadeh A et al (2021) Anaerobic membrane bioreactor for the production of bioH2: Electron flow, fouling modeling and kinetic study. Chem Eng J 426:130716. https://doi.org/10.1016/j.cej.2021.130716

Angikath F, Abdulrahman F, Khandavilli M et al (2021) Hydrogen evolution from hydrocarbon pyrolysis in a simulated liquid metal bubble reactor. Energy Fuel 35:14597–14609. https://doi.org/10.1021/acs.energyfuels.1c01880

Appel J, Schulz R (1998) Hydrogen metabolism in organisms with oxygenic photosynthesis: hydrogenases as important regulatory devices for a proper redox poising? J Photochem Photobiol B Biol 47:1–11. https://doi.org/10.1016/S1011-1344(98)00179-1

Argun H, Kargi F, Kapdan IK (2009) Effects of the substrate and cell concentration on biohydrogen production from ground wheat by combined dark and photo-fermentation. Int J Hydrog Energy 34:6181–6188. https://doi.org/10.1016/j.ijhydene.2009.05.130

Arias DM, Ortíz-Sánchez E, Okoye PU et al (2021) A review on cyanobacteria cultivation for carbohydrate-based biofuels: Cultivation aspects, polysaccharides accumulation strategies, and biofuels production scenarios. Sci Total Environ 794:148636. https://doi.org/10.1016/j.scitotenv.2021.148636

Arizzi M, Morra S, Gilardi G et al (2021) Improving sustainable hydrogen production from green waste: [FeFe]-hydrogenases quantitative gene expression RT-qPCR analysis in presence of autochthonous consortia. Biotechnol Biofuels 14:182. https://doi.org/10.1186/s13068-021-02028-3

Aydin MI, Karaca AE, Qureshy AMMI, Dincer I (2021) A comparative review on clean hydrogen production from wastewaters. J Environ Manag 279:111793. https://doi.org/10.1016/j.jenvman.2020.111793

Bakonyi P, Nemestóthy N, Simon V, Bélafi-Bakó K (2014) Fermentative hydrogen production in anaerobic membrane bioreactors: A review. Bioresour Technol 156:357–363. https://doi.org/10.1016/j.biortech.2014.01.079

Balderas-Hernandez VE, Landeros Maldonado KP, Sánchez A et al (2020) Improvement of hydrogen production by metabolic engineering of Escherichia coli: Modification on both the PTS system and central carbon metabolism. Int J Hydrog Energy 45:5687–5696. https://doi.org/10.1016/j.ijhydene.2019.01.162

Banerjee R (2006) Comparison of options for distributed generation in India. Energy Policy 34:101–111. https://doi.org/10.1016/j.enpol.2004.06.006

Banerjee M, Kumar A, Kumar H (1989) Factors regulating nitrogenase activity and hydrogen evolution in Azolla-anabaena symbiosis. Int J Hydrog Energy 14:871–879. https://doi.org/10.1016/0360-3199(89)90074-8
Barca C, Soric A, Ranava D et al (2015) Anaerobic biofilm reactors for dark fermentative hydrogen production from wastewater: A review. Bioresour Technol 185:386–398. https://doi.org/10.1016/j.biortech.2015.02.063
Barca C, Ranava D, Bauzan M et al (2016) Fermentative hydrogen production in an up-flow anaerobic biofilm reactor inoculated with a co-culture of Clostridium acetobutylicum and Desulfovibrio vulgaris. Bioresour Technol 221:526–533. https://doi.org/10.1016/j.biortech.2016.09.072
Beaton SE, Evans RM, Finney AJ et al (2018) The structure of hydrogenase-2 from Escherichia coli: implications for H2-driven proton pumping. Biochem J 475:1353–1370. https://doi.org/10.1042/BCJ20180053
Benoit SL, Maier RJ, Sawers RG, Greening C (2020) Molecular hydrogen metabolism: a widespread trait of pathogenic bacteria and protists. Microbiol Mol Biol Rev 84:e00092-19. https://doi.org/10.1128/MMBR.00092-19
Berman HM, Westbrook J, Feng Z et al (2000) The Protein Data Bank. Nucleic Acids Res 28:235–242. https://doi.org/10.1093/nar/28.1.235
Berney M, Greening C, Conrad R et al (2014a) An obligately aerobic soil bacterium activates fermentative hydrogen production to survive reductive stress during hypoxia. Proc Natl Acad Sci 111:11479–11484. https://doi.org/10.1073/pnas.1407034111
Berney M, Greening C, Hards K et al (2014b) Three different [NiFe] hydrogenases confer metabolic flexibility in the obligate aerobe Mycobacterium smegmatis. Environ Microbiol 16:318–330. https://doi.org/10.1111/1462-2920.12320
Bernstein FC, Koetzle TF, Williams GJ et al (1977) The Protein Data Bank: a computer-based archival file for macromolecular structures. J Mol Biol 112. https://doi.org/10.1016/s0022-2836(77)80200-3
Bowman L, Flanagan L, Fyfe PK et al (2014) How the structure of the large subunit controls function in an oxygen-tolerant [NiFe]-hydrogenase. Biochem J 458:449–458. https://doi.org/10.1042/BJ20131520
Brink LES, Tramper J, Luyben KCAM, Van 't Riet K (1988) Biocatalysis in organic media. Enzym Microb Technol 10:736–743. https://doi.org/10.1016/0141-0229(88)90118-4
Brugna-Guiral M, Tron P, Nitschke W et al (2003) [NiFe] hydrogenases from the hyperthermophilic bacterium Aquifex aeolicus: properties, function, and phylogenetics. Extremophiles 7:145–157. https://doi.org/10.1007/s00792-002-0306-3
Cai G, Jin B, Monis P, Saint C (2011) Metabolic flux network and analysis of fermentative hydrogen production. Biotechnol Adv 29:375–387. https://doi.org/10.1016/j.biotechadv.2011.02.001
Calusinska M, Happe T, Joris B, Wilmotte A (2010) The surprising diversity of clostridial hydrogenases: a comparative genomic perspective. Microbiology 156:1575–1588. https://doi.org/10.1099/mic.0.032771-0
Chen Y, Yin Y, Wang J (2021) Recent advance in inhibition of dark fermentative hydrogen production. Int J Hydrog Energy 46:5053–5073. https://doi.org/10.1016/j.ijhydene.2020.11.096
Christl S, Murgatroyd P, Gibson G, Cummings J (1992) Production, metabolism, and excretion of hydrogen in the large intestine. Gastroenterology 102:1269–1277. https://doi.org/10.1016/0016-5085(92)70022-4
Cornish AJ, Green R, Gärtner K et al (2015) Characterization of hydrogen metabolism in the multicellular green alga Volvox carteri. PLoS One 10:e0125324. https://doi.org/10.1371/journal.pone.0125324
Das D (2001) Hydrogen production by biological processes: a survey of literature. Int J Hydrog Energy 26:13–28. https://doi.org/10.1016/S0360-3199(00)00058-6
Das D (2009) Advances in biohydrogen production processes: An approach towards commercialization. Int J Hydrog Energy 34:7349–7357. https://doi.org/10.1016/j.ijhydene.2008.12.013

Das D, Khanna N, Dasgupta CN (2014) Biohydrogen production: fundamentals and technology advances. CRC, Boca Raton

da Silva SM, Venceslau SS, Fernandes CLV et al (2008) Hydrogen as an energy source for the human pathogen Bilophila wadsworthia. Antonie Van Leeuwenhoek 93:381–390. https://doi.org/10.1007/s10482-007-9215-x

de Vries W, van Berchum H, Stouthamer AH (1984) Localization of hydrogenase and nitrate reductase in Campylobacter sputorum subsp. bubulus. Antonie Van Leeuwenhoek 50:63–73. https://doi.org/10.1007/BF00404908

Dincer I (2012) Green methods for hydrogen production. Int J Hydrog Energy 37:1954–1971. https://doi.org/10.1016/j.ijhydene.2011.03.173

Doğan-Güner EM, Koku H (2022) Analysis of the carbon metabolism of Rhodopseudomonas palustris for biohydrogen production. Biotech Stud 31:1–9. https://doi.org/10.38042/biotechstudies.1062792

Du Y, Zou W, Zhang K et al (2020) Advances and applications of Clostridium co-culture systems in biotechnology. Front Microbiol 11:560223. https://doi.org/10.3389/fmicb.2020.560223

Dzulkarnain ELN, Audu JO, Wan Dagang WRZ, Abdul-Wahab MF (2022) Microbiomes of biohydrogen production from dark fermentation of industrial wastes: current trends, advanced tools and future outlook. Bioresour Bioprocess 9:16. https://doi.org/10.1186/s40643-022-00504-8

El-Dalatony MM, Zheng Y, Ji M-K et al (2020) Metabolic pathways for microalgal biohydrogen production: Current progress and future prospectives. Bioresour Technol 318:124253. https://doi.org/10.1016/j.biortech.2020.124253

ETEnergyWorld (2022) India's green hydrogen policy: unprecedented growth needed to achieve 2030 targets. ET EnergyWorld. https://energy.economictimes.indiatimes.com/news/renewable/indias-green-hydrogen-policy-unprecedented-growth-needed-to-achieve-2030-targets/90154361. Accessed 7 June 2022

Fakhimi N, Gonzalez-Ballester D, Fernández E et al (2020) Algae-bacteria consortia as a strategy to enhance H2 production. Cell 9:1353. https://doi.org/10.3390/cells9061353

Fontaine L, Meynial-Salles I, Girbal L et al (2002) Molecular characterization and transcriptional analysis of *adhE2*, the gene encoding the NADH-dependent aldehyde/alcohol dehydrogenase responsible for butanol production in alcohologenic cultures of *Clostridium acetobutylicum* ATCC 824. J Bacteriol 184:821–830. https://doi.org/10.1128/JB.184.3.821-830.2002

Fox JD, Kerby RL, Roberts GP, Ludden PW (1996) Characterization of the CO-induced, CO-tolerant hydrogenase from Rhodospirillum rubrum and the gene encoding the large subunit of the enzyme. J Bacteriol 178:1515–1524. https://doi.org/10.1128/jb.178.6.1515-1524.1996

François E, Dumas C, Gougeon RD et al (2021) Unexpected high production of biohydrogen from the endogenous fermentation of grape must deposits. Bioresour Technol 320:124334. https://doi.org/10.1016/j.biortech.2020.124334

Fu N, Peiris P, Markham J, Bavor J (2009) A novel co-culture process with Zymomonas mobilis and Pichia stipitis for efficient ethanol production on glucose/xylose mixtures. Enzym Microb Technol 45:210–217. https://doi.org/10.1016/j.enzmictec.2009.04.006

Galadima A, Muraza O (2015) From synthesis gas production to methanol synthesis and potential upgrade to gasoline range hydrocarbons: A review. J Nat Gas Sci Eng 25:303–316. https://doi.org/10.1016/j.jngse.2015.05.012

Gong C, Singh A, Singh P, Singh A (2021) Anaerobic digestion of agri-food wastes for generating biofuels. Indian J Microbiol 61:427–440. https://doi.org/10.1007/s12088-021-00977-9

González-Pajuelo M, Meynial-Salles I, Mendes F et al (2005) Metabolic engineering of Clostridium acetobutylicum for the industrial production of 1,3-propanediol from glycerol. Metab Eng 7:329–336. https://doi.org/10.1016/j.ymben.2005.06.001

Gorwa MF, Croux C, Soucaille P (1996) Molecular characterization and transcriptional analysis of the putative hydrogenase gene of Clostridium acetobutylicum ATCC 824. J Bacteriol 178:2668–2675. https://doi.org/10.1128/jb.178.9.2668-2675.1996

Greening C, Berney M, Hards K et al (2014) A soil actinobacterium scavenges atmospheric H2 using two membrane-associated, oxygen-dependent [NiFe] hydrogenases. Proc Natl Acad Sci USA 111:4257–4261. https://doi.org/10.1073/pnas.1320586111

Guo W-Q, Ren N-Q, Wang X-J et al (2008) Biohydrogen production from ethanol-type fermentation of molasses in an expanded granular sludge bed (EGSB) reactor. Int J Hydrog Energy 33:4981–4988. https://doi.org/10.1016/j.ijhydene.2008.05.033

Harris MA, Reddy CA (1977) Hydrogenase activity and the H2-fumarate electron transport system in Bacteroides fragilis. J Bacteriol 131:922–928. https://doi.org/10.1128/jb.131.3.922-928.1977

Harris SC, Devendran S, Méndez-García C et al (2018) Bile acid oxidation by Eggerthella lenta strains C592 and DSM 2243T. Gut Microbes 9:523–539. https://doi.org/10.1080/19490976.2018.1458180

Hu J (2021) Comparisons of biohydrogen production technologies and processes. In: Zhang Q, He C, Ren J, Goodsite ME (eds) Waste to renewable biohydrogen. Elsevier, London, pp 71–107

Hughes ER, Winter MG, Alves da Silva L et al (2021) Reshaping of bacterial molecular hydrogen metabolism contributes to the outgrowth of commensal E. coli during gut inflammation. eLife 10:e58609. https://doi.org/10.7554/eLife.58609

Köpke M, Straub M, Dürre P (2013) Clostridium difficile is an autotrophic bacterial pathogen. PLoS One 8:e62157. https://doi.org/10.1371/journal.pone.0062157

Kothari R (2017) A critical review on factors influencing fermentative hydrogen production. Front Biosci 22:1195–1220. https://doi.org/10.2741/4542

Kumar K, Roy K, Moholkar VS (2021) Mechanistic investigations in sonoenzymatic synthesis of n-butyl levulinate. Process Biochem 111:147–158. https://doi.org/10.1016/j.procbio.2021.09.005

Kumar K, Mohammad AS, Pragati et al (2022a) Computational investigations in inhibition of alcohol/aldehyde dehydrogenase in lignocellulosic hydrolysates. bioRxiv 2022.11.19.517192; https://doi.org/10.1101/2022.11.19.517192

Kumar K, Shah H, Moholkar VS (2022b) Genetic algorithm for optimization of fermentation processes of various enzyme productions. In: Optimization of sustainable enzymes production. Chapman and Hall/CRC, Boca Raton

Kumar K, Barbora L, Moholkar VS (2023a) Genomic insights into clostridia in bioenergy production: Comparison of metabolic capabilities and evolutionary relationships. Biotechnology and Bioengineering, 1–16. https://doi.org/10.1002/bit.28610

Kumar K, Patro P, Raut U et al (2023b) Elucidating the molecular mechanism of ultrasound-enhanced lipase-catalyzed biodiesel synthesis: a computational study. Biomass Conv Bioref. https://doi.org/10.1007/s13399-023-04742-4

Kumar K, Jadhav SM, Moholkar VS (2023c) Acetone-Butanol-Ethanol (ABE) fermentation with Clostridial Co-cultures for Enhanced Biobutanol Production. bioRxiv 2023.12.08.570763. https://doi.org/10.1101/2023.12.08.570763

Lamichhane-Khadka R, Benoit SL, Miller-Parks EF, Maier RJ (2015) Host hydrogen rather than that produced by the pathogen is important for Salmonella enterica serovar typhimurium virulence. Infect Immun 83:311–316. https://doi.org/10.1128/IAI.02611-14

Leskovac V (2003) Comprehensive enzyme kinetics. Springer, New York

Li Z, Gu J, Ding J et al (2021) Molecular mechanism of ethanol-H2 co-production fermentation in anaerobic acidogenesis: Challenges and perspectives. Biotechnol Adv 46:107679. https://doi.org/10.1016/j.biotechadv.2020.107679

Liu ZL, Moon J (2009) A novel NADPH-dependent aldehyde reductase gene from Saccharomyces cerevisiae NRRL Y-12632 involved in the detoxification of aldehyde inhibitors derived from lignocellulosic biomass conversion. Gene 446:1–10. https://doi.org/10.1016/j.gene.2009.06.018

Liu D, Sun Y, Li Y, Lu Y (2017) Perturbation of formate pathway and NADH pathway acting on the biohydrogen production. Sci Rep 7:9587. https://doi.org/10.1038/s41598-017-10191-7

Lukey MJ, Parkin A, Roessler MM et al (2010) How Escherichia coli is equipped to oxidize hydrogen under different redox conditions 2. J Biol Chem 285:3928–3938. https://doi.org/10.1074/jbc.M109.067751

Maier RJ, Fu C, Gilbert J et al (1996) Hydrogen uptake hydrogenase in Helicobacter pylori. FEMS Microbiol Lett 141:71–76. https://doi.org/10.1111/j.1574-6968.1996.tb08365.x
Maier L, Vyas R, Cordova CD et al (2013) Microbiota-derived hydrogen fuels Salmonella typhimurium invasion of the gut ecosystem. Cell Host Microbe 14:641–651. https://doi.org/10.1016/j.chom.2013.11.002
Malani RS, Umriwad SB, Kumar K et al (2019) Ultrasound-assisted enzymatic biodiesel production using blended feedstock of non-edible oils: Kinetic analysis. Energy Convers Manag 188:142–150. https://doi.org/10.1016/j.enconman.2019.03.052
Mathews J, Wang G (2009) Metabolic pathway engineering for enhanced biohydrogen production. Int J Hydrog Energy 34:7404–7416. https://doi.org/10.1016/j.ijhydene.2009.05.078
McDowall JS, Murphy BJ, Haumann M et al (2014) Bacterial formate hydrogenlyase complex. Proc Natl Acad Sci USA 111:E3948–E3956. https://doi.org/10.1073/pnas.1407927111
Menon NK, Chatelus CY, Dervartanian M et al (1994) Cloning, sequencing, and mutational analysis of the hyb operon encoding Escherichia coli hydrogenase 2. J Bacteriol 176:4416–4423. https://doi.org/10.1128/jb.176.14.4416-4423.1994
Meredith LK, Rao D, Bosak T et al (2014) Consumption of atmospheric hydrogen during the life cycle of soil-dwelling actinobacteria. Environ Microbiol Rep 6:226–238. https://doi.org/10.1111/1758-2229.12116
Meyer J (2007) [FeFe] hydrogenases and their evolution: a genomic perspective. Cell Mol Life Sci 64:1063. https://doi.org/10.1007/s00018-007-6477-4
Mishra J, Khurana S, Kumar N et al (2004) Molecular cloning, characterization, and overexpression of a novel [Fe]-hydrogenase isolated from a high rate of hydrogen producing Enterobacter cloacae IIT-BT 08. Biochem Biophys Res Commun 324:679–685. https://doi.org/10.1016/j.bbrc.2004.09.108
Miyake J, Matsunaga T, Pietro AS (2001) Biohydrogen II. Elsevier, Oxford
Morimoto K, Kimura T, Sakka K, Ohmiya K (2005) Overexpression of a hydrogenase gene in Clostridium paraputrificum to enhance hydrogen gas production. FEMS Microbiol Lett 246:229–234. https://doi.org/10.1016/j.femsle.2005.04.014
Morra S, Mongili B, Maurelli S et al (2016) Isolation and characterization of a new [FeFe]-hydrogenase from Clostridium perfringens. Biotechnol Appl Biochem 63:305–311. https://doi.org/10.1002/bab.1382
Myers MR, King GMY (2016) (2016) Isolation and characterization of Acidobacterium ailaaui sp. nov., a novel member of Acidobacteria subdivision 1, from a geothermally heated Hawaiian microbial mat. Int J Syst Evol Microbiol 66:5328–5335. https://doi.org/10.1099/ijsem.0.001516
Nelson WC, Stegen JC (2015) The reduced genomes of Parcubacteria (OD1) contain signatures of a symbiotic lifestyle. Front Microbiol 6. https://doi.org/10.3389/fmicb.2015.00713
Nixon JEJ, Field J, McArthur AG et al (2003) Iron-dependent hydrogenases of Entamoeba histolytica and Giardia lamblia: activity of the recombinant entamoebic enzyme and evidence for lateral gene transfer. Biol Bull 204:1–9. https://doi.org/10.2307/1543490
Oh Y-K, Lee Y, Choi E-H, Kim M-S (2008) Bioelectrocatalytic hydrogen production using Thiocapsa roseopersicina hydrogenase in two-compartment fuel cells. Int J Hydrog Energy 33:5218–5223. https://doi.org/10.1016/j.ijhydene.2008.05.015
Olson JW, Maier RJ (2002) Molecular hydrogen as an energy source for Helicobacter pylori. Science 298:1788–1790. https://doi.org/10.1126/science.1077123
Ooshima H, Takakuwa S, Katsuda T et al (1998) Production of hydrogen by a hydrogenase-deficient mutant of Rhodobacter capsulatus. J Ferment Bioeng 85:470–475. https://doi.org/10.1016/S0922-338X(98)80064-0
Pandey A, Chang J-S, Hallenbeck PC, Larroche C (2013) Biohydrogen, 1st edn. Elsevier, Amsterdam
Park SS, Decicco BTY (1974) Autotrophic growth with hydrogen of Mycobacterium gordonae and another scotochromogenic mycobacterium. Int J Syst Evol Microbiol 24:338–345. https://doi.org/10.1099/00207713-24-3-338

Pugazhendhi A, Kumar G, Sivagurunathan P (2019) Microbiome involved in anaerobic hydrogen producing granules: A mini review. Biotechnol Rep 21:e00301. https://doi.org/10.1016/j.btre.2018.e00301
Salimi F, Zhuang K, Mahadevan R (2010) Genome-scale metabolic modeling of a clostridial co-culture for consolidated bioprocessing. Biotechnol J 5:726–738. https://doi.org/10.1002/biot.201000159
Sarma S, Dubey VK, Moholkar VS (2016) Kinetic and thermodynamic analysis (with statistical optimization) of hydrogen production from crude glycerol using Clostridium pasteurianum. Int J Hydrog Energy 41:19972–19989. https://doi.org/10.1016/j.ijhydene.2016.08.204
Sarma S, Anand A, Dubey VK, Moholkar VS (2017) Metabolic flux network analysis of hydrogen production from crude glycerol by Clostridium pasteurianum. Bioresour Technol 242:169–177. https://doi.org/10.1016/j.biortech.2017.03.168
Sarma S, Ortega D, Minton NP et al (2019) Homologous overexpression of hydrogenase and glycerol dehydrogenase in Clostridium pasteurianum to enhance hydrogen production from crude glycerol. Bioresour Technol 284:168–177. https://doi.org/10.1016/j.biortech.2019.03.074
Schuchmann K, Müller V (2012) A bacterial electron-bifurcating hydrogenase. J Biol Chem 287:31165–31171. https://doi.org/10.1074/jbc.M112.395038
Schuchmann K, Müller V (2013) Direct and reversible hydrogenation of CO_2 to formate by a bacterial carbon dioxide reductase. Science 342:1382–1385. https://doi.org/10.1126/science.1244758
Schuchmann K, Chowdhury NP, Müller V (2018) Complex multimeric [FeFe] hydrogenases: biochemistry, physiology and new opportunities for the hydrogen economy. Front Microbiol 9
Seol E, Kim S, Raj SM, Park S (2008) Comparison of hydrogen-production capability of four different Enterobacteriaceae strains under growing and non-growing conditions. Int J Hydrog Energy 33:5169–5175. https://doi.org/10.1016/j.ijhydene.2008.05.007
Shima S, Thauer RK (2007) A third type of hydrogenase catalyzing H2 activation. Chem Rec 7:37–46. https://doi.org/10.1002/tcr.20111
Singh K (2022) Explained: What is the government's green hydrogen policy? The Indian Express. https://indianexpress.com/article/explained/explained-green-hydrogen-policy-7779679/. Accessed 7 June 2022
Singh L, Wahid ZA (2015) Methods for enhancing bio-hydrogen production from biological process: A review. J Ind Eng Chem 21:70–80. https://doi.org/10.1016/j.jiec.2014.05.035
Singh R, Kumar M, Mittal A, Mehta PK (2016) Microbial enzymes: industrial progress in 21st century. 3 Biotech 6:174. https://doi.org/10.1007/s13205-016-0485-8
Singh N, Kumar K, Goyal A, Moholkar VS (2022) Ultrasound-assisted biodiesel synthesis by in–situ transesterification of microalgal biomass: Optimization and kinetic analysis. Algal Res 61:102582. https://doi.org/10.1016/j.algal.2021.102582
Søndergaard D, Pedersen CNS, Greening C (2016) HydDB: A web tool for hydrogenase classification and analysis. Sci Rep 6:34212. https://doi.org/10.1038/srep34212
Takai K, Nakamura K, Suzuki K et al (2006) Ultramafics-Hydrothermalism-Hydrogenesis-HyperSLiME (UltraH3) linkage: a key insight into early microbial ecosystem in the Archean deep-sea hydrothermal systems. Paleontological Res 10:269–282. https://doi.org/10.2517/prpsj.10.269
Tian F, Toon OB, Pavlov AA, De Sterck H (2005) A hydrogen-rich early earth atmosphere. Science 308:1014–1017. https://doi.org/10.1126/science.1106983
Varjani S, Pandey A, Bhaskar T et al (2021) Biomass, biofuels, biochemicals: circular bioeconomy: technologies for biofuels and biochemicals. Elsevier, Amsterdam
Vignais PM, Billoud B (2007) Occurrence, classification, and biological function of hydrogenases: an overview. Chem Rev 107:4206–4272. https://doi.org/10.1021/cr050196r
Vignais PM, Billoud B, Meyer J (2001) Classification and phylogeny of hydrogenases. FEMS Microbiol Rev 25:455–501. https://doi.org/10.1111/j.1574-6976.2001.tb00587.x

Volbeda A, Amara P, Darnault C et al (2012) X-ray crystallographic and computational studies of the O2-tolerant [NiFe]-hydrogenase 1 from Escherichia coli. Proc Natl Acad Sci USA 109:5305–5310. https://doi.org/10.1073/pnas.1119806109

Vyas D (1995) Nitrogen fixation and hydrogen uptake in four cyanobacteria. Int J Hydrog Energy 20:163–168. https://doi.org/10.1016/0360-3199(94)E0008-M

Weerakoon DR, Borden NJ, Goodson CM et al (2009) The role of respiratory donor enzymes in Campylobacter jejuni host colonization and physiology. Microb Pathog 47:8–15. https://doi.org/10.1016/j.micpath.2009.04.009

Wolf PG, Biswas A, Morales SE et al (2016) H2 metabolism is widespread and diverse among human colonic microbes. Gut Microbes 7:235–245. https://doi.org/10.1080/19490976.2016.1182288

Xie B, Cheng J, Zhou J et al (2008) Cogeneration of hydrogen and methane from glucose to improve energy conversion efficiency. Int J Hydrog Energy 33:5006–5011. https://doi.org/10.1016/j.ijhydene.2008.07.048

Yan S, Dong D (2018) Improvement of caproic acid production in a Clostridium kluyveri H068 and Methanogen 166 co-culture fermentation system. AMB Express 8:175. https://doi.org/10.1186/s13568-018-0705-1

You L, Liu H, Blankenship RE, Tang YJ (2015) Using photosystem I as a reporter protein for 13C analysis in a coculture containing cyanobacterium and a heterotrophic bacterium. Anal Biochem 477:86–88. https://doi.org/10.1016/j.ab.2014.12.005

Younas M, Shafique S, Hafeez A et al (2022) An overview of hydrogen production: current status, potential, and challenges. Fuel 316:123317. https://doi.org/10.1016/j.fuel.2022.123317

Young JD (2014) 13C metabolic flux analysis of recombinant expression hosts. Curr Opin Biotechnol 30:238–245. https://doi.org/10.1016/j.copbio.2014.10.004

Zbell AL, Maier RJ (2009) Role of the Hya hydrogenase in recycling of anaerobically produced H2 in Salmonella enterica serovar typhimurium. Appl Environ Microbiol 75:1456–1459. https://doi.org/10.1128/AEM.02064-08

Chapter 3
Biohydrogen: Microbiology and Microbial Dynamics in Dark Fermentation

Reema Kumar, Satinder Kaur Brar, and Guneet Kaur

Abstract There is a global attention toward production of clean and green energy to reduce the dependency on fossil fuel-derived sources and its subsequent environmental impact. Biohydrogen, thus offers an attractive potential source of energy with both economic and environmental advantages. Hydrogen is produced biologically through light-driven and light-independent pathways. At present, light-independent process of dark fermentation appears more favorable due to low operational costs and high hydrogen yield, and thus, is advancing in research. Production involves a complex interplay of genetically and functionally diverse mixed microbial cultures with a multitude of organic substrates. The dominant types of microorganisms involved in these metabolic reactions function in mesophilic range and are both facultative and strict anaerobes, belonging to genera *Clostridium* and *Enterobacter*. However, with varying environmental and operational conditions, a shift occurs in the microbial community structure that further affects the yield of hydrogen produced. By understanding the syntrophic co-metabolism of hydrogen-producers and hydrogen-consumers, the bioreactor performance can be modified for scaling-up biohydrogen production in continuous systems. With modern molecular techniques and metagenomic studies, it is now becoming possible to study these reactor microbiomes to extract information about essential enzymes and respective microorganisms that may increase the hydrogen yield.

Keywords Clean energy · Biohydrogen · Dark fermentation · Bioreactors · Mixed-microbe consortium

---

R. Kumar · S. K. Brar · G. Kaur (✉)
Department of Civil Engineering, Lassonde School of Engineering, York University, Toronto, ON, Canada
e-mail: reemak15@yorku.ca; satinder.brar@lassonde.yorku.ca; guneetk@yorku.ca

C. R. Soccol et al. (eds.), *Biohydrogen - Advances and Processes*, Biofuel and Biorefinery Technologies 13, https://doi.org/10.1007/978-3-031-49818-3_3

3.1 Introduction

Ever since industrialization, fossil fuels have remained a key source of energy for the world. Along with being non-renewable resources, they also emit greenhouse gases upon utilization. A major constituent of these greenhouse gases includes CO_2. A tremendous increase is being observed in its emission which gravely impacts life and associated processes on earth, in addition to climate changes. The increasing human population has further burdened the resource demand leading to its depletion and aggravating pollution. Hence, the focus is being shifted to finding environment-friendly and renewable energy sources. Hydrogen shows huge potential as an alternative clean energy source, that is both renewable and green. It is essentially a carbon-neutral carrier as compared to fossil fuels with a 2.75 times higher yield of energy (Singh and Mahapatra 2019).

Different processes for hydrogen production are shown in Fig. 3.1. Conventional processes of hydrogen production include using natural gas by steam reforming, coal gasification, and water electrolysis. However, these are non-renewable and non-sustainable methods of hydrogen production (Das and Veziroğlu 2001). Biological processes offer a lesser energy-intensive alternative as these operate at ambient operational conditions and utilize renewable feedstocks such as: biomass, wastewater, agricultural waste and byproducts, lignocellulosic waste, food waste, sewage sludge, livestock effluent, etc. (Wang and Wan 2009). Hydrogen production is inherently observed in several organisms such as bacteria, green algae, blue-green algae, and higher plants (Ghirardi et al. 2007). Microbes exist in nature as dynamically changing communities with syntrophic interactions. Photosynthetic processes generate both oxygen and hydrogen at ambient temperature and pressure conditions by using practically non-exhaustive resources like sunlight and water. Fermentation processes utilize waste as a substrate to generate a range of products, including hydrogen. Presently a major problem with these biological processes is low hydrogen yield and higher production cost for commercial usage presently. However,

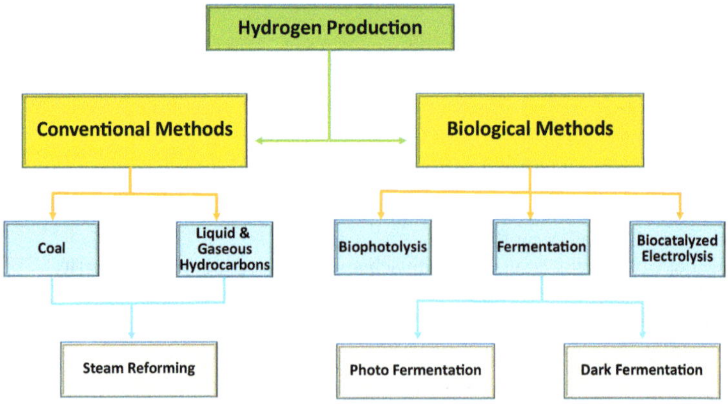

Fig. 3.1 Conventional and biological processes for hydrogen production

there are enhancement techniques to enhance hydrogen production rate and yield. These are physical, chemical, and thermochemical methods (Sarangi and Nanda 2020). With advancing molecular techniques, microbial community studies enrich the knowledge about hydrogen-producing microbes in reactors and their optimum operational requirements.

3.2 Methods of Biohydrogen Production

At present, there are five methods to produce biohydrogen: (1) biophotolysis (direct and indirect), (2) photofermentation, (3) dark fermentation, (4) two-stage integration of photo- and dark fermentation, and (5) bio catalyzed electrolysis. These biological processes involve the fundamental role of a hydrogen-producing enzyme catalyzing the following reaction:

$$2H^{+} + 2e^{-} \leftrightarrow H_2$$

These enzymes have metallo-clusters as active site. Four key enzymes known for carrying out these reactions are nitrogenase, Fe-hydrogenase, FeFe-hydrogenase, and NiFe-hydrogenase (Morra 2022; Hallenbeck and Benemann 2002). Nitrogenase enzyme is found in prokaryotic organisms for producing nitrogen and some amount of hydrogen under anaerobic conditions. It comprises of a two-part protein system that utilizes Mg-ATP and electrons derived from reduced flavodoxin or ferredoxin to reduce different kinds of substrates. However, nitrogenase catalyzed reaction yields a low turnover of hydrogen because of high energy input to drive the reaction. Hydrogenases are found in a variety of organisms like anaerobes, fungi, green algae, and protists (Dasgupta et al. 2010). Fe-hydrogenase, first isolated from *Clostridium pasteurianum* is a cytosolic enzyme that normally functions during fermentation reactions in anaerobic bacteria to remove reducing agents. This enzyme has a Fe-S center where one Fe atom complexes with CO and CN, and proton reduction follows an iron-hydride intermediate (Huang et al. 2019). The NiFe-hydrogenase is a heterodimer protein with a small and large subunit forming a unique metallocenter. It functions through reversible conversion and deriving electrons and protons from molecular hydrogen to be used through quinone pool to reduce NADP. NiFe-hydrogenase based hydrogen production yields higher amount of hydrogen as compared to Fe Hydrogenase/Fe-Fe hydrogenase (Ogata et al. 2016).

Biophotolysis

Direct biophotolysis is similar to the photosynthesis process found in plants and algae where solar energy is used to convert water into hydrogen and oxygen. Absorption of protons from direct sunlight leads to transfer of electrons to hydrogenases or nitrogenases for biohydrogen production. This is an attractive process due to easily available resources, primarily water. However, it poses inherent difficulty due to generation of possibly hazardous gas mixtures of oxygen and hydrogen along with inhibiting hydrogenase activity (Hallenbeck et al. 2009). This key problem gets bypassed in indirect biophotolysis in cyanobacteria by separation of oxygen and hydrogen gas evolution stages. Also, the hydrogen production takes place in heterocysts by nitrogenase which exhibits comparatively lower sensitivity to oxygen (Levin et al. 2004).

Photofermentation
Photofermentation is also a light-driven reaction to produce hydrogen by photosynthetic bacteria using organic acids as electron-donors under nitrogen-deficient condition (Levin et al. 2004). The released electrons are transported through electron carriers, and simultaneously pumps protons through the intracellular membrane. With a proton gradient in place, ATP synthase generates ATP which further transports electrons to ferredoxin, and thereby reducing protons to hydrogen (Reith et al. 2003).

Dark Fermentation
Dark fermentation reaction for hydrogen production is performed under anaerobic conditions using carbohydrate-rich substrates, majorly through catabolism of pyruvate. Acetyl CoA gets generated from pyruvate, from which ATP is derived, followed by hydrogen production from either formate or reduced ferredoxin. The enzymes involved are formate lyase and ferredoxin oxidoreductase, respectively (Manish and Banerjee 2008). A schematic diagram shows the process in dark fermentation in Fig. 3.2. Even though this process is comparatively cheaper and simpler than other processes, its drawback is the incomplete decomposition of the substrates as it is thermodynamically unfeasible and thus gives low hydrogen yields (Hallenbeck and Benemann 2002). But studies are being undertaken to enhance this process. Since hydrogen is produced as an intermediate metabolite to be used as an electron donor during methanogenesis, its production can be favored by arresting methanogenesis. According to studies, this could be achieved by regulating certain operational parameters of the bioreactors such as low pH, shorter hydraulic retention time (HRT), inhibiting hydrogen consumers using heat or chemical inhibitors (Chu et al. 2003; Zagrodnik and Laniecki 2015).

Two-Stage Integration of Photo- and Dark Fermentation
Integration of dark and photo-fermentation results in maximum yield of hydrogen. The energy barrier that prevents complete decomposition of substrates in dark fermentation can be overcome by combining photo-fermentation. The photosynthetic bacteria can further oxidize the products formed in dark fermentation which majorly include volatile fatty acids (Levin et al. 2004; Kim et al. 2006).

Biocatalyzed Electrolysis or Bioelectrohydrogenesis
Microbial aided electrolysis cells (MEC) or bioelectrochemically assisted microbial reactor (BEAMR) are used to perform the electrohydrogenesis, converting biodegradable material into hydrogen. It is another hybrid method for biohydrogen production that involves using electrical energy as the external energy instead of solar energy and is based on a microbial fuel cell concept coupled with anode-cathode reactions. Addition of electrical potential promotes reduction of protons to

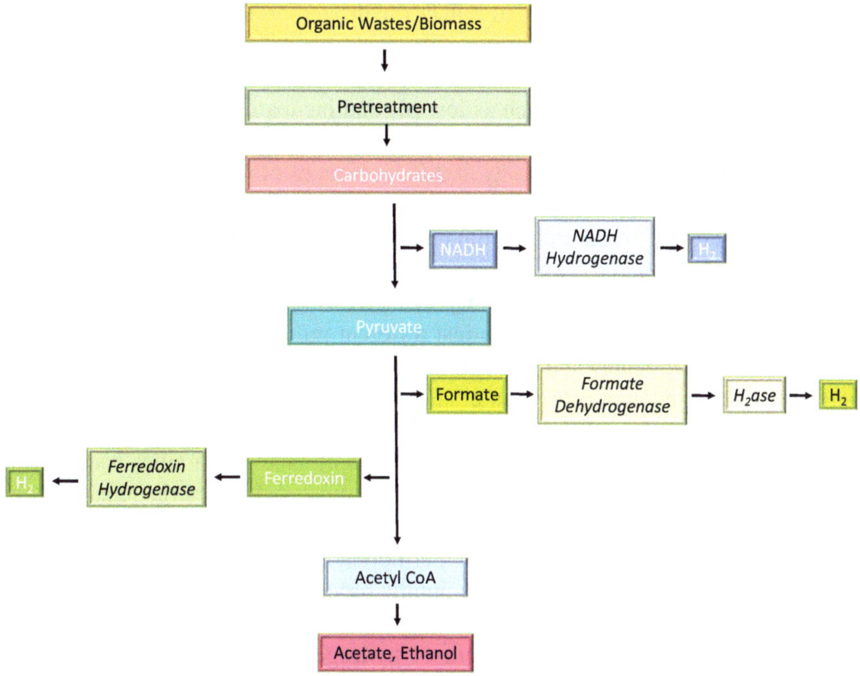

Fig. 3.2 Schematic diagram for hydrogen generation through various steps in dark fermentation of organic wastes or biomass

hydrogen by electrochemically active microorganisms (Parvanova-Mancheva et al. 2022). The electrogens degrade organic material and release electrons which then combine at the cathode. The applied electrical energy to the MEC leads to hydrogen production at the cathode when hydrogen ion reacts with electrons (Holladay et al. 2009).

3.3 Microorganisms Involved in Biohydrogen Production

Different biohydrogen production methods employ pure and mixed microbial cultures. Since fermentation technologies take precedence over other techniques, the microbial diversity in fermentative reactions has been extensively studied. Pure cultures have been mostly used for batch mode fermentation for study purposes. Consortia of mixed microbes are better suited for fermentation to produce biohydrogen as they have a wider range of optimum environmental conditions and eliminate the necessity of sterilized conditions. A mixed microbial consortium is also capable of decomposing different waste kinds than a pure culture as no single microbe possesses all possible hydrolytic machinery. However, along with hydrogen-producing microbes, a mixed culture would also comprise of hydrogen-consuming microbes that reduces hydrogen production yield by generation of other byproducts. Functioning of these hydrogen-consuming microbes, including methanogens, can be prevented by managing operational parameters and certain pretreatment methods such as low pH and incubation at high temperatures. The external parameters, nonetheless, require monitoring and optimization according to the shift of microbial diversity at different stages of reaction. For instance, in a study by Zhang et al. (2006), it was observed that hydraulic retention time (HRT) affects the microbial community in the reactor and as a result, affects the hydrogen yield. pH values around 5.5 promotes accumulation of butyric acid and abundance of bacteria from *Clostridiaceae* family that gives high hydrogen production rate (Alexandropoulou et al. 2018). With just different influent pH, the microbial communities develop uniquely and perform different fermentation types such as propionate, butyrate, or ethanol type (Ren et al. 2007). The type of substrate impacts the bacterial community structure where pentose-based substrates have high production rate compared to more complex sugars where major bacterial families become less active (Chatellard et al. 2016). So far, the isolated mesophilic bacteria mainly belong to genera *Enterobactericeae* and *Clostridiaceae*, while thermophilic bacteria belong to genus *Thermoanaerobacterium*. The thermophilic fermentation process is known to give higher hydrogen yield than mesophilic process where lactic acid-producing bacteria often become dominant and inhibit the activity of hydrogen-producing bacteria (Chen et al. 2022). Other genera of hydrogen producers include *Bacillus*, *Pseudomonas*, *Vibrio*, and *Aeromonas* (Hung et al. 2011). Some examples of different microbes involved in biohydrogen production are mentioned in Table 3.1.

3.4 Biohydrogen Production in Batch and Continuous Systems

The effectiveness of biohydrogen production process relies on reactor design, configuration, and operational conditions. Bioreactor's performance measurement can be assessed by its ability to function at lower hydraulic retention time (HRT) and reduced cell washout to achieve high production rate (Banu et al. 2021). Batch systems for biohydrogen production are majorly suitable for experimental studies or research purposes at laboratory scale to test effects of different substrate, inoculum, environmental conditions, etc., or in reactions with slow kinetics. Reactor needs to be operated continuously for commercial level and continuous biohydrogen production. The mode widely used bioreactor is continuous stirred-tank reactor (CSTR) (Li et al. 2010; Kim et al. 2011; Chang and Lin 2004).

Table 3.1 Types and examples of microorganisms functioning in different process of biohydrogen production

Process	Type of microorganism	Example of microorganisms	Reference
Biophotolysis	Cyanobacteria, green microalgae	*Anabaena, Gloebacter, Synechococcus, Aphanacapsa montana*	Yu and Takahashi (2007); Borodin et al. (2000)
Photo-fermentation	Green sulfur bacteria, purple sulfur bacteria, purple non-sulfur bacteria	*Chlorobium, Halobacterium, Rhodobater, Rhodopseudomonas palustris, Rhodospirillum rubrum*	Basak et al. (2014); Sağır and Hallenbeck (2019)
Dark fermentation	Anaerobic bacteria	*Alcaligenes eutrophus, Clostridium butyricum, Clostridium thermocellum, Clostridium tyrobutyricum, Clostridium bifermentans, Clostridium pasteurianum, Enterobacter aerogenes, Rumicococcus albus, Thermotoga neapolitana*	Ntaikou et al. (2008); Nguyen et al. (2010); Ren et al. (2009); Geng et al. (2010); Levin et al. (2006); Pattra et al. (2008)
Microbial electrolysis cell	Electrogens	*Ochrobactrum anthropic, Firmicutes thermincola, Klebsiella pneuminiae, Desulfobulbus propionnicus, Geobacter, Shewanella*	Zhao and Ci (2019); Chookaew et al. (2014)

3.5 Enhancement of Biohydrogen Production

Biological processes for hydrogen production fall under two major categories: light-dependent and light-independent pathways. The light-independent pathway of dark or anaerobic fermentation is more advantageous over other biological pathways. These advantages consist of no requirement of external light energy, anaerobic conditions preventing oxygen limitation problems, cost-effective, high cell-growth rate, use of organic waste as substrate, etc. However, the limitations to this process are low hydrogen yield, low production rate, and reactor instability to achieve commercial levels of continuous hydrogen production. These limitations arise because of the following major reasons: thermodynamic energy barrier for fermentative bacteria leading to incomplete decomposition of substrate, hydrogen consumption by hydrogen-consuming bacteria, solventogenesis, i.e., production of solvents (reduced products) in the stationary phase, and high substrate concentration due to continuous medium feeding (Hawkes et al. 2002; Singh and Wahid 2015). Additionally, other operational parameters such as pH, temperature, partial pressure of H_2/CO_2, hydraulic retention time, etc., affect the dark fermentation process. For instance, it is observed that hydrogen production rate increases with increased substrate degradation rate at high temperatures of 65–70 °C (Agler et al. 2011). Fundamental to fermentation reaction and environmental and operational parameters, there are techniques to enhance biohydrogen production: substrate pretreatment, immobilization, effluent recycling, bioaugmentation, sequential dark, and photo-fermentation.

Substrate Pre-treatment Processes
Substrate concentration and composition plays a key role in the fermentation process. Pre-treatment processes are often performed when the substrate's structural composition makes it harder to break down by fermenting microbes. Pre-treatment methods comprise of acid/alkali treatment, enzyme treatment, freezing and thawing, temperature changes, ultrasonication, etc. Ultrasonication method involves using ultrasound waves to enhance chemical reactions by breaking certain chemical bonds. The ultrasound waves create temperature and pressure changes, leading to changes in subsequent chemical reactions. In microwave pre-treatment, heat is generated using electromagnetic waves in the range of 300 MHz to 300 GHz. However, these treatments are done in addition to chemical treatments and need to be monitored to avoid overproduction of inhibiting compounds, while also raising costing concern to be used for commercial biohydrogen production (Bundhoo 2017). Depending on substrates, optimized acid or alkali treatments are carried out to initiate chemical breakdown and ease further degradation via fermentation (Prabakar et al. 2018). This shall also be achieved by enzymatic treatment by choosing the right enzyme for major compound to be broken down efficiently to enhance hydrolysis (Contreras-Dávila et al. 2017).

Immobilization
Biohydrogen production can be enhanced by cell immobilization in cases where suspended cells get washed out and the microbes are difficult to maintain in anaerobic conditions. This allows for maintaining a high cell density, thereby increasing hydrogen generation. It also prevents the possibility of contamination especially in case of pure cultures. Immobilization of cells is done by entrapment in biofilms, polymeric gel, etc., and surface adsorption on solids such as polyvinyl alcohol and alginate (Singh et al. 2013).

Effluent Recycling
Recycling of effluent helps improve hydrogen yield by increasing cell density and buffering activity. This prevents acid or alkali accumulation. This practice has mostly been performed in studies by using anaerobically digested wastewater in biohydrogen reactors and showed decrease in alkali accumulation, maintaining optimum thermophilic condition and thus increasing hydrogen production (Xie et al. 2014; Chen et al. 2012).

Bioaugmentation
Bioaugmentation is a process to add specific microbes that favor a certain reaction or maintain other factors at optimal levels. This approach has successfully been used previously in anaerobic digestion and acidogenic fermentation to produce biomethane and volatile fatty acids, respectively. Addition of acidogenic microbes for bioaugmentation favors acidogenesis over methanogenesis in the reactor (Goud et al. 2014; Medina-Morales et al. 2021).

Sequential Dark and Photo Fermentation Process

Dark and photo fermentation reactions, individually, have low yield of hydrogen generated. When performed sequentially, it allows for complete utilization of substrate and thus, increased hydrogen production. Firstly, the substrate is degraded in dark fermentation process and the resultant waste (organic acids) is subsequently used as a substrate in the second photo fermentation process to generate hydrogen (Redwood et al. 2009).

With revolution in modern molecular techniques, there is expansion of knowledge in microbial systematics and physiology as well as the knowledge about the lack of it. It is now clear that the abundance of microbes involved in dark fermentation is yet to be fully studied with respect to their metabolic functioning and interspecies networks. Therefore, mining of metagenomics data may provide insight into other potential enzymes or co-enzymes that might lead to better biohydrogen yield and production rates.

3.6 Conclusion and Future Perspectives

Climate change and global warming due to extensive greenhouse gas emission is an alarm for human population to shift toward renewable and environment friendly energy sources. Biohydrogen solves these problems by being a carbon neutral energy fuel that can be generated by using water or waste streams. The dark fermentation process gives a waste management alternative as well, while generating hydrogen. The functionally diverse microbial communities involved in dark fermentation break down complex organic wastes. According to several studies, waste streams from industries, agriculture, food, etc., can all be utilized for biohydrogen production. The use of mixed-microbial consortium negates the need of sanitizing practices involved with pure cultures and is more robust in processing complex substrates. The dark fermentation process can be optimized across various operational and environmental parameters required by the microbes for high efficiency. The current bottlenecks for biohydrogen production are low yield of biohydrogen and the production costs of the processes. At present, majority of studies are being done at laboratory scale and await its progression to commercial usage. In addition to process modifications and more efficient designs of the bioreactors, understanding the role and functional shifts of microbial population can further assist in metabolic engineering to achieve commercial level biohydrogen production rate and yield.

Acknowledgments Financial support from Natural Sciences and Engineering Research Council of Canada Collaborative Research and Training Experience Program (CREATE, CREAT/543238-2020 "Training in Applied Biotechnology for Environmental Sustainability (TABES)"), Discovery program (RGPIN-2020-06067, RGPIN-2021-03628), Coordination of Improvement of Higher Education Personnel—CAPES, project Print-BioADD, and James and Joanne Love Chair in Environmental Engineering are acknowledged.

References

Agler MT, Wrenn BA, Zinder SH, Angenent LT (2011) Waste to bioproduct conversion with undefined mixed cultures: the carboxylate platform. Trends Biotechnol 29(2):70–78. https://doi.org/10.1016/J.TIBTECH.2010.11.006

Alexandropoulou M, Antonopoulou G, Trably E, Carrere H, Lyberatos G (2018) Continuous biohydrogen production from a food industry waste: Influence of operational parameters and microbial community analysis. J Clean Prod 174:1054–1063. https://doi.org/10.1016/j.jclepro.2017.11.078

Banu JR et al (2021) A critical review on limitations and enhancement strategies associated with biohydrogen production. Int J Hydrog Energy 46(31):16565–16590. https://doi.org/10.1016/J.IJHYDENE.2021.01.075

Basak N, Jana AK, Das D, Saikia D (2014) Photofermentative molecular biohydrogen production by purple-non-sulfur (PNS) bacteria in various modes: The present progress and future perspective. Int J Hydrog Energy 39(13):6853–6871. https://doi.org/10.1016/J.IJHYDENE.2014.02.093

Borodin VB, Tsygankov AA, Rao KK, Hall DO (2000) Hydrogen production by Anabaena variabilis PK84 under simulated outdoor conditions. Biotechnol Bioeng 69:478–485. https://doi.org/10.1002/1097-0290

Bundhoo ZMA (2017) Effects of microwave and ultrasound irradiations on dark fermentative bio-hydrogen production from food and yard wastes. Int J Hydrog Energy 42(7):4040–4050. https://doi.org/10.1016/J.IJHYDENE.2016.10.149

Chang FY, Lin CY (2004) Biohydrogen production using an up-flow anaerobic sludge blanket reactor. Int J Hydrog Energy 29(1):33–39. https://doi.org/10.1016/S0360-3199(03)00082-X

Chatellard L, Trably E, Carrère H (2016) The type of carbohydrates specifically selects microbial community structures and fermentation patterns. Bioresour Technol 221:541–549. https://doi.org/10.1016/j.biortech.2016.09.084

Chen CC, Sen B, Chuang YS, Tsai CJ, Lay CH (2012) Effect of effluent recycle ratio in a continuous anaerobic biohydrogen production system. J Clean Prod 32:236–243. https://doi.org/10.1016/J.JCLEPRO.2012.04.006

Chen H et al (2022) Effects of temperature and total solid content on biohydrogen production from dark fermentation of rice straw: Performance and microbial community characteristics. Chemosphere 286:131655. https://doi.org/10.1016/j.chemosphere.2021.131655

Chookaew T, Prasertsan P, Ren ZJ (2014) Two-stage conversion of crude glycerol to energy using dark fermentation linked with microbial fuel cell or microbial electrolysis cell. N Biotechnol 31(2):179–184. https://doi.org/10.1016/J.NBT.2013.12.004

Chu CP et al (2003) Using filtrate of waste biosolids to effectively produce bio-hydrogen by anaerobic fermentation activated sludge microbiology view project photocatalytic oxidation view project using filtrate of waste biosolids to effectively produce bio-hydrogen by anaerobic fermentation. Water Res 37:2789–2793. https://doi.org/10.1016/S0043-1354(03)00004-6

Contreras-Dávila CA, Méndez-Acosta HO, Arellano-García L, Alatriste-Mondragón F, Razo-Flores E (2017) Continuous hydrogen production from enzymatic hydrolysate of Agave tequilana bagasse: Effect of the organic loading rate and reactor configuration. Chem Eng J 313:671–679. https://doi.org/10.1016/J.CEJ.2016.12.084

Das D, Veziroğlu TN (2001) Hydrogen production by biological processes: A survey of literature. Int J Hydrog Energy 26(1):13–28. https://doi.org/10.1016/S0360-3199(00)00058-6

Dasgupta CN et al (2010) Recent trends on the development of photobiological processes and photobioreactors for the improvement of hydrogen production. Int J Hydrog Energy 35(19):10218–10238. https://doi.org/10.1016/J.IJHYDENE.2010.06.029

Geng A, He Y, Qian C, Yan X, Zhou Z (2010) Effect of key factors on hydrogen production from cellulose in a co-culture of Clostridium thermocellum and Clostridium thermopalmarium. Bioresour Technol 101(11):4029–4033. https://doi.org/10.1016/J.BIORTECH.2010.01.042

Ghirardi ML, Posewitz MC, Maness P-C, Dubini A, Yu J, Seibert M (2007) Hydrogenases and hydrogen photoproduction in oxygenic photosynthetic organisms. Annu Rev Plant Biol 58(1):71–91. https://doi.org/10.1146/annurev.arplant.58.032806.103848

Goud RK, Sarkar O, Chiranjeevi P, Venkata Mohan S (2014) Bioaugmentation of potent acidogenic isolates: a strategy for enhancing biohydrogen production at elevated organic load. Bioresour Technol 165(C):223–232. https://doi.org/10.1016/J.BIORTECH.2014.03.049

Hallenbeck PC, Benemann JR (2002) Biological hydrogen production; fundamentals and limiting processes. Int J Hydrog Energy 27(11–12):1185–1193. https://doi.org/10.1016/S0360-3199(02)00131-3

Hallenbeck PC, Ghosh D, Skonieczny MT, Yargeau V (2009) Microbiological and engineering aspects of biohydrogen production. Indian J Microbiol 49(1):48–59. https://doi.org/10.1007/s12088-009-0010-4

Hawkes FR, Dinsdale R, Hawkes DL, Hussy I (2002) Sustainable fermentative hydrogen production: challenges for process optimisation. Int J Hydrog Energy 27(11–12):1339–1347. https://doi.org/10.1016/S0360-3199(02)00090-3

Holladay JD, Hu J, King DL, Wang Y (2009) An overview of hydrogen production technologies. Catal Today 139:244–260. https://doi.org/10.1016/j.cattod.2008.08.039

Huang G et al (2019) The atomic-resolution crystal structure of activated [Fe]-hydrogenase. Nat Catal 2(6):537–543. https://doi.org/10.1038/s41929-019-0289-4

Hung CH, Chang YT, Chang YJ (2011) Roles of microorganisms other than Clostridium and Enterobacter in anaerobic fermentative biohydrogen production systems—a review. Bioresour Technol 102(18):8437–8444. https://doi.org/10.1016/j.biortech.2011.02.084

Kim MS, Baek JS, Yun YS, Jun Sim S, Park S, Kim SC (2006) Hydrogen production from Chlamydomonas reinhardtii biomass using a two-step conversion process: Anaerobic conversion and photosynthetic fermentation. Int J Hydrog Energy 31(6):812–816. https://doi.org/10.1016/J.IJHYDENE.2005.06.009

Kim MS, Lee DY, Kim DH (2011) Continuous hydrogen production from tofu processing waste using anaerobic mixed microflora under thermophilic conditions. Int J Hydrog Energy 36(14):8712–8718. https://doi.org/10.1016/J.IJHYDENE.2010.06.040

Levin DB, Pitt L, Love M (2004) Biohydrogen production: Prospects and limitations to practical application. Int J Hydrog Energy 29(2):173–185. https://doi.org/10.1016/S0360-3199(03)00094-6

Levin DB, Islam R, Cicek N, Sparling R (2006) Hydrogen production by Clostridium thermocellum 27405 from cellulosic biomass substrates. Int J Hydrog Energy 31(11):1496–1503. https://doi.org/10.1016/J.IJHYDENE.2006.06.015

Li SL et al (2010) Effects of hydraulic retention time on anaerobic hydrogenation performance and microbial ecology of bioreactors fed with glucose–peptone and starch–peptone. Int J Hydrog Energy 35(1):61–70. https://doi.org/10.1016/J.IJHYDENE.2009.10.033

Manish S, Banerjee R (2008) Comparison of biohydrogen production processes. Int J Hydrog Energy 33(1):279–286. https://doi.org/10.1016/J.IJHYDENE.2007.07.026

Medina-Morales MA et al (2021) Biohydrogen production from thermochemically pretreated corncob using a mixed culture bioaugmented with Clostridium acetobutylicum. Int J Hydrog Energy 46(51):25974–25984. https://doi.org/10.1016/J.IJHYDENE.2021.04.046

Morra S (2022) Fantastic [FeFe]-hydrogenases and where to find them. Front Microbiol 13:601. https://doi.org/10.3389/FMICB.2022.853626/BIBTEX

Nguyen TAD, Kim KR, Kim MS, Sim SJ (2010) Thermophilic hydrogen fermentation from Korean rice straw by Thermotoga neapolitana. Int J Hydrog Energy 35(24):13392–13398. https://doi.org/10.1016/J.IJHYDENE.2009.11.112

Ntaikou I, Gavala HN, Kornaros M, Lyberatos G (2008) Hydrogen production from sugars and sweet sorghum biomass using Ruminococcus albus. Int J Hydrog Energy 33(4):1153–1163. https://doi.org/10.1016/J.IJHYDENE.2007.10.053

Ogata H, Lubitz W, Higuchi Y (2016) Structure and function of [NiFe] hydrogenases. J Biochem 160(5):251–258. https://doi.org/10.1093/JB/MVW048

Parvanova-Mancheva T, Vasileva E, Beschkov V (2022) Bio-hydrogen production through microbial electrolysis cells (review). J Chem Technol Metall 57:84–95

Pattra S, Sangyoka S, Boonmee M, Reungsang A (2008) Bio-hydrogen production from the fermentation of sugarcane bagasse hydrolysate by Clostridium butyricum. Int J Hydrog Energy 33(19):5256–5265. https://doi.org/10.1016/J.IJHYDENE.2008.05.008

Prabakar D et al (2018) Advanced biohydrogen production using pretreated industrial waste: Outlook and prospects. Renew Sustain Energy Rev 96:306–324. https://doi.org/10.1016/J.RSER.2018.08.006

Redwood MD, Paterson-Beedle M, MacAskie LE (2009) Integrating dark and light biohydrogen production strategies: Towards the hydrogen economy. Rev Environ Sci Biotechnol 8(2):149–185. https://doi.org/10.1007/S11157-008-9144-9/TABLES/2

Reith JH, Wijffels RH, Barten H (2003) Bio-methane & bio-hydrogen organic acids carbohydrates H2 + CO2 H2 + CO2 organic acids stage 1 stage 2 status and perspectives of biological methane and hydrogen production. Available: http://www.biohydrogen.nl. Accessed 7 Aug 2022 [Online]

Ren N, Xing D, Rittmann BE, Zhao L, Xie T, Zhao X (2007) Microbial community structure of ethanol type fermentation in bio-hydrogen production. Environ Microbiol 9(5):1112–1125. https://doi.org/10.1111/j.1462-2920.2006.01234.x

Ren Y, Wang J, Liu Z, Ren Y, Li G (2009) Hydrogen production from the monomeric sugars hydrolyzed from hemicellulose by Enterobacter aerogenes. Renew Energy 34(12):2774–2779. https://doi.org/10.1016/J.RENENE.2009.04.011

Sağır E, Hallenbeck PC (2019) Photofermentative hydrogen production. Biohydrogen 141–157. https://doi.org/10.1016/B978-0-444-64203-5.00006-X

Sarangi PK, Nanda S (2020) Biohydrogen production through dark fermentation. Chem Eng Technol 43(4):601–612. https://doi.org/10.1002/ceat.201900452

Singh L, Mahapatra DM (2019) Waste to sustainable energy, MFCs—prospects through prognosis. Waste Sustain Energy. https://doi.org/10.1201/9780429448799

Singh L, Wahid ZA (2015) Methods for enhancing bio-hydrogen production from biological process: A review. J Ind Eng Chem 21:70–80. https://doi.org/10.1016/J.JIEC.2014.05.035

Singh L, Siddiqui MF, Ahmad A, Rahim MHA, Sakinah M, Wahid ZA (2013) Application of polyethylene glycol immobilized Clostridium sp. LS2 for continuous hydrogen production from palm oil mill effluent in upflow anaerobic sludge blanket reactor. Biochem Eng J 70:158–165. https://doi.org/10.1016/J.BEJ.2012.10.010

Wang J, Wan W (2009) Factors influencing fermentative hydrogen production: A review. Int J Hydrog Energy 34(2):799–811. https://doi.org/10.1016/j.ijhydene.2008.11.015

Xie L, Dong N, Wang L, Zhou Q (2014) Thermophilic hydrogen production from starch wastewater using two-phase sequencing batch fermentation coupled with UASB methanogenic effluent recycling. Int J Hydrog Energy 39(36):20942–20949. https://doi.org/10.1016/J.IJHYDENE.2014.10.049

Yu J, Takahashi P (2007) Biophotolysis-based hydrogen production by cyanobacteria and green microalgae. Commun Curr Res Educ Top Trends Appl Microbiol 1:79–89

Zagrodnik R, Laniecki M (2015) The role of pH control on biohydrogen production by single stage hybrid dark- and photo-fermentation. Bioresour Technol 194:187–195. https://doi.org/10.1016/J.BIORTECH.2015.07.028

Zhang Z-P, Show K-Y, Tay J-H, Liang DT, Lee D-J, Jiang W-J (2006) Effect of hydraulic retention time on biohydrogen production and anaerobic microbial community. Process Biochem 41:2118–2123. https://doi.org/10.1016/j.procbio.2006.05.021

Zhao W, Ci S (2019) Nanomaterials as electrode materials of microbial electrolysis cells for hydrogen generation. In: Nanomaterials for the removal of pollutants and resource reutilization. Elsevier, Amsterdam, pp 213–242. https://doi.org/10.1016/B978-0-12-814837-2.00007-X

Chapter 4
Biohydrogen Production in Anaerobic Reactors

Reema Kumar, Rahul Saini, Satinder Kaur Brar, and Guneet Kaur

Abstract Over the past years, biohydrogen has emerged as one of the most appealing clean energy sources in order to eliminate the concerns of depleting fossil fuel reserves. Recently, anaerobic digestion or dark fermentation-based biohydrogen has gained lot of attention. Numerous techniques have been employed to increase the microbial-based biohydrogen production using renewable and sustainable substrates. With the aim to provide major insight of anaerobic-based biohydrogen production, this chapter summarizes the scope of anaerobic fermentation as well as different types of potential anaerobic bioreactors that can be employed for biohydrogen production. In addition, the effect of several operating parameters such as temperature, pH, hydraulic retention time, substrate, and type of microorganisms especially in mixed-microbe consortia are also discussed. The limitation and challenges associated with biohydrogen production as well as emerging strategies to increase the microbial growth and biohydrogen production rate are discussed in detail.

Keywords Biohydrogen · Dark fermentation · Bioreactors · Clean energy · Mixed-microbe consortia

R. Kumar · R. Saini · S. K. Brar (✉) · G. Kaur
Department of Civil Engineering, Lassonde School of Engineering, York University, Toronto, ON, Canada
e-mail: reemak15@yorku.ca; rahul53@yorku.ca; satinder.brar@lassonde.yorku.ca; guneetk@yorku.ca

C. R. Soccol et al. (eds.), *Biohydrogen - Advances and Processes*, Biofuel and Biorefinery Technologies 13, https://doi.org/10.1007/978-3-031-49818-3_4

4.1 Introduction

With increasing human population, there have been tremendous changes in the ways of generating or using energy sources. Post-industrialization, the dependency on fossil fuels for energy has been peaking. Their non-renewability and climate change effects is shifting the focus towards finding renewable and cleaner energy sources. Hydrogen is projected to be one such sustainable fuel that might solve all these problems and overcome the negative impact of using fossil fuels (Mona et al. 2020).

The conventional processes for hydrogen production rely on natural gas and coal (Das and Veziroğlu 2001). Whereas, the biological methods are sustainable, less energy-intensive, and require renewable and organic feedstock in processes such as dark fermentation. Since, hydrogen production is observed naturally in organisms like bacteria and algae, these organisms are being used to produce hydrogen by using organic waste streams in bioreactors. Due to the complexity of biological processes, different types of reactors are used to deal with different concerns that arise with varying parameters. Major studies on biohydrogen production now focus on increasing the yield and production rate to meet commercial level.

4.2 Biohydrogen Production Processes

The fundamental entity involved in the production processes of biohydrogen are enzymes which consist of metalloclusters in their active sites. These enzymes catalyze the following reaction:

$$2H^+ + 2e^- \leftrightarrow H_2$$

Broadly there are two types of biohydrogen-producing enzymes—nitrogenase and hydrogenase. Of these, hydrogenases are further divided into different subtypes based on the type of metallocluster present in their active site such as Fe-hydrogenase, Fe-Fe-hydrogenase, and Ni-Fe-hydrogenase (Vignais et al. 2006; Morra 2022). While nitrogenase is found in prokaryotic organisms, hydrogenase is more diversely present in organisms like protists, fungi, and green algae (Dasgupta et al. 2010).

Currently, there are five processes to produce hydrogen biologically: (1) biophotolysis, (2) photofermentation, (3) dark fermentation, (4) bio catalyzed electrolysis, and (5) two-stage integration of photo- and dark fermentation. Biophotolysis functions on the principle of photosynthesis where water is split into hydrogen and oxygen using sunlight that excites electrons which are then transferred to nitrogenases or hydrogenases as illustrated in Fig. 4.1. Cyanobacteria use indirect biophotolysis where there is a stage separation of evolution of hydrogen and oxygen gas. Hydrogen production occurs in heterocysts with absence of oxygen (Levin et al. 2004). Another light-driven process is photofermentation where photosynthetic bacteria

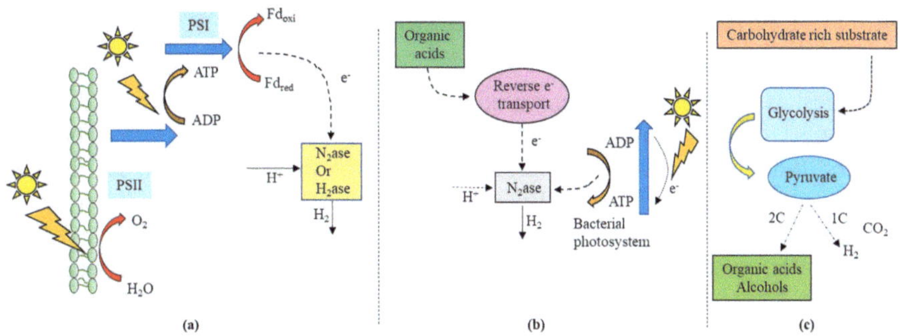

Fig. 4.1 Illustration of different biohydrogen production process (**a**) biophotolysis, (**b**) photofermentation, (**c**) dark fermentation (Aslam et al. 2018)

use organic acids as electron-donors. A proton gradient is generated, and further transfer of electrons to ferredoxin reduces protons to hydrogen (Reith et al. 2003). Bioelectrohydrogenesis is a hybrid method that uses microbial aided electrolysis cells (MEC) to convert biodegradable material into hydrogen. The microbes involved in this process are called electrogens (Parvanova-Mancheva et al. 2022). The dark fermentation process uses carbohydrate-rich substrates under anaerobic conditions to generate ATP and hydrogen by pyruvate catabolism pathway. However, the yield of hydrogen is lower because of incomplete degradation of the substrates. By integration of photo fermentation with dark fermentation, the drawback of incomplete decomposition gets overcome by photosynthetic bacteria which oxidize the partially degraded substrates further (Kim et al. 2006). As dark fermentation offers dual benefit of generating sustainable bioenergy and treatment of organic wastes, it is one method that is extensively being studied for increasing hydrogen yield.

4.3 Anaerobic Reactors and Its Type

Dark fermentation reactions are carried out in anaerobic high-speed reactors to achieve maximum hydrogen production rate. The objective is to sustain large volume of hydrogen-producing bacteria in the fermenter as films, granules, flocs, or membranes (Zhang et al. 2017). A well-designed reactor plays a pivotal role in effective sustenance of bacteria inside the reactor, and to provide optimized conditions like temperature, pH, substrate type, hydraulic retention time (HRT), etc. (Kumar et al. 2016). Four different types of these high-speed reactors are: continuous stirred tank reactor (CSTR), packed-bed reactor, up-flow anaerobic sludge blanket reactor (UASB), and membrane reactor.

4.3.1 Continuous Stirred Tank Reactor (CSTR)

The most extensively studied and used reactor for continuous production of hydrogen is CSTR where continuous addition of substrate along with product withdrawal occurs. It allows constant mixing of substrates and the microorganisms, thus the stirrer design and speed is one of the key parameters regulating the yield of hydrogen production. A well-designed stirrer in a CSTR ensures proper mixing, less shear tension, and good biomass transfer. Additionally, operational parameters like pH, temperature, and hydraulic retention time (HRT) are crucial for the reactor's performance. Shorter HRT (<6 h) has been found to enhance hydrogen production by limiting hydrogen-consuming bacteria and volatile fatty acid accumulation (Ri et al. 2019). However, cell washout in shorter HRT is a concerning limitation for CSTRs as retention of microbes in the fermentation mixture is necessary for reaction to continue. Studies suggest that this limitation can be overcome by flocculation, granulation, and cell immobilization for better retention of microbes and thus increased hydrogen production (Singh et al. 2013).

4.3.2 Packed-Bed Reactor

Packed-bed reactors (PBR) consist of a fermenter tank with a cell immobilization carrier for better retention of microbial biomass by biofilm formation or granulation. These reactors also offer better mixing with lesser shear than CSTRs. Even though these carriers are expensive, they are shown to increase hydrogen production and have lesser HRT (Sivagurunathan et al. 2017; Pugazhendhi et al. 2017). Different types of microorganism immobilization products studied that have shown high substrate conversion rate include polyester fiber beads, chlorinated polyethylene, bamboo stems, ceramic beads, scouring sponge pad, etc. (Karapinar et al. 2020; Zhao et al. 2019; Kirli and Karapinar 2018). The scouring sponge pad in PBR is known to increase hydrogen production rate by two times at a shorter HRT of 2 h when compared to an HRT of 13 h (Kirli and Karapinar 2018).

4.3.3 Up-Flow Anaerobic Sludge Blanket Reactor (UASB)

Similar to the packed-bed reactors, UASB is an enhancement to CSTR by addressing its limitation of cell washout. UASBs have an agglomeration of microbes in the form of granular beds in a suspension that reduces the cell washout. The higher biomass density reduces the inhibition by other compounds and the costs associated with mechanical agitation for better mixing (Muri et al. 2018). Studies employing UASBs for degrading organic effluents have shown increased hydrogen yield

(Castelló et al. 2009; Yang et al. 2006; Kongjan et al. 2019). The use of biofilm-induced carriers for biohydrogen production increases the production by 1.8 times (Hastuti et al. 2016). Their performance has limitation of higher HRT since lesser HRT is known to have a higher efficiency of substrate conversion.

4.3.4 Membrane Reactor (MBR)

Membrane bioreactors are hybrid systems that work by combining the principles behind activated sludge process and membrane filtration which enhances microbial retention. MBRs are broadly classified as aerobic and anaerobic; presence of aeration systems leads to higher biomass production in the former. Thus, the anaerobic system is slower in terms of its chemical oxygen demand (COD) reduction and thereby hydrogen production than aerobic MBR. The membrane materials used vary with pore size, and the commonly used ones are: hollow fiber, flat sheet, and multitubular. MBRs offer an advantage to other bioreactors as they have shown increased hydrogen production even at shorter HRT because of the membrane reducing cell-washout (Lee et al. 2008). Conventionally this is done by recycling of suspended cells by microfiltration or ultrafiltration. Major limitations of MBR are high operation cost and membrane fouling. One way of reducing membrane fouling is through gas bubbling (Yasin et al. 2014).

Examples of different reactor types, operational conditions, and hydrogen yield are mentioned in Table 4.1.

Table 4.1 Comparison between different types of bioreactors and their resultant hydrogen yield

Reactor	Substrate	Temperature (°C)	HRT	pH	Hydrogen yield	Reference
CSTR	Food waste	80	24 days	7	3.46 mol H_2/mol substrate	Saidi et al. (2018)
CSTR	Sugary waste	37	4 h	5.5	1.41 mol H_2/mol substrate	Hastuti et al. (2016)
PBR + Biofilm carrier	Glucose	37	5 h	3–5	0.92 mol H_2/mol substrate	Tomczak et al. (2018)
PBR + Bamboo stems	Cassava starch processing water	36	3–4 h	6	12.83 mol H_2/mol substrate	Andreani et al. (2015)
UASB	Sucrose	30	–	4.5–5	0.92 L H_2/h L	Muñoz-Páez et al. (2013)
UASB	Dairy	55	–	4	0.80 L H_2/h L	da Silva et al. (2019)

4.4 Parameters Affecting Hydrogen Production

An optimum hydrogen production requires minimum microbial washout, continuously stirred mixture, and lower HRT by maintaining optimum pH and temperature throughout the reaction (Carrillo-Reyes et al. 2016). Several studies had been carried out to understand the effects of differing operational parameters on the biohydrogen production.

4.4.1 pH and Temperature

The chemical reactions catalyzed by enzymes are pH and temperature sensitive; thus these two parameters are of critical importance in hydrogen production reactions catalyzed by hydrogenases and nitrogenases. Studies show that a pH range of 5.5–6.5 increases hydrogen production. Most organic waste streams have an initial pH of 7, which decreases as the organic acid production starts in the fermentation. This is maintained by adding alkali (Hu et al. 2018). The temperature within mesophilic range is mostly preferred due to its easier maintenance while hydrogen production is observed to be higher in thermophilic temperature ranges (Venetsaneas et al. 2009).

4.4.2 Type of Substrate

Appropriate substrate concentration as well as composition is essential to the hydrogen-producing microbes. Different composition of substrates can have different effects on the overall reaction. Nutrients like carbon, nitrogen, phosphate, and inorganic compounds are essential for cell activation and hydrogen generation. Carbon gets utilized for energy while nitrogen and phosphorus are essential to microbial growth, thus their ideal concentration can enhance the fermentation reaction (Lin and Lay 2004). In addition to simple sugars like glucose and sucrose, organic wastes from agriculture and industries, like food waste, molasses, and wheat straw are alternative carbon-rich substrates for anaerobic processes. They vary according to their constituents including carbohydrates, lipids, and proteins.

A substrate rich in carbohydrate is hydrolyzed by hydrolytic bacteria to form simple sugars which are further used by anaerobic bacteria to form hydrogen (Aceves-Lara et al. 2008). Lipid-rich substrates from dairy and food processing industries produce hydrogen at a slower rate due to inhibition by volatile fatty acids and low pH. The lipids are degraded by lipase enzyme produced by the bacteria. In protein-rich substrates, the proteins degrade into constituent polypeptides and then amino acids by the activity of protease enzyme. These amino acids are then converted to hydrogen, carbon dioxide, etc. (Carbonaro et al. 2012).

4.4.3 Hydraulic Retention Time

Hydraulic retention time (HRT) is the time duration where substrate is in contact with microbes for digestion to take place. The ideal HRT depends on the concentration of substrate and inoculum, reaction speed, temperature, etc. Shorter HRT is more effective for high hydrogen production rate by reducing cell washout and inhibiting methanogenic bacteria (Scoma et al. 2013).

4.4.4 Pure Culture or Microbial Consortia

Both pure and mixed microbial cultures can be employed for hydrogen production in bioreactors. However, a mixed-microbial consortium can be optimized to a wider range of environmental factors and negates the requirement of sterilized conditions as opposed to pure cultures. Hence, pure cultures have been extensively used for studying hydrogen production in batch mode. In dark fermentation reactions, anaerobic microbial consortium is generally taken from sewage or digestor sludge as an inoculum. The presence of both hydrogen-producing and hydrogen-consuming microbes requires certain pre-treatment processes to reduce the hydrogen-consumers. Pretreatment methods include chemicals, heat, freezing, thawing, aeration, ultrasound, and UV radiation (Poladyan et al. 2018).

4.5 Limitation of Biohydrogen Production and Emerging Strategies

Research on biohydrogen production has increased furthermore with the availability of genome sequencing technologies. The genetic and metabolic engineering has allowed knockout of hydrogen uptake genes, and overexpression of heterologous and homologous genes. For instance, hydrogen production was increased by three times, when NiFe-hydrogenase gene was overexpressed in *Pyrococcusfurious furiosus* (Wu et al. 2018). Similarly, Flanagan et al. (2016), modified the NiFe-hydrogenase genes using site-directed mutagenesis resulting in twofold increase in hydrogen production in *Escherichia coli*. Likewise, modification in nitrogenase enzymes had increased the biohydrogen production (Singh et al. 2021).

The scale-up of biohydrogen production has also been attempted. However, lack of substrate homogeneity in large-scale fermentations, incomplete conversion, lack of mixing, and inability to control pH causes variation in hydrogen production thus hindering the scaling-up process of biohydrogen production. In addition, anaerobic fermentation has also been reported to affect the biohydrogen-producing microorganisms and its titer. For instance, several compounds produced during anaerobic digestion such as alcohols and volatile fatty acids inhibit the biohydrogen-producing microorganism. Anaerobic microorganisms such as methanogens, sulfate-reducing bacteria, and nitrate-reducing bacteria tend to consume biohydrogen ultimately lowering the hydrogen generation rate (Aslam et al. 2018). Moreover, more than 65% of substrate is not biodegraded during anaerobic digestion and these metabolites remain in the effluents. Alternatively, anaerobic membrane bioreactor (AnMBR) tends to retain those metabolites and provide opportunity for microorganisms to convert them into biohydrogen. However, membrane bioreactor majorly suffers from membrane fouling as well as requiring huge capital investment.

In this sense, further studies are required to enhance the biohydrogen production with respect to current limitations as illustrated in Fig. 4.2. For instance, specific reactor should be designed depending on type of waste materials to vary and optimize the hydraulic and solid retention time. Similarly, utilization of different microbial consortia is another option to enhance the substrate digestion and biohydrogen production rate. However, the composition of consortia could change over time; hence designer consortia can be constructed to overcome this hurdle. It can have microorganisms which should be interdependent on each other, thus keeping each other in-check for longer period of time (Aslam et al. 2018). The process or reactor can be developed to produce both methane as well as hydrogen gas, which will ultimately aid in increasing the substrate conversion efficiency and reducing the production cost. Similarly, efforts on recycling the effluent after fermentation could also be performed in order to develop the circular bioeconomy and reduce the impact on environment and hydrogen production cost.

The following points can be considered to further investigate the biohydrogen production:

- Developing zero waste strategy and proper handling of waste produced.
- Characterization of different waste materials as feedstock and statistical optimization to develop model for predicting the potential effect of different feedstocks during scale-up.
- Production of co-products and their technoeconomic analysis to foresee its impact on production cost of biohydrogen.

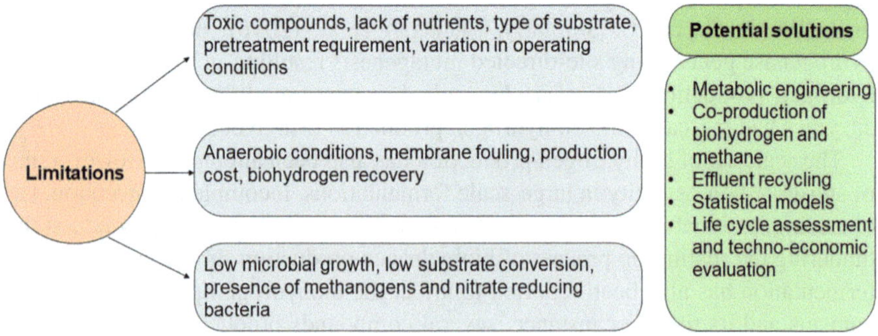

Fig. 4.2 Several limitations associated with biohydrogen production and potential solutions

- Life cycle assessment to identify the impact of biohydrogen production process on environmental system.

4.6 Conclusion

The anaerobic-based biohydrogen production holds an attractive future due to its potential to convert different substrates into biohydrogen with maximum efficiency and low pollution. The anaerobic bioreactors used for this process are being modified with advancements to their design to efficiently run the bioprocesses. The substantial improvement in biohydrogen yield and its productivity can be achieved by integrating the systems such as metabolic engineering and mixed-microbial consortium. Considering the current hurdles in terms of economic feasibility and practicality of anaerobic-based biohydrogen production at industrial scale, further studies will be required to refine the production process, reduce the energy consumption, increase microbial growth, and enhance substrate conversion.

Acknowledgment Financial support from Natural Sciences and Engineering Research Council of Canada Collaborative Research and Training Experience Program (CREATE, CREAT/543238-2020 "Training in Applied Biotechnology for Environmental Sustainability (TABES)"), Discovery program (RGPIN-2020-06067, RGPIN-2021-03628), and James and Joanne Love Chair in Environmental Engineering are acknowledged.

References

Aceves-Lara CA, Latrille E, Bernet N, Buffière P, Steyer JP (2008) A pseudo-stoichiometric dynamic model of anaerobic hydrogen production from molasses. Water Res [Internet] 42(10):2539–2550. Available from: https://www.sciencedirect.com/science/article/pii/S0043135408000808

Andreani CL, Torres DGB, Schultz L, de Carvalho KQ, Gomes SD (2015) Hydrogen production from cassava processing wastewater in an anaerobic fixed bed reactor with bamboo as a support material. Engenharia Agrícola 35(3):578–587

Aslam M, Ahmad R, Yasin M, Khan AL, Shahid MK, Hossain S et al (2018) Anaerobic membrane bioreactors for biohydrogen production: Recent developments, challenges and perspectives. Bioresour Technol 269:452–464

Carbonaro M, Maselli P, Nucara A (2012) Relationship between digestibility and secondary structure of raw and thermally treated legume proteins: a Fourier transform infrared (FT-IR) spectroscopic study. Amino Acids [Internet] 43(2):911–921. Available from: https://doi.org/10.1007/s00726-011-1151-4

Carrillo-Reyes J, Cortés-Carmona MA, Bárcenas-Ruiz CD, Razo-Flores E (2016) Cell wash-out enrichment increases the stability and performance of biohydrogen producing packed-bed reactors and the community transition along the operation time. Renew Energy 97:266–273

Castelló E, García y Santos C, Iglesias T, Paolino G, Wenzel J, Borzacconi L et al (2009) Feasibility of biohydrogen production from cheese whey using a UASB reactor: Links between microbial community and reactor performance. Int J Hydrog Energy 34(14):5674–5682

Das D, Veziroğlu TN (2001) Hydrogen production by biological processes: A survey of literature. Int J Hydrog Energy 26(1):13–28
Dasgupta CN, Jose Gilbert J, Lindblad P, Heidorn T, Borgvang SA, Skjanes K et al (2010) Recent trends on the development of photobiological processes and photobioreactors for the improvement of hydrogen production. Int J Hydrog Energy 35(19):10218–10238
da Silva AN, Macêdo WV, Sakamoto IK, Pereyra DDLAD, Mendes CO, Maintinguer SI et al (2019) Biohydrogen production from dairy industry wastewater in an anaerobic fluidized-bed reactor. Biomass Bioenergy [Internet] 120:257–264. Available from: https://www.sciencedirect.com/science/article/pii/S0961953418303210
Flanagan LA, Wright JJ, Roessler MM, Moir JW, Parkin A (2016) Re-engineering a NiFe hydrogenase to increase the H2 production bias while maintaining native levels of O2 tolerance. Chem Commun [Internet] 52(58):9133–9136. Available from: https://pubs.rsc.org/en/content/articlehtml/2016/cc/c6cc00515b. Cited 2023 Aug 22
Hastuti ZD, Chu CY, Rachman MA, Purwanto WW, Dewi EL, Lin CY (2016) Effect of concentration on biohydrogen production in a continuous stirred bioreactor using biofilm induced packed-carrier. Int J Hydrog Energy [Internet] 41(46):21649–21656. Available from: https://www.sciencedirect.com/science/article/pii/S0360319916310084
Hu BB, Li MY, Wang YT, Zhu MJ (2018) High-yield biohydrogen production from non-detoxified sugarcane bagasse: Fermentation strategy and mechanism. Chem Eng J 335:979–987
Karapinar I, Gokfiliz Yildiz P, Pamuk RT, Karaosmanoglu GF (2020) The effect of hydraulic retention time on thermophilic dark fermentative biohydrogen production in the continuously operated packed bed bioreactor. Int J Hydrog Energy [Internet] 45(5):3524–3531. Available from: https://www.sciencedirect.com/science/article/pii/S0360319918341995
Kim MS, Baek JS, Yun YS, Jun Sim S, Park S, Kim SC (2006) Hydrogen production from Chlamydomonas reinhardtii biomass using a two-step conversion process: Anaerobic conversion and photosynthetic fermentation. Int J Hydrog Energy [Internet] 31(6):812–816. Available from: https://koreauniv.pure.elsevier.com/en/publications/hydrogen-production-from-chlamydomonas-reinhardtii-biomass-using-. Cited 28 July 2022
Kirli B, Karapinar I (2018) The effect of HRT on biohydrogen production from acid hydrolyzed waste wheat in a continuously operated packed bed reactor. Int J Hydrog Energy [Internet] 43(23):10678–10685. Available from: https://www.sciencedirect.com/science/article/pii/S0360319918303185
Kongjan P, Inchan S, Chanthong S, Jariyaboon R, Reungsang A, O-Thong S. (2019) Hydrogen production from xylose by moderate thermophilic mixed cultures using granules and biofilm up-flow anaerobic reactors. Int J Hydrog Energy 44(6):3317–3324
Kumar G, Sivagurunathan P, Park JH, Park JH, Park HD, Yoon JJ et al (2016) HRT dependent performance and bacterial community population of granular hydrogen-producing mixed cultures fed with galactose. Bioresour Technol 206:188–194
Lee DY, Li YY, Noike T, Cha GC (2008) Behavior of extracellular polymers and bio-fouling during hydrogen fermentation with a membrane bioreactor. J Membr Sci 322(1):13–18
Levin DB, Pitt L, Love M (2004) Biohydrogen production: Prospects and limitations to practical application. Int J Hydrog Energy 29(2):173–185
Lin CY, Lay CH (2004) Effects of carbonate and phosphate concentrations on hydrogen production using anaerobic sewage sludge microflora. Int J Hydrog Energy 29(3):275–281
Mona S, Kumar SS, Kumar V, Parveen K, Saini N, Deepak B et al (2020) Green technology for sustainable biohydrogen production (waste to energy): A review. Sci Total Environ 728
Morra S (2022) Fantastic [FeFe]-hydrogenases and where to find them. Front Microbiol 13:601
Muñoz-Páez KM, Ruiz-Ordáz N, García-Mena J, Ponce-Noyola MT, Ramos-Valdivia AC, Robles-González IV et al (2013) Comparison of biohydrogen production in fluidized bed bioreactors at room temperature and 35 °C. Int J Hydrog Energy [Internet] 38(28):12570–12579. Available from: https://www.sciencedirect.com/science/article/pii/S0360319912022392
Muri P, Marinšek-Logar R, Djinović P, Pintar A (2018) Influence of support materials on continuous hydrogen production in anaerobic packed-bed reactor with immobilized hydrogen pro-

ducing bacteria at acidic conditions. Enzyme Microb Technol [Internet] 111:87–96. Available from: https://www.sciencedirect.com/science/article/pii/S0141022917301953

Parvanova-Mancheva T, Vasileva E, Beschkov V (2022) Bio-hydrogen production through microbial electrolysis cells (review). J Chem Technol Metall 57:84–95

Poladyan A, Trchounian K, Vassilian A, Trchounian A (2018) Hydrogen production by Escherichia coli using brewery waste: Optimal pretreatment of waste and role of different hydrogenases. Renew Energy 115:931–936

Pugazhendhi A, Anburajan P, Park JH, Kumar G, Sivagurunathan P, Kim SH (2017) Process performance of biohydrogen production using glucose at various HRTs and assessment of microbial dynamics variation via q-PCR. Int J Hydrog Energy 42(45):27550–27557

Reith JH, Wijffels RH, Barten H (2003) Bio-methane & bio-hydrogen organic acids carbohydrates H2 + CO2 H2 + CO2 organic acids stage 1 stage 2 status and perspectives of biological methane and hydrogen production. Available from: http://www.biohydrogen.nl. Cited 7 Aug 2022

Ri PC, Kim JS, Kim TR, Pang CH, Mun HG, Pak GC et al (2019) Effect of hydraulic retention time on the hydrogen production in a horizontal and vertical continuous stirred-tank reactor. Int J Hydrog Energy 44(33):17742–17749

Saidi R, Liebgott PP, Gannoun H, Ben Gaida L, Miladi B, Hamdi M et al (2018) Biohydrogen production from hyperthermophilic anaerobic digestion of fruit and vegetable wastes in seawater: Simplification of the culture medium of Thermotoga maritima. Waste Manag [Internet] 71:474–484. Available from: https://www.sciencedirect.com/science/article/pii/S0956053X17307110

Scoma A, Bertin L, Fava F (2013) Effect of hydraulic retention time on biohydrogen and volatile fatty acids production during acidogenic digestion of dephenolized olive mill wastewaters. Biomass Bioenergy 48:51–58

Singh L, Siddiqui MF, Ahmad A, Rahim MHA, Sakinah M, Wahid ZA (2013) Application of polyethylene glycol immobilized Clostridium sp. LS2 for continuous hydrogen production from palm oil mill effluent in upflow anaerobic sludge blanket reactor. Biochem Eng J 70:158–165

Singh T, Alhazmi A, Mohammad A, Srivastava N, Haque S, Sharma S et al (2021) Integrated biohydrogen production via lignocellulosic waste: Opportunity, challenges & future prospects. Bioresour Technol 338:125511

Sivagurunathan P, Anburajan P, Park JH, Kumar G, Park HD, Kim SH (2017) Mesophilic biogenic H2 production using galactose in a fixed bed reactor. Int J Hydrog Energy 42(6):3658–3666

Tomczak W, Ferrasse JH, Giudici-Orticoni MT, Soric A (2018) Effect of hydraulic retention time on a continuous biohydrogen production in a packed bed biofilm reactor with recirculation flow of the liquid phase. Int J Hydrog Energy [Internet] 43(41):18883–18895. Available from: https://www.sciencedirect.com/science/article/pii/S0360319918326351

Venetsaneas N, Antonopoulou G, Stamatelatou K, Kornaros M, Lyberatos G (2009) Using cheese whey for hydrogen and methane generation in a two-stage continuous process with alternative pH controlling approaches. Bioresour Technol 100(15):3713–3717

Vignais PM, Magnin JP, Willison JC (2006) Increasing biohydrogen production by metabolic engineering. Int J Hydrog Energy 31(11):1478–1483

Wu CH, Ponir CA, Haja DK, Adams MWW (2018) Improved production of the NiFe-hydrogenase from Pyrococcus furiosus by increased expression of maturation genes. Protein Eng Des Sel [Internet] 31(9):337–344. Available from: https://pubmed.ncbi.nlm.nih.gov/30358873/. Cited 22 Aug 2023

Yang H, Shao P, Lu T, Shen J, Wang D, Xu Z et al (2006) Continuous bio-hydrogen production from citric acid wastewater via facultative anaerobic bacteria. Int J Hydrog Energy 31(10):1306–1313

Yasin M, Park S, Jeong Y, Lee EY, Lee J, Chang IS (2014) Effect of internal pressure and gas/liquid interface area on the CO mass transfer coefficient using hollow fibre membranes as a high mass transfer gas diffusing system for microbial syngas fermentation. Bioresour Technol [Internet] 169:637–643. Available from: https://www.sciencedirect.com/science/article/pii/S0960852414009912

Zhang C, Kang X, Liang N, Abdullah A (2017) Improvement of biohydrogen production from dark fermentation by cocultures and activated carbon immobilization. Energy Fuels [Internet] 31(11):12217–12222. Available from: https://pubs.acs.org/doi/full/10.1021/acs.energyfuels.7b02035. Cited 20 Aug 2022

Zhao C, Zhang N, Zheng H, Zhu Q, Utsumi M, Yang Y (2019) Effective and long-term continuous bio-hydrogen production by optimizing fixed-bed material in the bioreactor. Process Biochem [Internet] 83:55–63. Available from: https://www.sciencedirect.com/science/article/pii/S1359511319301618

Chapter 5
Biohydrogen Production by Mono- Versus Co- and Mixed Cultures

A. A. Laikova, E. A. Zhuravleva, A. A. Kovalev, S. V. Shekhurdina, S. N. Parshina, and Yu. V. Litti

Abstract Acidogenesis and acetogenesis are the key steps in the formation of H_2 during the anaerobic decomposition of organic matter. Various groups of H_2-producing microorganisms, as well as aspects of their practical use for obtaining biohydrogen as an environmentally friendly fuel, have been studied. In terms of H_2 production, the most efficient acidogenic bacteria belong to the genera *Clostridium*, *Enterobacter* and *Escherichia*. They use different fermentation pathways, resulting in different H_2 yields and soluble metabolite products. Within the framework of the chapter, we summarized our current knowledge of the diversity of H_2-producing thermophilic, hyperthermophilic, mesophilic, and psychrophilic microorganisms and compared their metabolic pathways and cultivation conditions. Mono-, co-, and mixed cultures have different productivity depending on the environmental factors. In the case of co-cultivation, it is important to pay attention to symbiotic relationships of microorganisms, changes in microbial communities, and the development of predominant groups. Ways to enhance H_2 production by inhibiting hydrogen consumers and the use of bioaugmentation are discussed. Application of different methods of cultivation and their modifications in order to increase the yield of biohydrogen and the degree of decomposition of organic waste are included.

Keywords Biohydrogen · Dark fermentation · Hydrogen-producing bacteria · Co-culture · Mono-culture · Mixed culture · Bioaugmentation · Inoculum pretreatment

---

A. A. Laikova · E. A. Zhuravleva · S. V. Shekhurdina · S. N. Parshina · Y. V. Litti (✉)
Federal Research Center "Fundamentals of Biotechnology" of the Russian Academy of Sciences, Moscow, Russia

A. A. Kovalev
Federal State Budgetary Scientific Institution "Federal Scientific Agroengineering Center VIM", Moscow, Russia

C. R. Soccol et al. (eds.), *Biohydrogen - Advances and Processes*, Biofuel and Biorefinery Technologies 13, https://doi.org/10.1007/978-3-031-49818-3_5

5.1 Introduction

Hydrogen (H_2) is an environmentally friendly energy carrier that is considered as a future alternative to fossil fuels. Hydrogen has a high energy density, and its biological production is considered economically viable and environmentally friendly, since the only combustion waste is water, in contrast to industrial technologies for synthesis from fossil fuels (Ghimire et al. 2015). One of the environmentally friendly ways to produce hydrogen is dark fermentation (DF)—a shortened version of anaerobic methanogenic fermentation of organic matter. Many recent studies have been devoted to its fundamental study and practical optimization (Han et al. 2015; Wang et al. 2018; Ergal et al. 2020). This biological process for producing hydrogen (or biohydrogen) is the most optimal and efficient, since organic waste serves as a substrate for anaerobic fermentation, which contributes to its utilization simultaneously with the production of carbon-free biofuel (Yang and Wang 2019). The generation of gaseous hydrogen occurs at the stages of acidogenesis and acetogenesis due to the vital activity of the corresponding microbial groups. During DF, various soluble metabolite products (SMP) are co-produced—alcohols, volatile fatty acids, lactic acid, etc. (Sivaramakrishnan et al. 2021). These compounds can be further used by microorganisms to form biomethane, serve as an electron donor in denitrification, and find applications in various fields, including the production of solvents, adhesives, food additives, cosmetics and pharmaceuticals (Elbeshbishy et al. 2017). One of the most commonly used substrates for anaerobic fermentation is cellulose, the most abundant renewable natural resource in the world. Cellulose is a complex polymer that is difficult to access for microorganisms; a complex and diverse microbial community takes part in its hydrolysis and further chemical transformation (Liu et al. 2008).

The economic feasibility of dark fermentative biohydrogen production is essential for practical implementation of the technology. One of the important ways to optimize DF and achieve the highest biofuel yield is the search for and application of effective hydrogen-producing strains and their combinations. Microorganisms capable of producing hydrogen are widely distributed in nature. In particular, they are isolated from activated sludge, wastewater, anaerobic digestate, agricultural and domestic waste, animal manure (Łukajtis et al. 2018; Chen et al. 2021). Researchers are exploring the applications and benefits of using both pure and mixed cultures of hydrogen-producing microorganisms (Castelló et al. 2020). It is believed that mixed cultures are much more resistant to environmental stresses and more adapted to real substrates. Also, their use does not require sterile conditions and special pretreatment of the substrate (Fig. 5.1). The advantage of pure cultures is a higher stability and resilience of hydrogen production (Ozmihci and Kargi 2011; Castelló et al. 2020). In nature, hydrogen-producing bacteria usually live in the presence of other microorganisms, in particular, hydrogenotrophic methanogenic archaea, which consume the produced hydrogen. The removal of hydrogen from the medium makes it possible to maintain its production at a consistently high level, since an increase in the partial pressure of hydrogen inhibits its synthesis (Giovannini et al. 2016). The

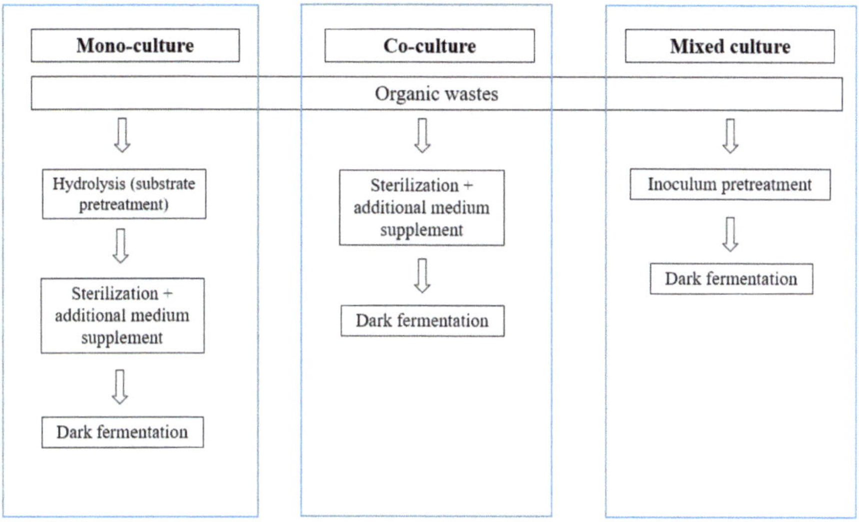

Fig. 5.1 Comparison of the mono-, co-, and mixed culture systems using complex organic substrate for dark fermentative hydrogen production

removal of hydrogen from the medium enhances the formation of other metabolite products, which also act as biotic inhibitors of hydrogen production. Thus, a low partial pressure of hydrogen is maintained in the medium, which allows maintaining the thermodynamic equilibrium and the anaerobic fermentation process (Ghimire et al. 2015; Chen et al. 2021). In laboratory experiments, various physical and chemical methods of preliminary treatment of the inoculum are used, taking into account the physiological and metabolic characteristics of hydrogen-consuming microorganisms for their successful inactivation (Łukajtis et al. 2018; Harirchi et al. 2022).

An important factor affecting the efficiency of hydrogen production is the availability of the substrate for microorganisms. Complex organic polymers are consumed by groups of hydrolytic microorganisms, while simple sugars are available to a wide range of fermentative bacteria (Geng et al. 2010). When using a mixed culture as an inoculum during dark fermentation, a change in the composition of the microbial community occurs, during which the main dominant groups of microorganisms are replaced. The change of the microbial community is influenced by various external factors: pH, cultivation temperature, substrate concentration, and the introduction of various materials to stimulate hydrogen production (Fang and Liu 2002; Kim et al. 2006). The researchers note that the use of mixed cultures of microorganisms obtained from natural sources makes it possible to effectively decompose the real complex substrate. The presence of various metabolic groups of microorganisms in the inoculum, compared to pure cultures, has many advantages: no sterilization requirements, simple preparation of the inoculum (minimum inoculum steps), the possibility of using mixed substrates, and the organization of a continuous process (Kleerebezem and van Loosdrecht 2007). However, the presence of hydrogen-consuming microorganisms (hydrogenotrophic methanogens, sulfate and nitrate reducers, homoacetogens) significantly reduces the final yield of biohydrogen and the efficiency of dark fermentation (Khan et al. 2018). The use of pure cultures for hydrogen production requires the expense of pretreating a complex organic substrate and converting it into available soluble sugars. Sterile conditions must also be maintained during pure or co-culture incubation (Wang et al. 2008). The advantage of pure cultures is a higher yield of pure hydrogen along with minimal formation of metabolite by-products (Pachapur et al. 2019). The co-cultivation model makes it possible to isolate the dominant and best-performing microorganisms for hydrogen production, while maintaining the stability of the microbiome and accelerating the consumption of the substrate. Co-cultivation of two microorganisms makes it possible to enter into a syntrophic relationship and complement each other's metabolic pathways to reduce the incubation time and increase the biohydrogen yield (Pachapur et al. 2015; Patel et al. 2014).

Enrichment of the natural microbial community with hydrogen-producing microorganisms can be considered a special type of mixed culture. This is a bioaugmentation technology that was developed to naturally intensify the production of biogas from anaerobic fermentation and is now used for both thermophilic and mesophilic modes. Reductant availability has been shown to be a regulating factor in biogas production (Kovács et al. 2013). The metabolic activity of hydrogenotrophic methanogens requires the presence of hydrogen to carry out redox reactions. Moreover, syntrophic relationships between acetogenic and methanogenic microorganisms are realized, in particular, due to interspecies electron transfer. The high rate of electron and hydrogen transfer ensures the rapid growth of methanogens (Bagi et al. 2007). In recent studies, bioaugmentation technology is used in two-stage anaerobic digestion, where biohydrogen production occurs in the first reactor, and the resulting effluent is processed by the methanogenic community in the second reactor (Cheng et al. 2020).

The use of monocultures, co-cultures, and mixed cultures for hydrogen production has its own advantages and disadvantages. In this chapter, a comparative analysis of cultures capable of dark fermentative hydrogen production, their metabolic capacity, the importance of preliminary treatment of the substrate and inoculum, as well as the interaction of microorganisms in the case of co- and mixed cultivation is evaluated. This helps to identify the most appropriate substrate ratios, their pretreatment methods, co-cultivation parameters, and evaluate the factors influencing hydrogen production. The possibilities of increasing the productivity of mixed cultures using the inhibition of hydrogen-consuming microorganisms and bioaugmentation technology are discussed. The most promising hydrogen-producing bacteria (HPB) used in experimental research and biotechnological practice are described. Important metabolic and physiological features of the main groups of HPB are analyzed. The key step in the study of mono-, co-, and mixed cultures of HPB is to explore the possibilities of increasing the efficiency of hydrogen production by improving metabolic characteristics, searching for optimal cultivation conditions, and increasing the availability of complex polymeric substrates. The main task is to find the most effective systems and conditions for hydrogen production, which will help develop new and optimize existing dark fermentation technologies using efficient hydrogen-producing strains.

5.2 Metabolism of Hydrogen-Producing Microorganisms

Hydrogen-producing microorganisms are widely distributed in natural and anthropogenic habitats, such as compost, soil, sewage sludge, organic waste, animal manure, hot springs, etc. These sources can be used to isolate new hydrogen producers (Wang and Yin 2017). The formation of hydrogen during dark fermentation is characterized by relatively high rates. However, the ratio of consumed substrate to released hydrogen is limited by the metabolic capabilities of bacteria. The theoretical limit is 4 mol of hydrogen per 1 mol of glucose, when only acetate is formed as a soluble by-product (Ergal et al. 2020).

The most effective hydrogen-producing microorganisms have been studied and described in detail. In the process of dark fermentation, as a rule, obligate and facultative anaerobes participate with an optimal growth temperature of 30–40 °C (mesophiles), 55–60 °C (thermophiles) and 70–80 °C (hyperthermophiles). Few studies have focused on psychrotrophs (4–25 °C). The optimal pH range for hydrogen production is 4–7 (Pachapur et al. 2015). Also, HPBs are broadly divided into three groups: spore-forming obligate anaerobes, non-spore-forming obligate anaerobes, and, finally, facultative anaerobes (Cabrol et al. 2017).

According to the literature data, bacterial strains of the genera *Clostridium*, *Enterobacter*, or *Escherichia*, which differ in the metabolic pathways of hydrogen production, are most often used for dark fermentative hydrogen production (Ren et al. 2022). Spore-forming obligate anaerobes are the main and most efficient hydrogen producers. They are often used as biocatalysts, to enrich bioreactors, or as a pure culture (Dahiya et al. 2021). For example, strictly anaerobic bacteria of the genus *Clostridium* produce hydrogen with high efficiency, yielding from 1.5 to 3 mol of hydrogen per mole of hexose (Masset et al. 2012; Cabrol et al. 2017). Oxidation of organic matter occurs with the formation of NADH and pyruvate, which is split into carbon dioxide and acetyl-CoA, which in turn is converted into other compounds during further reactions (Fig. 5.2). At a low partial pressure of hydrogen, the NADH molecule is oxidized by hydrogenases, with further proton reduction and hydrogen formation (Sikora et al. 2013; Haririchi et al. 2022). In addition to hydrogen, the soluble metabolite products of clostridial fermentation are ethanol and organic acids, acetate and butyrate (Moon et al. 2016). According to the literature, the end products of metabolism reduce the yield of hydrogen from the theoretical maximum and can also be toxic to the bacteria themselves (Xu et al. 2016; Łukajtis et al. 2018). Various types of microorganisms of the genus *Clostridium* exhibit variable metabolic pathways (Litti et al. 2023). Representatives of *Clostridium butyricum* form butyrate as the main product along with acetate and hydrogen. While during the DF of organic substances by *Clostridium articum*,

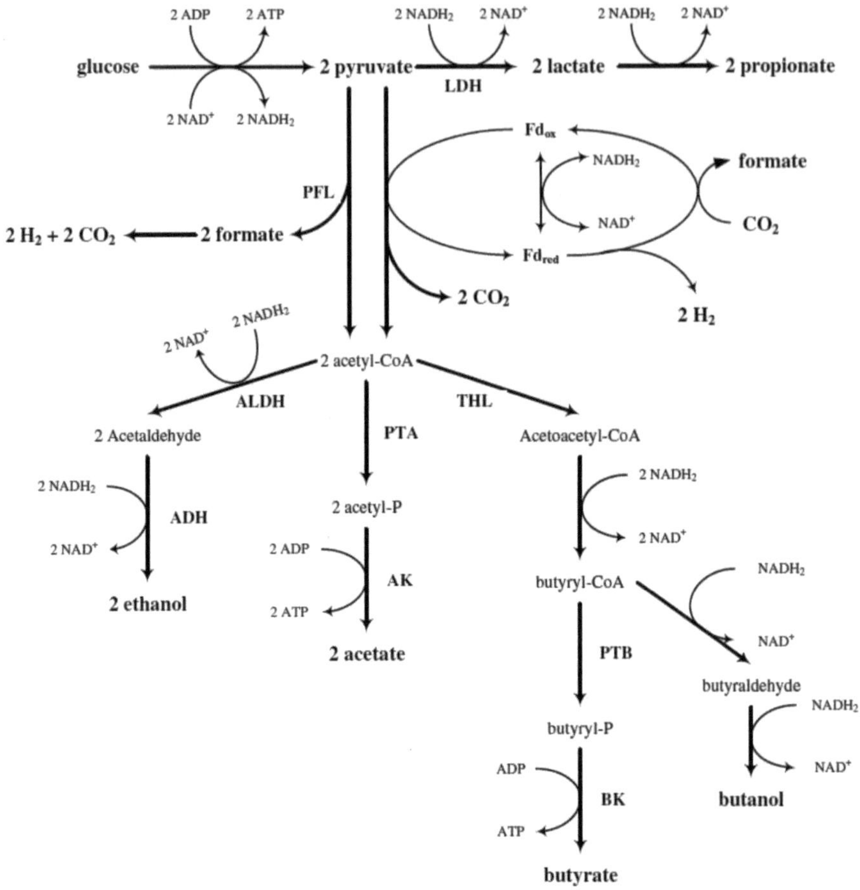

Fig. 5.2 Biochemical pathways of glucose fermentation by *Clostridium* species (Litti et al. 2023). Abbreviations: Fdox, the oxidized form of ferredoxin; Fdred, the reduced form of ferredoxin; PTA, phosphotransacetylase; PTB, phosphotransbutyrylase; LDH, lactate dehydrogenase; PFL, pyruvate-formate-lyase; HYD, hydrogenase; AK, acetate kinase; BK, butyrate kinase; ALDH, aldehyde dehydrogenase; ADH, alcohol dehydrogenase; THL, thiolase

hydrogen is consumed and propionate is formed. *Clostridium barkeri* produces lactate and ethanol and does not form hydrogen (Ghimire et al. 2015).

Sometimes the efficiency of hydrogen production is estimated by the molar ratio of butyrate and acetate, taking it as a quantitative characteristic of the process (Fuentes et al. 2021; Kim et al. 2006; Saady 2013). However, such an assessment of the process is not correct if homoacetogens are present in the community. Higher acetate concentrations do not always indicate efficient hydrogen yield, since homoacetogens (e.g., *Clostridium aceticum*) can convert CO_2 and H_2 to acetate or directly convert hexose to acetate (Guo et al. 2014a). By analyzing by-products, metabolic pathways and, therefore, the ability to produce biohydrogen can be determined. The relative abundance of representatives of the genus *Clostridium* in the reactor varies depending on the type and concentration of the substrate and cultivation conditions (Chojnacka et al. 2011). Microorganisms of the genus *Clostridium* are characterized by the production of hydrogen during the logarithmic phase of growth. During the stationary phase, metabolism switches to the production of VFAs. The sporulation ability of members of the genus *Clostridium* allows them to survive adverse conditions such as temperature or pH shocks. This gives significant advantages for use in biotechnological production. Often, *Clostridia* and other obligate anaerobes (for example, *Ruminococcus* sp., *Caldicellulosiruptor* sp.) are capable of utilizing a wide range of carbon substrates, including complex polymers such as cellulose (Nandi and Sengupta 1998). Microorganisms of this group efficiently produce hydrogen in pure culture, as part of microbial consortiums that decompose various raw materials, as well as during bioaugmentation (Łukajtis et al. 2018; Valdez-Vazquez et al. 2019). The efficiency of hydrogen production by clostridia depends on the volume of the inoculum (optimum is 10% of the substrate), temperature, pH, and cultivation time (Sivaramakrishnan et al. 2021).

The second group of hydrogen-producing microorganisms are non-spore-forming obligate anaerobic bacteria. They are considered auxiliary producers of hydrogen. The main representatives of this group are *Ethanoligenens harbinense*, *Acetanaerobacterium elongatum* (order *Clostridiales*), *Megasphaera* sp. (phylum *Firmicutes*), *Acidaminococcus* sp. (phylum *Firmicutes*) and *Prevotella* sp. (phylum *Bacteroidetes*). *Ethanoligenens harbinense* performs ethanol-type fermentation, in which, in addition to the main product of ethanol, hydrogen, acetate and carbon dioxide are also formed (Xu et al. 2016). The exact role of *Prevotella* sp. in anaerobic bioreactors is not completely clear. These bacteria are found during periods of active hydrogen production and can take part in the breakdown of complex substrates. *Megasphaera* sp. is capable of production of lactic acid and hydrogen and is found in bioreactors that treat wastewater. *Acidaminococcus* sp. participates in the biodegradation of amino acids with the formation of hydrogen. For these microorganisms, hydrogen is not the main metabolite product. Therefore, they are considered auxiliary producers of H_2 (Dahiya et al. 2021; Cabrol et al. 2017).

Among facultative anaerobes, there are also active hydrogen producers, for example, representatives of the genus *Enterobacter*—gram-negative non-spore-forming bacteria. They are more resistant to oxygen, which facilitates experimental work on dark fermentative H_2 research. For representatives of the genus *Enterobacter*, high partial pressure in the reactor does not affect the hydrogen yield during fermentation. The most active hydrogen producer in this group is *Enterobacter aerogenes* (Cabrol et al. 2017). Hydrogen production by enterobacteria in some cases exceeds the clostridial one. However, the theoretical hydrogen yield for facultative anaerobes is lower than for obligate ones—2 mol H_2/mol glucose instead of 4 mol (Argun et al. 2009; Patel et al. 2014; Tondro et al. 2020). Microorganisms of the genus *Enterobacter* carry out mixed acid fermentation with the production of hydrogen. The key enzyme in the process is pyruvate formate lyase, which reversibly converts pyruvate to formate and acetyl-CoA. During the reaction, a mixture of different organic acids (lactate, acetate, formate, propionate) is formed, and their ratio depends on the environmental conditions. For example, under conditions of low pH, lactate or propionate is more often formed instead of formate and acetate, which reduces oxidative stress (Ren et al. 2022). Hydrogen is formed upon decarboxylation of formate and further reduction of protons using the formate hydrogen lyase complex (McDowall et al. 2014). During mixed fermentation, *Enterobacter* sp. forms hydrogen from 2,3-butanediol and organic acids. However, due to the decomposition of formic acid to hydrogen and carbon dioxide, the hydrogen yield decreases compared to *Clostridium* sp. Usually, up to two formate molecules are formed from one glucose molecule during mixed acid fermentation, so the maximum hydrogen yield in this case is approximately 2 mol/mol of glucose (Łukajtis et al. 2018; Hallenbeck 2005). Representatives of the genus *Enterobacter* are found as native microbiota in communities of plant residues, eutrophic lakes, and bioreactors with a substrate rich in sugars (Kanso et al. 2011). *E. coli*, another efficient and widespread hydrogen producer, belongs to the same group. The main advantage of *E. coli* is higher metabolic activity and the ability to switch metabolic pathways. According to the literature data, the increase in the hydrogen productivity of *E. coli* due to genetic engineering is of great interest (Taifor et al. 2017). For the production of hydrogen under anaerobic conditions, *E. coli* uses different carbon sources: simple substrates—such as glycerol, glucose, galactose, gluconate, sucrose, xylose, fructose, and complex ones—industrial and agricultural waste. The theoretical yield of hydrogen for *E. coli* is 2 mol of hydrogen per 1 mol of glucose. The substrate is oxidized via different metabolic pathways that lead to the formation of pyruvate. Hydrogen is formed as a result of formate disproportionation, as in the case of *Enterobacter* sp. *E. coli* performs mixed acid fermentation, which produces a mixture of organic acids, alcohols, and hydrogen (Ren et al. 2022). *E. coli* is able to efficiently metabolize glycerol compared to other carbon sources. Glycerol is a source of reduced carbon, which makes it very promising for hydrogen production compared to other substrates. The efficiency of hydrogen production in *E. coli* is affected by the substrate concentration, pH range, kinetics, hydrogenase activity, and low CO_2 content. Bacterial activity is increased by the addition of nutrients and growth factors—peptone or yeast extract (Sivaramakrishnan et al. 2021). According to the literature data, the best hydrogen yield for *E. coli* is possible with co-cultivation with *E. aerogenes* (Maru et al. 2016; Sharma and Melkania 2017).

In addition to the two most well-known hydrogen producers described above, this group includes *Klebsiella* sp., *Citrobacter* sp., *Shewanella* sp., *Pseudomonas* sp., etc. They are reported to be capable of significant production of alcohols and biohydrogen in pure culture, but little is known about their metabolic activity in mixed cultures, continuous systems, when grown on real substrates. It is believed that hydrogen production in these microorganisms appeared during the evolution of metabolic pathways, including formate and pyruvate precursors (Cabrol et al. 2017). Sometimes, representatives of the *Enterobacteriaceae* family make up the dominant share in the hydrogen-producing community generating hydrogen from glycerol and are also found in batch DF systems with sucrose used as a substrate, while demonstrating weak hydrogen production (Cho and Lee 2011; Varrone et al. 2013). According to the literature data, the above-described bacteria are not obligate and effective hydrogen producers. However, under certain incubation conditions, efficient hydrogen production can be obtained, including in co-culture with other microorganisms.

The second way to classify HPBs is based on the optimal temperature ranges for their growth and metabolic activity. Depending on this, psychrophiles, mesophiles and thermophiles are distinguished. The most efficient work of the community is determined by the optimal temperature for the work of microbial enzymes. The yield of hydrogen at any temperature depends on the nature of the substrate and inoculum, while thermophilic hydrogen production gives a higher yield than mesophilic (Hallenbeck 2005). Mesophilic DF communities are more diverse than thermophilic ones, they can have a high potential yield of hydrogen, but the real hydrogen yield is relatively low due to the formation of other reduced products. Another factor influencing the decrease in the real hydrogen yield is the presence of many hydrogen-consuming microorganisms in the mesophilic community (Harirchi et al. 2022). Thus, hydrogen production under thermophilic conditions has many advantages: lower formation of by-products, low risk of contamination, a wide range of inexpensive consumable substrates, fast reactions, and a high yield of the target product. Most often, when working with thermophilic communities, an incubation temperature of 55 °C is used, since high temperatures inhibit the growth of many undesirable microorganisms. Thermophilic bacteria capable of DF are often isolated from hot springs and hydrotherms and have hydrolytic activity, which allows them to utilize lignocellulosic waste. Hydrolytic enzymes responsible for the degradation of lignocellulose are activated under thermophilic conditions (Kargi et al. 2012; Cappelletti et al. 2014; Nissilä et al. 2014). More readily biodegradable substrates decompose efficiently under both mesophilic and thermophilic conditions. However, the volumetric hydrogen production rate at high temperatures is low due to lower cell density, longer lag phase, and slow proliferation (Łukajtis et al. 2018). Under thermophilic conditions, the metabolism of hydrogen production most often shifts toward the acetate pathway (Sivaramakrishnan et al. 2021). The yield of hydrogen can be increased by pretreatment of the substrate or by co-cultivation. Thermophilic microorganisms are less sensitive to high hydrogen partial pressure compared to mesophilic ones, which also contributes to higher H_2 production efficiency. The main representatives of thermophilic hydrogen producers are microorganisms of the genera *Clostridium*, *Caldicellulosiruptor*, *Thermoanaerobacterium*, *Thermotoga*, and *Thermococcus* (Gupta et al. 2016). In a study by Yokoyama et al. (2007) under hyperthermophilic conditions, "peaks" of active hydrogen production were observed during DF of cow manure at 60 and 75 °C, and in the second case, the activity of the hydrogen-consuming microbiota was not recorded, and at 85 °C no hydrogen production was observed. Despite these advantages of thermophilic DF, hydrogen production by the mesophilic community (30–38 °C) is also widely used in practice, as it is considered to be more resistant to toxins and environmental changes, as well as more cost-effective (Labatut et al. 2014; Jung et al. 2021). In the literature, there are conflicting data on average optimal temperatures for the process of DF of both simple and complex substrates (Azbar et al. 2009; Ferreira et al. 2018). Under mesophilic conditions, representatives of the genus *Clostridium* and the family *Enterobacteriaceae* (genus

Enterobacter, *Klebsiella*, *Escherichia*) are most often used as pure cultures of HPB. However, in mixed cultures, microorganisms are often found that coexist with HPB, but are not capable of producing hydrogen (Fang et al. 2002).

Some studies on dark fermentative hydrogen production were carried out under psychrophilic conditions (4–20 °C). For example, a research was conducted on psychrophilic dark fermentation at 20 °C using cheese whey as a substrate. According to the results obtained, psychrophilic microorganisms can be used to produce hydrogen from real substrates. The highest yield was recorded for *Rahnella aquatilis*—from 1587.47 to 3087.57 mL H_2/g bacterial biomass (Dębowski et al. 2014). *Rahnella aquatilis* and *Clostridium algidixylanolyticum* strains were most effective at biohydrogen production, providing quite high concentrations of H_2 in biogas (32.61–43.21%) (Zieliński et al. 2017). In a study by Alvarado-Cuevas et al. (2015), hydrogen production was investigated using psychrophilic strains isolated from Antarctica and grown at 4 °C. Bacteria were incubated at a temperature of 25 °C; however, these conditions are close to psychrophilic. The maximum total hydrogen production and production rate of 253.3 mL and 16.64 mL/L/h, respectively, were attained by the GA051 strain, which is closely related to *Janthinobacterium agaricidamnosum*, whereas the maximum biohydrogen yield of 1.57 mol H_2/mol glucose was for the GA024 strain, closely related to *Polaromonas jejuensis* (Alvarado-Cuevas et al. 2015). The production of H_2 by a strain of *Bacteroides vulgatus* isolated from an inland water body at a depth of 40 ± 5 m has shown its feasibility during the fermentation of psychrophilic cattle slurry. 265.5 ± 31.2 mL biogas/g $COD_{removed}$ was yielded at 15 ± 1 °C, however H_2 concentration in resulting biogas was rather low (17.6 ± 1.4%) (Kazimierowicz et al. 2022). Based on these studies, it can be concluded that hydrogen-producing microorganisms can live and carry out DF at a wide range of temperatures.

Type of HPB culture (pure, mixed, co-culture), inoculum pretreatment method, incubation conditions are factors affecting hydrogen production. The most favorable combination of these factors during the experiment will lead to the best volumetric hydrogen yield. Below is an overview of experimental work on the production of hydrogen in the three proposed dark fermentation systems, a comparison of their performance and biotechnological potential.

5.3 Monocultures of Hydrogen-Producing Microorganisms

5.3.1 Simple Sugars

The use of pure bacterial cultures for dark fermentative hydrogen production is a very promising area for research and biotechnological applications. An analysis of the literature data shows differences in the efficiency of the process. Usually, when working with a pure culture, it is required to carry out preliminary preparation of the substrate and inoculum, sterilization and growth of the culture before the start of the experiment (Pachapur et al. 2019). Pretreatment of a complex substrate is important for the decomposition of complex carbohydrates into sugars (hexoses and pentoses) accessible to most HPB (Khamtib and Reungsang 2012). Pure cultures are generally characterized by a higher hydrogen yield and fewer metabolic by-products than mixed cultures. One of the big advantages is that the use of pure and co-cultures removes the limitation associated with hydrogen consumption by H_2-scavengers (Taherdanak et al. 2015). Simple sugars are the most common and available substrate for HPB. They are often used in experimental studies. A wide range of HPBs, differing in optimal temperature, pH, and substrate preference. They have been shown to produce biohydrogen from simple sugar with different H_2 production efficiency. For example, Ntaikou et al. (2009) studied glucose metabolism with hydrogen production in *Ruminococcus albus* monoculture. *R. albus* is a representative of the rumen microbiota and utilizes simple sugars—cellobiose, glucose, arabinose, and xylose through mixed acid fermentation (*Enterobacteriaceae* metabolic pathway). This genus is also known for its dominance in thermophilic DF reactors, treating carbohydrate-rich substrates (Litti et al. 2021). The result of the study was the creation of a kinetic model of glucose decomposition, interactions between metabolites were revealed, and the mechanism of hydrogen formation was studied (Ntaikou et al. 2009). *Bacillus* strain RTUA produced a maximum of 1.15 and 1.05 mol H_2 per mol glucose and xylose, respectively, in the work of Saleem et al. (2020). Another *Bacillus* strain, *B. firmus* NMBL-03, isolated from municipal sludge effectively utilizes a wide variety of carbohydrates, such as glucose, xylose, arabinose, lactose, sucrose, and starch. The highest volumetric hydrogen production rate (177.5 mL/L/h or 7.95 mmol H_2/L/h) and maximum H_2 yield (22.58 mmol H_2/L) was achieved using starch as the substrate. Butyrate and acetate were detected as the end soluble product in all the cases, while lactate was also detected in glucose fermentation (Sinha and Pandey 2014).

Clostridium butyricum NH-02 was shown to produce 1.90 mol-H_2/mol reducing sugar and was the first reported bio-H_2 producer within *C. butyricum* having a higher H_2 yield from maltose than that from glucose (Wang et al. 2020). Fermentative batch experiments show that the maximum hydrogen yield for *Clostridium butyricum* CGS2 (1.77 mmol/mmol glucose) is achieved at pH 6, whereas high hydrogen production with *Clostridium beijerinckii* L9 (1.72 mmol/mmol glucose) and *Clostridium tyrobutyricum* FYa102 (1.83 mmol/mmol glucose) could be achieved under uncontrolled pH conditions (initial pH of 6.4–6.6 and final pH of 4–4.2) (Liu et al. 2011).

Thermophilic HPBs produce H_2 from simple sugars at a relatively faster rate and generally at higher yields. Khamtib and Reungsang investigated the production of hydrogen by a strain of *Thermoanaerobacterium thermosaccharolyticum* KKU19 with xylose as a carbon source. Cultivation was carried out under thermophilic conditions (60 °C) at pH 6.5. The strain *T. thermosaccharolyticum* KKU19 was able to utilize various substrates, such as xylose, glucose, arabinose, xylan, while xylose showed the highest efficiency of hydrogen production—2.09 ± 0.02 mol H_2/mol xylose consumed and the rate of 3.95 ± 0.20 mmol H_2/L/h (Khamtib and Reungsang

2012). This is consistent with the findings of Litti et al. (2022) in which the newly isolated *T. thermosaccharolyticum* SP-H2 showed the highest H_2 production rate of 13.3 mL H_2/g COD/h from xylose along with 2.47 mmol H_2/L/h using cellobiose as a substrate (Litti et al. 2022). Singh et al. reported that *T. thermosaccharolyticum* TERI S7 was able to consume mono-, di-, and polysaccharides to produce hydrogen. As a substrate, a number of simple and complex carbohydrate substrates were used. Despite the fact that it was able to utilize almost all proposed substrates, the highest level of hydrogen production was observed when growing on sucrose, glucose, and xylose. The highest level of production of soluble metabolic products (ethanol, acetate, and butyrate) was observed during growth on xylose (Singh et al. 2014).

Sometimes, using polyextremophilic microorganisms as hydrogen producers, researchers can neglect the step of substrate sterilization and work in open systems, since cultivation conditions are the main limitation of the growth of undesirable microbiota (Ciranna et al. 2011). The culture of the thermophilic alkaliphilic bacterium *Thermobrachium celere* DSMZ 8682^T was used as the main hydrogen producer during growth on glucose. At the optimum level of pH 8.2 at the beginning of cultivation, the yield of hydrogen reached 3.36 mol H_2/mol glucose. Hyperthermophilic bacteria that use simple sugars as a substrate during dark fermentation also prevent microbial contamination and reduce substrate competition (Zeidan and van Niel 2010). In their work, the authors cultivated the *Caldicellulosiruptor owensensis* OL^T on glucose, xylose, their mixture, ribose, arabinose, etc. Hydrogen production by C. owenensis was the highest compared to the previously described thermophilic microorganisms when grown on simple sugar. Thus, replacement of yeast extract in the medium with a standard vitamin solution improved H_2 yield, which reached 100% of the theoretical maximum, i.e. 4 mol/mol hexose, on glucose. A very high H_2 yield of 3.85 ± 0.07 mol H_2/mol glucose and production rate of 51 mL/L/h were also obtained using the hyperthermophilic (80 °C optimum) bacterium *Thermotoga neapolitana* in the work of d'Ippolito et al. (2010).

5.3.2 Complex Substrates

Easily available substrates in the form of simple sugars are the most suitable for biohydrogen production using pure cultures. However, in order to increase the economic efficiency of the produced biohydrogen, it is of paramount importance to recover it from renewable raw materials, such as various organic wastes and sewage. Various wastewater and organic waste can be used as substrates for dark fermentation. Often, pretreatment, such as grinding or hydrolysis, is necessary to increase the availability of the substrate for HPB.

Heat pretreated and filtered cheese whey was utilized to produce hydrogen via dark fermentation by *Enterobacter aerogenes* 2822. The effect of change in total carbohydrate concentration in cheese whey (20–45 g/L), temperature (25–37 °C) and pH (5.5–7.5) was investigated on volumetric hydrogen production rate using the Box Behnken design. The highest hydrogen production rate of 24.7 mL/L/h was attained at an optimum concentration of 32.5 g/L cheese whey, 31 °C T and 6.5 pH, which was in good correlation with the predicted rate of 23.2 mL H_2/L/h (Rao and Basak 2021). When using *T. thermosaccharolyticum* SP-H2 as an inoculum, cheese whey and confectionery wastewater showed higher H_2 yield (152 mL H_2/g COD) and H_2 production rate (0.57 mmol H_2/L/h), respectively. The SP-H2 strain dominated the microbial community, despite the presence of indigenous microorganisms in real wastewater, so it can be considered as a promising HPB for the DF of non-sterile carbohydrate-rich wastewater under thermophilic conditions (Litti et al. 2022).

Clostridium sp. IODB-O3 was used by Patel et al. (2015) for biohydrogen production from low-cost agrowaste (wheat straw). The maximum H_2 was produced at 37 °C and pH 8.5. Maximum H_2 yield of 2.54 ± 0.2 mol-H_2/mol-reducing sugar from wheat straw pre-hydrolysate (WSPH) and 2.61 ± 0.1 mol-H_2/mol-reducing sugar from pretreated wheat straw enzymatic-hydrolysate (WSEH) was obtained at 37 °C and pH 8.5. Biomass pretreatment via steam-explosion generates an ample amount of WSPH which remains unutilized for bioethanol production due to non-availability of efficient C5-fermenting microorganisms. The authors suggested that the use of *Clostridium* sp. IODB-O3, efficient in WSPH utilization, will reduce the economic constraints for the entire cellulosic ethanol production process and also establish a sustainable biohydrogen production process (Patel et al. 2015).

Mahato and colleagues investigated the degradation of fruit waste by the *Clostridium* BOH3 strain. In the first experiment, the substrate was subjected to mechanical processing (grinding), microwave pretreatment or moist heat treatment to extract sugars and remove indigenous microbiota, as well as enzymatic hydrolysis (adding cellulase, amylase, xylanase, pectinase). The obtained hydrolyzate was used as a substrate for dark fermentation with the production of 2526–3118 mL H_2/L and 2.43–2.51 mol H_2/mol hexose. In the second case, the BOH3 strain was cultivated directly in heterogeneous media with the addition of fruit waste. It was shown that the strain is able to excrete saccharolytic and pectinolytic enzymes and directly ferment fruit waste with the formation of a sufficiently large volume of hydrogen (10,720 ± 137 mL/L—the maximum level after 60 h of cultivation). Along with the release of hydrogen, a decrease in the concentration of sugars in the substrate (decrease in total sugars) was shown (Mahato et al. 2020).

In the BioH_2 fermentation of pretreated and untreated wheat straw, rotten fruits and rotten vegetables, the cumulative H_2 production by *Bacillus* strains RTUA and RTUB was higher than the NaOH pretreated wheat straw, i.e., 156.4 mL/gVS, compared to 40–78 mL/gVS from rotten vegetables and rotten fruits. Owing to the diverse hydrolytic enzyme profile and high H_2 yield from complex substrates, RTUA and RTUB proved to be potential candidates for BioH_2 production from biomass (Saleem et al. 2020).

Hu and Zhu (2017) used acid hydrolysis for the pretreatment of sugarcane bagasse. The obtained hydrolyzate contains reducing sugars—glucose and xylose, available to a wider range of microorganisms capable of producing biohydrogen. In addition to reducing sugars, the hydrolyzate contains substances that can be toxic to bacteria (acetate, phenol, furan), which can be removed from the hydrolyzate, but this increases the cost of the process. *Thermoanaerobacterium thermosaccharolyticum* MJ1 isolated from waste paper performed better on non-detoxified sugarcane hydrolyzate compared to pure sugar (Hu and Zhu 2017). Similar results were obtained in a study by Brynjarsdottir et al. (2013), who investigated the properties of a new thermophilic strain, *Thermoanaerobacter* GHL15, when grown on sugars and lignocellulosic biomass. The work also explored various methods for the pretreatment of complex substrates—acid, alkaline, and enzymatic hydrolysis. Pretreatment of the complex substrate increased the amount of biohydrogen released compared to growth on simple sugars. Maximum hydrogen production (5.8–6.0 mmol H_2/g dw) was observed from Whatman paper (cellulose) hydrolysates although less hydrogen was produced by hydrolysates from other examined lignocellulosic materials (maximally 4.83 mmol H_2/g dw of grass hydrolysate). The hydrogen yields from all lignocellulosic hydrolysates were improved by acid and alkaline pretreatments, with maximum yields on grass, 7.6 mmol H_2/g dw. The authors also note that the choice of pretreatment method has a great influence on the yield of the target product, and this should be taken into account in further studies (Brynjarsdottir et al. 2013). However, these results are inconsistent with previous studies using *Caldicellulosireptor saccharolyticus*. Kadar et al. studied the production of hydrogen by an extremely thermophilic microorganism from the hydrolyzate of paper waste. The comparison was carried out when the strain was cultivated on glucose, xylose, and a mixture of these sugars. Raw paper sludge contains about 45% carbohydrates—xylan and cellulose. When cultivating on a medium containing glucose and xylose, the hydrogen production was 9–11 mmol/L/h, and on hydrolyzate—5–6 mmol/L/h. The lower hydrogen production from the paper sludge hydrolyzate was attributed to inhibitory substances. Nevertheless, the results obtained are comparable, and the use of paper sludge hydrolyzate is promising for the production of biohydrogen (Kádár et al. 2004).

The potential of a mesophilic fermentative hydrogen-producing soil bacterium, *Staphylococcus epidermidis* B-6, was investigated to produce hydrogen from rice straw. The maximum H_2 yield of 30 L/kg of rice straw was obtained from the hydrolysate prepared by treating 1% (w/v) rice straw with 0.5% (v/v) H_2SO_4. Thus, from this study, Mazumder et al. (2020) found that the isolated strain *Staphylococcus epidermidis* B-6 is a potential candidate for dark fermentative H_2 production and is used for the first time for the conversion of waste biomass into energy.

Lignocellulose has a stable ordered structure; therefore, it is often subjected to acid, alkali, or enzymatic hydrolysis in order to break it down to oligosaccharides and make it more accessible to hydrogen producers. These procedures are economically expensive, so to reduce the pretreatment steps, the researchers suggest using pure cultures of microorganisms that can degrade complex substrates. Gomez-Flores et al. studied hydrogen production by the hydrolytic strain *Clostridium termitidis* growing on cellulose under mesophilic conditions (Gomez-Flores et al. 2017). *C. termitidis* independently carried out all necessary biochemical steps: hydrolysis of polymers, fermentation of simple sugars (glucose), production of H_2, CO_2, and soluble metabolic products. *C. beijerinckii* has no hydrolytic activity compared to *C. termitidis* and can only degrade simple substrates. It was shown that the length of the lag phase in the case of monoculture and co-culture is approximately the same (about 17 days). However, in the case of co-cultivation, the cumulative yield of hydrogen was 30% higher, and the degree of cellulose decomposition was 15% higher than in the case of monoculture. There was a difference in the composition of the produced alcohols and VFAs. In the case of monocultures, the main fermentation products were acetate, lactate, formate, ethanol, and hydrogen, and in the case of co-culture, acetate, formate, ethanol, butyrate, and hydrogen. Despite the advantage of co-culture, monoculture of *C. termitidis* can be used to efficiently produce hydrogen from a complex substrate. However, the use of an additional acidogenic partner allows more complete decomposition of the substrate and more product (Gomez-Flores et al. 2017).

Another example of a thermophilic bacterium capable of utilizing complex biopolymers is *Thermoanaerobacterium thermosaccharolyticum* DD32, which showed the best result in hydrogen production from lignocellulose among monocultures of thermophilic bacteria—2.17 mol H_2/mol glucose (Sheng et al. 2015). In addition to the experiment using the model of cellulose, "Avicel" tests were carried out to obtain hydrogen from real lignocellulose substrates. The best hydrogen production was observed from corn stalks (6.38 mmol H_2/g) (Sheng et al. 2015). Some strains of thermophilic bacteria are capable of producing hydrogen from cellulose-containing substrates; however, in most studies, additional costs and time are required for strain isolation from a natural source and pre-cultivation.

Waste date seeds were valorized for biohydrogen production via consolidated bioprocessing by *Clostridium thermocellum* ATCC 27405. The effect of various surfactants (PEG1000, surfactin, Triton X-100) and sodium carbonate (buffering agent) on biohydrogen production from the acid pretreated substrate was examined. Among the various surfactants, the addition of Triton X-100 resulted in the maximum biohydrogen yield of 103.97 mmol/L and favored the production of ethanol and acetate as co-metabolites rather than butyrate. Addition of Na_2CO_3 enhanced the biohydrogen production by 33.16% and supported the glycolytic pathway and subsequent ethanol production rather than acetate/butyrate formation. The study revealed that the combined effects of the additives resulted in better substrate degradation, a favorable pH window and cell growth promotion, which ensured enhanced hydrogen productivity and yield (Rambabu et al. 2020).

The use of HPB monocultures has its own peculiarities, but under certain conditions it gives the best result. In most cases, the use of complex substrates for efficient dark fermentation requires pretreatment by physical, chemical, and biological methods, which increases the availability of microorganisms. Sometimes this step can be bypassed by using pure bacterial cultures capable of hydrolyzing polymers. However, based on the analysis of the literature, it can be concluded that such systems are claimed to be less effective. Some studies have shown that bacterial strains are better at using complex polymer hydrolysates and converting them into hydrogen and soluble metabolic products at a higher rate than simple sugars. The use of extremophilic bacteria for hydrogen production is very promising, since it allows:

1. Avoiding the cost of maintaining sterile cultivation conditions and reducing the risk of contamination
2. Increasing the yield of H_2, which is one of the determining factors for the practical implementation of dark fermentation technology

5.4 Co-cultivation of Hydrogen-Producing Microorganisms

Many researchers in the dark fermentation process note the effectiveness of systems based on two or more microbial cultures with matching metabolic capabilities. In nature, microorganisms live only as a community. Therefore, using complex trophic relationships with an efficient division of labor, it is possible to more efficiently utilize the substrate. This approach leads to improved production not only of hydrogen, but also of other dark fermentation products, such as butanol. The creation of co-cultivation systems makes it possible to form consortiums with specified characteristics, known as composition, and predictable metabolism (Hallenbeck and Ghosh 2009). The effectiveness of the use of co-cultures is also indicated by such features as increased resistance to oxygen (for strict anaerobes), the ability to expand the range of substrates, the variety of products obtained, and the synthesis of an intermediate that enhances growth (Cui et al. 2021). As in the case of monoculture, there are additional steps for growing cultures and sterilizing the substrate (Pachapur et al. 2019). However, co-cultures have been reported to produce more hydrogen and provide more complete substrate conversion, shorten fermentation time, and improve process stability compared to monocultures (Laxman Pachapur et al. 2015). Even the use of closely related microorganisms in co-culture, such as pure cultures of *Clostridium*, can provide a higher rate of H_2 production without adversely affecting H_2 yield (Masset et al. 2012).

The use of co-cultures is important in consolidated bioprocessing when symbiotic interactions are formed between two specific microorganisms (Nagarajan et al. 2019). Elsharnouby et al. (2013) distinguish three main types of co-cultures:

1. Co-culture of a strict anaerobe and a facultative anaerobe capable of producing hydrogen. The system is optimal for complex carbohydrate-rich substrates and allows you to maintain anoxic conditions in the environment without the use of additional reducing agents.
2. Co-culture of a microorganism capable of hydrolyzing composite substrates and an active hydrogen-producing acidogenic microorganism. Both microorganisms are capable of producing hydrogen, but the rapid consumption of the resulting sugars speeds up the process, improves the effect and economic benefits of the technology.
3. Co-culture of an acidophilic (aciduric) microorganism and a hydrogen producer, which reduces the consumption of alkali, thereby reducing or eliminating the need for a buffer to maintain a neutral or slightly acidic pH.

In most cases, bacteria that can hydrolyze complex polymers still don't produce much hydrogen. Therefore, combinations of cellulolytic and saccharolytic (acidogenic, solventogenic) HPB strains are used to achieve the best result (Gomez-Flores et al. 2017; Wen et al. 2019). Co-cultivation is also suitable for the efficient production of butanol through metabolic interactions between microorganisms (Cui et al. 2021). For a deeper understanding of the interaction of microorganisms, the researchers resorted to the creation of metabolic models based on the genome sequence, and compared systems containing monocultures of cellulolytic strains, and co-cultures from cellulolytic and acidogenic strains. Salimi et al. developed a model to analyze the integrated physiology of the co-cultivation of *Clostridium cellulolyticum* and *Clostridium acetobutylicum*. *C. cellulolyticum* possesses cellulosomes, multienzyme complexes that decompose cellulose to form glucose and soluble cellodextrins (cellobioses), and in addition produce simple metabolic intermediates. *C. acetobutylicum* decomposes the resulting sugars, as well as pyruvate, to form volatile fatty acids, alcohols, and hydrogen. In the presence of *C. acetobutylicum*, cellulose degradation, growth, and hydrogen production increased significantly, and cellulase activity improved. The simulation results showed that the elimination of cellobiose inhibition was not the main factor responsible for the improvement in cellulose utilization in co-cultivation. It has been suggested that *C. acetobutylicum* rapidly consumes cellobiose and produces pyruvate in return, which allows *C. cellulolyticum* to continue growing and hydrolyze cellulose during co-cultivation (Salimi et al. 2010). Significant amounts of butanol were also obtained during co-cultivation of *C. cellulolyticum* and *C. acetobutylicum* (Salimi and Mahadevan 2013). When *C. cellulolyticum* and *Citrobacter amalonaticus* were co-cultivated on corn straw, the bacteria consumed glucose and xylose formed during the hydrolysis of cellulose by *C. cellulolyticum*, which ensured complete utilization of the substrate. *Citrobacter amalonaticus* also converted formate, which is toxic to *Clostridium*, into hydrogen, which ensured its high yield of 51.9 L H_2/kg total solid (TS) (Zhang et al. 2016).

One approach is to isolate dominant hydrogen-producing species from mixed communities and co-cultivate them. Ozmihci and Kargi created a DF system with a combination of pure cultures of *Clostridium butyricum* and *Clostridium pasteurianum*, as these species traditionally predominate in anaerobic microbial communities. Cultivation was carried out under mesophilic conditions with periodic supply of the substrate; ground wheat starch served as a carbon source. This combination of *Clostridium* species has been shown to be effective for hydrogen production under optimal conditions (Ozmihci and Kargi 2011).

Encouraging results of co-cultivation under dark fermentation conditions have been shown not only for bacteria of the genus *Clostridium*. Valdez-Vazquez et al. studied the two-stage production of hydrogen, acetone, butanol, and ethanol by growing synthetic microbial consortia from representatives of the genera *Clostridium* and *Enterococcus* using wheat straw as a substrate. The first consortium was formed from three isolates of *Enterococcus* RP3, RP4, and RP7 to produce hydrogen from the xylan fraction of untreated (native) straw under mesophilic conditions (37 °C).

Due to the synergistic relationship between microorganisms of the genus *Enterococcus*, it was possible to achieve a hydrogen yield of about 79.5 mL H_2/g of added xylose from wheat straw. The second consortium, consisting of *Clostridium beijerinckii* 10132 and *Clostridium cellulovorans* 35296, efficiently converted cellulose into solvents (acetone-butanol-ethanol) (Valdez-Vazquez et al. 2015).

An increase in hydrogen production compared to pure cultures during co-cultivation of *Clostridium acetobutylicum* and *Enterobacter aerogenes* on pretreated sugarcane bagasse and rice straw was noted by Tondro et al. The substrates were preliminarily subjected to alkaline treatment and enzymatic hydrolysis, and then incubation was carried out with the participation of two microorganisms. Hydrogen production during co-cultivation increased 1.5 times compared with monoculture when cultivated on alkaline-pretreated rice straw and 3.4 times on sugarcane bagasse. A positive effect was observed in the production of metabolites: acetone, butanol, and ethanol. The formation of symbiotic relationships between bacteria is explained by the fact that the facultative anaerobic *E. aerogenes* consumes the remaining oxygen, maintaining the most favorable growth conditions for the obligate anaerobe *C. acetobutylicum*. This leads to a reduction in the duration of the process and an increase in the yield of products (Tondro et al. 2020). Yokoi et al. grew a mixed culture of *E. aerogenes* and *C. butyricum* on starch waste under mesophilic conditions. The authors attributed the significant production of hydrogen from starch (2 mol H_2/mol hexose) to the fact that the facultative anaerobic *E. aerogenes* maintained anaerobic conditions in the system that were optimal for dark fermentation (Yokoi et al. 2002). Another work by Kao et al. utilizing *C. butyricum* with *Rhodopseudomonas palustris* also showed the advantage of co-culture. In particular, the hydrogen yield of the co-culture produced 2.16 mol H_2/mol sucrose, *C. butyricum* 1.77 mol H_2/mol sucrose, and *R. palustris* 1.64 mol H_2/mol sucrose. However, hydrogen production ceased when the pH was lowered to 6.0, resulting from the accumulation of acidic products (Kao et al. 2016). Co-cultures of *C. butyricum* with *Sporolactobacillus vineae* increased the hydrogen production from 1.57 to 1.84 mol H_2/mol glucose, whereas lactic acid production did not increase in comparison with a single culture of *C. butyricum*. Interestingly, *C. butyricum* occupied more than 90% of the microbial composition; therefore, the microbial community analysis alone could not explain the difference in hydrogen production according to *S. vineae* addition. PICRUSt analysis showed that co-culture with *S. vineae* enhanced the expression of genes relating lactate to butyrate and the H_2 pathway, although *S. vineae* did not produce butyrate. Thus, enhanced H_2 production was associated with the synergy of *Clostridium* sp. and *Sporolactobacillus* sp. in co-cultivation in terms of regulation of the acidogenic pathway (Park et al. 2021).

Another example of an effective symbiotic relationship between two mesophilic bacteria, *Bacillus cereus* ATCC 14579^T, and *Brevundimonas naejangsanensis* BIO-TAS2-2^T in a DF system, was shown by Ma and Su (2019). Co-culturing system increased the substrate consumption rate by 34% and 70% and hydrogen production yield by 52% and 62% compared with the *B. cereus* and *B. naejangsanensis* fermentation, respectively. The analysis of soluble metabolites during the fermentation process indicated that the mixed acid fermentation conducted in two strains and co-culturing system was dominated by butyric acid type fermentation (Ma and Su 2019).

Co-cultures of thermophilic microorganisms are also capable of enhancing hydrogen production. Liu et al. investigated hydrogen production by *Clostridium thermocellum* JN4 and *Thermoanaerobacterium thermosaccharolyticum* GD17 from microcrystalline cellulose, filter paper and cellobiose at 60 °C. The monoculture of *C. thermocellum* JN4 produced 15 mmol H_2/L from microcrystalline cellulose, the production stopped after 60 h. While in co-cultivation, the yield was 55 mmol H_2/L, and production lasted up to 90 h. During the experiment on the cultivation of microorganisms using real lignocellulosic substrates (corn stalks and cob powder), similar results were obtained: during co-cultivation, the formation of hydrogen increased, and the composition and amount of volatile fatty acids and alcohols also differed (Liu et al. 2008). A thermophilic system consisting of two types of clostridia, cellulolytic *C. thermocellum* and non-cellulolytic *C. thermopalmarium*, growing on cellulose, was studied, and various factors influencing the production of hydrogen were analyzed. The inoculum ratio of the two bacterial species had little effect on hydrogen production: the authors attribute this to the low level of soluble sugars in the medium at the start of incubation. The maximum yield of hydrogen obtained was 1387 mL/L of culture under optimized conditions (Geng et al. 2010). Li and Liu used strains of *C. thermocellum* and *C. thermosaccharolyticum* to co-cultivate and enhance hydrogen production from corn waste under thermophilic conditions. The yield of hydrogen during co-cultivation was 68.2 mL/g waste, which is 94.1% higher than in the case of using *C. thermocellum* monoculture (Li and Liu 2012). Cheng and Zhu co-cultivated *C. thermocellum* and *Thermoanaerobacterium aotearoense* with sugarcane cake as a substrate. The substrate was previously subjected to alkaline hydrolysis. The synergism of *T. aotearoense* and *C. thermocellum* for hydrogen production was as follows. Pentoses are cellulose inhibitors that *C. thermocellum* possesses, while *T. aotearoense* is able to use xylose as a carbon source and also ferment hemicelluloses to produce hydrogen. The yield of hydrogen under optimized conditions was 50.05 ± 1.51 mmol H_2/L, which indicates the formation of synergistic relationships between microorganisms (Cheng and Zhu 2013). Similar results were obtained by Wang et al. in co-cultivation of *C. acetobutylicum* and *Ethanoligenens harbinense* with cellulose. *E. harbinense* was not capable of utilizing cellulose, but it quickly consumed the simple sugars formed during its hydrolysis, thereby maintaining thermodynamic equilibrium, which led to an increase in hydrogen production (Wang et al. 2008).

Co-cultivation of *Caldicellulosiruptor saccharolyticus* and *C. kristjanssonii* also resulted in a higher hydrogen yield compared to mono-cultivation (Zeidan et al. 2010). *C. kristjanssonii* has been shown to dominate when co-cultured using a mixture of glucose and xylose as a substrate, indicating that one or more compounds secreted by *C. saccharolyticus* may enhance the growth of the organism. This interaction may be a possible reason for the stable coexistence of the two strains even in the presence of only one growth-limiting substrate. Although the increase in hydrogen production from co-culture was small, its use still results in better conversion of various sugars and resistance to contamination (Zeidan et al. 2010).

A type of co-culture that combines an acidophilic hydrogen producer with an efficient hydrogen-producing bacterium also shows interesting results. Acidophilic bacteria are able to actively metabolize substrates and produce hydrogen at low pH and high concentrations of organic acids, which is inhibitory for most other bacteria. The use of acidophiles reduces the requirement for buffering (maintaining a neutral or slightly acidic pH) of the cultivation medium, which avoids the negative effects of chemicals on microorganisms. Thus, such a co-cultivation system makes it possible to obtain a high yield of biohydrogen even under conditions of acidification of the environment. Lu et al. evaluated the ability of a mixed culture under highly acidic conditions to produce hydrogen in a batch mode. *Candida maltosa* HY-35 is an acidophilic microorganism that can produce hydrogen at pH 1.3, while the facultative anaerobe *E. aerogenes* W-23 produces hydrogen at pH 4.0. Both the hydrogen yield and the rate of hydrogen production were significantly higher with co-culture than with monoculture. The co-cultivation of hydrogen-producing aciduric microorganisms and anaerobes with high hydrogen-producing activity allowed for a wider pH range for DF and avoided buffer dependence. In addition, *C. maltosa* consumed lactate, succinate, and citrate produced by *E. aerogenes* during fermentation, which slowed down the pH drop. Thus, a kind of proto-cooperative relationship was established between partners in co-cultivation (Lu et al. 2007).

The use of co-cultures can have not only advantages, but also disadvantages, in particular, lower growth rates and hydrogen production. This may be associated with competition for the substrate, different metabolic and physiological requirements, growth rates, hydrogen consumption by one of the strains. This was shown in the co-cultivation of *C. tyrobutyricum* F4 and *C. sporosphaeroides* F52, when hydrogen production during co-cultivation was lower than in the growth of pure cultures (Hsiao et al. 2009).

Some researchers note that co-cultures have an advantage over mono- and mixed cultures in technological processes (Elsharnouby et al. 2013). Co-cultivation makes it possible to exclude the influence of alien microbiota, provide more complete hydrolysis of polymers, and maintain anaerobic conditions and optimal acidity in the system (Ramírez-Morales et al. 2015). It is necessary to carefully study the metabolism of microorganisms involved in co-cultivation in order to predict the formation of symbiotic relationships between them and the possible effects based on this interaction.

5.5 Mixed Cultures of Hydrogen-Producing Microorganisms

The use of mixed cultures for dark fermentative hydrogen production is a simple and cost-effective approach, since it does not require the isolation of certain strains in pure culture, their cultivation and preliminary sterilization of the substrate. Hydrogen production when using mixed culture is affected by substrate concentration, pH, temperature, inoculum volume, nutritional requirements, additives, and redox potential (Mu et al. 2006; Pachapur et al. 2019). As mixed cultures, inoculums sampled from anthropogenic habitats—organic waste, sewage, compost, soil, animal manure—are usually used. The formation of a hydrogen-producing community in mixed cultures is due to natural factors and natural selection, due to which the microorganisms in them are resistant to many environmental factors and are able to utilize a wide range of substrates, which is also facilitated by symbiotic relationships and multiple trophic links (Kleerebezem and van Loosdrecht 2007; Baghchehsaraee et al. 2010). Mixed cultures are less susceptible to environmental changes and community fluctuations, making the hydrogen production process easier to control (Łukajtis et al. 2018). One of the disadvantages of technology is the lower purity of the gas compared to the gas produced by pure cultures (Ortigueira et al. 2015).

In mixed cultures, all stages of anaerobic degradation of organic matter occur: hydrolysis, acidogenesis, acetogenesis, and methanogenesis. The dominant microorganisms in dark fermentation communities are usually represented by the most efficient genus of hydrogen producers. For example, under mesophilic conditions at pH 5.5, representatives of the *Clostridium* genus dominate (Ghimire et al. 2015). Phylogenetic analysis showed that 64.6% of microorganisms in the community belonged to the three Clostridium species, 18.8% to the *Enterobacteriaceae* family (Fang et al. 2002). When working with waste activated sludge at 37 °C, a sharp numerical increase in the groups of hydrogen-producing microorganisms, the genera *Clostridium sensu stricto* and *Paraclostridium*, was observed in the community compared to the control (Yang and Wang 2020). Next-generation sequencing of cDNA also confirmed the dominance of the genera *Clostridium* and *Ruminiclostridium* at 37°C in the HPB community in a fluidized bed reactor inoculated with activated sludge and fed with xylose (Dessì et al. 2018).

Representatives of the genus *Thermoanaerobacterium* predominate in thermophilic communities. Nitipan et al. reported that the microbial community was dominated by *Thermoanaerobacterium* (~66%), *Clostridium* (~28%), and *Desulfotomaculum* spp. (~6%) in mixed culture dark thermophilic DF of palm oil mill effluent (Nitipan et al. 2014). The microbial community strongly depends on the composition of the DF substrate. Litti et al. (2021) showed that *Ruminococcus* (26–90%) and *Thermoanaerobacterium* (6–67%) dominated in DF of carbohydrate-rich substrates. DF of fats and proteins was characterized by higher microbial diversity. *Thermoanaerobacterium* (21%), *Thermobrachium* (19%), *Tepidiphilus* (16%), and *Acetomicrobium* (14%) dominated when using fats, while *Thermobrachium* (34%), *Acetomicrobium* (16%), and *Clostridium* sensu stricto 7 (12%) dominated when using proteins, as substrate. However, the dominance of the genus *Ruminococcus* was thought to be responsible for the highest hydrogen production according to correlation analysis (Litti et al. 2021). In a thermophilic batch reactor fed by rice straw, the predominance of hydrolytic and acidogenic bacteria *Thermoanaerobacterium* and *Clostridium* was also shown by Chen et al. (2012). It has also been shown that microorganisms that are not numerically dominant in the community can make a significant contribution to hydrogen production. For example, in a study by Rafrafi et al. *C. pasteurianum* was the dominant

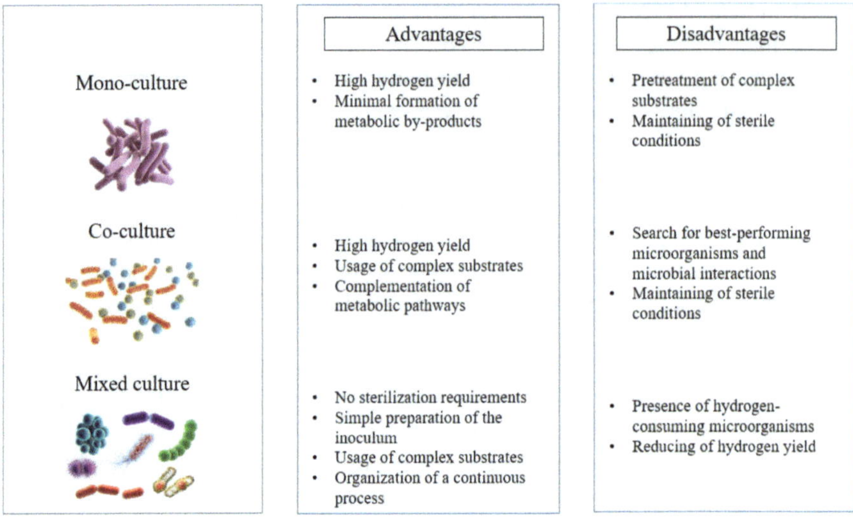

Fig. 5.3 Advantages and disadvantages of using mono-, co-, and mixed cultures in DF technology

hydrogen-producing bacterium, but the presence of *E. coli* switched metabolic pathways to the acetate and butyrate pathways of biohydrogen production (Rafrafi et al. 2013).

The disadvantage of using mixed cultures is the presence of hydrogen scavengers, the activity of which can significantly reduce the hydrogen yield. These include hydrogenotrophic methanogens, homoacetogens, nitrate reducers, sulfate reducers, and propionate-forming bacteria (Wong et al. 2014). Autotrophic sulfate and nitrate reducers use hydrogen as an electron donor when reducing sulfate to hydrogen sulfide or nitrate to ammonium (Khan et al. 2018). Propionic acid bacteria directly consume hydrogen to form propionate due to reverse hydrogenase activity (Cabrol et al. 2017). Lactic acid bacteria also have a significant effect on the system, competing for the substrate with hydrogen producers and forming lactic acid (Park et al. 2016).

In general, the use of mono-, co-, and mixed cultures in DF technology has both its advantages and disadvantages, which are briefly summarized in Fig. 5.3.

5.6 Methods of Inoculum Pretreatment

In order to increase hydrogen production, it is necessary to exclude these microorganisms from the community or inhibit their growth by pretreatment of the inoculum (Mohan et al. 2008; Gupta et al. 2016; Rafieenia et al. 2018; Soares et al. 2020). Most physical and chemical pretreatment methods are aimed at eliminating methanogenic archaea as the main hydrogen-consuming agents. Methods of pretreatment are not universal and can simultaneously affect non-target groups of microorganisms that are important for the production of biohydrogen. The applicability and effectiveness depend on the source of the inoculum. Table 5.1 summarizes the main inoculum pretreatment methods. Physical methods for pretreatment of the inoculum include heat shock, aeration, microwave irradiation, ultrasonication, infrared, ultraviolet and gamma irradiation, freezing and thawing. For chemical treatment, various chemical compounds and growth inhibitors of microorganisms that consume hydrogen are introduced. The main ones are 2-bromoethanesulfonate (BES), 2-bromoethanesulfonic acid (BESA), iodopropane, chloroform, and higher fatty acids (Kumar et al. 2015b). The heat-shock method (80–121 °C) is aimed at the destruction of non-spore-forming bacteria. It effectively acts on methanogens, lactic acid, acetic acid, propionic acid bacteria, but also on hydrogen-producing bacteria of the *Enterobacter* or *Citrobacter* genera. The above methods of pretreatment allow to increase the relative abundance of HPB in the microbial community and the production of hydrogen. Wang and Wan compared hydrogen production using untreated and pretreated inoculum in their study. The hydrogen yield in the experiment with heat-shock pretreated inoculum was more than three times higher than the hydrogen yield in the experiment with untreated inoculum, and the substrate to product conversion was 10% higher (Wang and Wan 2008). Taherdanak et al. used anaerobic sludge after heat-shock pretreatment (95 °C for 15 min) under anaerobic conditions to study the stimulation of hydrogen production (Taherdanak et al. 2016). Recent efforts by Mahata et al. (2020) were aimed at improvement of dark fermentative H_2 production from organic wastes using acidogenic mixed consortia. Using heat-shock pretreatment of anaerobically digested sludge and artificial intelligence (AI) based models, they obtained the maximum H_2 yield of 8.28 mol H_2/kg $COD_{removed}$ under the optimized condition of pH 6.75, groundnut de-oiled cake concentration 16.16 g/L and temperature 37.55 °C. H_2 yield was improved by 2.1-folds as compared to un-optimized condition (Mahata et al. 2020). The effect of waste frying oil (WFO) addition on H_2-producing and H_2-consuming metabolic pathways was studied by Rafieenia et al. (2018). Results revealed that low concentrations of WFO did not completely inhibit hydrogenotrophic methanogens. FBA showed that acetoclastic methanogens were more sensitive to WFO, in comparison with hydrogenotrophic methanogens.

It is reported that not only hydrogen-producing bacteria, but also sulfate reducers and VFA producers are able to survive heat shock. An increase in acid concentration, along with heat shock, leads to an almost complete suppression of the activity of methanogens, but not sulfate reducers (Zhu and Béland 2006). Freezing at –25 to –10 °C and thawing at 20–30 °C have a similar principle. Although it is used for selective enrichment with HPB, the hydrogen yield is often marginally increased. The hydrogen yield may even decrease compared to the untreated inoculum, which is associated with the survival of bacterial groups that adversely affect hydrogen production (Liu et al. 2009; Łukajtis et al. 2018). Chemical pretreatment methods are considered more efficient (Wang et al. 2019). The above chemical compounds are able to selectively block the metabolic pathways of methanogenic archaea. 2-BES and 2-BESA are chemical analogues of coenzyme M, which is involved in methanogenesis. They bind to the methyl group instead of CoM, thereby inhibiting or

Table 5.1 Inoculum pretreatment methods for dark fermentative hydrogen production using mixed cultures

Inoculum source	Pretreatment method	Pretreatment conditions	Dominating hydrogen producers after pretreatment	H_2 production and H_2 yield	Reference
Anaerobic sludge from wastewater treatment plant of baker yeast company	Heat	100 °C for 5 h	Not stated	H_2 production: 140–4500 mL H_2/L day H_2 yield: 20–646 mL H_2/g total sugar	Gorgec and Karapinar (2019)
Anaerobic sludge from wastewater treatment plant of baker yeast company	Heat	100 °C for 5 h	Not stated	H_2 production: 773.33–1754.76 mL H_2/L day H_2 yield: 0.6–1.6 mol H_2/mol total sugar	Kirli and Karapinar (2018)
Anaerobic sludge from a wastewater treatment plant	Heat	121 °C for 10 min	*Clostridium* sp.	H_2 production: 127.69 mL H_2/L H_2 yield: 18.6 mL H_2/g sugar	Moodley and Kana (2018)
Anaerobic sludge from an UASB reactor	Heat	90 °C for 30 min	*Clostridium* sp.	H_2 yield: 169.53–286.98 mL H_2/g biomass	Gonzales et al. (2019)
Anaerobic digester of the sewage treatment works	Base	5 M NaOH was used to set the inoculum pH value at 10	*Clostridium* sensu stricto 1	H_2 production: 140.75 ± 2.61 mL/L/h H_2 yield: 157.25 ± 7.62 mL/g of volatile solids	Luo et al. (2022)
Granular sludge from the methanogenic anaerobic reactor treating the effluent the vegetable oil industry	Heat	90 °C for 10 min	*Clostridium, Bacillus, Enterobacter*	H_2 production: 821.34 mL H_2 yield: 4.75 mmol H_2/g COD	Magrini et al. (2021)
Cow dung	Heat	100 °C, 20 min	Not stated	H_2 yield: 77.2 mL/g VS	Kumari and Das (2016)
Activated sludge from a domestic wastewater treatment plant	ORP	Oxidizing ORP control environment (−540 mV)	*Clostridium* sp., *Veillonellaceae, Enterobacteriaceae*	H_2 yield improved by 50–70%	Vesga-Baron et al. (2021)

Inoculum source	Pretreatment method	Pretreatment conditions	Dominating hydrogen producers after pretreatment	H_2 production and H_2 yield	Reference
Secondary settling tank of a local municipal wastewater treatment plant	Acid	Adjusting the pH of the sludge to 3.0 with 1 mol/L HCl and maintaining for 24 h	Not stated	H_2 production: 22.81 mmol H_2/g VSS H_2 yield: 1.51 mol/mol glucose consumed	Chang et al. (2011)
Secondary settling tank of a local municipal wastewater treatment plant	Base	Adjusting the pH of the sludge to 10.0 with 1 mol/L of NaOH and maintaining for 24 h	Not stated	H_2 production: 18.51 mmol H_2/g VSS H_2 yield: 1.34 mol H_2/mol glucose	Chang et al. (2011)
Anaerobic sludge	Heat	100 °C for 60 min	*Clostridium* sp.	H_2 yield: 4.62 mol H_2/mol sucrose	de Sá et al. (2013)
Rumen fluid	Acid	pH 3 for 24 h	*Clostridium* sp.	H_2 yield: 18.5 mmol/g cellulose	Ratti et al. (2014)
Anaerobic sludge	Ionizing irradiation	Irradiation at dose rate of 286 Gy/min	Not stated	H_2 production: 7.2 mL/h H_2 yield: 267.7 mL/g glucose	Yin et al. (2014)
Anaerobic mixed consortium was collected from the up-flow anaerobic sludge blanket reactor	Heat	85 °C for 30 min	Not stated	H_2 yield: 1.25 ± 0.01 mol H_2/mol carbohydrates	Mañunga et al. (2019)
Anaerobically digested sludge collected from a municipal sewage treatment plant	Gamma irradiation	5 kGy gamma irradiation	*Clostridium butyricum, Clostridium perfringens* ATCC 13124T (CP000246)	H_2 production: 300 mL/100 mL of mixture H_2 yield: 1.81 mol H_2/mol glucose	Yin and Wang (2016)

significantly reducing the formation of methane (Liu et al. 2011). Chloroform has a similar mechanism of action and can reduce the inhibitory activity of propionic acid bacteria, so it is used to treat the inoculum (Chang et al. 2011). However, due to its toxicity, chloroform can also affect hydrogen-producing bacteria (Hu and Chen 2007). The oxidation-reduction potential (ORP) control is a relatively new method to selectively enrich HPB. Vesga-Baron et al. (2021) studied the effect of ORP controlled by hydrogen peroxide and potassium ferricyanide on continuous hydrogen production and microbial diversity in an acidogenic sludge fed (HRT 12 h and pH 5.5) with glycerol. Results show that the more oxidizing ORP control environment (−540 mV) improves H_2 yield by 50–70% (0.31–0.51 mol H_2/mol glycerol) over non-ORP control conditions. Oxidizing ORP values were shown to enrich microorganisms of the genus *Clostridium*, which have been linked to high H_2 yields (Vesga-Baron et al. 2021).

Sometimes pretreatment of the inoculum, such as waste activated sludge, is required due to its complex floc structure. Pretreatment helps to achieve the optimal spatial arrangement of microorganisms, a better yield of biohydrogen and an increase in the availability of the substrate (Yang and Wang 2020). Acidity level (pH) and hydraulic retention time (HRT) also affect the activity of different groups of microorganisms. It is known that the optimal pH level for methanogenesis is close to neutral or slightly acidic values (Demirel and Scherer 2008), while hydrogen-producing microorganisms can grow and be metabolically active in a wider pH range. The growth rate of methanogens is on average lower than that of acidogenic microorganisms. By setting a lower pH and HRT, it is possible to create selective conditions and prevent the accumulation of methanogens (Hernández-Mendoza and Buitrón 2014; Nguyen et al. 2021; Kovalev et al. 2021). Thus, in the study by Hernández-Mendoza and Buitrón at pH 4.5 and a short HRT (5.5 h), it was possible to achieve a significant production of hydrogen from anaerobic methanogenic granules. However, some researchers believe that such methods of varying incubation conditions in many cases are not effective and do not significantly reduce the activity of methanogens (Wang et al. 2019). Moreover, the effectiveness of the pretreatment method may depend on the inoculum. For example, Mikheeva et al. (2021) reported that acid pretreatment was effective for thermophilic anaerobic sludge to suppress methanogenic activity, while heat treatment was more effective for mesophilic anaerobic sludge.

Despite the obvious disadvantages of the presence of hydrogen-consuming or competing microbiota in mixed cultures, some researchers report positive results of such cultivation. For example, Ohnishi et al. studied lactate-driven dark fermentation in the co-cultivation of *Megasphaera elsdenii* and several lactic acid bacteria (LAB). The yield of hydrogen was about half that of pure *M. elsdenii*, but the lactate-driven process reportedly produced hydrogen from lactate produced by lactic acid bacteria, as well as from less commonly used sugars such as xylose and cellobiose. Since LAB can ferment various substrates, the use of such a mixed culture is considered promising (Ohnishi et al. 2022). Moreover, lactate wastewater from the food industry can itself be a substrate for the production of biohydrogen. The highest hydrogen yield (0.85 mol H_2/mol lactate consumed) was achieved at

45 °C and an initial pH 8.5. At 35 °C, bacteria from the *Clostridium* genera were the most abundant at the end of the fermentation in the reactors that produced hydrogen, while at 45 °C *Sporanaerobacter*, *Clostridium,* and *Pseudomonas* were the most abundant (Ziara et al. 2019). Mthethwa et al. (2019) also reported better mixed culture performance obtained after heat pretreatment of anaerobic activated sludge compared to pure culture cultivation. The maximum hydrogen yield (HY) of 2.3 mol H_2/mol glucose was obtained at pH 5.5, fermentation time 6.7 h, substrate-to-biomass ratio (S/X) 0.5 g COD/g VSS, and 25 °C using a mixed culture system, including *Enterobacteriaceae*, *Gammaproteobacteria*, *Betaproteobacteria*, and *Clostridium histolyticum.*

Mixed microbial cultures are also used to ferment complex and untreated substrates, in particular various organic wastes. Such substrates do not require pretreatment, because the community contains microorganisms capable of synthesizing hydrolytic enzymes. Thus, in the work of Pérez-Rangel et al. inoculums obtained from different sources (anaerobic sludge, cow rumen, forest soil, native wheat straw microbiota) were used to obtain biohydrogen from untreated wheat straw. The highest rate of hydrogen production, as well as the specific H_2 yield, was recorded when native wheat straw microbiota was used as an inoculum. Researchers attribute this to the presence of hydrolytic fungi in the inoculum adapted to this type of substrate (Pérez-Rangel et al. 2015). Yang and Wang used digested anaerobic sludge to produce hydrogen from pasture ryegrass under mesophilic conditions, using several pretreatment methods (heat shock, acid, base, aeration, and gamma radiation) to eliminate methanogens and enrich the inoculum with hydrogen producers. Acid pretreatment was most efficient compared with other tested pretreatment methods, with relevant hydrogen yield of 64.4 mL/g dry grass and organics removal of 31.4%. *Clostridium* and *Enterococcus* were two dominant genera of HPB (Yang and Wang 2018).

Much more often when working with mixed cultures, a complex substrate, such as lignocellulosic biomass, is pretreated to improve its availability to microorganisms and increase bioconversion (Soares et al. 2020; Bundhoo 2019). It has also recently been shown that a new vortex layer pretreatment method can improve the performance of a dark fermentation system. Thorough disintegration, partial hydrolysis, and formation of iron nanoparticles have been reported to be major contributors to process improvement (Kovalev et al. 2022; Mikheeva et al. 2022). In their study, Phowan and Danvirutai used pre-acid hydrolyzed cassava pulp as a substrate. The inoculum was sewage sludge and anaerobic sludge; the maximum hydrogen yield was 342 mL H_2/g COD (Phowan and Danvirutai 2014). Mixed cultures are used to ferment not only complex natural substrates, but also simple sugars. Qiu et al. used digested manure and sewage sludge as an inoculum and xylose as a substrate for dark fermentation in the temperature range of 35–65 °C. The aim of the work was to identify ways of xylose utilization and succession of the microbial community depending on temperature. *Clostridium* species were dominant at 35–40 °C, while *Thermoanaerobacterium* dominated at 45–60 °C. 1.11 and 1.30 mol H_2/mol xylose consumed were obtained at 35 and 55 °C (Qiu et al. 2017). Sivagurunathan et al. used galactose in an up-flow anaerobic sludge blanket reactor as a substrate for dark fermentation when studying the effect of HRT (Sivagurunathan et al. 2016). Lee et al. studied hydrogen production from seacoast sludge at reduced pressure in a sucrose-fed reactor (Lee et al. 2012). These studies are productive and promising not only in terms of biotechnology, but also a fundamental understanding of the processes, along with work on the decomposition of real substrates.

5.7 Bioaugmentation Technology

Bioaugmentation in DF technology is the enrichment of indigenous microorganisms with one or more HPBs to enhance the ability of the microbial community to withstand environmental factors and achieve higher hydrogen production. Thus, the system is a mixed culture with a significantly dominant hydrogen-producing microorganism. Bioaugmentation technology is used in cases where it is necessary to quickly obtain a high hydrogen yield, shorten the lag phase, maintain community tolerance to high loads of organic matter, increase the substrate degrading capacity, and enhance the yield and formation rate of the target product (Okonkwo et al. 2020a; Soares et al. 2020) (Table 5.2).

The effect of gradually increasing glucose concentrations (from 5.6 to 111 mmol/L) on the fermentative H_2 production with and without bioaugmentation with *Thermotoga neapolitana* was studied by Okonkwo et al. They found that the rate of production and yield of H_2 was higher in bioaugmented cultures regardless of glucose concentration. Quantitative polymerase chain reaction targeting *T. neapolitana* hydA gene and MiSeq sequencing proved that *Thermotoga* was not only present in the augmented cultures but also the most abundant at the highest glucose concentrations (Okonkwo et al. 2020b). Bioaugmentation was also shown to enhance dark fermentative hydrogen production in cultures exposed to short-term temperature fluctuations. Synthetic mixed culture comprising of *Thermoanaerobacter thermohydrosulfuricus* (DSM-567), *Caldicellulosiruptor saccharolyticus* (DSM-8903), *Clostridium thermocellum* (DSM-1237), *Thermoanaerobacterium thermosaccharolyticum* (DSM-571), and *Thermotoga neapolitana* was added to the enrichment culture. This system was subjected to a 48-h downward or upward temperature fluctuation from 55 to 35 or 75 °C. Bioaugmentation has led to a significant improvement in H_2 production as reported by Okonkwo et al. (2020a).

Combined bioaugmentation by HPB and non-living additives can also be used. *Thermoanaerobacterium thermosaccharolyticum* MJ2 and biochar were used to enhance thermophilic hydrogen production from sugarcane bagasse. MJ2 bioaugmentation notably increased the hydrogen production by 95.31%, which was further significantly improved by 158.10% by adding biochar. The positive effect of combined bioaugmentation was attributed to hydrogenase and electron transport system activities stimulated by biochar (Huang et al. 2022).

It is reported that the bioaugmentation method makes it easier to utilize complex lignocellulosic substrates and their hydrolysates. In a study by Zhang et al. natural microbial communities were enriched with *T. thermosaccharolyticum* W16 to

Table 5.2 Effects of bioaugmentation on dark fermentative hydrogen production

Indigenous microbial community	Bioaugmenting microorganism	Improvements	Mechanism	Reference
Domestic sewage sludge	*Ruminiclostridium thermocellum* M3	Six-fold increase in H_2 yield	Higher cellulolytic activity and insensitivity to cellobiose inhibition	Sheng et al. (2021)
Rotten corn stover, cow dung compost, and digested sludge	*Thermoanaerobacterium thermosaccharolyticum* W16	Increase in the H_2 yield	Utilization of complex lignocellulosic substrates and their hydrolysates	Zhang et al. (2019)
Mixed culture obtained from a laboratory scale continuously stirred tank bioreactor	*Thermotoga neapolitana*	37.5% increase in H_2 yield	Usage of wide range of organic substrates and producing high hydrogen yields	Okonkwo et al. (2020b)
River sediment	*Thermoanaerobacterium thermosaccharolyticum* MJ2	Increase in the hydrogen production by 95.31%	Promotion of metabolic pathways related to substrate degradation and microbial metabolism	Huang et al. (2022)
Anaerobic sludge	*Clostridium sp.* TCW1	H_2 yield increased from 6 to 14 mol/kg-COD_{added} (+ 133.3%)	Efficient hydrolysis of cellulosic biomass	Kuo et al. (2012)
Anaerobic sludge	*Hydrogenispora ethanolica* LX-B	H_2 production increased from 0.57 to 1.78 mL/g VS/h	Altering the composition of the indigenous bacterial community and a shift in acidogenic metabolic pathway	Yang et al. (2016)
Sewage sludge	*Enterobacter cloacae* DSM 16657	H_2 yield increased from 220.6 to 236.5 mL/g-COD_{added}	Robustness and efficiency of hydrogen production	Kumar et al. (2015a)
Sewage sludge	*Escherichia coli* XL1-Blue	H_2 yield increased from 220.6 to 259.6 mL/g-COD_{added}	Robustness and efficiency of hydrogen production	Kumar et al. (2015a)
Anaerobic sludge from brewery industry	*Bacillus subtilis*	344% increase in H_2 production	High metabolic activity on organic wastes	Villanueva-Galindo, Moreno-Andrade (2021)
Cow dung compost	*Bacillus sp.* FS2011	16% increase in H_2 yield	High-efficiency hydrogen producing performance in a broad range of initial pH	Guo et al. (2014b)

improve hydrogen production from corn stover. The addition of even a small amount (5% of all microorganisms) to all seed sludge tested (rotten corn stover, cow dung compost, and digested sludge) has significantly increased hydrogen production from corn stover hydrolysate (Zhang et al. 2019). Sometimes, bioaugmentation with a cellulolytic strain makes it possible to avoid the stage of preliminary hydrolysis of the substrate. Sheng et al. enriched domestic sewage sludge with a strain of *Ruminiclostridium thermocellum* M3, which has cellulolytic activity and is insensitive to cellobiose inhibition. Several substrates with a high content of lignocellulose were used in the work, and a six-fold increase in hydrogen yield was shown for each of them. The hydrogen yield obtained by bioaugmentation exceeded that obtained by adding fungal cellulase (*Trichoderma viride*) to domestic sewage sludge (Sheng et al. 2021). However, positive interactions between the hydrolytic and fermentative members were only observed in co-culture tests when the lignocellulosic substrate was sterilized. In the presence of the host microbial community, the synergistic relationships of the bioaugmentation bacteria were lost, suggesting the importance of choosing the right dose of bioaugmentation bacteria that will allow them to thrive in the host environment (Valdez-Vazquez et al. 2019).

The benefits of bioaugmentation technology may be related to the symbiotic relationship between bacteria within a community. Kumar et al. enriched the industrial wastewater community with strains of facultative anaerobes *Escherichia coli* XL1-Blue and *Enterobacter cloacae* DSM 16657, resulting in improved biohydrogen production by strictly anaerobic microorganisms by removing trace oxygen in the system (Kumar et al. 2015a). Experiments comparing the production of biohydrogen by mixed and enriched cultures also confirm the effectiveness of bioaugmentation. Ortigueira et al. investigated hydrogen production from microalgae *Scenedesmus obliquus* biomass using a *C. butyricum* strain and heat-treated sewage sludge. The mesophilic mixed culture LE37 produced 34.8 mL H_2/g VS, while the one bioaugmented with *C. butyricum* strain produced 48.9 mL H_2/g VS, or 1.52 and 2.01 mol/mol of hexose, respectively (Ortigueira et al. 2015).

The study of bioaugmentation is a promising direction for further technological applications. However, it is still difficult to judge the universality of the method. The introduction of exogenous strains into an established community mitigates the instability of mixed cultures but causes strong competition for the substrate (Poirier et al. 2020). It is necessary to understand the specific metabolic interactions between the indigenous microbiota of the community and the introduced alien strain, as they affect the final yield of the product (Zhang et al. 2019). The added concentration of HPB (Villanueva-Galindo and Moreno-Andrade 2021) and HPB to indigenous community ratio (Sharma and Melkania, 2018) are critical parameters for the bioaugmentation process. Further research is required on the applicability of bioaugmentation to different types of inoculum and substrates, as well as the principles for choosing the dominant strain for enrichment.

5.8 Conclusions and Perspectives

The study of the applicability of pure, co-, and mixed cultures for the dark fermentative hydrogen production is an exciting area of modern science. Each of the considered systems has its own advantages and disadvantages, as well as areas for promising use in biotechnology. A wide variety of HPBs with different metabolic capabilities and physiological characteristics make it possible to obtain hydrogen from different substrates, at different temperatures, pH, and other technological parameters. The various inoculum and substrate pretreatment methods for maximizing biohydrogen yield discussed in this chapter help approach the scaling up of dark fermentation and its industrial application.

The most important challenges/issues are the availability and nature of the feedstock, the relationship between microorganisms and their metabolic capabilities, and cultivation conditions. To promote the industrial application of hydrogen production by HPB, more studies are needed from the following aspects:

1. More attention should be paid to the study of the metabolism of microbial communities in order to be able to stimulate the production of hydrogen by modifying metabolic pathways (Pachapur et al. 2015; Patel et al. 2014; Taifor et al. 2017), as well as to create consortiums suitable for the biodegradation of various substrates (Valdez-Vazquez et al. 2015). Based on the knowledge of the metabolic and physiological needs of HPBs, it is possible to create new strains with desired characteristics in order to use them for hydrogen production in a monoculture or for bioaugmentation of an indigenous microbial community.
2. It is promising to use bioaugmentation as a strategy aimed at obtaining hydrogen from complex raw substrates (Zhang et al. 2019). Sometimes, bioaugmentation with a cellulolytic strain makes it possible to avoid the stage of preliminary hydrolysis of the substrate (Sheng et al. 2021). Combined bioaugmentation by HPB and non-living additives such as biochar can also be beneficial (Huang et al. 2022).
3. For all areas of research, it is critical to work not only in the laboratory, but also to develop technologies on a larger scale in order to move to the production of large volumes of hydrogen for industry (Balachandar et al. 2020).

Acknowledgments This research was supported by the Ministry of Science and Higher Education of the Russian Federation, grant No. 075-15-2022-1225.

References

Alvarado-Cuevas ZD, López-Hidalgo AM, Ordoñez LG, Oceguera-Contreras E, Ornelas-Salas JT, De León-Rodríguez A (2015) Biohydrogen production using psychrophilic bacteria isolated from Antarctica. Int J Hydrog Energy 40(24):7586–7592. https://doi.org/10.1016/j.ijhydene.2014.10.063

Argun H, Kargi F, Kapdan IK (2009) Microbial culture selection for bio-hydrogen production from waste ground wheat by dark fermentation. Int J Hydrog Energy 34(5):2195–2200. https://doi.org/10.1016/j.ijhydene.2008.12.066

Azbar N, Dokgöz FT, Keskin T, Eltem R, Korkmaz KS, Gezgin Y et al (2009) Comparative evaluation of bio-hydrogen production from cheese whey wastewater under thermophilic and mesophilic anaerobic conditions. Int J Green Energy 6(2):192–200. https://doi.org/10.1080/15435070902785027

Baghchehsaraee B, Nakhla G, Karamanev D, Margaritis A (2010) Fermentative hydrogen production by diverse microflora. Int J Hydrog Energy 35(10):5021–5027. https://doi.org/10.1016/j.ijhydene.2009.08.072

Bagi Z, Ács N, Bálint B, Horváth L, Dobó K, Perei KR et al (2007) Biotechnological intensification of biogas production. Appl Microbiol Biotechnol 76(2):473–482. https://doi.org/10.1007/s00253-007-1009-6

Balachandar G, Varanasi JL, Singh V, Singh H, Das D (2020) Biological hydrogen production via dark fermentation: A holistic approach from lab-scale to pilot-scale. Int J Hydrog Energy 45(8):5202–5215. https://doi.org/10.1016/j.ijhydene.2019.09.006

Brynjarsdottir H, Scully SM, Orlygsson J (2013) Production of biohydrogen from sugars and lignocellulosic biomass using *Thermoanaerobacter* GHL15. Int J Hydrog Energy 38(34):14467–14475. https://doi.org/10.1016/j.ijhydene.2013.09.005

Bundhoo ZM (2019) Potential of bio-hydrogen production from dark fermentation of crop residues: a review. Int J Hydrog Energy 44(32):17346–17362. https://doi.org/10.1016/j.ijhydene.2018.11.098

Cabrol L, Marone A, Tapia-Venegas E, Steyer JP, Ruiz-Filippi G, Trably E (2017) Microbial ecology of fermentative hydrogen producing bioprocesses: useful insights for driving the ecosystem function. FEMS Microbiol Rev 41(2):158–181. https://doi.org/10.1093/femsre/fuw043

Cappelletti M, Zannoni D, Postec A, Ollivier B (2014) Members of the order *Thermotogales*: from microbiology to hydrogen production. In: Zannoni D, De Philippis R (eds) Microbial bioenergy: hydrogen production. Springer, Dordrecht, pp 197–224. https://doi.org/10.1007/978-94-017-8554-9_9

Castelló E, Ferraz-Junior ADN, Andreani C, del Pilar Anzola-Rojas M, Borzacconi L, Buitrón G et al (2020) Stability problems in the hydrogen production by dark fermentation: possible causes and solutions. Renew Sustain Energy Rev 119:109602. https://doi.org/10.1016/j.rser.2019.109602

Chang S, Li JZ, Liu F (2011) Evaluation of different pretreatment methods for preparing hydrogen-producing seed inocula from waste activated sludge. Renew Energy 36(5):1517–1522. https://doi.org/10.1016/j.renene.2010.11.023

Chen CC, Chuang YS, Lin CY, Lay CH, Sen B (2012) Thermophilic dark fermentation of untreated rice straw using mixed cultures for hydrogen production. Int J Hydrog Energy 37(20):15540–15546. https://doi.org/10.1016/j.ijhydene.2012.01.036

Chen Y, Yin Y, Wang J (2021) Recent advance in inhibition of dark fermentative hydrogen production. Int J Hydrog Energy 46(7):5053–5073. https://doi.org/10.1016/j.ijhydene.2020.11.096

Cheng J, Zhu M (2013) A novel anaerobic co-culture system for bio-hydrogen production from sugarcane bagasse. Bioresour Technol 144:623–631. https://doi.org/10.1016/j.biortech.2013.07.018

Cheng J, Li H, Ding L, Zhou J, Song W, Li YY, Lin R (2020) Improving hydrogen and methane co-generation in cascading dark fermentation and anaerobic digestion: The effect of magnetite nanoparticles on microbial electron transfer and syntrophism. Chem Eng J 397:125394. https://doi.org/10.1016/j.cej.2020.125394

Cho Y, Lee T (2011) Variations of hydrogen production and microbial community with heavy metals during fermentative hydrogen production. J Ind Eng Chem 17(2):340–345. https://doi.org/10.1016/j.jiec.2011.02.036

Chojnacka A, Błaszczyk MK, Szczęsny P, Nowak K, Sumińska M, Tomczyk-Żak K et al (2011) Comparative analysis of hydrogen-producing bacterial biofilms and granular sludge formed in continuous cultures of fermentative bacteria. Bioresour Technol 102(21):10057–10064. https://doi.org/10.1016/j.biortech.2011.08.063

Ciranna A, Santala V, Karp M (2011) Biohydrogen production in alkalithermophilic conditions: *Thermobrachium celere* as a case study. Bioresour Technol 102(18):8714–8722. https://doi.org/10.1016/j.biortech.2011.01.028

Cui Y, Yang KL, Zhou K (2021) Using co-culture to functionalize *Clostridium* fermentation. Trends Biotechnol 39(9):914–926. https://doi.org/10.1016/j.tibtech.2020.11.016

Dahiya S, Chatterjee S, Sarkar O, Mohan SV (2021) Renewable hydrogen production by dark-fermentation: Current status, challenges and perspectives. Bioresour Technol 321:124354. https://doi.org/10.1016/j.biortech.2020.124354

Dębowski M, Korzeniewska E, Filipkowska Z, Zieliński M, Kwiatkowski R (2014) Possibility of hydrogen production during cheese whey fermentation process by different strains of psychrophilic bacteria. Int J Hydrog Energy 39(5):1972–1978. https://doi.org/10.1016/j.ijhydene.2013.11.082

Demirel B, Scherer P (2008) The roles of acetotrophic and hydrogenotrophic methanogens during anaerobic conversion of biomass to methane: a review. Rev Environ Sci Biotechnol 7(2):173–190. https://doi.org/10.1007/s11157-008-9131-1

de Sá LRV, Cammarota MC, de Oliveira TC, Oliveira EMM, Matos A, Ferreira-Leitão VS (2013) Pentoses, hexoses and glycerin as substrates for biohydrogen production: An approach for Brazilian biofuel integration. Int J Hydrog Energy 38(7):2986–2997. https://doi.org/10.1016/j.ijhydene.2012.12.103

Dessì P, Porca E, Waters NR, Lakaniemi AM, Collins G, Lens PN (2018) Thermophilic versus mesophilic dark fermentation in xylose-fed fluidised bed reactors: Biohydrogen production and active microbial community. Int J Hydrog Energy 43(11):5473–5485. https://doi.org/10.1016/j.ijhydene.2018.01.158

d'Ippolito G, Dipasquale L, Vella FM, Romano I, Gambacorta A, Cutignano A, Fontana A (2010) Hydrogen metabolism in the extreme thermophile *Thermotoga neapolitana*. Int J Hydrog Energy 35(6):2290–2295. https://doi.org/10.1016/j.ijhydene.2009.12.044

Elbeshbishy E, Dhar BR, Nakhla G, Lee HS (2017) A critical review on inhibition of dark biohydrogen fermentation. Renew Sustain Energy Rev 79:656–668. https://doi.org/10.1016/j.rser.2017.05.075

Elsharnouby O, Hafez H, Nakhla G, El Naggar MH (2013) A critical literature review on biohydrogen production by pure cultures. Int J Hydrog Energy 38(12):4945–4966. https://doi.org/10.1016/j.ijhydene.2013.02.032

Ergal İ, Gräf O, Hasibar B, Steiner M, Vukotić S, Bochmann G et al (2020) Biohydrogen production beyond the Thauer limit by precision design of artificial microbial consortia. Commun Biol 3(1):1–12. https://doi.org/10.1038/s42003-020-01159-x

Fang HH, Liu H (2002) Effect of pH on hydrogen production from glucose by a mixed culture. Bioresour Technol 82(1):87–93. https://doi.org/10.1016/S0960-8524(01)00110-9

Fang H, Zhang T, Liu H (2002) Microbial diversity of a mesophilic hydrogen-producing sludge. Appl Microbiol Biotechnol 58(1):112–118. https://doi.org/10.1007/s00253-001-0865-8

Ferreira TB, Rego GC, Ramos LR, Soares LA, Sakamoto IK, de Oliveira LL et al (2018) Selection of metabolic pathways for continuous hydrogen production under thermophilic and mesophilic temperature conditions in anaerobic fluidized bed reactors. Int J Hydrog Energy 43(41):18908–18917. https://doi.org/10.1016/j.ijhydene.2018.08.177

Fuentes L, Palomo-Briones R, de Jesús Montoya-Rosales J, Braga L, Castelló E, Vesga A et al (2021) Knowing the enemy: homoacetogens in hydrogen production reactors. Appl Microbiol Biotechnol 105(23):8989–9002

Geng A, He Y, Qian C, Yan X, Zhou Z (2010) Effect of key factors on hydrogen production from cellulose in a co-culture of *Clostridium thermocellum* and *Clostridium thermopalmarium*. Bioresour Technol 101(11):4029–4033. https://doi.org/10.1016/j.biortech.2010.01.042

Ghimire A, Frunzo L, Pirozzi F, Trably E, Escudie R, Lens PN, Esposito G (2015) A review on dark fermentative biohydrogen production from organic biomass: process parameters and use of by-products. Appl Energy 144:73–95. https://doi.org/10.1016/j.apenergy.2015.01.045

Giovannini G, Donoso-Bravo A, Jeison D, Chamy R, Ruíz-Filippi G, Wouwer AV (2016) A review of the role of hydrogen in past and current modelling approaches to anaerobic digestion processes. Int J Hydrog Energy 41(39):17713–17722. https://doi.org/10.1016/j.ijhydene.2016.07.012

Gomez-Flores M, Nakhla G, Hafez H (2017) Hydrogen production and microbial kinetics of *Clostridium termitidis* in mono-culture and co-culture with *Clostridium beijerinckii* on cellulose. AMB Express 7(1):1–12. https://doi.org/10.1186/s13568-016-0256-2

Gonzales RR, Kim JS, Kim SH (2019) Optimization of dilute acid and enzymatic hydrolysis for dark fermentative hydrogen production from the empty fruit bunch of oil palm. Int J Hydrog Energy 44(4):2191–2202. https://doi.org/10.1016/j.ijhydene.2018.08.022

Gorgec FK, Karapinar I (2019) Biohydrogen production from hydrolyzed waste wheat by dark fermentation in a continuously operated packed bed reactor: the effect of hydraulic retention time. Int J Hydrog Energy 44(1):136–143. https://doi.org/10.1016/j.ijhydene.2018.08.155

Guo XM, Trably E, Latrille E, Carrere H, Steyer JP (2014a) Predictive and explicative models of fermentative hydrogen production from solid organic waste: role of butyrate and lactate pathways. Int J Hydrog Energy 39(14):7476–7485. https://doi.org/10.1016/j.ijhydene.2013.08.079

Guo YC, Dai Y, Bai YX, Li YH, Fan YT, Hou HW (2014b) Co-producing hydrogen and methane from higher-concentration of corn stalk by combining hydrogen fermentation and anaerobic digestion. Int J Hydrog Energy 39(26):14204–14211. https://doi.org/10.1016/j.ijhydene.2014.02.089

Gupta N, Pal M, Sachdeva M, Yadav M, Tiwari A (2016) Thermophilic biohydrogen production for commercial application: the whole picture. Int J Energy Res 40(2):127–145. https://doi.org/10.1002/er.3438

Hallenbeck PC (2005) Fundamentals of the fermentative production of hydrogen. Water Sci Technol 52(1–2):21–29. https://doi.org/10.2166/wst.2005.0494

Hallenbeck PC, Ghosh D (2009) Advances in fermentative biohydrogen production: the way forward? Trends Biotechnol 27(5):287–297. https://doi.org/10.1016/j.tibtech.2009.02.004

Han W, Liu DN, Shi YW, Tang JH, Li YF, Ren NQ (2015) Biohydrogen production from food waste hydrolysate using continuous mixed immobilized sludge reactors. Bioresour Technol 180:54–58. https://doi.org/10.1016/j.biortech.2014.12.067

Harirchi S, Wainaina S, Sar T, Nojoumi SA, Parchami M, Parchami M et al (2022) Microbiological insights into anaerobic digestion for biogas, hydrogen or volatile fatty acids (VFAs): a review. Bioengineered 13(3):6521–6557. https://doi.org/10.1080/21655979.2022.2035986

Hernández-Mendoza CE, Buitrón G (2014) Suppression of methanogenic activity in anaerobic granular biomass for hydrogen production. J Chem Technol Biotechnol 89(1):143–149. https://doi.org/10.1002/jctb.4143

Hsiao CL, Chang JJ, Wu JH, Chin WC, Wen FS, Huang CC et al (2009) *Clostridium* strain co-cultures for biohydrogen production enhancement from condensed molasses fermentation solubles. Int J Hydrog Energy 34(17):7173–7181. https://doi.org/10.1016/j.ijhydene.2009.06.028

Hu B, Chen S (2007) Pretreatment of methanogenic granules for immobilized hydrogen fermentation. Int J Hydrog Energy 32(15):3266–3273. https://doi.org/10.1016/j.ijhydene.2007.03.005

Hu BB, Zhu MJ (2017) Direct hydrogen production from dilute-acid pretreated sugarcane bagasse hydrolysate using the newly isolated *Thermoanaerobacterium thermosaccharolyticum* MJ1. Microb Cell Factories 16(1):1–11. https://doi.org/10.1186/s12934-017-0692-y

Huang JR, Chen X, Hu BB, Cheng JR, Zhu MJ (2022) Bioaugmentation combined with biochar to enhance thermophilic hydrogen production from sugarcane bagasse. Bioresour Technol 348:126790. https://doi.org/10.1016/j.biortech.2022.126790

Jung JH, Sim YB, Baik JH, Park JH, Kim SH (2021) High-rate mesophilic hydrogen production from food waste using hybrid immobilized microbiome. Bioresour Technol 320:124279. https://doi.org/10.1016/j.biortech.2020.124279

Kádár Z, Vrije TD, Noorden GEV, Budde MA, Szengyel Z, Réczey K, Claassen PA (2004) Yields from glucose, xylose, and paper sludge hydrolysate during hydrogen production by the extreme thermophile *Caldicellulosiruptor saccharolyticus*. In: Proceedings of the twenty-fifth symposium on biotechnology for fuels and chemicals held May 4–7, 2003, in Breckenridge, CO. Humana, Totowa, NJ, pp 497–508. https://doi.org/10.1385/ABAB:114:1-3:497

Kanso S, Dasri K, Tingthong S, Watanapokasin RY (2011) Diversity of cultivable hydrogen-producing bacteria isolated from agricultural soils, wastewater sludge and cow dung. Int J Hydrog Energy 36(14):8735–8742. https://doi.org/10.1016/j.ijhydene.2010.07.010

Kao PM, Hsu BM, Chang TY, Chiu YC, Tsai SH, Huang YL, Chang CM (2016) Biohydrogen production by *Clostridium butyricum* and *Rhodopseudomonas palustris* in Co-cultures. Int J Green Energy 13(7):715–719. https://doi.org/10.1080/15435075.2015.1088443

Kargi F, Eren NS, Ozmihci S (2012) Bio-hydrogen production from cheese whey powder (CWP) solution: comparison of thermophilic and mesophilic dark fermentations. Int J Hydrog Energy 37(10):8338–8342. https://doi.org/10.1016/j.ijhydene.2012.02.162

Kazimierowicz J, Dębowski M, Zieliński M (2022) Effectiveness of hydrogen production by *Bacteroides vulgatus* in psychrophilic fermentation of cattle slurry. Clean Technol 4(3):806–814. https://doi.org/10.3390/cleantechnol4030049

Khamtib S, Reungsang A (2012) Biohydrogen production from xylose by *Thermoanaerobacterium thermosaccharolyticum* KKU19 isolated from hot spring sediment. Int J Hydrog Energy 37(17):12219–12228. https://doi.org/10.1016/j.ijhydene.2012.06.038

Khan MA, Ngo HH, Guo W, Liu Y, Zhang X, Guo J et al (2018) Biohydrogen production from anaerobic digestion and its potential as renewable energy. Renew Energy 129:754–768. https://doi.org/10.1016/j.renene.2017.04.029

Kim SH, Han SK, Shin HS (2006) Effect of substrate concentration on hydrogen production and 16S rDNA-based analysis of the microbial community in a continuous fermenter. Process Biochem 41(1):199–207. https://doi.org/10.1016/j.procbio.2005.06.013

Kirli B, Karapinar I (2018) The effect of HRT on biohydrogen production from acid hydrolyzed waste wheat in a continuously operated packed bed reactor. Int J Hydrog Energy 43(23):10678–10685. https://doi.org/10.1016/j.ijhydene.2018.01.175

Kleerebezem R, van Loosdrecht MC (2007) Mixed culture biotechnology for bioenergy production. Curr Opin Biotechnol 18(3):207–212. https://doi.org/10.1016/j.copbio.2007.05.001

Kovács KL, Ács N, Kovács E, Wirth R, Rákhely G, Strang O et al (2013) Improvement of biogas production by bioaugmentation. Biomed Res Int. https://doi.org/10.1155/2013/482653

Kovalev AA, Kovalev DA, Nozhevnikova AN, Zhuravleva EA, Katraeva IV, Grigoriev VS, Litti YV (2021) Effect of low digestate recirculation ratio on biofuel and bioenergy recovery in a two-stage anaerobic digestion process. Int J Hydrog Energy 46(80):39688–39699. https://doi.org/10.1016/j.ijhydene.2021.09.239

Kovalev AA, Kovalev DA, Panchenko VA, Zhuravleva EA, Laikova AA, Shekhurdina SV et al (2022) Approbation of an innovative method of pretreatment of dark fermentation feedstocks. Int J Hydrog Energy. https://doi.org/10.1016/j.ijhydene.2022.08.051

Kumar G, Bakonyi P, Sivagurunathan P, Kim SH, Nemestóthy N, Bélafi-Bakó K, Lin CY (2015a) Enhanced biohydrogen production from beverage industrial wastewater using external nitrogen sources and bioaugmentation with facultative anaerobic strains. J Biosci Bioeng 120(2):155–160. https://doi.org/10.1016/j.jbiosc.2014.12.011

Kumar G, Bakonyi P, Periyasamy S, Kim SH, Nemestóthy N, Bélafi-Bakó K (2015b) Lignocellulose biohydrogen: practical challenges and recent progress. Renew Sustain Energy Rev 44:728–737. https://doi.org/10.1016/j.rser.2015.01.042

Kumari S, Das D (2016) Biologically pretreated sugarcane top as a potential raw material for the enhancement of gaseous energy recovery by two stage biohythane process. Bioresour Technol 218:1090–1097. https://doi.org/10.1016/j.biortech.2016.07.070

Kuo WC, Chao YC, Wang YC, Cheng SS (2012) Bioaugmentation strategies to improve cellulolytic and hydrogen producing characteristics in CSTR intermittent fed with vegetable kitchen waste and napiergrass. Energy Procedia 29:82–91. https://doi.org/10.1016/j.egypro.2012.09.011

Labatut RA, Angenent LT, Scott NR (2014) Conventional mesophilic vs. thermophilic anaerobic digestion: a trade-off between performance and stability? Water Res 53:249–258. https://doi.org/10.1016/j.watres.2014.01.035

Laxman Pachapur V, Jyoti Sarma S, Kaur Brar S, Le Bihan Y, Ricardo Soccol C, Buelna G, Verma M (2015) Co-culture strategies for increased biohydrogen production. Int J Energy Res 39(11):1479–1504. https://doi.org/10.1002/er.3364

Lee KS, Tseng TS, Liu YW, Hsiao YD (2012) Enhancing the performance of dark fermentative hydrogen production using a reduced pressure fermentation strategy. Int J Hydrog Energy 37(20):15556–15562. https://doi.org/10.1016/j.ijhydene.2012.04.039

Li Q, Liu CZ (2012) Co-culture of *Clostridium thermocellum* and *Clostridium thermosaccharolyticum* for enhancing hydrogen production via thermophilic fermentation of cornstalk waste. Int J Hydrog Energy 37(14):10648–10654. https://doi.org/10.1016/j.ijhydene.2012.04.115

Litti YV, Kovalev DA, Kovalev AA, Merkel AY, Vishnyakova AV, Russkova YI, Nozhevnikova AN (2021) Auto-selection of microorganisms of sewage sludge used as an inoculum for fermentative hydrogen production from different substrates. Int J Hydrog Energy 46(58):29834–29845. https://doi.org/10.1016/j.ijhydene.2021.06.174

Litti YV, Potekhina MA, Zhuravleva EA, Vishnyakova AV, Gruzdev DS, Kovalev AA et al (2022) Dark fermentative hydrogen production from simple sugars and various wastewaters by a newly isolated *Thermoanaerobacterium thermosaccharolyticum* SP-H2. Int J Hydrog Energy 47(58):24310–24327. https://doi.org/10.1016/j.ijhydene.2022.05.235

Litti YV, Khuraseva ND, Vishnyakova AV, Zhuravleva EA, Kovalev AA, Kovalev DA et al (2023) Comparative study on biohydrogen production by newly isolated Clostridium butyricum SP4 and Clostridium beijerinckii SP6. Int J Hydrog Energy. https://doi.org/10.1016/j.ijhydene.2023.03.424. Available online 17 April 2023. Corrected Proof

Liu Y, Yu P, Song X, Qu Y (2008) Hydrogen production from cellulose by co-culture of *Clostridium thermocellum* JN4 and *Thermoanaerobacterium thermosaccharolyticum* GD17. Int J Hydrog Energy 33(12):2927–2933. https://doi.org/10.1016/j.ijhydene.2008.04.004

Liu H, Wang G, Zhu D, Pan G (2009) Enrichment of the hydrogen-producing microbial community from marine intertidal sludge by different pretreatment methods. Int J Hydrog Energy 34(24):9696–9701. https://doi.org/10.1016/j.ijhydene.2009.10.025

Liu H, Wang J, Wang A, Chen J (2011) Chemical inhibitors of methanogenesis and putative applications. Appl Microbiol Biotechnol 89(5):1333–1340. https://doi.org/10.1007/s00253-010-3066-5

Lu W, Wen J, Chen Y, Sun B, Jia X, Liu M, Caiyin Q (2007) Synergistic effect of *Candida maltosa* HY-35 and *Enterobacter aerogenes* W-23 on hydrogen production. Int J Hydrog Energy 32(8):1059–1066. https://doi.org/10.1016/j.ijhydene.2006.07.010

Łukajtis R, Hołowacz I, Kucharska K, Glinka M, Rybarczyk P, Przyjazny A, Kamiński M (2018) Hydrogen production from biomass using dark fermentation. Renew Sustain Energy Rev 91:665–694. https://doi.org/10.1016/j.rser.2018.04.043

Luo L, Sriram S, Johnravindar D, Martin TLP, Wong JW, Pradhan N (2022) Effect of inoculum pretreatment on the microbial and metabolic dynamics of food waste dark fermentation. Bioresour Technol 358:127404. https://doi.org/10.1016/j.biortech.2022.127404

Ma H, Su H (2019) Effect of temperature on the fermentation of starch by two high efficient H2 producers. Renew Energy 138:964–970. https://doi.org/10.1016/j.renene.2019.01.126

Magrini FE, de Almeida GM, da Maia Soares D, Fuentes L, Ecthebehere C, Beal LL et al (2021) Effect of different heat treatments of inoculum on the production of hydrogen and volatile fatty acids by dark fermentation of sugarcane vinasse. Biomass Convers Biorefin 11(6):2443–2456. https://doi.org/10.1007/s13399-020-00687-0

Mahata C, Ray S, Das D (2020) Optimization of dark fermentative hydrogen production from organic wastes using acidogenic mixed consortia. Energy Convers Manag 219:113047. https://doi.org/10.1016/j.enconman.2020.113047

Mahato RK, Kumar D, Rajagopalan G (2020) Biohydrogen production from fruit waste by *Clostridium* strain BOH3. Renew Energy 153:1368–1377. https://doi.org/10.1016/j.renene.2020.02.092

Mañunga T, Barrios-Pérez JD, Zaiat M, Rodríguez-Victoria JA (2019) Evaluation of pretreatment methods and initial pH on mixed inoculum for fermentative hydrogen production from cassava wastewater. Biofuels 1–8. https://doi.org/10.1080/17597269.2019.1680041

Maru BT, López F, Kengen SWM, Constantí M, Medina F (2016) Dark fermentative hydrogen and ethanol production from biodiesel waste glycerol using a co-culture of *Escherichia coli* and *Enterobacter* sp. Fuel 186:375–384. https://doi.org/10.1016/j.fuel.2016.08.043

Masset J, Calusinska M, Hamilton C, Hiligsmann S, Joris B, Wilmotte A, Thonart P (2012) Fermentative hydrogen production from glucose and starch using pure strains and artificial co-cultures of *Clostridium* spp. Biotechnol Biofuels 5(1):1–15. https://doi.org/10.1186/1754-6834-5-35

Mazumder P, Nath D, Manhar AK, Gupta K, Saikia D, Mandal M (2020) Dark fermentative hydrogen production from lignocellulosic agro-waste by a newly isolated bacteria Staphylococcus epidermidis B-6. In: Kumar M, Munoz-Arriola F, Furumai H, Chaminda T (eds) Resilience, response, and risk in water systems. Springer, Singapore, pp 381–395. https://doi.org/10.1007/978-981-15-4668-6_21

McDowall JS, Murphy BJ, Haumann M, Palmer T, Armstrong FA, Sargent F (2014) Bacterial formate hydrogenlyase complex. Proc Natl Acad Sci USA 111(38):E3948–E3956. https://doi.org/10.1073/pnas.1407927111

Mikheeva ER, Katraeva IV, Kovalev AA, Kovalev DA, Nozhevnikova AN, Panchenko V et al (2021) The start-up of continuous biohydrogen production from cheese whey: Comparison of inoculum pretreatment methods and reactors with moving and fixed polyurethane carriers. Appl Sci 11(2):510. https://doi.org/10.3390/app11020510

Mikheeva ER, Katraeva IV, Kovalev AA, Kovalev DA, Litti YV (2022) Effects of pretreatment in a vortex layer apparatus on the properties of confectionery wastewater and its dark fermentation. Int J Hydrog Energy. https://doi.org/10.1016/j.ijhydene.2022.05.183

Mohan SV, Babu VL, Sarma PN (2008) Effect of various pretreatment methods on anaerobic mixed microflora to enhance biohydrogen production utilizing dairy wastewater as substrate. Bioresour Technol 99(1):59–67. https://doi.org/10.1016/j.biortech.2006.12.004

Moodley P, Kana EG (2018) Comparative study of three optimized acid-based pretreatments for sugar recovery from sugarcane leaf waste: a sustainable feedstock for biohydrogen production. Eng Sci Technol Int J 21(1):107–116. https://doi.org/10.1016/j.jestch.2017.11.010

Moon HG, Jang YS, Cho C, Lee J, Binkley R, Lee SY (2016) One hundred years of clostridial butanol fermentation. FEMS Microbiol Lett 363(3). https://doi.org/10.1093/femsle/fnw001

Mthethwa NP, Nasr M, Kiambi SL, Bux F, Kumari S (2019) Biohydrogen fermentation from *Pistia stratiotes* (aquatic weed) using mixed and pure bacterial cultures. Int J Hydrog Energy 44(33):17720–17731. https://doi.org/10.1016/j.ijhydene.2019.05.152

Mu Y, Zheng XJ, Yu HQ, Zhu RF (2006) Biological hydrogen production by anaerobic sludge at various temperatures. Int J Hydrog Energy 31(6):780–785. https://doi.org/10.1016/j.ijhydene.2005.06.016

Nagarajan D, Lee DJ, Chang JS (2019) Recent insights into consolidated bioprocessing for lignocellulosic biohydrogen production. Int J Hydrog Energy 44(28):14362–14379. https://doi.org/10.1016/j.ijhydene.2019.03.066

Nandi R, Sengupta S (1998) Microbial production of hydrogen: an overview. Crit Rev Microbiol 24(1):61–84. https://doi.org/10.1080/10408419891294181

Nguyen TT, Chu CY, Ou CM (2021) Pre-treatment study on two-stage biohydrogen and biomethane productions in a continuous co-digestion process from a mixture of swine manure and pineapple waste. Int J Hydrog Energy 46(20):11325–11336. https://doi.org/10.1016/j.ijhydene.2020.05.264

Nissilä ME, Lay CH, Puhakka JA (2014) Dark fermentative hydrogen production from lignocellulosic hydrolyzates—a review. Biomass Bioenergy 67:145–159. https://doi.org/10.1016/j.biombioe.2014.04.035

Nitipan S, Mamimin C, Intrasungkha N, Birkeland NK, Sompong O (2014) Microbial community analysis of thermophilic mixed culture sludge for biohydrogen production from

palm oil mill effluent. Int J Hydrog Energy 39(33):19285–19293. https://doi.org/10.1016/j.ijhydene.2014.05.139
Ntaikou I, Gavala HN, Lyberatos G (2009) Modeling of fermentative hydrogen production from the bacterium *Ruminococcus albus*: definition of metabolism and kinetics during growth on glucose. Int J Hydrog Energy 34(9):3697–3709. https://doi.org/10.1016/j.ijhydene.2009.02.057
Ohnishi A, Hasegawa Y, Fujimoto N, Suzuki M (2022) Biohydrogen production by mixed culture of *Megasphaera elsdenii* with lactic acid bacteria as lactate-driven dark fermentation. Bioresour Technol 343:126076. https://doi.org/10.1016/j.biortech.2021.126076
Okonkwo O, Escudie R, Bernet N, Mangayil R, Lakaniemi AM, Trably E (2020a) Bioaugmentation enhances dark fermentative hydrogen production in cultures exposed to short-term temperature fluctuations. Appl Microbiol Biotechnol 104(1):439–449. https://doi.org/10.1007/s00253-019-10203-8
Okonkwo O, Papirio S, Trably E, Escudie R, Lakaniemi AM, Esposito G (2020b) Enhancing thermophilic dark fermentative hydrogen production at high glucose concentrations via bioaugmentation with Thermotoga neapolitana. Int J Hydrog Energy 45(35):17241–17249. https://doi.org/10.1016/j.ijhydene.2020.04.231
Ortigueira J, Alves L, Gouveia L, Moura P (2015) Third generation biohydrogen production by *Clostridium butyricum* and adapted mixed cultures from *Scenedesmus obliquus* microalga biomass. Fuel 153:128–134. https://doi.org/10.1016/j.fuel.2015.02.093
Ozmihci S, Kargi F (2011) Dark fermentative bio-hydrogen production from waste wheat starch using co-culture with periodic feeding: effects of substrate loading rate. Int J Hydrog Energy 36(12):7089–7093. https://doi.org/10.1016/j.ijhydene.2011.03.071
Pachapur VL, Sarma SJ, Brar SK, Le Bihan Y, Buelna G, Verma M (2015) Biological hydrogen production using co-culture versus mono-culture system. Environ Technol Rev 4(1):55–70. https://doi.org/10.1080/21622515.2015.1068381
Pachapur VL, Kutty P, Pachapur P, Brar SK, Le Bihan Y, Galvez-Cloutier R, Buelna G (2019) Seed pretreatment for increased hydrogen production using mixed-culture systems with advantages over pure-culture systems. Energies 12(3):530. https://doi.org/10.3390/en12030530
Park JH, Lee SH, Ju HJ, Kim SH, Yoon JJ, Park HD (2016) Failure of biohydrogen production by low levels of substrate and lactic acid accumulation. Renew Energy 86:889–894. https://doi.org/10.1016/j.renene.2015.09.016
Park JH, Kim DH, Baik JH, Park JH, Yoon JJ, Lee CY, Kim SH (2021) Improvement in H_2 production from *Clostridium butyricum* by co-culture with *Sporolactobacillus vineae*. Fuel 285:119051. https://doi.org/10.1016/j.fuel.2020.119051
Patel SK, Kumar P, Mehariya S, Purohit HJ, Lee JK, Kalia VC (2014) Enhancement in hydrogen production by co-cultures of *Bacillus* and *Enterobacter*. Int J Hydrog Energy 39(27):14663–14668. https://doi.org/10.1016/j.ijhydene.2014.07.084
Patel AK, Debroy A, Sharma S, Saini R, Mathur A, Gupta R, Tuli DK (2015) Biohydrogen production from a novel alkalophilic isolate Clostridium sp. IODB-O3 Bioresour Technol 175:291–297. https://doi.org/10.1016/j.biortech.2014.10.110
Pérez-Rangel M, Quiroz-Figueroa FR, González-Castañeda J, Valdez-Vazquez I (2015) Microscopic analysis of wheat straw cell wall degradation by microbial consortia for hydrogen production. Int J Hydrog Energy 40(1):151–160. https://doi.org/10.1016/j.ijhydene.2014.10.050
Phowan P, Danvirutai P (2014) Hydrogen production from cassava pulp hydrolysate by mixed seed cultures: effects of initial pH, substrate and biomass concentrations. Biomass Bioenergy 64:1–10. https://doi.org/10.1016/j.biombioe.2014.03.057
Poirier S, Steyer JP, Bernet N, Trably E (2020) Mitigating the variability of hydrogen production in mixed culture through bioaugmentation with exogenous pure strains. Int J Hydrog Energy 45(4):2617–2626. https://doi.org/10.1016/j.ijhydene.2019.11.116
Qiu C, Yuan P, Sun L, Wang S, Lo S, Zhang D (2017) Effect of fermentation temperature on hydrogen production from xylose and the succession of hydrogen-producing microflora. J Chem Technol Biotechnol 92(8):1990–1997. https://doi.org/10.1002/jctb.5190
Rafieenia R, Lavagnolo MC, Pivato A (2018) Pre-treatment technologies for dark fermentative hydrogen production: current advances and future directions. Waste Manag 71:734–748. https://doi.org/10.1016/j.wasman.2017.05.024

Rafrafi Y, Trably E, Hamelin J, Latrille E, Meynial-Salles I, Benomar S et al (2013) Sub-dominant bacteria as keystone species in microbial communities producing bio-hydrogen. Int J Hydrog Energy 38(12):4975–4985. https://doi.org/10.1016/j.ijhydene.2013.02.008

Rambabu K, Show PL, Bharath G, Banat F, Naushad M, Chang JS (2020) Enhanced biohydrogen production from date seeds by *Clostridium thermocellum* ATCC 27405. Int J Hydrog Energy 45(42):22271–22280. https://doi.org/10.1016/j.ijhydene.2019.06.133

Ramírez-Morales JE, Tapia-Venegas E, Toledo-Alarcón J, Ruiz-Filippi G (2015) Simultaneous production and separation of biohydrogen in mixed culture systems by continuous dark fermentation. Water Sci Technol 71(9):1271–1285. https://doi.org/10.2166/wst.2015.104

Rao R, Basak N (2021) Optimization and modelling of dark fermentative hydrogen production from cheese whey by *Enterobacter aerogenes* 2822. Int J Hydrog Energy 46(2):1777–1800. https://doi.org/10.1016/j.ijhydene.2020.10.142

Ratti RP, Botta LS, Sakamoto IK, Silva EL, Varesche MBA (2014) Production of H2 from cellulose by rumen microorganisms: effects of inocula pre-treatment and enzymatic hydrolysis. Biotechnol Lett 36(3):537–546. https://doi.org/10.1007/s10529-013-1395-z

Ren Y, Si B, Liu Z, Jiang W, Zhang Y (2022) Promoting dark fermentation for biohydrogen production: Potential roles of iron-based additives. Int J Hydrog Energy 47(33:1499–1515. https://doi.org/10.1016/j.ijhydene.2021.10.137

Saady NMC (2013) Homoacetogenesis during hydrogen production by mixed cultures dark fermentation: unresolved challenge. Int J Hydrog Energy 38(30):13172–13191

Saleem A, Umar H, Shah TA, Tabassum R (2020) Fermentation of simple and complex substrates to biohydrogen using pure Bacillus cereus RTUA and RTUB strains. Environ Technol Innov 18:100704

Salimi F, Mahadevan R (2013) Characterizing metabolic interactions in a clostridial co-culture for consolidated bioprocessing. BMC Biotechnol 13(1):1–9. https://doi.org/10.1186/1472-6750-13-95

Salimi F, Zhuang K, Mahadevan R (2010) Genome-scale metabolic modeling of a clostridial coculture for consolidated bioprocessing. Biotechnol J 5(7):726–738. https://doi.org/10.1002/biot.201000159

Sharma P, Melkania U (2017) Biochar-enhanced hydrogen production from organic fraction of municipal solid waste using co-culture of *Enterobacter aerogenes* and *E. coli*. Int J Hydrog Energy 42(30):18865–18874. https://doi.org/10.1016/j.ijhydene.2017.06.171

Sharma P, Melkania U (2018) Effect of bioaugmentation on hydrogen production from organic fraction of municipal solid waste. Int J Hydrog Energy 43(15):7290–7298. https://doi.org/10.1016/j.ijhydene.2018.03.031

Sheng T, Gao L, Zhao L, Liu W, Wang A (2015) Direct hydrogen production from lignocellulose by the newly isolated *Thermoanaerobacterium thermosaccharolyticum* strain DD32. RSC Adv 5(121):99781–99788. https://doi.org/10.1039/C5RA20000H

Sheng T, Meng Q, Wen X, Sun C, Yang L, Li L (2021) Bioaugmentation with *Ruminiclostridium thermocellum* M3 to enhance thermophilic hydrogen production from agricultural solid waste. J Chem Technol Biotechnol 96(6):1623–1631. https://doi.org/10.1002/jctb.6682

Sikora A, Błaszczyk M, Jurkowski M, Zielenkiewicz U (2013) Lactic acid bacteria in hydrogen-producing consortia: on purpose or by coincidence? INTECH open science open minds, pp 488–514. https://doi.org/10.5772/50364

Singh S, Sarma PM, Lal B (2014) Biohydrogen production by *Thermoanaerobacterium thermosaccharolyticum* TERI S7 from oil reservoir flow pipeline. Int J Hydrog Energy 39(9):4206–4214. https://doi.org/10.1016/j.ijhydene.2013.12.179

Sinha P, Pandey A (2014) Biohydrogen production from various feedstocks by *Bacillus firmus* NMBL-03. Int J Hydrog Energy 39(14):7518–7525. https://doi.org/10.1016/j.ijhydene.2013.08.134

Sivagurunathan P, Anburajan P, Kumar G, Kim SH (2016) Effect of hydraulic retention time (HRT) on biohydrogen production from galactose in an up-flow anaerobic sludge blanket reactor. Int J Hydrog Energy 41(46):21670–21677. https://doi.org/10.1016/j.ijhydene.2016.06.047

Sivaramakrishnan R, Shanmugam S, Sekar M, Mathimani T, Incharoensakdi A, Kim SH et al (2021) Insights on biological hydrogen production routes and potential microorganisms for high hydrogen yield. Fuel 291:120136. https://doi.org/10.1016/j.fuel.2021.120136

Soares JF, Confortin TC, Todero I, Mayer FD, Mazutti MA (2020) Dark fermentative biohydrogen production from lignocellulosic biomass: technological challenges and future prospects. Renew Sustain Energy Rev 117:109484. https://doi.org/10.1016/j.rser.2019.109484

Taherdanak M, Zilouei H, Karimi K (2015) Investigating the effects of iron and nickel nanoparticles on dark hydrogen fermentation from starch using central composite design. Int J Hydrog Energy 40(38):12956–12963. https://doi.org/10.1016/j.ijhydene.2015.08.004

Taherdanak M, Zilouei H, Karimi K (2016) The effects of Fe^0 and Ni^0 nanoparticles versus Fe^{2+} and Ni^{2+} ions on dark hydrogen fermentation. Int J Hydrog Energy 41(1):167–173. https://doi.org/10.1016/j.ijhydene.2015.11.110

Taifor AF, Zakaria MR, Yusoff MZM, Toshinari M, Hassan MA, Shirai Y (2017) Elucidating substrate utilization in biohydrogen production from palm oil mill effluent by Escherichia coli. Int J Hydrog Energy 42(9):5812–5819. https://doi.org/10.1016/j.ijhydene.2016.11.188

Tondro H, Musivand S, Zilouei H, Bazarganipour M, Zargoosh K (2020) Biological production of hydrogen and acetone-butanol-ethanol from sugarcane bagasse and rice straw using co-culture of *Enterobacter aerogenes* and *Clostridium acetobutylicum*. Biomass Bioenergy 142:105818. https://doi.org/10.1016/j.biombioe.2020.105818

Valdez-Vazquez I, Pérez-Rangel M, Tapia A, Buitrón G, Molina C, Hernández G, Amaya-Delgado L (2015) Hydrogen and butanol production from native wheat straw by synthetic microbial consortia integrated by species of *Enterococcus* and *Clostridium*. Fuel 159:214–222. https://doi.org/10.1016/j.fuel.2015.06.052

Valdez-Vazquez I, Castillo-Rubio LG, Pérez-Rangel M, Sepúlveda-Gálvez A, Vargas A (2019) Enhanced hydrogen production from lignocellulosic substrates via bioaugmentation with *Clostridium* strains. Ind Crop Prod 137:105–111. https://doi.org/10.1016/j.indcrop.2019.05.023

Varrone C, Rosa S, Fiocchetti F, Giussani B, Izzo G, Massini G et al (2013) Enrichment of activated sludge for enhanced hydrogen production from crude glycerol. Int J Hydrog Energy 38(3):1319–1331. https://doi.org/10.1016/j.ijhydene.2012.11.069

Vesga-Baron A, Etchebehere C, Schiappacasse MC, Chamy R, Tapia-Venegas E (2021) Controlled oxidation-reduction potential on dark fermentative hydrogen production from glycerol: Impacts on metabolic pathways and microbial diversity of an acidogenic sludge. Int J Hydrog Energy 46(7):5074–5084. https://doi.org/10.1016/j.ijhydene.2020.11.028

Villanueva-Galindo E, Moreno-Andrade I (2021) Bioaugmentation on hydrogen production from food waste. Int J Hydrog Energy 46(51):25985–25994. https://doi.org/10.1016/j.ijhydene.2020.11.092

Wang J, Wan W (2008) Comparison of different pretreatment methods for enriching hydrogen-producing bacteria from digested sludge. Int J Hydrog Energy 33(12):2934–2941. https://doi.org/10.1016/j.ijhydene.2008.03.048

Wang J, Yin Y (2017) Principle and application of different pretreatment methods for enriching hydrogen-producing bacteria from mixed cultures. Int J Hydrog Energy 42(8):4804–4823. https://doi.org/10.1016/j.ijhydene.2017.01.135

Wang A, Ren N, Shi Y, Lee DJ (2008) Bioaugmented hydrogen production from microcrystalline cellulose using co-culture—*Clostridium acetobutylicum* X9 and *Ethanoigenens harbinense* B49. Int J Hydrog Energy 33(2):912–917. https://doi.org/10.1016/j.ijhydene.2007.10.017

Wang D, Duan Y, Yang Q, Liu Y, Ni BJ, Wang Q et al (2018) Free ammonia enhances dark fermentative hydrogen production from waste activated sludge. Water Res 133:272–281. https://doi.org/10.1016/j.watres.2018.01.051

Wang L, Trujillo S, Liu H (2019) Selective inhibition of methanogenesis by acetylene in single chamber microbial electrolysis cells. Bioresour Technol 274:557–560. https://doi.org/10.1016/j.biortech.2018.12.039

Wang Y, He L, Zhang Z, Zhao X, Qi N, Han T (2020) Efficiency enhancement of H_2 production by a newly isolated maltose-preferring fermentative bio-hydrogen producer of *Clostridium butyricum* NH-02. J Energy Storage 30:101426. https://doi.org/10.1016/j.est.2020.101426

Wen Z, Ledesma-Amaro R, Lin J, Jiang Y, Yang S (2019) Improved n-butanol production from Clostridium cellulovorans by integrated metabolic and evolutionary engineering. Appl Environ Microbiol 85(7):e02560-18. https://doi.org/10.1128/AEM.02560-18

Wong YM, Wu TY, Juan JC (2014) A review of sustainable hydrogen production using seed sludge via dark fermentation. Renew Sustain Energy Rev 34:471–482. https://doi.org/10.1016/j.rser.2014.03.008

Xu JF, Mi YT, Ren NQ (2016) Buffering action of acetate on hydrogen production by *Ethanoligenens harbinense* B49. Electron J Biotechnol 23:7–11. https://doi.org/10.1016/j.ejbt.2016.07.002

Yang G, Wang J (2018) Kinetics and microbial community analysis for hydrogen production using raw grass inoculated with different pretreated mixed culture. Bioresour Technol 247:954–962. https://doi.org/10.1016/j.biortech.2017.09.041

Yang G, Wang J (2019) Changes in microbial community structure during dark fermentative hydrogen production. Int J Hydrog Energy 44(47):25542–25550. https://doi.org/10.1016/j.ijhydene.2019.08.039

Yang G, Wang J (2020) Biohydrogen production from waste activated sludge pretreated by combining sodium citrate with ultrasonic: Energy conversion and microbial community. Energy Convers Manag 225:113436. https://doi.org/10.1016/j.enconman.2020.113436

Yang Z, Guo R, Shi X, He S, Wang L, Dai M et al (2016) Bioaugmentation of *Hydrogenispora ethanolica* LX-B affects hydrogen production through altering indigenous bacterial community structure. Bioresour Technol 211:319–326. https://doi.org/10.1016/j.biortech.2016.03.097

Yin Y, Wang J (2016) Changes in microbial community during biohydrogen production using gamma irradiated sludge as inoculum. Bioresour Technol 200:217–222. https://doi.org/10.1016/j.biortech.2015.10.027

Yin Y, Hu J, Wang J (2014) Enriching hydrogen-producing bacteria from digested sludge by different pretreatment methods. Int J Hydrog Energy 39(25):13550–13556. https://doi.org/10.1016/j.ijhydene.2014.01.145

Yokoi H, Maki R, Hirose J, Hayashi S (2002) Microbial production of hydrogen from starch-manufacturing wastes. Biomass Bioenergy 22(5):389–395. https://doi.org/10.1016/S0961-9534(02)00014-4

Yokoyama H, Waki M, Moriya N, Yasuda T, Tanaka Y, Haga K (2007) Effect of fermentation temperature on hydrogen production from cow waste slurry by using anaerobic microflora within the slurry. Appl Microbiol Biotechnol 74(2):474–483. https://doi.org/10.1007/s00253-006-0647-4

Zeidan AA, Van Niel EW (2010) A quantitative analysis of hydrogen production efficiency of the extreme thermophile *Caldicellulosiruptor owensensis* OL^T. Int J Hydrog Energy 35(3):1128–1137. https://doi.org/10.1016/j.ijhydene.2009.11.082

Zeidan AA, Rådström P, van Niel EW (2010) Stable coexistence of two *Caldicellulosiruptor* species in a de novo constructed hydrogen-producing co-culture. Microb Cell Factories 9(1):1–13. https://doi.org/10.1186/1475-2859-9-102

Zhang SC, Lai QH, Lu Y, Liu ZD, Wang TM, Zhang C, Xing XH (2016) Enhanced biohydrogen production from corn stover by the combination of *Clostridium cellulolyticum* and hydrogen fermentation bacteria. J Biosci Bioeng 122(4):482–487. https://doi.org/10.1016/j.jbiosc.2016.03.014

Zhang K, Cao GL, Ren NQ (2019) Bioaugmentation with *Thermoanaerobacterium thermosaccharolyticum* W16 to enhance thermophilic hydrogen production using corn stover hydrolysate. Int J Hydrog Energy 44(12):5821–5829. https://doi.org/10.1016/j.ijhydene.2019.01.045

Zhu H, Béland M (2006) Evaluation of alternative methods of preparing hydrogen producing seeds from digested wastewater sludge. Int J Hydrog Energy 31(14):1980–1988. https://doi.org/10.1016/j.ijhydene.2006.01.019

Ziara RM, Miller DN, Subbiah J, Dvorak BI (2019) Lactate wastewater dark fermentation: the effect of temperature and initial pH on biohydrogen production and microbial community. Int J Hydrog Energy 44(2):661–673. https://doi.org/10.1016/j.ijhydene.2018.11.045

Zieliński M, Korzeniewska E, Filipkowska Z, Dębowski M, Harnisz M, Kwiatkowski R (2017) Biohydrogen production at low load of organic matter by psychrophilic bacteria. Energy 134:1132–1139. https://doi.org/10.1016/j.energy.2017.05.119

Chapter 6
Biohydrogen Production by Photosynthetic Microorganisms

Jean Carlos Viccari Pereira, Rahul Saini, and Satinder Kaur Brar

Abstract Hydrogen is a promising green alternative to fossil fuel due to its low environmental impact and high energy density, which is nearly three times higher than current transportation fuels such as gasoline and diesel. Renewable means of hydrogen production can be achieved by several different microbial routes and a variety of phototrophic microorganisms are endowed with genes and proteins for biohydrogen production. In this context, research has been focused on improving photobioreactor performance for practical and commercial applications, specifically regarding the low yield associated with it compared to other biofuels. This chapter provides an overview of the different types of photosynthetic microorganisms involved in biohydrogen production, the substrates utilized, the various photobioreactor designs employed, and the role of genetic engineering in enhancing hydrogen production rates. Moreover, the hurdles and limitations encountered during hydrogen production are discussed, along with future perspectives for advancing biohydrogen production.

Keywords Biohydrogen · Biophotolysis · Photofermentation · Dark fermentation · Genetic engineering

6.1 Introduction

Over the past decades, the major source of energy has been derived from fossil fuels. However, they are a source of non-renewable energy associated with the increasing concentration of carbon dioxide in the air and tremendous environmental issues (Das and Veziroğlu 2001). Their depletion in the foreseeable future urges the requirement for new sustainable sources of energy.

J. C. V. Pereira · R. Saini · S. K. Brar (✉)
Department of Civil Engineering, Lassonde School of Engineering, York University, North York, Toronto, ON, Canada
e-mail: satinder.brar@lassonde.yorku.ca

C. R. Soccol et al. (eds.), *Biohydrogen - Advances and Processes*, Biofuel and Biorefinery Technologies 13, https://doi.org/10.1007/978-3-031-49818-3_6

In recent years, various studies have been conducted to find the best alternative to fossil fuels and mitigate the release of climate-changing gases and compounds harmful to human health. Photobiological hydrogen production has emerged as an attractive option which, in comparison with traditional methods, promises to generate clean and renewable energy from ample natural resources, such as water and sunlight (Melis and Happe 2001). The concept is based on the original reports of Gaffron and Rubin (1942) that some hydrogenase-containing green algae are capable of H_2 photoproduction as well as H_2 uptake in anaerobic conditions. Besides, existing research recognized hydrogen production from cyanobacteria in the 1890s (Hallenbeck 2012). Since these first reports, biohydrogen metabolism has been found in numerous microorganisms, including green algae/cyanobacteria photosynthetic bacteria, and dark fermentative bacteria. In a study conducted by Kossalbayev et al. (2020), biohydrogen production was successfully achieved using different strains of cyanobacteria. The research not only confirmed the ability of known cyanobacterial species to produce hydrogen but also discovered a new cyanobacterial species, namely *Desertifilum* species, which demonstrated the capability to generate hydrogen. Figure 6.1 illustrates the different routes for biohydrogen production.

In general, phototrophs convert light energy for H_2 production either with H_2O as the electron donor (autotrophic photobiological H_2 production) or with various organic substances such as saccharides, organic acids, etc. (heterotrophic photobiological H_2 production) (Bolatkhan et al. 2019). Hydrogen production via the photobiological process currently includes two fundamental approaches: biophotolysis of water by algae and cyanobacteria and photofermentation by photosynthetic bacteria. The first is an oxygenic process driven by hydrogenase enzymes, in which photosynthetic O_2 and H_2 gas occur simultaneously or are temporary and/or spatially separated (Ghirardi et al. 2014). In a non-oxygenic or photofermentation process, purple non-sulfur bacteria (PNSB) reduce H^+ ions to gaseous H_2 using the reducing power from the oxidation reaction of organic compounds and the light energy, in a reaction catalyzed by nitrogenase (Hallenbeck 2013). Since the early proposals of

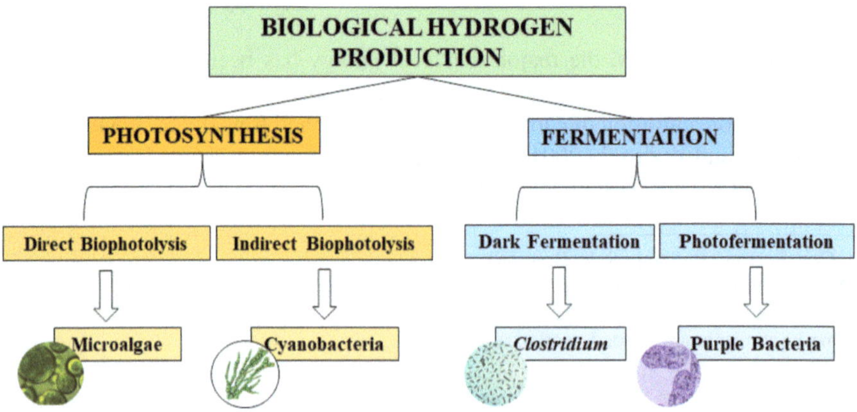

Fig. 6.1 Biohydrogen production routes. Reproduced with permission from Bolatkhan et al. (2019)

renewable photobiological H_2 energy production, there has been an increasing interest in this process with research focusing on the development of suitable photobioreactors, which are used for large-scale production of biohydrogen.

Various photobioreactors have been used for hydrogen production using photosynthetic microorganisms. Among them, the tubular, flat panel, and vertical-column types are widely used for hydrogen production (Dasgupta et al. 2010). New out-of-the-box geometries have also been investigated. However, although qualitatively feasible, the process requires additional research and development of the biological aspects for its commercial exploitation. The objective of this chapter is to provide an overview of the advancements achieved in hydrogen production via the photobiological process. Specifically, the chapter focuses on the utilization of green algae, cyanobacteria, and purple non-sulfur bacteria as potential sources of clean and renewable energy. The discussion will cover the current state of research, technological advancements in photobioreactors, and explore metabolic and genetic engineering approaches to enhance hydrogen production.

6.2 Photosynthetic Microorganisms for Biohydrogen Production

From the first findings of H_2 photoproduction in the green algae *Scenedesmus obliquus* (Gaffron and Rubin 1942), as well as the purple bacterium *Rhodospirillum rubrum* (Gest and Kamen 1949), a variety of microorganisms have been found capable of such metabolism. These are called phototrophs as they use light as their energy source. Phototrophs are generally classified into three types: aerobic green algae (eukaryotes); cyanobacteria (blue-green algae); and anaerobic photosynthetic bacteria (mostly Gram-negative prokaryotes). Table 6.1 lists some of these species and their hydrogen activity.

6.2.1 Green Algae

Green algae are unique eukaryotes that can grow in two distinctly different environments (aerobic and anaerobic) and photosynthetically generate molecular hydrogen (Melis and Happe 2001). These organisms have both PSI and PSII photosystems, structurally and functionally similar to the higher eukaryotes (Dasgupta et al. 2010).

They show a wide variation in biohydrogen production, depending on media and growth conditions. Generally, biohydrogen production from algae operates through indirect biophotolysis or direct biophotolysis. However, these processes require anaerobic conditions, given the acute oxygen sensitivity of the hydrogenase enzyme. A classic approach to overcome this limitation is sulfur deprivation. The lack of sulfur nutrients in green algae causes a substantial, but reversible inactivation of the

Table 6.1 Hydrogen producing microorganisms and their hydrogen activity

Species	Maximum photoproduction rate	Light intensity ($\mu E/m^2/s$)	H_2 evolution assay condition	References
Green algae				
Chlamydomonas reinhardtii	175 mL/L	30–40	Sulfur deprived cultures under constant argon flow	Laurinavichene et al. (2004)
Chlamydomonas moewusii	10.0 mL/L	150	Sulfur deprived cultures under constant argon flow	Winkler et al. (2002)
Cyanobacteria	32 μmol H_2[a]	200	Ar, $BG11_0$ medium using 0.5% fructose as a carbon source	Khetkorn et al. (2010)
Heterocystous				
Anabaena siamensis TISTR 8012				
Non-heterocystous				
Desertifilum sp. IPPAS B-1220	0.229 μmol H_2[a]	30	Ar, light/dark cycle	Kossalbayev et al. (2020)
Photosynthetic bacteria	1339 mL/L	7500 lx	Ar, different carbon and nitrogen sources	Assawamongkholsiri and Reungsang (2015)
Rhodobacter sp. KKU-PS				
Rhodobacter capsulatus	16.6 mL/L	Sunlight (+ initial artificial light)	Pilot-scale tubular photobioreactor using acetate as substrate	Boran et al. (2010)

[a]mg chl/a/h

O_2 evolution function and subsequent production of large amounts of pure H_2 in laboratory-scale experiments (Melis et al. 2000). Nevertheless, this approach is time-limited and a barrier for commercial applications. *Chlamydomonas reinhardtii* (Pinto et al. 2013) and *Chlamydomonas moewusii* (Das and Veziroğlu 2001) are the most promising strains to produce H_2 through direct biophotolysis when cultivated under sulfur starvation (Melis et al. 2000).

6.2.2 *Cyanobacteria*

Oxygenic prokaryotic cyanobacteria, also known as blue-green algae, form a large and diverse group, many of which can produce hydrogen. These bacteria are oxygenic phototrophs and therefore have type I and type II photosystems. In general, 14 cyanobacterial genera can perform H_2 production (Kufryk 2013) and are considered ideal candidates since they have simple nutritional requirements which are easily met by simple salt solutions (Lopes Pinto et al. 2002). They have mostly been used in direct and indirect biophotolysis (Demirbas 2009). The production is affected by intrinsic or environmental factors. The strain-type, the presence of uptake hydrogenase and molecular oxygen are considered intrinsic factors, while light, temperature, salinity, nutrient availability, and gaseous atmosphere can contribute to hydrogen synthesis (as environmental conditions) (Tiwari and Pandey 2012).

Cyanobacteria possess key enzymes such as nitrogenase and hydrogenase which are involved in hydrogen metabolism (Madamwar et al. 2000). For instance, the oxygen-sensitive nitrogenases play a pivotal role in the maintenance of the nitrogen cycle, conducting biological nitrogen fixation of N_2 to NH_3 and concomitantly evolving H_2 (Madamwar et al. 2000). On the other hand, hydrogenase normally catalyzes hydrogen uptake or hydrogen consuming reactions. This, however, depends on the type of microbial strain used (McIntosh et al. 2011).

Cyanobacteria developed unique adaptive mechanisms to protect the nitrogenase from inactivation by oxygen. For instance, they can be either unicellular or filamentous (heterocystous and non-heterocystous), and are present as cells of sizes varying between 0.5 and 100 μm and a broad spectrum of morphologies (McIntosh et al. 2011). Heterocystous cells are desirable for H_2 production since they can simultaneously carry out oxygenic photosynthesis and nitrogen fixation (H_2 evolution) (Hallenbeck 2012), whereas unicellular cyanobacteria typically separate oxygen evolution from nitrogenase activity by performing these functions during light and dark periods respectively (Allakhverdiev et al. 2010).

6.2.3 *Photosynthetic Bacteria*

With few exceptions, purple and some green bacteria constitute a group of gram-negative photosynthetic prokaryotes. Unlike algae and cyanobacteria, they contain a PSII-like reaction center that is incapable of water-splitting and cannot use water as an electron donor, therefore performing anoxygenic photosynthesis (Allakhverdiev et al. 2010). This is because the oxidant generated by the photochemical reaction of the PSII-like reaction center of PNSB is not strong enough to extract electrons from H_2O (Sakurai et al. 2013). Light energy is used for Adenosine triphosphate (ATP) generation, substantial amounts of which are needed to drive the nitrogenase-mediated hydrogen production.

However, the nitrogenase enzyme is expressed strictly in the light under anaerobic conditions and in the absence of ammonium. For this reason, photosynthetic bacteria produce hydrogen via photofermentation. Some of the common strains with high hydrogen production capability under photofermentation conditions include *Rhodobacter sphaeroides* (Assawamongkholsiri and Reungsang 2015), *Rhodopseudomonas palustris*, and *Rhodobacter capsulatus* (Montiel Corona et al. 2017; Muzziotti et al. 2016).

6.3 Types of Substrates

A wide variety of feedstocks have been explored for biohydrogen production such as wastewaters, agricultural biomass, and post-processing wastes rich in organic acids. Table 6.2 lists the numerous substrates employed for biohydrogen production. Normally, dark fermentation is coupled with photofermentative hydrogen production, so the volatile fatty acids presented in the effluent obtained from dark fermentation are used as the substrates to produce hydrogen by PNSB (Hallenbeck, 2013).

Mishra et al. (2016) reported a maximum hydrogen production rate of 26 mL H_2/h using *Rhodopseudomonas palustris* grown in dark fermentative pulp oil mill effluent. The authors observed that the dilution of the substrate by 50% improved light penetration and resulted in hydrogen yield enhancement from 2.86 to 3.064 mL H_2/mL. Similarly, Yang et al. (2015a, b) reported the same shielding effect on light penetration when using cornstalk hydrolysate as a substrate for *Rhodobacter sphaeroides* HY01. Therefore, the dark-photofermentation process often requires the dilution of effluents and/or higher light intensity (Tiang et al. 2020). In addition, pretreatment of fermentative effluents is also required due to their low pH, excess of fixed nitrogen, and high substrate concentration (Hallenbeck, 2013). High concentrations of nitrogen sources may lead to the accumulation of ammonia, which acts as an inhibitor for the photofermentation process (Tiang et al. 2020). To overcome this limitation, Cheng et al. (2015) treated residual dark fermentation solutions with zeolite for ammonia removal to further improve hydrogen gas production. The ammonia removal efficiencies using the modified zeolite treatment in the dark-fermentation effluents ranged from 81.3 to 91.9%. Apart from wastewater, agricultural biomass is another prospective feedstock. Lignocellulosic biomass is a byproduct of various processes found both in agriculture and forestry industries. The application of such waste for hydrogen production could help reduce the raw material cost of biohydrogen production. However, the recalcitrant nature of the lignocellulose complex makes it difficult to directly use as a substrate in biohydrogen production. For this reason, hydrolysis and pretreatment are needed to hydrolyze polymeric components of biomass. In a study utilizing the enzymatic hydrolysate of corn stalk pith as substrate, Jiang et al. (2016) obtained conversion efficiencies of 82–94%. On the other hand, hydrogen production decreased when initial substrate concentrations further increased up to 0.3 g/L. This finding is consistent with that of Yang et al. (2015a, b) who found a similar inhibition effect of hydrolysate when substrate concentration increased up to 15 g/L. These relationships may partly be explained by the fact that the hydrolysates generate fine particles, which have a considerable impact on light penetration into the hydrogen production medium.

Apart from the importance of pretreatment, another concern in hydrogen production from hydrolysate of agricultural biomass via photofermentation is the generation of inhibitive substances (Assawamongkholsiri and Reungsang 2015). Toxic compounds such as furan derivatives and phenolic compounds may be produced

Table 6.2 Hydrogen production from various substrates by different photosynthetic microorganisms

Substrate	Microorganism	Process conditions	Substrate conversion efficiency (%)	Hydrogen production rate	References
Tris-acetatephosphate	*Chlamydomonas reinhardtii* CC-124	Biophotolysis	ND	0.3 mL H_2/L	Tamburic et al. (2011)
$BG11_0$ medium	*Anabaena* PCC 7120 and AMC 414	Biophotolysis	0.042	14.9 mL H_2/L	Lindblad et al. (2002)
Palm oil mill effluent	*Rhodopseudomonas palustris*	Dark photofermentation	ND	26 mL H_2/L	Mishra et al. (2016)
Palm oil mill effluent	*Rhodopseudomonas palustris*	Photofermentation	ND	1.05 mL H_2/L	Jamil et al. (2009)
Crude glycerol	*Rhodopseudomonas palustris*	Dark photofermentation	80.21	6.42 mmol H_2/g COD	Chookaew et al. (2015)
Alanine in waste biomass	*Rhodopseudomonas palustris*	Dark photofermentation	25.1	418.6 mL H_2/g	Cheng et al. (2015)
Corn stalk hydrolysate	*Rhodobacter sphaeroides HY01*	Dark photofermentation	ND	74 mL H_2/L	Yang et al. (2015a)
Corn stalk pith	Photosynthetic microbial consortium consisted of *Rhodospirillum rubrum*, *Rhodopseudomonas capsulata*, *Rhodopseudomonas palustris*, *Rhodobacter sphaeroides*, and *Rhodobacter capsulatus*	Photofermentation	82–94	2.61 mol H_2/mol sugar consumed	Jiang et al. (2016)
Crude glycerol	*Rhodopseudomonas palustris*	Photofermentation	80–85	6.69 mol H_2/mol	Pott et al. (2013)

during the breakdown of lignocellulosic biomass. These molecules produce a strong oxidative effect on enzyme activity and causes DNA damage to cells, hence affecting the hydrogen yields (Ibraheem and Ndimba 2013). The tolerance of *Rhodopseudomonas palustris* CGA009 toward phenolic compound inhibition was tested at a concentration range of 0.054–0.143 g/L (Mabutyana and Pott 2021). The authors observed a 1.4-fold reduction in hydrogen production in vanillin-containing media (0.43 g/L). Therefore, attempts have been made to remove these inhibitors either by alkaline or bioelectrochemical detoxification methods (Mabutyana and Pott 2021). Since photofermentation is a nitrogenase-dependent process, some important inducers of this enzyme such as molybdenum (Mo) and iron (Fe) have been reported to enhance hydrogen production. Yang et al. (2015a, b) reported that an increase in Fe-nitrogenase concentration alone did not increase hydrogen production. However, the two nitrogenase systems functioned synergistically in *Rhodobacter capsulatus* YMP1 mutant strain and the maximum hydrogen production rate was enhanced by 24.1% compared with its wild-type strain. The results of these investigations show that the type of substrate as well as the presence of inducers or inhibitors can affect the hydrogen production rate. Hence, a natural progression of this work is optimization based on microbial response to a particular substrate.

6.4 Mechanisms of Biohydrogen Production

Biohydrogen production may be divided into four main groups: direct biophotolysis, indirect biophotolysis, photofermentation, and dark fermentation. Figure 6.2 describes the reaction mechanisms in each group.

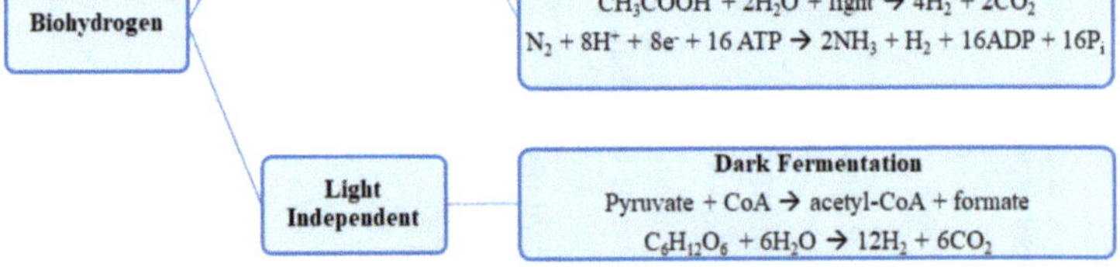

Fig. 6.2 Reaction mechanisms for the four types of biohydrogen production

6.4.1 *Direct and Indirect Biophotolysis*

Biophotolysis is the biological method of splitting water into O_2 and H_2, driven by sunlight as the energy source (Lam et al. 2019). This process relies on the same fundamentals of plant and algal photosynthesis, however leading to biohydrogen production instead of biomass accumulation (Das and Veziroğlu 2001). As discussed in previous reviews (Ghirardi et al. 2014; Goussias et al. 2002), the biohydrogen production of green algae and cyanobacteria depends on the photosystem II PSII-dependent pathway (direct biophotolysis) or PSII-independent pathway (indirect biophotolysis). In these photosystems, the light absorbed by antenna pigments of photosynthetic microorganisms is transferred to pigment-protein complexes called reaction centers (Antal et al. 2012), where a photosynthetic electron transfer chain is generated (Goussias et al. 2002). The PSII photosystem is at the start of this chain and uses light energy to generate the strongly oxidizing cation radical P_{680}, which catalyzes the water oxidation, thus providing the system with a source of electrons that is essentially unlimited (Goussias et al. 2002).

In direct biophotolysis, these electrons are transferred from PSII, via cytochrome *bsf* complex, and PSI to ferredoxin (Fd) (Lam et al. 2019), a protein that carries electrons to other reaction pathways outside the thylakoid (Bolatkhan et al. 2019). Under anaerobic conditions, electrons from reduced Fd are used for H_2 production by the [FeFe]- hydrogenase (green algae) enzyme (Bolatkhan et al. 2019), which reversibly facilitates the reduction of protons (H^+) to H_2, as represented in Fig. 6.3. For another hydrogen-evolving enzyme, the [Ni-Fe]-hydrogenase (cyanobacteria), nicotinamide adenine dinucleotide phosphate hydrogen (NAD(P)H) acts as the electron donor for H_2 production. NAD(P)H is produced either by light energy absorbed by PSI or anaerobic catabolism of organic substrates (Antal et al. 2012).

Although the process is attractive, as it just requires light and water for H_2 production, the main limitation of direct biophotolysis is the high sensitivity of most hydrogen-evolving enzymes to O_2 (Antal et al. 2012; Show et al. 2019). Different methods have been proposed to modify hydrogenases for enhanced O_2 tolerance, however, research on the subject has been mostly restricted to enzyme phenotypes

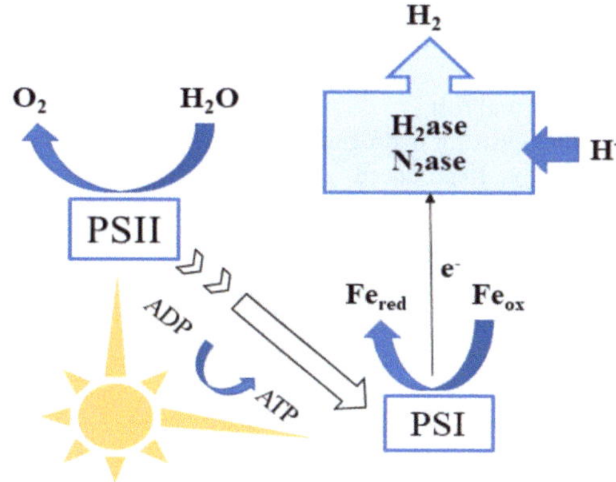

Fig. 6.3 Direct biophotolysis by green algae or cyanobacteria. Reproduced with permission from Bolatkhan et al. (2019)

with low hydrogen production (Fritsch et al. 2013; Lu and Koo 2019). The other approach, i.e., controlling the O_2 level, may involve the selective inactivation of PSII O_2 evolution activity by sulfur deprivation (Melis et al. 2000). For instance, sulfur deficiency, under constant illumination conditions, induces the down-regulation of PSII (cells are unable to replace photo-damaged D1 protein) and simultaneous starch accumulation (Antal et al. 2012). The decrease of O_2 evolution rate by PSII, to a level lower than the rate of respiration (compensation point), results in self-transition of cells into anaerobic conditions, following induction of hydrogenase activity and sustained H_2 production under constant illumination (Hallenbeck et al. 2018). In these conditions, the electron supply for hydrogenase activity comes either from the residual activity of PS II (primary source) and starch degradation (secondary source) (Touloupakis and Torzillo 2019). Since the process efficiency depends on the conversion of stored carbohydrates, it may be lengthy and the application for commercial purposes is not feasible (Show et al. 2019).

The limitations of direct biophotolysis led to the proposal of separating water splitting and hydrogen evolution reactions, in a two-stage light-driven process called indirect biophotolysis. During the first aerobic stage, green algae and cyanobacteria are allowed to grow under normal cultivation conditions, with the fixation of CO_2 into storage carbohydrates (e.g., starch in green algae, glycogen in cyanobacteria) (Bolatkhan et al. 2019). In the anaerobic stage, the fermentation of carbohydrates releases low potential electrons. These enter the electron chain at the plastoquinone pool (PQ) and are thereafter transferred to the hydrogenase, facilitating H_2 production (Ghirardi et al. 2014) (Fig. 6.4).

The main advantage of indirect biophotolysis is the minimal need for additional nutrients, however, the effectiveness of the process depends on intermediate steps and efficient temporal or spatial separation of H_2 and O_2 flows (Antal et al. 2012; Bolatkhan et al. 2019). For instance, nitrogen-fixing cyanobacteria spatially separate the processes of oxygen evolution and nitrogen fixation/hydrogen evolution by forming specific cells (heterocysts), segregating oxygenic photosynthetic activity in vegetative cells and nitrogenase activity in heterocysts with reduced gas diffusivity (Allakhverdiev et al. 2010).

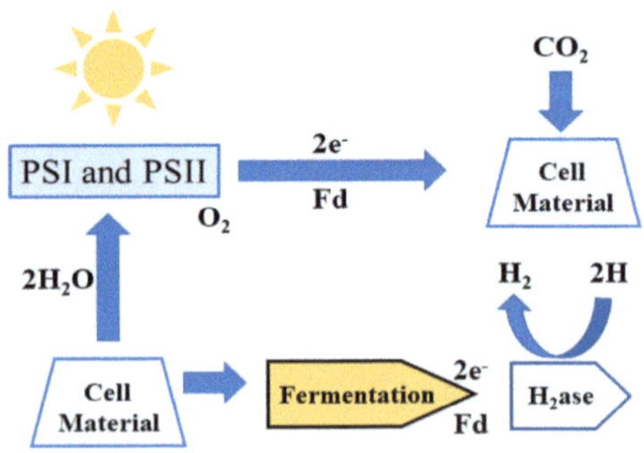

Fig. 6.4 Indirect biophotolysis for hydrogen production. Reproduced with permission from Bolatkhan et al. (2019)

Due to the fixed sustained availability of carbon sources during photosynthesis, cyanobacteria and algae are well-known for H_2 production via indirect biophotolysis.

6.4.2 *Photofermentation*

Unlike green algae and cyanobacteria, anoxygenic photosynthetic bacteria, such as purple and some green photosynthetic bacteria, only contain one photosystem (equivalent to the photosystem I in the green algae's photosynthesis) (Zhang and Zhang 2018) and use both light energy and reducing power from oxidation reaction of organic compounds to reduce H^+ to H_2.

In photofermentation, light energy drives electron flow through the photosynthetic apparatus, and the photo-phosphorylation process generates ATP for the dinitrogen fixation, accompanied by hydrogen evolution (Hallenbeck 2013). Hydrogen production is driven by the nitrogenase enzyme, whose physiological function is the biological reduction of dinitrogen to ammonia, with the release of one H_2 per N_2 reduced (Antal et al. 2012). Electrons for the reduction of N_2 and H^+ derive from the catabolism of exogenous inorganic or organic compounds, such as low-molecular-weight fatty acids. However, the substrate requirement is purely strain-specific (Gupta et al. 2013). Since both NH_3 and H_2 are products of nitrogenase, stressful concentrations of nitrogen are essential to increase the H_2/NH_3 product ratio. The main advantages of this method are the versatile metabolic capabilities of anoxygenic photosynthetic bacteria, and the lack of a PSII system, avoiding O_2-evolving reactions (Lazaro and Hallenbeck 2019). In addition, there is a possibility to couple the photofermentation with dark fermentation (Fig. 6.5). Thus potential environmental pollutants are reduced because waste streams from dark fermentation are reused in a two-stage process (Adessi and De Philippis 2012). On the other hand, the nitrogenase ATP-dependent activity requires highly reducing electrons and low ammonia concentrations.

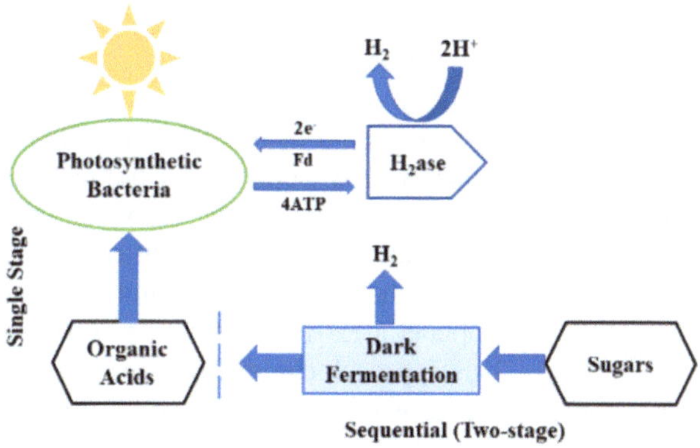

Fig. 6.5 Photofermentative hydrogen production

6.5 Photobioreactors

The general characteristics of photobioreactors (PBRs) are that it should be translucent to allow sufficient light penetration, impermeable to hydrogen and air, made from a durable unreactive material, easy to clean and sterilize, and a high surface-to-volume ratio (SVR) for better/wide distribution of light. They can be classified into open (raceway ponds, lagoons, and lakes) or closed (flat plate, tubular) systems.

Closed PBRs are generally preferred since they allow better control of parameters like temperature, nutrients, pH, the intensity of light, and oxygen production and facilitate the collection of produced hydrogen (Dasgupta et al. 2010). Closed PBRs can be categorized by their design (flat or tubular, horizontal, inclined, vertical or spiral, and manifold or serpentine) and mode of operation (batch, fed-batch, and continuous). Figure 6.6 illustrates the most common and some non-conventional designs of PBRs.

6.5.1 Tubular Reactors

Tubular PBRs consist of long translucent tubes, between 3 and 6 cm in diameter, that circulate liquid culture using mechanical or gas-lift pumps. Diverse geometries of tubular reactors have been designed, such as simple airlift or agitated bubble column (vertical type), horizontal or nearly horizontal tubular PBRs, and helical type PBRs (García-Galán et al. 2018).

In vertical type PBRs, gas is continuously supplied over the cultures for mixing purposes. Such reactors provide good mixing and gas-exchange properties, achieving efficient biomass growth and inhibiting the adhesion of cells on the surface (Hallenbeck 2009). However, one major drawback of the system is the dilution of produced hydrogen by mixing with inert gas, which raises the costs of gas separation. Compared with vertical-type configuration, horizontal or nearly horizontal tubular PBRs can obtain higher biomass concentration and hydrogen production because they receive a higher photon flux density. Also, this configuration reduces the shear stress on cells and low gas holdup (Boran et al. 2010). The disadvantages of using this type of reactor are the low gas exchange and sometimes an accumulation of biomass in the tubes (García-Galán et al. 2018). Moreover, temperature control is a challenge for this type of reactor.

Helical PBRs are made up of flexible materials into three-dimensional helical frameworks with the desired inclination. Cooling is achieved by an external heat exchanger (Dasgupta et al. 2010). Due to its high area to volume (A/V) ratio, a high hydrogen production rate has been reported (Assunção and Malcata 2020), however, it can also lead to the deposition of culture in the reactors' inner walls causing biofouling (Dasgupta et al. 2010). One way to scale up is to connect several tubes via manifolds. This approach can enable internal cooling of the reactor; however, the length of the tubes is limited because of the accumulation of gas (Wang 2011).

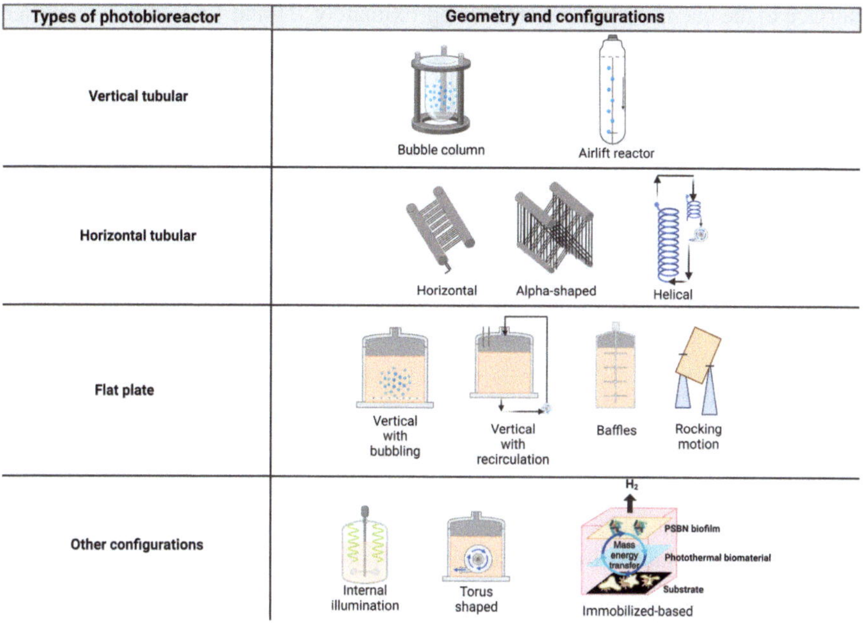

Fig. 6.6 Configuration of closed photobioreactor employed for biohydrogen production

6.5.2 Flat-Panel Reactors

Due to their high A/V ratio, the flat-panel configurations are a popular choice among closed PBRs. However, this high ratio is only tangible when their thickness ranges from 1 to 5 cm. They consist of two rectangular plates and can be made from transparent materials such as glass, plexiglass, and polycarbonate. The height and width can be varied to some extent, but in practice, flat panel reactors are smaller than 1 m (Uyar 2016).

In outdoor conditions, illuminated surfaces are typically orientated vertically or inclined in the direction of the sun, since PNSB mainly utilize longwave radiation which is more dominant in the morning and the evening. On the other hand, in indoor conditions, illumination can be provided from both sides to enhance light penetration into the inner parts of the culture medium (Assawamongkholsiri and Reungsang 2015). The spacing between the panels can be adjusted according to the targeted light and temperature distribution. Agitation can be provided by sparging (algae cultivation), magnetic stirring, impeller stirring, baffles, and a rocking motion. Preliminary work achieved better mixing by a rocking motion, using wave motion by positioning the reactor on the sea surface, possibly also including additional wind energy (Skjånes et al. 2016). However, one impediment of this technique may be the high power consumption.

In terms of required ground area, the flat-panel reactor has been reported to be more economically feasible than tubular reactors. Their ratio of illuminated reactor surface to the occupied ground area is approximately 1:1 and 1:8 in tubular and flat-panel reactors, respectively (Gebicki et al. 2010). In a comprehensive study investigating the performances of three different photobioreactors (tubular, bath, and flat panel) with hydrolyzed corncobs, Zhang et al. (2015) found the highest hydrogen yield and substrate conversion efficiency using the upflow baffle plate reactor.

Despite the advantages in terms of required ground area and easy construction, flat-panel PBRs have several drawbacks related to agitation methods. Mechanical restrictions arising from their design result in poor mixing and mass transfer rates. Therefore, the major shortcoming in scaling up is fouling. On the other hand, it is possible to arrange a set of reactors one behind another, at an appropriate distance, to increase the ground area productivity (Adessi and De Philippis 2012).

6.5.3 Internal-Illuminated Reactors

Since light is usually delivered externally via the surface of the PBR, the maximum size of a unit is limited by light penetration into the culture medium. In order to prevent the shaded and dark zones, their geometry must be very thin and long, and this may lead to insufficient gas transfer and mixing in the medium (Dasgupta et al. 2010). This impediment can be addressed by applying internal illumination systems, which brings the possibility of more compact reactors (Uyar 2016). Internally illuminated PBRs are generally classified into reactors with internal light guides or reactors with internal light sources. Firstly, optical fibers or optical plates are installed for solar light collection, transmission, and distribution of light energy more efficiently and evenly inside PBRs. Subsequently, an artificial light source is placed inside the reactor. Types of light sources include LED, tungsten filament lamps, halogen lamps, or metal-halide lamps. One limitation of solar energy is the varying sunlight intensity over time. Thus, a combination of internal light guides and artificial light sources enables a continuous supply of light to PBRs during cloudy weather or at night (Heining and Buchholz 2015). The main advantage over tubular and flat-plate PBRs is that light intensity and distribution are not governed by the concentration of cells and product, but they are controlled by the number and distribution of the optical fibers (Heining and Buchholz 2015).

Moreover, the scale-up of optical-fiber technology is challenging because of the high capital cost and the loss of light energy during transmission. Internal illumination via light sources offers lower light losses, but there are certain drawbacks associated with lower SVRs and higher volume losses (Ho et al. 2014). The small size and low heat dissipation of LEDs appear to be an interesting approach to significantly reducing construction costs and volume losses (Schulze et al. 2014).

6.5.4 Other Types of Photobioreactors

Apart from the aforementioned three main types, out-of-the-box geometries have also been investigated, such as alveolar panels or helical, alpha-shaped and torus-shaped photobioreactors (Dasgupta et al. 2010). Vertical alveolar panel PBRs consist of flat vessels and can be placed vertically or at an angle to the ground floor, to assure adequate light penetration (Assunção and Malcata 2020). However, they are prone to develop O_2 buildup partly due to their shape (Carvalho et al. 2006). Helical reactors are constructed with a helical shape in a series of coiled or looped tubes. Despite its notable SVR, high shear stress on microalgal cells is still a limitation (Carvalho et al. 2006). The alpha-shaped reactor is another hybrid type of tubular PBR, in which the culture is lifted 5 m by air to a receiver tank and flows down an inclined PVC tube. This configuration has all the requirements for hydrogen production but scaling up is only feasible with air replacement by inert gas (Dasgupta et al. 2010). The torus-type PBR was developed mainly for lab-scale studies on hydrogen production and the effect of mixing on phototrophic microorganisms (Pruvost et al. 2006). The culture is circulated in a torus-shape tank by the rotation of a marine impeller and the light is provided externally by tubes placed parallel to the illuminated surface. Moreover, the provided rotation allows good mixing with minimal shear stress. Nevertheless, it is unfortunate that torus-type PBR has been limited to lab-scale due to their alternative geometry, making it difficult to scale up. In addition, their active mixing generates a heterogeneous flow around the mechanical stirrer, hence, some shear stress can be generated (Assunção and Malcata 2020).

6.6 Role of Genetic Engineering

Recent genetic engineering efforts have been recognized to improve hydrogen production by phototrophic microorganisms. Several techniques based on this approach have been developed. A more detailed account of the main approaches to overcoming practical barriers encountered is provided in the following section.

6.6.1 Oxygen (O_2) Sensitivity

As previously stated, one of the greatest challenges in producing hydrogen using photobiological systems is the extreme sensitivity of the enzymes to oxygen. For this reason, attempts have been made at the molecular level, either physically blocking O_2 diffusion or chemically converting O_2 into harmless products (mainly H_2O) (Lu and Koo 2019). The first approach involves constricting the gas channels within the enzyme or externally protecting the enzyme with caging systems. Secondly, the conversion of O_2 into other innocuous molecules is executed by chemical units external to the enzymes. However, the success of these approaches primarily depends on blocking O_2 without restricting H_2 diffusion and hampering enzymatic activity.

Some alternative methods have been developed to induce anaerobic conditions, such as inactivation of PSII or increased O_2 uptake within the cell. As explained earlier, sulfur deprivation induces the selective inactivation of PS II; on the other hand, it also may cause cell inhibition growth. Alternatively, anaerobiosis can be achieved by using attenuated photosynthesis/respiration ratio (P/R ratio) and bringing it below 1 using DNA insertional mutagenesis. For instance, *Chlamydomonas reinhardtii* transformed with the *apr1* gene resulted in a low P/R ratio and steady biohydrogen production of 6.5 μmol H_2/mL after 6 h (Rühle et al. 2008). The authors observed twofold higher in vivo hydrogen production (72 nmol H_2/h) than previously reported for the sulfur-deprived *C. reinhardtii* wild type (36 nmol H_2/h). Another genetic engineering approach to induce anaerobiosis is by incorporating oxygen sequesters and decreasing its concentration within the cell. Genes encoding enzymes that remove O_2, such as oxidases and superoxide dismutases can be used for this purpose (Shestakov and Mikheeva 2006).

6.6.2 Proton Gradient

In addition to the O_2 sensitivity of hydrogenases, proton supply to the hydrogenase appears to be a critical issue. The transport of electrons from water to ferredoxin is accompanied by the formation of a proton gradient that drives the production of ATP, which is essential for carbon dioxide formation. However, the ATP requirement drops during H_2 evolution driven by hydrogenases, which impedes the dissipation of the proton gradient. Due to the inhibition of electron flow toward hydrogenase, the hydrogen production is reduced.

One possible strategy is the genetic insertion of an inducible, synthetically designed polypeptide proton channel into the thylakoid membrane of green algae. This will dissipate the proton gradient, avoid the ATPase channel, and promote the proton transfer toward hydrogenase (Lee and Greenbaum 2003). Previous studies have examined the feasibility of the proton-channel concept by making use of chemical proton uncouplers (such as carbonyl cyanide-*p*-trifluoromethoxyphenylhydrazone (FCCP)) to simulate the proton-shuttling effect of a proton channel across the algal thylakoid membrane (Cournac et al. 2002; Guan et al. 2004). The enhancing effect of proton uncouplers such as *m*-chlorophenylhydrazone (CCCP) and 2,4-dinitrophenol (DNF) was recently confirmed in the novel green microalgae *Parachlorella kessleri* RA-002 (Manoyan et al. 2019). The maximal H_2 yield of ~2.20 and 2.08 mmol/L was obtained in the presence of 15 μM CCCP and 50 μM DNF, respectively. The authors demonstrated that the dissipation of proton motive force across the thylakoid membrane directly increases H_2 yield.

6.6.3 Photosynthetic Efficiency

For enhanced biohydrogen production, photosynthetic efficiency can be increased by modifying the light-harvesting antenna complexes responsible for capturing solar energy. It is now understood that the size of the antenna of the photosynthetic apparatus is inversely proportional to the photosynthetic solar energy conversion efficiency (Bielczynski et al. 2020). Genetically truncated antenna sizes have demonstrated improvements in photosynthetic efficiency. The decrease in their size avoids dissipation and loss of excess photons during high light intensity conditions. For instance, mutant *Chlamydomonas reinhardtii* algae with a truncated chlorophyll antenna size resulted in increased photosynthetic efficiency of 3.2% (Polle et al. 2003).

In PNSB, genetic approaches were conducted to reduce the abundance of photosynthetic pigment. This procedure enabled the illumination of cells in the deeper parts of a reactor, thereby preventing the self-shade effect. Kondo et al. (2002) created a mutant of *Rhodobacter sphaeroides* RV by ultra violet (UV) radiation for the enhancement of hydrogen production under mass culture conditions. The reduction of the light-harvesting (LH1) complex enabled the production of 50% more hydrogen than the wild strain.

6.6.4 Competition for Electrons

The Calvin cycle is a significant competitor for the reducing power used by hydrogenases in algae and cyanobacteria. In this cycle, RuBisCO (the most abundant protein in the cell) plays a key role. To reduce this competition, cells were engineered to reduce RuBisCO levels. The hydrogen production of *Chlamydomonas* rubisco mutant Y67A was found to be 10–15 times greater than that of the wild type (Pinto et al. 2013).

Attempts have also been made via the deletion of the quinol oxidase gene in cyanobacteria (Cournac et al. 2002; Gutthann et al. 2007). The exceptionally high biohydrogen production is attributed to the absence of NADPH oxidation by NDH-1, which in turn results in the reduced rate of the Calvin cycle. Besides the respiratory activity, nitrate assimilation is another important electron sink that might compete with hydrogenase. For this reason, inhibition of the nitrate reductase gene in cyanobacteria also has a significant impact on biohydrogen production (Gutthann et al. 2007). The regulation of intracellular redox during the growth of photosynthetic bacteria with organic acids is achieved by reductant consumption by both nitrogenase and CO_2 fixation, via the Calvin cycle. For instance, Öztürk et al. (2012) inactivated CO_2 fixation by deleting one of the key enzymes (phosphoribulokinase) involved in this pathway, increasing the maximum rate of hydrogen production by 13.3% in *Rhodobacter capsulatus*.

6.6.5 Low Reductant Flux to the Hydrogenase

The presence of competitive electron pathways can reduce the reductant flow available for hydrogen production. Therefore, alterations in the metabolic pathways have also been made to redirect or increase the electron flux toward the hydrogen-producing enzyme. This occurred in a mutant of *Syncechocystis* PCC 6803 (M55) (a non-nitrogen fixing cyanobacteria) lacking the NADPH-dehydrogenase, that was found to accumulate NADPH and evolve significant amounts of hydrogen by the bidirectional NADP-dependent hydrogenase (Gutthann et al. 2007). Another example is a non-autotrophic mutant of *Rhodobacter capsulatus* named IR-4. When improving the synthesis of the D-malic enzyme and using D-malate as the sole carbon source, IR-4 showed an increased hydrogen production of 50% more than the wild type (Vignais et al. 2006).

6.7 Advantages and Disadvantages of Photobiological Hydrogen Production

The advantages and disadvantages of photobiological hydrogen production methods are outlined in Table 6.3.

Every method has its opportunities and limitations. However, they are all advantageous in comparison to conventional systems because biohydrogen can be produced naturally by microorganisms under ambient conditions of temperature and pressure. Moreover, it is also possible to couple the production of H_2 with other

Table 6.3 Comparison of photobiological hydrogen production methods

Process	Advantages	Disadvantages
Direct biophotolysis	Microorganisms produce H_2 gas from water and solar energy	High sensitivity of most hydrogen-evolving enzymes Possible generation of explosive hydrogen/oxygen mixtures
Indirect biophotolysis	Minimal need for additional nutrients	Temporal/spatial separation of H_2 and O_2 flows
	Nitrogen-fixing cyanobacteria produce H_2 from water and fix N_2 from the atmosphere	Energy loss in production and reuse of stored energy carrier
Photofermentation	Use of light energy to produce hydrogen from different substrates	Low light-conversion efficiency
	Photosynthetic bacteria can utilize a broad spectrum of light energy	The energetic requirement of the nitrogenase enzyme decreases the efficiency of H_2 production
	Can be coupled with dark fermentation to derive more hydrogen	Dilution of dark fermentative effluent is needed to remove inhibition

valuable metabolites like lactic acid, butyric acid, and acetic acid which are considered byproducts (Pandu and Joseph 2012).

Apart from the advantages, these processes also encounter many hurdles and deviations from the desired route. The major constraint of photobiological methods is the simultaneous release of oxygen along with hydrogen, which inhibits the oxygen-sensitive hydrogenase enzymes. This in turn challenges the scale-up of this process. In comparison, photofermentation has several advantages such as higher hydrogen yield efficiency, use of organic acid-rich spent effluent from the dark fermentation process as a viable feedstock, as well as high hydrogen purity (Budiman and Wu 2018). However, scale-up and economic feasibility remain major challenges due to the high cost of photobioreactors. The combination of dark-photo hydrogen fermentation in a two-stage process has considerable significance over all the aforementioned processes, as it considerably increases overall hydrogen yield.

6.8 Conclusion and Future Perspectives

This chapter set out to describe the current biohydrogen production systems along with a fundamental explanation of the photosynthetic microorganisms. At present time, a variety of geometries and immobilization matrices and methods are being evaluated for large-scale biohydrogen production. Moreover, efforts have also been made to reduce operational costs using waste-based feedstocks. However, marketable technologies also depend on the exploration of genetically engineered photosynthetic microorganisms to overcome practical barriers such as scale-up operation, oxygen sensitivity to hydrogenases, and light penetration.

By resolving these critical questions, the installation and maintenance costs associated with specially designed photobioreactors (PBRs) for biohydrogen production can be significantly reduced. Continued research on genetic interventions holds promise for overcoming the current challenges and limitations highlighted in this chapter. It is evident that photosynthetic microorganisms possess immense potential for sustainable biohydrogen production. However, achieving industrial-scale biohydrogen production will require the implementation of modern biotechnology. The resolution of these questions will therefore reduce the costs involved in installing and maintaining PBRs specially designed for biohydrogen production. Continued research on genetic interventions will provide solutions for a significant part of the current challenges and limitations described in this chapter. In summary, photosynthetic microorganisms have great potential for sustainable biohydrogen production but industrial scale biohydrogen production will be possible with the implementation of modern biotechnology.

Acknowledgments The authors are thankful for the support from Natural Sciences and Engineering Research Canada (Discovery Grant: 23541), Coordination of Improvement of Higher Education Personnel—CAPES, project Print-BioADD, and James and Joanne Love Chair in Environmental Engineering at York University, Canada.

References

Adessi A, De Philippis R (2012) Hydrogen production: Photofermentation. In: Hallenbeck PC (ed) Microbial technologies in advanced biofuels production. Springer, Boston, pp 53–75

Allakhverdiev SI, Thavasi V, Kreslavski VD et al (2010) Photosynthetic hydrogen production. J Photochem Photobiol C Photochem Rev 11:101–113

Antal T, Krendeleva T, Pashchenko V et al (2012) Photosynthetic hydrogen production: mechanisms and approaches. In: Azbar N, Levin DB (eds) State of the Art and progress in production of biohydrogen. Bentham Science, Sharjah, pp 25–54

Assawamongkholsiri T, Reungsang A (2015) Photo-fermentational hydrogen production of *Rhodobacter sp.* KKU-PS1 isolated from an UASB reactor. Electron J Biotechnol 18:221–230

Assunção J, Malcata FX (2020) Enclosed "non-conventional" photobioreactors for microalga production: A review. Algal Res 52:102107

Bielczynski LW, Schansker G, Croce R (2020) Consequences of the reduction of the Photosystem II antenna size on the light acclimation capacity of *Arabidopsis thaliana*. Plant Cell Environ 43:866–879

Bolatkhan K, Kossalbayev BD, Zayadan BK et al (2019) Hydrogen production from phototrophic microorganisms: Reality and perspectives. Int J Hydrog Energy 44:5799–5811

Boran E, Özgür E, van der Burg J et al (2010) Biological hydrogen production by *Rhodobacter capsulatus* in solar tubular photobioreactor. J Clean Prod 18:29–35

Budiman PM, Wu TY (2018) Role of chemicals addition in affecting biohydrogen production through photofermentation. Energy Convers Manag 165:509–527

Carvalho AP, Meireles LA, Malcata FX (2006) Microalgal reactors: a review of enclosed system designs and performances. Biotechnol Prog 22:1490–1506

Cheng J, Ding L, Xia A et al (2015) Hydrogen production using amino acids obtained by protein degradation in waste biomass by combined dark- and photo-fermentation. Bioresour Technol 179:13–19

Chookaew T, O-Thong S, Prasertsan P et al (2015) Biohydrogen production from crude glycerol by two-stage of dark and photo fermentation. Int J Hydrog Energy 40:7433–7438

Cournac L, Mus F, Bernard L et al (2002) Limiting steps of hydrogen production in *Chlamydomonas reinhardtii* and *Synechocystis* PCC 6803 as analysed by light-induced gas exchange transients. Int J Hydrog Energy 27:1229–1237

Das D, Veziroğlu TN (2001) Hydrogen production by biological processes: a survey of literature. Int J Hydrog Energy 26:13–28

Dasgupta CN, Jose Gilbert J, Lindblad P et al (2010) Recent trends on the development of photobiological processes and photobioreactors for the improvement of hydrogen production. Int J Hydrog Energy 35:10218–10238

Demirbas A (2009) Biohydrogen for future engine fuel demands. Springer, London

Fritsch J, Lenz O, Friedrich B (2013) Structure, function and biosynthesis of O_2-tolerant hydrogenases. Nat Rev Microbiol 11:106–114

Gaffron H, Rubin J (1942) Fermentative and photochemical production of hydrogen in algae. J Gen Physiol 26:219–240

García-Galán MJ, Gutiérrez R, Uggetti E et al (2018) Use of full-scale hybrid horizontal tubular photobioreactors to process agricultural runoff. Biosyst Eng 166:138–149

Gebicki J, Modigell M, Schumacher M et al (2010) Comparison of two reactor concepts for anoxygenic H_2 production by *Rhodobacter capsulatus*. J Clean Prod 18:36–42

Gest H, Kamen MD (1949) Photoproduction of molecular hydrogen by *Rhodospirillum rubrum*. Science 109:558–559

Ghirardi ML, King PW, Mulder DW et al (2014) Hydrogen production by water biophotolysis. In: Zannoni D, De Philippis R (eds) Microbial bioenergy: hydrogen production. Springer, Dordrecht, pp 101–135

Goussias C, Boussac A, Rutherford AW (2002) Photosystem II and photosynthetic oxidation of water: an overview. Philos Trans Roy Soc Lond Ser B Biol Sci 357:1369–1381

Guan Y, Zhang W, Deng M et al (2004) Significant enhancement of photobiological H_2 evolution by carbonylcyanide m-chlorophenylhydrazone in the marine green alga *Platymonas subcordiformis*. Biotechnol Lett 26:1031–1035

Gupta SK, Kumari S, Reddy K et al (2013) Trends in biohydrogen production: major challenges and state-of-the-art developments. Environ Technol 34:1653–1670

Gutthann F, Egert M, Marques A et al (2007) Inhibition of respiration and nitrate assimilation enhances photohydrogen evolution under low oxygen concentrations in *Synechocystis sp.* PCC 6803. Biochim Biophys Acta (BBA) Bioenerget 1767:161–169

Hallenbeck PC (2009) Fermentative hydrogen production: Principles, progress, and prognosis. Int J Hydrog Energy 34:7379–7389

Hallenbeck PC (2012) Hydrogen production by cyanobacteria. In: Hallenbeck PC (ed) Microbial technologies in advanced biofuels production. Springer, Boston, pp 15–28

Hallenbeck PC (2013) Photofermentative biohydrogen production. In: Pandey A, Chang JS, Hallenbeck PC, Larroche C (eds) Biohydrogen. Elsevier, Amsterdam, pp 145–159

Hallenbeck PC, Lazaro CZ, Sagir E (2018) Photosynthesis and hydrogen from photosynthetic microorganisms. In: Seibert M, Torzillo G (eds) Microalgal hydrogen production: Achievements and perspectives. Royal Society of Chemistry, London, pp 1–30

Heining M, Buchholz R (2015) Photobioreactors with internal illumination—A survey and comparison. Biotechnol J 10:1131–1137

Ho SH, Chan MC, Liu CC et al (2014) Enhancing lutein productivity of an indigenous microalga Scenedesmus obliquus FSP-3 using light-related strategies. Bioresour Technol 152:275–282

Ibraheem O, Ndimba BK (2013) Molecular adaptation mechanisms employed by ethanologenic bacteria in response to lignocellulose-derived inhibitory compounds. Int J Biol Sci 9:598–612

Jamil Z, Mohamad Annuar MS, Ibrahim S et al (2009) Optimization of phototrophic hydrogen production by *Rhodopseudomonas palustris* PBUM001 via statistical experimental design. Int J Hydrog Energy 34:7502–7512

Jiang D, Ge X, Zhang T et al (2016) Photo-fermentative hydrogen production from enzymatic hydrolysate of corn stalk pith with a photosynthetic consortium. Int J Hydrog Energy 41:16778–16785

Khetkorn W, Lindblad P, Incharoensakdi A (2010) Enhanced biohydrogen production by the N_2-fixing cyanobacterium *Anabaena siamensis* strain TISTR 8012. Int J Hydrog Energy 35:12767–12776

Kondo T, Arakawa M, Hirai T et al (2002) Enhancement of hydrogen production by a photosynthetic bacterium mutant with reduced pigment. J Biosci Bioeng 93:145–150

Kossalbayev BD, Tomo T, Zayadan BK et al (2020) Determination of the potential of cyanobacterial strains for hydrogen production. Int J Hydrog Energy 45:2627–2639

Kufryk G (2013) Advances in utilizing cyanobacteria for hydrogen production. Adv Microbiol 3:60–68

Lam MK, Loy ACM, Yusup S et al (2019) Biohydrogen production from algae. In: Pandey A, Mohan SV, Chang JS, Hallenbeck PC, Larroche C (eds) Biohydrogen. Elsevier, Amsterdam, pp 219–245

Laurinavichene T, Tolstygina I, Tsygankov A (2004) The effect of light intensity on hydrogen production by sulfur-deprived *Chlamydomonas reinhardtii*. J Biotechnol 114:143–151

Lazaro C, Hallenbeck PC (2019) Fundamentals of biohydrogen production. In: Pandey A, Mohan SV, Chang JS, Hallenbeck PC, Larroche C (eds) Biohydrogen. Elsevier, Amsterdam, pp 25–48

Lee JW, Greenbaum E (2003) A new oxygen sensitivity and its potential application in photosynthetic H_2 production. Appl Biochem Biotechnol 106:303–314

Lindblad P, Christensson K, Lindberg P et al (2002) Photoproduction of H_2 by wildtype *Anabaena* PCC 7120 and a hydrogen uptake deficient mutant: from laboratory experiments to outdoor culture. Int J Hydrog Energy 27:1271–1281

Lopes Pinto FA, Troshina O, Lindblad P (2002) A brief look at three decades of research on cyanobacterial hydrogen evolution. Int J Hydrog Energy 27:1209–1215

Lu Y, Koo J (2019) O_2 sensitivity and H_2 production activity of hydrogenases—a review. Biotechnol Bioeng 116:3124–3135

Mabutyana L, Pott RW (2021) Photo-fermentative hydrogen production by *Rhodopseudomonas palustris* CGA009 in the presence of inhibitory compounds. Int J Hydrog Energy 46:29088–29099
Madamwar D, Garg N, Shah V (2000) Cyanobacterial hydrogen production. World J Microbiol Biotechnol 16:757–767
Manoyan J, Gabrielyan L, Kozel N et al (2019) Regulation of biohydrogen production by protonophores in novel green microalgae *Parachlorella kessleri*. J Photochem Photobiol B Biol 199:111597
McIntosh CL, Germer F, Schulz R et al (2011) The [nife]-hydrogenase of the cyanobacterium *synechocystis sp.* pcc 6803 works bidirectionally with a bias to H_2 production. J Am Chem Soc 133:11308–11319
Melis A, Happe T (2001) Hydrogen production. Green algae as a source of energy. Plant Physiol 127:740–748
Melis A, Zhang L, Forestier M et al (2000) Sustained photobiological hydrogen gas production upon reversible inactivation of oxygen evolution in the green alga *Chlamydomonas reinhardtii*. Plant Physiol 122:127–136
Mishra P, Thakur S, Singh L et al (2016) Enhanced hydrogen production from palm oil mill effluent using two-stage sequential dark and photo fermentation. Int J Hydrog Energy 41:18431–18440
Montiel Corona V, le Borgne S, Revah S et al (2017) Effect of light-dark cycles on hydrogen and poly-β-hydroxybutyrate production by a photoheterotrophic culture and *Rhodobacter capsulatus* using a dark fermentation effluent as substrate. Bioresour Technol 226:238–246
Muzziotti D, Adessi A, Faraloni C et al (2016) H_2 production in *Rhodopseudomonas palustris* as a way to cope with high light intensities. Res Microbiol 167:350–356
Öztürk Y, Gökçe A, Peksel B et al (2012) Hydrogen production properties of *Rhodobacter capsulatus* with genetically modified redox balancing pathways. Int J Hydrog Energy 37:2014–2020
Pandu K, Joseph S (2012) Comparisons and limitations of biohydrogen production processes: a review. Int J Adv Eng Technol 2:342–356
Pinto TS, Malcata FX, Arrabaça JD et al (2013) Rubisco mutants of *Chlamydomonas reinhardtii* enhance photosynthetic hydrogen production. Appl Microbiol Biotechnol 97:5635–5643
Polle JEW, Kanakagiri SD, Melis A (2003) tla1, a DNA insertional transformant of the green alga *Chlamydomonas reinhardtii* with a truncated light-harvesting chlorophyll antenna size. Planta 217:49–59
Pott RWM, Howe CJ, Dennis JS et al (2013) Photofermentation of crude glycerol from biodiesel using *Rhodopseudomonas palustris*: Comparison with organic acids and the identification of inhibitory compounds. Bioresour Technol 130:725–730
Pruvost J, Pottier L, Legrand J (2006) Numerical investigation of hydrodynamic and mixing conditions in a torus photobioreactor. Chem Eng Sci 61:4476–4489
Rühle T, Hemschemeier A, Melis A et al (2008) A novel screening protocol for the isolation of hydrogen producing *Chlamydomonas reinhardtii* strains. BMC Plant Biol 8:107
Sakurai H, Masukawa H, Kitashima M et al (2013) Photobiological hydrogen production: Bioenergetics and challenges for its practical application. J Photochem Photobiol C Photochem Rev 17:1–25
Schulze PSC, Barreira LA, Pereira HGC et al (2014) Light emitting diodes (LEDs) applied to microalgal production. Trends Biotechnol 32:422–443
Shestakov SV, Mikheeva LE (2006) Genetic control of hydrogen metabolism in cyanobacteria. Russ J Genet 42:1272–1284
Show K, Yan Y, Zong C et al (2019) State of the art and challenges of biohydrogen from microalgae. Bioresour Technol 289:121747
Skjånes K, Andersen U, Heidorn T et al (2016) Design and construction of a photobioreactor for hydrogen production, including status in the field. J Appl Phycol 28:2205–2223
Tamburic B, Zemichael FW, Maitland GC et al (2011) Parameters affecting the growth and hydrogen production of the green alga *Chlamydomonas reinhardtii*. Int J Hydrog Energy 36:7872–7876

Tiang MF, Fitri Hanipa MA, Abdul PM et al (2020) Recent advanced biotechnological strategies to enhance photo-fermentative biohydrogen production by purple non-sulphur bacteria: An overview. Int J Hydrog Energy 45:13211–13230

Tiwari A, Pandey A (2012) Cyanobacterial hydrogen production—a step towards clean environment. Int J Hydrog Energy 37:139–150

Touloupakis E, Torzillo G (2019) Photobiological hydrogen production. In: Calise F, D'Accadia MD, Santarelli M, Lanzini A, Ferrero D (eds) Solar hydrogen production: processes, systems and technologies. Academic Press, Cambridge, pp 511–525

Uyar B (2016) Bioreactor design for photofermentative hydrogen production. Bioprocess Biosyst Eng 39:1331–1340

Vignais PM, Magnin JP, Willison JC (2006) Increasing biohydrogen production by metabolic engineering. Int J Hydrog Energy 31:1478–1483

Wang J (2011) Theory of flow distribution in manifolds. Chem Eng J 168:1331–1345

Winkler M, Hemschemeier A, Gotor C et al (2002) [Fe]-hydrogenases in green algae: photo-fermentation and hydrogen evolution under sulfur deprivation. Int J Hydrog Energy 27:1431–1439

Yang H, Shi B, Ma H et al (2015a) Enhanced hydrogen production from cornstalk by dark- and photo-fermentation with diluted alkali-cellulase two-step hydrolysis. Int J Hydrog Energy 40:12193–12200

Yang H, Zhang J, Wang X et al (2015b) Coexpression of Mo- and Fe-nitrogenase in *Rhodobacter capsulatus* enhanced its photosynthetic hydrogen production. Int J Hydrog Energy 40:927–934

Zhang Q, Zhang Z (2018) Biological hydrogen production from renewable resources by photo-fermentation. In: Li Y, Ge X (eds) Advances in bioenergy. Elsevier, Amsterdam, pp 137–160

Zhang Z, Wang Y, Hu J et al (2015) Influence of mixing method and hydraulic retention time on hydrogen production through photo-fermentation with mixed strains. Int J Hydrog Energy 40:6521–6529

Chapter 7
Metabolic Engineering and Synthetic Biology in Biohydrogen Production

Surjith Ramasamy and Kannan Pakshirajan

Abstract Hydrogen is a well-accepted clean fuel which is gradually replacing the conventional fossil fuels, and biohydrogen is a net zero fuel produced via biophotolysis, photofermentation, dark fermentation and microbial electrolysis cell. Commercialization of biohydrogen production is not yet realized due to its low yield and high production cost as compared to hydrogen produced from conventional methods. Modification of metabolic pathways, microorganisms and enzymes involved can lead to increase in biohydrogen yield and also eliminate the production of other undesired metabolites. Moreover, in order to overcome the disadvantages due to oxygen sensitivity of the hydrogenase and nitrogenase enzymes involved in hydrogen production, an in-depth analysis of the metabolic pathways is important. Metabolic flux balance analysis aids in understanding the complex metabolic pathways and to select specific genes for deletion in order to enhance hydrogen production. Flux models developed for hydrogen producing microorganisms can be used to accurately predict the hydrogen production, thereby minimizing the number of experiments required. This chapter presents an overview of metabolic engineering approach and synthetic biology tools developed to improve biohydrogen production.

Keywords Biohydrogen · Flux models · Metabolic engineering · Synthetic biology

---

S. Ramasamy · K. Pakshirajan (✉)
Department of Biosciences and Bioengineering, Indian Institute of Technology Guwahati, Guwahati, Assam, India
e-mail: pakshi@iitg.ac.in

C. R. Soccol et al. (eds.), *Biohydrogen - Advances and Processes*, Biofuel and Biorefinery Technologies 13, https://doi.org/10.1007/978-3-031-49818-3_7

7.1 Introduction

Hydrogen is one of the cleanest energy sources due to its high calorific value, and upon combustion it leaves no trace of pollutants (Hoffmann 2006). It is estimated that hydrogen will fulfil 8–10% of the world's energy demand by the year 2025 (Sharma et al. 2021). It is produced using different feedstock including fossil source, biomass and water. Most of the hydrogen in industry is produced through electrolysis, coal gasification, industrial oil and naphtha reforming, and steam reforming of natural gas in the ratio of 0.19:1.8:3:5 (Show et al. 2019). Hydrogen production from biomass as the feedstock consumes less energy and, therefore, it is also considered as the net neutral process (Hallenbeck 2004). Hydrogen is a high-energy (122 kJ/g) fuel and is used for different applications. Biohydrogen can be used directly in internal combustion engine of vehicles or used as fuel cell in automobile industry (Jo et al. 2010). Elemental hydrogen is universally available in organic and inorganic compounds, and these are widely used as the raw material for biohydrogen production; 1 g of hydrogen produces 34 kcal of energy whereas 1 g gasoline provides only 12 kcal. Combusted product of hydrogen is water and it does not release any oxides of nitrogen or sulphur when the hydrogen is combusted using air below 750 °C or under lean burn condition (Mahidhara et al. 2019; Lewis 2021).

Biological hydrogen production is mainly carried out through four methods, which are biophotolysis, photofermentation, dark fermentation and microbial electrolysis cell. Biophotolysis occurs in microalgae and cyanobacteria wherein photolysis of water takes place in the presence of sunlight to produce hydrogen; low conversion efficiency and inhibition of the hydrogenase enzyme due to oxygen are the major drawback of this method. In photofermentation, hydrogen is produced by nitrogenase in the absence of nitrogen or ammonium; however, low nitrogenase activity and inhibition of nitrogenase due to ammonium are the drawbacks. Hydrogen production by *Enterobacteriaceae* species and *Clostridium species* is advantageous due to high production rate but the formation of excessive amount of by-products is a major disadvantage of the dark fermentation process (Seol et al. 2012). Microalgae are alternative sources for hydrogen but their processing is challenging from an economic perspective (Pompelli et al. 2012). Microalgae are capable of converting 4–6% of light energy into biomass, and algae are capable of producing biohydrogen at high light intensity under anoxic environments (Zhan et al. 2017). Furthermore, microalgae have high growth rates, short life cycles and are capable of growing without pesticide and herbicide for biofuel production (Misra et al. 2016).

Several studies have been performed to increase the yield and productivity of biohydrogen via biological route. Genetic modification is the fastest way to improve the hydrogen production. Fermentative metabolism following the functional genomics approach further enhances the hydrogen yield. Overexpression and deletion of genes improve the hydrogen production (Seol et al. 2012; Wang et al. 2013). Metabolic engineering involves overexpression of existing product, synthesis of unnatural products and degradation of undesired products. Metabolic engineering eliminates the bottlenecks, increases the carbon flow, increases substrate utilization and induces the secretion of oxygen resistant enzymes for enhancing the hydrogen production (Kumar and Das 2001; Burgdorf et al. 2006). Knockout of a chromosomal gene is useful for metabolic engineering of the strains, and it is carried out using chemicals or ultraviolet radiation followed by screening of the mutants. Besides, certain advanced techniques are used for optimum modulation and deletion of genes rather than the time-consuming random mutagenesis approach.

This chapter deals with the genetic and metabolic engineering approaches in the biological route to produce biohydrogen, and improve its yield and productivity. Metabolic pathways, enzymes and metabolic flux analysis are discussed with respect to biohydrogen production.

7.2 Metabolic Engineering and Synthetic Engineering

Genetic engineering and synthetic biology approaches are applied to microorganisms for biohydrogen production. Genetically modified strains outperform wild strains with respect to hydrogen production, and for which metabolic and genetic engineering are the two valuable tools. Figure 7.1 shows various approaches followed in metabolic engineering and synthetic biology. In metabolic engineering, metabolic pathways are modified using recombinant technology for increasing the production of biofuel and other valuable bioproducts. Metabolic engineering includes the system biology through which systemic view of metabolic pathways and approaches aid in attaining optimal function. Total pathway shall be expressed through genes but the cost-effectiveness is determined by the titre value, rate of production and yield through optimization.

Metabolic engineering includes (a) defining and designing of the pathways which are involved in converting a substrate to product, (b) thermodynamic feasibility of the pathway, (c) estimation of pathway fluxes, (d) validation using radioactive tracer or metabolic flux, (e) use of genome model to target the specific gene, (f) optimal expression, (g) kinetic analysis of pathways and (h) inverse metabolic engineering for pathway optimization and evaluation of synthetic gene circuits. Flux

Synthetic biology	Metabolic engineering
• Genome	• Pathway and gene
• Promoter and Ribosomal binding site	• Overexpression
• Operon structure	• Deletion
• mRNA secondary structure	• Flux balance
• Transcription efficiency	• Modelling
• Translational rate	• Productivity
• mRNA stability	• Yield
	• Growth rate

Fig. 7.1 Synthetic biology and metabolic engineering approaches (Stephanopoulos 2012)

determination is the essential component of metabolic engineering along with reliable flux estimates such as measurement of tracer and assessment of acceptable confidence interval (Follstad et al. 1999; Antoniewicz et al. 2007). To overcome the time and resource demanding tracer method for determining flux, other methods were developed for evaluating the fluxes related to maximum growth. Genome scale models and linear optimization models are used for flux models for maximum growth (Varma and Palsson 1994). Pathway engineering in turn is classified into two types (1) native pathways such as amino acid pathway and (2) non-native pathways such as heterologous genes for amino acid synthesis.

Construction of first genetic counter and the genetic element to control the microbes have paved the way for synthetic engineering (Elowitz and Leibier 2000). Tremendous improvements are seen such as chemical synthesis of DNA to construction of the small genome. This has paved the way for synthetic cells for producing fuel and for environmental applications (Smith et al. 1989). Easy availability of synthetic DNA sequence has improved the scope of synthetic biology. Availability of toggle switch for controlling the gene expression for the product of interest and synthetic biology for pathway design have received tremendous interest due to industrial importance (Stephanopoulos 2012).

7.3 Metabolic Engineering and Synthetic Biology for Biohydrogen Production

7.3.1 Biophotolysis

In photolysis, water molecule is split into hydrogen and oxygen using solar energy in a biosystem. This is non-polluting and eco-friendly owing to the simple raw materials used and the products obtained in this process. Biophotolysis is divided into two types: direct and indirect photolysis. Figure 7.2 shows the mechanism involved in direct and indirect photolysis.

7.3.1.1 Direct Biophotolysis

In direct biophotolysis, Photosystems I and II absorb light energy to break water molecules into protons and electrons, and then hydrogen is produced with the help of hydrogenase enzyme (Eq. 7.1). Hydrogenase plays a vital role in converting the protons into hydrogen, but the enzyme is inhibited by high concentrations of hydrogen and oxygen, and it is applicable for both in direct and indirect biophotolysis (Li et al. 2022). The practical hydrogen yield efficiency is 2% whereas the theoretical yield efficiency is 80%, and the difference is due to the strong inhibitory effect of hydrogen and oxygen on the hydrogenase enzyme. Such kind of hydrogen production is common among microalgae such as *Chlamydomonas reinhardtii*.

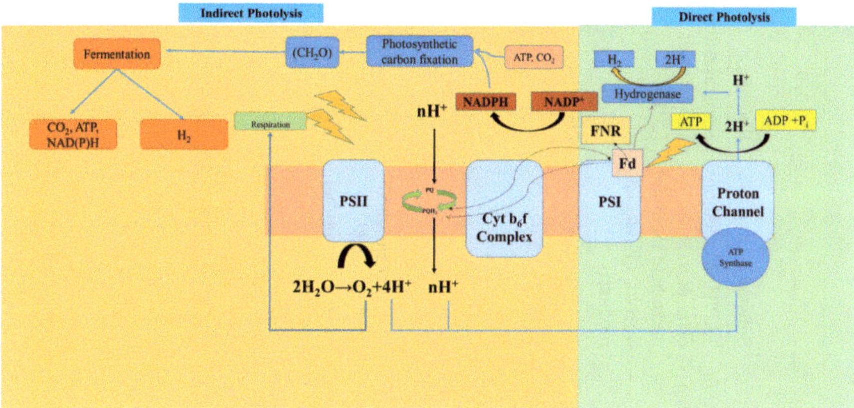

Fig. 7.2 Pictorial depiction of H_2 production through indirect biophotolysis and direct biophotolysis. Cyt b_6f, cytochrome b6f complex; Fd, Ferredoxin; FDP, Flavodiiron protein; FNR, Ferredoxin-NADP + reductase; H_2ase, hydrogenase; PC, plastocyanin; PQ, plastoquinones; PS, photosystem (Mahidhara et al. 2019)

$$2H_2O \xrightarrow{\text{Photon + Hydrogenase}} 2H_2 + O_2 \tag{7.1}$$

Light harvesting complex, electron transport system and hydrogenase enzyme mainly contribute towards hydrogen production by direct biophotolysis, and improvement in their functioning is key to increase the hydrogen production. Table 7.1 presents the literature on different metabolic and genetic engineering approaches used to improve the hydrogen production.

7.3.1.2 Indirect Biophotolysis

In indirect biophotolysis, electrons and protons are converted into carbohydrate in the light cycle (Eq. 7.2). During the dark cycle, reserved carbohydrate is transmuted to H_2 and CO_2 with the help of hydrogenase (Eq. 7.3) (Benemann 2000). These conversion reactions are observed in cyanobacteria, as follows:

$$6CO_2 + 6H_2O \xrightarrow{\text{Photon}} C_6H_{12}O_6 + 6O_2 \tag{7.2}$$

$$C_6H_{12}O_6 + 6H_2O \xrightarrow{\text{Hydrogenase}} 12H_2 + 6CO_2 \tag{7.3}$$

Hydrogenase enzymes in photolysis are highly sensitive to oxygen and less efficient in converting light, two main drawbacks in hydrogen production (Brentner et al. 2010; Lindblad et al. 2002; Zhang et al. 2002). Complexity of the algal hydrogenase has made it very difficult to engineer the enzyme for overcoming the oxygen sensitivity (Liebgott et al. 2011; Wu Chunfei et al. 2011). Investigations were done to

Table 7.1 Metabolic engineering strategies followed for improved biohydrogen production

S. no	Microorganism and strain	Metabolic engineering approach	Yield of hydrogen	Production rate	Substrate	Reference
	Biophotolysis and photofermentation					
1	*C. reinhardtii* 1. CC124 2. pgr5	1. Wild type 2. Knockout	1. 3.39 µmol/mL 2. 11.43 ± 0.5 µmol/mL	1. 2.05 ± 0.08 µmol/mg chl./h 2. 4.1 ± 0.21 µmol/mg chl./h	TAP/HSM medium (Tris-acetate phosphate medium/ Sueoka's High-Salt medium)	Khosravitabar and Hippler (2019)
2	*C. reinhardtii* 1. CC124 2. D240 3. D239-40 4. D240-41	1. Wild type 2. Double mutation of D1 complex 3. Double mutation of D1 complex 4. Double mutation of D1 complex	Total hydrogen produced 1. 175 mL/L 2. 165 mL/L 3. 490 mL/L 4. 388 mL/L	1. 50 mL/L/day 2. 70 mL/L/day 3. 140 mL/L/day 4. 95 mL/L/day	TAP medium	Oncel et al. (2014)
3	*Rhodobacter capsulatus* 1. MT1131 2. pHM14/FJ2 3. pFJ631/FJ2 4. pYO135/FJ2-R4 5. GK-32 6. KZ1 7. YO3 8. YO4 9. YO5	1. Wild type 2. cyt c2 3. cyt cy 4. cyt S-cyR35 5. cyt cbb_3^- 6. Qox^- 7. Hup^- 8. cyt cbb_3^-, Hup^- 9. Qox^-, Hup^-	1. 0.94 mL/mL of culture 2. 0.29 mL/mL of culture 3. 0.33 mL/mL of culture 4. 0.17 mL/mL of culture 5. 0.85 mL/mL of culture 6. 0.79 mL/mL of culture 7. 1.33 mL/mL of culture 8. 1.08 mL/mL of culture 9. 1.22 mL/mL of culture	1. 0.014 mL/mL of culture/ h 2. 0.014 mL/mL of culture/h 3. 0.0055 mL/mL of culture/h 4. 0.051 mL/mL of culture/h 5. 0.025 mL/mL of culture/h 6. 0.015 mL/mL of culture/h 7. 0.019 mL/mL of culture/h 8. 0.022 mL/mL of culture/h 9. 0.019 mL/mL of culture/h	(MPYE) enriched Medium and Sistrom's minimal medium	Öztürk et al. (2006)
4	1. *Rhodobacter sphaeroides* HY01 2. MHY01 3. MHY02 4. MHY03	1. Wild 2. hupSL::fdxN 3. hupSL::fdxN, △fdxN 4. △hupSL	1. 8445.7 mL of H_2/L 2. 9289.9 mL of H_2/L 3. 10058.8 mL of H_2/L 4. 9487.1 mL of H_2/L	1. 103.6 mL of H_2/L/h 2. 156.1 mL of H_2/L/h 3. 128.5 mL of H_2/L/h 4. 139.5 mL of H_2/L/h	Sistrom's succinate minimal (MedA) or MYPE	Ma et al. (2018)

S. no	Microorganism and strain	Metabolic engineering approach	Yield of hydrogen	Production rate	Substrate	Reference
	Biophotolysis and photofermentation					
5	*R. capsulatus* 1. MT1131 2 YO3 3. YO4 4. YO6 5. YO7 6. YO8	1. Wild 2. Hup^- 3. cyt cbb^-_3, Hup^- 4. PRK^- 5. Hup^-, PRK^- 6. cyt cbb^-_3, Hup^- PRK	1. 0.76 mL/mL culture 2. 1.27 mL/mL culture 3. 0.80 mL/mL culture 4. 0.61 mL/mL culture 5. 1.33 mL/mL culture 6. 0.90 mL/mL culture	1. 0.0075 mL/mL culture/h 2. 0.0181 mL/mL culture/h 3. 0.156 mL/mL culture/h 4. 0.0085 mL/mL culture/h 5. 0.0204 mL/mL culture/h 6. 0.129 mL/mL culture/h	MPYE medium (magnesium-calcium, peptone, yeast extract)	Öztürk et al. (2012)

S. no.	Microorganism and strains	Metabolic engineering approach	Yield of hydrogen (mol H_2/mol glucose)	Production rate (mmol H_2/g/h)	Substrate	Reference
	Dark fermentation					
1	*E. coli* 1. BW25113 2. SH1 (BW25113DhycA) 3. SH2 (SH1 DhyaAB) 4. SH3 (SH2 DhybBC) 5. SH4 (SH3 DldhA) 6. SH5 (SH4 DfrdAB)	1. Wild type 2. Inactivated negative regulator for FHL 3. Inactivated uptake hydrogenase 1 4. Inactivated uptake hydrogenase 2 5. Inactivated lactate dehydrogenase 6. Inactivated fumarate reductase	1. 1.20 2. 1.17 3. 1.37 4. 1.48 5. 1.61 6. 1.80	1. 30.6 2. 58.6 3. 46.2 4. 49.2 5. 45.6 6. 52.2	Luria Bertani (LB) medium	Kim et al. (2009)
2	*E. coli* 1. W3110 2. W3110 DldhA DfrdBC	1. Wild-type 2. Inactivated lactate dehydrogenase and fumarate dehydrogenase	1. 1.08 2. 1.82	1. 9.5 2. 13.4	Minimal medium containing sucrose	Yoshida et al. (2006)

(continued)

Table 7.1 (continued)

S. no.	Microorganism and strains	Metabolic engineering approach	Yield of hydrogen (mol H_2/mol glucose)	Production rate (mmol H_2/g/h)	Substrate	Reference
	Dark fermentation					
3	*E. coli* 1. BW25113/pCA24N 2. BW25113 DhyaB DhybC DhycA DfdoG/pCA24NFhlA	1. Wild type 2. Inactivated uptake hydrogenase and overexpressed FHL	1. 0.47 2. 0.70	1. 3.7 2. 12	LB media	Maeda et al. (2008)
4	*E. coli* 1. MC4100 2. HD705 3. FTD22 4. FTD67 5. FTD89 6. HD701, mutant	1. Wild type 2. Hyd-3; hycE knockout 3. Hyd-1; hyaB knockout 4. Hyd-2; hybC knockout 5. Hyd-1+Hyd-2; hyaB+hybC knockout 6. hycA knockout	1. 0.764 2. 0 3. 0.800 4. 1.024 5. 1.043 6. 0.737	NA	Nutrient broth	Redwood et al. (2008)
5	*E. coli* 1. BL21(DE3) pYdbK pAF 2. BL21(DE3) pA pET 3. BL21(DE3) pAF pET 4. BL21(DE3) ΔiscR pYdbK pAF 5.BL21(DE3) pA pYdbK 6. BL21(DE3) ΔiscR pA Pe	1. YdbK+CpFdx+hydA+hydF+hydG+hydE (overexpression) + MCS2 (Deletion) 2. hydA+hydF+hydG+hydE (overexpression) 3. CpFdx+hydA+hydF+hydG+hydE (overexpression) 4.YdbK+CpFdx+hydA+hydF+hydG+hydE iscR (overexpression) +MCS2 (Deletion) 5.YdbK+hydA+hydF+hydG+hydE (overexpression) 6.hydA+hydF+hydG+hydE (overexpression + iscR (Deletion)	1. 0.67 2. 0.01 3. 0.07 4. 1.46 5. 0.03 6. 0.04	NA	3-(*N*-Morpholino)-propanesulfonic acid (MOPS) minimal medium	Akhtar and Jones (2009)

S. no.	Microorganism and strains	Metabolic engineering approach	Yield of hydrogen (mol H_2/mol glucose)	Production rate (mmol H_2/g/h)	Substrate	Reference
	Dark fermentation					
6	*C. acetobutylicum* 1.DSM 792 2. DSM 792 [pSOS] 3. DSM 792 [pSOShydACa] 4. DSM 792 (pSOShydACb)	1. Wild type 2. thl promoter 3. hydA 4. hydA	1. 1.79 2. 1.77 3. 1.81 4. 1.80	1. 1810 mL H_2/L/h 2. 1740 mL H_2/L/h 3. 1650 mL H_2/L/h 4. 1710 mL H_2/L/h	Clostridial growth medium	Klein et al. (2010)
7	*C. paraputrificum* 1. M-21 2. M-21 pJIR751	1. Wild type 2. hydA	1. 1.4 2. 2.4	NA	Luria–Berrani (LB) medium	Morimoto et al. (2005)
8	*C. saccharoperbutylacetonicum* 1. N1-4 pNAK1, 2. N1-4 pAntihup	1. Wild type 2. hupCBA operon Knockout	NA	1. 500 mLH_2/L/h 2. 1546 mLH_2/L/h	TYA medium	Nakayama et al. (2008)
10	1. *C. tyrobutyricum* 2. PAK-Em	1. Wild type 2. ack Knockout	1. 1.44 2. 2.16	NA	Clostridial growth medium	Liu et al. (2006a)
11	*C. tyrobutyricum* 1. ATCC 25755 2. PAK-Em, mutant 3. ATCC 25755 PPTA-Em, mutant	1. Wild type 2. ack knockout 3. pta knockout	1. 1.35 2. 2.61 3. 1.08	NA	Clostridial growth medium	Liu et al. (2006b)

(continued)

Table 7.1 (continued)

S. no.	Microorganism and strains	Metabolic engineering approach	Yield of hydrogen (mol H_2/mol glucose)	Production rate (mmol H_2/g/h)	Substrate	Reference
	Dark fermentation					
12	*E. aerogenes* 1. IAM1183 2. IAM1183 A 3. IAM1183 O 4. IAM1183 AO	1. Wild type 2. hycA 3. hybO 4. hycA+hybO	1. 1.16 2. 1.20 3. 1.27 4. 1.36	1. 2321.86 mL/g/h 2. 2879 mL/g/h 3. 2747.20 mL/g/h 4. 3372.02 mL/g/h	Luria Broth	Lu et al. (2010)
13	*E. aerogenes* 1. IAM1183 2. IAM1183 Δldh, 3. IAM1183 (pCOM 10-fdh1)	1. Wild type 2. ldh knockout 3. ldh kockout +Fdh1 overexpression	1. 0.91 2. 1.01 3. 1.702	1. 0.141 L/L/h 2. 0.134 L/L/h 3. 0.110 L/L/h	Glucose 15.0 g, Tryptone 5.0 g, $(NH_4)_2SO_4$ 2.0 g, KH_2PO_4 14.0 g, $K_2HPO_4 \cdot 3H_2O$ 6.0 g, and $MgSO_4$ 0.2 g	Lu et al. (2010)
14	*E. aerogenes* 1. pMCL 2. pMCK	1. Wild type 2. ppk overexpression	1. 1.397 2. 1.504	1. 5.443 L/g 2. 5.809 L/g	Glucose 15.0 g, Tryptone 5.0 g, $(NH_4)_2SO_4$ 2.0 g, KH_2PO_4 14.0 g, $K_2HPO_4 \cdot 3H_2O$ 6.0 g, and $MgSO_4$ 0.2 g	Lu et al. (2011)
15	*E. cloacae* 1. IIT-BT-08 2. A3 3 DM11	1. Wild type 2. Alcohol dehydrogenase + butadienol dehydrogenase knockout 3. Alcohol dehydrogenase + butadienol dehydrogenase + acid blocking	1. 2.16 2. 1.65 3. 3.4	NA	MYG medium	Kumar et al. (2001)

decrease the oxygen level by reducing the photosynthesis rate and efficiency via sulphur deprivation which obviates the formation of sulphur containing polypeptides necessary in PS II reaction centre. The oxygen sensitivity could be avoided through this process, but the algae did not survive without photosynthesis (Esper et al. 2006).

In *Chlamydomonas reinhardtii*, inducible chloroplast gene expression was developed using the copper-sensitive cytochrome C_6 promoter and the nucleus-encoded Nac2 chloroplast protein which helps in accumulation of chloroplast *psbD* RNA and acts on its 5′ UTR(Untranslated region). *psbD* is expressed in copper-deficient condition but not under copper-rich conditions; in copper-deficient condition *psbD* encodes the D2 reaction centre polypeptide of photosystem II, increasing the oxygen concentration. Adding copper reduces the PSII, which in turn reduces the oxygen level and increases the hydrogenase activity (Surzycki et al. 2007). *Chlamydomonas reinhardtii* produces hydrogen by using [FeFe]-hydrogenase HYD1, after receiving electron from ferredoxin PetF. High amount of hydrogen is produced by *C. reinhardtii* in sulphur depleted conditions, and a high amount of electrons are produced from starch, which are then transferred to plastoquinone pool. Nitrogen starvation increases the starch and lipid content but the hydrogen production is delayed and the amount of hydrogen produced is less than that produced sulphur deficient condition (Philipps et al. 2012).

Size reduction of light harvesting complex (LHC) was carried out to increase the light harvesting efficiency. In *C. reinhardtii,* random mutagenesis was carried out to reduce the size of PSII, and it increased the hydrogen up to 300%, as compared to the original strain (Polle et al. 2002). Reduction in LHC was carried out using RNA interference technology and it showed increased quantum yield up to 1.8-fold and reduced the photoinhibition on PS II (Mussgnug et al. 2007). Deletion of TLA 1 gene in *C. reinhardtii*, *Scenedesmus obliquus*, and *Chlorella vulgaris* reduced the LHC number and improved the light harvesting efficiency (Mitra and Melis 2010). In *Chlamydomonas*, mutation was created on *stm6* gene, which reduced the movement of electrons in PSI and increased the H_2 production by five times compared to the native-type strain (Kruse et al. 2005).

Recent developments involve induction of hydrogen photo production in green algae with improved activity of oxygen sensitive [FeFe] hydrogenase located at the acceptor side of photosystem I to produce hydrogen for a prolonged period. Flavodiiron proteins (FDPs) were deleted in the *Chlamydomonas reinhardtii*, which is another competitor for the electrons involved in hydrogen production (Jokel et al. 2019). In *Chlorella* sp., hydA from the DT strain, in which the amino acid residues are modified (A1051, G113I, V273I and V265W), was inserted to prevent the oxygen at the active site. Even in the presence of 5% oxygen, seven times more hydrogen was produced than in the wild strain. In *C. reinhardtii*, genetic regulation of micro-RNA targeting PSII reaction protein expression was carried out using photon. The light-inducible heterodimer protein CRY2 and CIBI were mixed with VP16 (transcription activation domain) and GAL4 (DNA-binding domain), which in turn activated the target gene. Hydrogen production was enhanced by blue light-activated miRNA in the microalgae (Zhan et al. 2017; Wang et al. 2017).

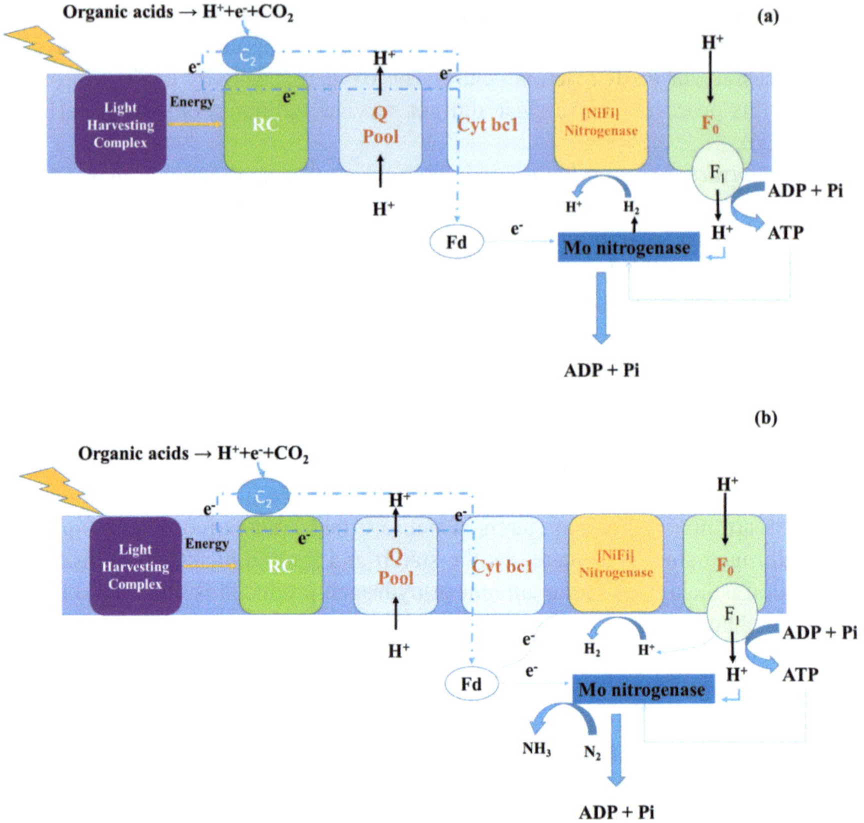

Fig. 7.3 Photofermentation mechanism in purple non-sulphur bacteria (**a**) nitrogen deficient condition. Cyt, Cytochrome; Fd, ferredoxin; Q, ubiquinone; RC, reaction centre (**b**) nitrogen-rich condition (Öztürk et al. 2006, 2012)

7.3.2 Photofermentation

In photofermentation, electrons are obtained from organic acids for hydrogen production by purple nonsulfur bacteria (PNSB). The electrons obtained from the organic sources are transported through membrane-bound transport mechanisms to oxidize ferredoxin, which in turn reduces nitrogen to ammonia in the presence of nitrogenase (Eq. 7.4). Hydrogen is produced in the absence of nitrogen because nitrogenase reduces the proton to hydrogen (Eq. 7.5). Figures 7.3 and 7.4 show the nitrogen production mechanism through photofermentation.

$$\text{Presence of nitrogen: } N_2 + 8H^+ + 8e^- + 16ATP \xrightarrow{\text{Light+Nitrogenase}} 2NH_3 + H_2 + 16ADP + 16P_i \quad (7.4)$$

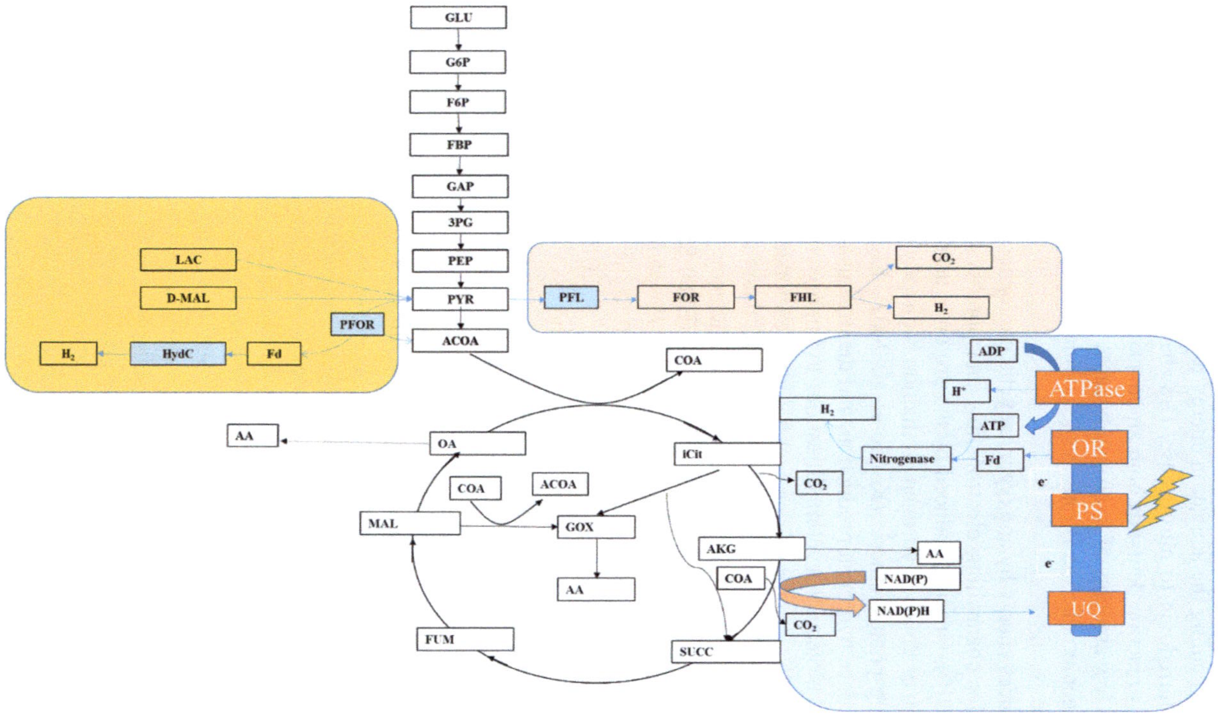

Fig. 7.4 Hydrogen production through photofermentation. AC, acetate; ACOA, acetyl-CoA; D-MAL, D-malate; Fd, ferredoxin; formate:hydrogen lyase; FHL; FOR, formate; GLC, glucose; FBP, fructose1,6biphosphate; FUM, fumarate; G6P, glucose6phosphate; GAP, glyceraldehyde 3 phosphate; Glu, glucose; GOX, glyoxalate; HydC, ferredoxin-dependent hydrogenase; LAC, lactate; N_2ase, nitrogenase; MAL, malate; OA, oxaloacetate; OR, oxidoreductase; PEP, phosphoenol pyruvate; PFL, pyruvate:formate lyase; PFOR, pyruvate:ferredoxinoxidoreductase; PG, phosphatidylglycerol; PRO, propionate; PYR, pyruvate; PS, photosystem; R5P, ribulose5phosphate; RNA, ribose nucleic acid; Ro5P, ribose5phosphate; SUCC, succinate; UQ, ubiquinone; X5P, xylulose 5 phosphate (Kim et al. 2009; Oh et al. 2011)

$$\text{Absence of nitrogen}: \ 8H^{+} + 8e^{-} + 16ATP \xrightarrow{\text{Light+Nitrogenase}} 4H_2 + 16ADP + 16P_i \quad (7.5)$$

A large amount of ATP is required in the process of nitrogen or hydrogen formation, i.e. ATP for overcoming the activation energy for nitrogenase and for transporting electron from ferredoxin to nitrogenase. Drawbacks of photofermentation are as follows: (1) this reaction demands high amounts of ATP for hydrogen production, (2) it produces low amounts of hydrogen because the turnover number of nitrogenase for hydrogen production is only 5/s (3) oxidation of final product (hydrogen) by hydrogenase

PNSB produces energy by anoxygenic cyclic photophosphorylation using complex proteins for channelling electron flow and photosystems (Oda et al. 2008; Huang et al. 2010). Gradient induced by the proton is used for ATP production in the photosystem of PNSB. Hydrogen production is directly proportional to the ATP produced during photosynthesis (Kondo et al. 2002). Light inhibition and shelf shading also reduce hydrogen production, and the organisms are subjected to various genetic and molecular level modifications to improve the light harvesting efficiency. Mutant *Rhodobacter sphaeroides* RV was developed by Kondo et al. (2002) by UV irradiation, which produced 150% higher H_2 than the native strain in a photobioreactor by increasing the light absorption efficiency. Mutant *R. sphaeroides* KCTC 12085 without light harvesting complex (B800-850) produced two-fold higher hydrogen than the wild strain. Mutant stain *R. sphaeroides* pRK puf DD13 produced higher amount of hydrogen as compared to the strain pRK puf ΔLM1 with peripheral light harvesting complex (Eltsova et al. 2010).

RegC/RegA activates the nitrogenase during photosynthesis, and prior to that cytochrome (cbb_3) and cyclic photosynthetic electron transfer chain in redox state activates the RegC/RegA in *Rhodobacter*. The hydrogen production in *Rhodobacter* with modified electron carrier cytochrome/cyt cbb_3 oxidase/quinol oxidase was evaluated. The hydrogen production was reduced by three- to fourfold in the *Rhodobacter* with modified electron transfer, whereas the organism without cbb_3 showed higher hydrogen production rate. Hup (hydrogenase deficient) showed increase in total hydrogen and its production rate. *R. capsulatus* was genetically modified to alter the electron carrier cytochromes or to remove cyt oxidase or quinol oxidase. Alteration in the specific gene reduced the H_2 production by three to four times, but the production rate increased. Reducing the activity of hydrogenase genes further improved the H_2 production efficiency (Schindel and Bauer 2016).

Nitrogenase produces hydrogen in the absence of nitrogen, whereas hydrogenase breaks the hydrogen for producing protons for other metabolic activities. The nitrogenase expression depends on the ammonium concentration and it is regulated in transcriptional and post-transcriptional levels. PSII like proteins (GlnB and GlnK) activate DraT which in turn inhibit the nitrogenase synthesis by ADP ribosylation; *NifA*, an activator of nitrogenase gene is also controlled by GlnB and GlnK (Oh et al. 2008). Deleting GlnB and GlnK in *R. capsulatus* increased the enzymatic activity and hydrogen production (150%) compared to the native strain even at high ammonium concentration (Drepper et al. 2003). A similar pattern was reported in the case of *R. sphaeroides* KCTC 12085 upon deletion of the PSII-like protein in the organism (Li et al. 2010). The glutamine synthase gene was deleted in *R. sphaeroides* 6016, which is responsible for ammonium synthesis, and due to which higher amount of hydrogen was obtained. Deleting the hydrogenase enzyme (Hup) in *Rhodopseudomonas palustris* increased the hydrogen production, and deletion of Hup and PHB synthase in *R. sphaeroides* KD131 yielded 2.5 times higher hydrogen per gram of biomass as compared to the wild strain (Oda et al. 2008; Koku et al. 2002).

The *Rhodobacter capsulatus* with hup (hydrogenase) knocked out was used for biohydrogen in small-scale (batch mode) and large-scale (continuous mode) reactor system. The hydrogen yield values were 34 and 31%, and its productivity were 0.49 and 0.26 mmol/L/h with the hydrogenase lacking strain and wild strain in the small-scale reactor. In the large-scale continuous photobioreactor, the hydrogen yield values were 48 and 46% and its productivity were 1.01 and 1.05 mmol/L/h with hydrogenase lacking strain and wild strain (Uyar et al. 2015). In *Rhodopseudomonas palustris*, the hydrogen produced did not depend on reduced carbon compounds but it depended on the biochemical pathway, and as a result both succinate and acetate produced the same amount of the hydrogen. Another important aspect is the amount of carbon flux towards the Calvin cycle, and deletion of the Calvin cycle increases the hydrogen production (McKinlay and Harwood 2011). Butyrate and succinate increased the hydrogen production by twofold, whereas acetate showed only 1.3-fold increase. This is because butyrate and succinate are processed through Calvin-Benson cycle. Elimination of cbbP (phosphoribulose kinase), which in turn arrested the Calvin-Benson cycle, showed lower hydrogen yield and productivity (Farmer et al. 2014). Additional mutation, i.e. DccoNOQP with cbbP, increased the hydrogen yield by 1.4% (Öztürk et al. 2012). Elimination of hydrogenase (hupR) gene and overexpression of nifA increased hydrogen production by 1.2 times in *R. capsulatus*. The double mutation of hvr A and hup AB in *R. capsulatus* enhanced hydrogen production by 44.5% than the wild-type strain (Cai and Wang 2014). Individual over expression of F1 or F0 operon in *R. sphaeroides* increased ATP content, activity of nitrogenase enzyme and hydrogen production, whereas expression of both F1 and F0 led to reduction in the activity of nitrogenase enzyme and hydrogen production (Zhang et al. 2017).

The enzyme hydrogenase are of three types: metal free hydrogenase, [Fe]-hydrogenase and [NiFe] hydrogenase in *R. sphaeroides, Ralstonia eutropha, Rhodobacter capsulatus, and Bradyhizobium legyminosarum.* [NiFe] hydrogenase enzyme is prominent one because the enzyme activity is unaffected in presence of O_2 and CO. This enzyme is further classified into membrane-associated respiratory uptake [NiFe] hydrogenase, cytoplasmic heterodimeric [NiFe] hydrogenase, cytoplasmic heterodimeric reversible [NiFe] hydrogenase and membrane associated H_2 evolving respiratory [NiFe] hydrogenase. Hydrogen production is carried out by membrane associated H_2 evolving respiratory [NiFe] hydrogenase in *R. sphaeroides*. Suicide vector was used in *R. sphaeroides* to knockout *hupSL* gene (hydrogenase), which resulted in enhanced hydrogen production (Penfold and Pemberton 1992).

Anabaena sp. PCC7120 with hupL⁻, hoxH⁻ mutation were assessed for the hydrogen production, hupL⁻ produced seven times higher amount of hydrogen, whereas in hoxH⁻ mutants hydrogen yield was slightly lower which was due to lower nitrogenase activity (Masukawa et al. 2002). The amount of the gas also affects hydrogen production: 100 % nitrogen inhibits hydrogen production, whereas hydrogen production is maximum at 10% hydrogen (Sasikala et al. 1993). Light also influences the hydrogen production: increase in light intensity increases the nitrogenase activity.

Mo-nitrogenase is the most studied nitrogenase, which has two parts: dinitrogenase (MoFe protein) and dinitrogenase reductase (Fe protein), and they are encoded as nifDK and nifH. The Fe Mo cofactor acts as the electron donor to dinitrogenase, which is important for fixing nitrogen and producing hydrogen. In *Rhodobacter capsulatus* a new nitrogen fixing gene has been discovered, which is rnfABCDGEH (Kars et al. 2006; Burgess et al. 2000). The main functions of NifH are transport of electrons to MoFe protein, synthesis of MoFe and maintaining the stability of the enzyme. Overexpression of the NifH may increase the electron flow to nitrogenase, increasing hydrogen production. Nitrogenase activity increased up to 1.5–2 times than in the wild strain of *R. capsulatus* (Schmehl et al. 1993).

*GLn*K and *Gln*B double mutant strains eliminated the ammonium regulated Mo-nitrogenase synthesis. NifL is also involved in oxygen regulation to repress the nitrogenase synthesis which was eliminated in a NifL mutant (Eydmann et al. 1995). Ferredoxin I donates electron to the nitrogenase and it is encoded by fdxN gene. Overexpression of the gene was carried out by inserting it in hupSL region in *Rhodobacter sphaeroides* HY01. The activity of fdxN and nitrogenase increased up to 177 and 61.7%, respectively. Use of 25 mM of acetate, 34 mM of butyrate and 6 mM of L-glutamate increased the hydrogen production up to 30%, 52.5% and 50.7%, respectively, in fdxN overexpressed *Rhodobacter sphaeroides*. Heat-resistant mutant *Rhodobacter sphaeroides* such as A52 and B41 were sequenced and they had 2137 and 2253 mutations as compared to the reference strain DSM1710. Mutant B41 had mutation in nifD, nifJ, glnD, nifB1, ccpA, hupD, dmsA, and cbbR1 genes, whereas A52 had mutations in nifB2, rnfF, nifJ, cbbO, anfH, amt, moeA, and hupD genes. Mutant B41 produced very high amount of hydrogen (Gökçe et al. 2017).

In photosynthesis, electron transfer between reaction centre and ubihydroquinone are important for energy production. Photochemical centre operates the ferrocytochromec-ubiquinoneoxidoreductase activity and cytochrome bc_1 complex operates the ubiquinol-ferricytochrome oxidoreductase activity. Deletion of cyt cbb_3 oxidase or quinol oxidase reduced the hydrogen production by three to four times, whereas hydrogen production increased in cbb_3 deleted *R. capsulatus* (Öztürk et al. 2006)

7.3.3 *Dark Fermentation*

In dark fermentation, microorganisms utilize carbon sources such as glucose, cellulose and hemicellulose for energy and electron to produce hydrogen without any light requirement (Oh et al. 2011; Moreno-Garcia et al. 2017). Three reactions in the Embden–Meyerhof–Parnas (EMP) pathway are involved in hydrogen production by dark fermentation route. Figure 7.5 shows the mechanism involved in dark fermentation for biohydrogen production.

1. In the EMP pathway, pyruvate is transmuted to acetyl-CoA and formate using pyruvate formate lyase (PFL). Hydrogen and carbon dioxide are produced by disintegrating formate by the enzyme formate:hydrogen lyase (FHL). *Escherichia coli* and *Enterobacteriaceae* produce hydrogen using this reaction (Hallenbeck 2005).
2. Pyruvate:ferredoxin oxidoreductase (PFOR) converts the pyruvate to acetyl-CoA and CO_2 in anoxic condition and Fd-dependent hydrogenase (HydA). Electron produced in the reaction is first transferred to Fdox and then combined with protons to form hydrogen by HydA. Majority of *Clostridium species* uses this reaction (Hallenbeck 2005).
3. NAD(P)H:ferredoxin oxidoreductase (NFOR) is reduced by NAD(P)H, and the electrons are coupled with protons to form hydrogen with the help of HydA. This

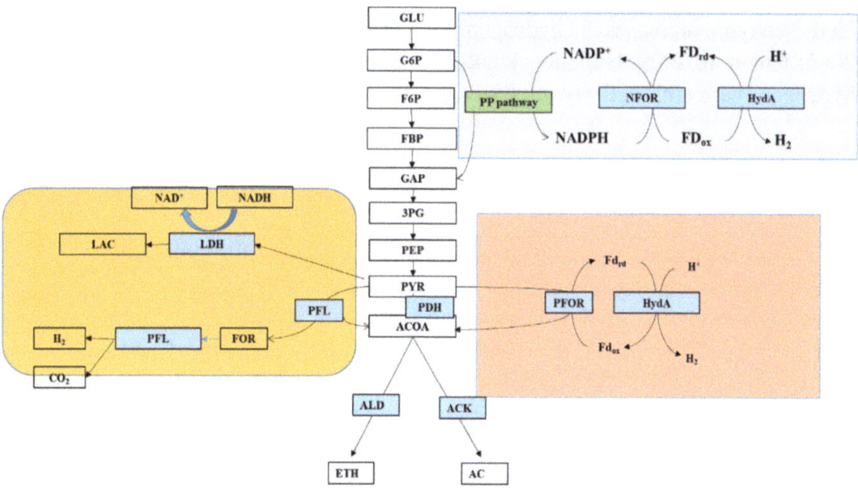

Fig. 7.5 Hydrogen production mechanism in dark fermentation. AC, acetate; ACOA, acetyl-CoA; ACK, acetate kinase; ALD, alcohol dehydrogenase; ETH, ethanol; Fdox, oxidized ferredoxin; Fdrd, reduced ferredoxin; FOR, formate; GLC, glucose; G3P, glyceraldehyde-3-phosphate; G6P, glucose-6-phosphate; HydA, ferredoxin-dependent hydrogenase; LAC, lactate; LDH, lactate dehydrogenase; PP pathway, pentose phosphate pathway; NFOR, NAD(P)H:ferredoxin oxidoreductase; PDH, pyruvate dehydrogenase; PFL, pyruvate:formate lyase; PFOR, pyruvate:ferredoxin oxidoreductase; PYR, pyruvate (Kim et al. 2009; Oh et al. 2011)

kind of hydrogen producing mechanism is common among thermophilic bacteria and in some *Clostridium species* (Wang et al. 2010).

Theoretical yield of hydrogen is 12 mol/mol of glucose ($C_6H_{12}O_6 + 6H_2O \rightarrow 12H_2 + 6CO_2$), but the experimental yield is only 33% which is due to the formation of metabolites such as alcohols and organic acids, which in turn inhibit the microbial growth (Moreno-Garcia et al. 2017; Levin et al. 2004; Lin and Chang 2004). Genetic engineering and synthetic biology have been applied for gene knockout, overexpression of genes, dual system and creation of new pathways to increase the hydrogen production (Goyal et al. 2013). Genetic engineering is advantageous only if the enzyme producing hydrogen is limited, else the gene modification may be detrimental to the microbes (Hallenbeck and Ghosh 2012). The theoretical yield of hydrogen in anaerobic organism which utilizes PFL and FHL is 2 mol/mol of glucose, whereas strict anaerobes which uses NFOR and PFOR yield 4 mol hydrogen/mol of glucose. Formation of alcohol and acid reduces the yield to 2 mol/mol of glucose, and metabolic engineering is done to attain the theoretical yield (Levin et al. 2004).

Deletion of the gene lactate dehydrogenase (ldh) increased the flux towards hydrogen production instead of lactate formation and this modification is favourable when the medium pH is low (Hallenbeck and Ghosh 2012). The enzyme fumarate reductase (frdBC) which converts phosphoenolpyruvate (PEP) to succinate also reduces the flux towards hydrogen production, and knockout of this enzyme improves hydrogen production. Hydrogenase enzymes are another bottleneck in hydrogen production as some of these enzymes such as Hyd-1 and Hyd-2 are capable of utilizing hydrogen through oxidation; on the other hand, increased activity of Hyd-3 enzyme increases the hydrogen production (Redwood et al. 2008; Hallenbeck 2005; Fan et al. 2009; Hallenbeck and Ghosh 2009). Similarly, increase in Formate hydrogen lyase (fhl) activity improves the hydrogen production.

7.3.3.1 *Escherichia coli*

E. coli has been widely used for heterologous expression for hydrogen production and other products due to complete genetic characterization. In *E. coli*, hydrogenases such as Hyd-1, Hyd-2, Hyd-3, and Hyd-4 are involved in hydrogen production and consumption, and the role of Hyd-4 is not known. Hyd-1 and Hyd-2 tend to consume hydrogen and knockout of Hyd-1 has no effect on the hydrogen yield, whereas Hyd-2 knockout increased the hydrogen yield (Redwood et al. 2008; Fan et al. 2009). Hyd-3 is involved in hydrogen production and removal of this enzyme leads to no hydrogen production. Contrastingly, other studies have shown no effect of Hyd 3 knockout on hydrogen production (Yoshida et al. 2006; Penfold and Macaskie 2004; Penfold et al. 2003). Knockout of hycA along with overexpressing of the fhlA enhanced H_2 production to 280%.

Alpha-formate dehydrogenase-N(fdnG) and alpha-formate dehydrogenase-O (fdoG) consume formate which reduces the hydrogen yield. Deletion of fdoG increased the hydrogen production whereas removal of fdnG had no effect on hydrogen production (Maeda et al. 2007). A high yield of hydrogen (3.12 mol/mol glucose) was obtained when hydrogenase gene (hydA) was deleted from *E. cloacae* IIT-BT-08 and *E.coli* BL-21 (Chittibabu et al. 2006). Similarly, adding hydrogenase (HupSL) from *Rhodobacter sphaeroides* in *E. coli* exhibited 20-fold increase in hydrogen (Lee et al. 2010).

Introduction of bidirectional hydrogenase (HoxEFUYH) enzyme from *Clostridium* and *Synechocystis* sp. shown 41-fold increase in hydrogen by inhibiting Hyd-1 and Hyd-2, but without disturbing Hyd-3 activity (Maeda et al. 2008). In *E. coli*, knockout of hycA gene increased FHL activity by two-fold and deletion of hya and hyb increased the hydrogen yield from 1.2 to 1.48 mol/mol of glucose. Furthermore, deletion of ldhA and frdAB genes increased the yield to 1.8 mol/mol of glucose at high partial pressure and 2.1 mol/mol of glucose at low partial pressure (Kim et al. 2009).

Immobilized *E. coli* SH5 with the genes hya, hyb, hycA, ldhA and frdAB knocked out shows high volumetric productivity of hydrogen (2.4 L H_2/L/h), and further the expression of fdhH and elimination of iscR increased the FHL activity (Seol et al. 2012). NiFe-hydrogenases are resistant to oxygen as compared to [FeFe]-hydrogenases. Expression of NiFe-hydrogenases1 in *E. coli* BL21 lacking hydrogen production capability yielded 12.5 mL of H_2/L/h. This hydrogenase is important to overcome the oxygen sensitivity (Seol et al. 2012). Ferredoxin-dependent hydrogenase, [4Fe-4S]-ferredoxin, NFORs from *Bacillus subtilis* and *Chlorobium tepidum* were incorporated in *E. coli* BL21(DE3) to form NAD(P)H-dependent hydrogen production pathway. The yield of the hydrogen was very low (0.04 mol H_2/mol glucose) as the NAD(P)H-dependent pathway is inhibited by hydrogen or the hydrogen produced was consumed by NFORs (Veit et al. 2008).

Transporter proteins play an important role in transporting hydrogenase enzyme, which either decreased or increased the hydrogen yield in the *E. coli*. Twin-arginine translocation (Tat) includes TatA, TatB, and TatC proteins for transporting the hydrogenase enzyme. Deletion of tatC and tatAE doubled the hydrogen production, due to inactivation of the hydrogenase enzyme (Penfold and Macaskie 2004; Gohlke et al. 2005). Elimination of lactate and succinate production pathways also increased the hydrogen yield. Knocking out the ldhA (lactate) and frdBC (succinate) pathways increased the hydrogen yield by two times, but it was 10% lower than the theoretical yield (Yoshida et al. 2006). Pyruvate dehydrogenase (aceE) and pyruvate oxidase (poxB) tend to consume pyruvate and deletion of aceE was found to increase the hydrogen production, whereas poxB did not show any effect on hydrogen production.

The following six proteins from different microbes were expressed in the *E. coli* to form a synthetic pathway for hydrogen production: YdbK (*E. coli*), [4Fe–4S]-ferredoxin (*Clostridium pasteurianum*), HydA, HydE, HydF, and HydG (*Clostridium acetobutylicum*). The synthetic pathway using the six proteins showed nine times higher hydrogen production as compared to the wild strain. Starch was used as the substrate which was metabolized by adding AmyE (*Bacillus subtilis*) protein into the synthetic pathway of the *E. coli* (Akhtar and Jones 2009). Knockout of formate transport (focA) increased formate concentration outside the cell instead of transporting into the cell, which in turn enhanced the hydrogen production (Fan et al. 2009). Deletion of ppc gene for phosphoenolpyruvate carboxylase reduced the succinate production and further inactivation of NarL increased the hydrogen yield by two times. NarL are repressor for FHL, PFL, and nickel transporters nik operon. Overexpression of selC gene for selenium metabolism, and fnr (transcription regulator) are found to increase the hydrogen yield (Fan et al. 2009).

7.3.3.2 *Clostridium*

Clostridium is a strict anaerobic bacteria, found in soil and intestine, and it is capable of producing hydrogen. Gene knockout of hydA in *Clostridium perfringens* resulted in no hydrogen production, which indicates that the specific gene is responsible for hydrogen production (Lecker et al. 2017). Increasing the expression of [FeFe]-hydrogenase in C. tyrobutyricum JM1 improved hydrogen production by up to 150% compared to the parental strain yield (Jo et al. 2010). Overexpression of hydA resulted in a 170% increase in hydrogen yield compared to the wild strain *Clostridium paraputrificum* M-21. Low lactic acid production and increased oxidation of NADH to NAD^+ were also noted when the hydA was over expressed (Morimoto et al. 2005). Downregulation of hydrogenase genes (hupCBA) in *Clostridium saccharoperbutylacetonicum* N1-4 showed a 310% increase in hydrogen production compared to the wild-type strain. Knockout of acetate kinase in *Clostridium tyrobutyricum* increased hydrogen production (Liu et al. 2006a; Hallenbeck and Ghosh 2009). Knockout of phosphotransacetylase gene (*pta*) in *Clostridium* reduced the hydrogen production (Liu et al. 2006a, b).

7.3.3.3 *Enterobacter aerogenes/E. cloacae*

Enterobacter aerogenes are capable of growing in both aerobic and anaerobic conditions. This bacterium has a high growth rate and hydrogen production rate. *Enterobacter* species are robust in producing very high hydrogen, and it grows along with other organisms such as *Clostridium species* and photofermentative organisms (Yokoi et al. 2001). Hydrogen production in the *Enterobacter aerogenes* involves two main reactions: pyruvate to formate and NADH oxidation (Lu et al. 2010; De Vos et al. 1983; Xia et al. 2013). FHL repressor (hycA) and hydrogenase uptake (hybO) were knocked out in the organism and this strategy was found to enhance the hydrogen production. Unblocking the FHL pathway also increased the production, whereas NADH pathway did not affect the production. Deletion of hycA and hybO decreased the growth rate (Zhao et al. 2009). Conversely, increasing the expression of heterologous gene (hydA) increased the hydrogen yield on glucose as the substrate. Overexpression of formate dehydrogenase (fdhF) and formate hydrogen lyase activator protein (fhlA) genes independently increased the hydrogen production (Lu et al. 2010).

E. aerogenes IAM1183 with FHL repressor (hycA) and hydrogenase (hybO) in different combinations were evaluated for hydrogen production. Strains without hycA produced 0.73 mol H_2/mol glucose, strains without hybO produced 0.78 mol H_2/mol glucose and strains with both the genes deleted yielded 0.83 mol H_2/mol glucose, whereas in the case of parental strain the hydrogen yield was 0.65 mol H_2/mol (Zhao et al. 2009). Deletion of NADH dependent lactate dehydrogenase and overexpression of NAD^+ dependent formate dehydrogenase increased the hydrogen production by 1.868 times in *E. aerogenes* IAM 1183 (Lu et al. 2010). *E. aerogenes* are capable of consuming high amount of NADH for lactate production, deletion of ldh gene resulted in higher hydrogen yield and reduced rate of production. Elimination of alcohol forming pathway in *E. aerogenes* IIT-BT reduced the hydrogen by enhancing the acid formation, whereas blocking the acid pathway increased the hydrogen by 150% (Kumar and Das 2001). Simultaneous overexpression of fdh1 and ldh knockout resulted in 1.86-fold higher hydrogen production by the organism (Lu et al. 2010). Diversion of NADH leads to hydrogen production; in general, NADH is utilized for the production of various organic acids such as ethanol, butanediol, lactic acid and butyric acid. Addition of NADH in the medium increased the hydrogen production from formate pathway and reduced hydrogen in NADH pathway, and the vice versa was observed when NAD^+ was added in the media (Zhao et al. 2009).

7.3.3.4 Extremophilic Bacterial Species

Formation of fewer biproducts, efficient utilization of substrate, high production rate and yield of hydrogen at high temperature are the advantages of utilizing thermophilic bacterial species for hydrogen production. Deleting the ldh gene in *Thermoanaerobacterium aotearoense* yielded zero lactate and 200% increase in hydrogen yield with a 250% increase in production rate compared to the wild strain (Xin et al. 2010).

Halanaerobium saccharolyticum subsp. *saccharolyticum* and *H. saccharolyticum* subsp. *senegalesis* were used to produce hydrogen by using glycerol as the substrate. Hydrogen production was higher in subspecies *senegalesis* than in the subspecies *saccharolyticum*, due to the formation of 1,3-propanediol by *saccharolyticum*. When glucose was used as the substrate, no 1,3-propanediol formation was observed (Kivistö et al. 2011).

Thermophilic bacteria are capable of yielding 3.3–4 mol of H_2/mole of glucose by using NFOR and PFOR pathway. Hydrogen production is thermodynamically not feasible by NAD(P)H but the activities of NFOR and Hyd A are maximum at higher temperature for yielding a high amount of hydrogen. Although the hydrogen yield is high, the cell density and volumetric productivity of hydrogen were less as compared to the mesophilic organisms (Hallenbeck and Ghosh 2009).

7.3.3.5 Substrate Utilization

Metabolic engineering of microorganisms is carried out to utilize a wide range of substrates, which also avoids the need for pre-treatment, thereby eliminating the release of furfurals or hydroxyl methyl furfurals that are known to inhibit hydrogen production by microorganisms. Substrates such as cellulose and hemicellulose can be utilized instead of costly glucose by metabolic engineering approaches. Extremophilic bacteria are capable of utilizing the polymeric substrate for hydrogen production, but the yield is very low, and therefore the scope for modifying these organisms is less (Geng et al. 2010). Zhang et al. (2017) used cell free system to produce hydrogen, in which 13 enzymes for hydrolysing cellulose and hydrogen production were used. Starch phosphorylase, phosphoglucomutase, 10 enzymes of the pentose phosphate pathway and enzymes for hydrogen and NADPH were involved in the system. This system produced 5.19 mol H_2/mol glucose equivalent of starch. Similarly, using 1 mol of cellobiose, 11.2 mol H_2 and 5.64 mol CO_2 were obtained using the cell free system (Zhang et al. 2017). Despite the high hydrogen yield, the enzymes' cost in the cell free system is a limiting factor.

7.4 Flux Modelling of Metabolic Pathways

Metabolic flux analysis (MFA) is an essential tool used to decide the significant metabolic or genetic modification for improving the hydrogen yield. Flux balance analysis (FBA) includes three major steps: (a) reconstruction of metabolic pathways and network, (b) creating a stoichiometric model and solving them through linear programming by defining the objective function. Metabolic reconstruction involves the interconnection of reactions, metabolites, and proteins to form a metabolic network. In the metabolic flux study, reactions are extracted from the Kyoto Encyclopedia of Genes and Genomes (KEGG) database and from the literature (Muthuraj et al. 2013). The reconstructed metabolic networks are represented in matrix (S) format. Reactants and products are presented with stoichiometric value in the appropriate place, the components in the reactions are presented in the row(m) and the column is for reaction(n). The negative value indicates substrate consumption and the positive value indicates the product formation. The vector (v) in Eq. (7.6) indicates the flux through the reactions, whereas x indicates the substrate concentration with respect to time. At steady state the accumulation is zero, so the systems mass balance is $dx/dt = 0$. Tools such as 'efm' or 'cobra' are used to develop the matrix and optimize the metabolic model. Linear programming in Matlab™ is used to solve the Eqs. (7.6), (7.7) and (7.8) by setting desired upper and lower bounds.

$$\frac{dx}{dt} = Sv \tag{7.6}$$

$$Sv \tag{7.7}$$

$$\begin{bmatrix} 1 & 0 & 0 \\ 0 & -1 & 0 \\ 0 & 1 & 0 \end{bmatrix} \begin{bmatrix} 0 \\ 0 \\ 1 \end{bmatrix} = \begin{bmatrix} 0 \\ 0 \\ 0 \end{bmatrix} \tag{7.8}$$

7.4.1 Photofermentation

A model was developed using 314 biochemical reactions for PNSB by Golomysova et al. (2010) to predict the hydrogen production rate for different carbon sources. The model prediction was 96.36% with lactate as the substrate, and the model accuracy reduced to 78.16% and 50.4%, when malate and acetate were considered as the substrate (Golomysova et al. 2010). Mutation was carried out on PNSB using the in silico model strain, which involved the genes for ribulose-1,5-bisphosphate carboxylase/oxygenase (RuBisCO), acetyl-CoA carboxylase, threonine:ammonia lyase, PFL, and pyruvate dehydrogenase that are known to positively influence the hydrogen production (Golomysova et al. 2010). Hädicke et al. (2011) made a framework using 143 reactions in the mutant without RuBisCO enzyme which showed a high yield of hydrogen on different substrates including succinate, malate, and fructose. MFA was further employed to analyse the photoproduction of hydrogen by *Synechocystis* sp. PCC6803 (Hädicke and Klamt 2011).

Metabolic flux balance analysis has been performed using 90 reactions and 56 compounds, and a maximum higher limit for hydrogen production was evaluated, which provided scope for improvement. MFA is used to determine feasible route for hydrogen production without any oxygen inhibition. Flux model was developed for *Rhodobacter sphaeroides* which had 183 reactions and 153 metabolites. Mass balance was performed by considering the reaction as pseudo steady state and the equations were solved using optimization. The predicted results showed a low specific rate of hydrogen production in the absence of light. The model can be developed further by including the light in the reaction (Chongcharoenthaweesuk et al. 2011).

In order to know the complex metabolic pathway of PNSB such as *Rhodospirillum (Rs.) rubrum*, *Rhodobacter (Rba.) sphaeroides* and *Rhodopseudomonas (Rps.) palustris* metabolic modelling was carried out (Hädicke et al. 2011). The model revealed metabolic constraints with respect to redox homeostasis in the PNSB. This model also showed a quantitative explanation for species-specific phenomena for photoheterotrophic growth of PNSB. Reproduction of quantitative wet lab data based on the model hypothesis reveals that a large number of catabolic pathways involved in PNSB are substrate specific. The model also explains quantitatively the need of CO_2 fixing pathways such as Calvin cycle for PNSB. This model further explains the pathways involved in redox metabolism and its effect on the Calvin cycle. The model also distinguishes acetate assimilation pathways in *R. sphaeroides and R. rubrum* without the involvement of isocitrate lyase. Thus, the theoretical basis of genetic modification in PNSB to improve hydrogen production was well demonstrated in this MFA and modelling study (Hädicke et al. 2011).

7.4.2 Dark Fermentation

A model with 34 reactions was developed for *C. butyricum* W5, which accurately predicted the experimental yield and production rate in two different media. MFA showed that the carbon was divided at three points (PEP, pyruvate and AcCoA), and it is highly influenced by the operational parameters such as pH and initial glucose concentration. Carbon flux towards oxaloacetate (OAA) increased, whereas flux towards pyruvate reduced at pH 7–6.5 at the PEP node; similar results were reported when the glucose concentration was increased from 4 to 8 g/L. Flux towards lactate reduced with a decrease in pH (7–6) or glucose concentration (10–4 g/L) in the pyruvate node. The division of carbon in the AcCoA node between acetate and butyrate was not affected by any change in pH or glucose concentration. The carbon fraction from AcCoA to ethanol at pH 7 was 3 times higher than at other pH values (Cai et al. 2010). Metabolic model was developed for *E. coli* using 27 reactions, and the model was accurate in predicting the hydrogen yield due to the wild strain (0.17 mol/mol of glucose) and mutant strain lacking lactate dehydrogenase (0.23 mol/mol of glucose) (Manish et al. 2007). Hydrogen production by *Citrobacter amalonaticus* Y19 using glucose as the substrate was studied. Batch studies were carried out at varying glucose concentration to obtain the mass balance and 81 biochemical reactions were used for the analysis. MetafluxNet™ was used for the analysis by setting specific growth rate as the objective, and the results obtained were closer to the experimental values. Maximum hydrogen was chosen as the objective, and it resulted in 8.7 mol-H_2/mol glucose and the pathway involved non-active NAD(P) linked hydrogenase and hydrogen production from NAD(P)H (Oh et al. 2008). Another model was developed for the hyperthermophilic archaeon *Thermotoga maritima* with 562 intercellular and 83 extracellular biochemical reactions. The model showed that in the absence of fructose bisphosphate aldolase and 2-dehydro-3-deoxy-phosphogluconate aldolase, and in presence of non-native NADPH:ferredoxin reductase, increase in hydrogen yield (7.6 mol of H_2 per glucose) could be obtained. Glucose flux was towards pentose phosphate pathway and hydrogen production was by hydrogenase enzyme using both Fd and NADH (Nogales et al. 2012).

Genome scale model has been used for evaluating hydrogen production in *E. coli* in which flux balance analysis, algorithm for blocking competitive pathway and manual selection were used. Eighty-one *E. coli* mutant possessing single gene deletion were evaluated for hydrogen production and biomass production in batch experiments and by FBA. The accuracy of MBA for gene essentiality prediction was 78 and 77% in glucose and galactose containing media, and 33% mutation predicted by the FBA revealed its positive effect on hydrogen production by the mutant strains. Flux balance analysis aids in understanding mainly the complex metabolic pathways and selecting specific genes for deletion to enhance the hydrogen production (Seppälä et al. 2013).

Fluxes that are difficult to solve through flux balance analysis are sorted through tracer technique. Radio isotope ^{13}C labelled substrates are introduced into the metabolic pathways to attain the steady state. Nuclear magnetic resonance spectroscopy (NMR), liquid chromatography–mass spectrometry (LC–MS), gas chromatography–mass spectrometry (GC–MS) and matrix assisted laser desorption ionization time-of-flight mass spectrometry (MALDI-TOF-MS) are employed to measure the tracer concentrations. Using ^{13}C radio labelled proteinogenic amino acids, the intracellular flux distribution for wild type and pyruvate kinase deficient *E. coli* were analysed (Emmerling et al. 2002). Flux distribution of *E. coli* K12 grown on different carbon source was estimated using radio labelled amino acids. Through radio labelling of amino acids, flux distribution in different pathways is analysed in *Actinobacillus succinogenes*. It was used to determine the change in flux due to perturbation in CO_2 and H_2 perturbations.

7.5 Conclusions

The synthetic biology and metabolic engineering discussed in this chapter deals with various methods for increasing the hydrogen production, substrate consumption and overcoming the oxygen inhibition of hydrogenase. Very few studies have been carried out using the microalgae due to its complex genetic and metabolic makeup. However, numerous studies have been reported on the *E. coli* for hydrogen production as its genetic constitutes are deeply explored. Since carbon dioxide, light and water are easily available substrates, biophotolysis as compared to dark fermentation is interesting for hydrogen production. Further research on bioreactor consideration, kinetics, low-cost feedstock, downstream processing and hydrogen storage are essential along with synthetic biology and metabolic engineering approaches for practical application of biohydrogen.

References

Akhtar MK, Jones PR (2009) Construction of a synthetic YdbK-dependent pyruvate:H2 pathway in Escherichia coli BL21(DE3). Metab Eng 11:139–147. https://doi.org/10.1016/j.ymben.2009.01.002

Antoniewicz MR, Kelleher JK, Stephanopoulos G (2007) Elementary metabolite units (EMU): A novel framework for modeling isotopic distributions. Metab Eng 9:68–86. https://doi.org/10.1016/j.ymben.2006.09.001

Benemann JR (2000) Hydrogen production by microalgae. J Appl Phycol 12:291–300. https://doi.org/10.1023/a:1008175112704

Brentner LB, Jordan PA, Zimmerman JB (2010) Challenges in developing biohydrogen as a sustainable energy source: Implications for a research agenda. Environ Sci Technol 44:2243–2254. https://doi.org/10.1021/es9030613

Burgdorf T, Lenz O, Buhrke T, Van Der Linden E, Jones AK, Albracht SPJ, Friedrich B (2006) [NiFe]-hydrogenases of Ralstonia eutropha H16: Modular enzymes for oxygen-tolerant biological hydrogen oxidation. J Mol Microbiol Biotechnol 10:181–196. https://doi.org/10.1159/000091564

Burgess SSO, Adams MA, Turner NC, Ward B (2000) Characterisation of hydrogen isotope profiles in an agroforestry system: Implications for tracing water sources of trees. Agric Water Manag 45:229–241. https://doi.org/10.1016/S0378-3774(00)00105-0

Cai J, Wang G (2014) Photo-biological hydrogen production by an acid tolerant mutant of Rhodovulum sulfidophilum P5 generated by transposon mutagenesis. Bioresour Technol 154:254–259. https://doi.org/10.1016/j.biortech.2013.12.057

Cai W, Wang F, van Veen A, Descorme C, Schuurman Y, Shen W, Mirodatos C (2010) Hydrogen production from ethanol steam reforming in a micro-channel reactor. Int J Hydrog Energy 35:1152–1159. https://doi.org/10.1016/j.ijhydene.2009.11.104

Chittibabu G, Nath K, Das D (2006) Feasibility studies on the fermentative hydrogen production by recombinant Escherichia coli BL-21. Process Biochem 41:682–688. https://doi.org/10.1016/j.procbio.2005.08.020

Chongcharoenthaweesuk P, Jiangdong Z, Mavituna F (2011) Metabolic flux balance analysis for biological hydrogen production by purple non-sulfur bacteria. Int Conf Food Eng Biotechnol 9:125–129. http://www.ipcbee.com/vol9/24-B059.pdf. Accessed 26 July 2022

De Vos P, Stevens P, De Ley J (1983) Hydrogen gas production from formate and glucose by different members of the Enterobacteriaceae. Biotechnol Lett 5:69–74. https://doi.org/10.1007/BF00132161

Drepper T, Groß S, Yakunin AF, Hallenbeck PC, Masepohl B, Klipp W (2003) Role of GlnB and GlnK in ammonium control of both nitrogenase systems in the phototrophic bacterium Rhodobacter capsulatus. Microbiology 149:2203–2212. https://doi.org/10.1099/mic.0.26235-0

Elowitz MB, Leibier S (2000) A synthetic oscillatory network of transcriptional regulators. Nature 403:335–338. https://doi.org/10.1038/35002125

Eltsova ZA, Vasilieva LG, Tsygankov AA (2010) Hydrogen production by recombinant strains of Rhodobacter sphaeroides using a modified photosynthetic apparatus. Appl Biochem Microbiol 46:487–491. https://doi.org/10.1134/S0003683810050042

Emmerling M, Dauner M, Ponti A, Fiaux J, Hochuli M, Szyperski T, Wüthrich K, Bailey JE, Sauer U (2002) Metabolic flux responses to pyruvate kinase knockout in Escherichia coli. J Bacteriol 184:152–164. https://doi.org/10.1128/JB.184.1.152-164.2002

Esper B, Badura A, Rögner M (2006) Photosynthesis as a power supply for (bio-)hydrogen production. Trends Plant Sci 11:543–549. https://doi.org/10.1016/j.tplants.2006.09.001

Eydmann T, Soderback E, Jones T, Hill S, Austin S, Dixon R (1995) Transcriptional activation of the nitrogenase promoter in vitro: Adenosine nucleotides are required for inhibition of NIFA activity by NIFL. J Bacteriol 177:1186–1195. https://doi.org/10.1128/jb.177.5.1186-1195.1995

Fan Z, Yuan L, Chatterjee R (2009) Increased hydrogen production by genetic engineering of Escherichia coli. PLoS One 4. https://doi.org/10.1371/journal.pone.0004432

Farmer RM, Laguna R, Panescu J, McCoy A, Logsdon B, Zianni M, Moskvin OV, Gomelsky M, Robert Tabita F (2014) Altered residues in key proteins influence the expression and activity of the nitrogenase complex in an adaptive CO2 fixation-deficient mutant strain of Rhodobacter sphaeroides. Microbiol (United Kingdom) 160:198–208. https://doi.org/10.1099/mic.0.073031-0

Follstad BD, Balcarcel RR, Stephanopoulos G, Wang DIC (1999) Metabolic flux analysis of hybridoma continuous culture steady state multiplicity. Biotechnol Bioeng 63:675–683. https://doi.org/10.1002/(SICI)1097-0290(19990620)63:6<675::AID-BIT5>3.0.CO;2-R

Geng A, He Y, Qian C, Yan X, Zhou Z (2010) Effect of key factors on hydrogen production from cellulose in a co-culture of Clostridium thermocellum and Clostridium thermopalmarium. Bioresour Technol 101:4029–4033. https://doi.org/10.1016/j.biortech.2010.01.042

Gohlke U, Pullan L, McDevitt CA, Porcelli I, De Leeuw E, Palmer T, Saibil HR, Berks BC (2005) The TatA component of the twin-arginine protein transport system forms channel complexes of variable diameter. Proc Natl Acad Sci U S A 102:10482–10486. https://doi.org/10.1073/pnas.0503558102

Gökçe A, Öztürk Y, Çakar ZP (2017) Comparative genomic analysis of two heat-resistant Rhodobacter capsulatus mutants with different hydrogen production levels reveals mutations related to hydrogen production. Int J Hydrog Energy 42:20529–20539. https://doi.org/10.1016/j.ijhydene.2017.07.005

Golomysova A, Gomelsky M, Ivanov PS (2010) Flux balance analysis of photoheterotrophic growth of purple nonsulfur bacteria relevant to biohydrogen production. Int J Hydrog Energy 35:12751–12760. https://doi.org/10.1016/j.ijhydene.2010.08.133

Goyal Y, Kumar M, Gayen K (2013) Metabolic engineering for enhanced hydrogen production: A review. Can J Microbiol 59:59–78. https://doi.org/10.1139/cjm-2012-0494

Hädicke O, Klamt S (2011) Computing complex metabolic intervention strategies using constrained minimal cut sets. Metab Eng 13:204–213. https://doi.org/10.1016/j.ymben.2010.12.004

Hädicke O, Grammel H, Klamt S (2011) Metabolic network modeling of redox balancing and biohydrogen production in purple nonsulfur bacteria. BMC Syst Biol 5. https://doi.org/10.1186/1752-0509-5-150

Hallenbeck PC (2004) Fundamentals and limiting processes of biological hydrogen production. Biohydrogen III Renew Energy Syst Biol Sol Energy Convers. https://doi.org/10.1016/B978-008044356-0/50008-7

Hallenbeck PC (2005) Fundamentals of the fermentative production of hydrogen. Water Sci Technol 52:21–29. https://doi.org/10.2166/wst.2005.0494

Hallenbeck PC, Ghosh D (2009) Advances in fermentative biohydrogen production: the way forward? Trends Biotechnol 27:287–297. https://doi.org/10.1016/j.tibtech.2009.02.004

Hallenbeck PC, Ghosh D (2012) Improvements in fermentative biological hydrogen production through metabolic engineering. J Environ Manag 95. https://doi.org/10.1016/j.jenvman.2010.07.021

Hoffmann P (2006) Tomorrow's energy: hydrogen, fuel cells, and the prospects for a cleaner planet. Cogener Distrib Gener J 21:76–77. https://doi.org/10.1080/15453660609509095

Huang JJ, Heiniger EK, McKinlay JB, Harwood CS (2010) Production of hydrogen gas from light and the inorganic electron donor thiosulfate by Rhodopseudomonas palustris. Appl Environ Microbiol 76:7717–7722. https://doi.org/10.1128/AEM.01143-10

Jo JH, Jeon CO, Lee SY, Lee DS, Park JM (2010) Molecular characterization and homologous overexpression of [FeFe]-hydrogenase in Clostridium tyrobutyricum JM1. Int J Hydrog Energy 35:1065–1073. https://doi.org/10.1016/J.IJHYDENE.2009.11.102. https://www.sciencedirect.com/science/article/pii/S0360319909018722. Accessed 26 July 2022

Jokel M, Nagy V, Tóth SZ, Kosourov S, Allahverdiyeva Y (2019) Elimination of the flavodiiron electron sink facilitates long-term H2 photoproduction in green algae. Biotechnol Biofuels 12:280. https://doi.org/10.1186/s13068-019-1618-1

Kars G, Gündüz U, Yücel M, Türker L, Eroglu I (2006) Hydrogen production and transcriptional analysis of nifD, nifK and hupS genes in Rhodobacter sphaeroides O.U.001 grown in media with different concentrations of molybdenum and iron. Int J Hydrog Energy 31:1536–1544. https://doi.org/10.1016/j.ijhydene.2006.06.021

Khosravitabar F, Hippler M (2019) A new approach for improving microalgal biohydrogen photoproduction based on safe & fast oxygen consumption. Int J Hydrog Energy 44:17835–17844. https://doi.org/10.1016/j.ijhydene.2019.05.038

Kim S, Seol E, Oh YK, Wang GY, Park S (2009) Hydrogen production and metabolic flux analysis of metabolically engineered Escherichia coli strains. Int J Hydrog Energy 34:7417–7427. https://doi.org/10.1016/j.ijhydene.2009.05.053

Kivistö A, Santala V, Karp M (2011) Closing the 1,3-propanediol route enhances hydrogen production from glycerol by Halanaerobium saccharolyticum subsp. saccharolyticum. Int J Hydrog Energy 36:7074–7080. https://doi.org/10.1016/j.ijhydene.2011.03.012

Klein M, Ansorge-Schumacher MB, Fritsch M, Hartmeier W (2010) Influence of hydrogenase overexpression on hydrogen production of Clostridium acetobutylicum DSM 792. Enzym Microb Technol 46:384–390. https://doi.org/10.1016/j.enzmictec.2009.12.015

Koku H, Erolu I, Gündüz U, Yücel M, Türker L (2002) Aspects of the metabolism of hydrogen production by Rhodobacter sphaeroides. Int J Hydrog Energy 27:1315–1329. https://doi.org/10.1016/S0360-3199(02)00127-1

Kondo T, Arakawa M, Hirai T, Wakayama T, Hara M, Miyake J (2002) Enhancement of hydrogen production by a photosynthetic bacterium mutant with reduced pigment. J Biosci Bioeng 93:145–150. https://doi.org/10.1016/S1389-1723(02)80006-8

Kruse O, Rupprecht J, Bader KP, Thomas-Hall S, Schenk PM, Finazzi G, Hankamer B (2005) Improved photobiological H2 production in engineered green algal cells. J Biol Chem 280:34170–34177. https://doi.org/10.1074/jbc.M503840200

Kumar N, Das D (2001) Continuous hydrogen production by immobilized Enterobacter cloacae IIT-BT 08 using lignocellulosic materials as solid matrices. Enzym Microb Technol 29:280–287. https://doi.org/10.1016/S0141-0229(01)00394-5

Kumar N, Ghosh A, Das D (2001) Redirection of biochemical pathways for the enhancement of H2 production by Enterobacter cloacae. Biotechnol Lett 23:537–541. https://doi.org/10.1023/A:1010334803961

Lecker B, Illi L, Lemmer A, Oechsner H (2017) Biological hydrogen methanation—A review. Bioresour Technol 245:1220–1228. https://doi.org/10.1016/j.biortech.2017.08.176

Levin DB, Pitt L, Love M (2004) Biohydrogen production: Prospects and limitations to practical application. Int J Hydrog Energy 29:173–185. https://doi.org/10.1016/S0360-3199(03)00094-6

Lewis AC (2021) Optimising air quality co-benefits in a hydrogen economy: a case for hydrogen-specific standards for NO_x emissions. Environ Sci Atmos 1:201–207. https://pubs.rsc.org/en/content/articlehtml/2021/ea/d1ea00037c. Accessed 8 May 2023

Li Y-S, Liang F-Y, Bux H, Feldhoff A, Yang W-S, Caro J (2010) Molecular sieve membrane: supported metal-organic framework with high hydrogen selectivity. Angew Chemie 122:558–561. https://doi.org/10.1002/ange.200905645

Lee, S. Y., Lee, H. J., Park, J. M., Lee, J. H., Park, J. S., Shin, H. S., ... & Min, J. (2010). Bacterial hydrogen production in recombinant *Escherichia coli* harboring a HupSL hydrogenase isolated from *Rhodobacter sphaeroides* under anaerobic dark culture. International Journal of Hydrogen Energy, 35(3), 1112–1116.

Liebgott PP, De Lacey AL, Burlat B, Cournac L, Richaud P, Brugna M, Fernandez VM, Guigliarelli B, Rousset M, Léger C, Dementin S (2011) Original design of an oxygen-tolerant [NiFe] hydrogenase: Major effect of a valine-to-cysteine mutation near the active site. J Am Chem Soc 133:986–997. https://doi.org/10.1021/ja108787s

Lin C-Y, Chang R-C (2004) Fermentative hydrogen production at ambient temperature. Int J Hydrog Energy 29:715–720. https://www.sciencedirect.com/science/article/pii/S0360319903002386. Accessed 9 May 2023

Lindblad P, Christensson K, Lindberg P, Fedorov A, Pinto F, Tsygankov A (2002) Photoproduction of H2 by wildtype Anabaena PCC 7120 and a hydrogen uptake deficient mutant: From laboratory experiments to outdoor culture. Int J Hydrog Energy 27:1271–1281. https://doi.org/10.1016/S0360-3199(02)00111-8

Liu X, Zhu Y, Yang ST (2006a) Butyric acid and hydrogen production by Clostridium tyrobutyricum ATCC 25755 and mutants. Enzym Microb Technol 38:521–528. https://doi.org/10.1016/j.enzmictec.2005.07.008

Liu D, Liu D, Zeng RJ, Angelidaki I (2006b) Hydrogen and methane production from household solid waste in the two-stage fermentation process. Water Res 40:2230–2236. https://doi.org/10.1016/j.watres.2006.03.029

Lu Y, Zhao H, Zhang C, Lai Q, Wu X, Xing XH (2010) Alteration of hydrogen metabolism of ldh-deleted Enterobacter aerogenes by overexpression of NAD(+)-dependent formate dehydrogenase. Appl Microbiol Biotechnol 86:255–262. https://doi.org/10.1007/s00253-009-2274-3

Lu Y, Wang L, Ma K, Li G, Zhang C, Zhao H, Lai Q, Li HP, Xing XH (2011) Characteristics of hydrogen production of an Enterobacter aerogenes mutant generated by a new atmospheric and room temperature plasma (ARTP). Biochem Eng J 55:17–22. https://doi.org/10.1016/j.bej.2011.02.020

Ma H, Zheng X, Yang H (2018) Enhancement on hydrogen production performance of Rhodobacter sphaeroides HY01 by overexpressing fdxN. Int J Hydrog Energy 43:17082–17090. https://doi.org/10.1016/j.ijhydene.2018.07.101

Maeda T, Sanchez-Torres V, Wood TK (2007) Enhanced hydrogen production from glucose by metabolically engineered Escherichia coli. Appl Microbiol Biotechnol 77:879–890. https://doi.org/10.1007/s00253-007-1217-0

Maeda T, Sanchez-Torres V, Wood TK (2008) Metabolic engineering to enhance bacterial hydrogen production. Microb Biotechnol 1:30–39. https://doi.org/10.1111/j.1751-7915.2007.00003.x

Mahidhara G, Burrow H, Sasikala C, Ramana CV (2019) Biological hydrogen production: molecular and electrolytic perspectives. World J Microbiol Biotechnol 35. https://doi.org/10.1007/s11274-019-2692-z

Manish S, Venkatesh KV, Banerjee R (2007) Metabolic flux analysis of biological hydrogen production by Escherichia coli. Int J Hydrog Energy 32:3820–3830. https://doi.org/10.1016/j.ijhydene.2007.03.033

Masukawa H, Mochimaru M, Sakurai H (2002) Disruption of the uptake hydrogenase gene, but not of the bidirectional hydrogenase gene, leads to enhanced photobiological hydrogen production by the nitrogen-fixing cyanobacterium Anabaena sp. PCC 7120. Appl Microbiol Biotechnol 58:618–624. https://doi.org/10.1007/s00253-002-0934-7

McKinlay JB, Harwood CS (2011) Calvin cycle flux, pathway constraints, and substrate oxidation state together determine the H2 biofuel yield in photoheterotrophic bacteria. MBio 2. https://doi.org/10.1128/mBio.00323-10

Misra N, Panda PK, Parida BK, Mishra BK (2016) Way forward to achieve sustainable and cost-effective biofuel production from microalgae: a review. Int J Environ Sci Technol 13:2735–2756. https://doi.org/10.1007/s13762-016-1020-5

Mitra M, Melis A (2010) Genetic and biochemical analysis of the TLA1 gene in Chlamydomonas reinhardtii. Planta 231:729–740. https://doi.org/10.1007/s00425-009-1083-3

Moreno-Garcia L, Adjallé K, Barnabé S, Raghavan GSV (2017) Microalgae biomass production for a biorefinery system: Recent advances and the way towards sustainability. Renew Sust Energ Rev 76:493–506. https://doi.org/10.1016/j.rser.2017.03.024

Morimoto K, Kimura T, Sakka K, Ohmiya K (2005) Overexpression of a hydrogenase gene in Clostridium paraputrificum to enhance hydrogen gas production. FEMS Microbiol Lett 246:229–234. https://doi.org/10.1016/j.femsle.2005.04.014

Mussgnug JH, Thomas-Hall S, Rupprecht J, Foo A, Klassen V, McDowall A, Schenk PM, Kruse O, Hankamer B (2007) Engineering photosynthetic light capture: Impacts on improved solar energy to biomass conversion. Plant Biotechnol J 5:802–814. https://doi.org/10.1111/j.1467-7652.2007.00285.x

Muthuraj M, Palabhanvi B, Misra S, Kumar V, Sivalingavasu K, Das D (2013) Flux balance analysis of Chlorella sp. FC2 IITG under photoautotrophic and heterotrophic growth conditions. Photosynth Res 118:167–179. https://doi.org/10.1007/s11120-013-9943-x

Nakayama SI, Kosaka T, Hirakawa H, Matsuura K, Yoshino S, Furukawa K (2008) Metabolic engineering for solvent productivity by downregulation of the hydrogenase gene cluster hupCBA in Clostridium saccharoperbutylacetonicum strain N1-4. Appl Microbiol Biotechnol 78:483–493. https://doi.org/10.1007/s00253-007-1323-z

Nogales J, Gudmundsson S, Knight EM, Palsson BO, Thiele I (2012) Detailing the optimality of photosynthesis in cyanobacteria through systems biology analysis. Proc Natl Acad Sci U S A 109:2678–2683. https://doi.org/10.1073/pnas.1117907109

Oda Y, Larimer FW, Chain PSG, Malfatti S, Shin MV, Vergez LM, Hauser L, Land ML, Braatsch S, Beatty JT, Pelletier DA, Schaefer AL, Harwood CS (2008) Multiple genome sequences reveal adaptations of a phototrophic bacterium to sediment microenvironments. Proc Natl Acad Sci U S A 105:18543–18548. https://doi.org/10.1073/pnas.0809160105

Oh YK, Kim HJ, Park S, Kim MS, Ryu DDY (2008) Metabolic-flux analysis of hydrogen production pathway in Citrobacter amalonaticus Y19. Int J Hydrog Energy 33:1471–1482. https://doi.org/10.1016/j.ijhydene.2007.09.032

Oh YK, Raj SM, Jung GY, Park S (2011) Current status of the metabolic engineering of microorganisms for biohydrogen production. Bioresour Technol 102:8357–8367. https://doi.org/10.1016/j.biortech.2011.04.054

Oncel SS, Kose A, Faraloni C, Imamoglu E, Elibol M, Torzillo G, Sukan FV (2014) Biohydrogen production using mutant strains of Chlamydomonas reinhardtii: The effects of light intensity and illumination patterns. Biochem Eng J 92:47–52. https://doi.org/10.1016/j.bej.2014.06.019

Öztürk Y, Yücel M, Daldal F, Mandaci S, Gündüz U, Türker L, Eroğlu I (2006) Hydrogen production by using Rhodobacter capsulatus mutants with genetically modified electron transfer chains. Int J Hydrog Energy 31:1545–1552. https://doi.org/10.1016/j.ijhydene.2006.06.042

Öztürk Y, Gökçe A, Peksel B, Gürgan M, Özgür E, Gündüz U, Eroğlu N, Yücel M (2012) Hydrogen production properties of Rhodobacter capsulatus with genetically modified redox balancing pathways. Int J Hydrog Energy 37:2014–2020. https://doi.org/10.1016/j.ijhydene.2011.06.078

Penfold DW, Macaskie LE (2004) Production of H2 from sucrose by Escherichia coli strains carrying the pUR400 plasmid, which encodes invertase activity. Biotechnol Lett 26:1879–1883. https://doi.org/10.1007/s10529-004-6035-1

Penfold RJ, Pemberton JM (1992) An improved suicide vector for construction of chromosomal insertion mutations in bacteria. Gene 118:145–146. https://doi.org/10.1016/0378-1119(92)90263-O

Penfold DW, Forster CF, Macaskie LE (2003) Increased hydrogen production by Escherichia coli strain HD701 in comparison with the wild-type parent strain MC4100. Enzym Microb Technol 33:185–189. https://doi.org/10.1016/S0141-0229(03)00115-7

Philipps G, Happe T, Hemschemeier A (2012) Nitrogen deprivation results in photosynthetic hydrogen production in Chlamydomonas reinhardtii. Planta 235:729–745. https://doi.org/10.1007/s00425-011-1537-2

Polle JEW, Kanakagiri S, Jin ES, Masuda T, Melis A (2002) Truncated chlorophyll antenna size of the photosystems—a practical method to improve microalgal productivity and hydrogen production in mass culture. Int J Hydrog Energy 27:1257–1264. https://doi.org/10.1016/S0360-3199(02)00116-7

Pompelli MF, França SC, Tigre RC, de Oliveira MT, Sacilot M, Pereira EC (2012) Spectrophotometric determinations of chloroplastidic pigments in acetone, ethanol and dimethylsulphoxide. Braz J Biosci 11:52–58

Redwood MD, Mikheenko IP, Sargent F, Macaskie LE (2008) Dissecting the roles of Escherichia coli hydrogenases in biohydrogen production. FEMS Microbiol Lett 278:48–55. https://doi.org/10.1111/j.1574-6968.2007.00966.x

Sasikala K, Ramana CV, Rao PR, Kovacs KL (1993) Anoxygenic phototrophic bacteria: physiology and advances in hydrogen production technology. Adv Appl Microbiol 38:211–295. https://doi.org/10.1016/S0065-2164(08)70217-X

Schindel HS, Bauer CE (2016) The RegA regulon exhibits variability in response to altered growth conditions and differs markedly between Rhodobacter species. Microb Genomics 2:e000081. https://doi.org/10.1099/mgen.0.000081

Schmehl M, Jahn A, Meyer zu Vilsendorf A, Hennecke S, Masepohl B, Schuppler M, Marxer M, Oelze J, Klipp W (1993) Identification of a new class of nitrogen fixation genes in Rhodobacter capsulatus: a putative membrane complex involved in electron transport to nitrogenase. Mol Gen Genet 241:602–615. https://doi.org/10.1007/BF00279903

Seol E, Jang Y, Kim S, Oh YK, Park S (2012) Engineering of formate-hydrogen lyase gene cluster for improved hydrogen production in Escherichia coli. Int J Hydrog Energy 37:15045–15051. https://doi.org/10.1016/j.ijhydene.2012.07.095

Seppälä JJ, Larjo A, Aho T, Yli-Harja O, Karp MT, Santala V (2013) Prospecting hydrogen production of Escherichia coli by metabolic network modeling. Int J Hydrog Energy 38:11780–11789. https://doi.org/10.1016/j.ijhydene.2013.07.002

Sharma S, Agarwal S, Jain A (2021) Significance of hydrogen as economic and environmentally friendly fuel. Energies 14(21):7389. https://www.mdpi.com/1346298. Accessed 9 May 2023

Show KY, Yan Y, Zong C, Guo N, Chang JS, Lee DJ (2019) State of the art and challenges of biohydrogen from microalgae. Bioresour Technol 289. https://doi.org/10.1016/j.biortech.2019.121747

Smith RG, Vanlerberghe GC, Stitt M, Turpin DH (1989) Short-term metabolite changes during transient ammonium assimilation by the N-limited green alga Selenastrum minutum. Plant Physiol 91:749–755. https://doi.org/10.1104/pp.91.2.749

Stephanopoulos G (2012) Synthetic biology and metabolic engineering. ACS Synth Biol 1:514–525. https://doi.org/10.1021/sb300094q

Surzycki R, Cournac L, Peltier G, Rochaix JD (2007) Potential for hydrogen production with inducible chloroplast gene expression in Chlamydomonas. Proc Natl Acad Sci U S A 104:17548–17553. https://doi.org/10.1073/pnas.0704205104

Uyar B, Gürgan M, Özgür E, Gündüz U, Yücel M, Eroglu I (2015) Hydrogen production by hup- mutant and wild-type strains of Rhodobacter capsulatus from dark fermentation effluent of sugar beet thick juice in batch and continuous photobioreactors. Bioprocess Biosyst Eng 38:1935–1942. https://doi.org/10.1007/s00449-015-1435-2

Varma A, Palsson BO (1994) Stoichiometric flux balance models quantitatively predict growth and metabolic by-product secretion in wild-type Escherichia coli W3110. Appl Environ Microbiol 60:3724–3731. https://doi.org/10.1128/aem.60.10.3724-3731.1994

Veit A, Akhtar MK, Mizutani T, Jones PR (2008) Constructing and testing the thermodynamic limits of synthetic NAD(P)H:H2 pathways. Microb Biotechnol 1:382–394. https://doi.org/10.1111/j.1751-7915.2008.00033.x

Wang YZ, Liao Q, Zhu X, Tian X, Zhang C (2010) Characteristics of hydrogen production and substrate consumption of Rhodopseudomonas palustris CQK 01 in an immobilized-cell photobioreactor. Bioresour Technol 101:4034–4041. https://doi.org/10.1016/j.biortech.2010.01.045

Wang S, Wang J, Xu L, Pi J, Zhang H, Yan Y (2013) Enhanced biohydrogen production by homologous over-expression of fnr, pncB, fdhF in Klebsiella sp. HQ-3, Shengwu Gongcheng Xuebao/Chinese. J Biotechnol 29:1278–1289. https://europepmc.org/article/med/24409691. Accessed 25 July 2022

Wang Y, Jiang X, Hu C, Sun T, Zeng Z, Cai X, Li H, Hu Z (2017) Optogenetic regulation of artificial microRNA improves H2 production in green alga Chlamydomonas reinhardtii. Biotechnol Biofuels 10. https://doi.org/10.1186/S13068-017-0941-7

Wu Chunfei C, Wang Leizhi L, Williams PT, Shi Jeffrey J, Huang J (2011) Hydrogen production from biomass gasification with Ni/MCM-41 catalysts: Influence of Ni content. Appl Catal B Environ 108–109:6–13. https://doi.org/10.1016/j.apcatb.2011.07.023

Xia Y, Fang HHP, Zhang T (2013) Recent studies on thermophilic anaerobic bioconversion of lignocellulosic biomass. RSC Adv 3:15528–15542. https://doi.org/10.1039/c3ra40866c

Xin L, Hong-ying H, Ke G, Ying-xue S (2010) Effects of different nitrogen and phosphorus concentrations on the growth, nutrient uptake, and lipid accumulation of a freshwater microalga Scenedesmus sp. Bioresour Technol 101:5494–5500. https://doi.org/10.1016/j.biortech.2010.02.016

Yokoi H, Aratake T, Hirose J, Hayashi S, Takasaki Y (2001) Simultaneous production of hydrogen and bioflocculant by Enterobacter sp. BY-29. World J Microbiol Biotechnol 17:609–613. https://doi.org/10.1023/A:1012463508364

Yoshida A, Nishimura T, Kawaguchi H, Inui M, Yukawa H (2006) Enhanced hydrogen production from glucose using ldh- and frd-inactivated Escherichia coli strains. Appl Microbiol Biotechnol 73:67–72. https://doi.org/10.1007/s00253-006-0456-9

Zhan J, Rong J, Wang Q (2017) Mixotrophic cultivation, a preferable microalgae cultivation mode for biomass/bioenergy production, and bioremediation, advances and prospect. Int J Hydrog Energy 42:8505–8517. https://doi.org/10.1016/j.ijhydene.2016.12.021

Zhang L, Happe T, Melis A (2002) Biochemical and morphological characterization of sulfur-deprived and H2-producing Chlamydomonas reinhardtii (green alga). Planta 214:552–561. https://doi.org/10.1007/s004250100660

Zhang Y, Hu J, Ma H, Yang H, Guo L (2017) Overexpressing atpXF enhanced photo-fermentative hydrogen production performance of Rhodobacter sphaeroides. Int J Hydrog Energy 42:9641–9649. https://doi.org/10.1016/j.ijhydene.2017.01.076

Zhao H, Ma K, Lu Y, Zhang C, Wang L, Xing XH (2009) Cloning and knockout of formate hydrogen lyase and H2-uptake hydrogenase genes in Enterobacter aerogenes for enhanced hydrogen production. Int J Hydrog Energy 34:186–194. https://doi.org/10.1016/j.ijhydene.2008.10.025

Chapter 8
Substrate Composition and Effects on Biohydrogen Production

A. A. Laikova, E. A. Zhuravleva, A. A. Kovalev, D. A. Kovalev, S. V. Shekhurdina, A. A. Ivanenko, and Yu. V. Litti

Abstract Dark fermentative biohydrogen production from various organic wastes has attracted increased interest in recent years due to the possibility of obtaining a carbon-neutral energy carrier along with partial stabilization of organic matter. This chapter compiles essential information on substrates of various origins that are of key importance for dark fermentative biohydrogen productivity. Simple sugars, industrial crops rich in carbohydrates, agricultural, industrial waste and wastewater are considered as the feedstocks. In addition, the role of the biochemical composition and concentration of the substrate are discussed. The diversity of native bacteria and the predominant groups of microorganisms inhabiting considered substrates of dark fermentation is studied. A significant point for increasing the efficiency of the dark fermentation process is physical, chemical, biological, and combined pretreatment of the feedstock in order to make it more accessible for microorganisms.

At the end of the chapter, the main points that should be addressed in further research are outlined.

Keywords Biohydrogen · Dark fermentation · Substrate · Pretreatment · Indigenous microbiota

---

A. A. Laikova · E. A. Zhuravleva · S. V. Shekhurdina · Y. V. Litti (✉)
Federal Research Center "Fundamentals of Biotechnology" of the Russian Academy of Sciences, Moscow, Russia

A. A. Kovalev · D. A. Kovalev
Federal State Budgetary Scientific Institution "Federal Scientific Agroengineering Center VIM", Moscow, Russia

A. A. Ivanenko
Department of Biology, Lomonosov Moscow State University, Moscow, Russia

C. R. Soccol et al. (eds.), *Biohydrogen - Advances and Processes*, Biofuel and Biorefinery Technologies 13, https://doi.org/10.1007/978-3-031-49818-3_8

8.1 Introduction

Currently, the development of inexpensive methods for producing hydrogen as an environmentally friendly fuel is being actively pursued. Molecular hydrogen is not as readily available in nature as fossil fuels or natural gas, but it can be produced from renewable resources such as water or biomass (Argun and Kargi 2011). Hydrogen production by natural gas reforming, water electrolysis, and other thermochemical methods requires high temperatures, high energy costs, and is not economically feasible. Dark fermentation (DF) is a natural microbial process of biodegradation of organic material; therefore, it does not require large economic and energy costs. In absence of light and oxygen acidogenic microorganisms ferment substrate to form organic acids, alcohols, and hydrogen (Akca et al. 2021). In this way, the two most important tasks are solved—partial stabilization of man-made waste and the generation of a clean energy carrier without additional burden on the environment. The disadvantages of this method include low yield, high process time, and the difficulty in maintaining the stability of large-scale reactor system (Levin and Chahine 2010).

To develop ways to optimize the DF technology, special attention should be paid to the selection of suitable substrates and its pretreatment methods for enhancing the utilization. The best substrates for DF are simple soluble sugars (monomers). However, it is obvious that their use is less economically attractive compared to waste organic feedstocks, such as industrial, agricultural, household and other wastes, which are powerful pollutants of the environment. The microbial community adapted to the biodegradation of complex substrates is often dominated by hydrolytic and acidogenic microorganisms of the genera *Clostridium* (strict anaerobes), *Enterobacter*, and *Escherichia* (facultative anaerobes), which are the most effective hydrogen producers (Ren et al. 2021). Biodegradability and carbohydrate content of the substrate are the key points when choosing feedstock for DF (Eker and Sarp 2017). Fundamental importance is the use of second-generation biomass sources, which are renewable and can be subjected to DF. As promising substrates in the production of biohydrogen, agricultural and food industry wastes rich in starch and cellulose (straw, manure, whey, distillery wastewater, molasses), pulp and paper industry wastes, biofuel production wastes and sewage sludge are used. Fast growing crops, industrial crops, and algae are also used in hydrogen production. Carbon dioxide formed during dark fermentation, along with hydrogen, can be taken up by plants and algae during photosynthesis (Liu et al. 2016; Łukajtis et al. 2018).

Many recent studies are devoted to the use of lignocellulose as a substrate for hydrogen production. It is a cheap renewable feedstock derived from plant components, having a complex structure, mainly consisting of three polymers (cellulose, hemicellulose, and lignin), and also containing ash and extractives. Lignocellulosic biomass includes agricultural and agro-industrial waste (sugarcane cake, stalks, cobs, bran, leaves, straw of corn, wheat, oats, and rice), forestry waste, wood, grass (Soares et al. 2020). The ratio of polymeric components in the feedstock used determines its availability for microorganisms. Complex substrates are available for microorganisms possessing a complex hydrolytic enzyme. The sugars resulting from hydrolysis are available to a wider range of microorganisms. The use of hydrolytic microorganisms as a pure culture or as a mixed culture makes it possible to bypass the expensive stages of substrate pretreatment such as hydrolysis and delignification (Elsharnouby et al. 2013). However, in many cases, researchers consider this stage to be important, since the inhibition of microbial hydrolysis by the reaction products is less and, in general, it affects hydrogen production positively (Ren et al. 2008; Soares et al. 2020). Pretreatment is important for the extraction of certain substances from hard-to-reach materials (Hassan et al. 2020). Some studies state that during microbial hydrolysis, hydrolytic bacteria subsequently grow on the resulting soluble products, thereby reducing the final yield of biohydrogen (Nissilä et al. 2014). Biohydrogen production is being actively studied using household and food waste, animal waste, organic substances that have a complex polymer composition which are difficult to ferment. The purpose of pretreatment is to break down long polymer chains into simple carbohydrates or shortened oligomers. Pretreatment methods are divided into physical, chemical, physico-chemical, and biological. They were developed to facilitate enzymatic reactions during the hydrolysis of cellulose and other polymers (Dou et al. 2019). The pretreatment process should be easy to use, have a low cost, stable efficiency when working with large volumes of biomass and lead to a high level of extraction of components of a complex substrate. Limiting substrate concentrations depend on the resistance of the selected strains or microbial community to inhibitors, the type of fermentation, and the overall effect of all toxins present (Akutsu et al. 2009; Gonzales et al. 2016). Incubation parameters such as temperature, pH, HRT affect the course of the dark fermentation process and the depth of substrate decomposition, so this aspect must be considered to identify the optimal conditions for the utilization of certain polymeric compounds (Wang and Wan 2009).

The biochemical composition of the substrate plays a key role in the formation of the microbial community that decomposes the substrate. Proteins, fats, and carbohydrates are found in almost all substrates of natural origin, but in different proportions (Hagos et al. 2017). For example, carbohydrate-rich feedstock includes food waste containing a large amount of easily degradable sugars, and vegetable waste, which has a more complex structure. Protein-rich substrates (manure, domestic wastewater) are energy rich, but the toxic effects of ammonia must be taken into account (Chen et al. 2021). Substrates rich in fats come mainly from the food industry and wastewater from slaughterhouses. Usually lipids are easily decomposed, however, at high concentrations, the process of anaerobic digestion is hindered. According to the literature data, by-products formed as a result of lipid fermentation can cause redox instability in the metabolism of microorganisms (Harirchi et al. 2022). When working with lipids, co-fermentation with carbohydrate-rich materials is optimal (Hagos et al. 2017). The use of various complex substrates for dark fermentation is promising in terms of obtaining environmentally friendly biofuels along with waste biodegradation. Recent studies are aimed at improving the availability of waste for microorganisms and optimizing the process by pre-preparing the substrate, choosing the optimal cultivation modes and strains of microorganisms. Particular attention is drawn to waste based on plant biomass containing lignocellulose and starch. Such wastes are generated in large volumes and are difficult to decompose, including for anaerobic digestion process. This chapter will review the most commonly used composite polymers for dark fermentation, both native and pretreated feedstocks rich in proteins, fats, or carbohydrates. The main groups of microorganisms that predominate in the degradation of different types of substrates and are the most promising in terms of hydrogen production will be emphasized Chemical analysis of substrates and their elemental composition will allow the development of suitable technological approaches for processing on an industrial scale. The classification of substrates according to their biochemical composition is important for assessing the biodegradability and availability of substrates. Comparative analysis of various chemical, physical, and biological methods of pretreatment will help to identify the most safe for microorganisms and at the same time, effective methods that allow decomposing the chemical structure of the substrate. Co-fermentation of a mixture of substrates with different biochemical characteristics is also a promising approach, since the method makes it possible to make difficult-to-decompose (troublesome) raw materials more accessible. To intensify the process of dark fermentation, it is important to understand the mechanism of the process, identify limiting stages, as well as stimulating or inhibiting agents. The chapter will show the influence of cultivation conditions on the efficiency of substrate decomposition and hydrogen production. The purpose is also to assess the biotechnological potential of using various wastes, primarily plant wastes, for microbial hydrogen production based on the available experimental work.

8.2 Substrates for Dark Fermentation

The spectrum of substrates that microorganisms ferment for the production of hydrogen is quite broad. The most accessible for consumption are simple mono- and disaccharides (glucose, xylose, lactose, etc.), such substrates do not require pretreatment, and it would yield higher amount of hydrogen during the fermentation process (Fang and Liu 2002; Chen and Lin 2003; Rosales-Colunga et al. 2012). Mono and disaccharides are often used as model substrates in dark fermentation. However, feedstocks in the form of polymers are more common in nature. Some polymers, such as starch, are easily degradable, while cellulose has a more rigid and stable structure due to hydrogen bonding between the monomers. Cellulose decomposition requires more severe conditions in order to change the chemical configuration of the substance (Łukajtis et al. 2018). Also, protein and lipid-rich waste can be used as a substrate for DF, however the complexities of biodegradation and the limiting steps of microbial metabolism must be taken into account (Table 8.1). In this part of the chapter, complex substrates will be considered, especially those containing plant tissues that can be used to produce hydrogen during DF.

Table 8.1 Substrates for dark fermentative hydrogen production, their advantages and existing challenges

Substrate type	Typical examples	Advantages	Challenges	References
Lignocellulosic waste	Bagasse, corn and wheat stalks, straw, wood chips	Minimizing of greenhouse gas emissions Energy-intensive and widely available substrate	Difficult hydrolysable components of substrate (e.g., lignin) Complex substrate structure Toxicity of several compounds Necessity of harsh pretreatment	Cha et al. (2013); Jiang et al. (2015); Wu et al. (2021)
Algal biomass	Micro- and macroalgae, e.g., *Scenedesmus obliquus*, *Scenedesmus quadricauda*, *Laminaria japonica*, *Pediastrum*	Easy to obtain in large quantities; Energy- and carbohydrate-rich substrate Simplicity for hydrolysis and fermentation	High initial cost investment Competition with H_2 production via biophotolysis or photo fermentation High substrate demand	Ferreira et al. (2013); Xia et al. (2013); Batista et al. (2014); Wu et al. (2020)
Other carbohydrate-rich waste	Kitchen waste, food by-products, cheese whey, effluents from biodiesel, dairy, brewery industry, and oil mills	High energy and VS content Simplicity of hydrolysis and fermentation Stable source of nutrients	Presence of pathogenic bacteria Requirement of co-fermentation Low content of nitrogen in some substrates Formation of toxins during fermentation	Ngo et al. (2011); Xiao et al. (2013); Noblecourt et al. (2018); Toledo-Alarcón et al. (2018)
Protein and lipid-rich substrates	Sewage sludge, food industry effluents	Proteins and lipids are favorable sources of energy Wide availability of substrates	Generally low H_2 production potential Complexity for biodegradation Limitations caused by microbial metabolism	Carbonaro et al. (2012); Xiao et al. (2014); Ali et al. (2017)

8.2.1 Lignocellulosic Waste

Lignocellulose is an important structural component of plants and a common energy-intensive substrate. Three types of fuel can be produced from lignocellulose by biochemical or thermochemical processing: electricity, liquid fuel (ethanol, pyrolysis oil), and gaseous fuel—hydrogen, methane (Menon and Rao 2012). The main source of lignocellulose in nature is plant residues (Bundhoo 2019). Lignocellulose is traditionally found in agricultural and industrial wastes such as bagasse, corn and wheat stalks, straw and wood chips. Due to its natural origin, the processing of lignocellulose provides energy security and minimizes greenhouse gas emissions, making it an attractive material for DF (Wu et al. 2021).

Lignocellulose consists of biopolymers linked by covalent and hydrogen bonds: cellulose and hemicellulose (up to 70–80% of the biomass). Cellulose and hemicellulose are chemically bonded to a sheath of lignin and pectin, a water-soluble, easily hydrolysable polymer. The ratio of components varies depending on the origin of the plant material (Kumar et al. 2015; Nagarajan et al. 2019). Cellulose and hemicellulose are common DF substrates, their monomers are hexoses and pentoses. Lignin is resistant to biodegradation, so its content in biomass reduces the efficiency of DF. Along with furan derivatives and phenolic compounds, lignin is able to inhibit the production of biohydrogen (Quéméneur et al. 2012a). As a result of cellulose biodegradation, glucose, cellobiose, and 5-hydroxymethylfurfural (5-HMF) are formed, which can suppress the activity of hydrogen-producing microorganisms (Lin et al. 2015). Inhibitors formed during the decomposition of lignocellulose have stronger effect on hydrogen production than on methane fermentation. Enrichment or adaptation of microorganisms to growth in the presence of inhibitors is a promising strategy for improving hydrogen production from lignocellulose (Sivagurunathan et al. 2017). Due to the complex structure of cellulose, various pre-treatment methods are often used to facilitate biodegradation, which will be discussed in more detail in the next part of this chapter. The main methods of lignocellulose pretreatment include physical grinding, chemical treatment (acidic, alkaline, oxidative), enzymatic hydrolysis, and combinations of the above methods. The described pretreatment methods are quite expensive, so alternative methods are followed (Zheng et al. 2017). Consolidated bioprocessing is a promising method for the production of biofuels from lignocellulose, combining the production of hydrolytic enzymes by certain bacteria, hydrolysis of pre-treated lignocellulose, and subsequent fermentation in a single volume. The released monomeric sugars are converted into hydrogen and biogas.

According to the literature, the use of thermophilic anaerobic microorganisms with a growth temperature above 50 °C is more effective than the use of mesophilic ones, due to an increase in the kinetics of the decomposition reaction, an increase in the synthesis of hydrolytic enzymes, and suppression of the growth of harmful microbiota (Ren et al. 2016; Jiang et al. 2020). In a study by Jiang et al., the optimum temperature for lignocellulose processing was shown to be around 55 °C. In the experiment, sewage sludge was used as an inoculum; the hydrogen yield was 15.4 mol H_2/kg cellulose and 2.77 mol H_2/mol hexose (Jiang et al. 2015). Under mesophilic and hyperthermophilic conditions, suppression of hydrogen production was observed. There are studies devoted to the biodegradation of untreated plant residues with the formation of hydrogen, for example, with the participation of thermophiles *Caldicellulosiruptor saccharolyticus* at a temperature of 70 °C (Ivanova et al. 2009). The yield of hydrogen during cultivation on the most effective substrate, wheat straw, was 3.8 mol H_2/mol glucose. According to the results obtained, plant biomass can be converted into hydrogen by microbial fermentation without resorting to costly pretreatment methods. However, with this approach, there is often a problem with a low degree of conversion of the substrate to the product. Wu et al. reviewed the prospects for using species of *Thermoanaerobacterium* for the biodegradation of lignocellulose. The results of these studies have shown that there are really promising *Thermoanaerobacterium* species. In addition, the characteristics of strains can be improved with the help of metabolic engineering. Among the three major components of lignocellulose, *Thermoanaerobacterium* is the best hemicellulose degrader, as they contain powerful xylanolytic enzymes. Hemicellulose consists of five types of monomers—pentoses or hexoses, including D-xylose, L-arabinose, D-galactose, D-mannose, and D-glucose. Xylan is the main component of hemicellulose, consisting of β-D-xylose residues (Wu et al. 2021).

According to the literature, there are works devoted to the artificial change in the metabolism of hydrogen-producing microorganisms in order to increase the effective direct conversion of lignocellulosic material into hydrogen. Cha et al. modified the *Caldicellulosiruptor bescii* strain by deleting the L-lactate dehydrogenase gene (Cha et al. 2013). The use of genetic engineering to control the metabolism of microorganisms and the development of multienzyme systems for the degradation of lignocellulose components can increase the yield of target products and reduce the cost of the entire process. Application of genetic engineering and development of multienzyme lignocellulose-degrading systems can increase the yield of target products and make the process more profitable.

8.2.2 Algal Biomass

Another source of feedstock, more accessible to microorganisms and suitable for DF, is the biomass of micro- and macroalgae. Algae are easy to obtain in large quantities and contain almost no lignin (Xia et al. 2015). Carbohydrates present in microalgae, including green algae and cyanobacteria, are simple polyglycans such as starch and glycogen, while macroalgae are composed of various organic acids and sugar alcohols such as glucuronic acid and mannitol. The content of carbohydrates in algae reaches 60–70%, proteins make up 15–50% of the biomass. Algae is a rich source of protein, which upon hydrolysis gives free amino acids, the fermentation of which is a potential source of energy (Kumar et al. 2018). The addition of amino acid fermenting clostridia to mixed or pure cultures ensures full utilization of microalgae biomass (Nagarajan et al. 2020). Different types of bacteria are capable of efficient hydrogen production from both raw and dried algal biomass (Batista et al. 2014). Ferreira et al. studied the production of hydrogen by *C. butyricum* culture from the acid hydrolyzate of the dried biomass of *Scenedesmus obliquus* under mesophilic conditions. The yield of hydrogen was 2.9 ± 0.3 mol H_2/mol sugars. However, the need to reduce energy costs when using such a hydrogen production route has been reported (Ferreira et al. 2013). According to the literature, untreated algae biomass is also used as a substrate for hydrogen production. Shi et al. investigated the production of hydrogen by anaerobic mixed cultures when grown on *Laminaria japonica*. The raw dried *Laminaria japonica* biomass was fermented at

35 ± 1 °C. After a series of batch experiments, where the maximum hydrogen production was 0.92 mol H_2/mol hexose, an experiment was carried out on continuous cultivation in an anaerobic sequencing batch reactor. After 30 days of incubation, a stable production of hydrogen of 0.79 ± 0.03 mol H_2/mol hexose was achieved. At present, there are many studies devoted to the production of hydrogen from the biomass of micro-, macro-, or mixtures of algae, since they are unpretentious in cultivation and quickly increase their biomass. The algae substrate is much more accessible than lignocellulose, but some studies resort to structure simplification by hydrolysis (Shi et al. 2011).

8.2.3 *Other Carbohydrate-Rich Waste*

Food waste is produced in huge quantities and is a rapidly degradable carbohydrate-rich substrate. According to studies, food waste is a promising substrate for the production of hydrogen and various biofuels (Dauptain et al. 2020; Sarangi and Nanda 2020). Typical examples of waste in this group are kitchen waste, food by-products, cheese whey (Toledo-Alarcón et al. 2018). Food waste has a high energy content and contains 85–95% volatile solids. The biochemical hydrogen potential varies depending on the nature of the substrate, but on average it is twice as high as the hydrogen potential of agricultural waste (Guo et al. 2010).

Wastes containing high amounts of carbohydrates, such as rice, flour, pasta, noodles and bread, are considered the most suitable for dark fermentative hydrogen production (Jarunglumlert et al. 2018). However, there are reports of technical limitations and low hydrogen yield from food waste fermentation. The main problem is the low content of nitrogen, a key element for the nutrition of microorganisms (Yun et al. 2018). Noblecourt et al. studied hydrogen production from the depackaging waste fraction of food waste. The results showed that the microbial consortium used (sewage sludge) has a great potential for converting lactate and carbohydrates into biohydrogen. In batch culture, lactate was used as the main carbon source and an overall yield of 0.4 mol H_2/mol lactate was obtained. Species of the *Clostridium* are most widely represented in the microbial community, mainly *C. butyricum*, which is the main producer of hydrogen from lactate (Noblecourt et al. 2018).

Digestion and pre-treatment of food waste can produce inhibitors such as furfural (Han et al. 2015). In the study, food waste was subjected to enzymatic hydrolysis before the experiment to release monosaccharides and facilitate the fermentation process. Fungal glucoamylases and proteases were used as the main hydrolysis enzymes; thermally treated anaerobic sludge served as an inoculum. According to the authors' findings, food waste hydrolyzate is a potential substrate for dark fermentation. The best yield of hydrogen from food waste was 39.14 mL H_2/g at a mass ratio of 4%. Dark fermentation of food waste is often carried out under mesophilic conditions at a temperature of 35–37 °C. Xiao et al. studied the production of hydrogen from synthetic food waste by certain types of bacteria. The most active hydrogen producers were *Escherichia cloacae* and *Enterobacter aerogenes*. The highest hydrogen yield was recorded for *E. cloacae*, 155.2 mL H_2/g VS (Xiao et al. 2013). Kim et al. showed the possibility of efficient hydrogen production from food waste under thermophilic conditions without inoculum using indigenous microbiota (Kim et al. 2011).

Effluents from biodiesel, dairy, brewery industry, and oil mills are suitable substrates for dark fermentation. Crude glycerol, a waste product of biodiesel fuel production, gives a high yield of hydrogen despite toxic impurities and moderate availability (Sarma et al. 2012). Hydrogen production from glycerol can be further improved by co-fermentation with the addition of glucose (Trchounian and Trchounian 2015). Ngo et al. studied the production of hydrogen from pure and crude glycerol by the extremely thermophilic bacterium *Thermotoga neapolitana* DSM 4359, whose genome contains sequences responsible for the metabolism and conversion of glycerol. The maximum yield of hydrogen was 1.97 ± 0.09 mol H_2/mol of consumed glycerol and was observed after 56 h of cultivation. The value obtained was almost 1.55 times higher than for crude glycerol waste (Ngo et al. 2011).

Cheese whey and cheese whey powder are widely used as substrates in DF (Kovalev et al. 2022a; Mikheeva et al. 2022a). Cheese whey powder is a highly concentrated and stable source of lactose, high in nitrogen. Kargi et al. studied the hydrogen production by dark fermentation of pretreated (protein precipitated and filtered) cheese whey powder under thermophilic conditions with anaerobic sludge as inoculum. The hydrogen yield was 1.03 mol H_2/mol glucose (Kargi et al. 2012). The fermentation of cheese whey by different types of psychrophilic bacteria at a temperature of 20 °C was studied. For *Rahnella aquatilis* as inoculum, the highest percentage of hydrogen in the biogas produced and the best efficiency of hydrogen production were recorded (Dębowski et al. 2014). By optimizing conditions, many substrates derived from food waste and food industry waste can be utilized during DF without any rigorous pretreatment.

Animal waste can be used as a substrate for dark fermentation. Among livestock waste, urine, solid manure, and wastewater (including liquid manure) are distinguished. Proper disposal and recycling are essential for such waste, as it pollutes the air and water. When using animal waste as a substrate for DF, it is necessary to pretreat not only to eliminate methanogenic activity, but also to remove pathogenic bacteria (Burton and Turner 2003; Toledo-Alarcón et al. 2018). Other types of pretreatments may be used to improve the availability of the substrate. The main problem of animal waste is the high content of nitrogen compounds, which during fermentation can be reduced to ammonium, which inhibits dark fermentation in certain concentrations. When dealing with animal waste, co-fermentation with glucose is used to optimize the C/N ratio, increase COD, and provide an additional, more available carbon source (Wu et al. 2009; Weide et al. 2020). Tang et al. investigated hydrogen production from cattle wastewater at 45 °C using sewage sludge as an inoculum. The hydrogen yield was 319 mL H_2/g COD (Tang et al. 2008). Wu et al. showed that when glucose is added to pig manure, a hydrogen yield of 1.18–1.63 mol H_2/mol glucose is observed (Wu et al. 2009).

8.2.4 Protein and Lipid-Rich Substrates

The potential for hydrogen production from carbohydrates is much higher than from lipids and proteins, so studies often estimate the productivity in terms of hydrogen yield per mole of hexose (Yun et al. 2018). However, it is worth paying attention to substrates rich in proteins and fats. Xiao et al. have investigated hydrogen production from protein-rich wastewater. Since the protein has a three-dimensional structure, hydrolysis by proteases to peptides and amino acids proceeds slowly and is a limiting step. Then, as a result of deamination reactions, hydrogen, VFA, and ammonium are formed. Unfolding the structure increases the availability of the protein to proteases. Wastewater rich in protein was pre-treated at pH 12 to break hydrogen and disulfide bonds to increase hydrogen production. Fermentation was carried out under neutral (control) and alkaline conditions at pH 7 and pH 10, respectively. Under alkaline conditions, the best hydrogen yield was observed, amounting to 205.2 mL H_2/g protein. Alkaline conditions contribute to the solubilization of sludge organic matter and the suppression of the activity of hydrogen-consuming microorganisms (Carbonaro et al. 2012; Xiao et al. 2014).

Wastewater rich in lipids is often generated in food production. The main components of lipids are triglycerides and long chain fatty acids (LCFAs). Hydrolytic bacteria break down lipids to form glycerol and long chain fatty acids (LCFA). LCFA are simple monomers and are more readily absorbed by bacteria. LCFA, sugars, and amino acids are decomposed into carbon dioxide, hydrogen gas, alcohols and organic acids at the stage of acidogenesis. Degradation of LCFA occurs via β-oxidation, in which coenzyme A is used to oxidize LCFA to acetate and hydrogen. Further decomposition during acetogenesis leads to the formation of acetic acid, carbon dioxide, hydrogen, and ammonia.

Lipids are a favorable source of energy, but the accumulation of LCFA prevents the formation of hydrogen, causes flotation and leaching of sludge from the anaerobic reactor (Rasit et al. 2015). Lipid hydrolysis is the rate-limiting step in the DF process. For efficient conversion of lipid-containing substrates, it is necessary to understand the kinetics of lipid hydrolysis. Ali et al. investigated the production of hydrogen and ethanol from lipid-rich food industry effluents in continuous fermentation at various hydraulic retention times (HRTs) (Ali et al. 2017). Wastewater containing residual sludge and no additional inoculum was added to the reactors. Monocultures of two different species of *Clostridium* sp. (EBR-02E-0045 and EBR-02E-0046) were found in the residual sludge. The best hydrogen production rate recorded was 80.3 ± 19.9 L H_2/day, at HRT of 12 h. The results of the aforementioned studies indicate that protein and lipid-rich wastes can be used as substrates for DF, but the complexities of biodegradation and the limiting steps of microbial metabolism must be taken into account.

Most natural, industrial, and anthropogenic wastes are an attractive and sustainable substrate for biofuel production. However, many polymers found in waste are complex and hard to reach for bacteria, so in many cases various pretreatment methods are employed.

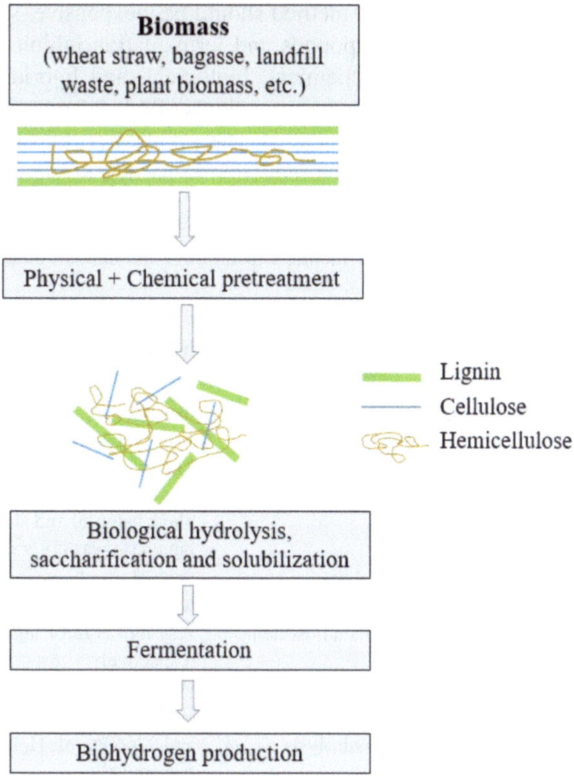

Fig. 8.1 Typical flowchart for improving dark fermentative hydrogen production based on pretreatment of lignocellulosic biomass

8.3 Methods of Substrate Pretreatment

The substrates must meet the nutritional needs of microorganisms, contain the components necessary for growth and stable processes in the microbial community (Hagos et al. 2017). Substrate pretreatment, especially in the case of lignocellulosic wastes, is important when working with monocultures of hydrogen-producing bacteria (HPB), since not all effective HPB are capable of hydrolyzing polymers (Pachapur et al. 2015). Some pretreatment methods also remove hydrogen-consuming microorganisms, which positively affect the final hydrogen yield (Guo et al. 2008). A typical flowchart for improving dark fermentative hydrogen production based on pretreatment is shown in Fig. 8.1.

8.3.1 Improving of Substrate Biodegradability

For some substrates, such as lignocellulose, pretreatment is almost always a mandatory step. The purpose of pretreatment is to reduce the degree of crystallization and polymerization and increase the availability of microfibrils for hydrolysis. The optimal pretreatment method should be inexpensive, safe, and not promote the formation of toxic compounds and fermentation inhibitors. Among the pretreatment methods, physical, chemical, biological, and hybrid methods are distinguished, which combine two or more methods (Table 8.2).

Physical methods, as a rule, are applied at the first stage of processing of the substrate and represent a reduction in particle size and degradation of the structure

Table 8.2 Pretreatment methods to improve the dark fermentative hydrogen production from various feedstocks

Feedstock	Pretreatment method	Improvements in H_2 production	References
Waste activated sludge	Freezing in the presence of nitrite	The maximum hydrogen yield increased from 7.96 to 19.40 mL/g VS	Liu et al. (2020)
Food waste	Addition of potassium ferrate	The maximum hydrogen yield was 173.5 mL/g (57.5 mL/g in blank sample) and shortening of lag phase was observed	Kuang et al. (2020)
Pine tree wood	Dilute acid pretreatment and pH adjustment	Peak H_2 production rate and yield were 1824 mL H_2/L/day and 1.27 mol H_2/mol sugar, respectively	Gonzales et al. (2017)
Empty fruit bunch of oil palm	Dilute acid and enzymatic hydrolysis	H_2 yield and H_2 production rate enhanced to 283.91 mL H_2/g TS and 3266.86 mL H_2/L/day, respectively	Gonzales et al. (2019)
Grain bamboo grass, oak, and camphor wood	Ammonia fiber explosion + NaOH	Promotion of bio-H_2 production (152.3, 99.1, and 76.9 mL/g TS, respectively)	Zhang et al. (2022)
Wheat flour	Grinding + acid hydrolysis	Enhancement of H_2 production rate and H_2 yield from 140 to 4500 mL H_2/L/day and from 20 to 646 mL H_2/g total sugar, respectively	Gorgec and Karapinar (2019)
Water hyacinth	Ultrasonic-assisted alkaline pretreatment	H_2 yield after pretreatment (110.4 mL/g VS) was about 4 times higher than without it (about 27 mL/g VS)	Tran and Nguyen (2022)
Garden wastes	Acid + enzymatic hydrolysis	2.7-fold increase in hydrogen yield when compared to the untreated garden wastes (35.9 mL H_2/g dry garden waste for untreated biomass)	Ramprakash and Incharoensakdi (2022)
Organic wastes (starchy wastewater and groundnut de-oiled cake)	Thermal pretreatment (121 °C, 15 psi for 15 min)	Increase in H_2 yield by 20% (maximum H_2 yield 12.05 mol H_2/kg COD)	Mahata et al. (2021)
Sugarcane bagasse	Milling + autoclave	H_2 production increased from 0.33 to 24.06 mmol H_2/L	Rabelo et al. (2018)

with the help of physical/mechanical force, which leads to an increase in the efficiency of further transformations.

Chemical pre-treatment of lignocellulose involves the use of acid, alkali, or oxidizing agent to dissolve and remove hemicellulose and lignin. Biological pretreatment involves the use of bacterial and fungal enzymes to degrade polymers. The most effective pre-treatment methods are physical and chemical, such as ammonia fiber expansion/explosion and steam explosion (Nagarajan et al. 2019; Zhao et al. 2020). Many studies on the DF of natural substrates point to the benefits of pre-treatment and increasing the efficiency of the process. Acidification of lignocellulosic waste with dilute hydrochloric acid increased hydrogen production from 10- to 136-fold compared to untreated waste (Chong et al. 2009). A study by He et al. examined the effect of hydrothermal treatment of rice straw on dark fermentative hydrogen production. It has been shown that the final hydrogen yield depends on the temperature and time of the hydrothermal treatment. The best hydrogen yield was 28 mL/g-VS, which is 93 times higher than control (He et al. 2014). Lopez-Hidalgo et al. treated wheat straw with sulfuric acid to increase the availability of sugars and additionally sterilized the substrate to work with a pure culture of *E. coli* strain WDHL. The yield of hydrogen was 3277.7 mL H_2/L reactor. In some cases, to improve the efficiency of the process, after acid hydrolysis, enzymatic hydrolysis is additionally carried out (Lopez-Hidalgo et al. 2017). Arreola-Vargas et al. compared the effectiveness of acid, alkaline, and combined methods of hydrolysis of lignocellulose with biological hydrolysis. Sequential combination of methods showed a higher molar yield of H_2 (2.39 mol H_2/mol reducing sugar) (Arreola-Vargas et al. 2015). An improvement in biohydrogen production after pre-treatment was observed for paper waste, lignocellulose-derived materials. Paper waste was subjected to acid hydrolysis to obtain various concentrations of glucose and studied the change in hydrogen production. The best hydrogen yield (140 mL H_2/g total sugars) was recorded at a sugar concentration of 3.84 g/L, and inhibition of the process was observed at concentrations above 18.9 g/L. Researchers attribute the inhibition of the process to a sharp accumulation of VFAs' and an increase in the concentration of toxic furan derivatives formed as a result of acid hydrolysis (Eker and Sarp 2017). Potassium ferrate (PF), a strong oxidant, was used by Kuang et al. (2020) to promote anaerobic dark fermentation of food waste. PF enhanced the hydrogen production from food waste, the maximal hydrogen yield was 173.5 mL/g, and the optimal PF dosage was 0.4 g/g total suspended solids. PF shortened the lag phase for hydrogen generation from 120 to 96 h.

The biological pretreatment of lignocellulose includes the precipitation of lignin by fungal or actinomycete enzymes (Soares et al. 2020). A promising strategy for the processing of lignocellulosic substrates, especially herbaceous ones, is bioaugmentation of the natural microbial community with hydrolytic hydrogen-producing strains' (Valdez-Vazquez et al. 2019). Many substrates are subjected to pre-treatment by hydrolysis to improve hydrogen production, and it's not restricted to lignocellulose. Substrates from microalgae are often subjected to pretreatment aimed at the destruction of the cell wall and the hydrolysis of carbohydrates. The choice of pre-treatment method depends on the class of algae and is associated with the polymer composition of cell walls (Montingelli et al. 2015). Mechanical forces are often used to degrade cell walls and hydrolyze carbohydrates: bead milling, high pressure homogenization, ultrasonication, microwave pretreatment, hydrodynamic cavitation, and pulsed electric fields. A common disadvantage of these methods is their high energy consumption. Bead milling, despite its simplicity, is an effective method and can destroy up to 95% of cells. High pressure homogenization has been successfully applied to algae with fragile cell walls, such as the red alga (*Rhodophyta*) *Porphyridium cruentum* and the cyanobacterium *Arthrospira platensis* (Safi et al. 2014). A new method of physical pretreatment using a vortex layer apparatus (VLA) showed promise. The mechanism for improving the production of H_2 is based on the disintegration of complex raw materials under the influence of physical impact, cavitation, and electrolysis occurring in the VLA (Mikheeva et al. 2022a, b; Kovalev et al. 2022b). The physical methods used for algae biomass pretreatment include thermal treatment options: high temperature thermal pretreatment, freeze drying, autoclaving, steam explosion. At high temperatures, the destruction of biomass occurs due to strong heating, and at low temperatures, due to the formation of ice crystals that destroy cells (Nagarajan et al. 2020). Chemical pretreatment methods usually involve treatment of the substrate with an alkali or acid, less commonly with organic or inorganic solvents. Despite low energy costs, chemical pretreatment methods have the disadvantage of forming inhibitory substances.

Many studies report a combination of thermal and chemical pretreatment methods. The heat treatment method liquefies the organic matter, while the acid/alkali chemical pretreatment promotes the hydrolysis of the substrate. In a study by Choi et al., the biodegradability of *Chlorella vulgaris* was enhanced by a combined hydrochloric acid pretreatment at 92 °C. The maximum hydrogen yield of the treated biomass was 48.4 mL H_2/g dry cells, and for the untreated biomass it was 8.1 mL H_2/g dry cells (Choi et al. 2016). Ammonia fiber explosion combined with NaOH pretreatment effects on the characteristics of grain bamboo grass (herbaceous), oak (hardwood), and camphor wood (softwood) were assessed by Zhang et al. (2022). Significant promotion of dark fermentative bio-H_2 production (152.3, 99.1, and 76.9 mL/g TS, respectively) and volatile acid production (4660.2, 3720.2, and 3496.2 mg/L, respectively) of *P. sinese*, oak, and camphor wood was observed. Giang et al. optimized the acid-thermal pre-treatment of *Chlorella* sp. biomass to increase hydrogen production. The hydrogen yield was 54.0 mL H_2/g volatile-solids (VS) after treatment with sulfuric acid and holding at 160 °C, which improved the process performance compared to pure acid hydrolysis (Giang et al. 2019).

8.3.2 Inhibition of Hydrogen-Consuming Microbiota

Substrates that are more accessible and richer in simple carbohydrates and peptides (household and food waste, sewage sludge) are pretreated to remove extraneous hydrogen-consuming microbiota, especially methanogenic archaea. The most commonly used methods are sterilization, heat shock, ultrasonic, acid, alkali treatment, or the addition of organic compounds (Cai et al. 2004). Dark fermentative hydrogen production was largely enhanced from waste activated sludge pretreated by freezing in the presence of nitrite. The maximum hydrogen yield increased 2.44 times from 7.96 to 19.40 mL/g VS. Mechanism explorations revealed that the proposed pre-treatment not only accelerated the disintegration of sludge but also promoted the proportion of biodegradable organics released, thereby provided more bio-available substrates for subsequent hydrogen production. Proposed pretreatment severely suppressed the sludge microorganisms responding to homoacetogenesis (−32.1%), methanogenesis (−58.4%), and sulfate-reducing process (−51.5%), inhibited the consumption of hydrogen (Liu et al. 2020). Guo et al. studied the effects of sterilization, ultrasonication, and microwave treatment on hydrogen production from waste sludge (Guo et al. 2008). The best hydrogen yield of 5.02 mL H_2/g TCOD was obtained by sterilizing the substrate. After microwave treatment, the hydrogen yield was lower (11.44 mL H_2/TCOD), ultrasonic treatment was the least effective; the hydrogen yield was 4.68 mL H_2/g TCOD. These results were probably due to the elimination of harmful bacteria at extremely high temperatures (121 °C) during sterilization. Some studies have shown the ineffectiveness of heat treatment for breaking down complex substrates and releasing soluble sugars. Heat treatment only affects the microbial community of the substrate, not always leading to an increase in biohydrogen yield (Dauptain et al. 2020). Biological enzymatic treatment is a rather mild method that effectively acts on insoluble and difficult to hydrolyze biopolymers, for example, alginan in the algae cell wall. During enzymatic process, complex carbohydrate components are hydrolyzed to simple sugars in one step. Various enzymes are used for dark fermentation. Cellobiohydrolase, hemicellulase, lysozymes, pectinases, chitinases are capable of effectively hydrolyzing the complex carbohydrates, whereas enzymes such as proteases and lipases tend to decompose proteins and lipids (Zabed et al. 2019). Special enzymes are often used to hydrolyze food waste. In a study by Han et al., glucoamylases and proteases derived from *Aspergillus oryzae* and *Aspergillus awamori* were used to hydrolyze kitchen waste. The best yield of hydrogen during dark fermentation was 219.91 mL H_2/g VS_{added} using anaerobic sludge as an inoculum (Han et al. 2015). Other methods, such as alkaline shock, are also used to pretreat food waste (Jang et al. 2015). The yield of hydrogen during the DF of food waste, after treatment with alkali at pH 9–13, was 17 L H_2/kg in batch mode. At high substrate pH (11–12), the dominant component of the microbial community was *Clostridium* sp. Thermal or chemical exposure destroys many non-spore-forming bacteria (for example, lactic acid bacteria that compete with acidogens for the substrate, fermenting carbohydrates). The composition of the microbial community of the substrate is important to consider, especially with animal waste containing pathogenic microbiota and various toxic substances. Waste of animal origin is characterized by a high content of hydrogen-consuming microorganisms, which must be suppressed during substrate pretreatment (Ghimire et al. 2015).

The control of animal manure is being strengthened, so it is important to make this substrate as convenient as possible for biodegradation and clean energy production. Xing et al. treated dairy manure with acid, alkali, and infrared radiation. A natural inoculum with hydrogen-producing bacteria was obtained from cattle feedlots. At optimal pH and substrate concentration, the maximum cumulative hydrogen yield from manure acid hydrolysate was 31.5 mL/g TVS (Xing et al. 2010). Different pre-treatment methods allow the substrate to be adapted for a more efficient DF and an increased yield of biohydrogen. Despite the above advantages, the use of some pretreatment methods can have negative consequences. For chemical and physicochemical methods, for example, the steam explosion process, the formation of DF inhibitors is typical. Inhibitors have different effects on the DF process, depending on the composition of the microbial community (Soares et al. 2020). Pure cultures are much more susceptible to the action of inhibitors of DF, mixed cultures are more resistant and adapted to the action of inhibitors, due to the presence of different groups of microorganisms capable of converting inhibitors into less toxic forms (Quéméneur et al. 2012a). According to the literature, phenol and furan compounds pose the greatest danger to DF, while organic acids (formic, levulinic, acetic) are less toxic. The main mechanism of furan inhibition is associated with interference with the synthesis of glycolytic enzymes and inhibition of biomass growth. Furan derivatives at a concentration of about 1 g/L suppress hydrogen production by 17.9–76.1%, but the effect strongly depends on the inoculum. For example, the anaerobic sludge community is able to degrade 5-hydroxymethyl furfural (Chen et al. 2021).

Sometimes the addition of furfural at certain concentrations changes the composition of the microbial community, shifting the ratio towards hydrogen-producing species and increasing hydrogen production by more than 40% (Liu et al. 2015). Cao et al. reported a synergistic inhibitory effect of several compounds (furfural, hydroxymethylfurfural (HMF), vanillin and syringaldehyde) on hydrogen production by *T. thermosaccharolyricum* W16 culture when grown on acid-pretreated hydrolysate of cellulose. Phenolic compounds, especially low molecular weight ones, damage the cell membrane, impairing permeability and are inhibiting at concentrations of 0.5 g/L (Cao et al. 2010). Some studies report that phenolic compounds are more toxic than furfural (Lin et al. 2015), but there is evidence to the contrary (Quéméneur et al. 2012a). To minimize the effect of DF inhibitors, physical, chemical, or biological detoxification methods are used. Biological detoxification involves the treatment of the hydrolysate with enzymes or fungi. Physical detoxification involves the use of evaporation techniques to remove volatile inhibitors from the hydrolysate. Chemical detoxification involves the use of alkali and other suitable chemicals to remove inhibitors (Bundhoo and Mohee 2016). Activated carbon is used to absorb inhibitors: vanillin, HMF, acetic acid, leading to an increase in hydrogen production compared to non-detoxified hydrolysate (Orozco et al. 2012; Cheng et al. 2015). DF inhibitors, in addition to those described above, include weak organic acids formed during pre-treatment and during fermentation itself. The source of organic acids is the majority of HPB, which, with substrate phosphorylation, form acids such as acetic, butyric, etc., capable of penetrating into the cell, acidifying the medium, and disrupting the proton gradient (Chen et al. 2021). The negative effect, the degree of formation and composition of acids depends on the metabolic pathway carried out by HPB. To minimize the formation of soluble metabolic products (SMP), it is recommended to maintain optimal pH, dilute the contents of the reactor, or use chloramphenicol to reduce ionic strength (Van Niel et al. 2003; Hwang et al. 2004). For VFAs and ethanol formed during fermentation, neutralization methods are not commercialized. At the moment they are only being developed and not applied in real conditions.

Pretreatment and saccharification is a key technology affecting the efficiency of biohydrogen production from lignocellulosic and other natural substrates. The positive effect is achieved due to the extraction of monosaccharides and inhibition of extraneous microbiota. The best results can be achieved using a combination of several pretreatment methods (Giang et al. 2019). It is important to continue research aimed at the conversion of feedstocks into a substrate suitable for DF. Various approaches should be explored to develop cost-optimal methods, including chemical, thermal, enzymatic, microbial, fungal, and multi-stage combined treatments (Urbaniec and Bakker 2015).

8.4 Factors Influencing Dark Fermentation Efficiency

Anaerobic digestion of substrates by pure and mixed cultures depends on environmental factors such as temperature, pH, and HRT that regulate the metabolic pathways of microorganisms, including HPB (Li and Fang 2007). A relationship has been established between factors such as temperature, pH, and HRT, which together affect the hydrogen production by microorganisms (Zagrodnik et al. 2022). The efficiency of the DF process is affected by the substrate concentration and the feeding rate. Temperature and pH are important parameters that determine the rate of substrate utilization, metabolic pathways for the synthesis and consumption of biohydrogen by microorganisms, and the formation of SMPs.

8.4.1 Temperature and pH

Dark fermentation is usually carried out under mesophilic (35–37 °C) or thermophilic (55–60 °C) conditions. According to literature, at neutral pH under thermophilic conditions, acetate is most actively formed as an SMP, and in mesophilic conditions, butyrate is mainly formed, with a decrease in pH to 5.5, the opposite results are observed (Shin et al. 2004; Valdez-Vazquez et al. 2005). The composition of SMP and the rate of substrate utilization depend on the chemical composition of the substrate. Thermophilic and extreme thermophilic conditions are considered to be better suited for hydrolysis of substrates. Higher temperatures increase the activity of enzymes responsible for hydrolysis, which increases the rate of the process (Guo et al. 2010). The advantage of thermophilic conditions is the lower solubility of gases in water, providing a higher yield of the product. Changing the incubation temperature affects the composition of the acidogenic microbial consortium. Upon transition from the mesophilic (35 °C) to the thermophilic temperature range (55 °C), inhibition of processes was observed, leading to a drop in the production of VFAs from 17 to 11 g/L during the fermentation of food waste (He et al. 2012). VFAs have a toxic effect on hydrogen production; a decrease in concentration stimulated acidogenesis (Harirchi et al. 2022). The difference between the optimal temperatures for biodegradation of different materials is related to the proportion of readily degradable compounds present in the composition of the substrate. For a higher yield of hydrogen during the DF of agricultural waste rich in lignocellulosic compounds, a high temperature is required to stimulate the hydrolysis process. On the contrary, temperatures in the mesophilic range are preferable for DF of readily biodegradable substrates (Ghimire et al. 2015). Luo et al. investigated the effect of pH and temperature on hydrogen production from cassava stillage vinasse. At a temperature of 60 °C and pH 6, optimal conditions were created for the production of hydrogen –67.8 mL H_2/g VS were obtained. Among VFAs, butyrate prevailed as the main metabolic product (Luo et al. 2010). The highest hydrogen production from cattle wastewater was 368 mL, the hydrogen yield was 319 mL H_2/g COD, and the butyrate to acetate ratio was 1.43. The process took place at pH 5.5 and a temperature of 45 °C. The presence of propionate and ethanol in the metabolic products indicates a decrease in hydrogen production, since the propionate and ethanol formation pathways suppress the hydrogen formation pathways (Tang et al. 2008). Homoacetogenesis and methanogenesis are strongly dependent on the initial pH of the system, despite the pretreatment of the inoculum. An acidic pH (below 6) suppresses methanogenic activity under both mesophilic and thermophilic conditions, and inhibition of the activity of hydrogen-consuming homoacetogens under thermophilic conditions is achieved by lowering the pH to 5.5 (Luo et al. 2011). Low pH (below 5) destabilizes the processes, as it limits the ability of bacteria to maintain intracellular pH at the desired level. On average, the optimal pH range for hydrogen production during DF roughly corresponds to the favorable pH range for bacterial growth, from 5 to 7. At lower pH values, such as 4, propionate and ethanol are produced (Li and Fang 2007; Łukajtis et al. 2018). Data shown is an estimate and optimum operating pH values will vary depending on type and source of pure or mixed culture. Some research suggests that the type of substrate used can affect optimal pH. A review by Guo et al. reported that a neutral pH is optimal for DF of animal waste, a pH of 6.5–7 for plant waste, and a pH of 5–6 is best for food waste (Guo et al. 2010). Liu and Shen studied the influence of several parameters on the production of hydrogen from starch, including pH in the range of 4–9. The pH 7 and 8 turned out to be optimal (103 and 120 mL H_2/g starch, respectively), and at pH 4 no hydrogen production was detected (Liu and Shen 2004). In the DF of brewery wastewater with a high organic content, the highest hydrogen yield was 1.41 mol H_2/mol hexose at pH 5.5 and a substrate concentration of 40 g total sugar/L (Chu et al. 2013). The optimal pH for DF of wastewater from a palm oil plant was 5.0, with pH 4 showing the lowest hydrogen production (Vijayaraghavan and Ahmad 2006). However, several papers report that a microbial community dominated by *Thermoanaerobacterium thermosaccharolyticum* can successfully produce hydrogen continuously at pH as low as 4 (Kovalev et al. 2021, 2022c), although the optimum pH is 7.5 (Litti et al. 2022).

Like temperature, the concentration of protons affects the metabolic pathways of hydrogen production. According to the literature, a neutral pH favors the acetate fermentation pathway, and an acidic pH favors switching to the butyrate pathway. In the work of Infantes et al., the effect of pH and temperature on hydrogen production from synthetic wastewater was investigated. The authors noted that the amount of hydrogen produced per unit of acetate produced increased as the pH increased. However, as the pH increased, the concentration of butyrate in the system increased. According to this study, pH is a more important parameter for acidogenesis than temperature (Infantes et al. 2011).

8.4.2 Hydraulic Retention Time

The HRT is an important parameter affecting the production of hydrogen in the continuous mode on an industrial scale. The rate of hydrogen production increases in a certain range of HRT values. After exceeding the optimal value, the rate of production decreases. The optimal HRT value depends on the type and biodegradability of the substrate used for DF. It is reported that the growth rate of hydrogenotrophic methanogens is lower than the growth rate of HPB; therefore, reduced HRT in reactors can be used to eliminate unwanted microbiota (Hawkes et al. 2007; Liu et al. 2014). HRT in the reactor is dynamically changed from long to short in order to adapt the culture to new conditions and gradually change the microbial population (Łukajtis et al. 2018). With a decrease in HRT, it is important to prevent the biomass of HPB from being washed out of the reactor. Thanwised et al. evaluated hydrogen production and COD removal from tapioca wastewater using anaerobic mixed cultures in an anaerobic baffled reactor. With a decrease in HRT from 24 to 6 h, an improvement in hydrogen production was observed (up to 883.19 ± 7.89 mL H_2/L/day), with an HRT of 3 h, a decrease in hydrogen production was observed. A similar trend was observed for the removal of COD (Thanwised et al. 2012). Mikheeva et al. found that to obtain the highest hydrogen yield and hydrogen production rate, HRTs of 4.5 and 2.02 days, respectively, should be maintained in an anaerobic filter and an anaerobic fluidized bed reactor (AFB) packed with a polyurethane carrier and fed with cheese whey (Mikheeva et al. 2021). Using confectionary wastewater as a substrate, the highest hydrogen yield of 44.73 mL/g COD and hydrogen production rate of 92.5 mL/(L day) was obtained in AFB at HRT of 7.5 days and 2.5 days, respectively (Mikheeva et al. 2022c). In the study by Kirli and Karapinar, the yield and production rate of hydrogen during cultivation in a packed bed reactor increased by almost 2 times when HRT was reduced from 13 to 2 h. Acid hydrolyzate of wheat waste was used as a substrate (Kirli and Karapinar 2018). The trend of decreasing HRT to increase hydrogen production is also true for thermophilic conditions. Ferreira et al. compared hydrogen production by reducing HRT from 8 to 1 h in mesophilic (30 °C) and thermophilic (55 °C) reactors fed with sugarcane juice. In the mesophilic reactor at the lowest HRT, hydrogen production was lower than at 8 h HRT, but the hydrogen production rate was higher. In the thermophilic reactor, the maximum hydrogen yield was 1.52 mol H_2/mol hexose and was observed at an HRT of 2 h (Ferreira et al. 2019). In a similar study by Rosa et al, with a gradual decrease in HRT, first an increase and then a decrease in hydrogen production was observed, which was explained by acclimatization of methanogenic archaea and stabilization of methane production (Rosa et al. 2014). At short HRTs, it is important to control the growth kinetics so that undesirable microorganisms such as propionic acid and methane producers cannot compete with the dilution effect.

8.4.3 Substrate Concentration

The parameters that are fundamentally important for the production of hydrogen include the concentration of the substrate. Substrate concentration affects the relationship between microorganisms in culture, the rate and efficiency of the process. According to the literature, the substrate concentration affects the production of hydrogen in both batch and continuous cultivation. A number of studies have shown that substrate concentration affects metabolic pathways and microbial community structures, and hydrogen production is inhibited by intermediates (Akutsu et al. 2009). Too high substrate concentrations lead to inhibition of DF and insufficient biohydrogen yield. For example, Eker and Sarp in their study used paper waste treated by acid hydrolysis as a raw material and tested different concentrations of the substrate (3.84–45.5 g/L) to obtain the best hydrogen yield. The best consumption of sugars occurred at low substrate concentrations—from 3.84 to 18.9 g/L. The maximum hydrogen yield was 1.01 mol H_2/mol sugar at a substrate concentration of 3.84 g/L, the highest cumulative hydrogen production was 125.89 mL and was observed at a substrate concentration of 18.9 g/L. At higher substrate concentrations, hydrogen production decreased. The authors attribute the inhibition of hydrogen production to the formation of a large amount of VFAs and furan derivatives resulting from hydrolysis (Eker and Sarp 2017). Using *Thermoanaerobacterium thermosaccharolyticum* SP-H2 as an inoculum, the highest concentrations of confectionery wastewater and cheese whey, at which the DF process took place, were 5915 and 7311 mg COD/L, respectively (Litti et al. 2022). Chu et al. investigated the effect of initial substrate concentration on anaerobic hydrogen production in a reactor inoculated with sewage sludge. Wastewater from a brewery with a high organic content was used as a substrate. The peak growth rate of the mixed culture was recorded at a concentration of 40 g total sugars/L, the most active hydrogen production of 28.3 L and the highest hydrogen yield of 2.00 mol H_2/mol hexose were observed at a substrate concentration of 60 g total sugars/L. At a higher concentration (100 g total sugars/L) of the substrate, inhibition of hydrogen production was observed (Chu et al. 2013). Gonzales et al. studied hydrogen production from three types of lignocellulosic biomass: palm fruits, rice husks, and pine wood. All feedstocks were previously treated with dilute acid to reduce the level of polymerization and increase the availability of substrates for microorganisms. Substrates with different concentrations of sugars (5, 10, 15, 20 g/L) were subjected to DF. For all types of waste, the optimal concentration of sugars for DF was 10 g/L, based on the analysis of yield and estimation of the rate of hydrogen production. At a sugar concentration of 5 g/L, the yield and production rate of hydrogen were the lowest. However, in the case of rice husks, hydrogen production at a substrate concentration of 10, 15, and 20 g/L was almost the same. The authors explain this by the lower content of inhibitory compounds in the rice husk hydrolyzate formed as a result of pretreatment compared to other substrates.

According to Gonzales et al., high concentrations of monomeric sugars do not suppress hydrogen production. However, with an increase in the concentration of monomeric sugars, the concentrations of VFAs with inhibitory properties increased (Gonzales et al. 2016). A similar effect is observed for the untreated substrate. In the work of Akutsu et al., starch was used as the initial substrate, but stable hydrogen production was possible only under thermophilic conditions. The maximum yield of hydrogen was 2.8 mol H_2/mol glucose at a starch concentration of 20 g/L. In the experiment, the hydrogen yields sharply decreased when the starch content exceeded 30 g/L, which is associated with the accumulation of SMP. The best results in terms of hydrogen yield were recorded at starch concentrations of 20 and 30 g/L. At starch concentrations above 50 g/L, increased formation of ethanol, formate and lactate was observed. A negative correlation was observed between the total concentration of undissociated acids and hydrogen production (Akutsu et al. 2009). In the work of Lazaro et al., it was not possible to establish a correlation between the concentration of the substrate (sugarcane vinasse) and the production of hydrogen under mesophilic conditions. However, an increase in the concentration of vinasse had a positive effect on the rate and potential of hydrogen production. Increasing the substrate concentration from 2 to 12 g/L caused a drop in the hydrogen yield from 2.31 to 0.44 mmol H_2/g COD under thermophilic conditions (55 °C). In fact, increasing the vinasse concentration had a negative effect on the hydrogen yield under thermophilic conditions. However, such an effect was not observed under mesophilic conditions; the hydrogen yield remained almost unchanged with similar changes in concentrations. The mesophilic microbial community was dominated by species of the genus *Clostridium*, while the thermophilic community was dominated by *Thermoanaerobacter*. Not all isolates belonging to the *Clostridium* genus contribute to high hydrogen yield, as a homoacetogenic pathway may take place (Lazaro et al. 2014). In most cases, the inhibition of hydrogen formation is associated with the rapid accumulation of VFAs in the system at a high substrate concentration. This kind of inhibition is observed for feedstocks rich in carbohydrates. This must be taken into account when choosing the optimal conditions for the biodegradation of natural substrates during DF.

8.5 Co-fermentation

The technology of co-digestion of two or more anaerobic substrates has long been studied and actively used in biotechnology. The co-fermentation technology is more efficient, since direct utilization of the mono-substrate is difficult due to nutrient imbalance, the absence of a diversified microbial community, and possible violations of the stability of operational factors. Anaerobic co-fermentation is widely used to increase the yield and production rate of biogas in anaerobic digesters. To optimize the co-digestion process, the decisive parameters are temperature, pH, HRT, chemical composition and biodegradability of the substrates, etc. (Hagos et al. 2017). The C/N ratio is an important factor in the availability of the substrate for microorganisms during DF. The range of carbon to nitrogen ratios is not universal and depends on the type and source of substrates. Kim et al. showed that the best hydrogen production from co-digestion of rice straw and sewage sludge was achieved at a C/N ratio of 25:1 (Kim et al. 2012). Sreela-or et al. noted that optimal hydrogen production by mixed cultures in batch co-digestion of food waste and anaerobic sludge occurs at a C/N ratio of 33:1 (Sreela-Or et al. 2011). A low C/N ratio is unfavorable for DF, so the substrates with different carbon and nitrogen contents are mixed to optimize the ratio. A low C/N ratio can cause an accumulation of ammonia, which inhibits the anaerobic fermentation process. A high C/N ratio leads to the accumulation of VFAs and the lack of nitrogen nutrition for microorganisms. Ding et al. co-fermented nitrogen-rich microalgae and carbon-rich macroalgae to optimize a two-step hydrogen and methane production process (Ding et al. 2016). At the stage of dark fermentation, co-fermentation of *Laminaria digitata* and microalgae contributed to the favorable course of hydrolysis and acidogenesis, resulting in a hydrogen yield of 94.5–97.0 mL/g VS. The co-fermentation results were 15.5–18.5% higher than for fermentations using only *L. digitata*. In the case of co-fermentation of two or more substrates, in order to achieve the maximum yield of biohydrogen, it is necessary to select the optimal ratio of substrate components. Cheng et al. studied hydrogen production from a mixture of food waste and sewage sludge. According to the literature, food waste and sewage sludge are most often used for co-fermentation to produce biohydrogen or biomethane, because they complement each other in terms of elemental composition, organic matter content, and microbiological characteristics. The optimal C/N ratio was 16.1, the maximum hydrogen production was 174.6 mL H_2/g VS, which is 49.9% higher than with monoenzymatic substrate fermentation (Cheng et al. 2016). Zhou et al. studied 21 mixtures of food waste (FW), primary sludge (PS), and waste activated sludge (WAS). The maximum yields of hydrogen during co-fermentation of FW+PS, FW+WAS, and FW+PS+WAS were obtained at volume ratios of 75:25, 75:25, and 80:15:5. The addition of sewage sludge made it possible to balance the content of carbon and nitrogen in the mixture for DF, since food waste is poor in nitrogen (Zhou et al. 2013). As a source of nitrogen in co-fermentation, animal waste can be used, for example, chicken manure. The co-fermentation of a mixture of wine stillage, sewage sludge, and chicken manure was studied by Sillero et al. In this study, wine stillage and sewage sludge were mixed in a 1:1 ratio and various concentrations of chicken manure were added. The best production of hydrogen was recorded at a ratio of sewage sludge, wine stillage and chicken manure of 49.5:49.5:1. With an increase in the concentration of chicken manure above 10 g/L, inhibition of the process was observed (Sillero et al. 2022).

Liu et al. investigated the production of hydrogen and methane by co-digesting corn stalks and sewage sludge in a 2:1 ratio (based on TS). An increase in the proportion of corn stalks led to an increase in the yield of target gases and the total energy of biogas. During co-fermentation of corn stalks and sewage sludge, the VFA composition differed from mono-fermentation (Liu et al. 2013).

In addition to food waste, crude glycerol, a product of biodiesel production, was used as a carbon source in co-fermentation. Rivero et al. investigated hydrogen production from a mixture of glycerol and sewage sludge under mesophilic conditions. The best operational condition was an OLR of 7.82 g COD/L/day, with removal of 93% COD and hydrogen and methane yields of 0.026 L H_2/g COD and 0.29 L CH_4/g COD, respectively (Rivero et al. 2014). Plant substrates are rich in carbohydrates and can be used in co-fermentation. Yang and Wang studied the co-fermentation of sewage sludge and perennial ryegrass in various ratios. The highest level of hydrogen production (60 mL H_2/g VS) was at a 30:70 ratio of sewage sludge and perennial ryegrass and was 5 times higher than hydrogen production from sewage sludge mono-fermentation (Yang and Wang 2017). The use of co-fermentation helps to improve the characteristics of the fermented material by creating a balance of nutrients, diluting toxins and inhibitors. This has a positive effect on biogas production and allows the simultaneous utilization of various substrates, reducing the burden on the environment and solving the problems of excessive generation of various wastes.

8.6 Indigenous Microbiota and Its Role in Dark Fermentation

To optimize the DF process, an external microbial inoculum is added. In the case of mixed culture, the inoculum can be pretreated to specifically select spore-forming HPB and limit hydrogen scavenging bacteria. Feedstock used for dark fermentative hydrogen production, such as animal waste, industrial wastewater, sewage sludge, etc. also have indigenous microbiota that can be both beneficial and detrimental to H_2 production. Pre-treatments affect both the structure of the feedstock and the microbial composition of the substrate, which can suppress some positive or negative interactions with exogenous bacteria, in particular by suppressing some non-spore-forming bacteria. Thus, pre-treatment of indigenous bacteria can positively or negatively affect the hydrogen yield. The effect of native bacteria on DF characteristics and influence of substrate pretreatment on native bacteria is not well understood. Few studies have addressed this issue. Favaro et al. (2013) reported a higher hydrogen yield for non-sterile organic fraction of municipal solid waste (OFMSW) after inoculum addition compared to sterile OFMSW and suggested that the positive interaction between native and exogenous bacteria was suppressed, likely resulting in less efficient hydrolysis. In contrast, Quéméneur et al. (2012b) reported a lower H_2 yield for non-sterile wheat straw compared to sterile wheat straw in a two-stage system (24 hours hydrolysis prior to DF). The authors attributed this difference to the production of other metabolites (acetate and lactate) rather than hydrogen. Andreani et al. (2019) reported a higher hydrogen yield from uninoculated cassava wastewater in the AnSBBR reactor. As a consequence, indigenous bacteria can positively or negatively influence DF by interacting with exogenous bacteria. Further selection of these microorganisms in order to enrich HPB microbial communities and avoid non-hydrogen producing pathways can be applied. Other authors have reported studies without a strict external inoculation requirement (for a pretreated substrate), such as Kim et al. (2009) with food waste (FW) as a substrate. These authors observed a low hydrogen yield for untreated FW (4 mL/g VS added) compared to FW pre-treated at 90 °C for 20 min, which had a higher hydrogen yield (97 mL/g VS added).

Some authors report that no inoculum nor pretreatment is required to achieve satisfactory dark fermentation performance from organic waste. For example, Dauptain et al. (2020) revealed a strong influence of the indigenous bacteria, which are as effective as thermally pretreated exogenous bacteria in producing H_2 and metabolites. High abundance in *Clostridiales* and/or *Enterobacteriales* in seven different organic substrates (Tunisian dates (pitted), corn silage, sorghum, microalgae (*Scenedesmus quadricauda,* and *Pediastrum*), sewage sludge, OFMSW, and FW) was associated with high H_2 yields (Dauptain et al. 2020). Similarly, the work by Li et al. (2012) showed that grass silage can become amenable to hydrogen fermentation by indigenous silage bacteria (including *Ruminobacillus xylanolyticum, Acetanaerobacterium elongatum,* and *Clostridium populeti*) through pH neutralization. Mushroom waste (MW) is a potential feedstock for DF. Indigenous HPB were successfully enriched from MW using thermophilic temperature (55 °C) for use as the seed inoculum for production of H_2 from MW without any external seeding (Lin et al. 2017). The indigenous microflora converted coffee silverskin into hydrogen without biomass pretreatment, reaching hydrogen yields of up to 24.1 mL H_2/g COD. Hydrogen yield was maintained at high biomass load (minimal dilution) during a semi-batch bioreactor operation. HPB affiliated to *Enterobacteriaceae* (27.5%) and *Clostridium* (5.1%) were selected from the indigenous microflora (Renaudie et al. 2022).

Results from literature indicate that indigenous bacteria can effectively compete with bioaugmented HPB cultures. Numerous factors affect the bio-augmentation, including a) the amount of time and existence of augmented biocatalyst b) specific targeting of the microbes for its desired function and c) enhancing the performance and yield (Goud et al. 2014). Apparently, due to all these critical factors, bioaugmented species have failed to compete with the indigenous microflora in a few cases (Bouchez et al. 2000).

8.7 Summary

Recycling organic waste and production of clean renewable energy is an urgent task. Based on the research pool analyzed, it should be noted that simple carbohydrate substrates are more suitable for microbial biohydrogen production. The main factor limiting the biodegradation of feedstocks is the complex biochemical structure. Many wastes that are generated in large quantities, such as lignocellulosic wastes, are difficult to access for microorganisms. Various pre-treatment technologies can increase bioavailability, facilitating the hydrolysis process and subsequent dark fermentation. An important task of microbiology and biotechnology is the creation of simple, low cost, and safe methods for pretreatment, that effectively reduce the degree of substrate polymerization and reduce the number of toxic by-products. Obtaining more versatile microbial consortia capable of utilizing a wide range of complex substrates and constructing strains with exclusive metabolic capabilities (for example, hydrolyzing complex polymeric bonds and more completely decomposing feedstocks into simple oligo- and monosaccharides) will make it possible to obtain high yields of biohydrogen, bypassing the stage of costly pretreatment. It is necessary to expand the use of the technology of co-fermentation of several substrates, as the co-fermentation process significantly increases the yield of the product due to the optimal balance of nutrients and microbial communities. Indigenous microbial community should also be taken into account to ensure efficient production of H_2 from a non-sterile feedstock. In this chapter, the main factors and parameters of incubation that affect the production of hydrogen during dark fermentation were discussed, and the importance of synergistic effects was revealed. To move to an industrial scale for the utilization of substrates and the production of renewable hydrogen, it is necessary to conduct more experiments in continuous mode in reactors and simulate real conditions.

Acknowledgements The work was partially supported by the Russian Science Foundation (grant no. 22-49-02002, https://rscf.ru/en/project/22-49-02002/.

References

Akca MS, Bostancı O, Aydin AK, Koyuncu I, Altinbas M (2021) BioH2 production from food waste by anaerobic membrane bioreactor. Int J Hydrog Energy 46(55):27941–27955. https://doi.org/10.1016/j.ijhydene.2021.06.042

Akutsu Y, Li YY, Harada H, Yu HQ (2009) Effects of temperature and substrate concentration on biological hydrogen production from starch. Int J Hydrog Energy 34(6):2558–2566. https://doi.org/10.1016/j.ijhydene.2009.01.048

Ali M, Danial A, Tawfik A (2017) Self-dark fermentation of lipids rich wastewater for 2-biofuels (H_2 and Et-OH) production. Process Saf Environ Prot 109:257–267. https://doi.org/10.1016/j.psep.2017.04.007

Andreani CL, Tonello TU, Mari AG, Leite LCC, Campaña HD, Lopes DD, Rodrigues JAD, Gomes SD (2019) Impact of operational conditions on development of the hydrogen-producing microbial consortium in an AnSBBR from cassava wastewater rich in lactic acid. Int J Hydrog Energy 44:1474–1482. https://doi.org/10.1016/j.ijhydene.2018.11.155

Argun H, Kargi F (2011) Bio-hydrogen production by different operational modes of dark and photo-fermentation: an overview. Int J Hydrog Energy 36(13):7443–7459. https://doi.org/10.1016/j.ijhydene.2011.03.116

Arreola-Vargas J, Razo-Flores E, Celis LB, Alatriste-Mondragón F (2015) Sequential hydrolysis of oat straw and hydrogen production from hydrolysates: role of hydrolysates constituents. Int J Hydrog Energy 40(34):10756–10765. https://doi.org/10.1016/j.ijhydene.2015.05.200

Batista AP, Moura P, Marques PA, Ortigueira J, Alves L, Gouveia L (2014) *Scenedesmus obliquus* as feedstock for biohydrogen production by *Enterobacter aerogenes* and *Clostridium butyricum*. Fuel 117:537–543. https://doi.org/10.1016/j.fuel.2013.09.077

Bouchez T, Patureau D, Dabert P, Juretschko S, Doré J, Delgenès P, Moletta R, Wagner M (2000) Ecological study of a bioaugmentation failure. Environ Microbiol 2(2):179–190. https://doi.org/10.1046/j.1462-2920.2000.00091.x

Bundhoo ZM (2019) Potential of bio-hydrogen production from dark fermentation of crop residues: a review. Int J Hydrog Energy 44(32):17346–17362. https://doi.org/10.1016/j.ijhydene.2018.11.098

Bundhoo MZ, Mohee R (2016) Inhibition of dark fermentative bio-hydrogen production: a review. Int J Hydrog Energy 41(16):6713–6733. https://doi.org/10.1016/j.ijhydene.2016.03.057

Burton CH, Turner C (2003) Manure management: Treatment strategies for sustainable agriculture. Editions Quae, Versailles

Cai M, Liu J, Wei Y (2004) Enhanced biohydrogen production from sewage sludge with alkaline pretreatment. Environ Sci Technol 38(11):3195–3202. https://doi.org/10.1021/es0349204

Cao GL, Ren NQ, Wang AJ, Guo WQ, Xu JF, Liu BF (2010) Effect of lignocellulose-derived inhibitors on growth and hydrogen production by *Thermoanaerobacterium thermosaccharolyticum* W16. Int J Hydrog Energy 35(24):13475–13480. https://doi.org/10.1016/j.ijhydene.2009.11.127

Carbonaro M, Maselli P, Nucara A (2012) Relationship between digestibility and secondary structure of raw and thermally treated legume proteins: a Fourier transform infrared (FT-IR) spectroscopic study. Amino Acids 43(2):911–921. https://doi.org/10.1007/s00726-011-1151-4

Cha M, Chung D, Elkins JG, Guss AM, Westpheling J (2013) Metabolic engineering of *Caldicellulosiruptor bescii* yields increased hydrogen production from lignocellulosic biomass. Biotechnol Biofuels 6(1):1–8. https://doi.org/10.1186/1754-6834-6-85

Chen CC, Lin CY (2003) Using sucrose as a substrate in an anaerobic hydrogen-producing reactor. Adv Environ Res 7(3):695–699. https://doi.org/10.1016/S1093-0191(02)00035-7

Chen Y, Yin Y, Wang J (2021) Recent advance in inhibition of dark fermentative hydrogen production. Int J Hydrog Energy 46(7):5053–5073. https://doi.org/10.1016/j.ijhydene.2020.11.096

Cheng J, Lin R, Song W, Xia A, Zhou J, Cen K (2015) Enhancement of fermentative hydrogen production from hydrolyzed water hyacinth with activated carbon detoxification and bacteria domestication. Int J Hydrog Energy 40(6):2545–2551. https://doi.org/10.1016/j.ijhydene.2014.12.097

Cheng J, Ding L, Lin R, Yue L, Liu J, Zhou J, Cen K (2016) Fermentative biohydrogen and biomethane co-production from mixture of food waste and sewage sludge: Effects of physiochemical properties and mix ratios on fermentation performance. Appl Energy 184:1–8. https://doi.org/10.1016/j.apenergy.2016.10.003

Choi JM, Han SK, Kim JT, Lee CY (2016) Optimization of combined (acid+ thermal) pretreatment for enhanced dark fermentative H_2 production from *Chlorella vulgaris* using response surface methodology. Int Biodeterior Biodegrad 108:191–197. https://doi.org/10.1016/j.ibiod.2015.06.013

Chong ML, Sabaratnam V, Shirai Y, Hassan MA (2009) Biohydrogen production from biomass and industrial wastes by dark fermentation. Int J Hydrog Energy 34(8):3277–3287. https://doi.org/10.1016/j.ijhydene.2009.02.010

Chu CY, Tung L, Lin CY (2013) Effect of substrate concentration and pH on biohydrogen production kinetics from food industry wastewater by mixed culture. Int J Hydrog Energy 38(35):15849–15855. https://doi.org/10.1016/j.ijhydene.2013.07.088

Dauptain K, Trably E, Santa-Catalina G, Bernet N, Carrere H (2020) Role of indigenous bacteria in dark fermentation of organic substrates. Bioresour Technol 313:123665. https://doi.org/10.1016/j.biortech.2020.123665

Dębowski M, Korzeniewska E, Filipkowska Z, Zieliński M, Kwiatkowski R (2014) Possibility of hydrogen production during cheese whey fermentation process by different strains of psychrophilic bacteria. Int J Hydrog Energy 39(5):1972–1978. https://doi.org/10.1016/j.ijhydene.2013.11.082

Ding L, Cheng J, Xia A, Jacob A, Voelklein M, Murphy JD (2016) Co-generation of biohydrogen and biomethane through two-stage batch co-fermentation of macro-and micro-algal biomass. Bioresour Technol 218:224–231. https://doi.org/10.1016/j.biortech.2016.06.092

Dou B, Zhang H, Song Y, Zhao L, Jiang B, He M et al (2019) Hydrogen production from the thermochemical conversion of biomass: issues and challenges. Sustain Energy Fuels 3(2):314–342. https://doi.org/10.1039/C8SE00535D

Eker S, Sarp M (2017) Hydrogen gas production from waste paper by dark fermentation: effects of initial substrate and biomass concentrations. Int J Hydrog Energy 42(4):2562–2568. https://doi.org/10.1016/j.ijhydene.2016.04.020

Elsharnouby O, Hafez H, Nakhla G, El Naggar MH (2013) A critical literature review on biohydrogen production by pure cultures. Int J Hydrog Energy 38(12):4945–4966. https://doi.org/10.1016/j.ijhydene.2013.02.032

Fang HH, Liu H (2002) Effect of pH on hydrogen production from glucose by a mixed culture. Bioresour Technol 82(1):87–93. https://doi.org/10.1016/S0960-8524(01)00110-9

Favaro L, Alibardi L, Lavagnolo MC, Casella S, Basaglia M (2013) Effects of inoculum and indigenous microflora on hydrogen production from the organic fraction of municipal solid waste. Int J Hydrog Energy 38:11774–11779. https://doi.org/10.1016/j.ijhydene.2013.06.137

Ferreira AF, Ortigueira J, Alves L, Gouveia L, Moura P, Silva CM (2013) Energy requirement and CO_2 emissions of bioH_2 production from microalgal biomass. Biomass Bioenergy 49:249–259. https://doi.org/10.1016/j.biombioe.2012.12.033

Ferreira TB, Rego GC, Ramos LR, de Menezes CA, Soares LA, Sakamoto IK et al (2019) HRT control as a strategy to enhance continuous hydrogen production from sugarcane juice under mesophilic and thermophilic conditions in AFBRs. Int J Hydrog Energy 44(36):19719–19729. https://doi.org/10.1016/j.ijhydene.2019.06.050

Ghimire A, Frunzo L, Pirozzi F, Trably E, Escudie R, Lens PN, Esposito G (2015) A review on dark fermentative biohydrogen production from organic biomass: process parameters and use of by-products. Appl Energy 144:73–95. https://doi.org/10.1016/j.apenergy.2015.01.045

Giang TT, Lunprom S, Liao Q, Reungsang A, Salakkam A (2019) Improvement of hydrogen production from *Chlorella* sp. biomass by acid-thermal pretreatment. PeerJ 7:e6637. https://doi.org/10.7717/peerj.6637

Gonzales RR, Sivagurunathan P, Parthiban A, Kim SH (2016) Optimization of substrate concentration of dilute acid hydrolyzate of lignocellulosic biomass in batch hydrogen production. Int Biodeterior Biodegrad 113:22–27. https://doi.org/10.1016/j.ibiod.2016.04.016

Gonzales RR, Kumar G, Sivagurunathan P, Kim SH (2017) Enhancement of hydrogen production by optimization of pH adjustment and separation conditions following dilute acid pretreatment of lignocellulosic biomass. Int J Hydrog Energy 42(45):27502–27511. https://doi.org/10.1016/j.ijhydene.2017.05.021

Gonzales RR, Kim JS, Kim SH (2019) Optimization of dilute acid and enzymatic hydrolysis for dark fermentative hydrogen production from the empty fruit bunch of oil palm. Int J Hydrog Energy 44(4):2191–2202. https://doi.org/10.1016/j.ijhydene.2018.08.022

Gorgec FK, Karapinar I (2019) Biohydrogen production from hydrolyzed waste wheat by dark fermentation in a continuously operated packed bed reactor: the effect of hydraulic retention time. Int J Hydrog Energy 44(1):136–143. https://doi.org/10.1016/j.ijhydene.2018.08.155

Goud RK, Sarkar O, Chiranjeevi P, Venkata Mohan S (2014) Bioaugmentation of potent acidogenic isolates: a strategy for enhancing biohydrogen production at elevated organic load. Bioresour Technol 165:223–232. https://doi.org/10.1016/j.biortech.2014.03.049
Guo L, Li XM, Bo X, Yang Q, Zeng GM, Liao DX, Liu JJ (2008) Impacts of sterilization, microwave and ultrasonication pretreatment on hydrogen producing using waste sludge. Bioresour Technol 99(9):3651–3658. https://doi.org/10.1016/j.biortech.2007.07.026
Guo XM, Trably E, Latrille E, Carrère H, Steyer JP (2010) Hydrogen production from agricultural waste by dark fermentation: a review. Int J Hydrog Energy 35(19):10660–10673. https://doi.org/10.1016/j.ijhydene.2010.03.008
Hagos K, Zong J, Li D, Liu C, Lu X (2017) Anaerobic co-digestion process for biogas production: Progress, challenges and perspectives. Renew Sustain Energy Rev 76:1485–1496. https://doi.org/10.1016/j.rser.2016.11.184
Han W, Ye M, Zhu AJ, Zhao HT, Li YF (2015) Batch dark fermentation from enzymatic hydrolyzed food waste for hydrogen production. Bioresour Technol 191:24–29. https://doi.org/10.1016/j.biortech.2015.04.120
Harirchi S, Wainaina S, Sar T, Nojoumi SA, Parchami M, Parchami M et al (2022) Microbiological insights into anaerobic digestion for biogas, hydrogen or volatile fatty acids (VFAs): a review. Bioengineered 13(3):6521–6557. https://doi.org/10.1080/21655979.2022.2035986
Hassan NS, Jalil AA, Vo DVN, Nabgan W (2020) An overview on the efficiency of biohydrogen production from cellulose. Biomass Convers Biorefin 1–23. https://doi.org/10.1007/s13399-020-01125-x
Hawkes FR, Hussy I, Kyazze G, Dinsdale R, Hawkes DL (2007) Continuous dark fermentative hydrogen production by mesophilic microflora: principles and progress. Int J Hydrog Energy 32(2):172–184. https://doi.org/10.1016/j.ijhydene.2006.08.014
He M, Sun Y, Zou D, Yuan H, Zhu B, Li X, Pang Y (2012) Influence of temperature on hydrolysis acidification of food waste. Procedia Environ Sci 16:85–94. https://doi.org/10.1016/j.proenv.2012.10.012
He L, Huang H, Lei Z, Liu C, Zhang Z (2014) Enhanced hydrogen production from anaerobic fermentation of rice straw pretreated by hydrothermal technology. Bioresour Technol 171:145–151. https://doi.org/10.1016/j.biortech.2014.08.049
Hwang MH, Jang NJ, Hyun SH, Kim IS (2004) Anaerobic bio-hydrogen production from ethanol fermentation: the role of pH. J Biotechnol 111(3):297–309. https://doi.org/10.1016/j.jbiotec.2004.04.024
Infantes D, Del Campo AG, Villaseñor J, Fernández FJ (2011) Influence of pH, temperature and volatile fatty acids on hydrogen production by acidogenic fermentation. Int J Hydrog Energy 36(24):15595–15601. https://doi.org/10.1016/j.ijhydene.2011.09.061
Ivanova G, Rákhely G, Kovács KL (2009) Thermophilic biohydrogen production from energy plants by *Caldicellulosiruptor saccharolyticus* and comparison with related studies. Int J Hydrog Energy 34(9):3659–3670. https://doi.org/10.1016/j.ijhydene.2009.02.082
Jang S, Kim DH, Yun YM, Lee MK, Moon C, Kang WS et al (2015) Hydrogen fermentation of food waste by alkali-shock pretreatment: microbial community analysis and limitation of continuous operation. Bioresour Technol 186:215–222. https://doi.org/10.1016/j.biortech.2015.03.031
Jarunglumlert T, Prommuak C, Putmai N, Pavasant P (2018) Scaling-up bio-hydrogen production from food waste: Feasibilities and challenges. Int J Hydrog Energy 43(2):634–648. https://doi.org/10.1016/j.ijhydene.2017.10.013
Jiang H, Gadow SI, Tanaka Y, Cheng J, Li YY (2015) Improved cellulose conversion to biohydrogen with thermophilic bacteria and characterization of microbial community in continuous bioreactor. Biomass Bioenergy 75:57–64. https://doi.org/10.1016/j.biombioe.2015.02.010
Jiang Y, Dong W, Xin F, Jiang M (2020) Designing synthetic microbial consortia for biofuel production. Trends Biotechnol 38(8):828–831. https://doi.org/10.1016/j.tibtech.2020.02.002
Kargi F, Eren NS, Ozmihci S (2012) Hydrogen gas production from cheese whey powder (CWP) solution by thermophilic dark fermentation. Int J Hydrog Energy 37(3):2260–2266. https://doi.org/10.1016/j.ijhydene.2011.11.018

Kim DH, Kim SH, Shin HS (2009) Hydrogen fermentation of food waste without inoculum addition. Enzym Microb Technol 45:181–187. https://doi.org/10.1016/j.enzmictec.2009.06.013

Kim DH, Wu J, Jeong KW, Kim MS, Shin HS (2011) Natural inducement of hydrogen from food waste by temperature control. Int J Hydrog Energy 36(17):10666–10673. https://doi.org/10.1016/j.ijhydene.2011.05.153

Kim M, Yang Y, Morikawa-Sakura MS, Wang Q, Lee MV, Lee DY et al (2012) Hydrogen production by anaerobic co-digestion of rice straw and sewage sludge. Int J Hydrog Energy 37(4):3142–3149. https://doi.org/10.1016/j.ijhydene.2011.10.116

Kirli B, Karapinar I (2018) The effect of HRT on biohydrogen production from acid hydrolyzed waste wheat in a continuously operated packed bed reactor. Int J Hydrog Energy 43(23):10678–10685. https://doi.org/10.1016/j.ijhydene.2018.01.175

Kovalev AA, Kovalev DA, Nozhevnikova AN, Zhuravleva EA, Katraeva IV, Grigoriev VS, Litti YV (2021) Effect of low digestate recirculation ratio on biofuel and bioenergy recovery in a two-stage anaerobic digestion process. Int J Hydrog Energy 46(80):39688–39699. https://doi.org/10.1016/j.ijhydene.2021.09.239

Kovalev AA, Mikheeva ER, Panchenko V, Katraeva IV, Kovalev DA, Zhuravleva EA, Litti YV (2022a) Optimization of energy production from two-stage mesophilic–thermophilic anaerobic digestion of cheese whey using a response surface methodology approach. Energies 15(23):8928. https://doi.org/10.1016/j.biortech.2015.01.092

Kovalev AA, Kovalev DA, Panchenko VA, Zhuravleva EA, Laikova AA, Shekhurdina SV et al (2022b) Approbation of an innovative method of pretreatment of dark fermentation feedstocks. Int J Hydrog Energy 47(78):33272–33281. https://doi.org/10.1016/j.ijhydene.2022.08.051

Kovalev AA, Kovalev DA, Zhuravleva EA, Katraeva IV, Panchenko V, Fiore U, Litti YV (2022c) Two-stage anaerobic digestion with direct electric stimulation of methanogenesis: The effect of a physical barrier to retain biomass on the surface of a carbon cloth-based biocathode. Renew Energy 181:966–977. https://doi.org/10.1016/j.renene.2021.09.097

Kuang Y, Zhao J, Gao Y, Lu C, Luo S, Sun Y, Zhang D (2020) Enhanced hydrogen production from food waste dark fermentation by potassium ferrate pretreatment. Environ Sci Pollut Res 27(15):18145–18156. https://doi.org/10.1007/s11356-020-08207-3

Kumar G, Bakonyi P, Periyasamy S, Kim SH, Nemestóthy N, Bélafi-Bakó K (2015) Lignocellulose biohydrogen: practical challenges and recent progress. Renew Sustain Energy Rev 44:728–737. https://doi.org/10.1016/j.rser.2015.01.042

Kumar G, Shobana S, Nagarajan D, Lee DJ, Lee KS, Lin CY et al (2018) Biomass based hydrogen production by dark fermentation—recent trends and opportunities for greener processes. Curr Opin Biotechnol 50:136–145. https://doi.org/10.1016/j.copbio.2017.12.024

Lazaro CZ, Perna V, Etchebehere C, Varesche MBA (2014) Sugarcane vinasse as substrate for fermentative hydrogen production: the effects of temperature and substrate concentration. Int J Hydrog Energy 39(12):6407–6418. https://doi.org/10.1016/j.ijhydene.2014.02.058

Levin DB, Chahine R (2010) Challenges for renewable hydrogen production from biomass. Int J Hydrog Energy 35(10):4962–4969. https://doi.org/10.1016/j.ijhydene.2009.08.067

Li C, Fang HH (2007) Fermentative hydrogen production from wastewater and solid wastes by mixed cultures. Crit Rev Environ Sci Technol 37(1):1–39. https://doi.org/10.1080/10643380600729071

Li YC, Nissilä ME, Wu SY, Lin CY, Puhakka JA (2012) Silage as source of bacteria and electrons for dark fermentative hydrogen production. Int J Hydrog Energy 37(20):15518–15524. https://doi.org/10.1016/j.ijhydene.2012.04.060

Lin R, Cheng J, Ding L, Song W, Zhou J, Cen K (2015) Inhibitory effects of furan derivatives and phenolic compounds on dark hydrogen fermentation. Bioresour Technol 196:250–255. https://doi.org/10.1016/j.biortech.2015.07.097

Lin CY, Lay CH, Sung IY, Sen B, Chen CC (2017) Anaerobic hydrogen production from unhydrolyzed mushroom farm waste by indigenous microbiota. J Biosci Bioeng 124(4):425–429. https://doi.org/10.1016/j.jbiosc.2017.05.001

Litti YV, Potekhina MA, Zhuravleva EA, Vishnyakova AV, Gruzdev DS, Kovalev AA et al (2022) Dark fermentative hydrogen production from simple sugars and various wastewaters by a newly isolated *Thermoanaerobacterium thermosaccharolyticum* SP-H2. Int J Hydrog Energy 47(58):24310–24327. https://doi.org/10.1016/j.ijhydene.2022.05.235

Liu G, Shen J (2004) Effects of culture and medium conditions on hydrogen production from starch using anaerobic bacteria. J Biosci Bioeng 98(4):251–256. https://doi.org/10.1016/S1389-1723(04)00277-4

Liu D, Li R, Ji M, Cai Y (2013) Enhanced hydrogen and methane production from sewage sludge by addition of cornstalk in two-stage fermentation process. Asian J Chem 25(12):6535–6539

Liu S, Zhu J, Chen M, Xin W, Yang Z, Kong L (2014) Hydrogen production via catalytic pyrolysis of biomass in a two-stage fixed bed reactor system. Int J Hydrog Energy 39(25):13128–13135. https://doi.org/10.1016/j.ijhydene.2014.06.158

Liu Z, Zhang C, Wang L, He J, Li B, Zhang Y, Xing XH (2015) Effects of furan derivatives on biohydrogen fermentation from wet steam-exploded cornstalk and its microbial community. Bioresour Technol 175:152–159. https://doi.org/10.1016/j.biortech.2014.10.067

Liu CM, Zheng JL, Wu SY, Chu CY (2016) Fermentative hydrogen production potential from washing wastewater of beverage production process. Int J Hydrog Energy 41(7):4466–4473. https://doi.org/10.1016/j.ijhydene.2015.08.079

Liu X, Dandan H, Wu Y, Xu Q, Wang D, Yang Q, Liu Y, Ni B, Wang Q, Li X (2020) Freezing in the presence of nitrite pretreatment enhances hydrogen production from dark fermentation of waste activated sludge. J Clean Prod 248:119305. https://doi.org/10.1016/j.jclepro.2019.119305

Lopez-Hidalgo AM, Sánchez A, De León-Rodríguez A (2017) Simultaneous production of bioethanol and biohydrogen by *Escherichia coli* WDHL using wheat straw hydrolysate as substrate. Fuel 188:19–27. https://doi.org/10.1016/j.fuel.2016.10.022

Łukajtis R, Hołowacz I, Kucharska K, Glinka M, Rybarczyk P, Przyjazny A, Kamiński M (2018) Hydrogen production from biomass using dark fermentation. Renew Sustain Energy Rev 91:665–694. https://doi.org/10.1016/j.rser.2018.04.043

Luo G, Xie L, Zou Z, Zhou Q, Wang JY (2010) Fermentative hydrogen production from cassava stillage by mixed anaerobic microflora: effects of temperature and pH. Appl Energy 87(12):3710–3717. https://doi.org/10.1016/j.apenergy.2010.07.004

Luo G, Karakashev D, Xie L, Zhou Q, Angelidaki I (2011) Long-term effect of inoculum pretreatment on fermentative hydrogen production by repeated batch cultivations: Homoacetogenesis and methanogenesis as competitors to hydrogen production. Biotechnol Bioeng 108(8):1816–1827. https://doi.org/10.1002/bit.23122

Mahata C, Dhar S, Ray S, Das D (2021) Effect of thermal pretreated organic wastes on the dark fermentative hydrogen production using mixed microbial consortia. Fuel 284:119062. https://doi.org/10.1016/j.fuel.2020.119062

Menon V, Rao M (2012) Trends in bioconversion of lignocellulose: biofuels, platform chemicals & biorefinery concept. Prog Energy Combust Sci 38(4):522–550. https://doi.org/10.1016/j.pecs.2012.02.002

Mikheeva ER, Katraeva IV, Kovalev AA, Kovalev DA, Nozhevnikova AN, Panchenko V et al (2021) The start-up of continuous biohydrogen production from cheese whey: Comparison of inoculum pretreatment methods and reactors with moving and fixed polyurethane carriers. Appl Sci 11(2):510. https://doi.org/10.3390/app11020510

Mikheeva ER, Katraeva IV, Kovalev AA, Biryuchkova PD, Zhuravleva EA, Vishnyakova AV, Litti YV (2022a) Pretreatment in vortex layer apparatus boosts dark fermentative hydrogen production from cheese whey. Fermentation 8(12):674. https://doi.org/10.3390/fermentation8120674

Mikheeva ER, Katraeva IV, Kovalev AA, Kovalev DA, Litti YV (2022b) Effects of pretreatment in a vortex layer apparatus on the properties of confectionery wastewater and its dark fermentation. Int J Hydrog Energy 47(55):23165–23174. https://doi.org/10.1016/j.ijhydene.2022.05.183

Mikheeva ER, Katraeva IV, Vorozhtsov DL, Kovalev DA, Kovalev AA, Grigoriev VS, Litti YV (2022c) Dark fermentative biohydrogen production from confectionery wastewater in continuous-flow reactors. Int J Hydrog Energy 47(53):22348–22358. https://doi.org/10.1016/j.ijhydene.2022.05.131

Montingelli ME, Tedesco S, Olabi AG (2015) Biogas production from algal biomass: A review. Renew Sustain Energy Rev 43:961–972. https://doi.org/10.1016/j.rser.2014.11.052

Nagarajan D, Lee DJ, Chang JS (2019) Recent insights into consolidated bioprocessing for lignocellulosic biohydrogen production. Int J Hydrog Energy 44(28):14362–14379. https://doi.org/10.1016/j.ijhydene.2019.03.066

Nagarajan D, Chang JS, Lee DJ (2020) Pretreatment of microalgal biomass for efficient biohydrogen production—Recent insights and future perspectives. Bioresour Technol 302:122871. https://doi.org/10.1016/j.biortech.2020.122871

Ngo TA, Kim MS, Sim SJ (2011) High-yield biohydrogen production from biodiesel manufacturing waste by *Thermotoga neapolitana*. Int J Hydrog Energy 36(10):5836–5842. https://doi.org/10.1016/j.ijhydene.2010.11.057

Nissilä ME, Lay CH, Puhakka JA (2014) Dark fermentative hydrogen production from lignocellulosic hydrolyzates—a review. Biomass Bioenergy 67:145–159. https://doi.org/10.1016/j.biombioe.2014.04.035

Noblecourt A, Christophe G, Larroche C, Fontanille P (2018) Hydrogen production by dark fermentation from pre-fermented depackaging food wastes. Bioresour Technol 247:864–870. https://doi.org/10.1016/j.biortech.2017.09.199

Orozco RL, Redwood MD, Leeke GA, Bahari A, Santos RCD, Macaskie LE (2012) Hydrothermal hydrolysis of starch with CO_2 and detoxification of the hydrolysates with activated carbon for bio-hydrogen fermentation. Int J Hydrog Energy 37(8):6545–6553. https://doi.org/10.1016/j.ijhydene.2012.01.047

Pachapur VL, Sarma SJ, Brar SK, Le Bihan Y, Buelna G, Verma M (2015) Biological hydrogen production using co-culture versus mono-culture system. Environ Technol Rev 4(1):55–70. https://doi.org/10.1080/21622515.2015.1068381

Quéméneur M, Hamelin J, Barakat A, Steyer JP, Carrère H, Trably E (2012a) Inhibition of fermentative hydrogen production by lignocellulose-derived compounds in mixed cultures. Int J Hydrog Energy 37(4):3150–3159. https://doi.org/10.1016/j.ijhydene.2011.11.033

Quéméneur M, Bittel M, Trably E, Dumas C, Fourage L, Ravot G, Steyer JP, Carrère H (2012b) Effect of enzyme addition on fermentative hydrogen production from wheat straw. Int J Hydrog Energy 37:10639–10647. https://doi.org/10.1016/j.ijhydene.2012.04.083

Rabelo CABS, Soares LA, Sakamoto IK, Silva EL, Varesche MBA (2018) Optimization of hydrogen and organic acids productions with autochthonous and allochthonous bacteria from sugarcane bagasse in batch reactors. J Environ Manag 223:952–963. https://doi.org/10.1016/j.jenvman.2018.07.015

Ramprakash B, Incharoensakdi A (2022) Dark fermentative hydrogen production from pretreated garden wastes by *Escherichia coli*. Fuel 310:122217. https://doi.org/10.1016/j.fuel.2021.122217

Rasit N, Idris A, Harun R, Ghani WAWAK (2015) Effects of lipid inhibition on biogas production of anaerobic digestion from oily effluents and sludges: An overview. Renew Sustain Energy Rev 45:351–358. https://doi.org/10.1016/j.rser.2015.01.066

Ren NQ, Guo WQ, Wang XJ, Xiang WS, Liu BF, Wang XZ et al (2008) Effects of different pretreatment methods on fermentation types and dominant bacteria for hydrogen production. Int J Hydrog Energy 33(16):4318–4324. https://doi.org/10.1016/j.ijhydene.2008.06.003

Ren NQ, Zhao L, Chen C, Guo WQ, Cao GL (2016) A review on bioconversion of lignocellulosic biomass to H_2: key challenges and new insights. Bioresour Technol 215:92–99. https://doi.org/10.1016/j.biortech.2016.03.124

Ren Y, Si B, Liu Z, Jiang W, Zhang Y (2021) Promoting dark fermentation for biohydrogen production: Potential roles of iron-based additives. Int J Hydrog Energy 47(3):1499–1515. https://doi.org/10.1016/j.ijhydene.2021.10.137

Renaudie M, Dumas C, Vuilleumier S, Ernst B (2022) New way of valorization of raw coffee silverskin: Biohydrogen and acetate production by dark fermentation without exogenous inoculum. Bioresour Technol Rep 17:100918. https://doi.org/10.1016/j.biteb.2021.100918

Rivero M, Solera R, Perez M (2014) Anaerobic mesophilic co-digestion of sewage sludge with glycerol: Enhanced biohydrogen production. Int J Hydrog Energy 39(6):2481–2488. https://doi.org/10.1016/j.ijhydene.2013.12.006

Rosa PRF, Santos SC, Sakamoto IK, Varesche MBA, Silva EL (2014) Hydrogen production from cheese whey with ethanol-type fermentation: effect of hydraulic retention time on the microbial community composition. Bioresour Technol 161:10–19. https://doi.org/10.1016/j.biortech.2014.03.020

Rosales-Colunga LM, Razo-Flores E, Rodríguez ADL (2012) Fermentation of lactose and its constituent sugars by *Escherichia coli* WDHL: impact on hydrogen production. Bioresour Technol 111:180–184. https://doi.org/10.1016/j.biortech.2012.01.175

Safi C, Charton M, Ursu AV, Laroche C, Zebib B, Pontalier PY, Vaca-Garcia C (2014) Release of hydro-soluble microalgal proteins using mechanical and chemical treatments. Algal Res 3:55–60. https://doi.org/10.1016/j.algal.2013.11.017

Sarangi PK, Nanda S (2020) Biohydrogen production through dark fermentation. Chem Eng Technol 43(4):601–612. https://doi.org/10.1002/ceat.201900452

Sarma SJ, Brar SK, Sydney EB, Le Bihan Y, Buelna G, Soccol CR (2012) Microbial hydrogen production by bioconversion of crude glycerol: A review. Int J Hydrog Energy 37(8):6473–6490. https://doi.org/10.1016/j.ijhydene.2012.01.050

Shi X, Jung KW, Kim DH, Ahn YT, Shin HS (2011) Direct fermentation of *Laminaria japonica* for biohydrogen production by anaerobic mixed cultures. Int J Hydrog Energy 36(10):5857–5864. https://doi.org/10.1016/j.ijhydene.2011.01.125

Shin HS, Youn JH, Kim SH (2004) Hydrogen production from food waste in anaerobic mesophilic and thermophilic acidogenesis. Int J Hydrog Energy 29(13):1355–1363. https://doi.org/10.1016/j.ijhydene.2003.09.011

Sillero L, Solera R, Perez M (2022) Anaerobic co-digestion of sewage sludge, wine vinasse and poultry manure for bio-hydrogen production. Int J Hydrog Energy. https://doi.org/10.1016/j.ijhydene.2021.11.032

Sivagurunathan P, Kumar G, Mudhoo A, Rene ER, Saratale GD, Kobayashi T et al (2017) Fermentative hydrogen production using lignocellulose biomass: An overview of pre-treatment methods, inhibitor effects and detoxification experiences. Renew Sustain Energy Rev 77:28–42. https://doi.org/10.1016/j.rser.2017.03.091

Soares JF, Confortin TC, Todero I, Mayer FD, Mazutti MA (2020) Dark fermentative biohydrogen production from lignocellulosic biomass: technological challenges and future prospects. Renew Sustain Energy Rev 117:109484. https://doi.org/10.1016/j.rser.2019.109484

Sreela-Or C, Plangklang P, Imai T, Reungsang A (2011) Co-digestion of food waste and sludge for hydrogen production by anaerobic mixed cultures: Statistical key factors optimization. Int J Hydrog Energy 36(21):14227–14237. https://doi.org/10.1016/j.ijhydene.2011.05.145

Tang GL, Huang J, Sun ZJ, Tang QQ, Yan CH, Liu GQ (2008) Biohydrogen production from cattle wastewater by enriched anaerobic mixed consortia: influence of fermentation temperature and pH. J Biosci Bioeng 106(1):80–87. https://doi.org/10.1263/jbb.106.80

Thanwised P, Wirojanagud W, Reungsang A (2012) Effect of hydraulic retention time on hydrogen production and chemical oxygen demand removal from tapioca wastewater using anaerobic mixed cultures in anaerobic baffled reactor (ABR). Int J Hydrog Energy 37(20):15503–15510. https://doi.org/10.1016/j.ijhydene.2012.02.068

Toledo-Alarcón J, Capson-Tojo G, Marone A, Paillet F, Júnior ADNF, Chatellard L et al (2018) Basics of bio-hydrogen production by dark fermentation. In: Liao Q, Chang J-s, Herrmann C, Xia A (eds) Bioreactors for microbial biomass and energy conversion. Springer, Singapore, pp 199–220. https://doi.org/10.1007/978-981-10-7677-0_6

Tran TTH, Nguyen PKT (2022) Enhanced hydrogen production from water hyacinth by a combination of ultrasonic-assisted alkaline pretreatment, dark fermentation, and microbial electrolysis cell. Bioresour Technol 357:127340. https://doi.org/10.1016/j.biortech.2022.127340

Trchounian K, Trchounian A (2015) Hydrogen production from glycerol by *Escherichia coli* and other bacteria: an overview and perspectives. Appl Energy 156:174–184. https://doi.org/10.1016/j.apenergy.2015.07.009

Urbaniec K, Bakker RR (2015) Biomass residues as raw material for dark hydrogen fermentation—A review. Int J Hydrog Energy 40(9):3648–3658. https://doi.org/10.1016/j.ijhydene.2015.01.073

Valdez-Vazquez I, Ríos-Leal E, Esparza-García F, Cecchi F, Poggi-Varaldo HM (2005) Semi-continuous solid substrate anaerobic reactors for H_2 production from organic waste: mesophilic versus thermophilic regime. Int J Hydrog Energy 30(13–14):1383–1391. https://doi.org/10.1016/j.ijhydene.2004.09.016

Valdez-Vazquez I, Castillo-Rubio LG, Pérez-Rangel M, Sepúlveda-Gálvez A, Vargas A (2019) Enhanced hydrogen production from lignocellulosic substrates via bioaugmentation with *Clostridium* strains. Ind Crop Prod 137:105–111. https://doi.org/10.1016/j.indcrop.2019.05.023

Van Niel EW, Claassen PA, Stams AJ (2003) Substrate and product inhibition of hydrogen production by the extreme thermophile, *Caldicellulosiruptor saccharolyticus*. Biotechnol Bioeng 81(3):255–262. https://doi.org/10.1002/bit.10463

Vijayaraghavan K, Ahmad D (2006) Biohydrogen generation from palm oil mill effluent using anaerobic contact filter. Int J Hydrog Energy 31(10):1284–1291. https://doi.org/10.1016/j.ijhydene.2005.12.002

Wang J, Wan W (2009) Factors influencing fermentative hydrogen production: a review. Int J Hydrog Energy 34(2):799–811. https://doi.org/10.1016/j.ijhydene.2008.11.015

Weide T, Hernandez Regalado RE, Brügging E, Wichern M, Wetter C (2020) Biohydrogen production via dark fermentation with pig manure and glucose using pH-dependent feeding. Chem Eng Technol 43(8):1578–1587. https://doi.org/10.1002/ceat.201900678

Wu X, Zhu J, Dong C, Miller C, Li Y, Wang L, Yao W (2009) Continuous biohydrogen production from liquid swine manure supplemented with glucose using an anaerobic sequencing batch reactor. Int J Hydrog Energy 34(16):6636–6645. https://doi.org/10.1016/j.ijhydene.2009.06.058

Wu H, Li J, Liao Q, Fu Q, Liu Z (2020) Enhanced biohydrogen and biomethane production from *Chlorella* sp. with hydrothermal treatment. Energy Convers Manag 205:112373. https://doi.org/10.1016/j.enconman.2019.112373

Wu M, Jiang Y, Liu Y, Mou L, Zhang W, Xin F, Jiang M (2021) Microbial application of thermophilic *Thermoanaerobacterium* species in lignocellulosic biorefinery. Appl Microbiol Biotechnol 105(14):5739–5749. https://doi.org/10.1007/s00253-021-11450-4

Xia A, Cheng J, Ding L, Lin R, Huang R, Zhou J, Cen K (2013) Improvement of the energy conversion efficiency of Chlorella pyrenoidosa biomass by a three-stage process comprising dark fermentation, photofermentation, and methanogenesis. Bioresour Technol 146:436–443. https://doi.org/10.1016/j.biortech.2013.07.077

Xia A, Cheng J, Song W, Su H, Ding L, Lin R et al (2015) Fermentative hydrogen production using algal biomass as feedstock. Renew Sustain Energy Rev 51:209–230. https://doi.org/10.1016/j.rser.2015.05.076

Xiao L, Deng Z, Fung KY, Ng KM (2013) Biohydrogen generation from anaerobic digestion of food waste. Int J Hydrog Energy 38(32):13907–13913. https://doi.org/10.1016/j.ijhydene.2013.08.072

Xiao N, Chen Y, Chen A, Feng L (2014) Enhanced bio-hydrogen production from protein wastewater by altering protein structure and amino acids acidification type. Sci Rep 4(1):1–9. https://doi.org/10.1038/srep03992

Xing Y, Li Z, Fan Y, Hou H (2010) Biohydrogen production from dairy manures with acidification pretreatment by anaerobic fermentation. Environ Sci Pollut Res 17(2):392–399. https://doi.org/10.1007/s11356-009-0187-4

Yang G, Wang J (2017) Co-fermentation of sewage sludge with ryegrass for enhancing hydrogen production: Performance evaluation and kinetic analysis. Bioresour Technol 243:1027–1036. https://doi.org/10.1016/j.biortech.2017.07.087

Yun YM, Lee MK, Im SW, Marone A, Trably E, Shin SR et al (2018) Biohydrogen production from food waste: current status, limitations, and future perspectives. Bioresour Technol 248:79–87. https://doi.org/10.1016/j.biortech.2017.06.107

Zabed HM, Akter S, Yun J, Zhang G, Awad FN, Qi X, Sahu JN (2019) Recent advances in biological pretreatment of microalgae and lignocellulosic biomass for biofuel production. Renew Sustain Energy Rev 105:105–128. https://doi.org/10.1016/j.rser.2019.01.048

Zagrodnik R, Duber A, Seifert K (2022) Dark-fermentative hydrogen production from synthetic lignocellulose hydrolysate by a mixed bacterial culture: the relationship between hydraulic retention time and pH conditions. Bioresour Technol 358:127309. https://doi.org/10.1016/j.biortech.2022.127309

Zhang J, Zhang W, Cai Z, Zhang J, Guan D, Ji D, Gao W (2022) Effect of ammonia fiber expansion combined with NaOH pretreatment on the resource efficiency of herbaceous and woody lignocellulosic biomass. ACS Omega 7(22):18761–18769. https://doi.org/10.1021/acsomega.2c01302

Zhao C, Shao Q, Chundawat SP (2020) Recent advances on ammonia-based pretreatments of lignocellulosic biomass. Bioresour Technol 298:122446. https://doi.org/10.1016/j.biortech.2019.122446

Zheng Y, Shi J, Tu M, Cheng YS (2017) Principles and development of lignocellulosic biomass pretreatment for biofuels. Adv Bioenergy 2:1–68. https://doi.org/10.1016/bs.aibe.2017.03.001

Zhou P, Elbeshbishy E, Nakhla G (2013) Optimization of biological hydrogen production for anaerobic co-digestion of food waste and wastewater biosolids. Bioresour Technol 130:710–718. https://doi.org/10.1016/j.biortech.2012.12.069

Chapter 9
Biomass Immobilization in Biohydrogen Production

E. R. Mikheeva, I. V. Katraeva, A. A. Kovalev, and Yu. V. Litti

Abstract In recent years, an enormous amount of research effort has been devoted to the study of immobilization methods and carrier materials for enhancement of the dark fermentation process. It is shown that immobilization has a complex multifactorial effect on the process of dark fermentation and opens up new technological possibilities, such as better acclimatization of hydrogen producers, a decrease in the lag phase, and an increase in biomass density, which provides greater resistance to inhibitory substances and shock loads. In addition, the use of carriers allows a wider variation of the main technological parameters, such as hydraulic retention time (HRT) and organic loading rate (OLR), and more flexible control. The chapter highlights the main immobilization methods, and their advantages and disadvantages. Parameters affecting attached growth systems, such as type of material, biofilm thickness, pH, substrate characteristics, are discussed. Particular attention is paid to materials that have a complex stimulating effect, such as conductive materials, as well as materials containing trace elements and nanoparticles that affect cell growth, metabolism, and enzyme activation. A comparative analysis of various types of biofilm reactors, their design, and operating modes is given. In conclusion, an assessment is given of the prospects for using immobilization to develop efficient, economical, and stable technologies for producing dark fermentative biohydrogen.

E. R. Mikheeva
Federal State Autonomous Educational Institution of Higher Education "Lobachevsky State University of Nizhni Novgorod", Nizhny Novgorod, Russia

I. V. Katraeva
Federal State Budgetary Educational Institution of Higher Education "Nizhny Novgorod State University of Architecture and Civil Engineering", Nizhny Novgorod, Russia

A. A. Kovalev
Federal State Budgetary Scientific Institution "Federal Scientific Agroengineering Center VIM", Moscow, Russia

Y. V. Litti (✉)
Federal Research Center "Fundamentals of Biotechnology" of the Russian Academy of Sciences, Moscow, Russia

C. R. Soccol et al. (eds.), *Biohydrogen - Advances and Processes*, Biofuel and Biorefinery Technologies 13, https://doi.org/10.1007/978-3-031-49818-3_9

Keywords Biohydrogen · Immobilization method · Immobilization material · Dark fermentation · Reactor configuration · Process parameters

9.1 Introduction

Cell immobilization technology has existed for many decades and is widely used in various production processes in the food, pharmaceutical, and wastewater treatment industries. In recent years there has been considerable interest in using microbial immobilization technology for the dark fermentative (DF) biohydrogen production process due to its advantages, such as high substrate conversion, high metabolic activity, shortened lag phase, increased cell density, ease of handling, reusability, better solid/liquid separation efficiency, and better operational stability, including the prevention of microbial wash-out, pH stability, and high biomass activity in continuous processes (Sekoai et al. 2018). It is noted that systems with immobilized biomass work well with substrates containing both easily biodegradable and recalcitrant compounds, and allow the dark fermentation process to be carried out both at much lower hydraulic retention time (HRT) and higher organic loading rate (OLR) (Mohan et al. 2008; Chen et al. 2021). A particular advantage of the use of immobilization is the ability to cultivate bacteria that have a low growth rate and have unique metabolic abilities, such as the decomposition of xenobiotics. In addition, the immobilization method to a large extent makes it possible to protect cells from environmental/chemical toxicity (Rai et al. 2022), to increase the survival of immobilized microorganisms under extreme conditions, after long-term storage, and under conditions of metabolic stress. It is noted that attached growth systems are more resistant to changes in process parameters and less energy intensive (Mohan et al. 2008; Kumar et al. 2016). The immobilized biomass technology can be used in various types of DF reactors, such as continuous stirred tank reactor, fluidized bed reactor, carrier-induced granular sludge bed reactor, packed bed reactor, fixed bed reactor, structured bed reactor, upflow anaerobic sludge bed reactor, and trickling biofilter. Various biological materials (alginate, agar, chitosan, cellulose, carrageenan, activated carbon (AC), etc.) or synthetic materials (acrylamide, polyurethane, polyaniline (PANI), polyvinyl alcohol (PVA), polyethylene glycol (PEG), etc.) can be used to immobilize microorganisms (Sekoai et al. 2018; Christopher et al. 2021).

In recent years, a large number of works has been directed to an in-depth study of immobilization methods and the selection of the best carrier materials to increase the productivity and efficiency of the DF process (Lu et al. 2018). The chapter provides information on various approaches to the implementation of DF technology with immobilization of microorganisms, parameters affecting the production of biohydrogen, types of reactors utilizing immobilized biomass, design, and operating modes, as well as prospects for further development and industrial application of technology.

9.2 Methods of Immobilization of Hydrogen-Producing Bacteria

There are two strategies for the immobilization of microorganisms: one is active (forced formation) and the other is passive (natural formation). The passive or self-immobilization approach has been marked by biofilm formation, while the active approach has been described as trapping the cultured cells in various polymers such as alginate, agar, polyethylene glycol, etc. (Güngörmüşler et al. 2021; Shuttleworth et al. 2014; Chen et al. 2021). For the production of biohydrogen by DF, four immobilization methods are mainly used: adsorption, entrapment, encapsulation, and polymer and nano-based immobilization (Fig. 9.1 modified from Sekoai et al. (2018)).

The results of some studies on the production of biohydrogen using various methods of hydrogen-producing bacteria (HPB) immobilization are presented in Table 9.1.

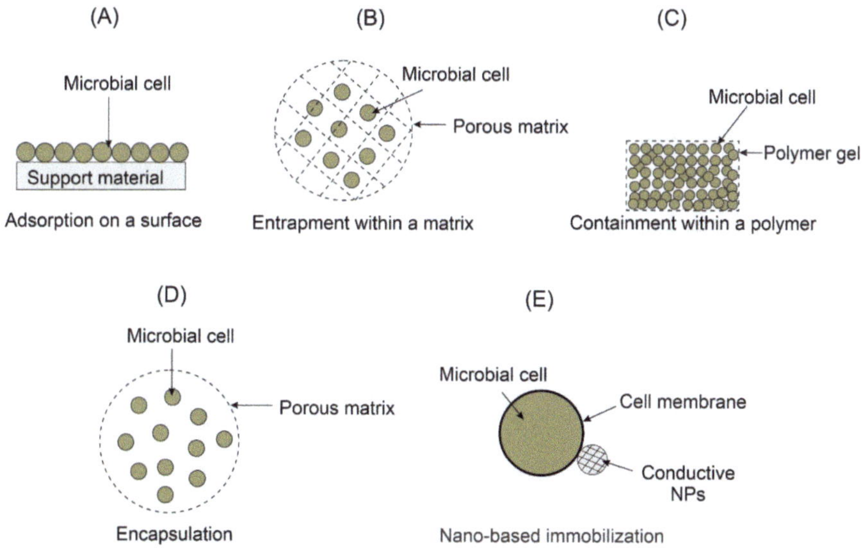

Fig. 9.1 Methods of immobilization of hydrogen-producing bacteria (HPB): (**a**) Adsorption on a surface, (**b**) Entrapment within a matrix, (**c**) Containment within a polymer, (**d**) Encapsulation, (**e**) Nano-based immobolization

Table 9.1 Types of immobilization techniques used in biohydrogen production process

Method	Microorganism	Carrier material	Substrate	Improvements	H_2 production rate	H_2 yield	Reference
Entrapment	Mixed culture	Alginate + chitosan + titanium oxide	Sucrose	Threefold H_2 production increase	21.3 mmol/L/h	–	(Wu et al. 2002)
Entrapment	Mixed culture	Calcium alginate + activated carbon	Sucrose	Twofold H_2 production increase	–	2.6 Mol H_2/Mol sucrose	(Wu et al. 2006)
Entrapment	*Clostridium LS2*	Polyethylene glycol	POME	H_2 increased at short HRT of 16 h and 90% substrate conversion efficiency achieved	–	0.35 L H_2/g Chemical oxygen demand (COD)	(Singh et al. 2013)
Entrapment	*Clostridium intestinale* URNW	Calcium-alginate	Cellobiose	Higher volumetric production rates and twice the H_2 molar yields compared to suspended cultures	–	3.57 Mol H_2/Mol cellobiose	(Güngörmüşler et al. 2021)
Entrapment	Mixed culture	PVA-alginate	Glycerol	Immobilization enhanced cumulative H_2 production and H_2 yield by 611.1 and 79.3%, respectively	–	0.52 Mol H_2/Mol glycerol	(Chen et al. 2021)
Adsorption	*Enterobacter aerogenes*	Carbon fiber and activated carbon	Glucose	23 and 34% enhancement of H_2 yield and production rate, respectively, compared to the free E. Aerogenes	2.48 L/L/h	2.56 Mol/Mol glucose	(Boshagh et al. 2019b)

Method	Microorganism	Carrier material	Substrate	Improvements	H_2 production rate	H_2 yield	Reference
Adsorption	Mixed culture (*Lachnospiraceae* was enriched)	Biochar	Pretreated sugarcane bagasse	Biochar boosted biohydrogen production by 317.1% through stimulating bacterial growth, improving critical enzymatic activities, manipulating the ratio of NADH/NAD+ and enhancing electron transfer efficiency of fermentation system	120.9 mL/L/h	2723.8 mL/L	(Bu et al. 2021)
Adsorption	Mixed culture	Activated carbon	Molasses	Maximum H_2 production at short HRT (5 h) and H_2 content of 62%, pH was stabilized at 4.60	12.27 mmol/L/h	–	(Wang et al. 2013)
Adsorption	Mixed culture	GAC	Sucrose	Repeated batch processes enhanced H_2 production	2.7 mmol/L/h	2.8 Mol H_2/Mol hexose	(Lutpi et al. 2016)
Adsorption	Mixed culture	Polyethylene terephthalate microplastics	Glucose	H_2 production rates were 21.6% (mesophilic) and 63.6% (thermophilic) higher than control	4.85 L/L/d (mesophilic) and 3.6 L/L/d (thermophilic)	–	(Ren et al. 2022)
Nano-based immobilization	Mixed culture	Silver nanoparticles	Glucose	Silver nanoparticles increased H_2 production	–	2.48 Mol H_2/Mol glucose	(Zhao et al. 2013)
Nano-based immobilization	Mixed culture	Nickel nanoparticles	Glucose	RSM was used to optimize nickel nanoparticles for maximum H_2 production	–	2.54 Mol H_2/Mol glucose	(Mullai et al. 2013)

(continued)

Table 9.1 (continued)

Method	Microorganism	Carrier material	Substrate	Improvements	H_2 production rate	H_2 yield	Reference
Nano-based immobilization	*Enterobacter aerogenes* ZJU1	Magnetite nanoparticles	Glucose	Increased ratio of NADH/NAD+ and the activities of hydrogenase and electron transport system (ETS) of E. Aerogenes ZJU1; hydrogen production increased by 21.1%	–	260.9 mL/g glucose	(Cheng et al. 2020)
Entrapment+ Nano-based immobilization	*Clostridium pasteurianum* CH5	K-carrageenan, alginic acid, and PVA + nano-metal nickel (npeni) and iron (npefe)	Glucose	H2 content in biogas was increased by 42–46.9%	–	Maximum 1.28 Mol H_2/Mol substrate	(Wannapokin et al. 2021)
Adsorption+ Nano-based immobilization	Mixed culture	Fe-modified zeolite	Glucose	Up to three times enhanced biohydrogen production	1893 mL/L	–	(Zhao et al. 2021)
Encapsulation	Mixed culture	Polyvinylidene fluoride	Hexanal	H_2 increased by 2.8 times	–	189 mL H_2/g COD	(Akinbomi et al. 2015)
Encapsulation	Mixed culture	Polyvinylidene fluoride	Octanol	H_2 increased by 2.25 times	–	198 mL H_2/g COD	(Akinbomi et al. 2015)
Encapsulation	Mixed culture	Polyvinylidene fluoride	Myrcene	H_2 increased by 1.3 times	–	179 mL H_2/g COD	(Akinbomi et al. 2015)

«–» not available, *POME* palm oil mill effluent, *GAC* granular activated carbon, *RME* rice mill effluent, *PVA* Polyvinyl alcohol, *RSM* response surface methodology

9.2.1 Immobilization Based on Adsorption and Attachment

Adsorption is one of the most commonly used methods for cell immobilization. During the adsorption process, bacterial cells are attached to the surface of an organic or inorganic support matrix, and there is a direct interaction between the carrier matrix and microbial cells inside the reactor medium. Adsorption process provides better mass transfer efficiency, stable biomass, better biomass retaining capacity, and substrate utilization under short HRT with enhanced hydrogen production performance. The choice and selection of matrix is crucial for cell adhesion and biofilm formation properties. The governing factors for cell adhesion on the support matrix include: cell surface charge of the bacteria, ionic nature of the matrix, electrostatic forces such as Van der Waals force, ionic, hydrophilic, and hydrophobic interactions (Kumar et al. 2016). The main requirements for the carrier matrix are: high porosity, large specific surface area, and high mechanical strength (Boshagh et al. 2019a; Liu et al. 2019).

Activated carbon (Jamali et al. 2021), biochar (Li et al. 2020a), alginate (Anjana and Kaushik 2014; Ma et al. 2017; Güngörmüşler et al. 2021), sponge (Kirli and Kapdan 2016), agar (Ishikawa et al. 2008), ceramics (Keskin et al. 2012), polyurethane foam (Mikheeva et al. 2021), lignocellulosic waste (Liu et al. 2019; Rai et al. 2022), microplastic (Ren et al. 2022), and carbon nanotubes (Boshagh et al. 2019a) are among the most studied biomass carrier materials for DF process. As examples of the effectiveness of immobilizing carrier materials for the production of biohydrogen, one can refer to the results of studies in which the production of biohydrogen using a porous glass carrier was two times higher than during DF with suspended biomass (Guevara-López and Buitrón 2015), and when using plastic sponges coated with metal grid, as an immobilizing material, the yield of hydrogen was 2.1 mol/mol glucose (Kirli and Kapdan 2016). The choice of carrier material is critical to obtaining stable hydrogen production performance with the adsorption-based immobilization method. Thus, weaker interaction forces between the adsorbent and bacterial cells for certain immobilizing materials can lead to washout of the active biomass, a decrease in the ability to form a biofilm, and failure of the system as a whole (Ma et al. 2017). For example, studies have shown that corn stalks have not only better physical and adsorption capacity, but also better biocompatibility due to the high content of hydroxyl groups compared to other adsorbent materials, maintaining high microbial activity and constant renewal of microbial cells were observed during their use, which made it possible to increase the yield of hydrogen by 29.3% with an increase in OLR from 10 g/L to 15 g/L (Ma et al. 2017). Poplar biomass has also been shown to be effective in immobilizing a well-balanced composition of HPB without washing out the high-yielding bacterium from the consortium, as is often the case with mixed cultures (Patel et al. 2021). The cell concentration during immobilized DF was about one order magnitude higher than that during planctonic cell DF when using the porous particulate carriers of activated carbon, bagasse, and brick for *Clostridium acetobutylicum* immobilization for co-production of hydrogen and butanol (Liu et al. 2019). Use of low-density polyethylene cylinders in an anaerobic structured-bed reactor (ASTBR) was successful in achieving 1.6 ± 0.7 L H_2 /L/d and an average hydrogen yield (HY) of 1.4 ± 0.7 mol H_2/mol of consumed lactose at OLR of 24 kg COD/m^3/d (chemical oxygen demand) and at HRT of 24 h during DF of cheese whey (Blanco et al. 2019). Aquarium biological sponge was used to immobilize heat-treated anaerobic sludge in a packed bed reactor to produce dark fermentative biohydrogen from hydrolyzed waste wheat with the highest hydrogen yield of 645.7 mL H_2/g total solids (TS) and hydrogen production rate of 2.51 L H_2/L/d (Gorgec and Karapinar 2019).

Nakatani et al. recently proposed a fast, easy method for immobilizing *Enterobacter aerogenes* by expressing AtaA, which encodes the adhesive protein of *Acinetobacter* sp. Tol 5. AtaA protein on the *E. aerogenes* cells carrying the AtaA gene was demonstrated by immunoblotting and flow cytometry. The AtaA-producing cells exhibited stronger adherence and auto-agglutination characteristics than wild-type cells, and were successfully immobilized (at approximately 2.5 mg/cm^3) on polyurethane foam. The AtaA production achieved simple and immediate immobilization of *E. aerogenes* on the foam, enabling repetitive and continuous hydrogen production (Nakatani et al. 2018).

9.2.2 Entrapment-Based Immobilization

Entrapment-based immobilization is one of the simplest and most common immobilization methods, in which cells of HPB are trapped inside a support matrix (Fig. 9.1). The carrier material creates a barrier around the cells, ensuring their protection and long-term use (Kumar et al. 2017). For an effective DF process, the carrier material must have good permeability, which will facilitate the diffusion of nutrients from the digestion medium to HPB. This method is implemented using various immobilizing materials such as agar, alginate, cellulose, carrageenan, polyacrylamide, polyurethane, polyvinyl, and polypropylene (Martins et al. 2013). However, at high concentrations of cations (K^+, Ca^{2+}, Mg^{2+}), alginate does not retain its structure (it liquefies). Therefore, it has been proposed to modify alginate granules to improve their mechanical strength. For example, a threefold increase in the production of biohydrogen was obtained after modification of alginate granules with chitosan and titanium oxide (Wu et al. 2006), and a twofold increase in the addition of calcium alginate granules with activated carbon (Wu et al. 2002). Singh et al. achieved an increase in biohydrogen yield during continuous DF (0.38 L H_2/g COD) at 16 h HRT by using polyethylene glycol to immobilize *Clostridium* LS2 cells (Singh et al. 2013). Immobilization with poly (vinyl alcohol)-alginate enhanced cumulative hydrogen production and hydrogen yield by 61.1 and 79.3%. In addition, immobilized microorganisms showed stronger tolerance to high substrate concentration and higher capability in glycerol utilization (Chen et al. 2021). The immobilization of anaerobic sludge in sodium alginate (SA) with polyaniline (PANI) nanoparticles (NPs) enhanced the degradation of organic matter in dairy wastewater into volatile fatty acids (VFAs) and improved biohydrogen yield.

Authors evidenced a direct positive relationship between PANI NPs concentration (5, 10, 20, 40 mg/L) in SA, degradation of organic matter, and bio-hydrogen production (Hellal et al. 2022). Ismail et al. reported an increase in the rate of biohydrogen production (2.1 L/L/day) from palm oil wastewater using polydimethylsiloxane-immobilized cells at a short HRT of 2 h (Ismail et al. 2011). In another study, a high hydrogen yield of 2.1 mol/mol glucose was obtained using alginate beads in which metal ions were incorporated (Kumar et al. 2016; Seelert et al. 2015). It was also shown that the additional introduction of metal cations, which play an important role for biohydrogen-producing enzymes, in particular dehydrogenase, contributes to better absorption of nutrients (Wong et al. 2014). The effects of various natural waste matrices (coir coir-CC, wood chips-WS) and low-cost matrices (foam-FM and alginate), as well as semiconducting TiO_2 nanoparticles incorporated in sodium alginate, were evaluated to increase the biohydrogen potential by Rai et al. Cell adsorption and cell entrapment methods were used for immobilization of *Bacillus licheniformis* AP1. Overall, the authors concluded that the cell absorption method was more efficient in comparison to the cell entrapment method (Rai et al. 2022).

9.2.3 Polymer-Based Immobilization

Polymer-based immobilization refers to a category of entrapment method of cell immobilization in which the stability of the immobilized cells has been improved by incorporating bacterial cells into a polymer matrix (5, 37–38). Matrices of polymeric materials that immobilize active biomass, in contrast to natural polysaccharides such as, for example, alginate, are more stable due to their hardness, strength, and non-biodegradable nature.

Examples of synthetic polymers used are polyacrylamide, polyurethane, polyvinyl alcohol, polyethylene glycol, polyvinylidene fluoride, polyethylene, polyvinyl chloride, etc. Synthetic polymers have been used in many studies to increase biohydrogen production. For example, Singh et al. obtained the maximum yield of biohydrogen (0.31 L H_2/g COD) using *Clostridium LS2* cells captured by polyethylene glycol gel (Singh et al. 2013). When comparing biological (ramie and loofah) and synthetic polymers (acrylic, polyethylene, and polyvinyl chloride), a twofold increase in the yield of biohydrogen was achieved using polyethylene (Wongthanate and Polprasert 2015). Barros and Silva, in their experiment on continuous dark fermentation to produce biohydrogen in anaerobic fluidized bed reactors (AFBR), evaluated three polymeric materials: polystyrene—R1, grounded tire—R2, and polyethylene terephthalate (PET)—R3. The best result was obtained using a grounded tire (60% of H_2 content in biogas and biohydrogen yield of 2.11 mol H_2/mol glucose) (Barros and Silva 2012). Using *Clostridium tyrobutyricum* JM1 cells immobilized on polyurethane foam, Jo et al. obtained the maximum biohydrogen production rate of 7.2 L H_2/L/d at HRT of 2 h with hydrogen content in biogas of 50% and a substrate decomposition efficiency of 97.4% (Jo et al. 2008). Mitchell et al. succeeded in increasing the yield of biohydrogen in a batch DF process using *Clostridium tyrobutyricum ATCC 25755* cells immobilized on a polyurethane matrix (Mitchell et al. 2009). Other synthetic immobilizing materials, such as polyacrylonitrile (PAN), have also been used to produce biohydrogen. Lie et al. obtained a good effect by immobilizing HPB cells on a polyacrylonitrile composite nanofiber mesh filled with carbon nanotubes (Li et al. 2014). Overall, the use of synthetic polymers could pave the way for large-scale biohydrogen production due to their versatility and mechanical stability.

9.2.4 Encapsulation-Based Immobilization

Encapsulation-based immobilization or microencapsulation involves the encapsulation of microorganisms in polymer microspheres ranging in size from 1 to 1000 μm (Fig. 9.1), while the polymer shell is a semi-permeable membrane that allows exchange with the medium. Unlike the other containment types of immobilization, i.e., entrapment, the encapsulated cells are retained inside the carrier material and the prevention of cell release into the medium aids in better downstream processing of the effluent (Kumar et al. 2016).

Encapsulation-based immobilization is a versatile technology that is used to encapsulate a wide range of products, such as pharmaceuticals, fragrances, volatile oils, herbal extracts, enzymes, etc. In recent decades, this technology has also been applied in the field of microbial cell immobilization due to its many advantages over other methods, such as higher cell loading capacity, increased cell survival, and high production rate of desired microbial products. Enclosing microbial cells inside a semipermeable polymeric matrix makes it possible to physically isolate the cells from the external environment, while maintaining a favorable internal microenvironment (Rathore et al. 2013).

Studies have shown that anaerobic bacteria captured by polyvinylidene fluoride provide a higher yield of bio-H_2 compared to non-encapsulated biomass during DF of hexanal, myrcene, and octanol (compounds present in fruits). The authors suggested that the encapsulation of HPB can effectively protect them from inhibitory compounds present in the reaction medium (Akinbomi et al. 2015).

A number of authors report interesting solutions for encapsulating HPB cells using TiO_2 (Stojkovic et al. 2015) and polyvinyl alcohol (Chen et al. 2021). Ta et al. encapsulated hydrogenic and methanogenic bacteria within κ-carrageenan and gelatin (2%/2% w/w) for use in a single-stage dark fermentation process to generate hydrogen-enriched methane. Adjusting the biomass ratio of the two encapsulated bacteria allowed control of the hydrogen produced, with hydrogen and methane generated at 64.6 and 395 mL/L/d, respectively (Ta et al. 2020). Recently, activated carbon, bagasse, and bricks have been used to encapsulate *Clostridium acetobutylicum*, increasing the concentration of cells in the medium and leading to higher glucose uptake (Liu et al. 2019).

9.2.5 Nano-Based Immobilization

In recent years, the growing interest among scientists in nanotechnology and its products with unique properties has led to an increase in the number of studies devoted to stimulating the DF process with nanoparticles and nanocomposites in order to increase biohydrogen yield (Kumar et al. 2015). A number of authors have noted the advantages of using nanoparticles as biocatalysts that contribute to changing biochemical pathways and subsequently positively affecting the final results of dark fermentation in terms of the resulting products (Shuttleworth et al. 2014; Dahiya et al. 2021; Pugazhendhi et al. 2019).

Nanosized particles (1–100 nm) have unique physicochemical properties, such as high activity, electrical conductivity, large reaction surface area, high selectivity, durability, and the ability to increase microbial growth during dark fermentation (Jadhav et al. 2021). It is noted that nanomaterials can act as oxygen scavengers and reduce the redox potential in enzymatic processes (Pugazhendhi et al. 2019). Figure 9.2 shows the nanomaterials that were used in the dark fermentation stimulation process (Pugazhendhi et al. 2019).

Nanomaterials strongly affect the metabolic activity of a microorganism, stimulating efficient electron (e^-) transfer to acceptors, increasing productivity and

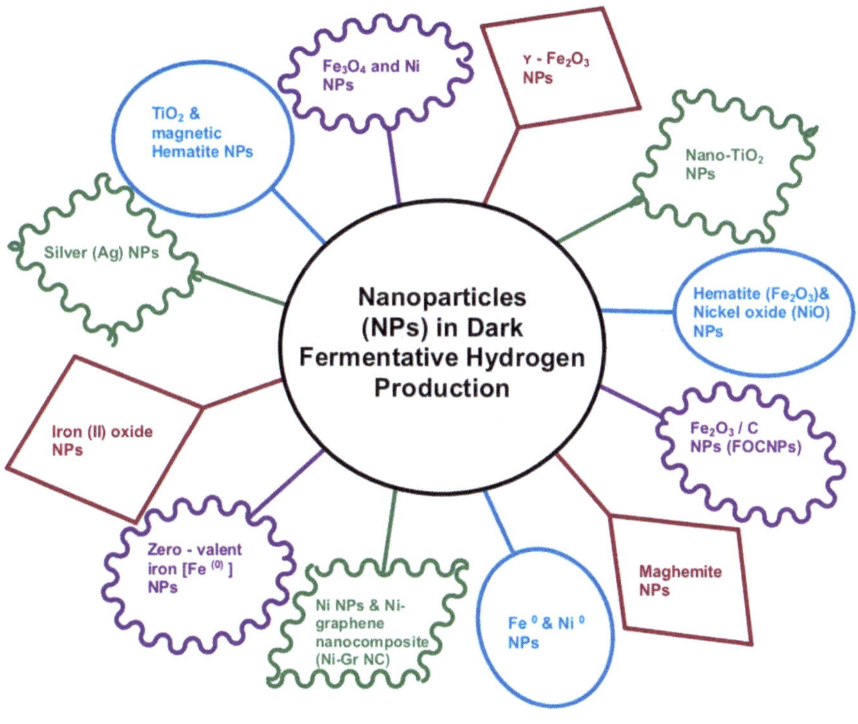

Fig. 9.2 Nanoparticles (NPs) in dark fermentative hydrogen production

biohydrogen yield (Patel et al. 2018). The presence of nanomaterials supports the dark enzymatic production of biohydrogen by creating anoxic conditions, enables rapid transfer of e^- from e^- donors to the corresponding e^- acceptors, and improves the reaction kinetics and thus biohydrogen yield (Sambusiti et al. 2015).

In general, the immobilization of microorganisms on nanomaterials supports the growth and metabolism of microorganisms, improves the thermal and pH stability, and reduces the lag phase of the microorganism and thereby significantly contributes to an increase in cumulative biohydrogen production (Kumar et al. 2019; Engliman et al. 2017). In addition, nanoparticles such as iron and nickel act as active site cofactors for enzymes (hydrogenases and nitrogenases), which greatly increases the biohydrogen yield (Shanmugam et al. 2020a).

The main disadvantages of the method today are the high cost of preparation, inhibition of processes at incorrectly selected concentrations, and the impossibility of reusing nanomaterials at each stage of the process. Although this problem can be solved thanks to the green synthesis of nanoparticles, nevertheless, it has its limitations (Srivastava et al. 2020). More detailed examples of the use of various nanomaterials to increase the yield of hydrogen during dark fermentation are presented in Sect. 9.3.

9.3 Parameters Affecting Biohydrogen Production in Attached Growth Systems

Biohydrogen production by immobilized HPB is affected by various factors, including carrier material, immobilization method, pH of the fermentation medium, carbon source used, hydraulic retention time, mass transfer pattern and reactor design, and finally the microorganisms used to produce hydrogen.

9.3.1 pH of the Fermentation Medium

pH is one of the most important parameters in the biohydrogen production process. It affects hydrogenase activity, metabolic activity, and substrate hydrolysis (De Gioannis et al. 2013). Numerous facultative organisms generate hydrogen by the oxidation of glucose to pyruvate by glycolysis. The end metabolite produced by the oxidation of pyruvate determines the yield of hydrogen (Antonopoulou et al. 2008). It is known that each enzyme has its own optimal pH range, at which its activity is maximum; therefore, the activity of hydrogenase, which is the key enzyme for the production of biohydrogen, is determined by the pH of the reaction medium (Dabrock et al. 1992).

An increase in the concentration of acid in the fermentation medium leads to a decrease in the pH of the medium. This, in turn, causes a shift in metabolism toward the conversion of acids to alcohols to reduce toxicity caused by acid accumulation (Khanal et al. 2004). In addition, the inhibitory effect occurs at low pH values, when non-polar non-dissociated acids present in the medium penetrate the cell walls. If the pH falls below 5.0, then hydrogen production decreases due to the formation of acidic metabolites that affect the ability of cells to maintain internal pH (Bowles and Ellefson 1985).

Many studies have evaluated the effect of pH on the performance of DF reactors with immobilized cultures. Wang et al. reported the maximum yield of biohydrogen (12.27 mmol/L/h) at pH 4.4–4.5 using activated carbon as a carrier material. It was also shown that pH and HRT are interdependent parameters, and a short HRT (5 h), along with a low pH of 4.4–4.5, inhibits the activity of biohydrogen consuming methanogenic archaea and, therefore, allows obtaining the maximum biohydrogen yield (Wang et al. 2013). A pH value of 6 was optimal for biofilm development on granular activated carbon (GAC) with a hydrogen production rate (HPR) of 2.8 mmol H_2/L/h during thermophilic biohydrogen production from palm oil mill effluent (POME) (Jamali et al. 2019). Keskin et al. achieved a fivefold increase in biohydrogen production while maintaining pH at 4.5–5.0 in a continuous dark fermentation process using ceramic balls as an immobilizing carrier (Keskin et al. 2012). Ren et al. reported successful continuous biohydrogen production using granular activated carbon (GAC) at pH 6 (Ren et al. 2010). Penniston et al. obtained a maximum hydrogen content in biogas of 35% with pH adjustment, while the peak hydrogen content observed without pH adjustment was 22%. The process was carried out using a mixed consortium of anaerobic bacteria immobilized with alginate. The authors also noted that the rate of pH change was lower for immobilized cells compared to suspended cells, indicating a better buffering capacity in the immobilized system under conditions without pH adjustment (Penniston and Gueguim Kana 2018). In general, studies have shown that the pH of the medium from 5.5 to 6.8 in many cases turned out to be suitable for obtaining a higher yield of hydrogen (Khan et al. 2018). However, other pH values in dark fermentation systems in the range of 4–9 have also been reported in the literature, depending on many factors including the type of carrier material, substrate used, microbial consortia, and process conditions (Wong et al. 2014).

9.3.2 Carbon Source

The substrate is one of the main cost items in the production of biohydrogen. Simple sugars such as glucose, maltose, xylose, sucrose, and other hexoses are widely used in research on the dark fermentation process with the production of biohydrogen using various immobilization methods (some examples are given in Table 9.1). However, these substrates are expensive and might escalate the process costs on a large scale (Ma et al. 2017).

Currently, many researchers are pursuing economically viable feedstocks such as biowaste effluents, with the criteria being availability, high nutrient content, and the ability to produce clean energy while reducing pollution (Sekoai et al. 2018). The chemical oxygen demand (COD) content plays an important role when using these substrates in the DF process (Li et al. 2020b). Many researchers are incorporating biowaste effluents in their biohydrogen augmentation studies using immobilized biocatalysts. Some examples are given in Table 9.1. POME is considered an ideal feedstock with a high COD value of 79,000 mg/L and a BOD of 49,000 mg/L (Jamali et al. 2019). In the study by Singh et al. *Clostridium butyricum EB6* was immobilized on PEG and used for hydrogen production from POME, and in this case a maximum hydrogen productivity of 22.7 mmol/L/h was obtained with a yield of 5.35 L H_2/L-POME (Litti et al. 2021). Also, carbon sources such as tequila vinasse (Arellano-García et al. 2021), cheese whey (Blanco et al. 2019; Mikheeva et al. 2021), hydrolyzed waste wheat (Gorgec and Karapinar 2019), molasses (Ferreira et al. 2020), oil palm trunk hydrolysate and slaughterhouse wastewater (Khamtib et al. 2021), confectionery wastewater (Litti et al. 2021; Mikheeva et al. 2022a; Mikheeva et al. 2022b), and sugar-beet processing wastewater (Litti et al. 2022) are gaining great popularity for the production of biohydrogen, including in systems with biomass immobilization.

9.3.3 *Properties of Immobilizing Materials*

In works with the immobilization of HPB, the importance of such a parameter as the size and porosity of carrier materials is noted. It is known that with a decrease in the size and density of particles in a fluidized bed, the fluid flow rate decreases, which makes it possible to reduce the energy consumption required for fluidization (Barca et al. 2015). An increase in the layer porosity to 91% provided continuous and cyclic production of hydrogen, while a lower layer porosity led to a reduction in the time of hydrogen production due to the accumulation of biomass, which was accompanied by a decrease in the working volume. At a layer porosity of 50 and 75%, the hydrogen yield reached its maximum value, but with time it decreased to zero values due to the accumulation of biomass in voids, which reduced the working volume of the reactor. The maximum productivity for hydrogen was 0.77, 0.80, and 1.11 L H_2/h at a layer porosity of 50, 75, and 91%, respectively (Fernandes et al. 2013).

Particle size in the immobilized system is an important factor in controlling the surface area, substrate diffusion limitation, and obtaining a sufficient void fraction for both gas extraction and uniform substrate-biomass contact. Gokfiliz and Karapinar showed that the yield of hydrogen with a polyester fiber diameter of 1.5 cm is 2.7 times higher than with a diameter of 3 cm (Gokfiliz and Karapinar 2017). Further work by the authors on the influence of the number of particles showed that the optimal number of polyester fiber particles in each bottle for fermentation is 240 per 200 mL of liquid medium (Gokfiliz-Yildiz and Karapinar 2018). Granules of biological mycelium that are formed spontaneously during the fermentation of *Aspergillus niger* Y3 were investigated as a carrier for the immobilization of *Clostridium* sp. T2 to improve hydrogen production. Batch fermentation tests have shown that the optimal dosage and size of mycelium granules for hydrogen production is 0.350 g per 150 mL of medium and 1.5 mm (Zhao et al. 2012a).

In some studies, such a parameter as the ratio of the diameter of the reactor (D) to the diameter of the carrier material (d) is controlled. It has been shown that D/d ratios above 10 improve the hydraulic performance of the immobilizing agent (Barca et al. 2015).

Surface hydrophobicity has been described as one of the most important factors influencing the bacterial adhesion process (Muri et al. 2018). For example, the Mutag BioChipTM carrier material was hydrophobic, while expanded clay and activated carbon (materials predominantly containing micro- and mesopores) were found to be hydrophilic. Moreover, Muri et al. suggested that the best adsorption is due to the numerous macropores on the surface of the Mutag BioChipTM, as these pores may be one of the decisive parameters influencing the binding of microorganisms to the surface (Muri et al. 2018). Basile et al. suggested that cell affinity could also be affected by charged functional chemical groups on the surfaces of the immobilizing carrier material (Basile et al. 2010). Since bacterial cells, as a rule, have an overall negative charge on the cell wall (Dickson and Koohmaraie 1989), their adsorption will be enhanced in the case of a positively charged surface.

When using positively charged carrier materials that differ in hydrophobicity and size, Muri et al. showed that the carrier material they used affected hydrogen production rate (HPR) and hydrogen yield (HY). Thus, 16.65 mmol H_2/L/h and 1.80 mol H_2/mol glucose, respectively, were achieved in the reactor with Mutag BioChipTM. In the expanded clay reactor, HPR and HY were 15.55 mmol H_2/L/h and 1.74 mol H_2/mol glucose, while in the system with activated carbon, HPR and HY were below 13.65 mmol H_2/L/h and 1.46 mol H_2/mol glucose. However, the authors concluded that the porosity and hydrophobicity of the carrier material did not affect the composition of H_2 in biogas, the content of which for three reactors with Mutag BioChipTM, polyethylene, expanded clay, and activated carbon was 52.3–53.2%, and the conversion of glucose was more than 98%. After 12 days of operation, the reactor filled with Mutag BioChipTM contained the highest amount of attached biomass (0.07 g VSS/g carrier) compared to expanded clay (0.04 g VSS/g carrier) and activated carbon (0.03 g VSS/g carrier). It is believed that this increased amount of biomass attached contributed to an increase in hydrogen production capacity (Muri et al. 2018).

The surface oxidation of carbon fiber (CF) with nitric acid and nitric and sulfuric acid mixture could increase the number of acidic functional groups on the CFs' surface and improve the wettability of the CFs, which may effectively improve the cell's immobilization. Thus, Boshagh et al. used CF and surface-modified CF for immobilization of *Enterobacter aerogenes*. Modifications of CF led to higher hydrogen yields of 2.56 and 2.11 mol/mol glucose compared to 1.83 mol/mol glucose in control treatment (CF) (Boshagh et al. 2019b).

It should be noted that most studies are carried out in a batch mode. When the DF process is carried out in continuous mode, another important parameter of the carrier material that is important is its stability. Immobilizing carriers based on sodium alginate are less stable in long-term tests compared to, for example, polyethylene glycol (PEG). Cells immobilized with PEG are more promising for the continuous production of hydrogen in the dark fermentation of real wastewater (Kumar et al. 2015).

The biomass carrier material can also influence the metabolic pathways of bacteria. Kerčmar and Pintar noted a decrease in the concentration of acetic acid and a transition from ethanol production to butyric acid production approximately on the fourth day of reactor operation when using polyurethane and activated carbon and on the sixth day when using polyethylene and expanded clay as a carrier material (Kerčmar and Pintar 2017).

9.3.4 Immobilizing Materials

Materials selected for immobilization should have the following properties: mechanical, chemical, and thermal stability, non-toxicity, economy, reusability, porous structure, high specific surface, and uniform permeability; they also should be resistant to fermentation products (Kumar et al. 2016; Jamali et al. 2016). The following were studied as immobilizing carriers: granular activated carbon (Jamali et al. 2016; Jamali et al. 2021), biochar (Li et al. 2020a; Bu et al. 2021), polyvinyl alcohol (PVA) (Chen et al. 2021), polyaniniline (Hellal et al. 2022), alginate (Ma et al. 2017; Ferreira et al. 2020; Güngörmüşler2021), agar (Ishikawa et al. 2008), polyester fibers (Gokfiliz and Karapinar 2017), expanded clay (De Amorim et al. 2009; Muri et al. 2018; Kerčmar and Pintar 2017), cellulose (Srivastava et al. 2019), ceramics (Keskin et al. 2012), polyurethane foam (Mikheeva et al. 2021; Mikheeva et al. 2022a; Mikheeva et al. 2022b), lignocellulose waste (Liu et al. 2019; Rai et al. 2022), microplastic (Ren et al. 2022), carbon nanotubes (Boshagh et al. 2019b), sponge (Kirli and Kapdan 2016), etc.

Some recent papers present valuable results of comparative studies of the effect of various materials on the dark fermentative production of biohydrogen. Solid matrices adsorbed by cells were characterized by the maximum biohydrogen potential: 62 mL H_2/30 mL (control), 138 mL H_2/30 mL (foam), 104 mL H_2/30 mL (coconut fiber), and 96 mL H_2/30 mL (wood chips). When cells were entrapped by TiO_2-NP with the addition of alginate, the biohydrogen potential also increased: 32 mL H_2/30 mL (control—0 mg TiO_2/L), 35 mL H_2/30 mL (200 mg TiO_2/L), 40 mL H_2/30 mL (400 mg TiO_2/L), 53 mL H_2/30 mL (600 mg TiO_2/L), 75 mL H_2/30 mL (800 mg TiO_2/L), and 93 mL H_2/30 mL (1000 mg TiO_2/L), respectively. Scanning electron microscopy (SEM) observation showed that the foam was the best carrier for high cell adhesion on its surface compared to other carriers due to the characteristics of the surface. Besides, a maximum yield of 2.07 mol/mol glucose was achieved using a foam carrier (Rai et al. 2022). Stainless steel scrubber, volcanic stone, tulle, polyester fiber, loofah sponge, plastic bath sponge, sea sponge, and aquarium biological sponges were used as support materials to investigate hydrogen production potentials of these particles. The highest hydrogen yield (1.96 mol H_2/mol glucose) and maximum hydrogen production rate (7.39 mL H_2/h) were obtained with polyester fiber particles (Gokfiliz and Karapinar 2017). Based on the immobilized mixed culture technology, both the hydrogen production and synergy mechanism between *Bacillus cereus* A1 and *Brevumdimonas naejangsanensis* B1 were investigated. In terms of hydrogen yield and the multi-cycle use of the carriers in batch fermentation, corn stalk as carrier was found to be a better candidate than fiber material (polyester fiber) and activated carbon (AC). The obtained cumulative hydrogen production was 2205 mL/L within 180 h, significantly higher than that of the suspended fermentation. The average cumulative hydrogen production and hydrogen yield were 1845 mL/L and 1.50 mol H_2/mol glucose for ten cycles of repeated fermentation batches, respectively, which was 62.5% higher than that of suspended fermentation (Ma et al. 2017).

The list of materials used for immobilization of HPB is quite extensive and is presented in Table 9.2. Also below is a brief overview of the recent works on the most widely used materials for immobilization of HPB.

9.3.4.1 Carbon

Carbon-based matrices such as conventional activated carbon (AC), carbon nanotubes, carbon cloth, carbon fiber, and biochar have been extensively studied for their application in the dark fermentative systems with attached growth (Boshagh et al. 2019a; Cheng et al. 2019). Adsorption of microbial cells on AC is a common method of immobilization, since AC has the properties of low toxicity, large surface area, and effective porosity. The porous structure of AC as an immobilizing material

Table 9.2 Classification of carrier materials for HPB immobilization

The nature of the carrier material	Example
Natural organic	Cellulose, dextran, agar, agarose, chitin, alginate, collagen, albumin, starch, chitosan, k-carrageenan, pectin, gelatin, lignocellulosic materials, sepharose, keratin, ferritin, wood chips, sea sponge, coconut coir
Synthetic organic	Polystyrol, polyacrylate, polymethacrylates, polyacrylamide, polyamides, vinyl and allyl polymers, polyvinyl alcohol, polyethylene glycol, polycarbamoylsulfonate, nylon, polyethylene, polypropylene, polyurethane, polyvinyl chloride, polyester, polyacrylonitrile, ion exchange resins, carbon fiber, carbon nanotube, nanoparticles, magnetic nanoparticles, nanofibers, nanocomposites, nanorods, ethylene-vinyl acetate copolymer, polyethylene–octene–elastomer, polymethyl methacrylate
Inorganic	Bentonite, alumina, activated carbon, silica gel, clay, zeolite, anthracite, porous glass, ceramics, brick dust, kaolinite, seabed, pumice, zirconia, diatomite, metal oxide nanoparticles, titanium dioxide, hydroxyapatite, sand, stainless steel scrubber, volcanic stone

helps to maintain the growth and viability of microbial cells and increase their density. In addition, carbon materials with conductive properties can enhance direct interspecies electron transfer (Cheng et al. 2022; Nozhevnikova et al. 2020). Zhang et al. showed that the use of AC as an immobilizing material for DF provides stable conditions for the rapid growth and metabolism of bacteria, and also increases the production of biohydrogen by 259% compared with control (Zhang et al. 2017). It should be kept in mind that the yield of hydrogen in the AC-immobilized system depends on its added amount. Adding an excess of AC can disrupt mass transfer processes. Studies have shown that the addition of AC in the amount of 200 g/L is optimal for increasing the yield of biohydrogen. A higher dose of AC limited the transfer of nutrients and VFAs, which led to a lower yield of hydrogen (300 g AC/L) and even impaired DF (400 g AC/L) (Zhang et al. 2017). Kerchmar et al. noted that the biofilm formed on the surface of AC contained more bacteria at the beginning of the process (which is comparable to polyethylene and expanded clay). AC is well known for its adsorption properties, which means that it can accumulate potential biochemically active compounds from the inoculum and cause a reduction in bacterial numbers or even result in a less diverse bacterial community (Kerčmar and Pintar 2017). Indeed, the microbial community on polyethylene and expanded clay carriers was more diverse than on AC and polyurethane substrates, suggesting selection of certain microbial species during DF. This is in agreement with some previous studies (Júnior et al. 2015) but differs from the observations published by Fernandes et al. (Fernandes et al. 2013). Granular activated carbon (GAC), immobilized with *Caldicellulosiruptor saccharolyticus* was suitable to carry out the DF process with POME as a substrate both under sterile and non-sterile conditions with high HPR and HY (Jamali et al. 2021).

Biochar has important characteristics such as electrical conductivity, functional groups, high surface area and high porosity, which makes it suitable for universal use in a wide range of applications. Li et al. showed that the addition of rice straw-derived biochar (3 g/L) increased H_2 production by 118.4% in ethanol-type and by 79.6% in butyrate-type DF due to its advantages of efficient cell immobilization and, thereby enhancing their growth and increasing substrate consumption (Li et al. 2020a). Sunyoto et al. showed that biochar addition increased the maximum hydrogen production rates by 32.5% and methane by 41.6%, improved hydrogen yield by 31% and methane 10%, and shortened the lag phases in the two phases by 36 and 41%, respectively. Biochar addition also enhanced volatile fatty acids (VFA) generation during hydrogen production and VFA degradation in methane production (Sunyoto et al. 2016). Increased hydrogen production upon addition of AC and biochar has also been observed in acetone-butanol-ethanol fermentation and other processes (Wu et al. 2019; Zhang et al. 2017).

High stability, high surface-to-volume ratio, good hydrophilicity and biocompatibility, and high biomass retention capacity are excellent properties of functionalized multi-walled carbon nanotube (MWCNT), making it very attractive for immobilization applications in DF systems. Boshagh et al. showed that immobilized *E. aerogenes* on 1.2 mg/mL MWCNT resulted in higher hydrogen yield (2.2 mol/mol glucose), hydrogen production rate (2.72 L/L/h), and glucose degradation efficiency (96.20%) and shorter lag phase (1 h) compared to the free *E. aerogenes* (Boshagh et al. 2019a).

9.3.4.2 Polymers

Natural biopolymers such as agarose, alginate, cellulose, collagen, keratin, and agar are widely used as immobilizing matrices in the DF process due to their high availability, low cost, non-toxicity, biocompatibility, and large surface area (Sekoai et al. 2020; Cheng et al. 2022). A study by Park et al. demonstrated that a reactor inoculated with alginate-immobilized sludge confirmed higher biofilm formation and higher biohydrogen production at a 3-h HRT by increasing hydrogen production and decreasing hydrogen consumption (Park et al. 2020). However, natural carriers such as alginate have the disadvantages of insufficient mechanical and chemical stability and reduced porosity. Therefore, they are unsuitable for long-term industrial use. To overcome the limitations of alginate matrices and ensure long-term use, alginate is mixed with other materials such as some synthetic polymers, activated carbon, and metal (Sekoai et al. 2020).

The mixture of PVA and sodium alginate was proved to have high activity and good mechanical stability. Yin et al. showed that sludge immobilized in PVA gels could remain active after ten repeated batch operations, which showed that sludge immobilization not only increased biohydrogen yield, but also ensured continuous biohydrogen production over a long period of operation (Yin et al. 2018). Dzul Rashidi et al. have recently reported that attaching microbial cells to GAC can provide a constant rate of biohydrogen production at a higher temperature, and further use of alginate as the immobilization material promotes the growth of HPB. GAC can act as a support for alginate and maintain pellet stability. However, it has been noted that the amount of alginate used to immobilize HPB should be controlled, as a higher concentration of alginate may prevent microbial cell growth and limit biohydrogen production (Dzul Rashidi et al. 2020). A new hybrid immobilization material with a combination of calcium alginate, activated carbon, silica, gel, and chitosan has also been developed and applied in the continuous production of biohydrogen. It was noted that a hybrid immobilization carrier with high efficiency and stability can be reused in a reactor (Sivagurunathan et al. 2014).

Hellal et al. showed that the immobilization of anaerobic sludge in sodium alginate with polyaniline nanoparticles (PANI NPs) increased the degradation of organic matter in dairy wastewater with the formation of VFA and biohydrogen. The authors revealed a direct relationship between the concentration of PANI NPs in sodium alginate, the degradation of dairy wastewater, and the formation of biohydrogen. The degradation of organic matter in terms of chemical oxygen demand was 75.5% at an NP concentration of 40 mg/L (Hellal et al. 2022).

9.3.4.3 Nanomaterials

Due to advances in nanotechnology, various types of nanomaterials with large surface area, high adsorption capacity, and high electrical conductivity, such as metal nanoparticles, metal oxide nanoparticles, nanocomposites and graphene-based nanomaterials, have been used in the DF process to improve the production of biohydrogen (Cheng et al. 2022). Nanomaterials can play a significant role in DF, acting as supporting matrices for microbial and enzyme immobilization, cofactors in the active site of hydrogenase, and enhancers of electron transfer between nanoparticles and enzymes. The end result is an increase in biohydrogen productivity (Elreedy et al. 2019; Sekoai et al. 2020; Srivastava et al. 2020). Seelert et al. showed that immobilization of *Clostridium beijerinckii* NCIMB8052 on magnetite nanoparticles reduced the lag phase of microbial growth and increased biohydrogen production (Seelert et al. 2015). In addition, nanomaterials can be used to immobilize cellulase enzymes to enhance the hydrolysis of lignocellulosic waste during pretreatment (Srivastava et al. 2017). Shanmugam et al. reported that *Trichoderma asperellum* laccase immobilized on chitosan-coated $Fe_3O_4@SiO_2$ nanoparticles performed better in the delignification of lignocellulosic biomass, with biohydrogen yield and production rates being 2.8 mol H_2/mol reducing sugar and 25 L H_2/L/d, respectively, in a continuous stirred tank reactor (Shanmugam et al. 2020a).

In the DF process, protons are reduced to molecular hydrogen under the action of Fe-Fe-hydrogenase and Ni-Fe-hydrogenase (Taherdanak et al. 2015). Therefore, the addition of certain amounts of iron and nickel nanoparticles promotes an increase in the activity of hydrogenases and an increase in the production of biohydrogen. Yu et al. (2014) showed that the hydrogen yield can be increased by 16% and the lag phase reduced by 36% by adding Fe^0 powder during dark fermentation of dehydrated sludge. Yang and Wang showed that biochar (600 mg/L) combined with Fe^0 NPs (400 mg/L) synergistically enhanced microbial activity and enriched much more *Clostridium* sp., which contributed to more efficient metabolic pathways for synergistically improving hydrogen productivity and substrate utilization. Hydrogen yield reached 50.6 mL/g dry grass when biochar+Fe^0 NPs were added, which was 89.8, 53.3, and 15.9% more than that for the control group, the individual biochar and individual Fe^0 NPs group, respectively. The addition of biochar+Fe^0 NPs also shortened the lag phase and enhanced the hydrolysis of grass, which accelerated the dark fermentation process (Yang and Wang 2018).

Taherdanak et al. evaluated the effect of Fe^{2+} and Ni^{2+} ions compared to Fe^0 and Ni^0 nanoparticles on the biohydrogen yield in the mesophilic DF process and indicated that the biohydrogen yield increased by 55, 37, and 15% under the action of Ni^{2+} ions, Fe^0 nanoparticles, and Fe^{2+} ions, respectively, due to a decrease in the formation of hydrogen inhibitors. In contrast, the addition of Ni^0 nanoparticles showed little effect on hydrogen yield in this study (Taherdanak et al. 2016). Sun et al. demonstrated that the addition of biochar and Ni^0 nanoparticles was found to significantly enhance the DF with the maximum potential hydrogen yield (1067 mL) and lagging phase time (4.78 h) being achieved. The acetate pathway was apparently facilitated by the addition of synergistic factors from kinetic study of major metabolites (Sun et al. 2019).

Carbon materials such as AC and biochar complement the nanoparticles, which can synergistically improve the activity of microorganisms and enzymes when the nanoparticles are added together (Sun et al. 2019). As expected, the co-supplementation of biochar and Fe^0 nanoparticles showed a synergistic effect on increasing biohydrogen production from grass fermentation due to their complementary functions and release of more Fe^{2+} as a result of microelectrolysis of Fe^0 and biochar nanoparticles (Taherdanak et al. 2015). The hydrogen yield was substantially increased up to 1.654, 1.908, and 2.187 mol/mol glucose with immobilization of HPB on graphene, hydroxyapatite, and graphene/hydroxyapatite nanoparticles during dark fermentation of black liquor from a paper-making industry (Tawfik et al. 2021).

Elreedy et al. found that higher biohydrogen production was observed with dual (α-Fe_2O_3 + NiO, α-Fe_2O_3 + ZnO, and NiO + ZnO) and multi-nanoparticles (α-Fe_2O_3 + NiO + ZnO) compared to individual nanoparticles by enhancing the growth of *Clostridium* sp. and the activity of alcohol dehydrogenase (ADH), aldehyde dehydrogenase (ALDH), and hydrogenase (hyd) (Elreedy et al. 2019). Zhao et al. obtained the maximum hydrogen yield of 2.48 mol/mol glucose at a silver NP concentration of 20 nmol/L at DF of glucose using mixed bacteria (Zhao et al. 2013).

In addition to the chemical nature of nanoparticles, their concentration during dark fermentation should also be taken into account. Mishra et al. found that the yield of biohydrogen can be increased by 1.51 and 1.61 times by adding 1.5 mg nickel (NiO)/L and 1.0 mg cobalt oxides (CoO)/L, respectively, in DF of POME. However, at an NP concentration of 3.0 mg of NiO and CoO per L in a modified Luria-Bertani (LB) medium, the authors showed a decrease in bacterial cell growth by 63 and 83% (Mishra et al. 2018). Table 9.3 presents the results of some studies on the effect of nanoparticles on the productivity of the DF process.

Currently, most of the nanoparticles are synthesized by chemical methods, which lead to high costs and harmful effects on the environment and human health. Therefore, in the future it is necessary to obtain cheaper and environmentally friendly nanomaterials obtained, for example, from microorganisms and plants.

9.3.5 Microorganisms

The production of biohydrogen by DF is carried out using various microorganisms, which are pure or mixed cultures. *Clostridium* and *Enterobacter* sp. are the most commonly used organisms in HPB immobilization studies (Kumar et al. 2016; Kumar et al. 2018a; Boshagh et al. 2019a; Boshagh et al. 2019b; Cheng et al. 2019; Cheng et al. 2020). *Clostridium* sp. are gram-positive, spore-forming, and rod-shaped obligate anaerobes (Kumar et al. 2018a). *Enterobacter* sp. are facultative, gram-negative, and rod-shaped microorganisms; however, they give lower yields

Table 9.3 Parameters of nanoparticles-assisted DF process

Nanoparticle carrier	Microorganism	Hydrogen yield	Hydrogen production rate	Reference
Chitosan coated $Fe_3O_4@SiO_2$ NPs for laccase immobilization	*Trichoderma asperellum*	2.8 Mol/Mol	25 L/L/d	(Shanmugam et al. 2020b)
Functionalized multi-walled carbon nanotube	*E. aerogenes*	2.2 Mol/Mol glucose	2.72 L/L/h	(Boshagh et al. 2019a)
Sodium alginate with polyaniline nanoparticles	From an upflow anaerobic sludge reactor treating industrial wastewater from a maize factory	54.5 mL/g VS	–	(Hellal et al. 2022)
Granular activated carbon (GAC)	From the sludge pit of palm oil mill effluent (POME)	1.24 Mol/ Mol sugar.	5.2 mmol/ L/h	(Jamali et al. 2019)
Graphene/ hydroxyapatite nanoparticle	Black liquor from a paper-making industry	Up to 2.187 Mol/ Mol glucose	–	(Tawfik et al. 2021)

compared to *Clostridium* sp. (Wang and Yin 2019). Recently, thermophilic *Thermoanaerobacterium* sp. have attracted increasing attention in consolidated bioprocessing, which can directly utilize lignocellulosic materials for biofuels production (Wu et al. 2021). A number of works have shown the high potential of *Thermoanaerobacterium* for use in DF-attached growth systems (Roy et al. 2014; Sompong et al. 2008; Zhao et al. 2012b; Jamali et al. 2019; Khamtib et al. 2021).

Several studies have shown that immobilization of pure cultures or defined microbial consortiums (DMC) for biohydrogen production could be a successful option for their retaining in the reactor space during long-term and continuous flow operation. The stable productivity of immobilized cells of *Clostridium intestinale* over 8 days with four changes of medium depicted that the immobilized cells can successfully yield higher hydrogen and lower soluble metabolites than suspended cells, suggesting a feasible process for future applications for bioH_2 production (Güngörmüşler et al. 2021). *Clostridium pasteurianum* immobilized in mixed k-carrageenan, alginic acid, and PVA granules could endure up to 14 days of operation (Wannapokin et al. 2021). Patel et al. demonstrates the successful continuous biogas production (40 days cultivation) with high H_2 content in it by immobilized DMC composed of *Klebsiella*, *Bacillus*, and *Proteus* spp. under non-sterile conditions (Patel et al. 2021). The use of expanded clay as a support material for *T. thermosaccharolyticum* KKU19 to produce hydrogen under non-sterile conditions was studied by Khamtib et al. Within 80 days of continuous flow cultivation, HPR of 6.82 L H_2/L/d, and an HY of 232 mL H_2/g-COD were obtained (Khamtib et al. 2021). Bao et al. conducted a study on dark fermentative biohydrogen production with a mixed culture of *Bacillus* sp. and *Brevumdimonas* sp. The results showed that the hydrogen yield was doubled in mixed culture compared to pure culture (Bao et al. 2012). In addition, studies have shown that the synergistic effect of mixed cultures can expand the type of substrate, improve the adaptability of the system to the environment, and increase the efficiency of fermentation (Smid and Lacroix 2013).

However, the majority of the studies utilizes mixed-cultures, obtained from various sludges, such as sludge pits (Jamali et al. 2019), activated sludge of wastewater treatment plant (wwtp) (Li et al. 2020b; Ren et al. 2022), granular sludge (Arellano-García et al. 2021), and methanogenic sludge (Ferreira et al. 2020; Gorgec and Karapinar 2019; Zhao et al. 2021). Mixed cultures are considered to be better suited than pure cultures (Elsharnouby et al. 2013), for the following reasons: (i) contamination levels are kept to a minimum, (ii) there is a high microbial diversity, (iii) there is mixed co-fermentation of microorganisms, (iv) ideal for continuous processing, and (v) the use of a variety of raw materials. Because pure cultures can become contaminated over time during fermentation, sterilization is again required, so the use of pure cultures in biohydrogen production becomes difficult for industrial applications (Ma et al. 2017).

Microbial biofilms are communities of microorganisms attached to surfaces and encased within a matrix of extracellular polymeric secretions that function as a cooperative consortium, which is increasingly being used in biological processes to generate renewable energy. Progressive formation of biofilm surface structures is observed during immobilization from single scattered microorganisms in thin continuous layers to complex forms of various sizes (1–3 mm long) and shapes (spherical to rod-shaped) embedded in the extracellular polymer matrix. Gaps, cracks, and pores on the surface of the substrates favored biomass immobilization (Nikolaev and Plakunov 2007; Vasudevan 2014). Kerčmar et al. noted that, in terms of the pleomorphic morphological appearance of bacterium-like structures and their rich distribution in biofilms on polyethylene and expanded clay, they differed from biofilms on the surface of polyurethane and activated carbon, which seemed to be poorer, which was confirmed during further assessments of the total number of viable bacteria in exfoliated biofilms. Biofilms on polyethylene and expanded clay substrates contained more bacteria (within 10^5 CFU/mL) during the entire immobilization process and also at the end compared to polyurethane (within 10^4 CFU/mL). The authors found that it takes about 7 days to achieve metabolic and microbiological stability of the process of immobilization of a bacterial consortium (Kerčmar and Pintar 2017).

Diversity of microbial communities may coincide with a change in the pH value in the reactor (Temudo et al. 2008). Kerchmar and Pintar showed that the isolated microbial communities at the steady-state stage in the case of polyethylene and expanded clay materials with wastewater pH values from 4.5 to 5.0 were more diverse and also more numerous compared to polyurethane and activated carbon with pH values ranging from 4.0 to 4.5. All bacteria isolated in this work could be divided into two phylogenetic groups: *Proteobacteria* and *Firmicutes*. *Proteobacteria* were present on all supporting materials, while *Firmicutes* were present only on polyethylene and expanded clay (Kerčmar and Pintar 2017). This is consistent with the work of Junior et al. where both phylogenetic groups were observed at the end of the DF process, and a higher proportion of *Firmicutes* was associated with higher hydrogen production in reactors filled with polyethylene and expanded clay (Júnior et al. 2015). The most frequently isolated from biofilms were *Klebsiella* sp., *Escherichia* sp., *Citrobacter* sp., *Acinetobacter* sp., and *Lactococcus* sp., while the first two genera (*Klebsiella* sp. and *Escherichia* sp.) were identified on all carrier materials. The spore-forming *Clostridium* sp. and *Bacillus* sp. were isolated only from biofilms formed on the surface of polyethylene and/or expanded clay carriers. Neither fungi nor yeasts were isolated in any of the bioreactors (Kerčmar and Pintar 2017). It is known that some of the bacteria can support the growth of HPB (Elsharnouby et al. 2015). *Klebsiella* sp. may support *Clostridium* sp. in competition with other undesirable bacteria for enhancing hydrogen production (Perna et al. 2013), with *Klebsiella* sp. reported to be hydrogen-producing bacteria themselves in numerous works (Niu et al. 2010; Zhang et al. 2021; Estevam et al. 2018; Song et al. 2018). The *Bacillus* strain is known for its multiple extracellular enzyme activities and can act as an oxygen consumer, creating an anaerobic environment for the *Clostridium* strain (Shanmugam et al. 2020b).

9.4 Biofilm Reactor Types, Design, and Operating Modes

To obtain biohydrogen by dark fermentation with immobilized microorganisms, reactors of various types are used, which can operate in batch or continuous modes (Fig. 9.3 modified from (Sekoai et al. 2018)). The choice of reactor mainly depends on the properties of the feedstock, as well as on the type and characteristics of the immobilized cells (Kumar et al. 2016).

Small volume batch reactors are successfully used for laboratory studies of dark fermentation to study the kinetics, process characteristics, mass transfer processes using various methods for immobilizing microbial cells, as well as the influence of various factors on the process stability and biohydrogen yield. However, for the production of hydrogen on an industrial scale, continuous bioreactors are recommended (Muri et al. 2018).

The selection of a bioreactor configuration for a continuous DF process is usually based on the analysis of data obtained during batch tests. The design of a continuously operating high-performance reactor with immobilized HPB cells should ensure a stable DF process for a long time with a high biohydrogen yield, efficient mass transfer in the reaction volume, a high biomass concentration, and operation at high organic loads.

Continuous stirred tank reactor (CSTR) is still a widely used type of reactor for model studies because it provides efficient mixing and homogeneity (Castelló et al. 2020; Arellano-García et al. 2021). However, CSTR cannot maintain high concentrations of biomass at low HRT, since it is constantly washed out from the reactor volume. Granular biomass reactors or fixed bed reactors such as upflow anaerobic sludge blanket bioreactor (UASB), anaerobic packed bed bioreactor (APBR),

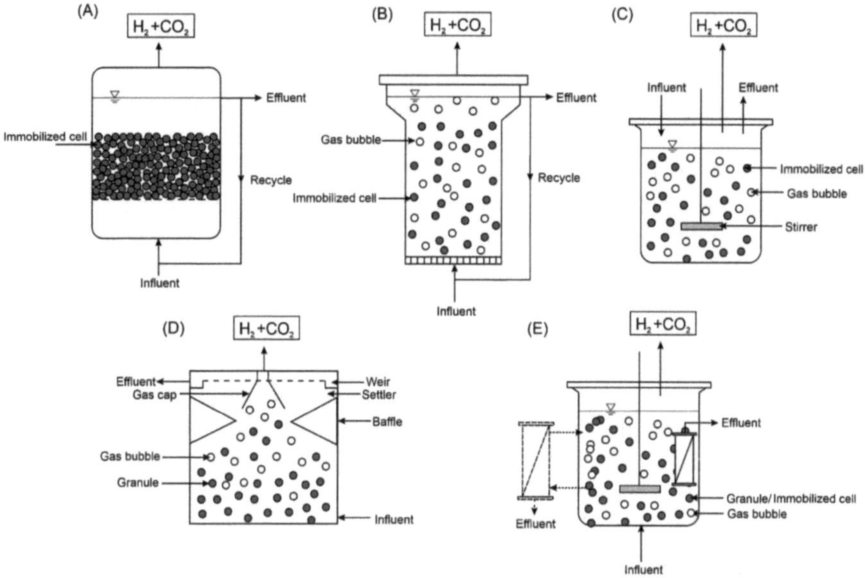

Fig. 9.3 Schematic diagrams of DF reactors with HPB immobilization: (**a**) anaerobic packed bed reactor (APBR), (**b**) anaerobic fluidized bed reactor (AFBR), (**c**) anaerobic stirred tank reactor (ASTR), (**d**) upflow anaerobic sludge bed reactor (UASB), (**e**) anaerobic membrane bioreactor with submerged or external membrane module (AMBR)

expanded granular sludge bed reactor (EGSB), and anaerobic fluidized bed reactor (AFBR) are more suitable systems for creating a high-performance dark fermentation process with the production of biohydrogen (Kumar et al. 2018b). Such reactors make it possible to achieve a high cell density due to immobilization, operate at an HRT that does not depend on the growth rate of bacteria, and provide greater resistance to inhibitory substances and shock loads (Łukajtis et al. 2018; Park et al. 2021).

Experiments show that the UASB is a promising reactor for dark fermentative biohydrogen production, which can work stably with spontaneous granulation of activated sludge microorganisms (Park et al. 2021). Fang et al. discovered for the first time that the hydrogen-producing microbial community could also form highly active granules. They found that the formation of granules with an average size of 1.6 mm occurred from the 15th day of a continuous dark fermentation process in a UASB reactor using sucrose as a substrate, the conversion of which was 97% (Fang et al. 2002).

A large number of studies on DF are carried out using simple design APBRs, which, thanks to various immobilizing materials, allow the biomass to be reliably retained in the system. For the immobilization of microbial cells, a wide variety of natural and synthetic materials are used, such as expanded clay, charcoal, porous ceramics, low-density polyethylene, polyurethane sponge, glass beads, Raschig rings, bamboo stalks, loofah, rice straw, aquarium biological sponge, etc. APBR is simple in design and highly efficient in hydrogen production. As disadvantages of APBR, insufficiently efficient mixing and possible problems with biogas removal, an increase in resistance to mass transfer, the development of a pH gradient, and an inhomogeneous distribution of microbial activity in the reactor volume are noted, to reduce which it is recommended to organize recirculation (Show et al. 2011; Karapinar et al. 2020).

Kongjan et al. compared the performance of UASB and APBR with a plastic carrier as an immobilizing material using xylose as a substrate in thermophilic mode (55 °C). At HRT of 4 h for UASB reactor, HY was 251 mL H_2/g xylose, HPR was 15.1 L/L/d; for APBR, HY was 221 mL H_2/g xylose, and HPR was 13.3 L/L/d. The granular reactor had more microbial diversity and more balance between economic efficiency in terms of hydrogen production rate and technical efficiency in terms of hydrogen yield, resulting in more reliable operation (Kongjan et al. 2019).

The process of biofilm formation in APBR is largely affected by the characteristics of immobilizing materials: hydrophobicity/hydrophilicity of the surface, the presence of a microporous structure, and the surface charge of the material. Surface hydrophobicity has been identified as one of the most important factors influencing the bacterial adhesion process (Sousa et al. 2009). Muri et al. conducted studies in three identical APBRs that differed in the carrier material: Mutag BioChipTM (reactor R1), expanded clay (R2), activated carbon (R3). Contact angle measurements showed that the Mutag BioChipTM carrier was hydrophobic, while the expanded clay and activated carbon were hydrophilic. Synthetic wastewater containing glucose as the main source of carbon was used as a substrate. The maximum hydrogen yield of 1.80 mol H_2/mol glucose was observed for R1, followed by 1.74 mol H_2/mol glucose in R2, and 1.46 mol H_2/mol glucose in R3. The main metabolic products were acetic and butyric acids. The data showed that in the reactors with the highest hydrogen yield (R1 and R2), the acetate/butyrate (HAc/HBu) ratio was 1.7 and 1.6, respectively, and in the reactor with the lowest hydrogen yield (R3), the resulting HAc/HBu ratio was 4.8. The experiment showed that the studied immobilizing materials affected the immobilization of hydrogen-producing bacteria and, consequently, the yield of hydrogen, as well as the composition of soluble metabolites (Muri et al. 2018). Due to its high porosity, an aquarium biological sponge was shown to be a feasible carrier material to produce biohydrogen from wheat powder in APBR (Gorgec and Karapinar 2019). A novel microorganism immobilization material, polyester fiber beads, was suggested by Karapinar et al. However, diffusion limitation of the substrate for the biomass within the particle was observed (Karapinar et al. 2020), which is one of the common drawbacks of packed-bed bioreactors.

To improve biohydrogen production, Zhao et al. proposed a hybrid bioreactor with integrated chlorinated polyethylene (CPE) fixed-bed and modified with iron zeolite as a microorganism nutrition carrier (MNC), aiming at enhancing and sustaining biohydrogen production during the anaerobic digestion process. The proposed mechanism revealed that the synergistic effect of the Fe-modified zeolite was based on supplying efficient micronutrients, facilitating enzyme and co-enzyme synthesis, inducing favorable pathways, promoting microbial immobilization, and supporting electron transfer between interconnected hydrogen-producing bacteria for superior and long-term biohydrogen conversion. Therefore, the developed hybrid-Fe bioreactor exhibits impressive potential as an effective and sustained strategy in biohydrogen production for practical applications. The hybrid reactor (CPE with Fe-modified zeolite) operated mainly along the acetate route and demonstrated high stability of the biohydrogen production process with a peak value of 1893.0 mL H_2/L during the 72-day experiment (Zhao et al. 2021) Expanded clay was successfully used by Khamtib et al. as a support material to immobilize *T. thermosaccharolyticum* KKU19 for the production of hydrogen from the co-digestion of oil palm trunk hydrolysate and slaughterhouse wastewater in a fixed-bed reactor. Hydraulic retention time of 6 h was optimal for hydrogen production with an HPR of 6.82 L H_2/L/d, and an HY of 231.99 mL H_2/g-COD under non-sterile conditions. Long-term operation results indicated that expanded clay can serve as an effective support material affording protection to microbial cells of *T. thermosaccharolyticum* KKU19 within fixed-bed reactors and can be used for long-term operation (Khamtib et al. 2021).

Anaerobic fluidized bed reactors (AFBR) for the DF process are also attracting the attention of scientists due to their excellent mass transfer characteristics and the provision of high biomass concentrations through immobilization, which can provide higher hydrogen production at low HRT values (Zhao et al. 2013). As immobilizing materials for AFBR, the following are used: PET, expanded clay, polystyrene, ethylene-vinyl acetate copolymer, activated carbon, modified alginate entrapment, and polyurethane (Kumar et al. 2016).

Start-up of continuous production of biohydrogen from cheese whey (CW) in an anaerobic filter (AF) and anaerobic fluidized bed (AFB) with a polyurethane carrier was studied by Mikheeva et al. Maximum specific yields of hydrogen, namely 178 mL/g chemical oxygen demand (COD) and 149 mL/g COD for AFB and AF, respectively, were obtained at the hydraulic retention time (HRT) of 4.5 days and organic loading rate (OLR) of 6.61 kg COD/m^3/d. At the same time, the maximum hydrogen production rates of 1.28 and 1.9 L/L/d for AF and AFB, respectively, were obtained at the HRT of 2.02 days and OLR of 14.88 kg COD/m^3/d. At the phylum level, the dominant taxa were *Firmicutes* (65% in AF and 60% in AFB), and at the genus level, *Lactobacillus* (40% in AF and 43% in AFB) and *Bifidobacterium* (24% in AF and 30% in AFB) (Mikheeva et al. 2021).

In their research, Chaves et al. evaluated the performance of AFBR for the production of hydrogen from sugarcane molasses with 25 g COD/L by dark fermentation. Shredded tires were used as an immobilizing material. The maximum hydrogen production rates (1.44 L H_2/h/L and the highest hydrogen yield (3.07 mol H_2/mol glucose) were achieved at HRT of 4 and 6 h, respectively. The highest COD removal (23.3 ± 8.5%) was achieved at HRT of 12 h, while the HRT of 6 h presented the maximum carbohydrate conversion of 70.1 ± 2.2%. Ethanol (44–67%) and acetic acid (18–38%) were the main metabolites produced, emphasizing the predominance of the ethanol-type fermentation pathway in the DF process (Chaves et al. 2021). To reduce start-up time and enhance hydrogen production efficiency, a sequential immobilization on a cube-shaped polyurethane sponge and deoxygenization strategy for hydrogen production was investigated in continuous-flow moving bed biofilm reactors. The pre-immobilization process accelerated the initial enrichment of HPB and promoted the biofilm formation, which contributes to higher hydrogen production efficiency compared to a non-immobilized reactor (Li et al. 2020b).

In recent years, researchers have shown interest in the anaerobic membrane bioreactor (AMBR) technology (Fig. 9.3E) for the implementation in the process of dark fermentation and offer new original solutions for its implementation. Renaudie et al. developed a liquid/gas membrane bioreactor (L/G MBR) for a continuous DF process, solving the problems of bacterial biomass immobilization and effective hydrogen removal. The hollow fiber module was used to continuously remove the resulting gases and ensure the development of the biofilm. The liquid phase circulated outside the hollow fibers, and the resulting gases diffused into the fiber (Fig. 9.4 modified from (Renaudie et al. 2021)). Unlike the solid/liquid membrane bioreactor, L/G MBR was not susceptible to membrane fouling because the biofilm does not restrict gas diffusion across the membrane (Renaudie et al. 2021).

MBR was repeatedly operated over a year under similar conditions (HRT = 12 h and OLR = 0.8 g-glucose/L-medium/h), and without further bacterial reseeding. Despite repeated draining and rinsing, consistent HY of 0.97 ± 0.09 mol H_2/mol glucose added, and HPR of 106.5 ± 10.6 mL H_2/L medium/h, were reached. Acetate, butyrate, and ethanol were the main metabolites produced. Different biogas extraction strategies showed that efficient in situ H_2 extraction is possible without sweeping gas into the lumen of the fibers, thus facilitating H_2 purification in an industrial setting (Renaudie et al. 2021).

Kim et al. conducted studies of DF with the production of biohydrogen in a bioreactor with a dynamic membrane (DMBR). Dynamic membranes are secondary membranes prior to the real membrane or support material and are formed when the filtered solution contains suspended solid particles such as microbial cells and flocs. Microbial aggregates are formed by extracellular polymeric substances that are secreted through the growth of microorganisms. The major hydrogen-producing bacterium was *Clostridium chromiireducens*, which was enriched up to 82% of the

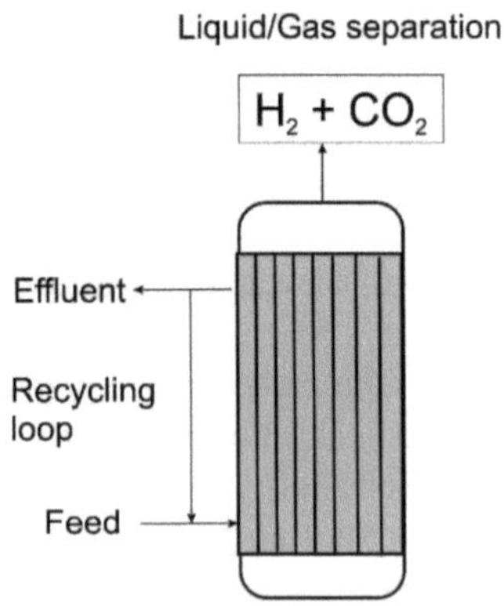

Fig. 9.4 Configurations of the MBR with in situ liquid/gas separation

total bacteria with a 2-h HRT, while *Enterobacter cloacae* was also observed in the range of 0.2–34.4% of total bacteria (Kim et al. 2021).

The characteristics of self-forming dynamic membranes are determined by the concentration, type, shape, molecular weight of the filter solution, and cross-flow rate (Park et al. 2021). Anburajan et al. applied a submerged dynamic filtration module that was used in the continuous operation. Peak H_2 yield was attained at 3 h HRT as 3.22 ± 0.22 mol H_2/mol glucose added and the HPR was attained at 2 h HRT as 54.07 ± 3.69 L H_2/L/d, respectively (Anburajan et al. 2019).

Table 9.4 shows the characteristics of the dark fermentative biohydrogen production for reactors of various types.

9.5 Future Outlook

The method of immobilization of hydrogen-producing bacterial cells can play a key role in the development of biohydrogen production technologies, providing a high yield, stability, and efficiency of the process when organizing industrial-scale production.

The immobilization method allows to increase the concentration of microorganisms in the reaction volume, improves the stability of the system and the efficiency of substrate conversion, prolongs the fermentation period, helps protect cells from toxic metabolites and toxic substances that can enter the reaction medium with the feedstock, and promotes the maximum yield of biohydrogen at high organic loads and low HRT.

For further development of DF systems with attached growth and overcoming the identified limitations, research should be directed to:

- development of more stable and inexpensive immobilizing materials;
- creation of hybrid synthetic and natural materials containing iron, nickel, and other elements in order to activate the enzyme system of HPB;
- improvement of the designs of anaerobic bioreactors that ensure stable operation and efficient mass transfer during long-term operation;
- effective suppression of hydrogen scavengers (homoacetogens, methanogens, etc.);
- improvement of symbiotic relationships between HPB and biofilm-forming microorganisms;
- carrying out further work to study the influence of the type, dosage, size, shape, and toxic effect of nanoparticles on the process of biohydrogen production, their influence on the structure, and dynamics of formation of the microbial community;
- expanding the range of organic waste used as carbon sources to reduce the cost of biohydrogen production technology and solve resource-saving problems;

Table 9.4 Parameters of the dark fermentative biohydrogen production with immobilization of HPB cells in reactors of various configurations (modified from Kumar et al. 2016; Park et al. 2021)

Reactor mode	Inoculum	Carrier material	Substrate	Retention time (h)	Temp. (°C)	Maximum H_2 yield	H_2 production rate (L/L/ d)	Reference
CSTR	*Clostridium* sp. T2	Mycelium pellet	Xylose	10	37	N.A.	1.62	(Zhao et al. 2012a)
ASTBR	Natural fermentation of synthetic wastewater	Low-density polyethylene cylinders	Reconstituted cheese whey powder	24	25	1.4 ± 0.7 Mol H_2/Mol of consumed lactose	1.6 ± 0.7	(Blanco et al. 2019)
APBR	Household solid waste sludge	Plastic carrier	Glucose	N.A.	70	2.21 Mol H_2/Mol glucose	0.074	(Zheng et al. 2008)
APBR	*Clostridium Tyrobutyricum* JM1	Poly urethane foam	Glucose	2.0	37	1.66 Mol H_2/Mol glucose	7.2	(Jo et al. 2008)
AF	Acid and heat pretreated anaerobic sludge	Poly urethane foam	Cheese whey	48	37	149 mL/g COD	1.28	(Mikheeva et al. 2021)
AFB	Acid and heat pretreated anaerobic sludge	Poly urethane foam	Cheese whey	48	37	178 mL/g COD	1.9	(Mikheeva et al. 2021)
APBR	Enriched cultures from xylose batch fermentation	Plastic carrier	Xylose	4.0	55	221 mL H_2/g xylose	13.3	(Kongjan et al. 2019)
AFBR	Sewage sludge	Granular activated carbon	Glucose	1.0	37	1.19 Mol H_2/Mol glucose	56.6	(Zhang et al. 2007)
UASB	Enriched cultures from xylose batch fermentation	Granulated activated sludge	Xylose	4.0	55	251 mL H_2/g xylose	15.1	(Kongjan et al. 2019)
L/G MBR	Fresh activated sludge introduced in a stirred tank reactor and went through a thermal treatment at 70 °C during 1 h	Cylindrical-shaped hollow-fiber membrane module	Glucose	12	37	0.97 ± 0.09 Mol-H_2/added Mol-glucose	106.5 ± 10.6 mL-H_2/L-medium/h	(Renaudie et al. 2021)

Reactor mode	Inoculum	Carrier material	Substrate	Retention time (h)	Temp. (°C)	Maximum H_2 yield	H_2 production rate (L/L/ d)	Reference
DMBR	Granular anaerobic sludge, collected from a beer factory wastewater treatment plant	Granulated activated sludge	Glucose	3.0	37	1.4 Mol H_2/Mol sugar	21.3	(Kim et al. 2021)

N.A.: not available

– development of interdisciplinary cooperation between biotechnologists, molecular biologists, biologists, process engineers, and chemists to improve the technology for obtaining dark fermentative biohydrogen and new discoveries.

Acknowledgments The research was supported by the Russian Science Foundation grant No. 21-79-10153, https://rscf.ru/project/21-79-10153. Litti Yu.V. was supported by the Ministry of Science and Higher Education of the Russian Federation.

References

Akinbomi J, Wikandari R, Taherzadeh MJ (2015) Enhanced fermentative hydrogen and methane production from an inhibitory fruit-flavored medium with membrane-encapsulated cells. Membranes 5(4):616–631

Anburajan P, Park JH, Pugazhendhi A, Kim JS, Kim SH (2019) Biohydrogen production from glucose using submerged dynamic filtration module: metabolic product distribution and flux-based analysis. Bioresour Technol 287:121445

Anjana K, Kaushik A (2014) Enhanced hydrogen production by immobilized cyanobacterium Lyngbya perelegans under varying anaerobic conditions. Biomass Bioenergy 63:54–57

Antonopoulou G, Stamatelatou K, Venetsaneas N, Kornaros M, Lyberatos G (2008) Biohydrogen and methane production from cheese whey in a two-stage anaerobic process. Ind Eng Chem Res 47(15):5227–5233

Arellano-García L, Velázquez-Fernández JB, Macías-Muro M, Marino-Marmolejo EN (2021) Continuous hydrogen production and microbial community profile in the dark fermentation of tequila vinasse: response to increasing loading rates and immobilization of biomass. Biochem Eng J 172:108049

Bao M, Su H, Tan T (2012) Biohydrogen production by dark fermentation of starch using mixed bacterial cultures of Bacillus sp. and Brevumdimonas sp. Energy Fuel 26(9):5872–5878

Barca C, Soric A, Ranava D, Giudici-Orticoni MT, Ferrasse JH (2015) Anaerobic biofilm reactors for dark fermentative hydrogen production from wastewater: a review. Bioresour Technol 185:386–398

Barros AR, Silva EL (2012) Hydrogen and ethanol production in anaerobic fluidized bed reactors: performance evaluation for three support materials under different operating conditions. Biochem Eng J 61:59–65

Basile MA, Dipasquale L, Gambacorta A, Vella MF, Calarco A, Cerruti P, d'Ayala GG (2010) The effect of the surface charge of hydrogel supports on thermophilic biohydrogen production. Bioresour Technol 101(12):4386–4394

Blanco VMC, Oliveira GHDD, Zaiat M (2019) Dark fermentative biohydrogen production from synthetic cheese whey in an anaerobic structured-bed reactor: performance evaluation and kinetic modeling. Renew Energy 139:1310–1319

Boshagh F, Rostami K, Moazami N (2019a) Biohydrogen production by immobilized Enterobacter aerogenes on functionalized multi-walled carbon nanotube. Int J Hydrog Energy 44(28):14395–14405

Boshagh F, Rostami K, Moazami N (2019b) Immobilization of Enterobacter aerogenes on carbon fiber and activated carbon to study hydrogen production enhancement. Biochem Eng J 144:64–72

Bowles LK, Ellefson W (1985) Effects of butanol on Clostridium acetobutylicum. Appl Environ Microbiol 50(5):1165–1170

Bu J, Wei HL, Wang YT, Cheng JR, Zhu MJ (2021) Biochar boosts dark fermentative H2 production from sugarcane bagasse by selective enrichment/colonization of functional bacteria and enhancing extracellular electron transfer. Water Res 202:117440

Castelló E, Ferraz-Junior ADN, Andreani C, del Pilar Anzola-Rojas M, Borzacconi L, Buitrón G, Etchebehere C (2020) Stability problems in the hydrogen production by dark fermentation: possible causes and solutions. Renew Sust Energ Rev 119:109602

Chaves TC, Gois GNSB, Peiter FS, Vich DV, de Amorim ELC (2021) Biohydrogen production in an AFBR using sugarcane molasses. Bioprocess Biosyst Eng 44(2):307–316

Chen Y, Yin Y, Wang J (2021) Comparison of fermentative hydrogen production from glycerol using immobilized and suspended mixed cultures. Int J Hydrog Energy 46(13):8986–8994

Cheng J, Li H, Zhang J, Ding L, Ye Q, Lin R (2019) Enhanced dark hydrogen fermentation of Enterobacter aerogenes/HoxEFUYH with carbon cloth. Int J Hydrog Energy 44(7):3560–3568

Cheng J, Li H, Ding L, Zhou J, Song W, Li YY, Lin R (2020) Improving hydrogen and methane co-generation in cascading dark fermentation and anaerobic digestion: the effect of magnetite nanoparticles on microbial electron transfer and syntrophism. Chem Eng J 397:125394

Cheng D, Ngo HH, Guo W, Chang SW, Nguyen DD, Deng L, Hoang NB (2022) Advanced strategies for enhancing dark fermentative biohydrogen production from biowaste towards sustainable environment. Bioresour Technol 351:127045

Christopher FC, Kumar PS, Vo DVN, Joshiba GJ (2021) A review on critical assessment of advanced bioreactor options for sustainable hydrogen production. Int J Hydrog Energy 46(10):7113–7136

Dabrock B, Bahl H, Gottschalk G (1992) Parameters affecting solvent production by Clostridium pasteurianum. Appl Environ Microbiol 58(4):1233–1239

Dahiya S, Chatterjee S, Sarkar O, Mohan SV (2021) Renewable hydrogen production by dark-fermentation: current status, challenges and perspectives. Bioresour Technol 321:124354

De Amorim ELC, Barros AR, Damianovic MHRZ, Silva EL (2009) Anaerobic fluidized bed reactor with expanded clay as support for hydrogen production through dark fermentation of glucose. Int J Hydrog Energy 34(2):783–790

De Gioannis G, Muntoni A, Polettini A, Pomi R (2013) A review of dark fermentative hydrogen production from biodegradable municipal waste fractions. Waste Manag 33(6):1345–1361

Dickson JS, Koohmaraie M (1989) Cell surface charge characteristics and their relationship to bacterial attachment to meat surfaces. Appl Environ Microbiol 55(4):832–836

Dzul Rashidi NF, Jamali NS, Mahamad SS, Ibrahim MF, Abdullah N, Ismail SF, Siajam SI (2020) Effects of Alginate and Chitosan on activated carbon as immobilisation beads in biohydrogen production. Processes 8(10):1254

Elreedy A, Fujii M, Koyama M, Nakasaki K, Tawfik A (2019) Enhanced fermentative hydrogen production from industrial wastewater using mixed culture bacteria incorporated with iron, nickel, and zinc-based nanoparticles. Water Res 151:349–361

Elsharnouby O, Hafez H, Nakhla G, El Naggar MH (2013) A critical literature review on biohydrogen production by pure cultures. Int J Hydrog Energy 38(12):4945–4966

Engliman NS, Abdul PM, Wu SY, Jahim JM (2017) Influence of iron (II) oxide nanoparticle on biohydrogen production in thermophilic mixed fermentation. Int J Hydrog Energy 42(45):27482–27493

Estevam A, Arantes MK, Andrigheto C, Fiorini A, da Silva EA, Alves HJ (2018) Production of biohydrogen from brewery wastewater using Klebsiella pneumoniae isolated from the environment. Int J Hydrog Energy 43(9):4276–4283

Fang HH, Liu H, Zhang T (2002) Characterization of a hydrogen-producing granular sludge. Biotechnol Bioeng 78(1):44–52

Fernandes BS, Saavedra NK, Maintinguer SI, Sette LD, Oliveira VD, Varesche MBA, Zaiat M (2013) The effect of biomass immobilization support material and bed porosity on hydrogen production in an upflow anaerobic packed-bed bioreactor. Appl Biochem Biotechnol 170(6):1348–1366

Ferreira TB, Rego GC, Ramos LR, de Menezes CA, Silva EL (2020) Improved dark fermentation of cane molasses in mesophilic and thermophilic anaerobic fluidized bed reactors by selecting operational conditions. Int J Energy Res 44(13):10442–10452

Gokfiliz P, Karapinar I (2017) The effect of support particle type on thermophilic hydrogen production by immobilized batch dark fermentation. Int J Hydrog Energy 42(4):2553–2561
Gokfiliz-Yildiz P, Karapinar I (2018) Optimization of particle number, substrate concentration and temperature of batch immobilized reactor system for biohydrogen production by dark fermentation. Int J Hydrog Energy 43(23):10655–10665
Gorgec FK, Karapinar I (2019) Biohydrogen production from hydrolyzed waste wheat by dark fermentation in a continuously operated packed bed reactor: the effect of hydraulic retention time. Int J Hydrog Energy 44(1):136–143
Guevara-López E, Buitrón G (2015) Evaluation of different support materials used with a photo-fermentative consortium for hydrogen production. Int J Hydrog Energy 40(48):17231–17238
Güngörmüşler M, Tamayol A, Levin DB (2021) Hydrogen production by immobilized cells of clostridium intestinale strain URNW using alginate beads. Appl Biochem Biotechnol 193(5):1558–1573
Hellal MS, Abou-Taleb EM, Rashad AM, Hassan GK (2022) Boosting biohydrogen production from dairy wastewater via sludge immobilized beads incorporated with polyaniline nanoparticles. Biomass Bioenergy 162:106499
Ishikawa M, Yamamura S, Ikeda R, Takamura Y, Sode K, Tamiya E, Tomiyama M (2008) Development of a compact stacked flatbed reactor with immobilized high-density bacteria for hydrogen production. Int J Hydrog Energy 33(5):1593–1597
Ismail I, Mohd AH, Nor'Aini AR, Chen SS (2011) Effect of retention time on biohydrogen production by microbial consortia immobilised in polydimethylsiloxane. Afr J Biotechnol 10(4):601–609
Jadhav P, Muhammad N, Bhuyar P, Krishnan S, Abd Razak AS, Zularisam AW, Nasrullah M (2021) A review on the impact of conductive nanoparticles (CNPs) in anaerobic digestion: applications and limitations. Environ Technol Innov 23:101526
Jamali NS, Jahim JM, Isahak WNRW (2016) Biofilm formation on granular activated carbon in xylose and glucose mixture for thermophilic biohydrogen production. Int J Hydrog Energy 41(46):21617–21627
Jamali NS, Rashidi NFD, Jahim JM, Sompong O, Jehlee A, Engliman NS (2019) Thermophilic biohydrogen production from palm oil mill effluent: effect of immobilized cells on granular activated carbon in fluidized bed reactor. Food Bioprod Process 117:231–240
Jamali NS, Jahim JM, Mumtaz T, Abdul PM (2021) Dark fermentation of palm oil mill effluent by Caldicellulosiruptor saccharolyticus immobilized on activated carbon for thermophilic biohydrogen production. Environm Technol Innov 22:101477
Jo JH, Lee DS, Park D, Park JM (2008) Biological hydrogen production by immobilized cells of clostridium tyrobutyricum JM1 isolated from a food waste treatment process. Bioresour Technol 99(14):6666–6672
Júnior ADNF, Etchebehere C, Zaiat M (2015) Mesophilic hydrogen production in acidogenic packed-bed reactors (APBR) using raw sugarcane vinasse as substrate: influence of support materials. Anaerobe 34:94–105
Karapinar I, Yildiz PG, Pamuk RT, Gorgec FK (2020) The effect of hydraulic retention time on thermophilic dark fermentative biohydrogen production in the continuously operated packed bed bioreactor. Int J Hydrog Energy 45(5):3524–3531
Kerčmar J, Pintar A (2017) Support material dictates the attached biomass characteristics during the immobilization process in anaerobic continuous-flow packed-bed bioreactor. Anaerobe 48:194–202
Keskin T, Giusti L, Azbar N (2012) Continuous biohydrogen production in immobilized biofilm system versus suspended cell culture. Int J Hydrog Energy 37(2):1418–1424
Khamtib S, Sittijunda S, Imai T, Reungsang A (2021) Co-digestion of oil palm trunk hydrolysate and slaughterhouse wastewater for biohydrogen production in a fixed bed reactor by immobilized Thermoanaerobacterium thermosaccharolyticum KKU19 on expanded clay. Front Energy Res 9:683989

Khan MA, Ngo HH, Guo W, Liu Y, Zhang X, Guo J, Wang J (2018) Biohydrogen production from anaerobic digestion and its potential as renewable energy. Renew Energy 129:754–768

Khanal SK, Chen WH, Li L, Sung S (2004) Biological hydrogen production: effects of pH and intermediate products. Int J Hydrog Energy 29(11):1123–1131

Kim DH, Park JH, Kim SH, Kumar G, Lee BD, Kumar S, Yoon JJ (2021) Shift of microbial community structure by substrate level in dynamic membrane bioreactor for biohydrogen production. Int J Energy Res 45(12):17408–17416

Kirli B, Kapdan IK (2016) Selection of microorganism immobilization particle for dark fermentative biohydrogen production by repeated batch operation. Renew Energy 87:697–702

Kongjan P, Inchan S, Chanthong S, Jariyaboon R, Reungsang A, Sompong O (2019) Hydrogen production from xylose by moderate thermophilic mixed cultures using granules and biofilm up-flow anaerobic reactors. Int J Hydrog Energy 44(6):3317–3324

Kumar G, Mudhoo A, Sivagurunathan P, Nagarajan D, Ghimire A, Lay CH, Chang JS (2016) Recent insights into the cell immobilization technology applied for dark fermentative hydrogen production. Bioresour Technol 219:725–737

Kumar G, Cho SK, Sivagurunathan P, Anburajan P, Mahapatra DM, Park JH, Pugazhendhi A (2018a) Insights into evolutionary trends in molecular biology tools in microbial screening for biohydrogen production through dark fermentation. Int J Hydrog Energy 43(43):19885–19901

Kumar G, Shobana S, Nagarajan D, Lee DJ, Lee KS, Lin CY, Chang JS (2018b) Biomass based hydrogen production by dark fermentation—recent trends and opportunities for greener processes. Curr Opin Biotechnol 50:136–145

Kumar AN, Bandarapu AK, Mohan SV (2019) Microbial electro-hydrolysis of sewage sludge for acidogenic production of biohydrogen and volatile fatty acids along with struvite. Chem Eng J 374:1264–1274

Li F, Arthur EE, La D, Li Q, Kim H (2014) Immobilization of CoCl2 (cobalt chloride) on PAN (polyacrylonitrile) composite nanofiber mesh filled with carbon nanotubes for hydrogen production from hydrolysis of NaBH4 (sodium borohydride). Energy 71:32–39

Li W, He L, Cheng C, Cao G, Ren N (2020a) Effects of biochar on ethanol-type and butyrate-type fermentative hydrogen productions. Bioresour Technol 306:123088

Li Z, Wang J, Feng K, Li Y, Ding J, Liu B, Xing D (2020b) Rapid recruitment of hydrogen-producing biofilms for hydrogen production in a moving bed biofilm reactor by a sequential immobilization and deoxygenization approach. Bioresour Technol 317:123979

Litti YV, Zhuravleva EA, Kovalev AA, Kovalev DA, Katraeva IV, Parshina SN (2021) Biohydrogen production from food processing wastewater by a newly isolated thermophilic bacterium. In: IOP conference series: earth and environmental science, vol 938(1). IOP Publishing, p 012017

Litti YV, Potekhina MA, Zhuravleva EA, Vishnyakova AV, Gruzdev DS, Kovalev AA, Parshina SN (2022) Dark fermentative hydrogen production from simple sugars and various wastewaters by a newly isolated Thermoanaerobacterium thermosaccharolyticum SP-H2. Int J Hydrog Energy 47(58):24310–24327

Liu J, Zhou W, Fan S, Qiu B, Wang Y, Xiao Z, Qin Y (2019) Coproduction of hydrogen and butanol by Clostridium acetobutylicum with the biofilm immobilized on porous particulate carriers. Int J Hydrog Energy 44(23):11617–11624

Lu Y, Ma L, Liang Y, Shan B, Chang J (2018) Enhancing nitrogen removal performance in a bioreactor using immobilized anaerobic ammonium oxidation sludge by polyvinyl alcohol-sodium alginate (PVA-SA). Pol J Environ Stud 27(2):773–778

Łukajtis R, Hołowacz I, Kucharska K, Glinka M, Rybarczyk P, Przyjazny A, Kamiński M (2018) Hydrogen production from biomass using dark fermentation. Renew Sust Energ Rev 91:665–694

Lutpi NA, Jahim JM, Mumtaz T, Harun S, Abdul PM (2016) Batch and continuous thermophilic hydrogen fermentation of sucrose using anaerobic sludge from palm oil mill effluent via immobilisation technique. Process Biochem 51(2):297–307

Ma Z, Li C, Su H (2017) Dark bio-hydrogen fermentation by an immobilized mixed culture of Bacillus cereus and Brevumdimonas naejangsanensis. Renew Energy 105:458–464
Martins SCS, Martins CM, Fiúza LMCG, Santaella ST (2013) Immobilization of microbial cells: a promising tool for treatment of toxic pollutants in industrial wastewater. Afr J Biotechnol 12(28):4412
Mikheeva ER, Katraeva IV, Kovalev AA, Kovalev DA, Nozhevnikova AN, Panchenko V, Litti YV (2021) The start-up of continuous biohydrogen production from cheese whey: comparison of inoculum pretreatment methods and reactors with moving and fixed polyurethane carriers. Appl Sci 11(2):510
Mikheeva ER, Katraeva IV, Kovalev AA, Kovalev DA, Litti YV (2022a) Effects of pretreatment in a vortex layer apparatus on the properties of confectionery wastewater and its dark fermentation. Int J Hydrog Energy 47(55):23165–23174
Mikheeva ER, Katraeva IV, Vorozhtsov DL, Kovalev DA, Kovalev AA, Grigoriev VS, Litti YV (2022b) Dark fermentative biohydrogen production from confectionery wastewater in continuous-flow reactors. Int J Hydrog Energy 47(53):22348–22358
Mishra P, Thakur S, Mahapatra DM, Ab Wahid Z, Liu H, Singh L (2018) Impacts of nano-metal oxides on hydrogen production in anaerobic digestion of palm oil mill effluent–a novel approach. Int J Hydrog Energy 43(5):2666–2676
Mitchell RJ, Kim JS, Jeon BS, Sang BI (2009) Continuous hydrogen and butyric acid fermentation by immobilized Clostridium tyrobutyricum ATCC 25755: effects of the glucose concentration and hydraulic retention time. Bioresour Technol 100(21):5352–5355
Mohan SV, Mohanakrishna G, Reddy SS, Raju BD, Rao KR, Sarma PN (2008) Self-immobilization of acidogenic mixed consortia on mesoporous material (SBA-15) and activated carbon to enhance fermentative hydrogen production. Int J Hydrog Energy 33(21):6133–6142
Mullai P, Yogeswari MK, Sridevi KJBT (2013) Optimisation and enhancement of biohydrogen production using nickel nanoparticles–a novel approach. Bioresour Technol 141:212–219
Muri P, Marinšek-Logar R, Djinović P, Pintar A (2018) Influence of support materials on continuous hydrogen production in anaerobic packed-bed reactor with immobilized hydrogen producing bacteria at acidic conditions. Enzym Microb Technol 111:87–96
Nakatani H, Ding N, Ohara Y, Hori K (2018) Immobilization of Enterobacter aerogenes by a trimeric autotransporter adhesin, AtaA, and its application to biohydrogen production. Catalysts 8(4):159
Nikolaev YA, Plakunov VK (2007) Biofilm—"City of microbes" or an analogue of multicellular organisms? Microbiology 76(2):125–138
Niu K, Zhang X, Tan WS, Zhu ML (2010) Characteristics of fermentative hydrogen production with Klebsiella pneumoniae ECU-15 isolated from anaerobic sewage sludge. Int J Hydrog Energy 35(1):71–80
Nozhevnikova AN, Russkova YI, Litti YV, Parshina SN, Zhuravleva EA, Nikitina AA (2020) Syntrophy and interspecies electron transfer in methanogenic microbial communities. Microbiology 89(2):129–147
Park JH, Sim YB, Kim JS, Kim SH (2020) Effects of alginate immobilization on dynamic membrane formation and H2 fermentation from galactose. Int J Hydrog Energy 45(10):5874–5880
Park JH, Chandrasekhar K, Jeon BH, Jang M, Liu Y, Kim SH (2021) State-of-the-art technologies for continuous high-rate biohydrogen production. Bioresour Technol 320:124304
Patel SK, Lee JK, Kalia VC (2018) Beyond the theoretical yields of dark-fermentative biohydrogen. Indian J Microbiol 58(4):529–530
Patel SK, Gupta RK, Das D, Lee JK, Kalia VC (2021) Continuous biohydrogen production from poplar biomass hydrolysate by a defined bacterial mixture immobilized on lignocellulosic materials under non-sterile conditions. J Clean Prod 287:125037
Penniston J, Gueguim Kana EB (2018) Impact of medium pH regulation on biohydrogen production in dark fermentation process using suspended and immobilized microbial cells. Biotechnol Biotechnol Equip 32(1):204–212

Perna V, Castelló E, Wenzel J, Zampol C, Lima DF, Borzacconi L, Etchebehere C (2013) Hydrogen production in an upflow anaerobic packed bed reactor used to treat cheese whey. Int J Hydrog Energy 38(1):54–62
Pugazhendhi A, Shobana S, Nguyen DD, Banu JR, Sivagurunathan P, Chang SW, Kumar G (2019) Application of nanotechnology (nanoparticles) in dark fermentative hydrogen production. Int J Hydrog Energy 44(3):1431–1440
Rai P, Pandey A, Pandey A (2022) Evaluation of low cost immobilized support matrices in augmentation of biohydrogen potential in dark fermentation process using B. licheniformis AP1. Fuel 310:122275
Rathore S, Desai PM, Liew CV, Chan LW, Heng PWS (2013) Microencapsulation of microbial cells. J Food Eng 116(2):369–381
Ren NQ, Tang J, Liu BF, Guo WQ (2010) Biological hydrogen production in continuous stirred tank reactor systems with suspended and attached microbial growth. Int J Hydrog Energy 35(7):2807–2813
Ren HY, Wang XW, Kong F, Zhao L, Xing D, Ren NQ, Liu BF (2022) Enhanced semi-continuous hydrogen production by addition of microplastics under mesophilic and thermophilic fermentation. Fuel 315:123247
Renaudie M, Clion V, Dumas C, Vuilleumier S, Ernst B (2021) Intensification and optimization of continuous hydrogen production by dark fermentation in a new design liquid/gas hollow fiber membrane bioreactor. Chem Eng J 416:129068
Roy S, Vishnuvardhan M, Das D (2014) Improvement of hydrogen production by newly isolated Thermoanaerobacterium thermosaccharolyticum IIT BT-ST1. Int J Hydrog Energy 39(14):7541–7552
Sambusiti C, Bellucci M, Zabaniotou A, Beneduce L, Monlau F (2015) Algae as promising feedstocks for fermentative biohydrogen production according to a biorefinery approach: a comprehensive review. Renew Sust Energ Rev 44:20–36
Seelert T, Ghosh D, Yargeau V (2015) Improving biohydrogen production using Clostridium beijerinckii immobilized with magnetite nanoparticles. Appl Microbiol Biotechnol 99(9):4107–4116
Sekoai PT, Awosusi AA, Yoro KO, Singo M, Olawale O, Ayeni AO, Bodunrin M, Daramola MO (2018) Microbial cell immobilization in biohydrogen production: a short overview. Crit Rev Biotechnol 38(2):157–171. https://doi.org/10.1080/07388551.2017.1312274
Sekoai PT, Daramola MO, Mogwase B, Engelbrecht N, Yoro KO, du Preez SP, Hlongwane GN (2020) Revising the dark fermentative H2 research and development scenario–an overview of the recent advances and emerging technological approaches. Biomass Bioenergy 140:105673
Shanmugam S, Hari A, Pandey A, Mathimani T, Felix L, Pugazhendhi A (2020a) Comprehensive review on the application of inorganic and organic nanoparticles for enhancing biohydrogen production. Fuel 270:117453
Shanmugam S, Krishnaswamy S, Chandrababu R, Veerabagu U, Pugazhendhi A, Mathimani T (2020b) Optimal immobilization of Trichoderma asperellum laccase on polymer coated Fe3O4@ SiO2 nanoparticles for enhanced biohydrogen production from delignified lignocellulosic biomass. Fuel 273:117777
Show KY, Lee DJ, Chang JS (2011) Bioreactor and process design for biohydrogen production. Bioresour Technol 102(18):8524–8533
Shuttleworth PS, Parker HL, Hunt AJ, Budarin VL, Matharu AS, Clark JH (2014) Applications of nanoparticles in biomass conversion to chemicals and fuels. Green Chem 16(2):573–584
Singh L, Siddiqui MF, Ahmad A, Rahim MHA, Sakinah M, Wahid ZA (2013) Biohydrogen production from palm oil mill effluent using immobilized mixed culture. J Ind Eng Chem 19(2):659–664
Sivagurunathan P, Kumar G, Sen B, Lin CY (2014) Development of a novel hybrid immobilization material (HY-IM) for fermentative biohydrogen production from beverage wastewater. J Chin Chem Soc 61(7):827–830
Smid EJ, Lacroix C (2013) Microbe–microbe interactions in mixed culture food fermentations. Curr Opin Biotechnol 24(2):148–154

Sompong O, Prasertsan P, Karakashev D, Angelidaki I (2008) High-rate continuous hydrogen production by Thermoanaerobacterium thermosaccharolyticum PSU-2 immobilized on heat-pretreated methanogenic granules. Int J Hydrog Energy 33(22):6498–6508

Song YH, Lu Y, Yu L (2018) Stoichiometry and thermodynamic analysis on biohydrogen production from xylose by klebsiella oxytoca GS-4-08. Energy Fuel 33(1):356–361

Sousa C, Teixeira P, Oliveira R (2009) Influence of surface properties on the adhesion of Staphylococcus epidermidis to acrylic and silicone. Int J Biomater 2009:718017

Srivastava N, Srivastava M, Manikanta A, Singh P, Ramteke PW, Mishra PK (2017) Nanomaterials for biofuel production using lignocellulosic waste. Environ Chem Lett 15(2):179–184

Srivastava N, Srivastava M, Malhotra BD, Gupta VK, Ramteke PW, Silva RN, Mishra PK (2019) Nanoengineered cellulosic biohydrogen production via dark fermentation: a novel approach. Biotechnol Adv 37(6):107384

Srivastava N, Srivastava M, Mishra PK, Kausar MA, Saeed M, Gupta VK, Ramteke PW (2020) Advances in nanomaterials induced biohydrogen production using waste biomass. Bioresour Technol 307:123094

Stojkovic D, Torzillo G, Faraloni C, Valant M (2015) Hydrogen production by sulfur-deprived TiO2-encapsulated Chlamydomonas reinhardtii cells. Int J Hydrog Energy 40(8):3201–3206

Sun Y, Yang G, Zhang J, Wen C, Sun Z (2019) Optimization and kinetic modeling of an enhanced bio-hydrogen fermentation with the addition of synergistic biochar and nickel nanoparticle. Int J Energy Res 43(2):983–999

Sunyoto NM, Zhu M, Zhang Z, Zhang D (2016) Effect of biochar addition on hydrogen and methane production in two-phase anaerobic digestion of aqueous carbohydrates food waste. Bioresour Technol 219:29–36

Ta DT, Lin CY, Ta TMN, Chu CY (2020) Biohythane production via single-stage anaerobic fermentation using entrapped hydrogenic and methanogenic bacteria. Bioresour Technol 300:122702

Taherdanak M, Zilouei H, Karimi K (2015) Investigating the effects of iron and nickel nanoparticles on dark hydrogen fermentation from starch using central composite design. Int J Hydrog Energy 40(38):12956–12963

Taherdanak M, Zilouei H, Karimi K (2016) The effects of Fe0 and Ni0 nanoparticles versus Fe2+ and Ni2+ ions on dark hydrogen fermentation. Int J Hydrog Energy 41(1):167–173

Tawfik A, Nasr M, Galal A, El-Qelish M, Yu Z, Hassan MA, Lee M (2021) Fermentation-based nanoparticle systems for sustainable conversion of black-liquor into biohydrogen. J Clean Prod 309:127349

Temudo MF, Muyzer G, Kleerebezem R, van Loosdrecht M (2008) Diversity of microbial communities in open mixed culture fermentations: impact of the pH and carbon source. Appl Microbiol Biotechnol 80(6):1121–1130

Vasudevan R (2014) Biofilms: microbial cities of scientific significance. J Microbiol Exp 1(3):00014

Wang J, Yin Y (2019) Progress in microbiology for fermentative hydrogen production from organic wastes. Crit Rev Environ Sci Technol 49(10):825–865

Wang B, Li Y, Ren N (2013) Biohydrogen from molasses with ethanol-type fermentation: effect of hydraulic retention time. Int J Hydrog Energy 38(11):4361–4367

Wannapokin A, Cheng YT, Wu SZ, Hsieh PH, Hung CH (2021) Potential of bio-hydrogen production by C. pasteurianum co-immobilized with selected nano-metal particles. Int J Hydrog Energy 46(20):11337–11344

Wong YM, Wu TY, Juan JC (2014) A review of sustainable hydrogen production using seed sludge via dark fermentation. Renew Sust Energ Rev 34:471–482

Wongthanate J, Polprasert C (2015) Immobilized biofilm in thermophilic biohydrogen production using synthetic versus biological materials. Braz Arch Biol Technol 58:124–130

Wu SY, Lin CN, Chang JS, Lee KS, Lin PJ (2002) Microbial hydrogen production with immobilized sewage sludge. Biotechnol Prog 18(5):921–926

Wu KJ, Chang JS, Chang C (2006) Biohydrogen production using suspended and immobilized mixed microflora. J Chin Inst Chem Eng 37(6):545

Wu J, Dong L, Zhou C, Liu B, Xing D, Feng L, Cao G (2019) Enhanced butanol-hydrogen coproduction by Clostridium beijerinckii with biochar as cell's carrier. Bioresour Technol 294:122141

Wu M, Jiang Y, Liu Y, Mou L, Zhang W, Xin F, Jiang M (2021) Microbial application of thermophilic Thermoanaerobacterium species in lignocellulosic biorefinery. Appl Microbiol Biotechnol 105(14):5739–5749

Yang G, Wang J (2018) Improving mechanisms of biohydrogen production from grass using zero-valent iron nanoparticles. Bioresour Technol 266:413–420

Yin Y, Zhuang S, Wang J (2018) Enhanced fermentative hydrogen production using gamma irradiated sludge immobilized in polyvinyl alcohol (pva) gels. Environ Prog Sustain Energy 37(3):1183–1190

Yu L, Jiang W, Yu Y, Sun C (2014) Effects of dilution ratio and Fe0 dosing on biohydrogen production from dewatered sludge by hydrothermal pretreatment. Environ Technol 35(24):3092–3104

Zhang ZP, Tay JH, Show KY, Yan R, Liang DT, Lee DJ, Jiang WJ (2007) Biohydrogen production in a granular activated carbon anaerobic fluidized bed reactor. Int J Hydrog Energy 32(2):185–191

Zhang C, Kang X, Liang N, Abdullah A (2017) Improvement of biohydrogen production from dark fermentation by cocultures and activated carbon immobilization. Energy Fuel 31(11):12217–12222

Zhang Q, Xu S, Li Y, Ding P, Zhang Y, Zhao P (2021) Green-synthesized nickel oxide nanoparticles enhances biohydrogen production of Klebsiella sp. WL1316 using lignocellulosic hydrolysate and its regulatory mechanism. Fuel 305:121585

Zhao L, Cao GL, Wang AJ, Guo WQ, Liu BF, Ren HY, Ma F (2012a) Enhanced bio-hydrogen production by immobilized Clostridium sp. T2 on a new biological carrier. Int J Hydrog Energy 37(1):162–166

Zhao L, Cao GL, Yao J, Ren HY, Ma F, Ren NQ, Wang AJ (2012b) Optimization of immobilization parameters of Thermoanaerobacterium thermosaccharolyticum W16 on a new carrier for enhanced hydrogen production. RSC Adv 2(19):7391–7395

Zhao W, Zhang Y, Du B, Wei D, Wei Q, Zhao Y (2013) Enhancement effect of silver nanoparticles on fermentative biohydrogen production using mixed bacteria. Bioresour Technol 142:240–245

Zhao C, Sharma A, Ma Q, Zhu Y, Li D, Liu Z, Yang Y (2021) A developed hybrid fixed-bed bioreactor with Fe-modified zeolite to enhance and sustain biohydrogen production. Sci Total Environ 758:143658

Zheng H, Zeng RJ, Angelidaki I (2008) Biohydrogen production from glucose in upflow biofilm reactors with plastic carriers under extreme thermophilic conditions (70 C). Biotechnol Bioeng 100(5):1034–1038

Chapter 10
Enhancement of Biohydrogen Production Using Chemical Additives and Nanoparticles

Anita Atukunda and Mahmoud Nasr

Abstract Biohydrogen is a promising fuel with the potential to satisfy the need for environment-friendly energy sources. The low efficiency of biohydrogen production is a bottleneck to the scale-up prospects for its production. Different enhancement strategies have been suggested to improve biohydrogen production efficiency. This chapter critically reviews some biohydrogen production processes, pretreatment technologies, chemicals (metal monomers, ions, and oxides, EDTA, vitamins, buffer solution, and surfactants), and nanoparticles used in the fermentation process. The impact of the additives on biohydrogen production and corresponding functional mechanisms were discussed. Furthermore, a bibliometric technique was used to fully explain the recent development, present trends, and prospective future of biohydrogen generation enhancement. From the available literature, there is a significant correlation between bio-H_2 production and the corresponding chemical additives and nanoparticles. Reagent supplements, nanomaterial, and pretreatment techniques have been used to improve the biodegradability of organic substrates for higher biohydrogen production rates. Finally, recommendations for future researchers, investors, and stakeholders are proposed to encourage renewable energy generation and sustainable production technologies.

Keywords Biofermentation pathways · Economics · H_2 yield · Inhibitory factors · Nanomaterial · Reagent additives

---

A. Atukunda
Environmental Engineering Department, Egypt-Japan University of Science and Technology (E-JUST), Alexandria, Egypt

M. Nasr (✉)
Environmental Engineering Department, Egypt-Japan University of Science and Technology (E-JUST), Alexandria, Egypt

Sanitary Engineering Department, Faculty of Engineering, Alexandria University, Alexandria, Egypt
e-mail: mahmoud.nasr@ejust.edu.eg

C. R. Soccol et al. (eds.), *Biohydrogen - Advances and Processes*, Biofuel and Biorefinery Technologies 13, https://doi.org/10.1007/978-3-031-49818-3_10

10.1 Hydrogen Production

Humanity has used carbon-based fuels like coal and oil to meet energy needs throughout the past century, releasing greenhouse gases (GHG) such as carbon dioxide into the atmosphere. Up to 50% of the available energy resources are expected to be consumed by 2050 (Bharathiraja et al. 2018). The GHG emissions peaked in 2018, and then grew by 6% in 2021, owing to a rise in CO_2 emissions from burning fossil fuels. This pattern occurs largely due to a resurgence in economic activity following the peak of the COVID-19 pandemic. Given this, production processes and technology utilizing renewable energy sources such as wind, solar, and biomass have attracted significant research interest. Among the energy fuels that could be obtained from renewable sources is hydrogen gas. Hydrogen is a promising fuel source with minimum (or no) GHG emissions. It is regarded as a viable green energy source capable of meeting the rising energy demand (Mthethwa et al. 2018). Hydrogen consumes oxygen when used as a fuel and only releases H_2O (Ali et al. 2021). It is more efficient than conventional fuels and a viable replacement for commercial projects like power plants, automobiles, and industrial processes (Qyyum et al. 2022). Moreover, H_2 has a minimum impact on the environment, and it positively contributes to greenization (Dincer and Acar 2015). Additionally, H_2 has a high mass-energy content of 141.9 MJ/kg, which is nearly 3–4 times higher than those of conventional hydrocarbon fuels such as coal (~33 MJ/kg), natural gas (~55 MJ/kg), and gasoline (~46 MJ/kg) (Elreedy et al. 2018). The amount of H_2 supply is very low compared with the global energy demand; however, this quantity is expected to increase more than twofold due to activities in the ammonia, petrochemical, fuel cell, and steel manufacturing industries (Staffell et al. 2019). Furthermore, H_2 will share more than half the global energy demand used for transport, specifically light-duty vehicles (Staffell et al. 2019). According to these predictions, the current infrastructures should be upgraded to connect to the green H_2 supply markets.

10.2 Bibliometric in Scopus (2011–2021) Related to Bio-H_2 Production

To assess the past and current trends of research on biohydrogen production and the associated enhancement patterns using chemical additives and nanoparticles, a bibliometric analysis was performed. Two main keyword searches were included (1) "biohydrogen" and "additives," and (2) "biohydrogen" and "nanoparticles," using the SCOPUS database in the last decade. The search queries applied were ((TITLE-ABS-KEY (("biohydrogen" and "additives")) and ((TITLE-ABS-KEY (("biohydrogen" and "nanoparticles")). The search revealed 56 articles for biohydrogen and additives and 128 documents for nanoparticles. These records were exported in a

CSV file format and analyzed using VOS viewer software version 1.6.17 (Dadebo et al. 2022).

The articles from the "biohydrogen and additives" search contained 135 keywords. A network diagram was constructed considering a threshold frequency of 3, revealing 58 keywords with strong links. The articles from the "biohydrogen and nanoparticles" search contained 1641 keywords. A network diagram was constructed considering a threshold frequency of 4, demonstrating strong connections among 78 keywords. The network diagrams and detailed explanation of the keyword links and clusters are presented in the relevant respective sections throughout this chapter.

Figure 10.1 shows the overlay visualization map representing the keyword co-occurrence in publications per year over the past decade. In this network, the size of bubbles increases for a higher occurrence of the specific keyword in research articles. The line connecting the keywords increases in thickness for strong relationships. The colors represent the publication year, with dark blue and green indicating keywords present in articles from earlier years and a light green to yellow indicating emerging or trending keywords.

In 2010, there is a co-occurrence of keywords like "biogas," "industrial wastes," "esters," and "environmental management," indicating that the concept of "renewable energy" was widely addressed. Industrial waste treatment has been considered

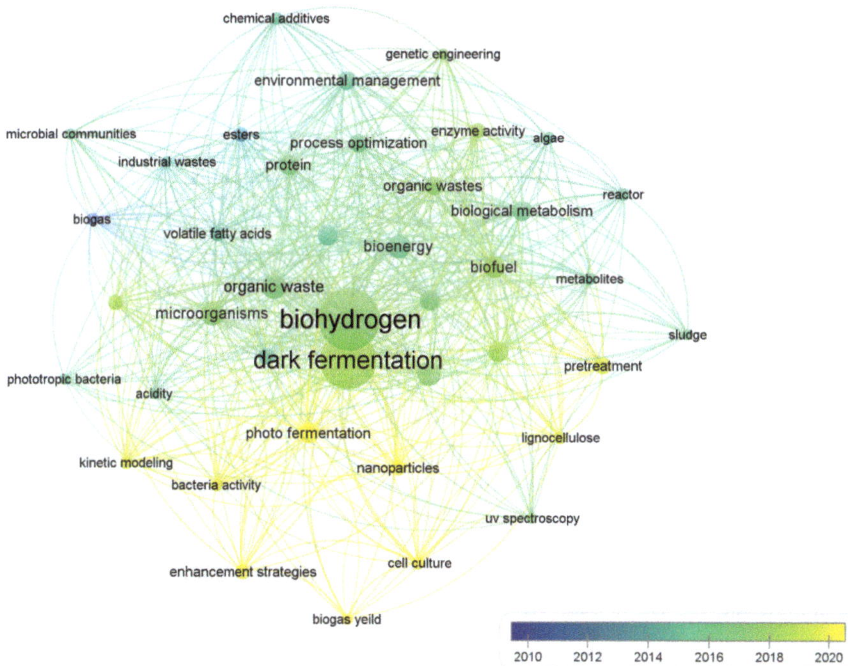

Fig. 10.1 Overlay visualization map of the top 58 keywords co-occurrences using search strings relevant to "biohydrogen" and "additives"

an environmental concern for the past century, and efforts to combat this, including anaerobic digestion, have been suggested. Due to the need to replace fossil fuels, biogas has also been greatly researched as a viable, sustainable, and environment-friendly option.

From 2014 to 2016, there is a high co-occurrence of keywords like "biohydrogen," "bioenergy," "process optimization," "dark fermentation," "organic wastes," etc. Biohydrogen production has been greatly championed as renewable energy due to its high calorific rates and nontoxic emissions during combustion (Mthethwa et al. 2018). Research into the use of dark fermentation for hydrogen production drastically increased mostly due to its economic advantages of coupling organic waste reduction with the recovery of renewable bioenergy (Akhlaghi and Najafpour-darzi 2020). Optimizing the bioreactor operating condition is vital to enhance bacterial activity and hydrogen production, making it an attractive alternative bioenergy.

Currently (2020 to date), there is emerging research in the field of "nanoparticles," "enhancement strategies," "biogas yield," "pretreatment," "lignocellulose," "photofermentation," "additives" "cell culture," etc. Using different techniques to optimize the effectiveness of the fermentation process and intensify the hydrogen yield is a trending concept. Various studies have explored different enhancement strategies, using additives such as nanoparticles and pretreatment techniques (like acid, alkali, and ultrasonic) to improve the organic matter solubility (Yang and Wang 2018a). Nanoparticles of metal additives have gained special attention such as Fe, Ni, and Ti, due to their positive effect on hydrolytic activity, electron transport, and microculture metabolism during dark and photofermentations (Cheng et al. 2022a; Arisht et al. 2021).

10.3 Methods and Technologies of H_2 Production

Hydrogen can be generated from biomass (organic wastes, algae), renewable energy (solar, wind), fossil fuel (coal, oil, natural gas), and nuclear energy via different routes (Sang et al. 2006). The production methods, with a wide range of techniques, can be grouped as biological, thermal, electrical, and photonic processes (Dincer and Acar 2015). Most technologies and methods involved in scaling up hydrogen production are relatively new and still under development.

Table 10.1 summarizes the hydrogen production methods, their primary source/feedstock, advantages, and disadvantages. The biological processes of hydrogen generation involve the use of organics as an energy source and the action of microorganisms, usually in the absence of oxygen. The main biological method is dark fermentation (Akhlaghi and Najafpour-darzi 2020). Biomass sources include organic waste and wastewater from agriculture, domestic, municipal, industrial, and lignocellulosic biomass. The biological method is the most attractive option for H_2 generation due to several fundamental benefits, including the simplicity of the production process, low energy requirements, economic viability, and accessibility of inexpensive feedstock (frequently obtained from wastes) (Saratale et al. 2013). The

Table 10.1 Comparison between different hydrogen production technologies

Methods	Brief description (feedstock/ source and mechanism)	Advantages	Disadvantages
Biological process			
Dark fermentation	• Biomass • Bioreactors are used to generate H_2 in the absence of light	• Well-established technology • Simple reactor technology • Low cost of materials • Relatively high production • Clean and sustainable technology	• Large amounts of byproducts (mainly volatile fatty acids) • Depends on organics' biodegradability • Low COD removal
Electrical process			
Electrolysis	• Water, urea • A direct current is used to split water molecules into O_2 and H_2 • Oxygen is produced at the anode: $2OH^- \rightarrow H_2O + \frac{1}{2}O_2 + 2e^-$ • H_2 is produced at the cathode: $2H_2O + 2e^- \rightarrow H_2 + 2OH^-$ • Overall reaction: $H_2O \rightarrow H_2 + \frac{1}{2}O_2$	• Possibility of treating wastewater that contain ammonia, urea, or hydrazine • Possibility of reducing COD and nitrogen • Cheap materials • Possible large-scale applications • No additional pollution • Integration with fuel cells	• Low system efficiency • High capital costs
Reverse electrodialysis	• A membrane-based technology (anion exchange membrane and cation exchange membrane) • Energy and hydrogen are produced due to a concentration gradient between two separate solutions (a diluted solution and a concentrated solution).	• Production of energy and hydrogen by only concentration gradient between two solutions (no special chemicals, or gas discharge are required)	• A narrow range of application • Still in the early research and development phase
Thermal process			
Thermolysis	• Thermochemical energy • Heat is the main reaction driver • Thermal decomposition reaction of water	• Reaction completion within a one step • Requires heat and water only • No electrodes are required	• Requires extremely high temperatures (>2700 °C) at 1 bar

(continued)

Table 10.1 (continued)

Methods	Brief description (feedstock/ source and mechanism)	Advantages	Disadvantages
Plasma decomposition	• Fossil fuels, hydrocarbons • Natural gas passes through a thermal plasma arc to generate H_2 • The decomposition of hydrogen-containing organic molecules	• Water decomposition in acousto-plasma discharge forms almost pure hydrogen (98%), with low concentrations of carbon oxides (5%–6%) • The decomposition of organic liquids forms solid particles • Can be used in waste management, such as healthcare waste and steel industry waste	• Technology is still in R&D[a]
Thermochemical water splitting	• Water molecules are split via a cyclic pathway (closed loop) consisting of a series of intermediate chemical reactions • A temperature range of 500–2000 °C is required	• Hydrogen-oxygen separation membranes are not required • No requirement for electricity • More economical in comparison to thermolysis	• Multistep process (two to four-step reactions)
Photonic process			
Photocatalysis	• Water • Water is split into H_2 in the presence of a photocatalyst $H_2O \xrightarrow{\text{Solar energy, photocatalyst}} H_2 + (1/2) O_2$	• Numerous new photocatalyst material studies • Availability of durable and efficient photocatalysts	• Low light-to-H_2 efficiency • Hazardous reagents and chemicals • May be required for advanced photocatalysis
PV electrolysis	• Water, solar • PV panels are used to generate the current to split water into H_2	• Readily available inputs (water and sun)	• Very expensive • The required cost is about 25 times higher than that of fossil fuel alternative

(continued)

Table 10.1 (continued)

Methods	Brief description (feedstock/ source and mechanism)	Advantages	Disadvantages
Photoelectrochemical	• Water • A photoelectrochemical cell produces H_2 along with photocurrent upon absorption of light	• Numerous new photocatalyst materials were developed • Durable and efficient photocatalysts available • Low operation temperature • Disinfection of wastewaters	• Low light-to-H_2 efficiency • Hazardous reagents may be required for the process
Hybrids			
Photofermentation (biological +photonic)	• Biomass and water • The fermentation process is activated by exposure to light	• Readily available waste streams as supply • Achieves higher substrate conversion than dark-fermentation	• Hydrogenase enzyme is highly sensitive to O_2 • Low light conversion efficiency • Metabolic and genetic engineering is required • Low volumetric production rate • Large surface area requirement
Biophotolysis (biological +photonic)	• Biomass and water • Bioreactors are used to generate H_2 and the process is activated by exposure to light	• Abundant supply (water) • Carbon-independent pathway • The only products are H_2 and O_2	• Separation of H_2 and O_2 is a complex step • Low conversion efficiencies • Large surface area requirement
Liquid gasification (biological + thermal)	• Water, chemicals, liquid fuels • Gasification beyond the critical pressure and critical temperature point of the liquid	• A better alternative to biomass thermal cracking	• High energy requirement • Challenges in scaling up • Needs higher temperatures for H_2 production • Produced gas has low H_2 content

(continued)

Table 10.1 (continued)

Methods	Brief description (feedstock/ source and mechanism)	Advantages	Disadvantages
Coal/ biomass gasification (biological + thermal + electrical)	• Biomass • Conversion of biomass into syngas, and further into H_2 • Temperature requirement of 500–900 °C	• Mitigating CO_2 emissions • Waste biomass can be used • Replacing fossil fuels with sustainable biomass fuel • Provides strategy for municipal solid waste management	• The thermal efficiency of the gasification process is typically low. • Seasonal availability and high handling costs (for coal) • Nontotal solid conversion and tars production • Fuel process limitations such as corrosion, pressure, resistance, and hydrogen aging
Microbial electrodialysis (electrical + biological)	• Liquid base with microorganisms • Energy and hydrogen are produced in the presence of electrolysis cells and microorganisms	• COD reduction with H_2 production is possible	• Still in the R&D phase • Detailed analysis of the system is still needed

[a]*R&D* Research and development

thermal and electrical methods involve heating renewable sources like water, biomass, and fossil fuels at high temperatures (over 500 °C). These processes, including gasification and pyrolysis, are used to release syngas ($H_2/CO/CO_2$). Solar irradiation is also used to maintain the photocatalytic water-splitting process, generating H_2 and O_2. This pathway is known as H_2 generation by photonics (solar-to-chemical energy conversion). A photoelectrochemical cell, a photocatalyst, or PV panels are used for water photosplitting to green hydrogen energy. Finally, the hybrid methods, e.g., electrical-thermal, photonic-biological, and electrical-photonic, have also been employed to generate hydrogen.

10.4 Biological Hydrogen Production: Concept and Methodology

In contrast to other methods, producing hydrogen using biological methods via the action of microorganisms requires less energy and it is considered more environmentally benign (Aydin et al. 2021). Agricultural waste, food waste, sewage sludge, and wastewater can be used as feedstocks to produce biohydrogen (Eun et al. 2021).

They provide a green and sustainable solution to the issues of energy scarcity, waste management, environmental pollution, and global warming (Cheng et al. 2022b). Biohydrogen production can be achieved by two major processes: light-dependent and light-independent. Biophotolysis and photofermentation are light-dependent processes, whereas dark anaerobic fermentation is light-independent.

10.4.1 Photofermentation and Biophotolysis

Photofermentation and biophotolysis are biochemical reactions used to generate H_2 by using photonic energy and water.

During photofermentation, purple nonsulfur bacteria (PNB) utilize light (either solar or artificial) as an energy source to convert organic acids into H_2, mainly under anaerobic conditions. When acetic acid is the representative volatile fatty acid (VFA) in the fermentation process, 4 mol of H_2 per 1 mol of CH_3COOH can be produced (Eq. 10.1) (Jat Baloch et al. 2022):

$$CH_3COOH + 2H_2O \xrightarrow{\text{Light}} 4H_2 + 2CO_2 \tag{10.1}$$

PNB use their single photosystem to produce adenosine triphosphate (ATP) molecules through cyclic photophosphorylation. As a result, energy electrons flow freely and they react with proton (H^+) to generate H_2 via the synthesis and activity of nitrogenase enzymes (Eqs. 10.1 and 10.2) (Hu et al. 2018):

$$N_2 + 8H^+ + 8e^- + 16ATP \rightarrow 2NH_3 + H_2 + 16ADP + 16Pi \tag{10.2}$$

$$2H^+ + 2e^- + 4ATP \rightarrow H_2 + 4ADP + 4P \tag{10.3}$$

The rate of H_2 synthesis in photofermentation is influenced by light intensity, wavelength, and illumination pattern (Boran et al. 2012). Low light intensity might cause poor effectiveness of PNB and insufficient light transformation for biohydrogen production. Due to cell shadowing, light penetration inside the photoreactor can be limited, thereby reducing the light utilization rate and thus biohydrogen production (Cheng et al. 2022a). A maximum light intensity of 150–200 W/m^2 (5000–7000 lux) could be favorable for bio-H_2 production by photofermenting bacteria (Su et al. 2009; Zhao and Chen 2011; Zhang et al. 2021), whereas, a low light intensity of about 10.5 W/m^2 could also be employed (Liu et al. 2010).

Biophotolysis generates hydrogen from water splitting using sunlight radiation as an energy source. This water dissociation process occurs in biological systems. The method has picked researchers' interest because of the potential to generate clean fuels from only water and light (Sang et al. 2006). Hydrogen production via biophotolysis can be mediated by green algae (*Chlamydomonas reinhardtii*) or cyanobacteria (*Anabaena cylindrical*). Both nitrogenase and hydrogenase enzyme systems can be responsible for this light-driven splitting of water into H_2 and O_2 (Sun et al. 2019).

Biophotolysis can be further classified into direct and indirect. In direct biophotolysis, hydrogen formation is catalyzed by both the photosynthesis and hydrogenase enzymes, maintaining solar energy to chemical energy conversion (Akhlaghi and Najafpour-darzi 2020). This step depends on photosystems (both PSI and PSII) under anaerobic conditions (Aydin et al. 2021). With the help of the light energy captured by the PSI (700 nm) and/or PSII (680 nm), water is oxidized into protons, electrons/reducing equivalents, and O_2. This reaction is followed by protons reduction via the ferredoxin (Fd) enzyme to form H_2, giving an overall reaction Eq. 10.4:

$$2H_2O \xrightarrow{\text{PS,Hydrogenase,light}} 2H_2 + O_2 \tag{10.4}$$

In indirect biophotolysis, photosynthesis is separated from the hydrogenase enzyme reactions (Sun et al. 2019). This effectively limits O_2 generation in PSII, thereby avoiding the inhibition of oxygen-sensitive enzymes (Sang et al. 2006). Separating the release of O_2 and H_2 could also facilitate the H_2 purification process, where CO_2 can be easily eliminated from the gas mixture. This pathway is shown in Eqs. 10.5 and 10.6:

$$6CO_2 + 6H_2O \xrightarrow{\text{PS,light}} C_6H_{12}O_6 + 6O_2 \tag{10.5}$$

$$C_6H_{12}O_6 + 6H_2O \xrightarrow{\text{hydrogenase}} 12H_2 + 6CO_2 \tag{10.6}$$

In addition to renewable resources (sunlight and water) utilization, biophotolysis has advantages like being cost-effective (biocatalysts are inexpensive), high purity hydrogen generation, and no additional carbon or nutrient supplements, making the process "low-to-net zero carbon" (Shanmugam et al. 2020). Nonetheless, these photoprocesses have major setbacks regarding the application (Aydin et al. 2021), such as

- the high sensitivity of hydrogenase enzyme to oxygen; hence, oxygen-resistant hydrogenase should be developed,
- the light conversion efficiency of the process can be lower than 2%, especially when treating concentrated wastewater with darker colors,
- several factors such as light intensity, wavelength, and illumination pattern should be controlled and optimized to have high-purity H_2.

10.4.2 Dark Fermentation

Dark fermentation uses hydrogen-producing microorganisms under anaerobic conditions to produce bioenergy from organic-rich substrates (Akhlaghi and Najafpour-darzi 2020). The main outputs of this process by several heterotrophic bacteria are H_2, CO_2, VFAs, and alcohols. The most common microbial communities responsible for bio-H_2 production under dark conditions include facultative anaerobes (*Citrobacter* sp., *Escherichia coli*, and *Enterobacter*), and obligate anaerobes (rumen microbes, *Clostridia*, and methylotrophic methanogens) (Argun and Kargi 2011). The process is affected by medium pH, fermentation time, substrate composition, microculture, and gas partial pressure (Akhlaghi and Najafpour-darzi 2020). The amount of H_2 generated depends on soluble metabolite formation, following either the acetate or butyrate pathway. Considering the acetic metabolic route, each mole of glucose is accompanied by 4 moles of H_2 (Eq. 10.7), whereas 2 moles of H_2 can be acquired for a final product of butyric acid (Eq. 10.8) (Nasr et al. 2013).

$$C_6H_{12}O_6 + 2H_2O \rightarrow 2C_2H_4O_2 + 2CO_2 + 4H_2 \quad (10.7)$$

$$C_6H_{12}O_6 \rightarrow C_4H_8O_2 + 2CO_2 + 2H_2 \quad (10.8)$$

In dark fermentation, bacteria use various substrates (e.g., glucose) to synthesize pyruvate and initiate the acidogenic glycolysis pathway to generate H_2. In this process, NAD^+ is reduced to nicotinamide adenine dinucleotide (NADH) for generating adenosine triphosphate (ATP), the energy required for microorganisms' activity during anaerobic conditions (Sinha and Pandey 2011). Secondly, pyruvate is converted into acetyl-CoA. This may happen via two pathways, depending on the species (Łukajtis et al. 2018):

1. Pyruvate dehydrogenase pathway: pyruvate can be oxidized in a reaction catalyzed by pyruvate:ferredoxin oxido-reductase (PFOR) to produce acetyl-CoA, CO_2, and reduced ferredoxin. This reduced ferredoxin can be utilized used as a substrate (a direct electron donor) by hydrogenases, catalyzing the synthesis of H_2. This pathway is also supported by NADH:ferredoxin oxido-reductase (NFOR).
2. Formate hydrogen lyase by facultative anaerobes: another way to convert pyruvate to acetyl-CoA is by producing formate, which is mediated by pyruvate formate lyase (PFL). In the presence of NiFe-hydrogenases enzymes, formate is transformed into H_2 and CO_2.

Dark fermentation has challenges, such as insufficient hydrogen yields and slow substrate conversion rates (Sang et al. 2006). However, it is operated with lower costs compared with biophotolysis and photofermentation for biohydrogen generation because of less electricity consumption. Currently, several studies are investigating advanced strategies to improve biohydrogen production through metabolic and genetic engineering, optimizing the dark fermentation process (Cheng et al. 2022a, b).

10.5 Sludge Pretreatment to Enhance Hydrogen Production

Recently, most researchers have focused on modifying the sludge composition to overcome the issues related to low bio-H_2 yields (Cheng et al. 2022b). This approach is known as sludge pretreatment. For instance, waste-activated sludge is widely used as a source of mixed culture in biohydrogen production because it is readily available and contains high organic contents. Proteins and carbohydrates represent about 72% of waste-activated sludge's total chemical oxygen demand (Wang et al. 2014). However, the microbial cell is protected by extracellular polymeric substances (EPS) and outside membrane (Li et al. 2022), hindering the transfer of soluble organics into the sludge matrix. Additionally, sludge is considered a mixed culture inoculum that contains various microbial species, such as acetogens, methanogens, and sulfate-reducing bacteria, in addition to the hydrogen producers. Some of these microorganisms consume H_2 as an electron donor, inhibiting H_2 accumulation (Cheng et al. 2022b). Therefore, sludge pretreatment is one of the efficient methods used to deactivate the hydrogen consumers, improving biohydrogen production. The common chemical, physical, and biological techniques used to boost sludge biodegradability and eliminate the undesired hydrogen-consuming microorganisms while preserving the hydrogen-producing bacteria can be described as follows (Carrère et al. 2010):

10.5.1 Chemical Pretreatment

Using chemicals and reagents to improve biohydrogen production has received increasing attention (Budiman and Wu 2018). The basis of the technique is to increase the organics' solubility by breaking down the complex structure of the cells, releasing more simple and soluble organics efficiently (Wacławek et al. 2018). The most commonly used methods for sludge pretreatment are alkaline, acid, and oxidation.

10.5.1.1 Alkaline Pretreatment

The often-used alkaline reagents in sludge solubilization are NaOH, KOH, and $Ca(OH)_2$ (Carrère et al. 2010). Alkaline conditions reportedly increase the solubilization of protein and carbohydrates in the sewage sludge (Liu et al. 2011), making them available for microorganisms in the fermentation process.

Taheri et al. (2018) reviewed hydrogen production under dark fermentation through the acetate-butyrate and acetate-ethanol metabolic pathway with anaerobic sludge pretreatment by 10 M NaOH and KOH. The highest biohydrogen production obtained was 2.6 ± 0.5 L and 2.2 ± 0.7 L for NaOH and KOH pretreatments, respectively.

Similarly, Kim and Choi (2013) investigated the effects of alkalization with NaOH on biohydrogen production from sewage sludge. Using NaOH of 2 g/L sludge effectively decomposes the sludge flocs and cell walls. The biohydrogen yield increased by about 8.2 times at pH 12 in 24 hr. It is attributed to the increased COD solubilization rate, thus contributing to more substrate utilization by anaerobes.

10.5.1.2 Acid Pretreatment

Acid pretreatment is another chemical method used in sludge solubilization. Its primary mechanism, comparable to alkaline pretreatment, is to disrupt the sludge flocs and cell walls, resulting in more easily degradable forms and soluble phases in the sludge matrix.

Amin et al. (2016) studied biohydrogen production using anaerobically digested sludge pretreated with 95% H_2SO_4. The pretreatment was supposed to selectively enrich the H_2-producing bacteria and inhibit the growth of nonspore-forming methanogenic bacteria. Kim et al. (2013) pretreated raw sewage with 72% H_2SO_4 and found that hydrogen yield was increased up to ~2 times. Free nitrous acid (FNA) was also employed in acid pretreatment to accelerate sludge solubilization and inhibit hydrogen consumers, showing a 2.21-folds increase in H_2 yield (Niu et al. 2019). However, some factors should be considered for applying the acid pretreatment process (Sang et al. 2006): (1) limit the release of some inhibitory by-products such as furfural compounds, (2) minimize loss of fermentable sugar, (3) alkaline reagent is needed to avoid VFAs accumulation in the medium, and (4) equipment protection from corrosion.

10.5.1.3 Oxidation

Chemical methods using strong oxidants have been also used in sludge pretreatment, where some oxidants generate more eco-friendly degradation products and residues compared with the alkaline and acid pretreatments. Common oxidants used include ozone and hydrogen peroxide (Wang et al. 2014). Peracetic acid has also been effectively used as an oxidant, increasing H_2 productivity via dark fermentation using waste-activated sludge by 10.9 times (Li et al. 2022). The mechanism used to enhance H_2 productivity under the oxidation-based condition is comparable with the application of other chemical methods. As such, the aims of these techniques are to disintegrate sludge flocs into smaller pieces and transform the complex compounds (proteins, carbohydrates, and lipids) into soluble organic forms. However, unlike other chemical methods, oxidation requires a substantial energy input (Carrère et al. 2010).

10.5.2 *Physical Pretreatment*

Thermal, ultrasonic, and microwave radiation processes are some physically-assisted forms of sludge pretreatment. These methods attempt to improve sludge solubilization by fragmenting the sludge flocs, suggesting that microbial culture gets access to the intracellular protein and carbohydrate contents.

10.5.2.1 Thermal

Heat is one of the commonly employed methods for sludge pretreatment. For enhancing biohydrogen production, the temperature range can vary from 60 to 270 °C, depending on the thermal treatment duration (Wacławek et al. 2018). At high temperatures (over 180 °C), the enhancement mechanisms rely on cleaving the chemical bonds in the cell membrane and EPS, making the soluble organic components accessible to fermentative bacteria. In contrast, at low temperatures (50–100 °C), sludge heating encourages macromolecule deflocculation and improves the hydrolase activities (Carrère et al. 2010). Like oxidation, thermal pretreatment also has a high energy requirement.

10.5.3 *Microwave*

Sludge pretreatment using microwave radiation has been recently applied to enhance the fermentative H_2 production process because it provides energy-saving fast heating rates (Elagroudy et al. 2020). The microwave input is used to reduce the sludge particle size in short processing periods, causing the discharge of intracellular and extracellular organics (Siddique et al. 2017). A hydrogen yield of 144.3 mL/g-corn stalk was observed after the pretreatment of cow dung compost by microwave irradiation, which elevated about 99.6% to that of the control (Song et al. 2012). Their study verified the role of microwave irradiation to suppress hydrogen-consuming bacterial activity and preserve hydrogen-producing bacteria. This advantage ensures the dominance of H_2-generating spore-forming anaerobes in the fermentation medium with no generation of secondary waste. Microwave radiation of Petha waste at 320 W for 5 min was used to maximize the hydrogen yield to 14 mmol H_2/mol soluble sugar (Singhal and Singh 2014). This condition was suitable for encouraging the hydrogen-producing bacterial activity.

10.5.3.1 Ultrasonic

Ultrasonic technique has received growing attention due to its effectiveness and nonsecondary pollution production. Its primary working mechanism is to induce cavitation, which causes cell membranes and sludge flocs to break down and release intercellular organics into the liquid phase (Wacławek et al. 2018). Ultrasonic pretreatment can be combined with other thermal and alkali methods to enhance the solubilization and disintegration of the sludge flocs (Sang et al. 2006). Integrated alkalization and ultrasonication were associated with peaking the hydrogen yield to 13.8 mL H_2/g $VSS_{consumed}$ using a mixture of food waste and sewage sludge (Kim and Choi 2013).

10.5.4 Biological Pretreatment

The essence of biological pretreatment involves the inoculation and/or isolation of specific enzymes and bacterial strains to enhance sludge fermentation. Isolation of acid-producing bacteria improves anaerobic degradation and cumulative biogas production. Biological pretreatment success is attributed to the improvement in the hydrolysis and acidogenesis processes (Ai et al. 2018). The isolated strain of bacteria or enzyme acts as a biocatalyst in the sludge disintegration. The biological process performance depends on various factors, such as treatment time, temperature, and stirring rate. These factors influence sludge solubilization significantly and thus affect hydrogen production.

10.6 Chemical Additives to Improve H_2 Production

During dark fermentation, pyruvate is generally broken down to produce acetyl-CoA, which is transformed into organic metabolites like ethanol and acetic or butyric acid with the simultaneous oxidation of NADH (i.e., $NADH + H^+ \rightarrow H_2 + NAD^+$) and formation of ATP (Łukajtis et al. 2018). Several enzymes facilitate the process, such as the pyruvate ferredoxin oxidoreductase (PFOR) pathway, pyruvate-formate lyase (PFL) pathway, [NiFe]-hydrogenases, and [FeFe]-hydrogenases, speeding up the biochemical reactions.

Presently, using additives to enhance biohydrogen production has gained the interest of researchers (Yang and Wang 2018a). The overlay visualization map (see Fig. 10.1) shows an increased trend in research focusing on the application of nanoparticle techniques for enhancing biohydrogen (articles published from 2020 to date). Just like pretreatment, the primary function of additives is to improve microbial development and increase the metabolism of H_2-producing bacteria along with the catalytic performance of key enzymes (Ren et al. 2021). Unlike other approaches to improve biohydrogen production, such as cofermentation, process

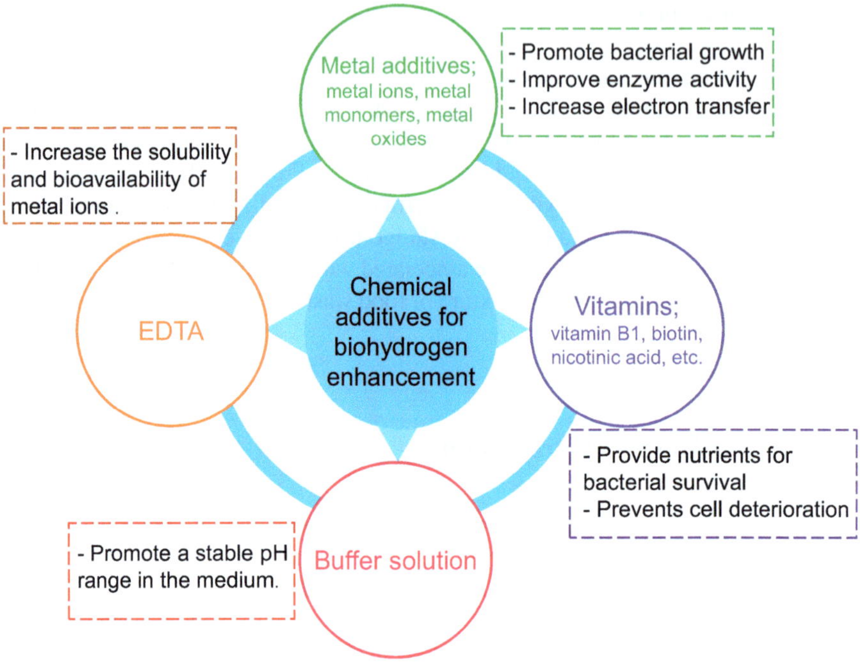

Fig. 10.2 Types of chemical additives and their enhancement mechanism

optimization, and pretreatment, adding additives has the advantages of reusability and cost-effectiveness (Yang and Wang 2018b). Figure 10.2 shows the additives used in the enhancement of biohydrogen and the associated benefits.

10.6.1 Metal Additives

Metal additives are among the most frequently used chemicals. Studies have shown that trace metals play a significant role in anaerobic fermentation (Ren et al. 2021). For the convenience of discussion, the metal additives are further divided into monomers, metal ions, and metal oxide. Each of these subcategories uses different enhancement mechanisms to impact hydrogen production.

10.6.1.1 Metal Monomers

The frequently used metal monomers include iron (Fe), Nickel (Ni), Gold (Au), Silver (Ag), Molybdenum (Mo), and Copper (Cu). Fe and Ni are the most attractive metals for improving bio-H_2 fermentation. They could aid in electron transfer and enable the synthesis of enzymes. Based on a review by Sun et al. (2019), different metal monomers, under specific conditions, have recorded enhanced hydrogen yield from 10% to 110%, compared with the control test without metal addition (Sun et al. 2019). Their mechanism of improvement is based on the facilitation of electron transport between donors and acceptors. Specifically, iron monomer dissolves in the solution to release iron ions (Fe^{2+}), facilitating the synthesis of NiFe and FeFe hydrogenase and iron-sulfur protein in ferredoxin. Zero valent iron (Fe^{0}) also assists in reducing the oxidative-reductive potential (ORP) of the system (Yang and Wang 2018a; Yu et al. 2014). This condition creates a favorable ecosystem for the dominance of fermentative bacteria like *Enterobacter* sp. and *Clostridium* sp. Yu et al. (2014) found that a zero-valent iron dosage of 550 mg/L resulted in a 16% increase in hydrogen yield and shortened lag time by 36%.

Similarly, zero valent nickel addition to a fermentation system improved hydrogen production (Elreedy et al. 2017). The study found that adding 60 mg/L of Ni to the industrial effluents increased the hydrogen yield by 23%. However, inhibition of hydrogen production was recorded at a dosage above 60 mg/L.

Mo and Fe metal elements make up part of the nitrogenase enzyme; accordingly, the enzyme can be classified into Mo-nitrogenase and Fe-nitrogenase, with the Mo-nitrogenase being considered the most effective nitrogenase enzyme (Al-Mohammedawi and Znad 2020). Hence, adding Mo to the fermentation medium will affect the photofermentation process.

Al-Mohammedawi and Znad (2020) showed the impact of Fe, Mo, and EDTA concentrations on photofermentative hydrogen production and bacterial growth. The results indicated that the maximum hydrogen production (105 mL) was obtained at a Mo dosage of 14 μM, equivalent to an increase of 27% than the control. The bacterial growth curves for the control run and all Mo-supplemented effluents were approximately similar, meaning that Mo addition did not greatly affect microbe growth. The lack of Mo concentration did not significantly interfere with bacterial growth (Al-Mohammedawi and Znad 2020).

Similarly, Eroglu et al. (2011) investigated the effects of Fe and Mo on photofermentative hydrogen from olive mill wastewater by *Rhodobacter sphaeroides* OU 001. They observed that compared to the control (40 mL), the maximum cumulative hydrogen achieved under Mo supplementation was only 62 mL compared to 125 mL under Fe supplementation.

10.6.1.2 Metal Ions

Fe^{2+}/Fe^{3+} and Ni^{2+} are widely employed metal ions for enhancing anaerobic fermentation. They are relatively cost-effective compared with other metal ions and constitute part of the hydrogenase and ferredoxin enzymes (Yang and Wang 2018a). The authors provide a summary of the usage of different metal ions as additives during the dark fermentation of substrates, such as glucose, sucrose, food waste, and organic-rich wastewater (Sun et al. 2019). The results show that adding different doses of metal ions, especially Fe^{2+} and Ni^{2+}, increased biohydrogen production by up to 500%.

The enhancement mechanism of Fe^{2+} and Ni^{2+} in the fermentation system is responsible for accelerating electron transfer between NiFe and FeFe hydrogenase in the bacteria cells (*Bacillus* sp. and *Clostridium* sp.) (Reddy et al. 2017). Other metal ions, such as calcium (Ca^{2+}), Magnesium (Mg^{2+}), and Sodium (Na^{+}), could increase cell biomass, indirectly affecting hydrogen production via dark fermentation (Lee et al. 2004). Lee et al. (2004) found that adding 5.4 mg/L of Ca^{2+} to the influent resulted in a 3.5-fold increase in biomass concentration, accompanied by a twofold increase in H_2 yield and a 64% increase in H_2 content.

Ferredoxin, an electron carrier for nitrogenase (the functional enzyme of H_2 production in anoxygenic phototrophic bacteria), includes Fe (Kumari et al. 2017). Hydrogen production by *Rhodobacter sphaeroides* using sodium lactate as substrate increased by adding ferrous ion concentrations of Fe^{2+} (0–3.2 mg/L) (Zhu et al. 2007). In this photoheterotrophic hydrogen production process, the highest H_2 production rate was 0.15 mL/h/mg dry weight at Fe^{2+} of 2.4 mg/L with substrate and light conversion efficiencies of 28.2% and 1.9%, respectively. Increasing the Fe^{2+} concentration over 3.2 mg/L negatively affected H_2 productivity probably due to the opportunity of nonphysiological impact, e.g., coagulation effect.

10.6.1.3 Metal Oxides

Similar to metal monomers and ions, the addition of metal oxides efficiently helps microbes transfer electrons between ferredoxin and hydrogenase and thereby induces a more efficient metabolic pathway in dark fermentation, providing essential elements for bacterial growth. Some metal oxides used in hydrogen fermentation include Fe_2O_3, Fe_3O_4, NiO, and TiO_2 (Cos et al. 2022). Though an excessive dosage of metal additives causes toxicity to fermentative bacteria and a decrease in the enzymatic activity field (Yang and Wang 2018b), they have the advantage of simple usability and low energy consumption, making them more practical than alternative methods from an application standpoint (Sharmila et al. 2021).

10.6.2 Ethylenediaminetetraacetic Acid (EDTA)

Metal additives, either individually or as cofactors, are used to improve the hydrogenase and nitrogenase performances during dark fermentation. However, the overdosing of these additives could inhibit the nitrogenase enzyme activity for *R. faecalis* RLD-53 (Ren et al. 2012) and *Rhodobacter sphaeroides* 158 DSM (Al-Mohammedawi and Znad 2020) in photofermentation. Adding ethylenediaminetetraacetic acid (EDTA) to the fermentation medium could increase the solubility and bioavailability of the metal ions in the medium for easy uptake into the bacterial cells (Vintiloiu et al. 2013; Lu et al. 2022).

At low concentrations of 0.2 g/L, EDTA enhanced biohydrogen production by 131% from the fermentation of brewery and restaurant effluents (Al-Mohammedawi and Znad 2020). However, EDTA acted as a chelating agent at higher concentrations, lessening metal ion availability in the culture medium. Similarly, a 40% increase in biohydrogen yield was obtained due to the improvement of nitrogenase activity when 0.3 g/L EDTA was added to the fermentation system. Moreover, the cosupplement of EDTA (0.3 g/L) and Fe^{2+} (2.4 mg/L) further increased biohydrogen yield and nitrogenase activity by 42% and 297%, respectively, than the addition of Fe^{2+} alone (Ren et al. 2012).

10.6.3 Buffer Solution

A buffer solution is added to fermentation systems to maintain a stable pH range suitable for the metabolic process. During the anaerobic fermentation process, the bacterial community activity becomes highly deteriorated by pH instability (Liu et al. 2011). A drop or increase in pH could destroy the metabolism environment for the bacteria, leading to a decline in hydrogen production. A review summary by Budiman and Wu (Budiman and Wu 2018) found that the commonly used buffer solution in biohydrogen fermentation is the potassium phosphate buffer.

Lu et al. (2020) studied the effect of buffer solution Na_2HPO_4/NaH_2PO_4 on photofermentation biohydrogen production. They observed a maximum hydrogen content of 53.88% at pH 6 in 48 hr. It exceeded the hydrogen content from the control. Similarly, Guo et al. (2020) investigated the impact of phosphate buffer addition to the fermentation of cornstalk and found that the cumulative hydrogen production was maximized to 569.6 mL at the initial pH value of 6.0. In contrast, the maximum hydrogen production rate of 23.96 mL/h was achieved at the buffer's initial pH value of 6.5. With an initial pH value above 6.0, the alkaline environment would inhibit the hydrogen production ability of photosynthetic bacteria.

10.6.4 Vitamins

Vitamins are chemical additives used to provide essential nutrients for bacterial growth and metabolism (Zhang et al. 2021). However, because bacteria cannot produce vitamins through their cells, they must be introduced using an outside source. Several fermentation systems record the addition of vitamin solutions consisting of compounds such as vitamin B_1, biotin, nicotinic acid, thiamine hydrochloride, p-aminobenzoic acid, pyridoxamine hydrochloride, calcium pantothenate, and vitamin B_{12}.

Li et al. (2011) investigated the effect of vitamin supplementation on *Rhodobacter sphaeroides* ZX-5 cell growth and hydrogen production, revealing that the presence of nicotinic acid was directly related to the electron transfer process. During photofermentation, nicotinic acid (representing a precursor of $NAD^+/NADH$) accelerated phototrophic hydrogen synthesis. Vitamins strengthen the photosynthetic bacterial system, preventing cellular deterioration and ensuring a longer lifespan for microorganisms (Budiman and Wu 2018). Vitamins may not directly affect biohydrogen production but are essential for the proper functioning of the microorganism in the medium. Vitamin B_1 and biotin had no direct effect on hydrogen production, but their deficiency in the medium significantly reduces hydrogen production (Li et al. 2011).

10.6.5 Defoamers

Defoamers have been recently reported to enhance biohydrogen production during lignocellulose photofermentation. As a highly abundant organic matter, lignocellulose-based materials have been processed to generate value-added products such as bio-H_2 (Arevalo-gallegos et al. 2017). However, foam formation is a major challenge detected during the fermentation of carbohydrate-rich materials (Wang et al. 2022).

Wang et al. (2022) studied the enhancement effect of defoamer additives on the photofermentation biohydrogen production process. The results depicted an increased hydrogen yield for samples spiked with defoamer compared to the control with no defoamer addition. At the optimum defoamer dosage of 0.125 mL/L, the cumulative hydrogen yield improved by 23% (363 mL). However, the maximum foam reduction of 55% was achieved at 1 mL/L defoamer in 36 hr. The defoamer did not appear to cause direct physiological toxicity to the fermentation bacteria culture.

Of all the additives-associated approaches, chemical supplementation is considered to be the most attractive and practical one because of its operational simplicity, i.e., no further modifications are needed, and it has a relatively low energy requirement (Liu et al. 2021).

10.7 Nanoparticles (NP) Addition to Improve Bio-H_2 Production

Nanotechnology has played an essential role, as a novel, simple, and cost-effective approach, in increasing biohydrogen production (Bosu and Rajamohan 2022; Dikshit et al. 2022). It involves creating and using materials at the size scale of intracellular structures and molecules; typically, material size is below 100 nm (Kumar et al. 2019).

A network visualization map, showing the keyword co-occurrence for publications about biohydrogen production and nanoparticles, was created (Fig. 10.3). The map shows three clusters distinguished by colors:

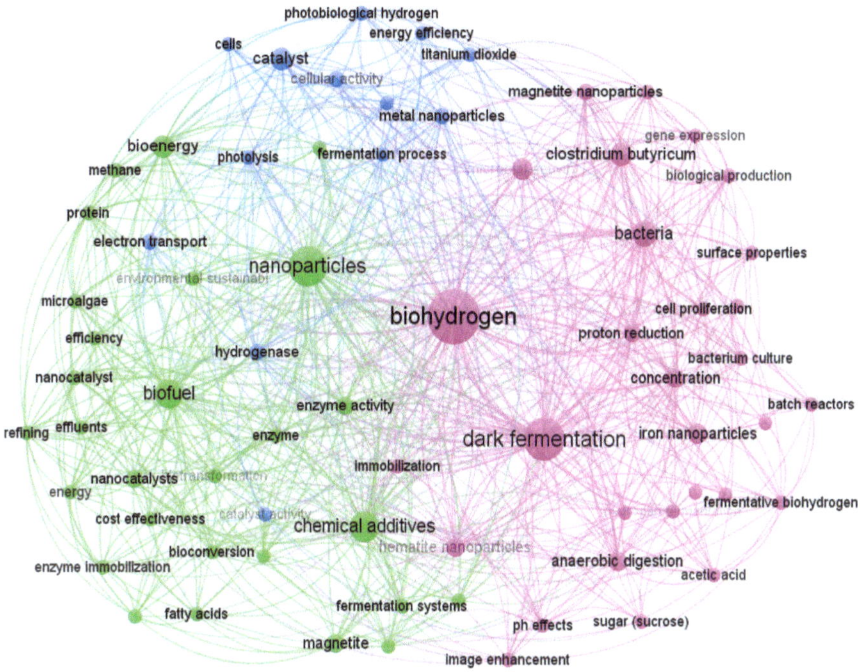

Fig. 10.3 Network visualization map of the top 78 keywords co-occurrences using search strings related to "biohydrogen" and "nanoparticles"

Cluster 1 (Green): nanoparticles, chemical additives, enzyme activity, electron transport, hydrogenase, enzyme immobilization.

This cluster represents the major benefits associated with the utilization of "nanoparticles" and "chemical additives," denoting "biofuel/bioenergy" generation, "cost-effectiveness," and "process efficiency." Nanocatalyst addition improves enzyme activity, electron transport, and bioconversion of complex proteins and carbohydrates into simple fatty acids (Zhao and Chen 2011). These mechanisms are important to accelerate the electron transfer between ferredoxin (Fd) and the active enzymes in fermentation bacteria, thus improving the hydrogen production rate (Zhang et al. 2021).

Cluster 2 (Purple): biohydrogen, dark fermentation, bacteria, pH, sugar.

This cluster denotes a co-occurrence of keywords related to the "biohydrogen" production via "fermentation," such as "dark fermentation," "photofermentation," "anaerobic digestion," "batch reactor," "bacteria," "pH," "sugar," "proton reduction," and "*Clostridium butyricum.*" Some of these keywords are operational conditions; for example, pH is a critical factor used to maintain the microorganisms' activity. Also, sugar/glucose is an energy source utilized for bacterial metabolism (Liu et al. 2011). Specifically, a strong occurrence of *Clostridium butyricum* sp. is identified. This gram-positive bacillus is often isolated in sewage sludge for bio-H_2 production (Reddy et al. 2017). "Iron nanoparticles" and "magnetite nanoparticles" are also included in this cluster because several researchers have integrated metal ions to improve the electron transport between the enzyme and microbes (Zhang et al. 2021).

Cluster 3 (Blue): photofermentation, photolysis, photocatalysts.

The cluster comprises multiple terms related to photofermentation. "Photolysis" is decomposition that occurs by the action of light. This process can be divided into:

Direct photolysis: In the PSII photosystem, the light energy (at 680 nm) could split water molecules into oxygen, protons, and electrons ($2H_2O \xrightarrow{\text{light energy}} O_2 + 4H^+ + 4e^-$).

Indirect photolysis: light energy is involved in producing O_2 and carbohydrate ($6CO_2 + 6H_2O \xrightarrow{\text{light energy}} C_6H_{12}O_6 + 6O_2$) followed by carbohydrate conversion to H_2 ($C_6H_{12}O_6 + 6H_2O \xrightarrow{\text{light energy}} 12H_2 + 6CO_2$).

Biophotolysis is also one of the methods used for hydrogen production via water splitting into hydrogen and oxygen (Sang et al. 2006). However, this process could be incompletely maintained due to "Hydrogenase" sensitivity to oxygen (Sun et al. 2019). Other relevant terms in the cluster are "catalyst," "Titanium oxide," and "electron process." As such, photocatalysts are used to facilitate electron transfer in the biophotolysis process. One of the most promising photocatalysts used for producing hydrogen is TiO_2 due to its attractive properties regarding chemical stability, corrosion resistance, nontoxicity, and low-cost (Ni et al. 2007).

Biohydrogen yields and production rates depend on organics' microorganism metabolism integrated with cellulolytic enzymes. Moreover, most enhancement techniques focus on improving enzyme production and hydrolytic activity (Lu et al. 2022); (Saha et al. 2021). High electroconductivity, large surface area, and the elevated surface-to-volume ratio of nanoparticles support the fermentation system's metabolic process at an intracellular and micro level (Bosu and Rajamohan 2022). Figure 10.4 shows the different categories of nanoparticles researchers use to improve biohydrogen production.

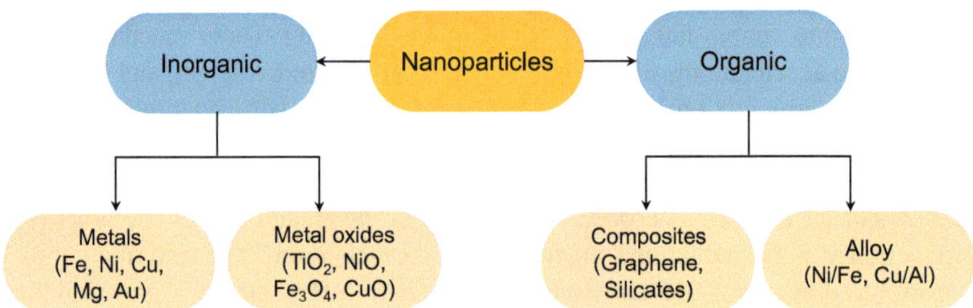

Fig. 10.4 Categorization of nanoparticles used in biohydrogen production

10.7.1 Inorganic Nanoparticles

10.7.1.1 In Photofermentation

Metal ions and oxides can be used to improve biohydrogen production in photofermentation systems. Many purple phototrophic bacteria generate hydrogen, and the reaction is catalyzed by nitrogenase. Since metal ions are components of nitrogenase, they are affiliated with ferredoxin's (Fd) function as an electron transporter, combining the electrons with the protons to produce molecular hydrogen in the system (Zhang et al. 2021; Hay et al. 2012). Therefore, adding trace metal elements like Fe, Mg, Ni, Ti, or their oxides improves enzyme activity.

A study by Dolly et al. (2015) compared the improvement of biohydrogen production by coculturing *Rhodobacter sphaeroides* NMBL-02 and *Escherichia coli* NMBL-04 using Fe NP and Fe in bulk form. The results of biohydrogen yield derived from the Fe NP were higher than the associated yield in the Fe bulk medium by 1.2 times (Dolly et al. 2015). The greater affinity of Fe NP toward H_2 productivity was due to the easy transfer of electrons to their respective acceptors, enhancing the microorganisms' metabolism.

Zhao and Chen (2011) demonstrated that adding 100 mg/L of TiO_2 NP to a photofermentation system increased biohydrogen production from waste sludge by 46.1%. This enhancement was attributed to the better decomposition of protein and polysaccharide in the waste sludge to simple compounds, improving nitrogenase activity and photosynthetic bacteria growth.

The use of metal oxide NP could transfer photo-generated electrons to the enzyme system of the purple sulfur photosynthetic bacterium effectively, thereby promoting photohydrogen production. Moreover, the TiO_2 NP in the photofermentative system could purify wastewater by producing hydroxyl radicals, improving light penetration (Cheng et al. 2022a; Liu et al. 2017). Thus, integrating the NP in the photofermentation system is a promising method for enhancing biohydrogen production.

10.7.1.2 In Dark Fermentation

During dark fermentation, organics are converted to molecular hydrogen with the catalysis of hydrogenase. Furthermore, adding Fe and Ni metal ions and oxides is especially reported to enhance the activity of [NiFe]- and [FeFe]-hydrogenases (Reddy et al. 2017).

At optimized conditions, adding Fe NP showed a 1.85-fold increase in hydrogen productivity and accelerated the fermentation process with a significantly shortened lag time. The lag phase showed a 96% rapid reaction (from 6.86 h in the control sample to 0.25 h under optimum conditions). This pattern is attributed to the facilitated hydrolysis phase and metabolic process after NP addition (Arisht et al. 2021).

Nasr et al. (2015) indicated that adding maghemite (Fe_2O_3) nanoparticles improved the hydrolysis step and enhanced starch conversion into biohydrogen. The hydrogen yield and production rate were higher in the reactor containing maghemite NP (0.90 mol-H_2/mol-glucose and 16.80 L-H_2/d) as compared to the control (0.57 mol-H_2/mol-glucose and 11.67 L-H_2/d).

Similarly, Reddy et al. (2017) investigated the effect of Fe and magnetite (Fe_3O_4) NP supplementation for biohydrogen production from dark fermentation of sugarcane bagasse. Adding 200 mg/L of Fe^{2+} and magnetite NP improved the biohydrogen yield to 1.154 and 1.211 mol/mol glucose, respectively, while the control achieved 0.714 mol/mol glucose. The results also suggest improved bioactivity in the reactor with magnetite NP. The electron-equivalent balance indicated that the microorganism activity in the reactor with magnetite NP transferred more electrons to protons, hence the higher biogas yields.

Zhang et al. (2022) used calcium ferrite NP ($CaFe_2O_4$ NPs) to enhance H_2 yield via dark fermentation. The highest H_2 yield of 250.1 ± 6.5 mL/g glucose was achieved at 100 mg/L $CaFe_2O_4$ NPs. Further, an increase in NPs concentration above 100 mg/L slightly lowered biogas yield. The $CaFe_2O_4$ NP released Ca^{2+} and Fe^{2+} in the fermentation medium, promoting sludge granulation and microbial activity. $CaFe_2O_4$ NPs appear to have increased soluble metabolite formation, specifically butyric acid. This finding indicates an improved metabolic pathway for H_2 production. Most microbial populations were related to *Firmicutes* (71.22%) in the NP-supplemented sample.

10.7.2 Organic Nanoparticles

Organic nanomaterials commonly used in bio-H_2 fermentation include metal alloys, graphene, and silica-based materials. Graphene is a bonded carbon material with a single layer of atoms organized in a two-dimensional (2D) honeycomb (hexagonal) lattice nanostructure (Gadipelli and Guo 2015). Graphene has unique physicochemical properties such as high specific surface area, and improved mechanical strength, and electrical and thermal conductivities, facilitating the hydrogenase enzyme's electron transfer (Elreedy et al. 2017).

10.7.2.1 In Photofermentation and Biophotolysis

One of the most well-known and promising photocatalysts for producing hydrogen is TiO_2, due to its chemical stability, corrosion resistance, nontoxicity, and low-cost (Ni et al. 2007). TiO_2 acts as a semiconductor by offering a large surface for interactions between protons and electrons. The main principle of photocatalytic hydrogen production is proton reduction; thus, nanoscale semiconductors can improve reactions because they have a greater reactive surface area (Ni et al. 2007). Synthesizing

TiO_2 mesoporous composites has also been performed to offer a larger surface area compared with the sole TiO_2 material.

Hakamizadeh et al. (2014) synthesized nanocomposite using Platinum (Pt), TiO_2, and activated carbon (AC) to improve the photocatalytic production of hydrogen gas via water splitting. The study compared the performance of TiO_2/AC, Pt/TiO_2, and Pt/TiO_2/AC composites on hydrogen yield. The study found that the average rate of H_2 gas produced using Pt/TiO_2/AC composite was about 20-times and 75-times higher than that of TiO_2/AC and the control (commercial photocatalyst), respectively. As such, combining both AC and Pt with TiO_2 elevated hydrogen evolution. For instance, AC could control the agglomeration and dispersion of TiO_2 particles with smaller sizes and greater surface area, increasing the photocatalytic activity (Liu et al. 2007).

Liu et al. (2017) performed photofermentation in batch culture with TiO_2, ZnO, and silicon carbide (SiC) nanoparticles addition and studied hydrogen production. The findings from the three nanoparticles showed improved hydrogen production and metabolic performance of *Rhodopseudomonas* sp. nov. strain A7 at their optimal conditions. The organic SiC NP outperformed its counterparts, recording the maximum hydrogen volume of 2.99 mol-H_2/mol-acetate compared to 2.81 and 2.64 mol-H_2/mol-acetate for TiO_2 and ZnO, respectively.

10.7.2.2 In Dark Fermentation

Elreedy et al. (2017) recorded substantial improvements of 23% and 105% in hydrogen production from mono-ethylene glycol (MEG)-laden wastewater at an optimum dosage of 60 mg/L for Ni NP and Ni/graphene nanocomposites, respectively, compared with the control without NP addition. The H_2 production results from the Ni/graphene composite increased by 4.5-fold compared to the Ni NP.

A study by Tawfik et al. (2021) revealed that the activities of protein and carbohydrate-degrading enzymes such as protease, α-amylase, xylanase, and CM-cellulase were highly improved after immobilization on graphene and hydroxyapatite nanoparticles during black liquor fermentation. An economic assessment of the study encouraged the utilization of graphene nanoparticles to recover the cash investment in a short payback time (i.e., 5.92 years). The scenario of NP implementation was considered a more economically and highly sustainable.

10.7.3 Summary of the Mechanism of Nanoparticles

All the mentioned studies showed that the improvement of hydrogen production by nanomaterial supplementation could be attributed to the following:

- NPs influence the hydrogenase and nitrogenase activities by facilitating the electron transfer between ferredoxin (Fd) and the active enzymes in the fermentation culture.
- NPs could penetrate the bacterial cell due to their microstructure and high surface area. This property enables them to alter the electron transfer rate by enhancing the Fd activity.
- Adding NP could increase the production of acetate and butyric in dark fermentation, indicating an improved metabolic pathway.
- NP encourages the growth of dominant hydrogen-producing bacteria.

10.8 Recommendations

Several studies have investigated the addition of various chemicals and nanoparticles to the fermentation medium for enhancing biohydrogen production. Despite the developments, several topics are still to be addressed:

Optimization of nanoparticles dosage: more research should focus on identifying the optimal dosage of nanoparticles to increase the bioavailability of metallic ions without causing adverse toxicity to fermentation microbes. The hydrogenases and nitrogenases utilize metallic ions at an optimal level to maintain their catalytic function. The optimum nanoparticles dosage is also required to promote the hydrolytic enzyme activity and generation of fermenting microbes (H_2 producers).

Continuous H_2 production: The addition of chemicals and nanomaterials has been done mostly at the lab scale in batch mode. There is a need to run more continuous and combined fermentation systems. These studies should overcome the technical barriers associated with large-scale application and process commercialization. The continuous mode of operation should also define the shift in the bacterial community and fermentation dynamics due to chemical addition.

Sludge management: Microorganism immobilization onto nanomaterials creates focused sites with active microbes/enzymes and bacterial colonies, increasing operational efficiency. However, the toxicology of the immobilized nanoparticles and residue chemicals needs further research to inform the possibility of sludge recyclability. This sludge management possibility could reduce operation costs and encourage the H_2 fermentation sustainability approach.

Economic evaluation: More studies should explore the profitability of enhancing H_2 productivity from nanoparticles and chemicals supplementation, regarding the initial investment, operating cost, and payback period.

10.9 Conclusions and Perspectives

The pressing need to switch to an environment-friendly energy resource has encouraged more researchers to enhance the performances of the H_2-producing microbes. There are various ways to produce hydrogen, with photo/biophotolysis and dark fermentation being the most environmentally benign and long-lasting methods. Several strategies have been considered to overcome some challenges, such as the higher expenses and lower H_2 yields, associated with inadequate operating conditions. The current chapter succeeded in representing the sludge pretreatment methods and chemicals and nanoparticle supplementation approaches recently used in the H_2 fermentation processes. The chapter comprehensively addressed the role of chemical additives, like metal ions, oxides, and monomers, to efficiently facilitate the electron transfer between the ferredoxin and the enzyme and promote cell growth. Chemicals like EDTA support biomass hydrolysis and solubility, vitamins provide bacteria nutrients, and buffered solutions are added to retard a decrease in pH, thus creating a suitable environment for bacterial growth. Nanosized additives accelerate electron transfer and intracellular metabolism, showing improved biohydrogen yields with more profits and shorter payback periods during implementation. Chemical additives could be more feasible from an application viewpoint than other intensification approaches, owing to their easy operation and low energy consumption. However, determining the optimal additive dosages and analyzing their impact on bacterial community structure and dynamics in the fermentation residue should be studied as part of future research. The associated environmental impact and economic feasibility of such chemicals- and nanoparticles-assisted H_2 fermentation routes should also be comprehensively explored.

Acknowledgments The second author would like to acknowledge Nasr Academy for Sustainable Environment (NASE).

References

Ai S, Liu H, Wu M, Zeng G, Yang C (2018) Roles of acid-producing bacteria in anaerobic digestion of waste activated sludge. Front Environ Sci 12:1–11

Akhlaghi N, Najafpour-darzi G (2020) A comprehensive review on biological hydrogen production. Int J Hydrog Energy 45:22492–22512

Ali M, Elsaid K, Wilberforce T, Kamil M, Taha E, Olabi A (2021) Environmental aspects of fuel cells: a review. Sci Total Environ 752:141803

Al-Mohammedawi HH, Znad H (2020) Impact of metal ions and EDTA on photofermentative hydrogen production by *Rhodobacter sphaeroides* using a mixture of pre-treated brewery and restaurant effluents. Biomass Bioenergy 134:105482

Amin MM, Bina B, Taheri E, Fatehizadeh A, Ghasemian M (2016) Stoichiometry evaluation of biohydrogen production from various carbohydrates. Environ Sci Pollut Res 23:20915–20921

Arevalo-gallegos A, Ahmad Z, Asgher M, Parra-saldivar R, Iqbal HMN (2017) Lignocellulose: a sustainable material to produce value-added products with a zero waste approach – a review. Int J Biol Macromol 99:308–318

Argun H, Kargi F (2011) Bio-hydrogen production by different operational modes of dark and photo-fermentation: an overview. Int J Hydrog Energy 36:7443–7459

Arisht SN, Roslan R, Gie GA, Mahmod SS (2021) Effect of nano zero-valent iron (nZVI) on biohydrogen production in anaerobic fermentation of oil palm frond juice using *Clostridium butyricum* JKT37. Biomass Bioenergy 154:106270

Aydin MI, Karaca AE, Qureshy AMMI, Dincer I (2021) A comparative review on clean hydrogen production from wastewaters. J Environ Manag 279:111793

Bharathiraja B, Sudharsana T, Jayamuthunagai J, Praveenkumar R, Chozhavendhan S, Iyyappan J (2018) Biogas production – a review on composition, fuel properties, feed stock and principles of anaerobic digestion biogas production – a review on composition, fuel properties, feed stock and principles of anaerobic digestion. Renew Sust Energ Rev 90:570–582

Boran E, Özgür E, Yücel M, Gündüz U, Eroglu I (2012) Biohydrogen production by Rhodobacter capsulatus in solar tubular photobioreactor on thick juice dark fermenter effluent. J Clean Prod 31:150–157

Bosu S, Rajamohan N (2022) Nanotechnology approach for enhancement in biohydrogen production- review on applications of nanocatalyst and life cycle assessment. Fuel 323:124351

Budiman PM, Wu TY (2018) Role of chemicals addition in affecting biohydrogen production through photofermentation. Energy Convers Manag 165:509–527

Carrère H, Dumas C, Battimelli A, Batstone DJ, Delgenès JP, Steyer JP, Ferrer I (2010) Pretreatment methods to improve sludge anaerobic degradability: a review. J Hazard Mater 183:1–15

Cheng D, Hao H, Guo W, Woong S, Duc D (2022a) Enhanced photo-fermentative biohydrogen production from biowastes: an overview. Bioresour Technol 357:127341

Cheng D, Hao H, Guo W, Woong S, Duc D (2022b) Advanced strategies for enhancing dark fermentative biohydrogen production from biowaste towards sustainable environment. Bioresour Technol 351:127045

Cos B, Kantu A, Ozkaya B (2022) Screening of biohydrogen production based on dark fermentation in the presence of nano-sized Fe_2O_3 doped metal oxide additives. Int J Hydrog Energy 7:1–14

Dadebo D, Ibrahim MG, Fujii M, Nasr M (2022) Transition towards sustainable carwash wastewater management: trends and enabling technologies at global scale. Sustainability 14:5652

Dikshit PK, Jamnal N, Singh S, Kumar J (2022) Advancement of nanoparticles in gaseous biofuel production. In: Bio-clean energy technologies, pp 227–248

Dincer I, Acar C (2015) Review and evaluation of hydrogen production methods for better sustainability. Int J Hydrog Energy 40:11094–11111

Dolly S, Pandey A, Pandey BK, Gopal R (2015) Process parameter optimization and enhancement of photo-biohydrogen production by mixed culture of *Rhodobacter sphaeroides* NMBL-02 and *Escherichia coli* NMBL-04 using Fe-nanoparticle. Int J Hydrog Energy 40(46):16010–16020

Elagroudy S, Radwan AG, Banadda N, Mostafa NG, Owusu PA, Janajreh I (2020) Mathematical models comparison of biogas production from anaerobic digestion of microwave pretreated mixed sludge. Renew Energy 155:1009–1020

Elreedy A, Ibrahim E, Hassan N, El-dissouky A, Fujii M, Yoshimura C, Tawfik A (2017) Nickel-graphene nanocomposite as a novel supplement for enhancement of biohydrogen production from industrial wastewater containing mono-ethylene glycol. Energy Convers Manag 140:133–144

Elreedy A, Fujii M, Tawfik A (2018) Psychrophilic hydrogen production from petrochemical wastewater via anaerobic sequencing batch reactor: techno-economic assessment and kinetic modelling. Int J Hydrog Energy 44:5189–5202

Eroglu E, Gunduz U, Yucel M, Eroglu I (2011) Effect of iron and molybdenum addition on photofermentative hydrogen production from olive mill wastewater. Int J Hydrog Energy 36:5895–5903

Eun J, Shafiq I, Hussain M, Shiung S, Hoon G, Park Y (2021) A review on integrated thermochemical hydrogen production from water. Int J Hydrog Energy 47:4346–4356

Gadipelli S, Guo ZX (2015) Graphene-based materials: synthesis and gas sorption, storage and separation. J Prog Mater Sci 69:1–60

Guo S, Lu C, Wang K, Wang J, Zhang Z, Jing Y, Zhang Q (2020) Enhancement of pH values stability and photo-fermentation biohydrogen production by phosphate buffer. Bioengineered 11:291–300

Hakamizadeh M, Afshar S, Tadjarodi A (2014) Improving hydrogen production via water splitting over Pt/TiO_2/activated carbon nanocomposite. Int J Hydrog Energy 39:7262–7269

Hay JXW, Wu TY, Juan JC (2012) Biohydrogen production through photo fermentation or dark fermentation using waste as a substrate: overview, economics, and future prospects of hydrogen usage. Biofuels Bioprod Biorefin 6:246–256

Hu C, Choy SY, Giannis A (2018) Evaluation of lighting systems, carbon sources, and bacteria cultures on photofermentative hydrogen production. Appl Biochem Biotechnol 185:257–269

Jat Baloch MY, Talpur SA, Iqbal J, Munir M, Baidya P, Talpur HA (2022) Process design for biohydrogen production from waste materials and its application. Sustain Environ 7:47

Kim S, Choi K (2013) Biological hydrogen production by anaerobic digestion of food waste and sewage sludge treated using various pretreatment technologies. Biodegradation 24:753

Kim M, Liu C, Noh J, Yang Y, Oh S, Shimizu K, Lee D, Zhang Z (2013) Hydrogen and methane production from untreated rice straw and raw sewage sludge under thermophilic anaerobic conditions. Int J Hydrog Energy 38:8648–8656

Kumar G, Mathimani T, Rene ER (2019) Application of nanotechnology in dark fermentation for enhanced biohydrogen production using inorganic nanoparticles. Int J Hydrog Energy 44:13106–13113

Kumari S, Nasr M, Kumar S (2017) Technological advances in biohydrogen production from microalgae. In: Algal biofuels: recent advances and future prospects, pp 347–360

Lee KS, Lo YS, Lo YC, Lin PJ, Chang JS (2004) Operation strategies for biohydrogen production with a high-rate anaerobic granular sludge bed bioreactor. Enzym Microb Technol 35:605–612

Li X, Shi H, Wang Y, Zhang S, Chu J, Zhang M, Huang M, Zhuang Y (2011) Effects of vitamins (nicotinic acid, vitamin B1 and biotin) on phototrophic hydrogen production by *Rhodobacter sphaeroides* ZX-5. Int J Hydrog Energy 36:9620–9625

Li C, Liu X, Du M, Yang J, Lu Q, Fu Q, He D (2022) Peracetic acid promotes biohydrogen production from anaerobic dark fermentation of waste activated sludge. Sci Total Environ 844:156991

Liu SX, Chen XY, Chen X (2007) A TiO_2/AC composite photocatalyst with high activity and easy separation prepared by a hydrothermal method. J Hazard Mater 143:257–263

Liu BF, Ren NQ, Xie GJ, Ding J, Guo WQ, Xing DF (2010) Enhanced bio-hydrogen production by the combination of dark- and photo-fermentation in batch culture. Bioresour Technol 101:5325–5329

Liu H, Wang J, Liu X, Fu B, Chen J, Yu H (2011) Acidogenic fermentation of proteinaceous sewage sludge: effect of pH. Water Res 6:1–8

Liu B, Jin Y, Wang Z, Xing D, Ma C (2017) Enhanced photo-fermentative hydrogen production of *Rhodopseudomonas* sp. nov. strain A7 by the addition of TiO_2, ZnO and SiC nanoparticles. Int J Hydrog Energy 42:18279–18287

Liu Y, Liu J, He H, Yang S, Wang Y, Hu J, Jin H, Cui T, Yang G, Sun Y (2021) A review of enhancement of biohydrogen productions by chemical addition using a supervised machine learning method. Energies 14:5916

Lu C, Tahir N, Li W, Zhang Z, Jiang D, Guo S, Wang J, Wang K, Zhang Q (2020) Enhanced buffer capacity of fermentation broth and biohydrogen production from corn stalk with Na_2HPO_4/NaH_2PO_4. Bioresour Technol 313:123783

Lu B, Xia D, Zhao S, Yang Z, Chen Z, Huang S, Li S, Zhang Z (2022) The influence mechanism of ethylenediaminetetraacetic acid (EDTA) and ferrous iron on the bioavailability of Fe in the process of low rank coal fermentation. Biochem Eng J 185:108520

Łukajtis R, Hołowacz I, Kucharska K, Glinka M, Rybarczyk P, Przyjazny A, Kamiński M (2018) Hydrogen production from biomass using dark fermentation. Renew Sust Energ Rev 91:665–694

Mthethwa NP, Nasr M, Bux F, Kumari S (2018) Utilization of *Pistia stratiotes* (aquatic weed) for fermentative biohydrogen: electron-equivalent balance, stoichiometry, and cost estimation. Int J Hydrog Energy 43:8243–8255

Nasr M, Tawfik A, Ookawara S, Suzuki M (2013) Biological hydrogen production from starch wastewater using a novel up-flow anaerobic staged reactor. Bioresources 8:4951–4968

Nasr M, Tawfik A, Ookawara S, Suzuki M, Kumari S, Bux F (2015) Continuous biohydrogen production from starch wastewater via sequential dark-photo fermentation with emphasize on maghemite nanoparticles. J Ind Eng Chem 21:500–506

Ni M, Leung MK, Leung DY, Sumathy K (2007) A review and recent developments in photocatalytic water-splitting using TiO_2 for hydrogen production. Renew Sust Energ Rev 11:401–425

Niu Q, Xu Q, Wang Y, Wang D, Liu X, Liu Y, Wang Q, Ni B, Yang Q, Li X, Li H (2019) Enhanced hydrogen accumulation from waste activated sludge by combining ultrasonic and free nitrous acid pretreatment: performance, mechanism, and implication. Bioresour Technol 285:121363

Qyyum MA, Ismail S, Ni SQ, Ihsanullah I, Ahmad R, Khan A, Tawfik A, Nizami AS, Lee M (2022) Harvesting biohydrogen from industrial wastewater: production potential, pilot-scale bioreactors, commercialization status, techno-economics, and policy analysis. J Clean Prod 340:130809

Reddy K, Nasr M, Kumari S, Kumar S, Gupta SK, Enitan AM, Bux F (2017) Biohydrogen production from sugarcane bagasse hydrolysate: effects of pH, S/X, Fe^{2+}, and magnetite nanoparticles. Environ Sci Pollut Res 24:8790–8804

Ren HY, Liu BF, Ding J, Nan J, Xie GJ, Zhao L, Chen MG, Ren NQ (2012) Enhanced photo-hydrogen production of *Rhodopseudomonas faecalis* RLD-53 by EDTA addition. Int J Hydrog Energy 37:8277–8281

Ren Y, Si B, Liu Z, Jiang W (2021) Promoting dark fermentation for biohydrogen production: potential roles of iron-based additives. Int J Hydrog Energy 47:1499–1515

Saha R, Bhattacharya D, Mukhopadhyay M (2021) Enhanced production of biohydrogen from lignocellulosic feedstocks using microorganisms: a comprehensive review. Energy Convers Manag X 13:100153

Sang BI, Kim DK, Lee YJ, Chung YC (2006) Biohydrogen production from organic wastes. Amer Chemical Soc

Saratale GD, Saratale RG, Chang J (2013) Biohydrogen from renewable resources. In: Biohydrogen. Elsevier

Shanmugam S, Hari A, Pandey A, Mathimani T (2020) Comprehensive review on the application of inorganic and organic nanoparticles for enhancing biohydrogen production. Fuel 270:117453

Sharmila VG, Tamilarasan K, Kumar MD, Kumar G, Varjani S, Kumar SA, Banu JR (2021) Trends in dark biohydrogen production strategy and linkages with transition towards low carbon economy: an outlook, cost-effectiveness, bottlenecks and future scope. Int J Hydrog Energy 47:15309–15332

Siddique MNI, Munaim MSA, Zularisam AW (2017) The combined effect of ultrasonic and microwave pre-treatment on bio-methane generation from co-digestion of petrochemical wastewater. J Clean Prod 145:303–309

Singhal Y, Singh R (2014) Effect of microwave pretreatment of mixed culture on biohydrogen production from waste of sweet produced from *Benincasa hispida*. Int J Hydrog Energy 39:7534–7540

Sinha P, Pandey A (2011) An evaluative report and challenges for fermentative biohydrogen production. Int J Hydrog Energy 36:7460–7478

Song ZX, Wang ZY, Wu LY, Fan YT, Hou HW (2012) Effect of microwave irradiation pretreatment of cow dung compost on bio-hydrogen process from corn stalk by dark fermentation. Int J Hydrog Energy 37:6554–6561

Staffell I, Scamman D, Abad V, Balcombe P, Dodds PE, Ekins P, Ward KR (2019) The role of hydrogen and fuel cells in the global energy system. Energy Environ Sci 12:463–491

Su H, Cheng J, Zhou J, Song W, Cen K (2009) Combination of dark- and photo-fermentation to enhance hydrogen production and energy conversion efficiency. Int J Hydrog Energy 34:8846–8853

Sun Y, He J, Yang G, Sun G (2019) A review of the enhancement of bio-hydrogen generation by chemicals addition. Catalysts 9:353

Taheri E, Amin M, Pourzamani H, Fatehizadeh A, Ghasemian M, Bina B (2018) Comparison of acetate-butyrate and acetate-ethanol metabolic pathway in biohydrogen production. J Med Signals Sensors 8:101

Tawfik A, Nasr M, Galal A, El-Qelish M, Yu Z, Hassan MA, Salah HA, Hasanin MS, Meng F, Bokhari A, Qyyum MA, Lee M (2021) Fermentation-based nanoparticle systems for sustainable conversion of black-liquor into biohydrogen. J Clean Prod 309:127349

Vintiloiu A, Boxriker M, Lemmer A, Oechsner H, Jungbluth T, Mathies E, Ramhold D (2013) Effect of ethylenediaminetetraacetic acid (EDTA) on the bioavailability of trace elements during anaerobic digestion. Chem Eng J 223:436–441

Wacławek S, Grübel K, Silvestri D, Padil VVT, Wacławek M, Černík M, Varma RS (2018) Disintegration of wastewater activated sludge (WAS) for improved biogas production. Energies 12:1–16

Wang L, Liu W, Kang L, Yang C, Zhou A, Wang A (2014) Enhanced biohydrogen production from waste activated sludge in combined strategy of chemical pretreatment and microbial electrolysis. Int J Hydrog Energy 39:11913–11919

Wang K, Lu C, Zhang H, Guo S, Ru G, Wang J, Hu J, Zhang N, Zhang Q (2022) Enhancement effect of defoamer additives on photo-fermentation biohydrogen production process. Bioresour Technol 352:127070

Yang G, Wang J (2018a) Improving mechanisms of biohydrogen production from grass using zero-valent iron nanoparticles. Bioresour Technol 266:413–420

Yang G, Wang J (2018b) Various additives for improving dark fermentative hydrogen production: a review. Renew Sust Energ Rev 95:130–146

Yu L, Jiang W, Yu Y, Sun C (2014) Effects of dilution ratio and Fe dosing on biohydrogen production from dewatered sludge by hydrothermal pretreatment. Environ Technol 35:3330

Zhang Q, Zhu S, Zhang Z, Zhang H, Xia C (2021) Enhancement strategies for photo-fermentative biohydrogen production: a review. Bioresour Technol 340:125601

Zhang J, Zhang H, Zhang J, Zhou C, Pei Y, Zang L (2022) Improved biohydrogen evolution through calcium ferrite nanoparticles assisted dark fermentation. Bioresour Technol 361:127676

Zhao Y, Chen Y (2011) Nano-TiO_2 enhanced photofermentative hydrogen produced from the dark fermentation liquid of waste activated sludge. Environ Sci Technol 45:8589–8595

Zhu H, Fang HHP, Zhang T, Beaudette LA (2007) Effect of ferrous ion on photo heterotrophic hydrogen production by *Rhodobacter sphaeroides*. Int J Hydrog Energy 32(17):4112–4118

Chapter 11
Biohydrogen Production from Agroindustrial Wastes: Pre-Treatment, Process Engineering and Techno-Economic Analysis

Arindam Sinharoy and Kannan Pakshirajan

Abstract The depletion of fossil fuel sources of energy and environmental pollution due to their burning for energy application are two major problems faced by the present world, which have led to alternative sources of energy to feed the needs of continuously growing population and industries. The use of alternative renewable energy including energy derived from renewable biomass is one of the solutions proposed by researchers. This bioenergy includes ethanol, butanol, biodiesel, methane, hydrogen, etc., among which hydrogen is one of the best candidate to replace the traditional fossil fuel energy with high energy content. The recent focus on hydrogen production is towards use of waste resources which not only reduces the process cost but provides for alternative ways for waste management. The agriculture industry offers a good choice for this strategy as it generates large quantities of residues every year that are primarily disposed of by burning, dumping or unplanned landfilling. However, the vast diversity of agroindustrial waste and their compositional variation can be challenging for biohydrogen production. This chapter discusses the suitability of agroindustrial waste for biohydrogen production, the processes and important factors involved, including different improvement strategies. The key challenges and future research directions to overcome the same are also presented.

A. Sinharoy
Department of Microbiology, University of Galway, Galway, Ireland

K. Pakshirajan (✉)
Department of Biosciences and Bioengineering, Indian Institute of Technology Guwahati, Guwahati, Assam, India
e-mail: pakshi@iitg.ac.in

C. R. Soccol et al. (eds.), *Biohydrogen - Advances and Processes*, Biofuel and Biorefinery Technologies 13, https://doi.org/10.1007/978-3-031-49818-3_11

Graphical Abstract

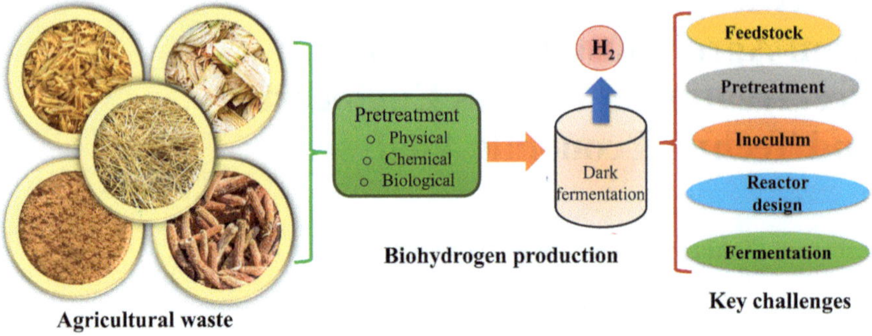

Keywords Biohydrogen · Agroindustrial waste · Lignocellulosic biomass · Waste to energy · Biofuel

11.1 Introduction

Numerous studies have been carried out to explore renewable energy sources and to make the process sustainable and economical for production of biofuels, chemicals and other useful products (Panagiotopoulos et al. 2015). Biomass provides enormous potential for production of renewable and clean energy through various thermo-chemical and biological methods (Zheng et al. 2021). The biomass-based energy sources include hydrogen, methane, ethanol, butanol, propanol, biodiesel, etc. (Roy et al. 2016). Based on the type of biomass used, biofuels can be categorized into four groups: first-, second-, third- and fourth-generation biofuels (Hirani et al. 2018). First-generation biofuels used sugars, starch, oils and fats from primarily food crops which potentially created conflict with the food/feed resources (Paul et al. 2020). In the second generation, these resources were replaced with waste (mainly lignocellulosic biomass) such as agricultural waste, forest residue and yard waste (grass clippings). In addition to avoiding conflict with food sources, use of waste biomass further helped in land management. The third-generation biofuels are mainly produced from algal biomass as the substrate (Singh et al. 2011).

Among the different biofuels, hydrogen is highly advantageous owing to its very high energy value (122 MJ/kg) and capability to burn without any pollution problems (Chatellard et al. 2017). Furthermore, utilization of waste resources for hydrogen production is feasible by using both thermo-chemical and biological routes (Dou et al. 2019; Sinharoy et al. 2019). The utilization of waste biomass for biohydrogen production can be beneficial as it helps in resource recovery from waste and prevents environmental degradation by safe disposal of the wastes (Sinharoy et al. 2020a). Although, thermo-chemical methods such as biomass gasification, pyrolysis and steam reforming of natural gases are efficient in terms of biomass conversion to energy (Paul et al. 2022), they involve highly energy intensive and costly process, perishable catalysts, which limit their application at large scale (Kumar et al. 2019; Singh et al. 2022). Compared with thermo-chemical methods, biological methods are inexpensive, environment-friendly and sustainable for converting waste biomass to hydrogen (Kamaraj et al. 2020).

Large quantities of waste are generated by agricultural industry every year (Zheng et al. 2021). Management of such large amount of waste is a challenging task for any industry. Some of these agricultural wastes are utilized as animal feed, or disposed of by either burning, composting, landfilling or land spreading (Sinharoy et al. 2020b; Sinharoy and Lens 2020, 2022; Bhatia et al. 2022). The industrial refuse of agricultural products processing also generates large quantities of organic rich waste which needs safe disposal. Such waste produced from agricultural industry can be used for bioenergy, including hydrogen production (Singh et al. 2022; Kamaraj et al. 2020). For example, the worldwide biohydrogen potential of two of the major crop residues, i.e., rice and wheat straw are estimated to be 58,002 and 34,680 Mm^3/year, respectively, which corresponds to energetic potentials of 623 and 373 PJ/year. These values are for untreated residues and treatment of these residues prior to their use in biohydrogen production can significantly improve the hydrogen yield and consequently their energy potential (Zheng et al. 2021; Bundhoo 2019).

Different biological methods, including photo-fermentation, dark fermentation and microbial electrolysis cells, have been used to convert agroindustrial waste to biohydrogen (Ndayisenga et al. 2021). However, most of the studies on hydrogen production from agricultural waste have utilized dark fermentation as the preferred method with a limited number of studies on photo-fermentation.

11.2 Biohydrogen Production from Agroindustrial Waste

Agroindustrial wastes can be broadly categorized into two groups, agriculture residues, which remain in the field after harvesting crops, and industrial residues, which are mainly the remains of industrial processing of agricultural products. The agriculture residues include field residues such as crop stalk, stem, husk, leaves, etc. which contribute relatively a larger fraction of the crops. On the other hand, agroindustrial refuse are fruit and vegetable peels, oil cakes, bagasse, etc. The organic content in such waste is high and can serve as a suitable substrate for biohydrogen production. Table 11.1 provides an overview of biohydrogen production from different agricultural waste along with the experimental conditions followed. Figure 11.1 shows the biohydrogen production using agricultural waste including information on different feedstock, pre-treatment and fermentation strategies.

Table 11.1 Biohydrogen production from different agroindustrial waste

Type of waste	Bioreactor	Inoculum	pH	Temp. (°C)	HRT (h)	Substrate concentration	Hydrogen production/yield	Reference
Rice straw hydrolysate	Continuously external circulating bioreactor	Heat-treated sludge	5.5–6.8	37	2, 4 and 8	20 g/L total sugar	Yield = 0.72 and 1.02 Mol/Mol hexose Rate = 5.52 and 16.32 L/L/d	(Liu et al. 2014)
Rice straw	Continuous stirred tank reactors (semi-continuous mode)	Pig manure	5.0	55	NA	6% total solids	Yield = 63.6 mL/g VS	(Chen et al. 2022)
Rice straw (ethanol organosolv pre-treated)	Batch serum bottles	*Enterobacter aerogenes*	5.8	37	NA	Not mentioned	Yield = 19.73 mL/g straw	(Asadi and Zilouei 2017)
Rice bran hydrolysate	Batch duran bottles	*Clostridium saccharoper-butylacetonicum*	6.0	30	NA	~30 g/L	Yield = 3.37 Mol/ Mol sugar Productivity = 7.58 mmol/(L.h)	(Dada et al. 2013)
Wheat straw hydrolysate	UASB	Heat-treated granular sludge	4.9	70	1	3.9 g/L total sugar	Yield = 212 mL/g sugars Rate = 821.4 mL/L.d	(Kongjan and Angelidaki 2010)
Wheat straw hydrolysate	Batch shake flask (under light condition)	*Rhodobacter capsulatus*	7.0	30	NA	Not mentioned	254 mL/L	(Mirza et al. 2013)
Wheat straw (acid-treated)	Batch serum bottles	Cow dung compost	6.5	36	NA	15 g/L	Yield = 68.1 mL/g VS Rate = 10.14 mL/g VS.h	(Fan et al. 2006)
Waste ground wheat hydrolysate	Batch serum bottles	Heat-treated anaerobic sludge	6.8	37	NA	10 g/L total sugar	Yield = 1.46 Mol/ Mol glucose	(Sagnak et al. 2011)

Type of waste	Bioreactor	Inoculum	pH	Temp. (°C)	HRT (h)	Substrate concentration	Hydrogen production/yield	Reference
Corn stalk	Batch sealed bottles	Photosynthetic bacteria HAU-M1	6.0	30	NA	5 g corn stalk (per 120 mL media)	Yield = 132.69 mL/g corn stalk	(Lu et al. 2020)
Corn stalk	Batch glass bottles	Anaerobic sludge (pre-treated at 100 °C for 15 min)	7.0	36	NA	10 g/L	Yield = 141.29 mL/g corn stalk Rate = 12.31 mL/g.h	(Ma et al. 2011)
Corn stalk	Batch glass bottles	Cow dung composts	7.0	36	NA	15 g/L	Yield = 149.69 mL/g VS Rate = 7.6 mL/h	(Zhang et al. 2007)
Sugarcane bagasse hydrolysate	Batch serum bottle	Elephant dung	6.5	37	NA	Not mentioned	Yield = 0.84 Mol/Mol total sugar	(Fangkum and Reungsang 2011)
Sugarcane bagasse	Batch glass bottles	Anaerobic sludge (pre-treated at 100 °C for 15 min)	7.0	55	NA	Not mentioned	Yield = 4.89 mmol/g bagasse	(Tian et al. 2015)
Sugarcane stillage	Batch glass bottles	Cow dung composts	5.5–7.5	55	6–7.5 h	84.2–110 kg COD/m^3/d	Rate = 2107 mL/L.d	(Fuess et al. 2016)
Cotton stalk hydrolysate	Batch serum bottle	Enterobacter cloacae WL1318 (recombinant)	–	37	NA	40 g/L total reducing sugar	Cumulative hydrogen production 1700 mL/L	(Zhang et al. 2020b)
Cotton plant wastes (pre-treated)	Batch serum bottle	Heat treated wastewater treatment sludge	5.5	37	NA	0.5 g of substrate per bottle	Yield = 17 and 15.2 mL/g VS	(Ghasemian et al. 2016)
Cotton stalk hydrolysate	Batch flask	*Klebsiella* sp. WL1316	8.0	37	NA	40 g/L total sugar	Yield = 1.44 Mol/ Mol sugar	(Li et al. 2018)
Barley straw (ozone treated)	Batch serum bottles	Mixture of cow manure and pond sediment (heat treated)	6.0	35	NA	2 g per bottle (dry weight basis)	139.2–186.8 mL of cumulative hydrogen	(Wu et al. 2013b)

(continued)

Table 11.1 (continued)

Type of waste	Bioreactor	Inoculum	pH	Temp. (°C)	HRT (h)	Substrate concentration	Hydrogen production/yield	Reference
Waste barley hydrolysate	Batch serum bottles	*Rhodobacter sphaeroides*	7.0	29	NA	11 g/L total sugar	Productivity = 0.4 L H_2/L culture	(Kars and Ceylan 2013)
Oat straw hydrolysate	Anaerobic sequencing batch reactors	Anaerobic granular sludge from full-scale UASB	5.5	35	24 h (cycle time of 12 h)	5 g COD/L	Yield = 2 Mol/Mol sugar Rate = 278 mL/L/h	(Arreola-Vargas et al. 2013)
Sweet sorghum Stover	Batch polyacrylic bioreactor	Heat-treated sewage sludge	7.0	37	NA	Not mentioned	Yield = 2.8 Mol/Mol reducing sugar Rate = 25 L/L.d	(Shanmugam et al. 2020)

NA Not applicable

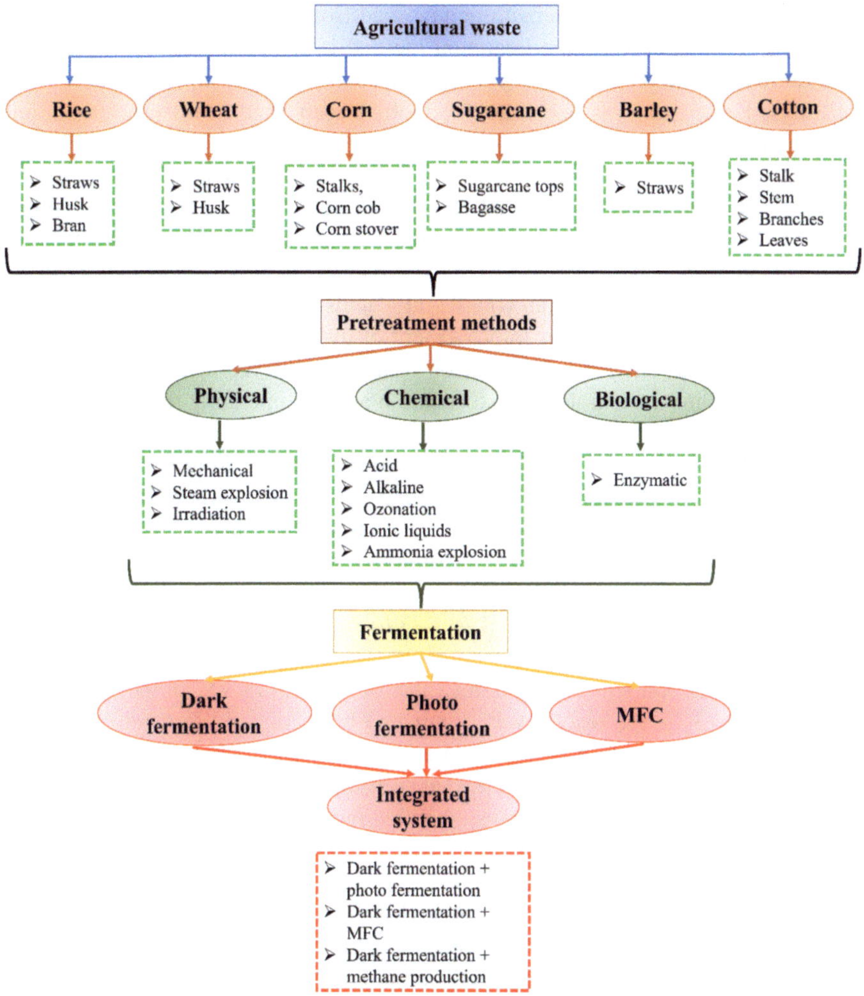

Fig. 11.1 Overview of biohydrogen production using agricultural waste: different feedstock, pretreatment methods and fermentation strategies

11.2.1 Rice

Rice is one of the major food crops as it is consumed by almost half of world population as its staple food. Rice is the main crop in south and south-east Asian countries such as China, India, Indonesia, Bangladesh, Vietnam, Thailand, Myanmar, etc. Rice straw and rice husk are the main agricultural residues from the rice crop. The rice straws consist of 39.2%, 23.5% and 36.1% (w/w) cellulose, hemicellulose and lignin, respectively (El-Tayeb et al. 2012).

There are many reports on hydrogen derived from rice crop waste as substrate. However, in most cases, a pre-treatment method is applied to make the rice crop residue more biodegradable for hydrogen production. Liu et al. (2013) used acid-hydrolysed rice straw as substrate for biohydrogen production. The maximum hydrogen yield obtained was 0.44 mol/mol sugar with 82% substrate utilization under batch mode. Further addition of food industry wastewater to rice straw hydrolysate resulted in 1.5 times increase in the hydrogen yield (0.69 mol/mol sugar) during continuous mode. The microbial community analysis revealed that *Clostridium pasteurianum* strain was predominant in the bioreactor, which favoured the production of hydrogen from volatile fatty acids formed as the major intermediate during the process. Further improvement of hydrogen production using a continuously external circulating bioreactor could be achieved with hydrogen yield values of 0.72 and 1.02 mol/mol hexose at 8 h and 4 h operating time, respectively (Liu et al. 2014).

In a study by Chen et al. (2022), biohydrogen production from rice straw using a stirred tank reactor was evaluated under semi-continuous mode of operation. The effect of temperature (37 °C and 55 °C) and total solids content (3%–12%) on biohydrogen production was studied. The authors reported that the biohydrogen production was higher under thermophilic condition (0.46–63.60 mL/g VS) than in mesophilic condition (0.19–2.13 mL/g VS) for all the three solid concentrations. Among the three solid content, a moderate solid concentration (6%) showed best hydrogen production with a maximum value of 63.6 mL/g VS. In the case of low solid content (3%), the reactor was dominated by lactic acid fermenting bacteria which caused acidification of the reactor and lowered the hydrogen production. Whereas, methane production increased at high solid content (12%) due to the presence of excess organic compounds, thereby reducing the hydrogen content in the biogas.

Co-digestion of rice straw with an activated sewage sludge improved the hydrogen yield on substrate (Alemahdi et al. 2015). The process improvement strategy included studying the effect of different inoculum treatment condition, pH, substrate to sludge ratio and substrate concentration. The effect of duration of heat treatment and temperature applied to the inoculum were studied to inactivate the methanogens present in the inoculum sludge. The best pre-treatment condition was found to be 100 °C for 60 min which resulted in a hydrogen yield of 14.22 mL/g VS. Following optimization of the other process parameters employing response surface methodology, a maximum biohydrogen yield of 14.70 mL/g VS was obtained at initial pH of 5.01 and substrate to biomass ratio of 4.54:1 and 2–20 mm size of rice straw.

Combination of two or more pre-treatment methods for improving biohydrogen production from rice straw is also reported in the literature. In a study by Kannah et al. (2019), combined dispersion and thermochemical disintegration was followed for treating rice straw. The results showed that 20.9% disintegration could be achieved at optimum conditions of pH 10, 80 °C, 120 rpm agitation and a reaction time of 30 min. The biohydrogen production improved significantly using the pre-treated rice straw (129 mL/g COD) as compared with untreated rice straw (8 mL/g

COD). In another strategy, organosolv treatment of rice straw with ethanol was carried out to improve the biohydrogen production by *Enterobacter aerogenes* (Asadi and Zilouei 2017). The effect of different process parameters, such as temperature, reaction time and ethanol concentration on pre-treatment efficiency were examined in this study. Maximum hydrogen yield of 19.73 mL/g of rice straw was obtained under pre-treatment conditions of 45% ethanol (v/v), 180 °C temperature and 30 min reaction time.

11.2.2 Wheat

Wheat is another grain which is equally popular as rice as a food source throughout the world. Production wise China is the leading producer (134.3 million metric tonnes) of wheat, followed by India (107.6 million metric tonnes), Russia (85.9 million metric tonnes), United States (49.7 million metric tonnes) and Canada (35.2 million metric tonnes) (values are for the year 2020) (FAO 2022). Wastes produced from wheat are mainly in the form of straw and husk which are generated during its processing. The wheat straws contain 32.9% cellulose, 24% hemicellulose and 8.9% lignin (by weight) (Martin et al. 2012).

Considering the amount of wheat production, large quantities of waste are generated that can be used for the production of hydrogen. For example, Fan et al. (2006) examined acid-treated wheat straw for biohydrogen production using cow dung digestate as the inoculum. The hydrogen content in the biogas produced from wheat straw was high (52%) without any significant methane production. The maximum hydrogen yield and production rate values were 68.1 mL/g VS and 10.14 mL/g VS.h, respectively. Another study examined three different bioreactors, namely continuous stirred tank reactor (CSTR), upflow anaerobic sludge bed (UASB) reactor and anaerobic filter (AF) reactor, for biohydrogen production from wheat straw hydrolysate under thermophilic (70 °C) condition (Kongjan and Angelidaki 2010). Among the three reactors, UASB showed best performance with hydrogen yield of 212 mL/g sugars under optimum HRT of 1 day. Reduction in HRT to 0.5 day showed detrimental effect on the hydrogen production performance.

Pre-treatment of wheat straw enhances the biohydrogen production potential as demonstrated by several studies in the literature (Fig. 11.1). For example, ozonation to improve hydrogen production from wheat straw was found effective by removing lignin fraction from the lignocellulosic biomass (Wu et al. 2013a). Ozone treatment enhanced the hydrogen production by 107%–158%, and the maximum H_2 production was obtained with 45 min ozonation. Reilly et al. (2014) investigated the effect of calcium hydroxide treatment on wheat straw for biohydrogen production (Reilly et al. 2014). In this study, to keep the pre-treatment process economical, the treatment was carried out under ambient temperature (20 °C) over different reaction times ranging from 2–12 days. The average hydrogen production was 58.78 mL/g VS which is 35.8% higher than the value obtained using untreated wheat biomass. Another similar study used calcium peroxide and calcium oxide for pre-treatment of wheat straw (Zhang et al. 2020a). Although both the treatments were found to improve the hydrogen production, the performance due to calcium oxide was slightly better with very high hydrogen production of 144 mL/g total solids under optimum conditions of 121 °C temperature and 6% CaO (w/v).

Wheat straw was utilized for simultaneous production of biohydrogen, bioethanol and methane under a biorefinery concept (Kaparaju et al. 2009). Wheat straw was treated with combined hydrothermal process and enzymatic hydrolysis. Cellulose obtained from the wheat straw was used for ethanol (0.41 g/g glucose) and biohydrogen (178.0 mL/g sugars) production. The effluents from the ethanol and hydrogen production processes were utilized for methane production which yielded 0.324 and 0.381 m^3 methane/kg volatile solids, respectively. In another study, bioethanol and biohydrogen production from wheat straw hydrolysate by *Escherichia coli* WDHL in a single reactor was carried out (Lopez-Hidalgo et al. 2017). The ethanol and hydrogen production values were 0.050 g/L/h and 18 mL/L/h under the optimum conditions of pH 8.2, 31 °C temperature and 23.4 g/L total reducing sugar.

11.2.3 Corn

Corn is another major crop, and the United States, China, Brazil, Argentina and Ukraine are the top five producers. In addition to being a staple food for large numbers of people throughout the world, corn is used as animal feed and to produce corn oil, corn starch and corn syrup. The on-field waste generated from corn crop is in the form of corn stalks, corn cob and corn stover. The cellulose, hemicellulose and lignin content in corn stalks are approximately 61.2%, 19.3% and 6.9% (w/w), respectively (El-Tayeb et al. 2012). On the other hand, corn cob contains approximately 36% cellulose, 26% hemicelluloses and 17% lignin.

There are a significant number of studies that explored corn stalks as the substrate for biohydrogen production. The effect of substrate concentration and initial pH on hydrogen production from corn stalk was examined by Ma et al. (2011). The substrate concentration was found to significantly affect the biohydrogen production and 10 g/L was found to be the optimum concentration. The maximum hydrogen yield and production rate values were 141.29 mL/g corn stalk and 12.31 mL/g/h, respectively. A novel strain, *Bacillus licheniformis* MSU AGM 2, was isolated from paper mill effluent treating reactor for biohydrogen production from corn stalk (Bala-Amutha and Murugesan 2013). The maximum hydrogen yield obtained was 82.5 mL/g of delignified corn stalk (treated with 2% NaOH). The cellulose utilization was 80.56% after 6 days of incubation.

The effect of iron addition on photo-fermentative hydrogen production from corn stalk was examined by Lu et al. (2022). The biohydrogen production by the phototrophic bacteria HAU-M1 improved by 19.98% with the addition of iron in the media. The maximum hydrogen yield obtained was 70.25 mL/g corn stalk with 2500 μmol/L iron addition, whereas the maximum hydrogen production rate was

17.82 mL/h at a slightly lower iron concentration of 2000 μmol/L. In another study using the same bacterial strain, photo-fermentation of corn stalk resulted in a biohydrogen yield of 132.69 mL/g corn stalk along with nearly 54% hydrogen content in the biogas (Lu et al. 2020). The buffering capacity of the medium was maintained by adding a buffer solution of Na_2HPO_4/NaH_2PO_4 of varying pH (4.8–7.0). The buffer addition improved hydrogen production from corn stalk by preventing pH reduction due to acids production. The pH 6.0 was found to be optimum in this study.

As with the other lignocellulosic biomass, pre-treatment of corn stalk is also reported to improve the biohydrogen production. Acid pre-treatment followed by enzymatic hydrolysis is reported to enhance hydrogen yield from corn stalk (Wang et al. 2010). The optimum pre-treatment conditions were acid (HCl 0.6% w/w) treatment at 90 °C for 2 h followed by enzymatic hydrolysis (cellulase from *Trichoderma viride*) for 72 h at 50 °C and pH 4.8. The maximum hydrogen yield obtained from the pre-treated corn stalk was 126.22 mL/g substrate along with a hydrogen production rate of 9.58 mL/g/h at a substrate concentration of 20 g/L.

In a different strategy involving pre-treatment and two-stage fermentation (dark fermentation and photo-fermentation; Fig. 11.2), maximum hydrogen yield from corn stalk was reported (Zhang et al. 2017). The dark fermentation was carried out using cow dung and photo-fermentation using a recombinant photosynthetic bacterium, *Rhodobacter capsulatus*. Corn stalk was treated using 0.75% NaOH, 12 IU cellulase/g corn stalk and 2400 IU hemicellulase/g corn stalk under different temperature and treatment time. The maximum hydrogen yield obtained was 156.7 mL/g corn stalk during dark fermentation along with 2827.5 mL/L of hydrogen during photo-fermentation using the effluent from the previous stage. Biohydrogen yield using the two-stage approach was 439.4 mL/g corn stalk and COD removal was 90.6%.

11.2.4 Sugarcane

Sugarcane is another major crop accounting for 79% of the total worldwide sugar production. Sugarcane is largely produced in tropical and subtropical regions of the world. The largest producer of sugarcane is Brazil (29.61 million metric tonnes), followed by India, European Union, China and Thailand. Processing of sugarcane crop produces wastes like plant tops and bagasse. The sugarcane tops are one of the most abundantly found crop residue from sugarcane cultivation and composed of 40% cellulose, 30% hemicellulose and 30% lignin (Khaire et al. 2021). The cellulose, hemicellulose and lignin content in bagasse are 40%–45%, 30%–35% and 20%–30%, respectively (Rai et al. 2014).

The waste generated from sugarcane crop can be utilized for biohydrogen production by different processes. However, it is important to optimize process parameters and apply different pre-treatment strategies to enhance hydrogen yield on the biomass. For example, a study examined the effect of substrate concentration, substrate:buffer ratio and inoculum:substrate ratio on biohydrogen production from

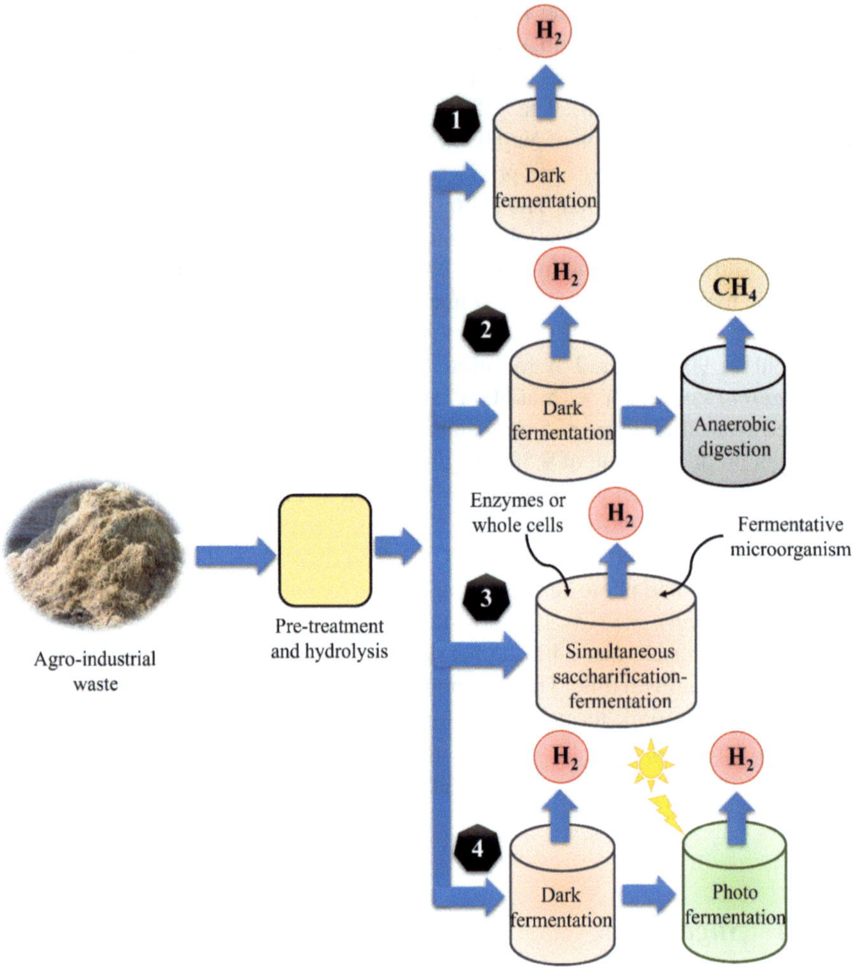

Fig. 11.2 Biohydrogen production strategies from agricultural wastes reported in the literature. (1): single-stage dark fermentation, (2): two-stage dark fermentation and methane production, (3): single-stage simultaneous saccharification and fermentation and (4): two-stage dark fermentation and photo-fermentation

acid-hydrolysed sugarcane bagasse (Sangyoka et al. 2016). The maximum hydrogen production of 6980 mL/L was obtained under the optimum conditions 22.77 g/L total sugar, substrate:buffer ratio 4.31 and inoculum:substrate ratio 0.31.

A two-stage approach was evaluated for hydrogen production from sugarcane bagasse by integrating dark fermentation (using *Enterobacter aerogenes* MTCC 2822) followed by photo-fermentation (Fig. 11.2) (using *Rhodopseudomonas* BHU 01) (Rai et al. 2014). Sugarcane bagasse was hydrolysed with H_2SO_4 followed by furfural removal by adsorption onto Amberlite XAD-4 resin column prior to the dark-fermentative hydrogen production step. A simultaneous saccharification-fermentation step (Fig. 11.2) was also carried out by adding *Cellulomonas fimi* which is capable of releasing sugars from cellulosic residues, and its performance was compared with that of the single dark-fermentation step. The results revealed that cumulative hydrogen production was higher for dark fermentation (1000 mL/L) than for simultaneous saccharification-fermentation (613 mL/L). The hydrogen production by *Rhodopseudomonas* sp. during photo-fermentation was 755 and 351 mL/L using the effluents from dark fermentation and simultaneous saccharification-fermentation, respectively.

Combined production of hydrogen and methane in a two-stage approach (Fig. 11.2) using sugarcane bagasse along with water hyacinth was examined using a continuous stirred tank reactor (Kumari and Das 2019). Water hyacinth was added to improve the nitrogen and micronutrient content in the media which is beneficial in terms of utilizing the waste biomass rather than the use of costly chemicals. The optimum HRT for the hydrogenogenic and methanogenic stages were 8 h and 10 days, respectively. The maximum hydrogen and methane yield values were 303 L/g COD and 142 mL/g COD, respectively.

A biorefinery approach with stepwise production of multiple products such as nanocellulose, lignin and biohydrogen from sugarcane bagasse was undertaken by Katakojwala and Mohan (2022). Acid treatment of bagasse resulted in production of the same amount (0.15 g/g sugarcane bagasse) of nanocellulose and lignin from the biomass. The amount of biohydrogen produced was 0.15 L/g COD by dark fermentation of delignified biomass.

11.2.5 Cotton

Cotton is the most widely used natural fibre and encompasses nearly 40% of the worldwide fibre market. Cotton is grown as short shrub in the tropical and subtropical climatic regions including America, Africa and Asia. As per the data provided by United Nations Food and Agriculture Organization, the largest producer of cotton is China, followed by India, United States, Brazil and Pakistan. After cotton is harvested, a large amount of lignocellulosic waste namely, stalk, stem, branches and leaves remain unutilized which can be converted to biohydrogen. The general composition of cotton stalk is 58.5% cellulose, 14.4% hemicellulose and 21.5% lignin (Sadh et al. 2018).

A study investigated biogas and biohydrogen production from cotton plant wastes and examined the effect of different pre-treatment conditions on the process efficiency (Ghasemian et al. 2016). The plant stalks were treated with ammonia solution and the boll (i.e. stem and branches) were treated with sodium hydroxide. Maximum biogas yields of 246 and 219 mL/g VS were obtained, respectively, from bolls pre-treated by sodium hydroxide (8% by weight) at 100 °C for 10 min and stalks pre-treated by ammonia solution (4% by weight) at 80 °C for 12 h. The biogas production values were, respectively, 178% and 171% higher than the values obtained with untreated biomass. The maximum hydrogen yields for the same pre-treatment conditions were 17 and 15.2 mL/g VS. In another study, Li et al. (2018)

investigated biohydrogen production from cotton stalk hydrolysate using *Klebsiella* sp. WL1316. The maximum daily and cumulative hydrogen production values were 937.0 mL/L.d and 2908.5 mL/L, respectively, and the final yield of biohydrogen was 1.44 mol/mol of sugar consumed.

Different pre-treatment strategies for improving biohydrogen production from cotton stalk by *Clostridium* sp. strain, isolated from a domestic wastewater sludge digester, were examined by Yoruklu et al. (2020). Among the three treatment methods (acid, alkaline and autoclaving), acid treatment using HCl (1% by volume) was found to be the best with a maximum hydrogen yield of 33 mL/g VS, which is attributed to almost two times higher sugar concentration in the acid-treated hydrolysate compared with the other treatment methods.

A recent study found that a genetically engineered strain of *Enterobacter cloacae* WL1318 with overexpression of formate hydrogen lyase activator (*fhlA*) gene showed increased production of hydrogen from cotton stalk hydrolysate (Zhang et al. 2020b). The recombinant strain produced ~1700 mL/L of cumulative hydrogen which was 23.5% higher than that of the control (wild strain). The glucose and xylose consumption also improved by 12% and 3.4%, respectively, due to the engineered strain. The better performance by the recombinant strain is due to the overexpression of *fhlA* gene, inhibition of metabolic pathways (including enzymes and metabolites) responsible for the synthesis of VFAs and alcohol and enhanced metabolic flux towards pyruvate node and formate splitting branch in the recombinant strain.

11.2.6 Barley

Barley is one of the major cereal grains, mainly cultivated in temperate climatic areas of the world, such as Russia, Germany, France, Ukraine, Australia, Canada, Spain, etc. Barely is the fourth largest produced grain, just after rice, wheat and corn. The majority of barley (~70%) is grown to be used in animal feed whereas the rest is used as substrate for beer and some other alcoholic beverages fermentation. Barley straw is the main on-field waste generated from the barley crop, and it contains around 33.8% cellulose, 21.9% hemicellulose and 13.8% lignin (Nigam 2017), which can serve as substrate for biofuel production including hydrogen.

Biohydrogen production from barley straw hydrolysate by sequential dark and photo-fermentation processes (Fig. 11.2) was carried out using *Caldicellusiruptor saccharolyticus* and two purple non-sulphur bacteria (*Rhodobacter capsulatus* DSM1710 and *Rhodobacter capsulatus* YO3), respectively (Özgür and Peksel 2013). The study observed that among the different media components, iron/molybdenum addition had significant effect on the biohydrogen yield, whereas yeast extract did not show any effect. The maximum hydrogen productivity was 0.58 mmol/L/h using *R. capsulatus* YO3 strain with added Fe/Mo to the media.

Wu et al. (2013b) studied the effect of ozone treatment on barley straw for hydrogen production by dark fermentation. To further improve the hydrogen production efficiency, an enzyme mixture containing cellulase, endoxylanase and β-glucosidase enzymes was added during dark fermentation to stimulate simultaneous saccharification and fermentation of the substrate. Ozonation for different durations (15–90 min) improved biohydrogen production from barley straw by 94%–166% and 45 min was found to the optimum.

In another study, along with hydrogen, 5-aminolevulinic acid was produced from barley waste hydrolysate by *Rhodobacter sphaeroides* O.U.001 (Kars and Ceylan 2013). The results showed that increase in total sugar concentration in the fermentation media improved the biohydrogen accumulation with a maximum hydrogen production of 0.4 L/L at 11 g/L of total sugar concentration. The maximum 5-aminolevulinic acid production reached 67.4 μM with the same initial total sugar concentration but with vitamin B_{12} (2.5 g/L) and levulinic acid (3.5 g/L) addition to the media.

11.3 Techno-Economic Analysis of Biohydrogen Production from Agricultural Waste

Techno-economic analysis is important to evaluate the economics of an industrial process, product or service. In general, this analysis is carried out based on estimated different costs (installation and operating) associated with the process and revenue potential of the process, depending upon the technical and financial inputs. The analysis is important to understand the feasibility of any process in terms of its economic viability for successful implementation at a large scale. Hence, feasibility of biohydrogen production from different agroindustrial wastes should be studied using this tool in order to ensure that the process can be realized. There are quite a few studies on techno-economic analysis of biohydrogen production from agroindustrial waste, which are discussed in this section.

Li et al. (2012) evaluated the potential of agricultural waste and beverage industry wastewater for commercial production of biohydrogen using techno-economic analysis. A reference model was developed using the data obtained from lab and pilot scale process with the help of Aspen Plus™ software program. The feasibility of using these waste substrates for biohydrogen production was established with this model which revealed that in order to maximize the profit from agricultural waste an annual return rate of 39% with the Aspen Plus™ model is necessary. In case of beverage industry wastewater, this annual return rate should be much higher at 60% with the same model. The estimated annual revenue from agricultural waste and beverage wastewater-based hydrogen production were estimated to be 2,001,000 USD/year and 2,658,000 USD/year using the Aspen simulation. Moreover, the optimum plant sizes for hydrogen production from both these substrates were estimated as 300.57 and 52.51 m^3, respectively.

Zúñiga et al. (2016) studied the techno-economic sensitivity of hydrogen production from empty palm fruit bunches under Colombian conditions to evaluate the feasibility of a future palm-based biorefinery. The key parameters, namely, net present value, payback period and economic potentials of the process were calculated. The economic analysis was carried out with the assumption of a maximum stable hydrogen production capacity of 8747 tonnes/year from the empty palm branches. The results showed that the raw materials cost is the most critical parameter influencing the plant economics and an increase of 50% of raw materials cost can affect the process profitability. The variable operating cost has a critical value of 1500 US dollars/tonnes of raw materials and even a small increase to it can cause significant effect on the payback period of the plant. Hence, to ensure consistent supply of raw materials and to maintain the production cost, it is recommended to associate the biohydrogen production plant with a palm oil production plant.

Lucas et al. (2015) examined the techno-economic feasibility of biohydrogen production from cassava processing wastewater. The study found that the potential energy recovery from cassava wastewater due to hydrogen production can reach up to 0.59×10^9 MJ/year which is much higher than for dairy or citrus wastewater. The high production is due to the presence of high organic matter (8 g/l of COD) in cassava wastewater in comparison to the other two types of wastewater. The economic analysis revealed that the biohydrogen production can generate a maximum economic yield of USD 0.009 cents/L of wastewater processed. Moreover, a further increase in revenue generation could be achieved (up to USD 0.61 cents/L of wastewater) if a sequential hydrogen and methane production strategy is implemented. The energy recovery from the wastewater could also be improved in such combined hydrogen and methane production plant with values as high as 10.48 kJ/g of COD consumed. The study further observed that nutritional supplementation to the wastewater may improve hydrogen yield, but it would significantly increase the operation cost to USD 22.3 cents/L of wastewater. Hence, according to this study, media supplementation is not an economically viable strategy for biohydrogen production from cassava wastewater.

The techno-economic analysis can also be used to identify the most suitable substrate for biohydrogen production. For instance, Panagiotopoulos et al. (2015) used technical suitability mapping to select most economically suitable feedstock among different agricultural wastes for biohydrogen production. The model for this study was developed by taking into account hydrogen yield potential, sugar mobilization efficiency, fermentability, and co-product yield and value. It was observed that sugar rich substrates were most suited for biohydrogen production compared to agricultural residues, and sugar beet juice, sweet sorghum juice and potato steam peels were the most preferred. Among the agricultural residues, such as straw, stalks, stem, etc., barley straw showed the maximum potential followed by miscanthus, sweet sorghum bagasse, carrot press cake, wheat straw and wheat bran. This study showed that hydrogen production potential can vary even though the substrate composition is almost similar. Hence, the model provided an easy tool to differentiate between different substrates and for selecting the most suitable substrate for hydrogen production. However, the drawback of this model is that it considers only substrate and no other factors that can influence biohydrogen production. Hence, further research efforts towards developing a more robust model need to be undertaken.

11.4 Challenges and Future Research Perspectives

Despite its huge potential, there exists no full-scale biohydrogen production plant which can utilize agricultural waste as feedstock. This is mainly attributed to low conversion efficiency of the biological hydrogen production along with high operational cost of the process. Poor yield of hydrogen on the substrate is another challenge faced by the biological processes which limits its commercial applicability. Due to the biochemical and thermodynamic limitations of the biological process, the maximum moles of hydrogen produced per mol of substrate is only 12 which further reduces to only 4 and 2 when acetic acid and butyric acid are the by-products. Moreover, under experimental conditions, the theoretical yield values are not achievable and an observed yield value in the range 1–2.5 mol H_2 per mole hexose is most common due to partial oxidation of the substrate for cellular growth and maintenance, which reduces the electron flow towards hydrogen production.

Dark fermentation is the main route to biohydrogen production, but it is hindered by many other factors. For instance, the lignin content in the agricultural waste makes the biomass structure difficult to degrade by the fermentative microorganisms, which is one of the major challenges. In addition, different pre-treatment methods applied to remove lignin creates more recalcitrant compounds (e.g. furfural, hydroxymethylfurfural, etc.) which need to be removed from the biomass hydrolysate before the substrate can be successfully utilized for biohydrogen production. Another challenge is that mixed microbial consortia employed in dark fermentation generally tends to produce methane from organic substrate instead of hydrogen. In order to prevent methanogenesis, either pre-treatment of the inoculum or controlling the reactor operating conditions (pH, temperature, HRT, etc.) to inhibit methanogenic microorganisms needs to be carried out (Sinharoy and Pakshirajan 2021). For example, low pH condition, high temperature and short HRT are some of the desired operating conditions for preventing methanogenic activity in anaerobic reactors (Chatellard et al. 2017). Moreover, acid accumulation during continuous bioreactor operation, particularly at high organic loading conditions can inhibit hydrogen production by shifting the optimum pH to highly acidic pH condition. Apart from these factors, there is a lack of scientific information on techno-economic feasibility of different agricultural wastes for biohydrogen production. In the absence of such scientific data, it is difficult to make any informed decision towards successful commercialization of this process. To summarize, different challenges in biohydrogen production processes are its operational cost, performance, scale-up issues, process improvement strategies, etc. (Kamaraj et al. 2020). In addition, storage and transportation issues, environmental issues, national plan and policy, rules and legislation, and public awareness are few other major concerns in implementing hydrogen as a future fuel.

Future research should be focused towards addressing the aforementioned challenges (Fig. 11.3). Some key recommendations are as follows. Optimization of process parameters, including environmental conditions (pH, temperature, agitation, etc.), organic loading and substrate to microorganism ratio to improve hydrogen

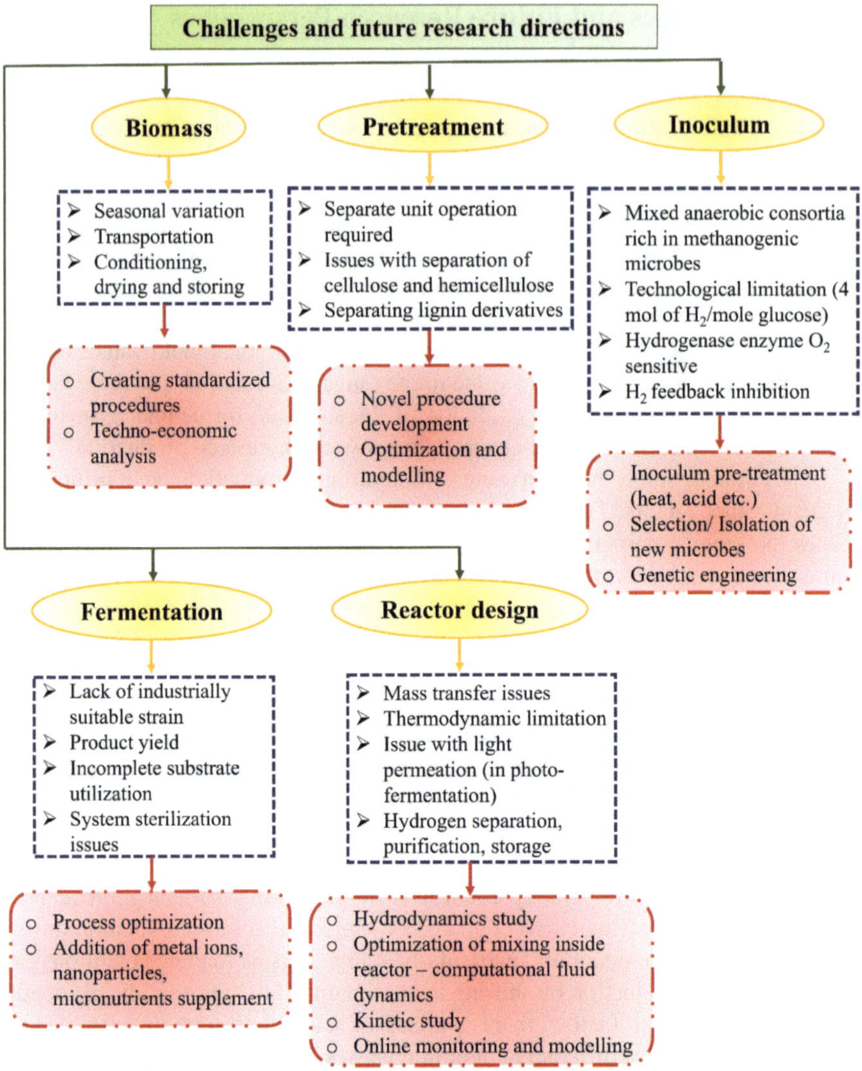

Fig. 11.3 Challenges and future research directions to overcome challenges in biohydrogen production from agricultural waste

production from agricultural waste, needs to be taken up. Other process intensification strategies include addition of nutrients, metal ions, nanoparticles, biochar, activated carbon, etc. to the fermentation media.

From the microbial perspective, selective enrichment of hydrogen-producing microorganisms by various methods such as long-term enrichment study or providing ideal environmental condition to shape the microbial community is important. With the advancement of genetic engineering and molecular biology tools, better understanding of metabolic pathways, key enzymes and co-factors involved in the hydrogen production process became possible. Improving hydrogen production and yield through metabolic engineering with such available information are proving helpful. Furthermore, these methods are also useful for identifying microorganisms responsible for biohydrogen production among the mixed microbial consortium, which provide necessary impetus for process intensification.

There are many recent studies on novel bioreactors for enhancing biohydrogen production from agricultural waste. Such novel reactors can prevent acidification of the rector and provide ideal environmental conditions for improving hydrogen production. Continuous monitoring of different bioreactor parameters is another important area of research for achieving high performance. Many advanced tools for monitoring and modelling reactor performance, including artificial neural network based methods, are emerging in the recent times that provides for process upgradation.

Another strategy to improve the process economics is development of novel approaches involving diversification of the product range under the concept of biorefinery. Biorefinery mimics the traditional Petro-chemical industry by simultaneously producing multiple products, including energy from biomass. In addition, integration of dark-fermentative biohydrogen production with other associated processes such as photo-fermentation, methane production and microbial electrochemical cells, can improve the energy conversion efficiency. Such integrated processes utilize effluent (mainly VFAs) from the previous dark-fermentation stage as the substrate for improving the production of hydrogen or even methane. Among these three processes, integration of the methanogenic step seems to be feasible for agricultural waste due to better understanding of the process involved as well as the technological advances already made in biogas production.

11.5 Conclusions

Biohydrogen production from agricultural waste has great prospect due to their abundance and long-term availability of the substrate. From a waste management perspective, production of hydrogen is ideally suited for treating agricultural residue rather than unsafely disposing them through burning, land-spreading or landfilling. As seen from the detailed literature survey, dark fermentation is the preferred method for biohydrogen production from agricultural waste. Different process parameters such as pH, temperature, organic load, HRT, agitation, biomass and inoculum pre-treatment etc. can significantly affect the process efficiency as well as the economics. However, biohydrogen production using dark fermentation is generally limited by low yield of the process, which is quite challenging for its commercial success. Many studies have focused on improving biohydrogen productivity by using various strategies including nutrient supplementation, application of novel bioreactor designs and developing integrated biorefinery. Another main challenge is development of industrial-scale systems that are economically viable and technologically sustainable, and towards which scientific efforts need to be directed.

References

Alemahdi N, Man HC, Nasirian N, Yang Y (2015) Enhanced mesophilic bio-hydrogen production of raw rice straw and activated sewage sludge by co-digestion. Int J Hydrog Energy 40(46):16033–16044. https://doi.org/10.1016/j.ijhydene.2015.08.106

Arreola-Vargas J, Celis LB, Buitrón G, Razo-Flores E, Alatriste-Mondragón F (2013) Hydrogen production from acid and enzymatic oat straw hydrolysates in an anaerobic sequencing batch reactor: performance and microbial population analysis. Int J Hydrog Energy 38(32):13884–13894. https://doi.org/10.1016/j.ijhydene.2013.08.065

Asadi N, Zilouei H (2017) Optimization of organosolv pretreatment of rice straw for enhanced biohydrogen production using *Enterobacter aerogenes*. Bioresour Technol 227:335–344. https://doi.org/10.1016/j.biortech.2016.12.073

Bala-Amutha K, Murugesan AG (2013) Biohydrogen production using corn stalk employing *Bacillus licheniformis* MSU AGM 2 strain. Renew Energy 50:621–627. https://doi.org/10.1016/j.renene.2012.07.033

Bhatia L, Sarangi PK, Singh AK, Prakash A, Shadangi KP (2022) Lignocellulosic waste biomass for biohydrogen production: future challenges and bio-economic perspectives. Biofuels Bioprod Biorefin 16(3):838–858. https://doi.org/10.1002/bbb.2329

Bundhoo ZM (2019) Potential of bio-hydrogen production from dark fermentation of crop residues: a review. Int J Hydrog Energy 44(32):17346–17362. https://doi.org/10.1016/j.ijhydene.2018.11.098

Chatellard L, Marone A, Carrère H, Trably E (2017) Trends and challenges in biohydrogen production from agricultural waste. In: Singh A, Rathore D (eds) Biohydrogen production: sustainability of current technology and future perspective. Springer, New Delhi, pp 69–95. https://doi.org/10.1007/978-81-322-3577-4_4

Chen H, Wu J, Huang R, Zhang W, He W, Deng Z, Han Y, Xiao B, Luo H, Qu W (2022) Effects of temperature and total solid content on biohydrogen production from dark fermentation of rice straw: performance and microbial community characteristics. Chemosphere 286:131655. https://doi.org/10.1016/j.chemosphere.2021.131655

Dada O, Yusoff WMW, Kalil MS (2013) Biohydrogen production from ricebran using *Clostridium saccharoperbutylacetonicum* N1-4. Int J Hydrog Energy 38(35):15063–15073. https://doi.org/10.1016/j.ijhydene.2013.07.048

Dou B, Zhang H, Song Y, Zhao L, Jiang B, He M, Ruan C, Chen H, Xu Y (2019) Hydrogen production from the thermochemical conversion of biomass: issues and challenges. Sustain Energy Fuels 3(2):314–342. https://doi.org/10.1039/C8SE00535D

El-Tayeb TS, Abdelhafez AA, Ali SH, Ramadan EM (2012) Effect of acid hydrolysis and fungal biotreatment on agro-industrial wastes for obtainment of free sugars for bioethanol production. Braz J Microbiol 43:1523–1535. https://doi.org/10.1590/S1517-83822012000400037

Fan YT, Zhang YH, Zhang SF, Hou HW, Ren BZ (2006) Efficient conversion of wheat straw wastes into biohydrogen gas by cow dung compost. Bioresour Technol 97(3):500–505. https://doi.org/10.1016/j.biortech.2005.02.049

Fangkum A, Reungsang A (2011) Biohydrogen production from sugarcane bagasse hydrolysate by elephant dung: effects of initial pH and substrate concentration. Int J Hydrog Energy 36(14):8687–8696. https://doi.org/10.1016/j.ijhydene.2010.05.119

FAO (2022). https://www.fao.org/faostat/en/#data/QCL. Accessed 19 July 2022

Fuess LT, Kiyuna LSM, Garcia ML, Zaiat M (2016) Operational strategies for long-term biohydrogen production from sugarcane stillage in a continuous acidogenic packed-bed reactor. Int J Hydrog Energy 41(19):8132–8145. https://doi.org/10.1016/j.ijhydene.2015.10.143

Ghasemian M, Zilouei H, Asadinezhad A (2016) Enhanced biogas and biohydrogen production from cotton plant wastes using alkaline pretreatment. Energy Fuel 30(12):10484–10493. https://doi.org/10.1021/acs.energyfuels.6b01999

Hirani AH, Javed N, Asif M, Basu SK, Kumar A (2018) A review on first-and second-generation biofuel productions. In: Kumar A, Ogita S, Yau YY (eds) Biofuels: greenhouse gas mitigation and global warming. Springer, New Delhi, pp 141–154. https://doi.org/10.1007/978-81-322-3763-1_8

Kamaraj M, Ramachandran KK, Aravind J (2020) Biohydrogen production from waste materials: benefits and challenges. Int J Environ Sci Technol 17(1):559–576. https://doi.org/10.1007/s13762-019-02577-z

Kannah RY, Kavitha S, Sivashanmugham P, Kumar G, Nguyen DD, Chang SW, Banu JR (2019) Biohydrogen production from rice straw: effect of combinative pretreatment, modelling assessment and energy balance consideration. Int J Hydrog Energy 44(4):2203–2215. https://doi.org/10.1016/j.ijhydene.2018.07.201

Kaparaju P, Serrano M, Thomsen AB, Kongjan P, Angelidaki I (2009) Bioethanol, biohydrogen and biogas production from wheat straw in a biorefinery concept. Bioresour Technol 100(9):2562–2568. https://doi.org/10.1016/j.biortech.2008.11.011

Kars G, Ceylan A (2013) Biohydrogen and 5-aminolevulinic acid production from waste barley by *Rhodobacter sphaeroides* OU 001 in a biorefinery concept. Int J Hydrog Energy 38(14):5573–5579. https://doi.org/10.1016/j.ijhydene.2013.03.013

Katakojwala R, Mohan SV (2022) Multi-product biorefinery with sugarcane bagasse: process development for nanocellulose, lignin and biohydrogen production and lifecycle analysis. Chem Eng J 446:137233. https://doi.org/10.1016/j.cej.2022.137233

Khaire KC, Moholkar VS, Goyal A (2021) Bioconversion of sugarcane tops to bioethanol and other value added products: an overview. Mater Sci Energy Technol 4:54–68. https://doi.org/10.1016/j.mset.2020.12.004

Kongjan P, Angelidaki I (2010) Extreme thermophilic biohydrogen production from wheat straw hydrolysate using mixed culture fermentation: effect of reactor configuration. Bioresour Technol 101(20):7789–7796. https://doi.org/10.1016/j.biortech.2010.05.024

Kumar M, Oyedun AO, Kumar A (2019) A comparative analysis of hydrogen production from the thermochemical conversion of algal biomass. Int J Hydrog Energy 44(21):10384–10397. https://doi.org/10.1016/j.ijhydene.2019.02.220

Kumari S, Das D (2019) Biohythane production from sugarcane bagasse and water hyacinth: a way towards promising green energy production. J Clean Prod 207:689–701. https://doi.org/10.1016/j.jclepro.2018.10.050

Li YC, Liu YF, Chu CY, Chang PL, Hsu CW, Lin PJ, Wu SY (2012) Techno-economic evaluation of biohydrogen production from wastewater and agricultural waste. Int J Hydrog Energy 37(20):15704–15710. https://doi.org/10.1016/j.ijhydene.2012.05.043

Li Y, Zhang Q, Deng L, Liu Z, Jiang H, Wang F (2018) Biohydrogen production from fermentation of cotton stalk hydrolysate by *Klebsiella* sp. WL1316 newly isolated from wild carp (*Cyprinus carpio* L.) of the Tarim River basin. Appl Microbiol Biotechnol 102(9):4231–4242. https://doi.org/10.1007/s00253-018-8882-z

Liu CM, Chu CY, Lee WY, Li YC, Wu SY, Chou YP (2013) Biohydrogen production evaluation from rice straw hydrolysate by concentrated acid pre-treatment in both batch and continuous systems. Int J Hydrog Energy 38(35):15823–15829. https://doi.org/10.1016/j.ijhydene.2013.07.055

Liu CM, Wu SY, Chu CY, Chou YP (2014) Biohydrogen production from rice straw hydrolyzate in a continuously external circulating bioreactor. Int J Hydrog Energy 39(33):19317–19322. https://doi.org/10.1016/j.ijhydene.2014.05.175

Lopez-Hidalgo AM, Sánchez A, De León-Rodríguez A (2017) Simultaneous production of bioethanol and biohydrogen by *Escherichia coli* WDHL using wheat straw hydrolysate as substrate. Fuel 188:19–27. https://doi.org/10.1016/j.fuel.2016.10.022

Lu C, Tahir N, Li W, Zhang Z, Jiang D, Guo S, Wang J, Wang K, Zhang Q (2020) Enhanced buffer capacity of fermentation broth and biohydrogen production from corn stalk with Na_2HPO_4/NaH_2PO_4. Bioresour Technol 313:123783. https://doi.org/10.1016/j.biortech.2020.123783

Lu C, Jiang D, Jing Y, Zhang Z, Liang X, Yue J, Li Y, Zhang H, Zhang Y, Wang K, Zhang Q (2022) Enhancing photo-fermentation biohydrogen production from corn stalk by iron ion. Bioresour Technol 345:126457. https://doi.org/10.1016/j.biortech.2021.126457

Lucas SDM, Peixoto G, Mockaitis G, Zaiat M, Gomes SD (2015) Energy recovery from agro-industrial wastewaters through biohydrogen production: kinetic evaluation and technological feasibility. Renew Energy 75:496–504. https://doi.org/10.1016/j.renene.2014.10.025

Ma S, Wang H, Wang Y, Bu H, Bai J (2011) Bio-hydrogen production from cornstalk wastes by orthogonal design method. Renew Energy 36(2):709–713. https://doi.org/10.1016/j.renene.2010.08.019

Martin JGP, Porto E, Corrêa CB, Alencar SM, Gloria EM, Cabral ISR, Aquino LM (2012) Antimicrobial potential and chemical composition of agro-industrial wastes. J Nat Prod 5:27–36

Mirza SS, Qazi JI, Zhao Q, Chen S (2013) Photo-biohydrogen production potential of *Rhodobacter capsulatus*-PK from wheat straw. Biotechnol Biofuels 6(1):1–12. https://doi.org/10.1186/1754-6834-6-144

Ndayisenga F, Yu Z, Zheng J, Wang B, Liang H, Phulpoto IA, Habiyakare T, Zhou D (2021) Microbial electrohydrogenesis cell and dark fermentation integrated system enhances biohydrogen production from lignocellulosic agricultural wastes: substrate pretreatment towards optimization. Renew Sust Energy Rev 145:111078. https://doi.org/10.1016/j.rser.2021.111078

Nigam PS (2017) An overview: recycling of solid barley waste generated as a by-product in distillery and brewery. Waste Manag 62:255–261. https://doi.org/10.1016/j.wasman.2017.02.018

Özgür E, Peksel B (2013) Biohydrogen production from barley straw hydrolysate through sequential dark and photofermentation. J Clean Prod 52:14–20. https://doi.org/10.1016/j.jclepro.2013.02.035

Panagiotopoulos IA, Karaoglanoglou LS, Koullas DP, Bakker RR, Claassen PA, Koukios EG (2015) Technical suitability mapping of feedstocks for biological hydrogen production. J Clean Prod 102:521–528. https://doi.org/10.1016/j.jclepro.2015.04.055

Paul T, Sinharoy A, Pakshirajan K, Pugazhenthi G (2020) Lipid-rich bacterial biomass production using refinery wastewater in a bubble column bioreactor for bio-oil conversion by hydrothermal liquefaction. J Water Process Eng 37:101462. https://doi.org/10.1016/j.jwpe.2020.101462

Paul T, Sinharoy A, Baskaran D, Pakshirajan K, Pugazhenthi G, Lens PNL (2022) Bio-oil production from oleaginous microorganisms using hydrothermal liquefaction: a biorefinery approach. Crit Rev Environ Sci Technol 52(3):356–394. https://doi.org/10.1080/10643389.2020.1820803

Rai PK, Singh SP, Asthana RK, Singh S (2014) Biohydrogen production from sugarcane bagasse by integrating dark-and photo-fermentation. Bioresour Technol 152:140–146. https://doi.org/10.1016/j.biortech.2013.10.117

Reilly M, Dinsdale R, Guwy A (2014) Mesophilic biohydrogen production from calcium hydroxide treated wheat straw. Int J Hydrog Energy 39(30):16891–16901. https://doi.org/10.1016/j.ijhydene.2014.08.069

Roy AS, Chingkheihunba A, Pakshirajan K (2016) An overview of production, properties, and uses of biodiesel from vegetable oil. In: Soccol C, Brar S, Faulds C, Ramos L (eds) Green fuels technology. Green energy and technology. Springer, Cham, pp 83–105. https://doi.org/10.1007/978-3-319-30205-8_4

Sadh PK, Duhan S, Duhan JS (2018) Agro-industrial wastes and their utilization using solid state fermentation: a review. Bioresour Bioprocess 5(1):1–15. https://doi.org/10.1186/s40643-017-0187-z

Sagnak R, Kargi F, Kapdan IK (2011) Bio-hydrogen production from acid hydrolyzed waste ground wheat by dark fermentation. Int J Hydrog Energy 36(20):12803–12809. https://doi.org/10.1016/j.ijhydene.2011.07.076

Sangyoka S, Reungsang A, Lin CY (2016) Optimization of biohydrogen production from sugarcane bagasse by mixed cultures using a statistical method. Sustain Environ Res 26(5):235–242. https://doi.org/10.1016/j.serj.2016.05.001

Shanmugam S, Krishnaswamy S, Chandrababu R, Veerabagu U, Pugazhendhi A, Mathimani T (2020) Optimal immobilization of *Trichoderma asperellum* laccase on polymer coated Fe_3O_4@SiO_2 nanoparticles for enhanced biohydrogen production from delignified lignocellulosic biomass. Fuel 273:117777. https://doi.org/10.1016/j.fuel.2020.117777

Singh A, Olsen SI, Nigam PS (2011) A viable technology to generate third-generation biofuel. J Chem Technol Biotechnol 86(11):1349–1353. https://doi.org/10.1002/jctb.2666

Singh H, Tomar S, Qureshi KA, Jaremko M, Rai PK (2022) Recent advances in biomass pretreatment technologies for biohydrogen production. Energies 15(3):999. https://doi.org/10.3390/en15030999

Sinharoy A, Lens PNL (2020) Biological removal of selenate and selenite from wastewater: options for selenium recovery as nanoparticles. Curr Pollut Rep 6(3):230–249. https://doi.org/10.1007/s40726-020-00146-4

Sinharoy A, Lens PNL (2022) Biological selenate and selenite reduction by waste activated sludge using hydrogen as electron donor. J Environ Manag 319:115745. https://doi.org/10.1016/j.jenvman.2022.115745

Sinharoy A, Pakshirajan K (2021) Methane free biohydrogen production from carbon monoxide using a continuously operated moving bed biofilm reactor. Int J Hydrog Energy 46(1):306–313. https://doi.org/10.1016/j.ijhydene.2020.09.250

Sinharoy A, Baskaran D, Pakshirajan K (2019) Sustainable biohydrogen production by dark fermentation using carbon monoxide as the sole carbon and energy source. Int J Hydrog Energy 44(26):13114–13125. https://doi.org/10.1016/j.ijhydene.2019.03.130

Sinharoy A, Kumar M, Pakshirajan K (2020a) An overview of bioreactor configurations and operational strategies for dark fermentative biohydrogen production. In: Yousuf A, Mahapatra DM (eds) Singh L. Elsevier, Bioreactors - sustainable design and industrial applications in mitigation of GHG emissions, pp 249–288. https://doi.org/10.1016/B978-0-12-821264-6.00014-0

Sinharoy A, Pakshirajan K, Lens PNL (2020b) Biological sulfate reduction using gaseous substrates to treat acid mine drainage. Curr Pollut Rep 6(4):328–344. https://doi.org/10.1007/s40726-020-00160-6

Tian QQ, Liang L, Zhu MJ (2015) Enhanced biohydrogen production from sugarcane bagasse by *clostridium thermocellum* supplemented with $CaCO_3$. Bioresour Technol 197:422–428. https://doi.org/10.1016/j.biortech.2015.08.111

Wang Y, Wang H, Feng X, Wang X, Huang J (2010) Biohydrogen production from cornstalk wastes by anaerobic fermentation with activated sludge. Int J Hydrog Energy 35(7):3092–3099. https://doi.org/10.1016/j.ijhydene.2009.07.024

Wu J, Upreti S, Ein-Mozaffari F (2013a) Ozone pretreatment of wheat straw for enhanced biohydrogen production. Int J Hydrog Energy 38(25):10270–10276. https://doi.org/10.1016/j.ijhydene.2013.06.063

Wu J, Ein-Mozaffari F, Upreti S (2013b) Effect of ozone pretreatment on hydrogen production from barley straw. Bioresour Technol 144:344–349. https://doi.org/10.1016/j.biortech.2013.07.001

Yoruklu HC, Koroglu EO, Ozdemir OK, Demir A, Ozkaya B (2020) Bioenergy production from cotton straws using different pretreatment methods. Int J Hydrog Energy 45(60):34720–34729. https://doi.org/10.1016/j.ijhydene.2020.02.104

Zhang ML, Fan YT, Xing Y, Pan CM, Zhang GS, Lay JJ (2007) Enhanced biohydrogen production from cornstalk wastes with acidification pretreatment by mixed anaerobic cultures. Biomass Bioenergy 31(4):250–254. https://doi.org/10.1016/j.biombioe.2006.08.004

Zhang Y, Li Q, Wang X, Yang H, Guo L (2017) Enhanced biohydrogen production from cornstalk through a two-step fermentation: dark fermentation and photofermentation. Int J Energy Res 41(15):2491–2501. https://doi.org/10.1002/er.3810

Zhang J, Kong C, Yang M, Zang L (2020a) Comparison of calcium oxide and calcium peroxide pretreatments of wheat straw for improving biohydrogen production. ACS Omega 5(16):9151–9161. https://doi.org/10.1021/acsomega.9b04368

Zhang Q, You S, Li Y, Qu X, Jiang H (2020b) Enhanced biohydrogen production from cotton stalk hydrolysate of *Enterobacter cloacae* WL1318 by overexpression of the formate hydrogen lyase activator gene. Biotechnol Biofuels 13(1):1–11. https://doi.org/10.1186/s13068-020-01733-9

Zheng Y, Zhang Q, Zhang Z, Jing Y, Hu J, He C, Lu C (2021) A review on biological recycling in agricultural waste-based biohydrogen production: recent developments. Bioresour Technol 347:126595. https://doi.org/10.1016/j.biortech.2021.126595

Zúñiga DP, Barrios EL, Peralta-Ruiz YY, González-Delgado AD (2016) Techno-economic sensitivity of bio-hydrogen production from empty palm fruit bunches under Colombian conditions. Chem Eng Trans 52:1117–1122. https://doi.org/10.3303/CET1652187

Chapter 12
Bioprocessing of Domestic Wastewater and Sewage Sludge for Biohydrogen Production: Different Routes and Pretreatment Strategies

Arindam Sinharoy and Kannan Pakshirajan

Abstract Hydrogen is a clean and non-polluting source of energy, which has received a lot of interest in the recent times. However, for its sustainable production and positive impact on the environment, it is necessary to utilize renewable feedstock as substrate. Municipal wastewater is one such alternative substrate that can be utilized for production of hydrogen through biological route. Large quantities of municipal wastewater are generated worldwide which require treatment. Combining municipal wastewater treatment process with biohydrogen production will help in recovering energy and value addition to the treatment process. Municipal wastewater is a mixture of domestic grey water from kitchen, wastewater from toilets, and sometimes can include industrial wastewater as well. The organic content of the municipal wastewater is generally lower than many other industrial wastewaters and it ranges in between 250 and 100 mg/L of COD. Municipal wastewater is also characterized by low dissolved and suspended solids, making it ideal for easy microbial degradation without any pre-treatment. There are many previous studies that explored biohydrogen production from municipal wastewater, including up scaling the process to industrial level. This chapter aims to provide information from the literature and case studies on biohydrogen production from municipal wastewater including bioreactor designs. Further, it describes the potential use of microbial electrolysis cell (MEC) for biohydrogen production from sewage. It also provides a holistic view on the effect of important process parameters, challenges for its commercialization and future research directions in the field.

A. Sinharoy
Department of Microbiology, University of Galway, Galway, Ireland

K. Pakshirajan (✉)
Department of Biosciences and Bioengineering, Indian Institute of Technology Guwahati, Guwahati, Assam, India
e-mail: pakshi@iitg.ac.in

C. R. Soccol et al. (eds.), *Biohydrogen - Advances and Processes*, Biofuel and Biorefinery Technologies 13, https://doi.org/10.1007/978-3-031-49818-3_12

Graphical Abstract

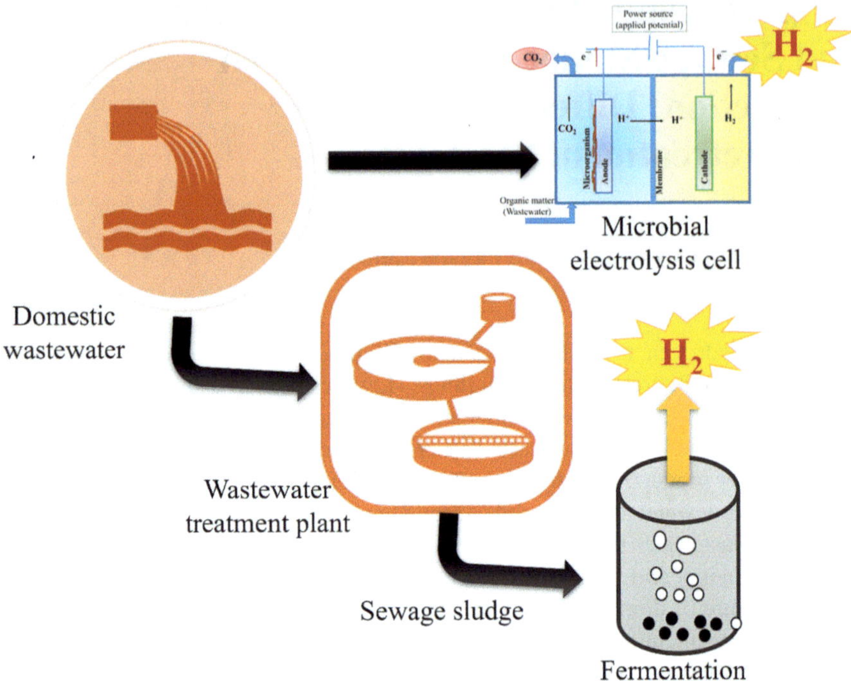

Keywords Biohydrogen · Dark fermentation · Waste to energy · Biological wastewater treatment · Municipal wastewater

12.1 Introduction

The widespread problem of water pollution caused by release of untreated wastewater from various industries and household is causing havoc to our health and the environment (Sinharoy et al. 2020a; Sinharoy and Lens 2020). Such unsafe wastewater disposal practices not only contaminate the surface water bodies, but also seeps into the groundwater table, causing long term hazardous impact (Cheng et al. 2020). Furthermore, the drinkable water sources are limited to less than 1% of the earth's freshwater reserve available for human use (Kanaujiya et al. 2019). This challenge of supplying safe drinking water to large population will only get worse by 2050, when the global demand for freshwater is expected to increase by one-third of what it is now (Darwall et al. 2018). Clean water supply is essential for all types of anthropogenic activities, including domestic activities, agriculture and animal farming, power production, etc. (Xiao et al. 2021).

Ensuring sufficient water supply with acceptable quality is critical for maintaining livelihoods, sustaining a healthy ecosystem and, above all, achieving sustainable developmental goals (Chrispim et al. 2019; Sinharoy and Pakshirajan 2019; Qadir et al. 2020).

Among the different wastewater sources, municipal or urban wastewater is one of the largest and common type of wastewater generated throughout the world (Kehrein et al. 2020). It is often termed as sewage, and among others it contains faecal matter, urine and laundry waste, which can cause immediate health human hazards when discharged into the environment without proper treatment (Dwivedi et al. 2018). Although, wastewater treatment plants for treating sewage are well established in developed countries, it is still a major problem in many of the developing countries (Chrispim et al. 2019; Lu et al. 2019). Other issue with sewage treatment plants is unprofessional management by unskilled workforce which can render the whole treatment system useless. Moreover, unscientific disposal of chemicals and pharmaceuticals through domestic wastewater can create further problem to the sewage treatment system (Kanaujiya et al. 2019). These chemicals are often nonbiodegradable, recalcitrant in nature and persist in the environment for years. Bacteria and viruses released along with bodily excretions by ill and infected persons also end up in the wastewater treatment plants and can pose harm to other humans (Dwivedi et al. 2018). Hence, developing a proper and safe sewage disposal system is essential for not only protecting human health but also for the benefit of the whole environment. Table 12.1. provides typical characteristics of domestic wastewater and sewage sludge as reported in different literature.

The conventional municipal wastewater treatment plants are composed of a combination of physico-chemical and biological systems (Fig. 12.1.). The most common biological treatment of domestic wastewater involves an activated sludge process, which is an aerobic treatment system. Anaerobic treatment is also used in many cases as biological treatment, either as the main biological treatment system or as an additional treatment where sewage sludge generated from the aerobic system is further processed to recover energy and reduce the sludge volume (Hreiz et al. 2015). The anaerobic system produces biogas (mainly methane) which can add value to the wastewater treatment system by generating energy (Chan et al. 2009). The organic matter present in domestic wastewater is converted by the anaerobic microorganism to methane (Kehrein et al. 2020). The same anaerobic system can also be adopted for the production of hydrogen by selecting suitable inoculum and optimizing the operating conditions (Roy et al. 2016; Sinharoy et al. 2020b). Hydrogen is more attractive than methane or any other biofuels due to its very high energy content, and it is considered as a clean fuel because it does not generate any pollutants (CO_2, SO_x, NO_x, etc.) upon combustion. Figure 12.2. depicts integration of hydrogen production as part of biological municipal wastewater treatment. Both wastewater and sewage sludge can be utilized as substrates for biohydrogen production.

Table 12.1 Characteristics of domestic wastewater and sewage sludge

Parameter	Amounts											
	Domestic wastewater						Sewage sludge					
pH	7.4	7.2	7.64	7.4	7.38	7.4	6.8	NM	7.3	7.1	6.3	6.61
COD (mg/L)	104.4	286.5	705.43	185	418.36	430	35,200	40,260	44,200	38,800	17,800	75,940
Suspended solids (mg/L)	NM	75.8	NM	NM	164.6	NM	NM	NM	NM	NM	NM	NM
Total nitrogen (mg/L)	NM	NM	185.88	35	47.15	60	1040	NM	1400	1200	1800	3310
Ammonia nitrogen (mg/L)	NM	38.2	131.92	32	33.01	47	NM	NM	NM	NM	NM	NM
Phosphate (mg/L)	7.9	NM	16.98	2.5	11.43	6.9	850	NM	NM	NM	NM	1300
Total solids (mg/L)	NM	NM	NM	NM	NM	NM	35,400	30,500	30,500	26,300	44,000	52,500
Volatile solids (mg/L)	NM	NM	NM	NM	NM	NM	26,100	26,840	20,700	19,700	32,200	39,300
Reference	Jwa et al. (2019)	Wang et al. (2006)	Yang et al. (2021)	Liu et al. (2010)	Valipour et al. (2015)	Hülsen et al. (2016)	Fountoulakis et al. (2010)	Caporgno et al. (2015)	Maragkaki et al. (2018)	Maragkaki et al. (2017)	Lee et al. (2014)	Liu et al. (2021)

NM Not mentioned

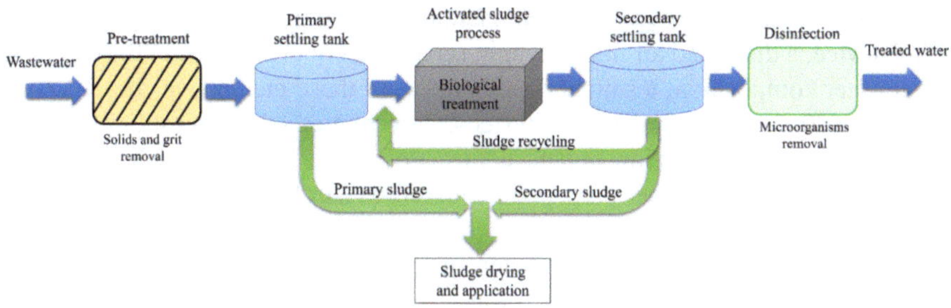

Fig. 12.1 Conventional activated sludge process based domestic wastewater treatment system

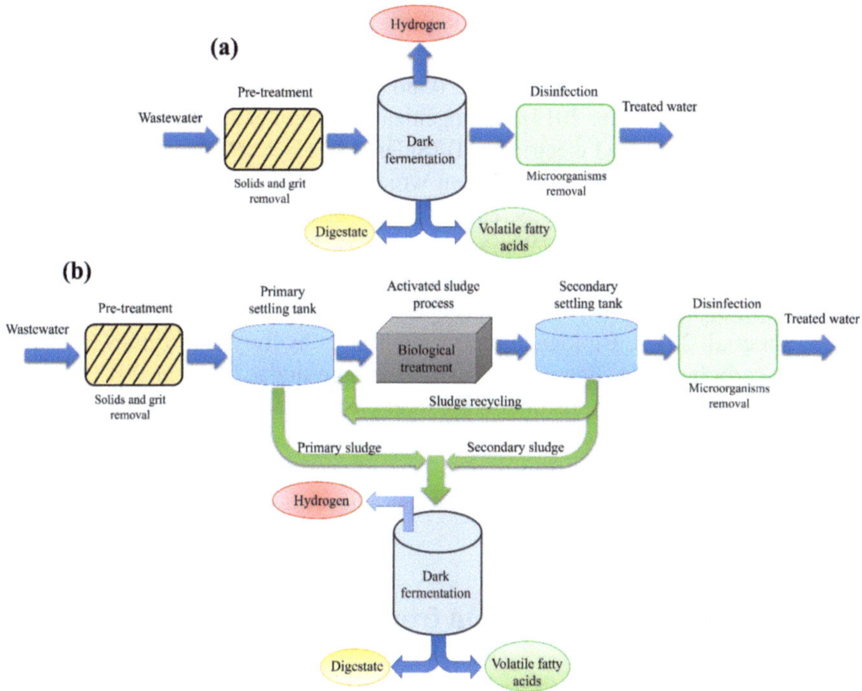

Fig. 12.2 (**a**) Domestic wastewater treatment by dark fermentation and (**b**) integration of dark fermentative hydrogen production with activated sludge process based domestic wastewater treatment

12.2 Pretreatment of Domestic Wastewater and Sewage Sludge

Primary treatment of domestic wastewater involves removal of oil, grease and various large-sized solid particles. Removal of these insoluble matter is necessary as they can interfere with the actual wastewater treatment by clogging and wearing due to the deposition of such materials (Sonune and Ghate 2004). The primary treatment methods in an actual sewage treatment plant include an oil and grease trap, a skimming device, a dissolved air flotation unit, screens, a grit chamber, etc. Based on the wastewater composition, a suitable combination of these methods is chosen (Zhou and Smith 2001). Other than these physico-chemical methods, domestic wastewater treatment is simple due to low organic content compared to more complex wastewaters such as those from paper-pulp, petroleum refinery, or food processing industries.

In a conventional domestic wastewater treatment plant, the sewage sludge produced is treated to reduce its water content and volume by applying techniques such as thickening, stabilization, (aerobic or anaerobic treatment), conditioning (applying inorganic or organic flocculants), dewatering and drying. However, biogas or hydrogen production from sewage sludge necessitates pretreatment to improve its biodegradability. For this purpose, many different pretreatment methods have been applied to sewage sludge, which include mechanical, thermal, chemical and biological methods (Zhen et al. 2017; Yao et al. 2018). Examples of mechanical treatment of sewage sludge are microwave (Guo et al. 2008), ultrasonication, gamma irradiation (Yin and Wang 2015), homogenization (Nabi et al. 2020) and electrokinetic degradation (pulsed electric field) (Salerno et al. 2009). Heating is the most common method of sludge pretreatment which involves variation of temperature. Besides, sludges can be treated by repetitive freezing and thawing cycles for enhancing biohydrogen production (Wang et al. 2003). Chemical treatments used for degrading sewage sludge are acid (Assawamongkholsiri et al. 2013), alkaline (Wei et al. 2010), ozonation and oxidation (Fenton and persulfate oxidation) (Tony et al. 2008; Shi et al. 2015). Biological pretreatment uses hydrolytic enzymes (amylase, protease, carbohydrases, lipases, etc.) to treat sewage sludge (Yu et al. 2013; Nguyen et al. 2021). Biological pretreatments are rarely used due to their high cost and need for large quantities of enzymes as compared with chemical agents to treat the same volume of sludge. Figure 12.3. provides a broad overview of the different methods used for pretreatment of domestic wastewater and sewage sludge.

12.3 Biohydrogen Production from Municipal Wastewater

Municipal wastewater can be utilized for hydrogen production as the organic content present in municipal wastewater can serve as nutrient for the hydrogen producing organisms (Sinharoy et al. 2020b). There are various methods of biological production of hydrogen which use different classes of microorganisms (Usman et al. 2019). These methods are dark fermentation, photo fermentation, direct biophotolysis, indirect biophotolysis and microbial electrochemical cells. Among these methods, dark fermentation and microbial electrochemical cells are widely reported in the literature on biohydrogen production from municipal wastewater. Moreover, in the case of microbial electrochemical cells, pilot scale studies have also been demonstrated, indicating suitability of this technology for treating real wastewater and value addition in the form of energy recovery from waste (Table 12.2.).

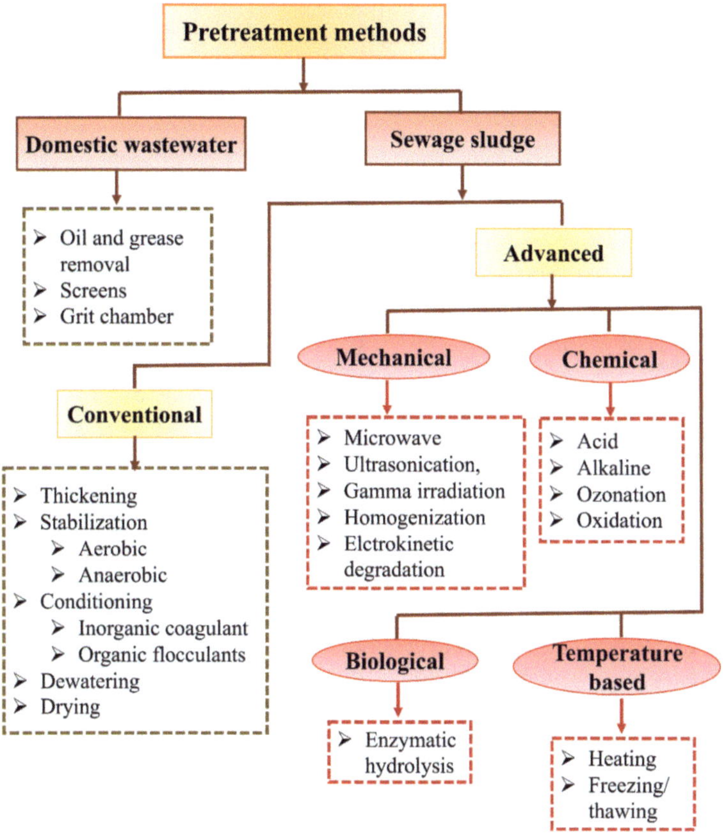

Fig. 12.3 Pretreatment methods for domestic wastewater and sewage sludge

Prakash et al. (2018) utilized domestic wastewater along with biodiesel industry effluent for biohydrogen production by a co-digestion strategy. The two bacterial strains used in this study were *Bacillus thuringiensis* strain EGU45 and *Bacillus amyloliquefaciens* strain CD16. During the batch study, the hydrogen yields were 55–110 L/L crude glycerol and 70–100 L/L crude glycerol for *B. thuringiensis* and *B. amyloliquefaciens*, respectively. Continuous biohydrogen production using bacteria immobilized on dried coconut coir packed into a PVC column was evaluated, which showed increase in the hydrogen production with maximum values of 120 and 100 L of hydrogen/L crude glycerol for *B. amyloliquefaciens* and *B. thuringiensis*, respectively.

In another study, co-digestion of press mud (sugar industry waste) and sewage as the substrate was studied for biohydrogen production using a UASB reactor under continuous mode of operation (Radjaram and Saravanane 2011). The substrate was prepared by mixing press mud with water to sewage at ratios of 1:7.5, 1:10 and 1:12.5. The specific hydrogen production rate of 187 mL/g VS along with biohydrogen production rate of 7960 mL/day could be achieved at press mud to sewage ratio

Table 12.2 Selected studies on biohydrogen production from municipal wastewater by microbial electrolysis cell

Type of wastewater	MEC setup	pH	Temp. (°C)	HRT (h)	Substrate concentration	Hydrogen production/ yield	Energy recovery (%)	Reference
Domestic	Bioelectrochemically assisted microbial reactor with packed bed (graphite granules) type anode (batch)	7.1	30	NA	204–481 mg/L COD	Yield = 154 mL/ g COD	26	Ditzig et al. (2007)
Domestic	Single chamber (continuous)	6.7	30	6 h	OLR = 642–836 mg/L.d Influent COD = 321–418 mg/L	Productivity = 0.3 L/L.d	38	Escapa et al. (2012)
Domestic	Single chamber (continuous)	7.0	25	10 h	Influent COD = 250–300 mg/L and 100–180 mg/L	Productivity = 1.0–2.6 L/L.d	23	Gil-Carrera et al. (2013a)
Domestic	Single chamber, semi-pilot modular MEC (continuous)	6.8–7.3	20	4 h	54–118.2 mg/L COD	Productivity = 0.05 L/L.d	719	Gil-Carrera et al. (2013b)
Domestic	Single chamber, pilot scale (continuous)	7.0	13.5–21	24 h	OLR = 0.14 kg COD/m^3/day	Productivity = 0.015 L/L.d	55	Heidrich et al. (2013)
Domestic	Single chamber, pilot scale (continuous)	7.0	1–22	24 h	147–1976 mg/L influent COD	Productivity = 0.007 L/L.d Rate = 0.6 L/day	41	Heidrich et al. (2014)
Domestic	Double chamber	6.5–7.5	25–35	30 h	OLR = 0.5 g COD/L/d	Not mentioned	11	Brown et al. (2014)
Domestic	Single chamber, cassette-style designed MEC	7.0–8.5	11.4 ± 2.5	5 h	Average COD of 347 mg/L	124 and 521 mL/d	21	Cotterill et al. (2017)

Type of wastewater	MEC setup	pH	Temp. (°C)	HRT (h)	Substrate concentration	Hydrogen production/yield	Energy recovery (%)	Reference
Domestic	Single chamber	NA	30 ± 1	NA	Total COD of 345 ± 47 mg/L	0.28 ± 0.04 $m^3/m^3/d$	64	Cusick et al. (2010)
Domestic	Single chamber, cassette-style designed MEC	NA	NA	0.015–18 days	4.5 g/L COD	0.011–0.015 $m^3/m^3/d$	2.9–131	Leicester et al. (2020)
Domestic	Tubular microbial electrolysis cell	7.4	30	24 h	OLR = 0.27–0.8 g COD/l/d	Rate = 0.18 ± 0.03 $m^3/m^3/d$	151.9	Jwa et al. (2019)

NA not available

of 1:10 and a fixed hydraulic retention time (HRT) of 30 h. The COD and VS removal under the same conditions were 61% and 59%, respectively. The optimum pH range for biohydrogen production was 5–6, which could be easily maintained during the co-digestion process.

12.3.1 Biohydrogen Production from Municipal Wastewater by Microbial Electrolysis Cells

Table 12.2. presents selected studies on biohydrogen production using microbial electrochemical cells. Microbial electrolysis cells are an advancement of bioelectrochemical cells and can provide alternative solutions to wastewater treatment along with energy generation in the form of hydrogen. This technology has the potential to replace conventional domestic wastewater treatment systems such as activated sludge process and anaerobic digestion (Rozendal et al. 2008; Arun et al. 2020). The working principle of the microbial electrolysis cell (MEC) system is opposite to the microbial fuel cell (MFC) where electric potential is applied to generate the desired product (hydrogen) (Escapa et al. 2016). In this system, hydrogen is produced in the cathode chamber by reducing protons (that are produced in the anode chamber and transported to cathode side thorough a proton selective membrane) in the presence of either inorganic catalysts or electroactive bacteria and low voltages (Neto et al. 2018).

Figure 12.4. is a schematic representation of a MEC. Unlike MFCs, as the electron flow is not spontaneous from anode chamber to cathode chamber in MEC, an external electric potential is applied between anode and cathode (Escapa et al. 2016; Katuri et al. 2019). The applied electric voltage depends on the organic compound (substrate) used in the anode chamber. In general, the potential applied is small and cost wise much more sustainable than employing expensive catalyst (e.g. platinum

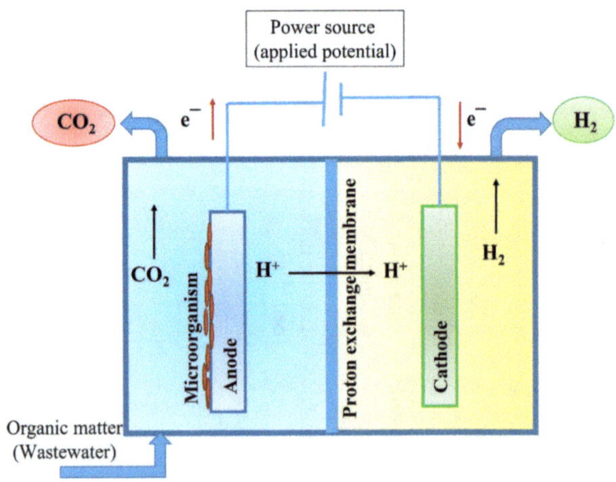

Fig. 12.4 Schematic representation of a microbial electrolysis cell

based catalyst) used in traditional electrolysis for hydrogen production (Escapa et al. 2016). Due to these reasons, MECs have captured the attention of scientists and industry alike for wastewater treatment and more particularly for the management of a large volume of urban wastewater. In the following section, selected studies on biohydrogen production using domestic wastewater are discussed in detail.

A tubular microbial electrolysis cell (MEC) equipped with a robust membrane electrode assembly (MEA) was used for biohydrogen production from domestic wastewater (Jwa et al. 2019). The study compared two different electrode assembly, one by directly coating a catalyst (Pt) on a self-supporting tubular membrane and the other consisted of a conventional cathode physically combined to a membrane. The novel electrode with catalyst coating showed better performance compared to the conventional one with 0.18 ± 0.03 $m^3/m^3/d$ of hydrogen production rate, 34 ± 2 mA of current generation. Performance of a microbial electrolysis cell fed with domestic wastewater was examined for hydrogen production, COD removal and energy consumption at different organic loading rates and applied voltages (Escapa et al. 2012). The maximum hydrogen production rate achieved in the MEC system was 0.3 L/L/d at HRT of 3–6 h and the hydrogen production followed Monod type kinetics. The hydrogen production highly depended on the wastewater strength with a saturated hydrogen production rate at an organic loading rate (OLR) above 2000 mg/L/d.

Biohydrogen production from domestic wastewater by *Enterobacter aerogenes* in a microbial electrolysis cell was studied with three different volumes (300–500 mL) of anode chambers (Afify et al. 2017). The maximum biohydrogen production of 316 cm^3 with *E. aerogenes* at 500 mL anode volume and 0.4 V power supply was achieved, which was almost three times higher than the value obtained without microbial addition. In another study, biohydrogen production from domestic wastewater was studied using a bioelectrochemical reactor in which performance of a commonly used carbon electrode was compared with that of a granular graphite packed-bed electrode (Ditzig et al. 2007). The results showed that the bioelectrochemical reactor containing granular graphite packed-bed electrode performed better in terms of both wastewater treatment and hydrogen production. The treatment efficiency in terms of COD, BOD and dissolved organics removal reached as high as 87–100%, and all these parameters were below the permissible level, following treatment using the bioelectrochemical reactor. The maximum hydrogen yield obtained in this study was 0.0125 mg H_2/mg COD.

A pilot-scale microbial electrolysis cell (MEC) was operated on site in Northern England to treat raw domestic wastewater to produce hydrogen (Heidrich et al. 2013). The system produced hydrogen (100 ± 6.4%) of very high purity at a volumetric COD loading rate of 0.14 $kg/m^3/d$ over 90 days of operation. The hydrogen production in the MEC reactor was 0.015 L/L/d which was equivalent to 70% recovery of net electrical input. Despite certain disadvantages such as low COD removal efficiency and inability to recover the entire electrical energy invested in the system, the MEC showed good potential in treating low strength wastewater along with hydrogen generation.

Another similar study examined long term (12 months) operation of a 100 L microbial electrolysis cell (MEC) for hydrogen production with raw domestic wastewater at ambient temperature (1 °C to 22 °C) (Heidrich et al. 2014). The system produced an average hydrogen production rate of 0.6 L/day; however, the volumetric gas production reduced over time. The energy recovery by the system was ~48.7% of the electrical energy input. The main problems encountered by the system was inconsistent removal of COD with poor effluent standard, poor pumping system and large overpotential.

Scale-up potential of MEC system for biohydrogen production using domestic wastewater was carried out by Escapa et al. (2015). The two membrane-less MEC units used in this study achieved 92% COD removal efficiency along with rapid hydrogen generation within the first 48 h of operation. The cumulative hydrogen production reached 223 mL and 208 mL in the two experimental setups. However, performance of the MECs declined during continuous operation and the COD removal efficiency was below 10%. This poor performance during continuous mode of operation was attributed to the insufficient capacity of the anodic chamber, poor mixing inside the anode chamber and ineffective hydrogen management on the cathode side.

In a study to improve the scalability of MEC system for domestic wastewater treatment along with biohydrogen production at ambient temperature in the European countries, Cotterill et al. (2017) used six 1 m^2 anodes of cassette-type design. The MEC system showed low internal resistance and produced 0.8 L/d hydrogen (93% purity) at ambient winter temperatures (11.4 ± 2.5 °C) and at 5 h HRT. The COD was around 63.5% with an effluent COD concentration of 124.7 mg/L which was in agreement with the European standard.

Microbial community plays an important role in any biological process and the role of biofilm in microbial electrolysis cell (MEC) is well-known. Analysis of microbial community from two pilot scale MECs operated continuously to treat domestic wastewater from a single source for more than 6 months was examined by Cotterill et al. (2018). The authors observed no significant difference in the microbial community present in the two MECs, and the electrogenic bacteria *Geobacter* dominated in the two systems. The study demonstrated that the MEC systems work well with most types of domestic wastewater, provided the microbial community present in the anode is maintained the same. Furthermore, the study suggested that influent wastewater flow rate need to be optimized for achieving maximum performance of the system.

12.3.2 Factors Influencing Performance of Microbial Electrolysis Cells

MECs use an ion exchange membrane which separate the anode and cathode chambers. The main advantage of this membrane is that it allows for altering cathode chamber for its maximum performance, without affecting the functioning of microbial community in the anode chamber (Escapa et al. 2016; Katuri et al. 2019). The membrane presence allows relatively pure hydrogen to be obtained from the cathodic side (Escapa et al. 2016). However, ion exchange membrane can hinder the flow of electrons from one chamber to another, thus increasing the internal resistance and deteriorating the reactor performance (Tartakovsky et al. 2009). Possible solutions to address this problem are the use of specifically designed catholyte solution that prevents pH increase and show better electrical conductivity, provision for recirculating the anolyte. However, these strategies may increase the operating cost of the system and recirculation of anolyte can reduce the purity of hydrogen produced (Escapa et al. 2015). Membrane fouling is another major concern that reduces the flow of protons through the membrane and ultimately affects hydrogen generation (Escapa et al. 2016).

Another strategy to avoid ion exchange membrane is the use of membrane-free MEC system, also known as single chamber MECs. The theoretical value of hydrogen production in this single chamber MECs is two times more than that in conventional MECs, but such high hydrogen yield is not reported thus far (Gil-Carrera et al. 2013a). This is due to excess potential in the single chamber system (Escapa et al. 2012). Another major problem with the membrane-free system is that the hydrogen produced is consumed in several other reactions. Hence, rather than obtaining pure hydrogen as desired, several unwanted by-products are produced in this system (Escapa et al. 2016; Katuri et al. 2019). For example, the hydrogen formed can get converted to acetate or methane in the presence of homoacetogenic and methanogenic bacteria (Escapa et al. 2016; Sinharoy et al. 2022). The acetate produced can also be oxidized at anode, resulting in a cyclic process (known as hydrogen recycling) without producing the desired product. Moreover, the hydrogen itself can be oxidized in the anode and consume electrical energy which will inadvertently reduce the energy efficiency of the system (Escapa et al. 2016).

Another technical challenge with biohydrogen production using MECs is that it is difficult to contain hydrogen in closed space and preventing its loss from the system (Moreno et al. 2015). Moreover, it is problematic to store large volume of hydrogen produced during large-scale real wastewater treatment operation due to the need for storing it under pressurized condition and explosive nature of hydrogen. There is always some net loss of hydrogen during its compression and storage (Escapa et al. 2016). Another challenge with the application of MECs for domestic wastewater treatment is with variation in wastewater characteristics. It is difficult to treat high strength wastewater using MECs as it creates problem of methane production and excess biomass (biofilm) growth which prevent effective electron transfer in the system. Also, as the treatment efficiency is quite low in MECs, treatment of high strength wastewater often fails to meet the effluent standard required for safe discharge of sewage (Katuri et al. 2019). High amounts of ammonia and phosphorus present in the wastewater are also difficult to remove in MECs, which require additional treatment steps for their removal from the MEC effluent (Escapa et al. 2015). The lack of essential nutrients in the wastewater to support growth of desired microorganisms in the system is another setback, which requires supplementation of the wastewater with additional nutrients (Moreno et al. 2015). This nutrient supplementation further increases the operating cost of the system and also creates additional operational complexity. Keeping all these factors in mind, rather than complete replacement of the conventional systems for domestic wastewater treatment with MECs, it can be integrated with already established biological treatment systems, which will ensure both better effluent standards and energy generation in the form of hydrogen (Fig. 12.5.). This strategy is also useful for integrating MECs with anaerobic digestion or dark fermentation systems, wherein the MECs can utilize the VFAs and other organics present in the effluent of anaerobic treatment in the previous step for H_2 production (Escapa et al. 2015, 2016). MECs can also be used for pretreatment of domestic wastewater, prior to aerobic treatment (e.g. activated sludge process) (Fig. 12.5.c). In this strategy the organic load can be reduced in MECs which will help in achieving better effluent quality in the subsequent aerobic treatment in comparison to the situation where only a single ASP system is used.

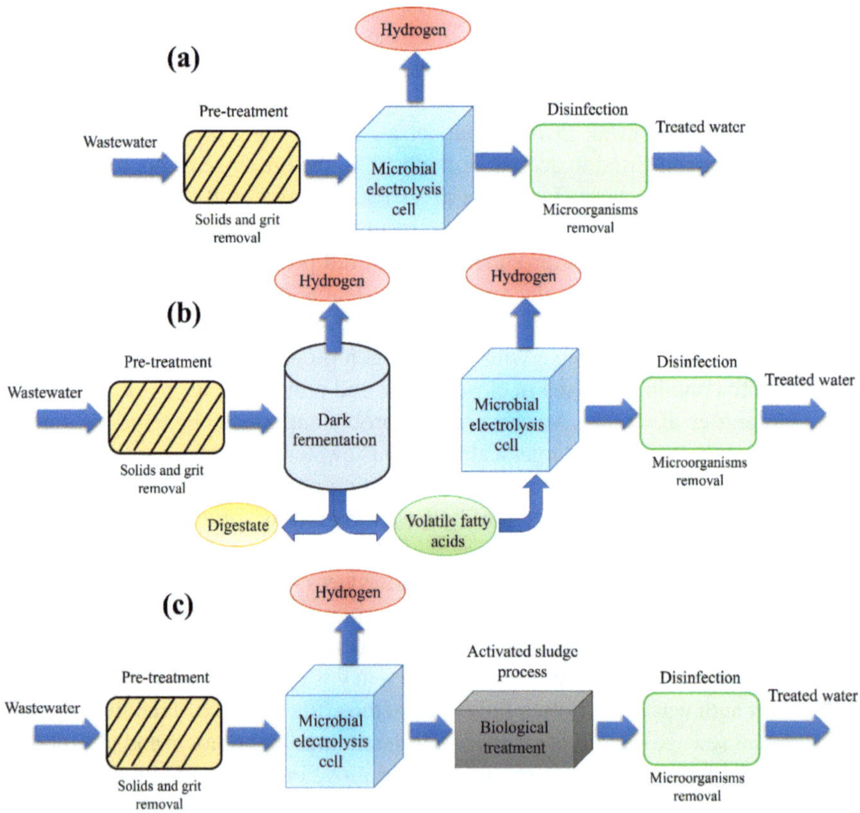

Fig. 12.5 (**a**) Microbial electrolysis cell based domestic wastewater treatment, (**b**) integration of dark fermentation based domestic wastewater treatment with microbial electrolysis cell and (**c**) integration of microbial electrolysis cell with activated sludge process

12.4 Biohydrogen Production from Municipal Sewage Sludge

Sewage sludge is a by-product of wastewater treatment, and it includes both primary and secondary sludge. The quantum of primary and secondary sludge generated are estimated to be 110–170 and 70–100 kg/mega litre of wastewater treated, respectively (Pilli et al. 2015). Common methods for the disposal of sewage sludge are land filling, land spreading, composting, anaerobic digestion, recycling as a construction material and incineration (Fytili and Zabaniotou 2008; Kirchmann et al. 2017). This sewage sludge can be utilized by hydrogenogenic bacteria to produce hydrogen along with volatile fatty acids such as acetic acid, butyric acid, etc. (Sinharoy et al. 2020b). The process of hydrogen production from sewage sludge is mainly through dark fermentation (Table 12.3.). In this process, the steps involved are hydrolysis, hydrogen and acid production and methanogesis. During the hydrolysis step, the larger molecular weight compounds such as starch, fiber, proteins are broken down into smaller compounds such as glucose, sucrose, amino acids, fatty acids, etc. In the next step, these low molecular weight compounds are fermented by homoacetogenic bacteria to produce hydrogen, carbon dioxide, acetic acid, butyric acid and propionic acid. In the final step, the methanogenic microorganisms convert the products formed in the previous step to methane by either hydrogenotrophic or acetoclastic route (Wong et al. 2014). If hydrogen is the desired product, methanogens need to be inhibited by either adding methanogenic inhibitor, pre-treatment of the sludge (inoculum) or proper choice of operational parameters such as pH, temperature, etc. Numerous studies have reported hydrogen production from sewage sludge by co-digestion, pretreatment, process optimization, etc. In the following section, some selected studies are discussed in detail (Yao et al. 2018; Wong et al. 2014).

Co-digestion of sewage sludge with grass residue (3:7 ratio VS basis) for hydrogen production was studied at different substrate concentrations (5–80 g/L based on VS) (Yang and Wang 2019). The hydrogen production improved with increase in substrate concentration till 10 g/L value, and above which the hydrogen production declined due to overloading of the substrate in the system and accumulation of metabolites such as acetate, butyrate, formate, etc. which reduced the pH inside the reactor and deteriorated the hydrogen production. Maximum biohydrogen yield of 45.6 mL/g VS was obtained at an optimum substrate concentration of 10 g/L VS. The effect of different organic loading rates (OLR) and HRTs on biohydrogen production by co-digestion of sewage sludge and crude glycerol in a two-stage process was examined (Rivero et al. 2014). Although HRT did not have any effect on the hydrogen production in this study, OLR significantly affected the reactor performance. The maximum hydrogen and methane yields were 0.026 L/g COD and 0.29 L/g COD, respectively, along with 93% COD removal at optimum 7.82 g COD/L/d OLR.

A study examined poultry manure addition (10–30 g/L) on biohydrogen production by co-digestion of sewage sludge and wine vinasse (equal amount) under

Table 12.3 Selected studies on biohydrogen production from sewage sludge by dark fermentation

Type of waste	Bioreactor	pH	Temp. (°C)	HRT (h)	Substrate concentration	Hydrogen production/Yield	Reference
Sewage sludge (alkaline treated)	Serum bottles	11.0	36 ± 1	NA	50 mL sludge	Yield = 16.6 mL/g of dry solids	Cai et al. (2004)
Food waste and sewage sludge	Wheaton media lab bottles	6.0	35	NA	1–28 g COD/L	Yield = 122.9 mL/g carbohydrate Rate = 111.2 mL/g VS/h	Kim et al. (2004)
Fallen leaves and sewage sludge	Serum bottles	7.0	37	NA	10 g VS/L	Yield = 37.8 mL/g VS	Yang et al. (2019)
Sewage sludge and grass residue	Serum bottles	7.0	37	NA	10 g VS/L	Yield = 45.6 mL/g VS	Yang and Wang (2019)
Sewage sludge (combined sodium citrate and thermal treated)	Serum bottles	7.0	37	NA	50 m of sludge	Yield = 28.6 mL/g VS	Yang and Wang (2021)
Sewage sludge with glycerol	Semi-continuous stirred tank reactor	5–6	-	6–10 days	OLR = 17.90 and 7.82 g COD/L.d	Yield = 0.032 L/g COD	Rivero et al. (2014)
Food waste, sewage sludge and landfill leachate	Serum bottles	7.81	36	NA	15.7 g/L VS	H_2 production potential = 193.85 mL/gVS H_2 production rate = 94.35 mL/g VS.h	Li et al. (2008)
Tofu residue and sewage sludge	Sealed bottles	7.0	35	NA	Not mentioned	Yield = 1.48 Mol/Mol hexose Rate = 161 mL/L/h	Kim et al. (2011)
Rice straw and activated sewage sludge	Serum bottles	5.01	35	NA	Substrate to biomass ratio of 4.54:1	Yield = 14.54 ± 0.29 mL/g VS	Alemahdi et al. (2015)
Press mud and sewage	UASB	5–6	30–38	30 h	OLR = 8–20 g/L/day	H_2 production = 7960 mL/day	Radjaram and Saravanane (2011)
Flower wastes and sewage sludge	Serum bottles	7.1	37	NA	1 g VS per bottle	Yield = 39 mL/g VS	Yang and Wang (2018)

Type of waste	Bioreactor	pH	Temp. (°C)	HRT (h)	Substrate concentration	Hydrogen production/Yield	Reference
Waste activated sludge (treated with gamma irradiation)	Sealed Erlenmeyer flasks	7.0	36	NA	Not mentioned	H_2 production 6.64 mL/h	Yin and Wang (2015)
Sewage sludge (alkali and ultrasonication treated)	Semi-continuous stirred tank reactor	10.0	37	NA	OLR = 2.2 ± 0.2 kg VS/m^3.Day	Yield = 0.35 mmol/g VS	El-Qelish et al. (2020)

NA not available

thermophilic condition (55 °C) (Sillero et al. 2022). Maximum hydrogen production was obtained with 10% poultry manure which resulted in C/N ratio of 27. The hydrogen yield was 22.34 mL/g soluble COD and 27.1 mL/g VS at the optimum substrate concentration. Increase in poultry manure addition to 20% reduced the hydrogen yield by 18.2% and 27.6%. Kim et al. (2011) studied hydrogen production by co-digestion of pretreated tofu waste and sewage sludge. The addition of sewage sludge improved hydrogen production rate and yield to 161 mL/L/h and 1.48 mol/mol hexose, respectively. The increase in hydrogen production was attributed to the availability of additional micronutrients (e.g. minerals, vitamins and metals) in pretreated tofu waste, which were required to digest the sewage sludge.

Pretreatment of sewage sludge by different methods such as thermal treatment, alkaline treatment, acid treatment, sonication, etc. have been examined for enhancing hydrogen production from sewage sludge. For example, alkaline treatment of sludge improved biohydrogen production from 9.1 mL/g of dry solids in untreated sludge to 16.6 mL/g of DS in alkaline treated sludge (pH 11.0) (Cai et al. 2004). Results showed that protein present in the sewage sludge acts as the main substrate for biohydrogen production, which were released due to the breakdown of sludge by alkaline treatment. Another study compared four different pretreatments, namely, acid, alkaline, thermal and ultrasonic treatment of sewage sludge to enhance hydrogen production (Xiao and Liu 2009). Although, all the four methods could enhance hydrogen production, the alkaline treatment proved to be the best (11.68 mL H_2/g VS at pH 11.5) which was followed by thermal treatment (8.62 mL/g VS). The pretreatments led to breakdown of sludge flocs and microbial structure which resulted in increase in soluble COD concentration and decreased volatile solids and dry solids content. In another study, a combination of alkali and ultrasonication treatment of sewage sludge was examined for their effect on biohydrogen production (El-Qelish et al. 2020). The results showed that hydrogen production improved by almost seven times to reach 0.35 mmol/g VS due to pretreatment of the sludge. Furthermore, the hydrogen yield due to the combined pretreatment was three times higher than with only alkali treatment. In a recent study, a novel strategy using sodium citrate in combination with thermal treatment was applied to treat sewage sludge for biohydrogen production (Yang and Wang 2021). The use of sodium citrate, a cation binding agent, was aimed at removing the extracellular polymeric substance (EPS) layer from sludge flocs for improving its thermal degradation efficiency. The study found that the combined treatment of sewage sludge produced maximum hydrogen (28.6 mL/g VS) which was 73–346.9% higher than the value obtained using untreated sludge or sludge treated with either of these methods. Thus, the combined treatment showed a synergistic effect on the sludge degradation which resulted in 107.5 times increase in the soluble organic content, thereby enabling the microorganisms present to efficiently utilize the substrate and produce high amount of hydrogen.

12.4.1 *Factors Influencing Biohydrogen Production from Sewage Sludge by Dark Fermentation*

Activated sludge contains microbes that have rigid cell wall and form floc type structure. As a result, it is difficult to degrade organics present in sewage sludge during the hydrolysis step of dark fermentation or anaerobic digestion, which hinders the hydrogen production from the sludge. Moreover, the unutilized sludge poses secondary pollution problems to the environment, thus rendering the waste treatment ineffective. Due to these reasons many different pretreatment technologies are developed to disintegrate and hydrolyse the sewage sludge. Another advantage of pretreatment is that it can prevent proliferation of unwanted methanogenic bacteria required for enhancing hydrogen production. Hydrogen producing microorganisms are hardy, form endospores and able to withstand extreme environmental conditions such as high temperature, low pH, high alkalinity, etc. Hence, pretreatment can selectively enrich hydrogenogenic microorganisms while inhibiting sensitive methanogens (Sinharoy and Pakshirajan 2021). Examples of different sludge pretreatment methods that are commonly reported in the literature are heat treatment, acid treatment, alkaline treatment, ultrasonication, irradiation, microwave, etc. (Yao et al. 2018). Although these methods are effective in enhancing hydrogen production, all these methods have their own disadvantages as well as advantages. Some of the common drawbacks of these methods include the use of costly chemicals, high energy requirement and use of sophisticated instrument, which limit their application in large scale. Hence, appropriate cost benefit analysis should be performed before selecting any of the methods to clearly understand if any economic benefits could be derived from such pretreatment.

Effect of different operational conditions such as pH, temperature, oxidation reduction potential (ORP), carbon to nitrogen ratio, substrate concentration, hydraulic retention time (HRT), solids retention time (SRT), etc. in the anaerobic system is crucial to achieve maximum hydrogen production performance (Sinharoy et al. 2019, 2020b). The optimum pH for hydrogen production is around 6.5, and hence most of the studies have used a near neutral pH for biohydrogen production from sewage sludge. The reason for the choice of pH 6.5 is hydrogenase, which is the key enzyme responsible for hydrogen production that has an optimum activity at pH 6.5. However, biohydrogen production at acidic or alkaline pH has also been reported. But in general, low pH value can harm hydrogenogenic organisms, and at slightly high pH values methanogenic microorganisms thrive. Temperature is another important factor which influences the biohydrogen yield and production of byproducts from sewage sludge. Both mesophilic (30–45 °C) and thermophilic (50–65 °C) temperature conditions have been reported for biohydrogen production. Higher temperature often improves the hydrogen production rate and prevents methanogenic activity (Ferreira et al. 2018). HRT is the reaction time provided to microorganisms to degrade the organics to desired products. Together with SRT, HRT depends on the wastewater flowrate, reactor volume and the design of the anaerobic system used. Both these parameters are much akin to the other environmental conditions in shaping the microbial community inside the reactor which affect the process performance.

Initial concentration of substrate is another major factor that influences hydrogen production and yield. In continuous system, it is usually referred as organic loading rate. There is no fixed optimum organic loading rate for hydrogen production as it depends on many other factors. However, it has often been observed that reactor operation fais at a very high loading condition due to accumulation of acid inside the reactor (Yao et al. 2018). Hence, a moderate optimum organic loading condition need to be determined for each system. The carbon to nitrogen ratio is another important parameter for optimization. Microorganisms utilize carbon to the extent of 25 to 30 times higher than nitrogen and due to which a ratio of 25–30: 1 of carbon and nitrogen is optimum for better performance (Balat and Balat 2009). A low C/N ratio accelerates cell growth but not hydrogen production, whereas a high C/N ratio inhibits cell growth and hinders hydrogen production. Hence, an optimum C/N ratio is crucial for enhancing hydrogen production from sewage sludge (Yao et al. 2018). Due to this reason, many studies have focused on co-digestion strategy to optimise C/N ratio by adding a nitrogen rich substrate, such as grass clippings, algae, plant biomass to carbon rich sewage sludge (Yang and Wang 2019). Supplementation of metal ions (e.g. Fe, Ni, Mg) during dark fermentation is shown to improve hydrogen production (Elreedy et al. 2019; Yin and Wang 2021). These metal ions play an important role by either forming the core of metalo-enzymes such as hydrogenase or act as enzyme co-factor which are involved in the metabolic pathway of hydrogenogenic microorganism (Sinharoy and Pakshirajan 2020). Addition of metal nanoparticles, even at small dose have shown positive effect on hydrogen production due to similar role in improving the enzyme activity (Sinharoy and Pakshirajan 2020; Sinharoy et al. 2020c; Yang et al. 2022).

12.5 Conclusions and Perspectives

Hydrogen is one of the most important energy carrier and clean source for energy in the future, and due to which there is a huge demand for hydrogen worldwide. The demand is expected to further increase as the world is transitioning from a fossil fuel based economy to green fuel based economy. Hence, there is an urgent need to explore new sources and find alternative avenues for hydrogen production. In this regard, the large amount of waste generated can serve as an alternative source for producing hydrogen through different biological processes. Municipal wastewater (sewage) and byproduct of wastewater treatment - sewage sludge - have attracted interest among the researchers. Among the available biological methods, microbial electrolysis cell has gained importance in generation of hydrogen by simultaneously treating municipal wastewater. Despite known as a new technology in wastewater treatment, it has been successfully tested at pilot scale level with an immense potential for industrial scale installation to treat domestic wastewater and produce hydrogen. However, as a prerequisite for its successful implementation at large-scale, factors including cost, use of membrane or membrane-less system, over-potential and hydrogen recycling, need to be addressed. In case of sewage sludge, dark fermentation is found to be the most suited for biohydrogen production. Important issues affecting its performance are operating parameters, sludge pretreatment, inoculum selection and its pretreatment, nutrient supplementation and choice of reactor system, which need to be optimized for achieving maximum hydrogen production performance. Hydrogen production from domestic wastewater and sewage sludge has immense potential, and future research need to be directed towards making the process economically viable and technologically feasible for its large-scale implementation.

References

Afify A, Gwad AE, Rahman AE (2017) Using *Enterobacter aerogenes* DSM 30053 for biohydrogen production by microbial electrolysis cells from domestic wastewater. J Agri Chem Biotechnol 8(6):167–171. https://doi.org/10.21608/JACB.2017.38674

Alemahdi N, Man HC, Nasirian N, Yang Y (2015) Enhanced mesophilic bio-hydrogen production of raw rice straw and activated sewage sludge by co-digestion. Int J Hydrog Energy 40(46):16033–16044. https://doi.org/10.1016/j.ijhydene.2015.08.106

Arun S, Sinharoy A, Pakshirajan K, Lens PNL (2020) Algae based microbial fuel cells for wastewater treatment and recovery of value-added products. Renew Sust Energ Rev 132:110041. https://doi.org/10.1016/j.rser.2020.110041

Assawamongkholsiri T, Reungsang A, Pattra S (2013) Effect of acid, heat and combined acid-heat pretreatments of anaerobic sludge on hydrogen production by anaerobic mixed cultures. Int J Hydrog Energy 38(14):6146–6153. https://doi.org/10.1016/j.ijhydene.2012.12.138

Balat M, Balat H (2009) Biogas as a renewable energy source—a review. Energy Sources Part A 31(14):1280–1293. https://doi.org/10.1080/15567030802089565

Brown RK, Harnisch F, Wirth S, Wahlandt H, Dockhorn T, Dichtl N, Schröder U (2014) Evaluating the effects of scaling up on the performance of bioelectrochemical systems using a technical scale microbial electrolysis cell. Bioresour Technol 163:206–213. https://doi.org/10.1016/j.biortech.2014.04.044

Cai M, Liu J, Wei Y (2004) Enhanced biohydrogen production from sewage sludge with alkaline pretreatment. Environ Sci Technol 38(11):3195–3202. https://doi.org/10.1021/es0349204

Caporgno MP, Trobajo R, Caiola N, Ibáñez C, Fabregat A, Bengoa C (2015) Biogas production from sewage sludge and microalgae co-digestion under mesophilic and thermophilic conditions. Renew Energy 75:374–380. https://doi.org/10.1016/j.renene.2014.10.019

Chan YJ, Chong MF, Law CL, Hassell DG (2009) A review on anaerobic–aerobic treatment of industrial and municipal wastewater. Chem Eng J 155(1–2):1–18. https://doi.org/10.1016/j.cej.2009.06.041

Cheng D, Ngo HH, Guo W, Chang SW, Nguyen DD, Liu Y, Wei Q, Wei D (2020) A critical review on antibiotics and hormones in swine wastewater: water pollution problems and control approaches. J Hazard Mater 387:121682. https://doi.org/10.1016/j.jhazmat.2019.121682

Chrispim MC, Scholz M, Nolasco MA (2019) Phosphorus recovery from municipal wastewater treatment: critical review of challenges and opportunities for developing countries. J Environ Manag 248:109268. https://doi.org/10.1016/j.jenvman.2019.109268

Cotterill SE, Dolfing J, Jones C, Curtis TP, Heidrich ES (2017) Low temperature domestic wastewater treatment in a microbial electrolysis cell with 1 m^2 anodes: towards system scale-up. Fuel Cells 17(5):584–592. https://doi.org/10.1002/fuce.201700034

Cotterill SE, Dolfing J, Curtis TP, Heidrich ES (2018) Community assembly in wastewater-fed pilot-scale microbial electrolysis cells. Front Energy Res 6:98. https://doi.org/10.3389/fenrg.2018.00098

Cusick RD, Kiely PD, Logan BE (2010) A monetary comparison of energy recovered from microbial fuel cells and microbial electrolysis cells fed winery or domestic wastewaters. Int J Hydrog Energy 35(17):8855–8861. https://doi.org/10.1016/j.ijhydene.2010.06.077

Darwall W, Bremerich V, De Wever A, Dell AI, Freyhof J, Gessner MO, Grossart HP, Harrison I, Irvine K, Jähnig SC, Weyl O (2018) The alliance for freshwater life: a global call to unite efforts for freshwater biodiversity science and conservation. Aquat Conserv: Mar Freshw Ecosyst 28(4):1015–1022. https://doi.org/10.1002/aqc.2958

Ditzig J, Liu H, Logan BE (2007) Production of hydrogen from domestic wastewater using a bioelectrochemically assisted microbial reactor (BEAMR). Int J Hydrog Energy 32(13):2296–2304. https://doi.org/10.1016/j.ijhydene.2007.02.035

Dwivedi S, Mishra S, Tripathi RD (2018) Ganga water pollution: a potential health threat to inhabitants of ganga basin. Environ Int 117:327–338. https://doi.org/10.1016/j.envint.2018.05.015

El-Qelish M, Chatterjee P, Dessì P, Kokko M, El-Gohary F, Abo-Aly M, Rintala J (2020) Biohydrogen production from sewage sludge: screening for pretreatments and semi-continuous reactor operation. Waste Biomass Valor 11(8):4225–4234. https://doi.org/10.1007/s12649-019-00743-5

Elreedy A, Fujii M, Koyama M, Nakasaki K, Tawfik A (2019) Enhanced fermentative hydrogen production from industrial wastewater using mixed culture bacteria incorporated with iron, nickel, and zinc-based nanoparticles. Water Res 151:349–361. https://doi.org/10.1016/j.watres.2018.12.043

Escapa A, Gil-Carrera L, García V, Morán A (2012) Performance of a continuous flow microbial electrolysis cell (MEC) fed with domestic wastewater. Bioresour Technol 117:55–62. https://doi.org/10.1016/j.biortech.2012.04.060

Escapa A, San-Martín MI, Mateos R, Morán A (2015) Scaling-up of membrane-less microbial electrolysis cells (MECs) for domestic wastewater treatment: bottlenecks and limitations. Bioresour Technol 180:72–78. https://doi.org/10.1016/j.biortech.2014.12.096

Escapa A, Mateos R, Martínez EJ, Blanes J (2016) Microbial electrolysis cells: an emerging technology for wastewater treatment and energy recovery. From laboratory to pilot plant and beyond. Renew Sust Energ Rev 55:942–956. https://doi.org/10.1016/j.rser.2015.11.029

Ferreira TB, Rego GC, Ramos LR, Soares LA, Sakamoto IK, de Oliveira LL, Varesche MBA, Silva EL (2018) Selection of metabolic pathways for continuous hydrogen production under thermophilic and mesophilic temperature conditions in anaerobic fluidized bed reactors. Int J Hydrog Energy 43(41):18908–18917. https://doi.org/10.1016/j.ijhydene.2018.08.177

Fountoulakis MS, Petousi I, Manios T (2010) Co-digestion of sewage sludge with glycerol to boost biogas production. Waste Manag 30(10):1849–1853. https://doi.org/10.1016/j.wasman.2010.04.011

Fytili D, Zabaniotou A (2008) Utilization of sewage sludge in EU application of old and new methods—a review. Renew Sust Energ Rev 12(1):116–140. https://doi.org/10.1016/j.rser.2006.05.014

Gil-Carrera L, Escapa A, Mehta P, Santoyo G, Guiot SR, Morán A, Tartakovsky B (2013a) Microbial electrolysis cell scale-up for combined wastewater treatment and hydrogen production. Bioresour Technol 130:584–591. https://doi.org/10.1016/j.biortech.2012.12.062

Gil-Carrera L, Escapa A, Moreno R, Morán A (2013b) Reduced energy consumption during low strength domestic wastewater treatment in a semi-pilot tubular microbial electrolysis cell. J Environ Manag 122:1–7. https://doi.org/10.1016/j.jenvman.2013.03.001

Guo L, Li XM, Bo X, Yang Q, Zeng GM, Liao DX, Liu JJ (2008) Impacts of sterilization, microwave and ultrasonication pretreatment on hydrogen producing using waste sludge. Bioresour Technol 99(9):3651–3658. https://doi.org/10.1016/j.biortech.2007.07.026

Heidrich ES, Dolfing J, Scott K, Edwards SR, Jones C, Curtis TP (2013) Production of hydrogen from domestic wastewater in a pilot-scale microbial electrolysis cell. Appl Microbiol Biotechnol 97(15):6979–6989. https://doi.org/10.1007/s00253-012-4456-7

Heidrich ES, Edwards SR, Dolfing J, Cotterill SE, Curtis TP (2014) Performance of a pilot scale microbial electrolysis cell fed on domestic wastewater at ambient temperatures for a 12 month period. Bioresour Technol 173:87–95. https://doi.org/10.1016/j.biortech.2014.09.083

Hreiz R, Latifi MA, Roche N (2015) Optimal design and operation of activated sludge processes: state-of-the-art. Chem Eng J 281:900–920. https://doi.org/10.1016/j.cej.2015.06.125

Hülsen T, Barry EM, Lu Y, Puyol D, Keller J, Batstone DJ (2016) Domestic wastewater treatment with purple phototrophic bacteria using a novel continuous photo anaerobic membrane bioreactor. Water Res 100:486–495. https://doi.org/10.1016/j.watres.2016.04.061

Jwa E, Yun YM, Kim H, Jeong N, Park SC, Nam JY (2019) Domestic wastewater treatment in a tubular microbial electrolysis cell with a membrane electrode assembly. Int J Hydrog Energy 44(2):652–660. https://doi.org/10.1016/j.ijhydene.2018.11.036

Kanaujiya DK, Paul T, Sinharoy A, Pakshirajan K (2019) Biological treatment processes for the removal of organic micropollutants from wastewater: a review. Curr Pollution Rep 5(3):112–128. https://doi.org/10.1007/s40726-019-00110-x

Katuri KP, Ali M, Saikaly PE (2019) The role of microbial electrolysis cell in urban wastewater treatment: integration options, challenges, and prospects. Curr Opin Biotechnol 57:101–110. https://doi.org/10.1016/j.copbio.2019.03.007

Kehrein P, Van Loosdrecht M, Osseweijer P, Garfí M, Dewulf J, Posada J (2020) A critical review of resource recovery from municipal wastewater treatment plants–market supply potentials, technologies and bottlenecks. Environ Sci Water Res Technol 6(4):877–910. https://doi.org/10.1039/C9EW00905A

Kim SH, Han SK, Shin HS (2004) Feasibility of biohydrogen production by anaerobic co-digestion of food waste and sewage sludge. Int J Hydrog Energy 29(15):1607–1616. https://doi.org/10.1016/j.ijhydene.2004.02.018

Kim DH, Lee DY, Kim MS (2011) Enhanced biohydrogen production from tofu residue by acid/base pretreatment and sewage sludge addition. Int J Hydrog Energy 36(21):13922–13927. https://doi.org/10.1016/j.ijhydene.2011.03.085

Kirchmann H, Börjesson G, Kätterer T, Cohen Y (2017) From agricultural use of sewage sludge to nutrient extraction: a soil science outlook. Ambio 46(2):143–154. https://doi.org/10.1007/s13280-016-0816-3

Lee JY, Yun J, Kim TG, Wee D, Cho KS (2014) Two-stage biogas production by co-digesting molasses wastewater and sewage sludge. Bioprocess Biosyst Eng 37(12):2401–2413. https://doi.org/10.1007/s00449-014-1217-2

Leicester DD, Amezaga JM, Moore A, Heidrich ES (2020) Optimising the hydraulic retention time in a pilot-scale microbial electrolysis cell to achieve high volumetric treatment rates using concentrated domestic wastewater. Molecules 25(12):2945. https://doi.org/10.3390/molecules25122945

Li M, Zhao Y, Guo Q, Qian X, Niu D (2008) Bio-hydrogen production from food waste and sewage sludge in the presence of aged refuse excavated from refuse landfill. Renew Energy 33(12):2573–2579. https://doi.org/10.1016/j.renene.2008.02.018

Liu Y, Shi H, Xia L, Shi H, Shen T, Wang Z, Wang G, Wang Y (2010) Study of operational conditions of simultaneous nitrification and denitrification in a carrousel oxidation ditch for domestic wastewater treatment. Bioresour Technol 101(3):901–906. https://doi.org/10.1016/j.biortech.2009.09.015

Liu X, Wang Q, Tang Y, Pavlostathis SG (2021) A comparative study on biogas production, energy balance, and nutrients conversion with inter-stage hydrothermal treatment of sewage sludge. Appl Energy 288:116669. https://doi.org/10.1016/j.apenergy.2021.116669

Lu JY, Wang XM, Liu HQ, Yu HQ, Li WW (2019) Optimizing operation of municipal wastewater treatment plants in China: the remaining barriers and future implications. Environ Int 129:273–278. https://doi.org/10.1016/j.envint.2019.05.057

Maragkaki AE, Fountoulakis M, Gypakis A, Kyriakou A, Lasaridi K, Manios T (2017) Pilot-scale anaerobic co-digestion of sewage sludge with agro-industrial by-products for increased biogas production of existing digesters at wastewater treatment plants. Waste Manag 59:362–370. https://doi.org/10.1016/j.wasman.2016.10.043

Maragkaki AE, Fountoulakis M, Kyriakou A, Lasaridi K, Manios T (2018) Boosting biogas production from sewage sludge by adding small amount of agro-industrial by-products and food waste residues. Waste Manag 71:605–611. https://doi.org/10.1016/j.wasman.2017.04.024

Moreno R, Escapa A, Cara J, Carracedo B, Gómez X (2015) A two-stage process for hydrogen production from cheese whey: integration of dark fermentation and biocatalyzed electrolysis. Int J Hydrog Energy 40(1):168–175. https://doi.org/10.1016/j.ijhydene.2014.10.120

Nabi M, Zhang G, Li F, Zhang P, Wu Y, Tao X, Bao S, Wang S, Chen N, Ye J, Dai J (2020) Enhancement of high pressure homogenization pretreatment on biogas production from sewage sludge: a review. Desalin Water Treat 175:341–351. https://doi.org/10.5004/dwt.2020.24670

Neto SA, Reginatto V, De Andrade AR (2018) Microbial fuel cells and wastewater treatment. In: Martínez-Huitle CA, Rodrigo MA, Scialdone O (eds) Electrochemical water and wastewater treatment. Butterworth-Heinemann, Oxford, pp 305–331. https://doi.org/10.1016/B978-0-12-813160-2.00012-2

Nguyen VK, Chaudhary DK, Dahal RH, Trinh NH, Kim J, Chang SW, Hong Y, La DD, Nguyen XC, Ngo HH, Nguyen DD (2021) Review on pretreatment techniques to improve anaerobic digestion of sewage sludge. Fuel 285:119105. https://doi.org/10.1016/j.fuel.2020.119105

Pilli S, Yan S, Tyagi RD, Surampalli RY (2015) Thermal pretreatment of sewage sludge to enhance anaerobic digestion: a review. Crit Rev Environ Sci Technol 45(6):669–702. https://doi.org/10.1080/10643389.2013.876527

Prakash J, Sharma R, Patel SK, Kim IW, Kalia VC (2018) Bio-hydrogen production by co-digestion of domestic wastewater and biodiesel industry effluent. PloS One 13(7):e0199059. https://doi.org/10.1371/journal.pone.0199059

Qadir M, Drechsel P, Jiménez Cisneros B, Kim Y, Pramanik A, Mehta P, Olaniyan O (2020) Global and regional potential of wastewater as a water, nutrient and energy source. Nat Res Forum 44(1):40–51. https://doi.org/10.1111/1477-8947.12187

Radjaram B, Saravanane R (2011) Assessment of optimum dilution ratio for biohydrogen production by anaerobic co-digestion of press mud with sewage and water. Bioresour Technol 102(3):2773–2780. https://doi.org/10.1016/j.biortech.2010.11.075

Rivero M, Solera R, Perez M (2014) Anaerobic mesophilic co-digestion of sewage sludge with glycerol: enhanced biohydrogen production. Int J Hydrog Energy 39(6):2481–2488. https://doi.org/10.1016/j.ijhydene.2013.12.006

Roy AS, Chingkheihunba A, Pakshirajan K (2016) An overview of production, properties, and uses of biodiesel from vegetable oil. In: Soccol C, Brar S, Faulds C, Ramos L (eds) Green fuels technology. Green energy and technology. Springer, Cham, pp 83–105. https://doi.org/10.1007/978-3-319-30205-8_4

Rozendal RA, Hamelers HV, Rabaey K, Keller J, Buisman CJ (2008) Towards practical implementation of bioelectrochemical wastewater treatment. Trends Biotechnol 26(8):450–459. https://doi.org/10.1016/j.tibtech.2008.04.008

Salerno MB, Lee HS, Parameswaran P, Rittmann BE (2009) Using a pulsed electric field as a pretreatment for improved biosolids digestion and methanogenesis. Water Environ Res 81(8):831–839. https://doi.org/10.2175/106143009X407366

Shi Y, Yang J, Yu W, Zhang S, Liang S, Song J, Xu Q, Ye N, He S, Yang C, Hu J (2015) Synergetic conditioning of sewage sludge via Fe^{2+}/persulfate and skeleton builder: effect on sludge characteristics and dewaterability. Chem Eng J 270:572–581. https://doi.org/10.1016/j.cej.2015.01.122

Sillero L, Solera R, Perez M (2022) Anaerobic co-digestion of sewage sludge, wine vinasse and poultry manure for bio-hydrogen production. Int J Hydrog Energy 47(6):3667–3678. https://doi.org/10.1016/j.ijhydene.2021.11.032

Sinharoy A, Lens PNL (2020) Biological removal of selenate and selenite from wastewater: options for selenium recovery as nanoparticles. Curr Pollution Rep 6(3):230–249. https://doi.org/10.1007/s40726-020-00146-4

Sinharoy A, Pakshirajan K (2019) Heavy metal sequestration by sulfate reduction using carbon monoxide as the sole carbon and energy source. Process Biochem 82:135–143. https://doi.org/10.1016/j.procbio.2019.04.002

Sinharoy A, Pakshirajan K (2020) A novel application of biologically synthesized nanoparticles for enhanced biohydrogen production and carbon monoxide bioconversion. Renew Energy 147:864–873. https://doi.org/10.1016/j.renene.2019.09.027

Sinharoy A, Pakshirajan K (2021) Methane free biohydrogen production from carbon monoxide using a continuously operated moving bed biofilm reactor. Int J Hydrog Energy 46(1):306–313. https://doi.org/10.1016/j.ijhydene.2020.09.250

Sinharoy A, Baskaran D, Pakshirajan K (2019) Sustainable biohydrogen production by dark fermentation using carbon monoxide as the sole carbon and energy source. Int J Hydrog Energy 44(26):13114–13125. https://doi.org/10.1016/j.ijhydene.2019.03.130

Sinharoy A, Pakshirajan K, Lens PNL (2020a) Biological sulfate reduction using gaseous substrates to treat acid mine drainage. Curr Pollution Rep 6(4):328–344. https://doi.org/10.1007/s40726-020-00160-6

Sinharoy A, Kumar M, Pakshirajan K (2020b) An overview of bioreactor configurations and operational strategies for dark fermentative biohydrogen production. In: Singh L, Yousuf A, Mahapatra DM (eds) Bioreactors - sustainable design and industrial applications in mitigation of GHG emissions. Elsevier, Amsterdam, pp 249–288. https://doi.org/10.1016/B978-0-12-821264-6.00014-0

Sinharoy A, Baskaran D, Pakshirajan K (2020c) Process integration and artificial neural network modelling of biological sulfate reduction using a carbon monoxide fed gas lift bioreactor. Chem Eng J 391:123518. https://doi.org/10.1016/j.cej.2019.123518

Sinharoy A, Pakshirajan K, Lens PNL (2022) Syngas fermentation for bioenergy production: advances in bioreactor systems. In: Sinharoy A, Lens PNL (eds) Renewable energy technologies for energy efficient sustainable development. Applied environmental science and engineering for a sustainable future. Springer, Cham, pp 325–358. https://doi.org/10.1007/978-3-030-87633-3_13

Sonune A, Ghate R (2004) Developments in wastewater treatment methods. Desalination 167:55–63. https://doi.org/10.1016/j.desal.2004.06.113

Tartakovsky B, Manuel MF, Wang H, Guiot SR (2009) High rate membrane-less microbial electrolysis cell for continuous hydrogen production. Int J Hydrog Energy 34(2):672–677. https://doi.org/10.1016/j.ijhydene.2008.11.003

Tony MA, Zhao YQ, Fu JF, Tayeb AM (2008) Conditioning of aluminium-based water treatment sludge with Fenton's reagent: effectiveness and optimising study to improve dewaterability. Chemosphere 72(4):673–677. https://doi.org/10.1016/j.chemosphere.2008.03.032

Usman TM, Banu JR, Gunasekaran M, Kumar G (2019) Biohydrogen production from industrial wastewater: an overview. Bioresour Technol Rep 7:100287. https://doi.org/10.1016/j.biteb.2019.100287

Valipour A, Raman VK, Ahn YH (2015) Effectiveness of domestic wastewater treatment using a bio-hedge water hyacinth wetland system. Water 7(1):329–347. https://doi.org/10.3390/w7010329

Wang CC, Chang CW, Chu CP, Lee DJ, Chang B, Liao CS (2003) Producing hydrogen from wastewater sludge by *clostridium bifermentans*. J Biotechnol 102(1):83–92. https://doi.org/10.1016/S0168-1656(03)00007-5

Wang C, Li J, Wang B, Zhang G (2006) Development of an empirical model for domestic wastewater treatment by biological aerated filter. Process Biochem 41(4):778–782. https://doi.org/10.1016/j.procbio.2005.09.015

Wei S, Xiao B, Liu J (2010) Impact of alkali and heat pretreatment on the pathway of hydrogen production from sewage sludge. Chin Sci Bull 55(8):777–786. https://doi.org/10.1007/s11434-009-0591-7

Wong YM, Wu TY, Juan JC (2014) A review of sustainable hydrogen production using seed sludge via dark fermentation. Renew Sust Energ Rev 34:471–482. https://doi.org/10.1016/j.rser.2014.03.008

Xiao B, Liu J (2009) Effects of various pretreatments on biohydrogen production from sewage sludge. Chin Sci Bull 54(12):2038–2044. https://doi.org/10.1007/s11434-009-0100-z

Xiao L, Liu J, Ge J (2021) Dynamic game in agriculture and industry cross-sectoral water pollution governance in developing countries. Agric Water Manag 243:106417. https://doi.org/10.1016/j.agwat.2020.106417

Yang G, Wang J (2018) Synergistic biohydrogen production from flower wastes and sewage sludge. Energy Fuel 32(6):6879–6886. https://doi.org/10.1021/acs.energyfuels.8b01122

Yang G, Wang J (2019) Biohydrogen production by co-fermentation of sewage sludge and grass residue: effect of various substrate concentrations. Fuel 237:1203–1208. https://doi.org/10.1016/j.fuel.2018.10.026

Yang G, Wang J (2021) Enhancing biohydrogen production from disintegrated sewage sludge by combined sodium citrate-thermal pretreatment. J Clean Prod 312:127756. https://doi.org/10.1016/j.jclepro.2021.127756

Yang G, Hu Y, Wang J (2019) Biohydrogen production from co-fermentation of fallen leaves and sewage sludge. Bioresour Technol 285:121342. https://doi.org/10.1016/j.biortech.2019.121342

Yang S, Zheng Y, Mao Y, Xu L, Jin Z, Zhao M, Kong H, Huang X, Zheng X (2021) Domestic wastewater treatment for single household via novel subsurface wastewater infiltration systems (SWISs) with Nii Mi process: performance and microbial community. J Clean Prod 279:123434. https://doi.org/10.1016/j.jclepro.2020.123434

Yang J, Zhang H, Liu H, Zhang J, Pei Y, Zang L (2022) Unraveling the roles of lanthanum-iron oxide nanoparticles in biohydrogen production. Bioresour Technol 351:127027. https://doi.org/10.1016/j.biortech.2022.127027

Yao Z, Su W, Wu D, Tang J, Wu W, Liu J, Han W (2018) A state-of-the-art review of biohydrogen producing from sewage sludge. Int J Energy Res 42:4301–4312. https://doi.org/10.1002/er.4188

Yin Y, Wang J (2015) Biohydrogen production using waste activated sludge disintegrated by gamma irradiation. Appl Energy 155:434–439. https://doi.org/10.1016/j.apenergy.2015.05.105

Yin Y, Wang J (2021) Mechanisms of enhanced hydrogen production from sewage sludge by ferrous ion: insights into functional genes and metabolic pathways. Bioresour Technol 321:124435. https://doi.org/10.1016/j.biortech.2020.124435

Yu S, Zhang G, Li J, Zhao Z, Kang X (2013) Effect of endogenous hydrolytic enzymes pretreatment on the anaerobic digestion of sludge. Bioresour Technol 146:758–761. https://doi.org/10.1016/j.biortech.2013.07.087

Zhen G, Lu X, Kato H, Zhao Y, Li YY (2017) Overview of pretreatment strategies for enhancing sewage sludge disintegration and subsequent anaerobic digestion: current advances, full-scale application and future perspectives. Renew Sust Energ Rev 69:559–577. https://doi.org/10.1016/j.rser.2016.11.187

Zhou H, Smith DW (2001) Advanced technologies in water and wastewater treatment. Can J Civ Eng 28(S1):49–66. https://doi.org/10.1139/l00-091

Chapter 13
Feedstocks and Cases III: Biohydrogen From Algal Biomass (Macro and Microalgae)

Prashant Bhagwat, Kugenthiren Permaul, Suren Singh, and Santhosh Pillai

Abstract The rapid depletion of fossil fuels along with the negative effects of their use in industries and combustion engines has necessitated the development of more sustainable energy sources such as biofuel. Biohydrogen gas, alongside other biofuels such as biodiesel, bioethanol, and biomethane, is gaining momentum as a clean energy option, owing to its high energy density, pollutant-free combustion products, and renewable nature. Though domestic and agricultural wastes, plants, and microbes are all useful sources of renewable biomasses for biofuel production, algal biomass has emerged as a potential raw material in this area. Algal biomass, which has predominantly been investigated for biodiesel production, is an ideal substrate for biohydrogen production. Furthermore, algae-based biohydrogen production has been described as a highly appealing strategy that can provide bioenergy sustainability through the generation of safe, clean energy with net-zero carbon dioxide emissions. This chapter, therefore, investigates the different types of micro- and macroalgae and their biochemical composition for biohydrogen production. Furthermore, the mechanisms of direct and indirect bio-photolysis, photo fermentation, and dark fermentation for biohydrogen production are described. The algal cultivation and harvesting techniques, pretreatment methods, and strategies to improve biomass and biohydrogen production are also briefly discussed.

Keywords Algae · Biofuel · Biohydrogen · Dark fermentation · Fossil fuels · Pretreatment methods

---

P. Bhagwat · K. Permaul · S. Singh · S. Pillai (✉)
Department of Biotechnology and Food Science, Faculty of Applied Sciences, Durban University of Technology, Durban, South Africa
e-mail: santhoshk@dut.ac.za

C. R. Soccol et al. (eds.), *Biohydrogen - Advances and Processes*, Biofuel and Biorefinery Technologies 13, https://doi.org/10.1007/978-3-031-49818-3_13

13.1 Introduction

Climate change and environmental pollution have become a global issue, particularly for scientists and policymakers in energy economics and environmental sustainability, due to their various detrimental impacts (Chien et al. 2022). There has been a substantial increase in greenhouse gas emissions over the last 150 years due to human activity, which poses a threat to various global economies. The majority of greenhouse gas emissions from human activity could be attributed to the use of fossil fuels for electricity, heat generation, and transportation (Yuan et al. 2022). Therefore, limiting the combustion of fossil fuels, largely through reducing their consumption and replacing them with biofuels, is an environmentally friendly solution. In particular, it is anticipated that the use of biofuels would surpass that of fossil fuels, enhancing both environmental sustainability and the world economy (Fan and Hajiyeva 2022).

Biofuels are sustainable, renewable products made from eco-friendly biosources that can meet the demand for petroleum-derived goods while utilizing biowaste and minimizing the environmental impact (Kushwaha et al. 2022). Based on feedstock consumption, biofuels can be organized into the first, second, and third generations. Crops high in sugar (fruits, molasses, palm juice, sugar beet, sugar cane, and sugar juice), crops high in starch (barley, cassava, corn, potato, rice, sweet sorghum, and wheat), cellulosic biomasses (agricultural residues, forest waste, and waste fibers), and edible oils (from palm, rapeseed, soybean, and sunflower) are the recurrent feedstocks targeted in the first-generation biofuel production. However, there are several issues with first-generation biofuels; they compete with human staple foods for resources, use agricultural land, freshwater, and manure, which increases the price of food crops and negatively impacts the environment (Goswami et al. 2022). The concept of utilizing bio-waste for fuel production negates these problems to a greater extent. Thus, the feedstocks utilized for second-generation biofuel production include industrial bio-waste (corn cobs, fruit peelings, pulp waste, sawdust, and sugarcane bagasse), lignocellulosic waste (Alfalfa, Camelina, Jatropha, poplar, switch grass, and willow), municipal and domestic waste (food waste, and organic waste) as an alternative to food crops. These feedstocks also contribute to reducing bio-waste accumulation in the environment without increasing the cost of food crops or causing food shortages (Singh et al. 2022). However, both first- and second-generation biofuels majorly rely on agriculture, which necessitates substantial reserves of cultivable land, major human resource engagement, irrigation systems, and is significantly influenced by fluctuations in rainfall patterns (Hirani et al. 2018).

As a potential third-generation feedstock for biofuel production, algae have received growing global attention in recent years due to their significant advantages, including fast growth rates and high oil contents, the ability to grow in inapplicable water resources such as saline water and wastewater, the use of wasteland and marginal land unsuitable for agriculture, and no conflicts with food or feed production (Goswami et al. 2022). In addition, algae are an excellent alternative for third-generation biofuel production due to their ability to thrive in challenging conditions with limited nutrients, and year-round production rate (Sinha and Sengupta 2019). Most importantly, their high cellular carbohydrate content is advantageous because many algal cells contain easily fermentable sugars with minimal hemicellulose and little or no lignin, which favours biofuel production. Despite these appealing characteristics, biofuel generation from microalgae is still being researched and has yet to be commercialized due to three major obstacles: high production costs, technological constraints, and environmental challenges (Abbasi et al. 2021).

13.1.1 Algae

Algae are typically photosynthetic plant-like organisms (unicellular to multicellular forms) but do not have roots, stems and leaves. There are almost 300,000 algal species distributed throughout the world in freshwater, seawater, and wastewater. Algae are classified into two groups based on their size, viz., microalgae, which are microscopic single-celled organisms that are typically less than 1 mm in size and macroalgae, known as 'seaweeds' (Bhateria and Dhaka 2014; Rajauria and Yuan 2021). Microalgae are photosynthetic microbes that can grow quickly and survive in severe environments due to their unicellular or simple multicellular structure. Cyanobacteria (Cyanophyceae) are examples of prokaryotic, while green algae (Chlorophyta) and diatoms (Bacillariophyta) are examples of eukaryotic microalgae. Microalgae are more effective at converting solar energy into biomass than land-based plants due to their basic cellular structure and the fact that they are immersed in an aquatic ecosystem where they have easy access to water, CO_2, and other nutrients (John et al. 2011). Compared to terrestrial plants, microalgae are better at capturing CO_2, and under ideal conditions, 1.83 kg of CO_2 are transformed into 1 kg of dry algal biomass (Karpagam et al. 2021). On the other hand, macroalgae are a broad group of photosynthetic eukaryotic marine organisms existing in three main categories; green, brown, and red algae (Aguilar-Toalá et al. 2021). In contrast to unicellular microalgae, macroalgal species are multicellular and have plant-like characteristics, however, they do not have roots, stems and leaves and well-defined vascular tissues (Rajauria and Yuan 2021). Macroalgae typically consist of the thallus, the lamina, the sorus, the stipe, and the fucus. The key components of importance for biomass to energy conversion are the stipe and blade, which are collectively referred to as the frond (Sudhakar et al. 2018). They are common in coastal areas, particularly near-shore coastal waters, where an appropriate substrate for attachment is present.

China is the biggest microalgae producer, accounting for more than 60% of global production (Chen et al. 2016). Until recently, microalgae production was done on a small scale in facilities with less than 10 ha of surface area, but now large-scale facilities with 200 ha or more of surface area are in operation, capable of producing up to 3000 tonnes/year of microalgae, with *Chlorella, Dunaliella, Haematococcus*, and *Spirulina* being the most cultivated genera (Acién et al. 2017). Approximately 80% of the microalgae produced is used as human food, medicine, or nutritional supplements, with the remaining being used as animal feed additives, fertilizers, and bioenergy production (Sudhakar et al. 2019). As a result of its numerous industrial applications, it has been determined that systematic algal cultivation is the only way to meet current and future demand, which will lead to the development of large-scale sustainable algal biorefineries.

According to data from the Food and Agriculture Organization of the United Nations (FAO), the macroalgae aquaculture business continues to grow in size and value on an annual basis. In 2016, the world produced almost 30 million tonnes of macroalgae worth USD11.6 billion (Ferdouse et al. 2018). Asia is the largest producer of macroalgae in the world, with China topping the list with 14 million tonnes corresponding to USD 8.6 billion; whereas Africa comes second with around 140,000 tonnes, followed by America in the third position with 15,634 tonnes (Kostas et al. 2021). Despite this, production is still considered to be in its infancy in Africa, Europe, and America. This is due to the fact that the cultivation methods used successfully in Asia are not necessarily viable or profitable elsewhere (Buschmann et al. 2017).

13.1.2 The Biochemical Composition of Algae

The biochemical composition of micro- and macroalgae varies considerably depending on the species. However, the algal composition has been observed to be influenced by cultivation conditions, such as the growth medium's composition and salinity. Thus, the upstream processes could influence the algal composition to some extent and in natural aquatic habitats, the biochemical makeup of the algae changes substantially at different seasons (Zhou et al. 2022). Therefore, it is also critical to ascertain the composition of algal biomass to understand how the underlying structure of the biomass influences biofuel production and how the biomass could be transformed into accessible forms by different pretreatment methods. A summary of the organic content (carbohydrates, lipids, and proteins) of some micro- and macroalgae has been presented in Table 13.1.

The algal biomass can be easily fermented to produce biofuels or converted into high-value compounds since they have abundant carbohydrates and little or no lignin. Reports suggest that algal biomass could be fermented to yield biohydrogen (Pathy et al. 2022), biomethane (Darko et al. 2022), bioethanol, and biobutanol (Elkatory et al. 2022), making it a promising candidate for biofuel production. In addition to biofuel, algal polysaccharides can be converted into economically valuable chemicals such as levulinic acid and 5-hydroxymethylfurfural (HMF) (Zhou et al. 2022). The carbohydrate content of micro- and macroalgae varies from 15 to 50% (Chew et al. 2017). Microalgae carbohydrates include starch, cellulose, and other complex polysaccharides. The major component of the microalgal cell wall is cellulose, whereas starch is accumulated intracellularly. In addition to cellulose and starch, microalgae cells include additional polysaccharides comprising heteropolymers of glucose, galactose, and xylose (De Jesus Raposo et al. 2015). In contrast, macroalgae include not only cellulose and starch, but also more complex

Table 13.1 Biochemical composition of different micro- and macroalgae

	Carbohydrate (%)	Lipid (%)	Protein (%)	Reference
Microalgae				
Acutodesmus dimorphus	38.60	18.80	28.10	Tibbetts et al. (2015)
Anabaena cylindrica	25–30	4–7	43–56	de Farias Silva et al. (2019)
Arthrospira platensis	15.60	8.10	62.90	Aouir et al. (2017)
Botryococcus braunii	18.50	34.40	39.90	Tibbetts et al. (2015)
Chlamydomonas reinhardtii	17.00	21.00	48.00	de Farias Silva et al. (2019)
Chlorella pyrenoidosa	26.00	2.00	57.00	de Farias Silva et al. (2019)
Chlorella sorokiniana	26.80	22–24	40.50	Sajjadi et al. (2018)
Chlorella sp.	56.00	19.00	22.00	Hossain and Mahlia (2019)
Chlorella sp. FC2IITG	18–46	15–54	22–41	Muthuraj et al. (2013)
Chlorella vulgaris	38.00	23.14	15.20	Dong et al. (2016)
Chlorella vulgaris	12–17	14–22	51–58	de Farias Silva et al. (2019)
Chlorella vulgaris	18.35	23.57	ND	Rinanti and Purwadi (2019)
Coelastrum sp.	30.00	30.40	28.10	Mehrabadi et al. (2017)
Desmodesmus sp.	23.00	40.20	27.10	Mehrabadi et al. (2017)
Dunaliella salina	69.70	9.10	5.40	Efremenko et al. (2012)
Dunaliella tertiolecta	46.50	18.00	31.30	Efremenko et al. (2012)
Dunaliella tertiolecta	50.60	23.50	8.30	Efremenko et al. (2012)
Isochrisis galbana	16.98	17.16	26.99	Tokuşoglu and Üunal (2003)
Micractinium pusillum	29.20	46.30	16.80	Mehrabadi et al. (2017)
Mucidosphaerium pulchellum	27.80	48.20	13.50	Mehrabadi et al. (2017)
Nannochloropsis granulata	36.20	23.60	33.50	Tibbetts et al. (2015)
Nannochloropsis spp.	40.40	21.00	22.20	Rebolloso-Fuentes et al. (2001)
Nannochloropsis granulata	10.89	57.33	9.4	Dong et al. (2016)
Neochloris oleoabundans	37.80	15.40	30.10	Tibbetts et al. (2015)
Nostoc commune	34–56	22.00	20–43	Sajjadi et al. (2018)
Pediastrum boryanum	28.50	22.10	47.40	Mehrabadi et al. (2017)
Phaeodactylum tricornutum	25.20	18.20	39.60	Tibbetts et al. (2015)
Porphyridium aerugineum	45.80	13.70	31.60	Tibbetts et al. (2015)
Porphyridium cruentum	40–57	9–14	28–39	de Farias Silva et al. (2019)
Rhodomonas sp.	9.00	15.00	74.00	Hossain and Mahlia (2019)
Scendesmus protuberans	21–29	18–29	25–45	Sajjadi et al. (2018)
Scenedemsus sp.	32.7–41	17–24	29–37	Sajjadi et al. (2018)
Scenedesmus dimorphus	21–52	16–40	8–18	de Farias Silva et al. (2019)
Scenedesmus obliquus UTEX 393	55.40	22.60	20.20	Singh et al. (2019)

(continued)

Table 13.1 (continued)

	Carbohydrate (%)	Lipid (%)	Protein (%)	Reference
Scenedesmus sp.	18.00	12.00	56.00	Hossain and Mahlia (2019)
Scenedesmus acutus	39.42	38.55	7.80	Dong et al. (2016)
Spirogyra sp.	20.00	16.00	55.00	Hossain and Mahlia (2019)
Spirulina maxima	13–16	6–7	60–71	de Farias Silva et al. (2019)
Spirulina platensis	60.00	8.00	12.00	Hossain and Mahlia (2019)
Spirulina platensis	8–14	4–9	46–63	de Farias Silva et al. (2019)
Spirulina platensis	15.81	8.03	61.32	Tokuşoglu and Üunal (2003)
Spirulina sp.	22.20	14.20	55.80	Tibbetts et al. (2015)
Synechococcus sp.	15.00	11.00	63.00	de Farias Silva et al. (2019)
Tetraselmis chuii	25.00	12.30	46.50	Tibbetts et al. (2015)
Tetraselmis maculata	15.00	3.00	52.00	de Farias Silva et al. (2019)
Macroalgae				
Acanthophora spicifera	12.40	11.00	12.60	Suganya et al. (2016)
Azolla filiculoides[a]	41.00	ND	20.00	Miranda et al. (2016)
Caulerpa cupressoides	51.75	10.97	7.43	Suganya et al. (2016)
Caulerpa cupressoides	26.00	7.90	3.90	Dromard et al. (2017)
Ceramium cf. nitens	51.00	2.70	10.60	Dromard et al. (2017)
Cladophora fascicularis	49.50	15.70	15.53	Suganya et al. (2016)
Codium decorticatum	50.63	9.00	6.08	Suganya et al. (2016)
Codium tomentosum	32.80	3.60	18.80	Rodrigues et al. (2015)
Eichhornia crassipes	49.98	2.20	8.20	Suleiman et al. (2019)
Gracilaria gracilis	46.60	0.60	20.20	Rodrigues et al. (2015)
Grateloupia turuturu.	60.40	2.60	22.90	Denis et al. (2010)
Ipomoea aquatica	22.12	ND	26.45	Adelakun et al. (2016)
Landoltia punctata	24.50	ND	16.20	Chen et al. (2012)
Lemna minor[a]	20.30	8.70	32.30	Ge et al. (2012)
Nymphaea lotus	34.92	ND	17.23	Adelakun et al. (2016)
Osmundea pinnatifida	32.40	0.90	23.80	Rodrigues et al. (2015)
Pistia stratioties	25.62	ND	23.27	Adelakun et al. (2016)
Saccorhiza polyschides	45.60	1.10	14.44	Rodrigues et al. (2015)
Salvinia molesta	43.88	1.49	9.19	Syaichurrozi et al. (2019)
Sargassum muticum	49.30	1.45	16.90	Rodrigues et al. (2015)
Ulva lactuca	29.60	1.40	24.90	Dromard et al. (2017)
Ulva rigida	65.93	3.10	26.00	Balar et al. (2019)

ND: Not determined
[a]Aquatic plants

polysaccharides such as alginate, laminarin, and fucoidan (brown seaweed), agar (red seaweed), and ulvan (green seaweed), which are made up of galactose, glucose, mannose, rhamnose and xylose (Cesário et al. 2018). Thus, saccharification of algal biomass yields a large number of monosaccharides which can be used to create a variety of bioproducts.

In the past few decades, a lot of research has focused on using lipid-rich algae species as potential substrates for the production of biodiesel (Yaashikaa et al. 2022). Microalgae, which typically have a higher lipid concentration of around 20–50% as compared to macroalgae comprising ~5%, offer a superior option for the generation of biofuel (Zhou et al. 2022). Furthermore, fatty acids (the primary components of polar and neutral lipids) with carbon numbers ranging from 12 to 22 are promising for biofuel synthesis, whereas microalgal lipids include considerable amounts of such neutral and polar lipids, substantially bolstering their viability (Sajjadi et al. 2018).

Proteins are also one of the principal compositions of micro- and macroalgae, accounting for 50–70% and 10–30%, respectively, and can be hydrolyzed to amino acids. As a result, both microalgae and macroalgae are essential sources of amino acids and proteins, which are used in animal and human nutrition (Zhou et al. 2022). In addition, algal biomass contains a variety of pigments, including chlorophylls, sterols, and carotenoids, as well as volatile organic compounds (VOCs) and polyunsaturated fatty acids (PUFAs), which have bioactive properties and can be exploited in nutraceuticals or medicine (Khoo et al. 2019; Sun et al. 2018).

13.2 Algal Biorefinery

The algal biomass has been commercially utilized as an essential source for human nutrition, animal feed, cosmetics, bio-fertilizers, pigments, nutraceuticals, and biofuels (Fig. 13.1). For instance, the use of algal biofuels and bioactive compounds in a variety of commercial sectors, such as transportation, chemical, biomedical, pharmaceutical, cosmeceutical, and nutraceutical industries, has increased demand for the most appropriate and standardized methods to produce these bioproducts in a sophisticated and cost-effective manner, paving the way for the establishment of algal biorefinery systems in the coming future. In this regard, the goal of algal-based biorefineries is to transform algal biomass in a sustainable way to create a variety of industrial products that will advance the bioeconomy.

The market price of algal biomass is entirely determined by its application, such as human consumption (€100/kg), aquatic and animal feed (€5–20/kg), chemicals (€1–5/kg), and biofuels (€0.40/kg) (Katiyar et al. 2021). A study evaluating the economic feasibility of an algae production process based on a biorefinery approach concluded that the yield of products per kg of macroalgae cellulosic residue ($0.056/kg) were 0.24 kg of anhydrous ethanol, 0.41 kg of L-Lactic acid, and 0.28 kg of fertilizer which could generate the value of $0.80/kg, $2.49/kg, and $0.24/kg, respectively (Wong et al. 2022).

The whole algal cells can be converted through thermochemical processes like pyrolysis, hydrothermal liquefaction (HTL), and gasification to produce a variety of fuels. However, the composition of the resultant products is frequently complex and has several limitations for usage as a direct replacement for fossil fuels.

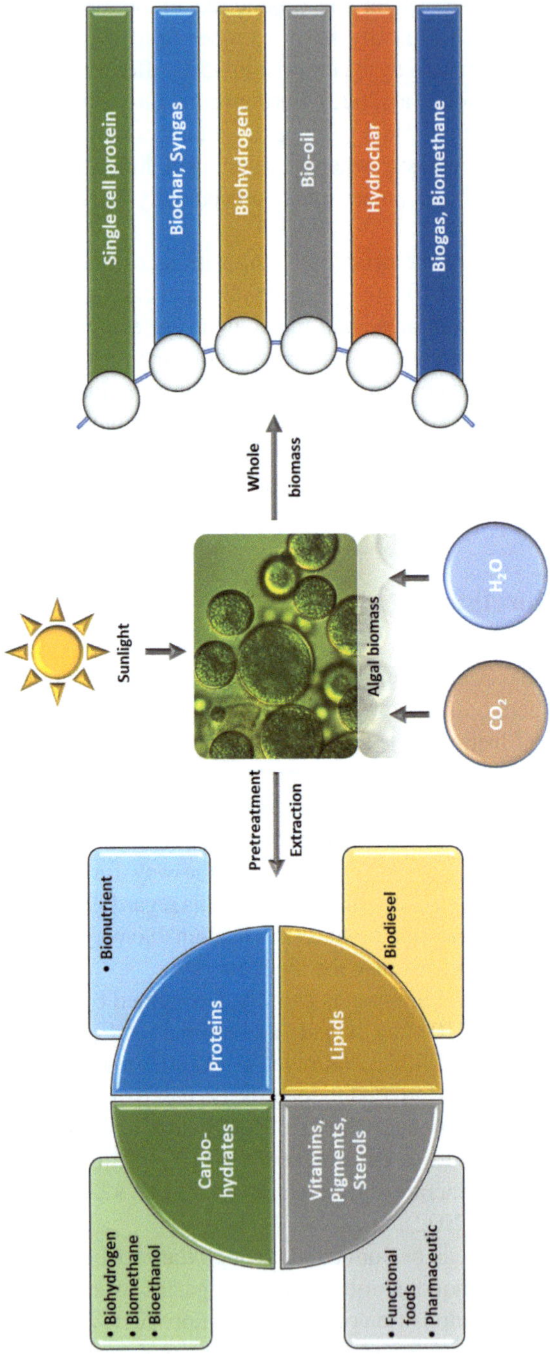

Fig. 13.1 Algal biorefinery concept

Furthermore, it has been demonstrated that the market value of algal biofuels (€0.30/kg) is the lowest, whereas the application of algae in cosmetics, food sources, nutraceuticals, and chemicals has a superior economic value (Zhou et al. 2022). It is therefore necessary for various algal components to be fractionated or transformed in a sequential manner in order to achieve products with a simple composition and high selectivity. Some high-value fatty acids in algal lipids, for example, can be recovered selectively and routed to biodiesel production. At the same time, carbohydrates can be transformed to biofuels such as biohydrogen, biomethane and bioethanol and proteins can be used to harvest nitrogen-rich compounds like amino acids.

13.3 Biohydrogen

Transportation fuel accounts for 33% of total global energy consumption, whereas the combustion of 1 L of diesel results in the production of 2.9 kg of greenhouse gases (Ağbulut and Sarıdemir 2021). Hydrogen is regarded as the most promising cleaner substitute for fossil fuels since its combustion does not produce any harmful by-products. Unlike other fossil fuels, hydrogen as fuel emits no carbon into the atmosphere, and it has a specific energy content of 142 MJ kg^{-1} which is higher than methane, natural gas, and gasoline having a specific energy content of 56 MJ kg^{-1}, 54 MJ kg^{-1}, and 47 MJ kg^{-1}, respectively (Acar and Dincer 2019). However, the conventional hydrogen production techniques are not sustainable as the majority of hydrogen produced currently is reliant on fossil fuels and chemical processes. Hydrogen generated from photosynthetic microorganisms and biomass, on the other hand, is regarded as a sustainable energy source (Kim et al. 2021). In recent decades, microalgae have been employed to produce biohydrogen more sustainably with the help of special metallocluster enzymes. Additionally, algae contain a variety of fermentable carbohydrates that are ideal for hydrogen generation by fermentation. Algae are capable of producing biohydrogen in a sustainable manner due to their faster growth rates and the lack of lignin (which eliminates the need for chemical pretreatment procedures) (Kumari et al. 2017; Pathy et al. 2022).

Algal H_2 production can be accomplished either by direct/indirect photolysis, depending on the amount of sunlight available, or through the fermentation and thermochemical processes as a result of biomass conversion. The biological synthesis of hydrogen using algae involves two main approaches: (1) a light-dependent process involving photolysis, and (2) a light-independent mechanism in which algal intracellular macromolecules act as a substrate for fermentation (Pathy et al. 2022; Reddy et al. 2017). The following section provides detailed mechanisms of the four basic methods by which algae produce biohydrogen (Fig. 13.2).

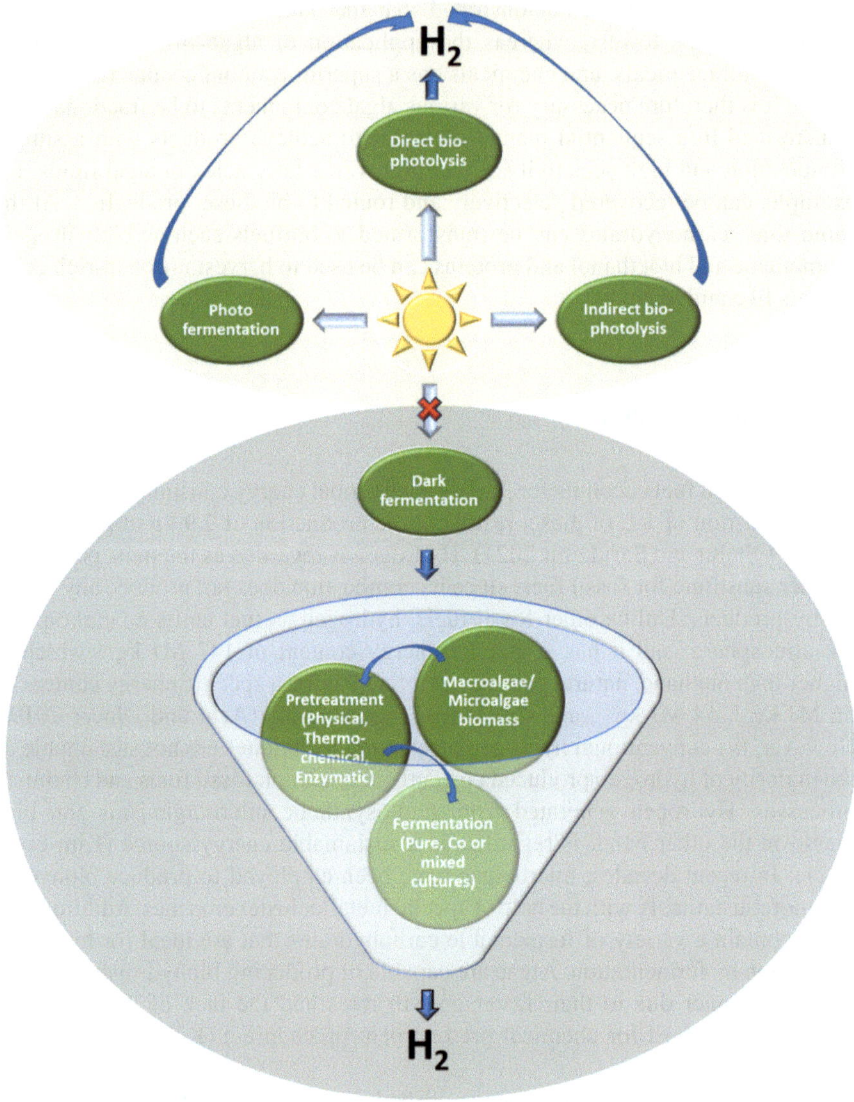

Fig. 13.2 Hydrogen production processes using algae

13.3.1 Direct Biophotolysis

Microalgae can convert solar energy into chemical energy in a direct biophotolysis system (direct biophotolysis from water). In summary, direct biophotolysis is a transitional reaction that occurs for a short amount of time in green microalgae. The light energy fixed by PSI and PSII reduces H_2O (Eq. 13.1a) by transferring electrons to ferredoxin, thereby initiating oxygenic photosynthesis where electrons are moved to Fd to [Fe-Fe]-hydrogenase which produces the molecular H_2 (Eq. 13.1b) (Benemann 2000).

$$2H_2O + \text{light energy} = 4H^+ + 4^{e-} + O_2 \tag{13.1a}$$

$$4H^+ + 4^{e-} + Fe_{red} + \text{Hydrogenase} = 2H_2 \tag{13.1b}$$

This method is highly advantageous because it does not require additional nutritional (carbon) sources or ATP for process stability and can be regulated by solar energy. It is an enzyme-dependent process in which [Fe-Fe]-hydrogenase is O_2-sensitive and acts effectively in low O_2 or anaerobic conditions (Goswami et al. 2020). Since hydrogenase reactions are O_2 sensitive and numerous uncontrolled pathways are involved, this process remains a long-term process for practical application, as it has been for the past 50 years (Li et al. 2022).

13.3.2 Indirect Biophotolysis

Indirect photolysis is a two-step process that overcomes the limitations of direct photolytic H_2 production (Jiménez-Llanos et al. 2020). The microalgae first fix CO_2 and transform it into internal storage materials and metabolites (Eq. 13.2a), which lowers the in-situ O_2 levels and halts the electron transport chains. In the second stage, electrons created by biomolecule catabolism (glycolytic and citric acid pathways) are transferred to plastoquinone, which is then passed to the [Fe-Fe]-hydrogenase enzyme for H_2 production (Eq. 13.2a). Blue-green algae (cyanobacteria) are promising in this regard (Sharma and Arya 2017). It is also noted that the sulphur-depleted medium inhibits the oxygenic photosynthesis pathway in PSII and lowers the level of O_2, thereby decreasing mitochondrial respiration and inducing anaerobic conditions that favor in-situ indirect biophotolytic H_2 synthesis (Goswami et al. 2020).

$$12H_2O + 6CO_2 + \text{light energy} = C_6H_{12}O_6 + 6O_2 + 6H_2O \tag{13.2a}$$

$$C_6H_{12}O_6 + 12H_2O + Fe_{red} + \text{Hydrogenase} = 12H_2 + 6CO_2 + 6H_2O \tag{13.2b}$$

13.3.3 Photo Fermentation

The hydrogen can be created biologically by light-driven fermentation of organic substrates (Li et al. 2022). Photo fermentation occurs only in the presence of light, where it helps microalgae to manufacture energy through photosynthesis, avoiding pyruvate breakdown and the limitations of respiratory ATP generation (Bhatia et al. 2021). Using light energy, electrons are extracted through photo-catabolism of organic substrate coupled with oxidative carbon metabolism (Eq. 13.3). The efficiency estimations of photoconversion to H_2 show values close to 100%, but these evaluations typically ignore the energy content of the organic substrates (Hallenbeck and Benemann 2002). Therefore, the practical photosynthesis efficiency is substantially lower under ideal (low) light conditions. Although hydrogenases are found in photosynthetic microorganisms, the photoproduction of molecular H_2 by these microorganisms is mostly through nitrogenase, as evidenced in purple non-sulfur bacteria (Maeda 2022).

$$CH_3COOH + 2H_2O + \text{light energy} = 4H_2 + 2CO_2 \tag{13.3}$$

13.3.4 Dark Fermentation

The generation of biohydrogen from organic substrates is of interest as a viable technology for the development of sustainable energy. In contrast to the preceding methods, which entirely rely on sunlight for energy, H_2 can be produced by anaerobic fermentation via dark conversion of organic substrates in the absence of sunlight (Li et al. 2022). Generally, dark fermentation involves three metabolic steps in an oxygen-free environment such as hydrolysis, acidogenesis, and acetogenesis. All three phases involve the breakdown of organic biomass and their conversion to volatile fatty acids (VFAs), followed by the synthesis of molecular H_2. The primary microbial species or cultures that create hydrogen from the carbohydrate-rich substrate are *Clostridium*, *Enterobacter*, and *Bacillus*. However, many a time, mixed cultures or anaerobic digestate of different types of sludge has also been used for biohydrogen production (Kumar et al. 2022). The anaerobic degradation of glucose into acetic acid with the end product H_2 is shown in Eq. 13.4.

$$C_6H_{12}O_6 + 2H_2O = 2CH_3COOH + 4H_2 + 2CO_2 \tag{13.4}$$

Pretreatment of biomass by physical, thermal, chemical and biological methods before anaerobic fermentation has been demonstrated to be effective in improving H_2 production (Mandotra et al. 2021).

13.4 Cultivation and Harvesting of Algae

13.4.1 Algae Cultivation Systems

Cost-effective cultivation and harvesting technologies with lower energy input are vital for improving the economic feasibility of algae in the bioenergy sector (Kim et al. 2013). Due to competitive growth from contaminating bacteria and fungus, a substantial share of cultivation and harvesting issues are noticed in microalgae rather than macroalgae, necessitating more specialized cultivation approaches.

The groups Bacillariophyceae (including the diatoms), Chlorophyceae (green algae), Chrysophyceae (including golden algae) and Cyanophyceae (blue-green algae) are the most sought-after microalgae for industrial use, while genera like *Spirulina*, *Chlorella*, and *Dunaliella* are of particular interest as a source of fuels, high-value chemicals, and bioactive compounds (Costa et al. 2019; Rajkumar et al. 2013). Land-based ponds, nearshore coastal farms, and offshore farms are the three most prevalent farming methods utilized in microalgal production (Roesijadi et al. 2010). However, various culture conditions such as light, temperature, pH, salinity, O_2, CO_2, nutritional stress, toxic compounds, pathogens, and competition all have a significant effect on the optimal biomass production of microalgae (Lu et al. 2020). Furthermore, pollution from the atmosphere and other land surfaces is a significant drawback of the open cultivation technique. This may be overcome by using enclosed bioreactors, which are often carried out in glass, plastic, or other transparent bags, plates, or tubes, where microalgae are provided with appropriate light, nutrients, and CO_2 (Amaral et al. 2020). However, they are constrained by their small operational size and high maintenance and construction costs.

Unlike microalgae production, macroalgae cultivation is relatively simple with low maintenance involved. More than 90% of the macroalgae consumed come through diverse farming operations, with the remainder coming from natural sources (Ghadiryanfar et al. 2016). For example, the majority of macroalgal biomass comes from *Kappaphycus*, *Eucheuma*, *Laminaria*, *Gracilaria*, *Undaria*, *Porphyra*, *Sargassum*, and *Ulva*, which drives the world markets for seaweed-derived hydrocolloid polymers and other high-value products for food, feed, and agriculture applications (Laurens et al. 2020). There are two types of macroalgae cultivation methods such as extensive and intensive culture methods. In extensive macroalgae cultivation, seaweeds are cultivated in natural water bodies using only naturally available light, heat, energy, and nutrients (Zollmann et al. 2019). For instance, large-scale commercial mariculture also falls into this category. On the other hand, intensive algal culture methods involve the cultivation of one or more algal species in small natural water bodies such as lagoons, lakes, and ponds utilizing artificial or natural light, nutrients, and plant hormones. It also includes the use of organic and inorganic fertilizers, as well as agronomic procedures including managing light and water movements, weeding, and minimizing epiphyte development (Cole et al. 2016). Although macroalgae farming presents fewer challenges than microalgae farming, several complications arise, including epiphytic and fouling algae, grazing animals, infections, and the detrimental impact of macroalgae on benthic ecosystems (Hurtado et al. 2019).

13.4.2 Algae Harvesting Systems

Algal biomass harvesting costs form a significant component of its total production cost. In certain circumstances, harvesting expenses have been estimated to reach close to 30% of the overall production cost of biofuel and other value-added products from algae (Ramachandra and Saranya 2022). Furthermore, microalgae harvesting is time-consuming, energy-intensive and inefficient due to its small size, low density, and low concentration in the growth medium (Roy and Mohanty 2019). Microalgae are harvested using a variety of methods, including centrifugation, flocculation, filtration, gravity sedimentation, thermal dewatering and ultrasonication, depending upon the type of algae and the desired product (Ferreira et al. 2020; Lam et al. 2019). Sedimentation by gravity is a relatively simple but time-consuming method while, extracting algal biomass exclusively by centrifugation, filtration, thermal dewatering or ultrasonication is still far from practical applicability due to high energy consumption and operational costs. Bulk harvesting technologies, on the other hand, such as flocculation, provide an alternate strategy for collecting algal biomass with less energy input and at a fair cost (Lam et al. 2019; Zeng et al. 2016). Integrating flocculation with other harvesting techniques might enhance the process and lower operational expenses.

Harvesting of macroalgae can be done by manual or mechanized methods due to their large sizes which are less tedious, cost-effective and efficient as compared to microalgae harvesting. However, handheld harvesting is a time-consuming procedure that involves hand-picking or cutting with hooks, nets, and sickles, while mechanical harvesting methods such as dredging cutters, revolving blades, or collection with mesh conveyors, boats or tractors minimize seaweed harvest expenses (Alam et al. 2021; Kumar et al. 2022). Although mechanical harvesting of macroalgae is less expensive, manual harvesting has been demonstrated to perform better in terms of biomass quality and uniformity, since it allows for more onsite contamination removal (Ghadiryanfar et al. 2016). Due to the non-selective nature of mechanical harvesting, it also eliminates significant non-target species, including fish, frogs, snails, and nonspecific plants other than macroalgal biomass. Hence, there is a need to improve macroalgal harvesting efficiency by combining mechanized methods with manual processing, which would result in getting better quality biomass as well as products. Furthermore, before subjecting macroalgal material to biofuel processing or bioproducts recovery, further treatments such as washing, cleaning, dewatering, drying, and size reduction is required. Similarly, both micro- and macroalgae require rigorous pretreatment procedures (physical, thermal, chemical and biological) for superior biofuel production (Kumar et al. 2022; Mandotra et al. 2021).

13.5 Pretreatment of Algal Biomass for Biohydrogen Production

Considering that carbohydrates are the primary substrate for hydrogen synthesis, which is mainly found in the cell walls of algae, their breakdown is essential in the release of polysaccharides. Pretreatment improves the recovery of simple carbohydrates, making it a crucial factor in hydrogen production (Alam et al. 2021; Goswami et al. 2020), and theoretically increases the fermentative H_2 production by 50–70% (Shobana et al. 2017). The pretreatment processes can be broadly classified into four categories such as physical, thermal, chemical and biological methods (Kumar et al. 2022, Mandotra et al. 2021).

The common physical methods of algal biomass pretreatment include, bead-beating, ultrasonication, high-pressure homogenization, microwave, and electroporation (Nagarajan et al. 2020). These physical pretreatment methods essentially alter the structure of the biomass using mechanical shear without adding any biological or chemical reagents. This is a critical stage since it influences the final particle size, degree of polymerization, and cellulose crystallinity (Alam et al. 2021). Alternatively, thermal pretreatment involves the direct application of temperatures ranging from 50 to 250 °C on biomass via heat exchanges to destroy the hydrogen bonds that maintain mechanical strength, resulting in cell wall disintegration and organic matter release (Thompson et al. 2019). The thermal treatment can be further categorized into three types: low-temperature thermal treatment (below 100 °C), hydrothermal pretreatment (100–150 °C) and steam explosion thermal pretreatment (above 150 °C). Thermal pretreatment is often seen as a simple method that can be carried out with standard equipment; however, it can result in the production of toxic compounds such as 5-HMF, furfural, levulinic acid which further act as inhibitors during enzyme treatment and biohydrogen production (Alam et al. 2021; Sulfahri et al. 2020).

Chemical pretreatment involves the use of a variety of chemicals, including acids, alkali, carbon dioxide, organic solvents, and other similar chemicals. The primary goal of this pretreatment is to increase the structural destabilization of algal biomass by depolymerization and increase the crystallinity of the cellulose component (Cheng et al. 2019). The chemical pretreatment approach has some advantages over the mechanical and thermal methods; nevertheless, it has several downsides as well. For example, it induces corrosion in bioreactors, and the use of acid or base interferes with biomolecule yield, generating toxic compounds, and increasing operational costs (Rodriguez et al. 2015). Biological pretreatment, on the other hand, is the safest and greenest processing phase, which consists of using microorganisms or their enzymes that are capable of breaking down complex macromolecules. Generally, a cocktail of enzymes is utilized in the process of treating algal biomass, and it operates under extremely specific conditions. Though the cell wall of algae is made up of cellulose, hemicelluloses, pectin, and algaenan, the cell wall components vary depending on factors such as algal strain, nutrient availability, and growth conditions (Anto et al. 2020). Therefore, enzymes such as amylases, cellobiohydrolase, cellulases, chitinases, hemicellulases, proteases, pectinases, lipases, and lysozymes are used depending upon the algal strain to be used for pretreatment (Nagarajan et al. 2020). Despite its efficacy, enzymatic pretreatment of algal biomass has a few limitations, including the expensive cost of the enzymes and their selective and specialized mode of action; nonetheless, they generate more hydrolyzable sugars and no toxic compounds (Mandotra et al. 2021).

Combined pretreatment is complicated, yet it has been shown to be promising in increasing hydrogen production from algal biomass. As an alternative to common pretreatment methods, other pretreatment combinations are being extensively investigated such as combinations of steam heating/hydrothermal heating and acid treatment (Cheng et al. 2019); microwave and acid treatment (Xia et al. 2013); homogenization, bead beating, thermal-acid treatment (Pinto et al. 2018); surfactant aided microwave treatment (Kumar et al. 2019); hydrothermal and enzymatic method (Ding et al. 2020), all showed better hydrogen yields than individual pre-treatment methods (Table 13.2).

13.6 Strategies to Improve Algal Biomass and Biohydrogen Production

Since the favourable possibilities of cultivating algal cells for biofuel production have been extensively discussed in the literature, advancements in research and development have pushed this industry one step closer to scaling-up and commercialization. However, the cultivation of algae is one of the most expensive procedures involved in the generation of biohydrogen. Unfortunately, several life cycle analyses (LCA) of algal biofuel production indicated a negative energy balance in their evaluations, particularly when algal cells were cultured in closed-photobioreactors (Lam et al. 2019). The economic viability of algal biohydrogen production could be achieved by using strategies such as optimizing the medium composition and physicochemical parameters, in addition to genetic engineering alterations (Fayyaz et al. 2020; Kumar Sharma et al. 2021; Kumari et al. 2015). Additionally, it is crucial to integrate processes into a biorefinery method, such as combining bioenergy with useful co-products (Kumar et al. 2022).

The effectiveness of biohydrogen synthesis from algae may be greatly impacted by operating parameters such as light intensity, temperature, pH, substrate availability, and conversion efficiency in addition to the type of organism used to produce H_2. An adequate quantity of light is required for algae growth and the subsequent synthesis of hydrogen or biomass. The reactors used for this purpose should be constructed to expose the organisms to optimal light conditions. At greater intensities, the light conversion efficiency might be poor due to the light saturation effect. As a result of high absorption, excess photons might be emitted as heat which could induce photoinhibition in the higher layers of algae (Degrenne et al. 2010; Mandotra et al. 2021).

For biohydrogen production, temperature may be the most important factor determining the composition of microorganisms as well as impacting metabolic pathways. Algal species can endure a broad range of temperatures between 20 and 25 °C, notably for mesophilic species, and they can even withstand temperatures below 15 °C, however this results in restricted growth. Temperatures above the optimum, typically lead to cell death (Yaashikaa et al. 2022). Whereas, the organisms used in fermentation methods varied from mesophilic to thermophilic, where thermophiles showed higher H_2 yields (Kumar et al. 2018). The pH also plays a detrimental factor in the growth of algae; however, it is highly species-specific

Table 13.2 Pretreatment methods and hydrogen yields from different micro- and macroalgae

	Pretreatment conditions/Inoculum	Hydrogen yield	Reference
Microalgae			
Chlamydomonas mexicana	Ultrasonication treatment 40 kHz, 450 W, 50 °C, 30 min/*Clostridium acetobutylicum*	2.9 mL/g carbohydrate	El-Dalatony et al. (2022)
Chlorella sp.	Hydrothermal treatment/Heat-treated anaerobic sludge	8.29 mL/g VS	Wu et al. (2020)
Dunaliella primolecta	Lipid-extracted biomass/*Thermococcus eurythermalis* A501	192.35 mL/g VS	Chen et al. (2020)
Dunaliella tertiolecta	Lipid-extracted biomass/*Thermococcus eurythermalis* A501	183.02 mL/g VS	Chen et al. (2020)
Microcystis aeruginosa	Steam heating and acid treatment 135 °C, 15 min, 2% H_2SO_4/Heat-treated anaerobic sludge	18.63 mL/g TVS	Cheng et al. (2019)
Microcystis wesenbergii	Hydrothermal heating and acid treatment 135 °C, 15 min, 2% H_2SO_4/ Heat-treated anaerobic sludge	24.96 mL/g TVS	Cheng et al. (2019)
Nannochloropsis oceanica	Acid treatment 0–2.0% H_2SO_4; microwave treatment 80–180 °C, 5–25 min/Anaerobic sludge	2–39 mL/g VS	Xia et al. (2013)
Scenedesmus obtusiusculus	Thermal acid treatment 160 °C, 3 h, 3% HCl/Granular sludge	48.08 mL/g-VS	Rincón-Pérez et al. (2020)
Scenedesmus obliquus	Autoclaving 121 °C, 15 min, 1.4 bar/*Enterobacter aerogenes*	40.9 mL/g biomass	Batista et al. (2018)
Spirogyra sp.	Homogenization, bead beating, thermal-acid treatment 121 °C, 1 N H_2SO_4/g biomass, 1 h/*Clostridium butyricum*	146.3 mL/g biomass	Pinto et al. (2018)
Macroalgae			
Chaetomorpha antennina	Microwave treatment 0.36 kW, 15 min/ Heat-treated anaerobic sludge	63 mL/g COD	Kumar et al. (2019)
Chaetomorpha antennina	Surfactant aided microwave treatment 0.36 kW, 15 min/Heat-treated anaerobic sludge	74.5 mL/g COD	Kumar et al. (2019)
Gelidium amansii	Thermal treatment 159 °C, 2% H_2SO_4/ Heat-treated anaerobic sludge	37.0 mL/g dry biomass	Park et al. (2013)
Kappaphyccus alvarezzi	Enzymatic treatment Celluclast 60 U/g algae, 50 °C/*C. Beijerinckii* Br21	23.8 mmol/g	Rodrigues et al. (2019)
Laminaria digitata	Hydrothermal pretreatment 140 °C, 20 min, pH 4.5, glucoamylase 0.05 g/g VS/sludge from industrial digester	55.2 mL/g VS	Ding et al. (2020)
Laminaria digitata	Enzymatic treatment glucoamylase 0.05 g/g VS/sludge from industrial digester	44.0 mL/g VS	Ding et al. (2020)
Laminaria japonica	Thermal treatment 170 °C, 20 min/ Heat-treated anaerobic sludge	109.6 mL/g COD	Jung et al. (2011)

(continued)

Table 13.2 (continued)

	Pretreatment conditions/Inoculum	Hydrogen yield	Reference
Laminaria japonica	Thermal treatment 121 °C, 30 min/ Heat-treated anaerobic sludge	66.68 mL/g biomass	Liu and Wang (2014)
Laminaria japonica	Alkaline treatment NaOH 1.0 Mol/L, 30 min/Heat-treated anaerobic sludge	15.00 mL/g biomass	Liu and Wang (2014)
Laminaria japonica	Acid treatment HCl 1.0 Mol/L, 30 min/ Heat-treated anaerobic sludge	43.65 mL/g biomass	Liu and Wang (2014)
Laminaria japonica	Ultrasonic sonic treatment, 20 kHZ/ Heat-treated anaerobic sludge	23.56 mL/g biomass	Liu and Wang (2014)
Saccharina japonica	Thermal acid treatment 121 °C, 2% H_2SO_4, 30 min, 5 kGy ionizing irradiation digested sludge from WWTP, batch	38.1 mL/g VS	Yin and Wang (2019)
Saccharina japonica	Thermal alkaline treatment 121 °C, 2% NaOH, 5 kGy ionizing irradiation digested sludge from WWTP	42 mL/g VS	Yin and Wang (2019)
Saccharina latissima	Hydrothermal treatment 140 °C, 20 min/Heat-treated digestate from biogas plant	10.7 mL/g VS	Lin et al. (2019)
Ulva reticulata	Alkaline treatment (pH 7–12), microwave and H_2O_2 combination/ Anaerobic digested sludge	87.5 mL/g COD	Dinesh Kumar et al. (2020)

VS volatile solids, *TVS* total volatile solids, *COD* chemical oxygen demand, *kGy* kilogray, *WWTP* wastewater treatment plant

(Surendhiran and Sirajunnisa 2019). Similarly, the pH of the fermentation processes of biohydrogen production is also crucial, and the optimal pH range is considered to be between 4.5 and 8.0. It influences the specific activity of the enzymes involved, the hydrolysis of complicated substrates, the diversity of microorganisms, and their metabolic pathways (Pathy et al. 2022; Vargas et al. 2020).

Carbohydrate is an essential component in the fermentative production of H_2. Many mono- and polysaccharides have been identified as potential substrates for biohydrogen synthesis, glucose being the most preferred (Pathy et al. 2022). The build-up of various macromolecules such as carbohydrates, lipids and proteins are dependent on medium composition and other environmental factors. However, several micronutrients are also essential for algal growth. Iron and manganese, for example, are necessary for minute amounts (2.5 to 30 ppm). On the other hand, trace elements such as boron, cobalt, copper, molybdenum, nickel, and zinc, are reportedly added in extremely small concentrations ranging from 2.5 to 4.5 ppm (Banerjee et al. 2020). In addition, iron and nickel are active core constituents of the enzymes [FeFe]-hydrogenase and [NiFe]-hydrogenase, which catalyzes the oxidative generation of H_2 and their content in a medium might alter the enzyme activity (Pathy et al. 2022). Nutrient stress is also one of the variables responsible for significant changes in the biochemical composition of algal biomass. The carbon concentration mechanism (CCM) improves carbohydrate content in algal biomass under CO_2 deprivation by allowing algal cells to absorb and retain inorganic carbon from their extracellular environment (Banerjee et al. 2020). On the other hand, nitrogen (Huang et al. 2019) and phosphate (Labry et al. 2020) deficiency have previously been documented to enhance the carbohydrate contents in algal biomass. However, many reports suggest that nutrient starvation also ends up distorting photosynthetic efficiency, subsequently resulting in reduced chlorophyll content affecting overall algal biomass after repeated cultivations (Banerjee et al. 2020; Takahashi et al. 2020). These limitations might be effectively addressed by using genetic manipulation approaches in which focused enzyme alteration can boost oxygen tolerance ability and glucose accumulation without sacrificing algal biomass levels (Goswami et al. 2020; Mandotra et al. 2021; Surendhiran and Sirajunnisa 2019).

The alteration of the algal genome using novel genome editing technologies including CRISPR/Cas9, RNAi, TALENs, and ZNFs could be utilized to increase the quality and quantity of a variety of products. This has also allowed scientists to create transgenic algal strains with improved hydrogen production ability and biomass yield (Banerjee et al. 2020; Fayyaz et al. 2020; Li et al. 2022). Various gene modifications in algae have been previously studied for enhanced production of hydrogen such as blocking of proton translocation across the thylakoid membrane (Lee and Greenbaum 2003), knockdown of OEE gene (Li et al. 2015), expression of hydrogenase CaI gene from *Clostridium* (Noone et al. 2017), knockdown of SIR1 and IFR1 genes (Venkanna 2018), blocking of electron flow between PS II and PS I (Ban et al. 2018), site-directed mutagenesis and homologous transformation of hydrogenase (Yang et al. 2019). Similarly, other studies reported over expression of carbon fixing RuBisCo gene (Liu et al. 2010), transcription factor PSR1 for starch (Bajhaiya et al. 2015) and isoamylase gene in the algal system (Banerjee et al. 2020) for overproduction of fermentable carbohydrates. In the last decade, numerous studies reported successful algal genome modification; nonetheless, the durability of genetically modified (GM) algal strains and their usage in an open environment remain a major challenge. As a result, before releasing the GM organism for commercialization, a major hazard evaluation undergoing the long-term evaluation of the influence of transgenic microalgal strains on the surrounding environment is crucial. This would ascertain that the genetically modified algal strains do not pose a risk to the ecology (Surendhiran and Sirajunnisa 2019).

13.7 Conclusion and Future Prospects

Since fossil fuels cannot meet expanding energy demand and are also responsible for rising greenhouse gas emissions, it is imperative to develop alternative energy sources that are clean, renewable, and environmentally friendly. Hydrogen shows promise as a future fuel and renewable energy source despite the fact that the current manufacturing of hydrogen via traditional practices is still expensive, energy-intensive, and unsustainable. An excellent alternative to the conventional hydrogen production techniques is the synthesis of biohydrogen from organic biomass. Although algae-derived hydrogen is touted as a clean energy source, the cost of production is still high when compared to other alternative fuel options. Recent studies have focused on developing alternative techniques for enhancing process efficiency; however, the cost of operation and production could be reduced without compromising the sustainability of the process. The remarks listed below are crucial in this regard.

(a) The selection of algal strain is the most important aspect of determining the fate of algal biohydrogen production. To overcome the constraints of light-dependent hydrogen synthesis and dark fermentation, algal species having hydrogenases with low oxygen sensitivity and high carbohydrate content should be selected.
(b) One of the major requirements for hydrogen production is the availability of inexpensive algal biomass. To cut costs, wastewater may be used to produce algal biomass on an industrial scale. However, due to contamination issues in wastewater-based units, flat-plate photo-bioreactor (PBR) and open raceway pond (OPR) are also economically feasible choices. As a result, prior to selecting the kind of manufacturing method and conducting scale-up research, a thorough assessment should be conducted.
(c) Separating microalgal biomass from liquid culture is a difficult and costly process. Thermal dewatering techniques can be substituted by a low-cost solar-powered drier, which might be used in tropical areas. Drying technologies based on other natural renewable energy sources, on the other hand, remain undeveloped and should be researched.
(d) The selection of a fast-growing algae strain may be the best step toward enhancing production efficiency. To some extent, physicochemical optimization can enhance the production; however, genetically modified algae that alleviate the oxygen sensitivity issue of hydrogenase and/or have superior metabolic pathways for biomass (carbohydrate) accumulation may result in increased biohydrogen generation.
(e) Pretreatment procedures for releasing polysaccharides from harvested algal biomass have a direct influence on biohydrogen production. The combination of different pretreatment procedures is an appealing strategy that should be investigated for industrial-scale production.
(f) Individual procedures (light-dependent hydrogen synthesis and dark fermentation) were shown to be inefficient in terms of hydrogen production. As a result, several integrated ways must be created in order to boost efficiency. Furthermore, the total time spent in the complete hydrogen production process is critical to its economics. Hence, a study into this issue is required to minimize the lag time in the process.
(g) Many experiments are conducted in a laboratory size, but further studies on a bigger scale are required to understand the obstacles associated with commercializing biohydrogen production. In addition, algal biomass should be utilized from a biorefinery perspective as this will lead to the development of sustainable processes.

Acknowledgements This study is part of a multilateral BRICS Joint Science and Technology Research Collaboration project supported by the National Research Foundation (NRF) of South Africa, under grant numbers (UID 150503 and 116019), also with support of the Coordination of Improvement of Higher Education Personnel—CAPES of Brazil, project Print-BioADD.

References

Abbasi M, Pishvaee MS, Mohseni S (2021) Third-generation biofuel supply chain: a comprehensive review and future research directions. J Clean Prod 323:129100. https://doi.org/10.1016/j.jclepro.2021.129100

Acar C, Dincer I (2019) Review and evaluation of hydrogen production options for better environment. J Clean Prod 218:835–849. https://doi.org/10.1016/j.jclepro.2019.02.046

Acién F, Molina E, Fernández-Sevilla J, Barbosa M, Gouveia L, Sepúlveda C et al (2017) Economics of microalgae production. Microalgae-based biofuels and bioproducts. Elsevier, Amsterdam, pp 485–503

Adelakun K, Kehinde A, Amali R, Ogundiwin D, Omotayo O (2016) Nutritional and phytochemical quality of some tropical aquatic plants. Poultry Fisheries Wildlife Sci 04:1000164. https://doi.org/10.4172/2375-446X.1000164

Ağbulut Ü, Sarıdemir S (2021) A general view to converting fossil fuels to cleaner energy source by adding nanoparticles. Int J Ambient Energy 42(13):1569–1574. https://doi.org/10.1080/01430750.2018.1563822

Aguilar-Toalá JE, Hall FG, Urbizo-Reyes U, Liceaga AM (2021) Chapter 18 - sustainable, alternative sources of bioactive peptides. In: Toldrá F, Wu J (eds) Biologically active peptides. Academic Press, Cambridge, pp 427–453

Alam SN, Khalid Z, Guldhe A, Singh B, Korstad J (2021) Harvesting and pretreatment techniques of aquatic macrophytes and macroalgae for production of biofuels. Environ Sustain 4(2):299–316. https://doi.org/10.1007/s42398-021-00178-6

Amaral MS, Loures CC, Naves FL, Samanamud GL, Silva MB, Prata AM (2020) Microalgae cultivation in Photobioreactors aiming at biodiesel production. Biomass, IntechOpen

Anto S, Mukherjee SS, Muthappa R, Mathimani T, Deviram G, Kumar SS et al (2020) Algae as green energy reserve: technological outlook on biofuel production. Chemosphere 242:125079. https://doi.org/10.1016/j.chemosphere.2019.125079

Aouir A, Amiali M, Bitam A, Benchabane A, Raghavan VG (2017) Comparison of the biochemical composition of different *Arthrospira platensis* strains from Algeria, Chad and the USA. J Food Measurement Characterization 11(2):913–923. https://doi.org/10.1007/s11694-016-9463-4

Bajhaiya AK, Dean AP, Zeef LAH, Webster RE, Pittman JK (2015) PSR1 is a global transcriptional regulator of phosphorus deficiency responses and carbon storage metabolism in *Chlamydomonas reinhardtii*. Plant Physiol 170(3):1216–1234. https://doi.org/10.1104/pp.15.01907

Balar N, Sharnagat P, Kumari P, Mantri VA (2019) Variation in the proximate composition of edible marine macroalga *Ulva rigida* collected from different coastal zones of India. J Food Sci Technol 56(10):4749–4755. https://doi.org/10.1007/s13197-019-03929-w

Ban S, Lin W, Wu F, Luo J (2018) Algal-bacterial cooperation improves algal photolysis-mediated hydrogen production. Bioresour Technol 251:350–357. https://doi.org/10.1016/j.biortech.2017.12.072

Banerjee S, Banerjee S, Ghosh AK, Das D (2020) Maneuvering the genetic and metabolic pathway for improving biofuel production in algae: present status and future prospective. Renew Sust Energ Rev 133:110155. https://doi.org/10.1016/j.rser.2020.110155

Batista AP, Gouveia L, Marques PASS (2018) Fermentative hydrogen production from microalgal biomass by a single strain of bacterium *Enterobacter aerogenes* – effect of operational

conditions and fermentation kinetics. Renew Energy 119:203–209. https://doi.org/10.1016/j.renene.2017.12.017
Benemann JR (2000) Hydrogen production by microalgae. J Appl Phycol 12(3):291–300. https://doi.org/10.1023/A:1008175112704
Bhateria R, Dhaka R (2014) Algae as biofuel. Biofuels 5(6):607–631. https://doi.org/10.1080/17597269.2014.1003701
Bhatia SK, Mehariya S, Bhatia RK, Kumar M, Pugazhendhi A, Awasthi MK et al (2021) Wastewater based microalgal biorefinery for bioenergy production: Progress and challenges. Sci Total Environ 751:141599. https://doi.org/10.1016/j.scitotenv.2020.141599
Buschmann AH, Camus C, Infante J, Neori A, Israel Á, Hernández-González MC et al (2017) Seaweed production: overview of the global state of exploitation, farming and emerging research activity. Eur J Phycol 52(4):391–406. https://doi.org/10.1080/09670262.2017.1365175
Cesário MT, da Fonseca MMR, Marques MM, de Almeida MCMD (2018) Marine algal carbohydrates as carbon sources for the production of biochemicals and biomaterials. Biotechnol Adv 36(3):798–817. https://doi.org/10.1016/j.biotechadv.2018.02.006
Chen Q, Jin Y, Zhang G, Fang Y, Xiao Y, Zhao H (2012) Improving production of bioethanol from duckweed (*Landoltia punctata*) by pectinase pretreatment. Energies 5(8):3019–3032. https://doi.org/10.3390/en5083019
Chen J, Wang Y, Benemann JR, Zhang X, Hu H, Qin S (2016) Microalgal industry in China: challenges and prospects. J Appl Phycol 28(2):715–725. https://doi.org/10.1007/s10811-015-0720-4
Chen S, Qu D, Xiao X, Miao X (2020) Biohydrogen production with lipid-extracted *Dunaliella* biomass and a new strain of hyper-thermophilic archaeon *Thermococcus eurythermalis* A501. Int J Hydrog Energy 45(23):12721–12730. https://doi.org/10.1016/j.ijhydene.2020.03.010
Cheng J, Yue L, Ding L, Li Y-Y, Ye Q, Zhou J et al (2019) Improving fermentative hydrogen and methane production from an algal bloom through hydrothermal/steam acid pretreatment. Int J Hydrog Energy 44(12):5812–5820. https://doi.org/10.1016/j.ijhydene.2019.01.046
Chew KW, Yap JY, Show PL, Suan NH, Juan JC, Ling TC et al (2017) Microalgae biorefinery: high value products perspectives. Bioresour Technol 229:53–62. https://doi.org/10.1016/j.biortech.2017.01.006
Chien F, Hsu C-C, Ozturk I, Sharif A, Sadiq M (2022) The role of renewable energy and urbanization towards greenhouse gas emission in top Asian countries: evidence from advance panel estimations. Renew Energy 186:207–216. https://doi.org/10.1016/j.renene.2021.12.118
Cole AJ, Neveux N, Whelan A, Morton J, Vis M, de Nys R et al (2016) Adding value to the treatment of municipal wastewater through the intensive production of freshwater macroalgae. Algal Res 20:100–109. https://doi.org/10.1016/j.algal.2016.09.026
Costa JAV, Freitas BCB, Lisboa CR, Santos TD, Brusch LRF, de Morais MG (2019) Microalgal biorefinery from CO2 and the effects under the blue economy. Renew Sust Energ Rev 99:58–65. https://doi.org/10.1016/j.rser.2018.08.009
Darko CNS, Agyei-Tuffour B, Faloye DF, Goosen NJ, Nyankson E, Dodoo-Arhin D (2022) Biomethane production from residual algae biomass (*Ecklonia maxima*): effects of inoculum acclimatization on yield. Waste Biomass Valorization 13(1):497–509. https://doi.org/10.1007/s12649-021-01497-9
de Farias Silva CE, Sforza E, Bertucco A (2019) Chapter 3 - enhancing carbohydrate productivity in photosynthetic microorganism production: a comparison between cyanobacteria and microalgae and the effect of cultivation systems. In: Hosseini M (ed) Advances in feedstock conversion Technologies for Alternative Fuels and Bioproducts. Woodhead Publishing, Cambridge, pp 37–67
De Jesus Raposo MF, De Morais AMB, De Morais RMSC (2015) Marine polysaccharides from algae with potential biomedical applications. Mar Drugs 13(5):2967–3028. https://doi.org/10.3390/md13052967
Degrenne B, Pruvost J, Christophe G, Cornet JF, Cogne G, Legrand J (2010) Investigation of the combined effects of acetate and photobioreactor illuminated fraction in the induction

of anoxia for hydrogen production by *Chlamydomonas reinhardtii.* Int J Hydrog Energy 35(19):10741–10749. https://doi.org/10.1016/j.ijhydene.2010.02.067

Denis C, Morançais M, Li M, Deniaud E, Gaudin P, Wielgosz-Collin G et al (2010) Study of the chemical composition of edible red macroalgae *Grateloupia turuturu* from Brittany (France). Food Chem 119(3):913–917. https://doi.org/10.1016/j.foodchem.2009.07.047

Dinesh Kumar M, Yukesh Kannah R, Kumar G, Sivashanmugam P, Rajesh Banu J (2020) A novel energetically efficient combinative microwave pretreatment for achieving profitable hydrogen production from marine macro algae (*Ulva reticulate*). Bioresour Technol 301:122759. https://doi.org/10.1016/j.biortech.2020.122759

Ding L, Cheng J, Lin R, Deng C, Zhou J, Murphy JD (2020) Improving biohydrogen and biomethane co-production via two-stage dark fermentation and anaerobic digestion of the pretreated seaweed *Laminaria digitata.* J Clean Prod 251:119666. https://doi.org/10.1016/j.jclepro.2019.119666

Dong T, Van Wychen S, Nagle N, Pienkos PT, Laurens LM (2016) Impact of biochemical composition on susceptibility of algal biomass to acid-catalyzed pretreatment for sugar and lipid recovery. Algal Res 18:69–77. https://doi.org/10.1016/j.algal.2016.06.004

Dromard CR, Bouchon-Navaro Y, Harmelin-Vivien M, Bouchon C (2017) The nutritional quality of non-calcified macroalgae in Guadeloupe (Lesser Antilles) evaluated by their biochemical composition. Gulf Caribbean Res 28(1):GCFI1–GCFI6. https://doi.org/10.18785/gcr.2801.04

Efremenko EN, Nikolskaya AB, Lyagin IV, Senko OV, Makhlis TA, Stepanov NA et al (2012) Production of biofuels from pretreated microalgae biomass by anaerobic fermentation with immobilized *clostridium acetobutylicum* cells. Bioresour Technol 114:342–348. https://doi.org/10.1016/j.biortech.2012.03.049

El-Dalatony MM, Basak B, Kurade MB, Roh H-S, Jang M, Jeon B-H (2022) Effect of sonication pretreatment on hydrogen and acetone-butanol-ethanol coproduction from *Chlamydomonas mexicana* biomass using *clostridium acetobutylicum.* J Environ Chem Eng 10(3):107600. https://doi.org/10.1016/j.jece.2022.107600

Elkatory MR, Hassaan MA, El Nemr A (2022) Chapter 12 - algal biomass for bioethanol and biobutanol production. In: El-Sheekh M, Abomohra AE-F (eds) Handbook of algal biofuels. Elsevier, Cambridge, pp 251–279

Fan Q, Hajiyeva AM (2022) Nexus between energy efficiency finance and renewable energy development: empirical evidence from G-7 economies. Renew Energy 195:1077–1086. https://doi.org/10.1016/j.renene.2022.06.113

Fayyaz M, Chew KW, Show PL, Ling TC, Ng IS, Chang J-S (2020) Genetic engineering of microalgae for enhanced biorefinery capabilities. Biotechnol Adv 43:107554. https://doi.org/10.1016/j.biotechadv.2020.107554

Ferdouse F, Holdt SL, Smith R, Murúa P, Yang Z (2018) The global status of seaweed production, trade and utilization. Globefish Res Programme 124:I. https://www.proquest.com/scholarly-journals/global-status-seaweed-production-trade/docview/2164110004/se-2

Ferreira J, de Assis LR, de Sousa Oliveira AP, de Siqueira CJ, Calijuri ML (2020) Innovative microalgae biomass harvesting methods: technical feasibility and life cycle analysis. Sci Total Environ 746:140939. https://doi.org/10.1016/j.scitotenv.2020.140939

Ge X, Zhang N, Phillips GC, Xu J (2012) Growing *Lemna minor* in agricultural wastewater and converting the duckweed biomass to ethanol. Bioresour Technol 124:485–488. https://doi.org/10.1016/j.biortech.2012.08.050

Ghadiryanfar M, Rosentrater KA, Keyhani A, Omid M (2016) A review of macroalgae production, with potential applications in biofuels and bioenergy. Renew Sust Energ Rev 54:473–481. https://doi.org/10.1016/j.rser.2015.10.022

Goswami RK, Mehariya S, Karthikeyan OP, Verma P (2020) Advanced microalgae-based renewable biohydrogen production systems: a review. Bioresour Technol 320:124301. https://doi.org/10.1016/j.biortech.2020.124301

Goswami L, Kayalvizhi R, Dikshit PK, Sherpa KC, Roy S, Kushwaha A et al (2022) A critical review on prospects of bio-refinery products from second and third generation biomasses. Chem Eng J 448:137677. https://doi.org/10.1016/j.cej.2022.137677

Hallenbeck PC, Benemann JR (2002) Biological hydrogen production; fundamentals and limiting processes. Int J Hydrog Energy 27(11):1185–1193. https://doi.org/10.1016/S0360-3199(02)00131-3

Hirani AH, Javed N, Asif M, Basu SK, Kumar A (2018) A review on first-and second-generation biofuel productions. Biofuels: greenhouse gas mitigation and global warming. Springer, Cham, pp 141–154

Hossain N, Mahlia TMI (2019) Progress in physicochemical parameters of microalgae cultivation for biofuel production. Crit Rev Biotechnol 39(6):835–859. https://doi.org/10.1080/07388551.2019.1624945

Huang Y, Hong J, Liang C, Li G, Shen L, Zhang H et al (2019) Nitrogen limitation affects the sinking property of *Microcystis* by promoting carbohydrate accumulation. Chemosphere 221:665–671. https://doi.org/10.1016/j.chemosphere.2019.01.080

Hurtado AQ, Neish IC, Critchley AT (2019) Phyconomy: the extensive cultivation of seaweeds, their sustainability and economic value, with particular reference to important lessons to be learned and transferred from the practice of eucheumatoid farming. Phycologia 58(5):472–483. https://doi.org/10.1080/00318884.2019.1625632

Jiménez-Llanos J, Ramírez-Carmona M, Rendón-Castrillón L, Ocampo-López C (2020) Sustainable biohydrogen production by *chlorella* sp. microalgae: a review. Int J Hydrog Energy 45(15):8310–8328. https://doi.org/10.1016/j.ijhydene.2020.01.059

John RP, Anisha G, Nampoothiri KM, Pandey A (2011) Micro and macroalgal biomass: a renewable source for bioethanol. Bioresour Technol 102(1):186–193

Jung K-W, Kim D-H, Shin H-S (2011) Fermentative hydrogen production from *Laminaria japonica* and optimization of thermal pretreatment conditions. Bioresour Technol 102(3):2745–2750. https://doi.org/10.1016/j.biortech.2010.11.042

Karpagam R, Jawaharraj K, Gnanam R (2021) Review on integrated biofuel production from microalgal biomass through the outset of transesterification route: a cascade approach for sustainable bioenergy. Sci Total Environ 766:144236. https://doi.org/10.1016/j.scitotenv.2020.144236

Katiyar R, Banerjee S, Arora A (2021) Recent advances in the integrated biorefinery concept for the valorization of algal biomass through sustainable routes. Biofuels Bioprod Biorefin 15(3):879–898. https://doi.org/10.1002/bbb.2187

Khoo KS, Lee SY, Ooi CW, Fu X, Miao X, Ling TC et al (2019) Recent advances in biorefinery of astaxanthin from *Haematococcus pluvialis*. Bioresour Technol 288:121606. https://doi.org/10.1016/j.biortech.2019.121606

Kim J, Yoo G, Lee H, Lim J, Kim K, Kim CW et al (2013) Methods of downstream processing for the production of biodiesel from microalgae. Biotechnol Adv 31(6):862–876. https://doi.org/10.1016/j.biotechadv.2013.04.006

Kim S-H, Kumar G, Chen W-H, Khanal SK (2021) Renewable hydrogen production from biomass and wastes (ReBioH2-2020). Bioresour Technol 331:125024. https://doi.org/10.1016/j.biortech.2021.125024

Kostas ET, Adams JMM, Ruiz HA, Durán-Jiménez G, Lye GJ (2021) Macroalgal biorefinery concepts for the circular bioeconomy: a review on biotechnological developments and future perspectives. Renew Sust Energ Rev 151:111553. https://doi.org/10.1016/j.rser.2021.111553

Kumar Sharma A, Kumar Ghodke P, Manna S, Chen W-H (2021) Emerging technologies for sustainable production of biohydrogen production from microalgae: a state-of-the-art review of upstream and downstream processes. Bioresour Technol 342:126057. https://doi.org/10.1016/j.biortech.2021.126057

Kumar G, Shobana S, Nagarajan D, Lee D-J, Lee K-S, Lin C-Y et al (2018) Biomass based hydrogen production by dark fermentation—recent trends and opportunities for greener processes. Curr Opin Biotechnol 50:136–145. https://doi.org/10.1016/j.copbio.2017.12.024

Kumar D, Eswari AP, Park J-H, Adishkumar S, Banu JR (2019) Biohydrogen generation from macroalgal biomass, *Chaetomorpha antennina* through surfactant aided microwave disintegration. Front Energy Res 7:78. https://doi.org/10.3389/fenrg.2019.00078

Kumar MD, Kavitha S, Tyagi VK, Rajkumar M, Bhatia SK, Kumar G et al (2022) Macroalgae-derived biohydrogen production: biorefinery and circular bioeconomy. Biomass Conversion Biorefinery 12(3):769–791. https://doi.org/10.1007/s13399-020-01187-x

Kumari S, Singh P, Gupta SK, Kumar S (2015) Genetic engineering tools for enhancing lipid production in microalgae. In: Singh B, Bauddh K, Bux F (eds) Algae and environmental sustainability. Springer India, New Delhi, pp 119–127

Kumari S, Nasr M, Kumar S (2017) Technological advances in biohydrogen production from microalgae. In: Gupta SK, Malik A, Bux F (eds) Algal biofuels: recent advances and future prospects. Springer International Publishing, Cham, pp 347–360

Kushwaha A, Goswami S, Sultana A, Katiyar NK, Athar M, Dubey L et al (2022) Waste biomass to biobutanol: recent trends and advancements. In: Waste-to-energy approaches towards zero waste. Elsevier, Cambridge, pp 393–423. https://doi.org/10.1016/B978-0-323-85387-3.00004-5

Labry C, Delmas D, Moriceau B, Gallinari M, Quere J, Youenou A (2020) Effect of P depletion on the functional pools of diatom carbohydrates, and their utilization by bacterial communities. Mar Ecol Prog Ser 641:49–62. https://doi.org/10.3354/meps13297

Lam MK, Khoo CG, Lee KT (2019) Chapter 19 - scale-up and commercialization of algal cultivation and biofuels production. In: Pandey A, Chang J-S, Soccol CR, Lee D-J, Chisti Y (eds) Biofuels from algae, 2nd edn. Elsevier, Cambridge, pp 475–506

Laurens LML, Lane M, Nelson RS (2020) Sustainable seaweed biotechnology solutions for carbon capture, composition, and deconstruction. Trends Biotechnol 38(11):1232–1244. https://doi.org/10.1016/j.tibtech.2020.03.015

Lee JW, Greenbaum E (2003) A new oxygen sensitivity and its potential application in photosynthetic H2 production. In: Davison BH, Lee JW, Finkelstein M, McMillan JD (eds) Biotechnology for fuels and chemicals: the twenty-fourth symposium. Humana Press, Totowa, NJ, pp 303–313

Li H, Zhang L, Shu L, Zhuang X, Liu Y, Chen J et al (2015) Sustainable photosynthetic H2-production mediated by artificial miRNA silencing of OEE2 gene in green alga *Chlamydomonas reinhardtii*. Int J Hydrog Energy 40(16):5609–5616. https://doi.org/10.1016/j.ijhydene.2015.02.073

Li S, Li F, Zhu X, Liao Q, Chang J-S, Ho S-H (2022) Biohydrogen production from microalgae for environmental sustainability. Chemosphere 291:132717. https://doi.org/10.1016/j.chemosphere.2021.132717

Lin R, Deng C, Ding L, Bose A, Murphy JD (2019) Improving gaseous biofuel production from seaweed *Saccharina latissima*: the effect of hydrothermal pretreatment on energy efficiency. Energy Convers Manag 196:1385–1394. https://doi.org/10.1016/j.enconman.2019.06.044

Liu H, Wang G (2014) Fermentative hydrogen production from macro-algae *Laminaria japonica* using anaerobic mixed bacteria. Int J Hydrog Energy 39(17):9012–9017. https://doi.org/10.1016/j.ijhydene.2014.03.244

Liu C, Young AL, Starling-Windhof A, Bracher A, Saschenbrecker S, Rao BV et al (2010) Coupled chaperone action in folding and assembly of hexadecameric Rubisco. Nature 463(7278):197–202. https://doi.org/10.1038/nature08651

Lu W, Alam MA, Liu S, Xu J, Saldivar RP (2020) Critical processes and variables in microalgae biomass production coupled with bioremediation of nutrients and CO2 from livestock farms: a review. Sci Total Environ 716:135247. https://doi.org/10.1016/j.scitotenv.2019.135247

Maeda I (2022) Potential of phototrophic purple nonsulfur bacteria to fix nitrogen in rice fields. Microorganisms 10(1):28. https://doi.org/10.3390/microorganisms10010028

Mandotra SK, Sharma C, Srivastava N, Ahluwalia AS, Ramteke PW (2021) Current prospects and future developments in algal bio-hydrogen production: a review. Biomass Conversion Biorefinery 13:8575–8592. https://doi.org/10.1007/s13399-021-01414-z

Mehrabadi A, Farid MM, Craggs R (2017) Potential of five different isolated colonial algal species for wastewater treatment and biomass energy production. Algal Res 21:1–8. https://doi.org/10.1016/j.algal.2016.11.002

Miranda AF, Biswas B, Ramkumar N, Singh R, Kumar J, James A et al (2016) Aquatic plant *Azolla* as the universal feedstock for biofuel production. Biotechnol Biofuels 9(1):221. https://doi.org/10.1186/s13068-016-0628-5

Muthuraj M, Palabhanvi B, Misra S, Kumar V, Sivalingavasu K, Das D (2013) Flux balance analysis of *chlorella* sp. FC2 IITG under photoautotrophic and heterotrophic growth conditions. Photosynth Res 118(1):167–179. https://doi.org/10.1007/s11120-013-9943-x

Nagarajan D, Chang J-S, Lee D-J (2020) Pretreatment of microalgal biomass for efficient biohydrogen production – recent insights and future perspectives. Bioresour Technol 302:122871. https://doi.org/10.1016/j.biortech.2020.122871

Noone S, Ratcliff K, Davis R, Subramanian V, Meuser J, Posewitz MC et al (2017) Expression of a clostridial [FeFe]-hydrogenase in *Chlamydomonas reinhardtii* prolongs photoproduction of hydrogen from water splitting. Algal Res 22:116–121. https://doi.org/10.1016/j.algal.2016.12.014

Park J-H, Cheon H-C, Yoon J-J, Park H-D, Kim S-H (2013) Optimization of batch dilute-acid hydrolysis for biohydrogen production from red algal biomass. Int J Hydrog Energy 38(14):6130–6136. https://doi.org/10.1016/j.ijhydene.2013.01.050

Pathy A, Nageshwari K, Ramaraj R, Pragas Maniam G, Govindan N, Balasubramanian P (2022) Biohydrogen production using algae: potentiality, economics and challenges. Bioresour Technol 360:127514. https://doi.org/10.1016/j.biortech.2022.127514

Pinto T, Gouveia L, Ortigueira J, Saratale GD, Moura P (2018) Enhancement of fermentative hydrogen production from *spirogyra* sp. by increased carbohydrate accumulation and selection of the biomass pretreatment under a biorefinery model. J Biosci Bioeng 126(2):226–234. https://doi.org/10.1016/j.jbiosc.2018.02.017

Rajauria G, Yuan YV (2021) Recent advances in micro- and macroalgal processing: food and health perspectives. Wiley, New York

Rajkumar R, Yaakob Z, Takriff MS (2013) Potential of micro and macro algae for biofuel production: a brief review. Bioresources 9(1):1606–1633. https://ojs.cnr.ncsu.edu/index.php/BioRes/article/view/4049

Ramachandra TV, Saranya G (2022) Sustainable bioeconomy prospects of diatom biorefineries in the Indian west coast. Renew Sust Energ Rev 161:112399. https://doi.org/10.1016/j.rser.2022.112399

Rebolloso-Fuentes MM, Navarro-Pérez A, García-Camacho F, Ramos-Miras J, Guil-Guerrero J (2001) Biomass nutrient profiles of the microalga *Nannochloropsis*. J Agric Food Chem 49(6):2966–2972. https://doi.org/10.1021/jf0010376

Reddy K, Nasr M, Kumari S, Kumar S, Gupta SK, Enitan AM et al (2017) Biohydrogen production from sugarcane bagasse hydrolysate: effects of pH, S/X, Fe2+, and magnetite nanoparticles. Environ Sci Pollut Res 24(9):8790–8804. https://doi.org/10.1007/s11356-017-8560-1

Rinanti A, Purwadi R (2019) Increasing carbohydrate and lipid productivity in tropical microalgae biomass as a sustainable biofuel feed stock. Energy Procedia 158:1215–1222. https://doi.org/10.1016/j.egypro.2019.01.310

Rincón-Pérez J, Razo-Flores E, Morales M, Alatriste-Mondragón F, Celis LB (2020) Improving the biodegradability of *Scenedesmus obtusiusculus* by thermochemical pretreatment to produce hydrogen and methane. Bioenergy Res 13(2):477–486. https://doi.org/10.1007/s12155-019-10067-w

Rodrigues D, Freitas AC, Pereira L, Rocha-Santos TA, Vasconcelos MW, Roriz M et al (2015) Chemical composition of red, brown and green macroalgae from Buarcos bay in central west coast of Portugal. Food Chem 183:197–207. https://doi.org/10.1016/j.foodchem.2015.03.057

Rodrigues EL, Fonseca BC, Gelli VC, Carli S, Meleiro LP, Furriel RPM et al (2019) Enzymatically and/or thermally treated macroalgae biomass as feedstock for fermentative H 2 production. Matéria (Rio de Janeiro) 24:2. https://doi.org/10.1590/S1517-707620190002.0678

Rodriguez C, Alaswad A, Mooney J, Prescott T, Olabi AG (2015) Pre-treatment techniques used for anaerobic digestion of algae. Fuel Process Technol 138:765–779. https://doi.org/10.1016/j.fuproc.2015.06.027

Roesijadi G, Jones SB, Snowden-Swan LJ, Zhu Y (2010) Macroalgae as a biomass feedstock: a preliminary analysis. Pacific Northwest National Lab.(PNNL), Richland

Roy M, Mohanty K (2019) A comprehensive review on microalgal harvesting strategies: current status and future prospects. Algal Res 44:101683. https://doi.org/10.1016/j.algal.2019.101683

Sajjadi B, Chen W-Y, Raman AAA, Ibrahim S (2018) Microalgae lipid and biomass for biofuel production: a comprehensive review on lipid enhancement strategies and their effects on fatty acid composition. Renew Sust Energ Rev 97:200–232. https://doi.org/10.1016/j.rser.2018.07.050

Sharma A, Arya SK (2017) Hydrogen from algal biomass: a review of production process. Biotechnol Rep 15:63–69. https://doi.org/10.1016/j.btre.2017.06.001

Shobana S, Saratale GD, Pugazhendhi A, Arvindnarayan S, Periyasamy S, Kumar G et al (2017) Fermentative hydrogen production from mixed and pure microalgae biomass: key challenges and possible opportunities. Int J Hydrog Energy 42(42):26440–26453. https://doi.org/10.1016/j.ijhydene.2017.07.050

Singh H, Varanasi JL, Banerjee S, Das D (2019) Production of carbohydrate enrich microalgal biomass as a bioenergy feedstock. Energy 188:116039. https://doi.org/10.1016/j.energy.2019.116039

Singh A, Kushwaha A, Sen S, Goswami S, Katiyar S, Kumar A et al (2022) Recent advancement in microwave-assisted pyrolysis for biooil production. In: Waste-to-energy approaches towards zero waste. Elsevier, Cambridge, pp 197–219

Sinha A, Sengupta T (2019) Impact of natural resource rents on human development: what is the role of globalization in Asia Pacific countries? Resour Policy 63:101413. https://doi.org/10.1016/j.resourpol.2019.101413

Sudhakar K, Mamat R, Samykano M, Azmi W, Ishak W, Yusaf T (2018) An overview of marine macroalgae as bioresource. Renew Sust Energ Rev 91:165–179. https://doi.org/10.1016/j.rser.2018.03.100

Sudhakar M, Kumar BR, Mathimani T, Arunkumar K (2019) A review on bioenergy and bioactive compounds from microalgae and macroalgae-sustainable energy perspective. J Clean Prod 228:1320–1333. https://doi.org/10.1016/j.jclepro.2019.04.287

Suganya T, Varman M, Masjuki H, Renganathan S (2016) Macroalgae and microalgae as a potential source for commercial applications along with biofuels production: a biorefinery approach. Renew Sust Energ Rev 55:909–941. https://doi.org/10.1016/j.rser.2015.11.026

Suleiman M, Khadija AY, Nasiru Y, Garba AA, Alhassan M, Bello HJ (2019) Proximate, minerals and anti-nutritional composition of water hyacinth (*Eichhornia crassipes*) grass. Earthline J Chem Sci 3(1):51–59. https://doi.org/10.34198/ejcs.3120.5159

Sulfahri MS, Langford A, Tassakka ACMAR (2020) Ozonolysis as an effective pretreatment strategy for bioethanol production from marine algae. Bioenergy Res 13(4):1269–1279. https://doi.org/10.1007/s12155-020-10131-w

Sun X-M, Geng L-J, Ren L-J, Ji X-J, Hao N, Chen K-Q et al (2018) Influence of oxygen on the biosynthesis of polyunsaturated fatty acids in microalgae. Bioresour Technol 250:868–876. https://doi.org/10.1016/j.biortech.2017.11.005

Surendhiran D, Sirajunnisa AR (2019) Chapter 18 - role of genetic engineering in bioethanol production from algae. In: Ray RC, Ramachandran S (eds) Bioethanol production from food crops. Academic Press, Cambridge, pp 361–381

Syaichurrozi I, Villta PK, Nabilah N, Rusdi R (2019) Effect of sulfuric acid pretreatment on biogas production from *Salvinia molesta*. J Environ Chem Eng 7(1):102857. https://doi.org/10.1016/j.jece.2018.102857

Takahashi M, Kumari P, Li C, Mikami K (2020) Low temperature causes discoloration by repressing growth and nitrogen transporter gene expression in the edible red alga *Pyropia yezoensis*. Mar Environ Res 159:105004. https://doi.org/10.1016/j.marenvres.2020.105004

Thompson TM, Young BR, Baroutian S (2019) Advances in the pretreatment of brown macroalgae for biogas production. Fuel Process Technol 195:106151. https://doi.org/10.1016/j.fuproc.2019.106151

Tibbetts SM, Milley JE, Lall SP (2015) Chemical composition and nutritional properties of freshwater and marine microalgal biomass cultured in photobioreactors. J Appl Phycol 27(3):1109–1119. https://doi.org/10.1007/s10811-014-0428-x

Tokuşoglu Ö, Üunal M (2003) Biomass nutrient profiles of three microalgae: *Spirulina platensis*, *Chlorella vulgaris*, and *Isochrisis galbana*. J Food Sci 68(4):1144–1148. https://doi.org/10.1111/j.1365-2621.2003.tb09615.x

Vargas SR, Zaiat M, Calijuri MC (2020) Influence of culture age, ammonium and organic carbon in hydrogen production and nutrient removal by *anabaena* sp. in nitrogen-limited cultures. Int J Hydrog Energy 45(55):30222–30231. https://doi.org/10.1016/j.ijhydene.2020.08.072

Venkanna D (2018) Synthetic rewiring of *Chlamydomonas reinhardtii* to improve biological H2 production. Universität Bielefeld, Bielefeld, p 126

Wong KH, Tan IS, Foo HCY, Chin LM, Cheah JRN, Sia JK et al (2022) Third-generation bioethanol and L-lactic acid production from red macroalgae cellulosic residue: prospects of industry 5.0 algae. Energy Convers Manag 253:115155. https://doi.org/10.1016/j.enconman.2021.115155

Wu H, Li J, Liao Q, Fu Q, Liu Z (2020) Enhanced biohydrogen and biomethane production from *chlorella* sp. with hydrothermal treatment. Energy Convers Manag 205:112373. https://doi.org/10.1016/j.enconman.2019.112373

Xia A, Cheng J, Lin R, Lu H, Zhou J, Cen K (2013) Comparison in dark hydrogen fermentation followed by photo hydrogen fermentation and methanogenesis between protein and carbohydrate compositions in *Nannochloropsis oceanica* biomass. Bioresour Technol 138:204–213. https://doi.org/10.1016/j.biortech.2013.03.171

Yaashikaa PR, Keerthana Devi M, Senthil Kumar P, Pandian E (2022) A review on biodiesel production by algal biomass: outlook on lifecycle assessment and techno-economic analysis. Fuel 324:124774. https://doi.org/10.1016/j.fuel.2022.124774

Yang D-W, Syn J-W, Hsieh C-H, Huang C-C, Chien L-F (2019) Genetically engineered hydrogenases promote biophotocatalysis-mediated H2 production in the green alga *chlorella* sp. DT. Int J Hydrog Energy 44(5):2533–2545. https://doi.org/10.1016/j.ijhydene.2018.11.088

Yin Y, Wang J (2019) Hydrogen production and energy recovery from macroalgae *Saccharina japonica* by different pretreatment methods. Renew Energy 141:1–8. https://doi.org/10.1016/j.renene.2019.03.139

Yuan M, Barron AR, Selin NE, Picciano PD, Metz LE, Reilly JM et al (2022) Meeting US greenhouse gas emissions goals with the international air pollution provision of the clean air act. Environ Res Lett 17(5):054019. https://doi.org/10.1088/1748-9326/ac6227

Zeng X, Guo X, Su G, Danquah MK, Chen XD, Lin L et al (2016) Harvesting of microalgal biomass. In: Bux F, Chisti Y (eds) Algae biotechnology: products and processes. Springer International Publishing, Cham, pp 77–89

Zhou Y, Liu L, Li M, Hu C (2022) Algal biomass valorisation to high-value chemicals and bioproducts: recent advances, opportunities and challenges. Bioresour Technol 344:126371. https://doi.org/10.1016/j.biortech.2021.126371

Zollmann M, Robin A, Prabhu M, Polikovsky M, Gillis A, Greiserman S et al (2019) Green technology in green macroalgal biorefineries. Phycologia 58(5):516–534. https://doi.org/10.1080/00318884.2019.1640516

Chapter 14
Biohythane Production

Adriane Bianchi Pedroni Medeiros, Marcela Moreira Albuquerque, Sabrina Vieira, Walter José Martinez Burgos, Júlio Cesar de Carvalho, and Carlos Ricardo Soccol

Abstract Biofuels such as biodiesel and bioethanol have been considered good alternatives to fossil fuels in transportation. The main advantage of biofuels is their environment-friendly status, as they are obtained from renewable sources and are less polluting fuels. Among the types of biofuels, gaseous biofuels have also been attracting attention. Biohythane is a blend of biomethane and biohydrogen with the potential to substitute natural gas and hythane from fossil sources. Biohythane is produced via the anaerobic digestion of wastewaters and various organic solid wastes. Thus, being a biotechnological process that involves several microorganisms and substrates, the operating conditions of the process and reactor configuration must be adequate, aiming for the best yields of biogas production. In general, the costs related to anaerobic digestion processes are attractive and arouse research and the development of technologies to produce biohythane.

Keywords Biomethane · Biohydrogen · Biofuel · Agro-industrial residues · Wastewaters

14.1 Introduction

The growing concern about the consequences of the use of fossil fuels, such as the greenhouse effect, global warming, and the scarcity of these fuels, directs the search for alternative energy sources. In the context of sustainability, biofuels are considered viable alternative fuels to fossil fuels due to the lower pollutant emissions and their production from renewable organic matter, especially agro-industrial residues rich in carbohydrates. Gaseous fuels are less common vehicular fuels compared to

A. B. P. Medeiros (✉) · M. M. Albuquerque · S. Vieira · W. J. M. Burgos · J. C. de Carvalho · C. R. Soccol
Federal University of Paraná, Department of Bioprocess Engineering and Biotechnology, Centro Politécnico, Curitiba, Paraná, Brazil
e-mail: adrianebpm@ufpr.br

C. R. Soccol et al. (eds.), *Biohydrogen - Advances and Processes*, Biofuel and Biorefinery Technologies 13, https://doi.org/10.1007/978-3-031-49818-3_14

more energy-dense liquid fuels. To overcome problems associated with storage, which limits the performance in terms of vehicle range (Bauer and Forest 2001), a gaseous mixture of methane and hydrogen has been an considered an interesting alternative (Ta et al. 2020).

The term "hythane" designates the gaseous blend composed of methane and about 10–25% (v/v) of hydrogen (Pasupuleti and Venkata Mohan 2015; Sun et al. 2019). A higher percentage of hydrogen in the gaseous mixture results in a better quality fuel with environmental and energy benefits. The burning of hydrogen results in only water and has a high calorific value. Besides reducing CO_2 and NO_x emissions in the atmosphere, studies showed that the presence of hydrogen in the biogas also enhances its combustion rate and increases the efficiency of methane-powered vehicles (Feng et al. 2022). Currently, hythane is mainly obtained from fossil sources. When produced via biotechnological processes, hythane is called *biohythane*. Biohythane is gaining more attention due to its superior characteristics compared to other biofuels. According to (Abdur Rawoof et al. 2021) the production of a gaseous mixture (biohydrogen and biomethane) can improve the energy efficiency of the fuel and reduce the emissions of pollutants such as CO_2 and NO_x that are released into the atmosphere. Furthermore, biohythane does not require special storage conditions or significant modifications to the ignition systems commonly used in the combustion of compressed natural gas (Shanmugam et al. 2021b).

Biomethane and biohydrogen can be produced from anaerobic digestion (AD) of wastewaters and solid waste and has been the subject of many studies. The most common process for obtaining biohythane is via two-stage anaerobic digestion. The anaerobic digestion of renewable organic materials takes place in two distinct stages, the dark fermentation and the methanogenic phase, which result in the production of hydrogen (10–15%) and methane (50–55%), as well as CO_2 (30–40%) (Sasidhar et al. 2022). The composition of biohythane can vary depending on the final application. If used as an automotive fuel, the CO_2 content may be higher than for use in electricity generation. However, the complexity of operating two-stage reactors and the investment cost still limits this type of process. Biohythane production can also take place in a single-stage anaerobic digester (Pasupuleti and Venkata Mohan 2015; Ta et al. 2020).

In this chapter, some biotechnological aspects of the production of biohythane as a potential fuel are addressed. Various organic wastes and microorganisms have been studied and proposed to produce biohythane mainly by two-stage anaerobic digestion. Table 14.1 presents data from various studies on biohythane production using different substrates, microorganisms, types of reactors, and information about process conditions, biogas yield, and hydraulic retention time. In addition, the advantages and economic aspects of biohythane as a promising alternative fuel were briefly discussed.

Table 14.1 Biohythane ($H_2 + CH_4$) production from different substrates

Substrate	Inoculum	Reactors (H_2 and CH_4)	Temperature and pH	HRT[a] H_2	HRTa H_4	H_2 Yield	CH_4 Yield	Predominant microorganisms at H_2 production	Predominant microorganisms at CH_4 production	Reference
Wheat straw hydrolysate	Methanogenic granules from a potato-processing wastewater treatment plant and digested manure from a thermophilic pilot scale plant	UASB	70 and 55 °C, pH was not controlled during the process	1 day	5.2 days	89 mL-H_2/g VS	307 mL CH_4/g VS	*Thermoanaerobacter wiegelii*, *Caldanaerobacter subterraneus*, *Caloramator fervidus*	*Methanosarcina mazei*, *Methanothermobacter defluvii*, *Thermoanaerobacterium* sp., *C. Roseum, and C. Isatidis*	Kongjan et al. (2011)
Vinasse	Thermophilic anaerobic digested manure from biogas plant	CSTR	55 °C, pH 6.0	3 days	12 days	69 mL H_2/g VS	348 mL CH_4/g VS	*Thermoanaerobacterium thermosaccharolyticum*	*Clostridium thermocellum*, *Methanosarcina acetivorans*	Luo et al. (2011)
Desugared molasse	Effluent obtained from the thermophilic biogas plant	UASB	H_2 production: 55 °C, pH 5.2–5.8 CH_4 production:7.1–7.8	16 h	2.5 days	263 mL H_2/g VS	379 mL CH_4/g VS	*Thermoanaerobacterium thermosaccharolyticum*, *Thermoanaerobacterium aciditolerans*	*Methanosarcina mazei*	Kongjan et al. (2013)
Palm oil mill effluent (POME)	Sludge from a palm oil mill biogas plant	ASBR and UASB	55 °C and 34 °C, pH 7.3–7.5	2 days	15 days	210 mL H_2 /g COD	315 mL CH_4/g COD	*Thermoanaerobacterium thermosaccharolyticum*	*Metanosarcina* sp., *Methanoculleus* sp	Mamimin et al. (2015)

(continued)

Table 14.1 (continued)

Substrate	Inoculum	Reactors (H_2 and CH_4)	Temperature and pH	HRT[a] H_2	HRTa H_4	H_2 Yield	CH_4 Yield	Predominant microorganisms at H_2 production	Predominant microorganisms at CH_4 production	Reference
Palm oil mill effluent (POME)	Anaerobic sludge from the bottom of the anaerobic pond	UASB and CSTR	55 °C, 37 °C, pH 5.5 and 7.5	2 days	5 days	205 L H_2 /kg COD	310 L CH_4 /kg COD	*Thermoanaerobacterium*	*Methanobrevibacter* sp.	Krishnan et al. (2016)
Sugarcane top	Acidogenic and methanogenic mixed consortia	Double jacketed reactor with a working volume of 500 mL	37 °C, pH 6.5 and 7.5	3.38 h	2.9 days	87.40 mL/g VS	180.86 mL/g VS	[b]	[b]	Kumari and Das (2016)
Palm oil mill effluent (POME)	Sludge for hydrogen production collected from a palm oil mill biogas plant	CSTR and UASB	55 °C, pH 5.5 35 °C, pH 7.5	2 days	15 days	135 mL H_2/g VS	414 mL/g VS	*Thermoanaerobacterium* sp	*Methanosarcina* sp. and *Methanoculleus* sp.	O-Thong et al. (2016)
Chlorella sp. biomass	Anaerobic sludge from palm oil mill wastewater treatment plant	Solid-state anaerobic digestion, 1 L reactor	55 °C, pH 5.5 and 7.0	5 days	30 days	190 mL H_2/g VS	230.1 mL CH_4/g VS	*Clostridium* sp	*Clostridium* sp. e *Ruminococcus* sp., *Methanoculleus* sp. e *Methanothermobacter* sp.	Jehlee et al. (2019)
Algal biomass	Anaerobic sludge	100 mL serum bottle	37 °C, pH 7.0	24 h	12 days	116.1 mmol/L	38.88 mmol/L	*Clostridium* spp	Methanogenic consortium	Mishra et al. (2017)

Substrate	Inoculum	Reactors (H_2 and CH_4)	Temperature and pH	HRT[a] H_2	HRTa H_4	H_2 Yield	CH_4 Yield	Predominant microorganisms at H_2 production	Predominant microorganisms at CH_4 production	Reference
Palm oil factory waste (oil palm frond OPF)	Anaerobic sludge collecting from the biogas plant	1-L Pyrex glass bottles	55 °C, pH 3.9	14 days	50 days	88.5 mL H_2/gVS	305.4 mL CH_4/gVS	*Clostridium* sp.	*Methanosphaera* sp.	Mamimin et al. (2019a)
Palm oil mill effluent	Hydrogen producing inoculum	500 mL serum bottles	55 °C, pH was not controlled during the process	4 days	45 days	139 mL H_2/gVS	454 mL CH_4/gVS	*Clostridium, bacillus, Thermoanaerobacterium* sp.	*Methanosarcina, Methanoculleus, Methanomassiliicoccus*	Mamimin et al. (2019b)
Palm oil mill effluent	Sludge from a biohydrogen production reactor	ASBR and UASB	55 °C, pH 6.6 and 7.5	2 days	10 days	73 mL/g-COD)	342 mL/g-COD)	*Thermoanaerobacterium* sp.	*Methanoculleus* sp., *Methanosarcina* sp., and *Methanbrevibacter* sp.	Seengenyoung et al. (2019)
Pineapple peel waste juice	From hydrogen production digester, the biogas digesters of the pig farm	$CSABR_H$ and $CSABR_C$	37 °C, pH 5.5 and 7.0	4 h	3 days	599.0 ± 77.8 mL/LR-d	174.6 ± 9.4 mL/LR-d	[b]	[b]	Chu et al. (2020)
Cornstalk	Seed sludge from an anaerobic digester	CSTR and UASB	37 ± 1 °C, pH 5.53 ± 0.20;	48 h	24 h	25.02 ± 0.90 L/kg TS	95.38 ± 3.48 L/kg TS	[b]	[b]	Li et al. (2020)
Tofu processing residue	Anaerobic digested sludge	500 mL glass bottles as digesters	37 °C, initial pH was adjusted to 6 ± 0.2 the second pH was adjusted to 7 ± 0.1	36 h	13 days	30.7 mL/g-VS	59.8 mL/g-VS	*Mobilitalea* and *Clostridium* sensu stricto	*Methanoculleus thermophilus*	Ali et al. (2022)

(continued)

Table 14.1 (continued)

Substrate	Inoculum	Reactors (H_2 and CH_4)	Temperature and pH	HRT[a] H_2	HRTa H_4	H_2 Yield	CH_4 Yield	Predominant microorganisms at H_2 production	Predominant microorganisms at CH_4 production	Reference
Rice straw and pig manure	Digestate from a mesophilic anaerobic digester	CSTR	55 ± 1,0 °C (stage 1) and 37 ± 1,0 °C (stage 2) pH 5.67–6.05 (stage 1), 5.87–6.11 (stage 2)	3 days	27 days	73.09 ± 3.03 mL/g VS	235.81 ± 9.30 mL/g VS	*Clostridium*	*Methanobacterium* e *Methanosaeta*	Liu et al. (2022)

[a]HRT: Hydraulic retention time

[b] Not specified in the study

14.2 Biohythane Production Process

Biohythane can be obtained from several substrates, including those composed of cellulose, such as lignocellulosic agricultural residues and algae biomass. Considering the complex structure of three-dimensional polysaccharides, which gives them crystallinity and low solubility, this kind of substrates need pretreatment to facilitate the accessibility of microorganisms to fermentable sugars (Ahmad et al. 2018). In addition, the presence of lignin in the fermentation process interferes with biohythane yields, as it generates compounds that are toxic and can inhibit microbial action (Shanmugam et al. 2021b). A pretreatment aims to reduce crystalline, partially solubilize lignin bonds, or break the carbohydrate-lignin bond (Kumar et al. 2022).

Pretreatment methods are classified as physical, chemical, and biological. Different pretreatment strategies based on these categories are proposed in the literature, all to break down the recalcitrant heteropolymeric structure (cellulose, hemicellulose, and lignin) into molecules accessible for enzymatic hydrolysis and other biorefinery processes. Acid, alkaline, explosion-current, microwave, and ionizing radiation pretreatments require specific and resistant equipment, have high energy consumption, and generate inhibitory compounds that affect hydrolysis and fermentation. Chemical pretreatment uses concentrated acids along with high temperatures and pressures and ends up generating large amounts of fermentation-inhibiting molecules (such as furfural and hydroxymethylfurfural). Alkaline treatments are milder and result in lower inhibitor content, but achieve lower yields (Vieira et al. 2020). Biological pretreatment is a promising technique for the production of biofuels from lignocellulosic biomass: it uses lignocellulolytic enzymes or microorganisms, does not require chemicals, and does not release toxic compounds to the environment (Sindhu et al. 2016). In a study using biological pretreatment in the sugarcane top, there was maximum lignin removal of 60.4% w/w after 21 days of incubation at 28 °C in static conditions. The total recovery of gaseous energy was 37.7%, 54% higher than that of the untreated energy (Kumari and Das 2016).

On a commercial scale, hydrothermal pretreatment is reported as efficient (Kongjan et al. 2011). Water in the liquid phase or vapor phase under high pressure penetrates the biomass, hydrates cellulose, and removes most hemicelluloses and a small part of lignin. It is a relatively gentle pretreatment method that requires no catalyst and does not cause significant corrosion problems (Jönsson and Martín 2016). The pretreatment methodology varies according to the characteristics of the substrate.

To obtain biohythane, a combination of hydrogen and methane is required, with a final hydrogen composition of approximately 10–25% (Sun et al. 2019). The bioprocess for obtaining biohythane is conducted by combining dark fermentation and anaerobic digestion in bioreactors with various configurations, in one phase or two. In single-stage biohythane production systems, co-production of H_2 and CH_4 occurs simultaneously by the simultaneous cultivation of fermentative and methanogenic microbial consortia in a single reactor configuration with controlled adequate operating conditions (Sasidhar et al. 2022). However, in this single-stage configuration, the production of H_2 becomes the limiting step, because its rate influences the composition of the gas and the yield of the biohythane (Abdur Rawoof et al. 2021).

In most cases, fermentation is conducted in two stages, where H_2 is produced in the first stage, and the resulting mixture of volatile fatty acids undergoes methanogenic digestion in the second stage, generating CH_4 (Shanmugam et al. 2021a). In this process, two separate reactors are connected in series, separated from the acidogenesis and methanogenesis phases, operating in different configurations of temperature, acidity, and hydraulic retention time, optimizing the growth of microbial communities of interest in each reactor. Acidogenic bacteria convert organic substrates mainly into hydrogen, organic acids, solvents, and CO_2 in the first stage under mildly acidic conditions (between 5.0 and 6.0) and shorter hydraulic retention times from 1 to 3 days. Methane-producing microorganisms in the second phase have a lower growth rate, which requires a longer hydraulic retention time (10–20 days) and neutral acidity (between 7.0 and 8.0) for the conversion of organic acids and solvents contained in the effluent of the first reactor. In a study of sequential hydrogen and methane generation from POME, it was found that there was a 34% increase in energy yield, compared with single-stage methane production, and 90% in single-stage hydrogen production (Mamimin et al. 2015). The two-stage process is based on the differences between acidogenic and methanogenic microorganisms, such as physiology, nutritional needs, growth kinetics, and sensitivity to environmental conditions. However, system operations are not mutually exclusive, there is a degree of dependency. Hydrolysis and acidogenesis occur in the first stage of the anaerobic digestion process. During hydrolysis, fermentative bacteria produce enzymes that break down complex carbohydrates into glucose monomers, amino acid proteins, and lipids into long-chain volatile fatty acids or/and glycerol. The first stage of acidogenesis involves the conversion of these organic compounds into H_2, CO_2, volatile fatty acids, alcohols, and lactate (Lay et al. 2020). H_2 yields are generally lower (Table 14.1) because, in the conventional fermentation process, only about 7.5 to 15% of the energy contained in organic waste is converted into hydrogen, the rest of the energy is in volatile fatty acids. About 65% of the energy contained in organic residues remains in the liquid, mainly in butyric acid, acetic acid, lactic acid, and propionic acid (Mamimin et al. 2015). These acids can still be converted into other products by methanogenesis in anaerobic digestion, so the second stage consists of the conversion of non-gaseous products into CH_4.

Bioreactors can be optimized by adjusting the necessary microbial fermentation conditions, thus ensuring the required H_2/CH_4 ratio within the preferred range. The most widely used bioreactors for biohythane production are type continuous stirred-tank reactor (CSTR) and upflow anaerobic sludge blanket (UASB). CSTR bioreactors operate in a steady state with a continuous flow of reagents and products. They have optimal temperature control, continuous operation, and high methane recovery. However, hydrogen and methane yield are mainly affected by environmental factors such as hydraulic retention time, pH, and temperature (Krishnan et al. 2016; Ghimire et al. 2022).

The UASB reactors are specifically designed for anaerobic digestion and can handle high organic load rate, and short hydraulic retention time. This type of reactor is indicated for liquid substrates with total solids below 2% in hydrogen and methane production (Abdur Rawoof et al. 2021).

It is also reported the use of anaerobic fluidized bed bioreactors (AFBRs), as they allow the retention of methanogenic arches (which grow more slowly) in a fixed bed with inert material. The AFBR has characteristics of supporting high load capacity, concentrated biomass, and resistance to hydraulic or organic shocks, enabling greater treatment efficiency and without the need for mechanical mixing (Yeshanew et al. 2016).

14.3 Substrates Used for the Production of Biohythane

The main criteria used in choosing the substrates used in the production of biofuels are their availability, cost, carbohydrate content, and biodegradability (Shanmugam et al. 2021a). Thus, due to its wide availability and low cost, the use of biomass and organic waste from agricultural processing, industrial processing waste, organic fractions of urban solid waste, livestock effluents and algae biomass have been commonly used in the production of biohythane in two stages, combining economic viability with effluent treatment, mitigation of greenhouse gases and renewable energy generation (Kongjan et al. 2013; Si et al. 2016; Shanmugam et al. 2021a).

14.3.1 Agricultural Processing Waste

The wide production of lignocellulosic residues annually around the world exceeds 200 billion tons, so this cheap and uninterrupted availability, together with characteristics such as the large presence of fermentable sugars and the non-competition with food resources has increased the demand for these residues as substrates to produce biofuels such as biohythane (Ren et al. 2009). In addition, lignocellulosic biomass has a large volume of carbon and nitrogen, ensuring an adequate ratio for the production of hydrogen and methane (Ramos and Silva 2018). Corn, sugar cane, rice, cassava, sorghum, sugar beet, wheat, peanut husks, and hazelnut husks fall into this category of agricultural base residues (Meena et al. 2020). Many types of agricultural residues have been used in the production of biohythane, such as wheat residues (Kongjan et al. 2011), sugar cane (Kumari and Das 2016), and corn (Li et al. 2020) and others.

In their work, Kongjan et al. (2011) demonstrated the two-step production of biohythane under thermophilic conditions (70 and 55 °C) using wheat straw hydrolysate generated by hydrothermal pretreatment of wheat straw containing mainly hemicellulose. The authors achieved specific hydrogen yields of 89 mL-H_2/g-VS (volatile solids) (190 mL H_2 /g-sugars) and methane 307 mL-CH_4/g-VS. The sugarcane top pre-treated for maximum delignification was used by Kumari and Das (2016) for the production of biohythane, demonstrating in their study that the use of this step in the substrate can considerably increase the production of hydrogen and methane. The study was carried out in bioreactors using pre-treated sugarcane tops, with a temperature of 37 °C and pH 6.5 and 7.5, culminating in the production of 77.2 L H_2/kg-VS and 157 L CH_4/kg-VS when the biomass used went through pre-treatment and delignification. On the other hand, without pre-treatment, 55 L H_2 /kg-VS and 84 L CH_4/kg-VS were obtained. The use of corn stalk for biohythane production is also presented as a viable alternative to the use of these residues. Li et al. (2020) conducted their studies at a temperature of 37 °C and pH of 5.53, obtaining a yield of H_2 of 25.02 L/kg-TS (total solids) and CH_4 of 95.38 L/kg-TS. The higher yields obtained by (Kongjan et al. 2011) can be explained by the use of higher temperatures in the production of biohythane, since higher temperatures are thermodynamically more favorable for this type of fermentation (Li et al. 2020).

14.3.2 Industrial Processing Waste

Among the organic residues from industrial processing that have been used in the production of biohythane, we can mention the effluents from the ethanol industry (Luo et al. 2011) dairy (Cota-Navarro et al. 2011), alcoholic beverage processing (Buitrón et al. 2014), palm oil (Mamimin et al. 2015, 2019b; Krishnan et al. 2016; O-Thong et al. 2016; Seengenyoung et al. 2019), effluents from tofu processing (Ali et al. 2022) and others.

Vinasse is an industrial effluent resulting from the distillation of sugar-rich biomass, which presents itself as a serious environmental problem due to its incorrect use as a fertilizer, leading to intense pollution of soils and water bodies (Christofoletti et al. 2013).

Due to this great availability and polluting capacity, alternatives have been sought for the treatment of this effluent, being used by Luo et al. (2011) to produce biohythane in two stages at a temperature of 55 °C and pH 6.0, leading to a hydrogen yield of 69 ml-H_2 /gVS and a methane yield of 348 ml-CH_4/ gVS.

The production of biohythane from tequila vinasse with high concentrations of organic matter was demonstrated by Buitrón et al. (2014), the study was conducted at a temperature and pH of 35 °C and 5.5 and 6.8 respectively, showing a yield of 190 mL H_2 /g-COD of hydrogen and 257.9 mL CH_4/gCOD of methane. Whey, also an abundant residue from the dairy industry, consists mainly of mineral salts, phosphorus, nitrogen, proteins and lactose as the main sugar, reaching up to about 70% of its constitution(Castelló et al. 2009). The high concentration of lactose sugar in this effluent makes it suitable for biohythane production, which was tested by Cota-Navarro et al. (2011) under controlled temperature and pH of 25–30° and 7.5–7.8 respectively, leading to the production of 28 L^{-1} d^{-1} hydrogen and 5 L^{-1} d^{-1} methane. Palm oil factories (POME), which are abundant mainly on the Asian continent, dispose of a high amount of waste, being used by several works for the production of biohythane. This residue is extremely problematic environmentally due to a large amount of total solids, reaching (35–42 g/L) suspended solids (8.5–12 g/L), total nitrogen (0.83–0.92 g/L) and total phosphorus (0.097–0.125 g/L) (O-Thong et al. 2008). Palm oil showed high yields of hydrogen and methane under similar production parameters in two stages. O-Thong et al. (2008) investigated the biohythane production from palm oil effluent in two stages, at temperatures of 35 °C and 55 °C and pH of 5.5 and 7.5 respectively, obtaining a hydrogen yield of 135 mL of H_2/gVS and methane of 414 mL of CH_4/gVS. Mamimin et al. (2019a, b) also investigated the production of biohythane in two stages using palm oil effluent, at a temperature of 55 °C and uncontrolled pH during the process. However, supplementing the medium with trace metals to increase the thermophilic hydrogen fermentation and methane production, it was obtained a hydrogen yield of 139 mL H_2/gVS, and methane of 454 mL/gVS, with yields increasing by 90.4% and 21% respectively (Mamimin et al. 2019b).

Approximately 14 million tons of tofu processing waste (okara) are generated annually. Thus, this residue has received greater attention regarding its use in the biofuel industry (Ali et al. 2022) used tofu processing residues for the production of biohythane in two steps. The conditions used in the process were 37 °C with pH 6.0 and 7.0, culminating in a hydrogen yield of 30.7 mL/g-VS and methane of 29.9 mL/g-VS. The low productivity of hydrogen and methane in this study may be the result of many factors, including the non-use of pre-treatment of the residue, which has low water solubility. Ali et al. (2022) performed the ultrasonic and acid pretreatment of this residue, obtaining a hydrogen yield of 42.5 ml/g-VS and methane of 80.4 ml/g-VS.

14.3.3 Municipal Solid Waste

The use of food waste, an organic fraction of urban solid waste, can be used for biofuels production due to its uninterrupted availability, a large amount of available energy and for being a renewable energy source uniting the production of a valuable compound to the waste treatment (Shanmugam et al. 2021a). A waste mixture consisting of 79% vegetables and fruits, 5% pasta and rice, 6.0% bread, 8.0% meat and fish and 2.0% dairy products were used in the two-step production of biohythane with a temperature of 55 °C in the first step and 37 °C in the second step, not carrying out the pH adjustment, resulting in a production of hydrogen of 391.7 L H_2 /m^3/d and methane of 2021.6 L CH_4 /m^3/d (Yeshanew et al. 2016). Chu et al. (2020) used pineapple peel for the production of biohythane in two stages, with a temperature of 55 °C in the two reactors and pH of 5.5 and 7.0 respectively, culminating in a hydrogen production rate of 599.0 mL/LR/ d and methane of 174.6 mL/L R/ d.

14.3.4 Livestock Effluents

Effluents from livestock are an intense environmental and human health concern due to the presence of strong odor, pathogens and the generation of greenhouse gases (GHG), in addition to having an acidic pH and a high volume of proteins, lipids, chemical oxygen demand (COD), total solids (TS), volatile solids (VS) and microorganisms, which makes this residue highly polluting, requiring alternative ways to treat and dispose of it (Bernet and Béline 2009). However, the high content of nitrogen, ammonia and sulfate in the manure causes the inhibition of hydrogen production (Elbeshbishy et al. 2017), which can be circumvented by mixing the manure with lignocellulosic materials, culminating in a carbon/hydrogen ratio ideal and increasing buffering, thus allowing the use of manure compounds for the biohythane production (Keskin et al. 2019).

In this context, Nguyen et al. (2022) investigated the biohythane production by mixing swine manure and pineapple residues in a single fermentation stage in a two-chamber digester, with a temperature of 37 °C and a pH of 5.5, obtaining a hydrogen production of 1240 mL/L.d and methane of 812 mL/L.d. Also noteworthy was the removal of COD from the residue, carbohydrates and ammonia, reaching percentages of 50–75.2, 80.3–98.5 and 77.5–80.3% respectively. The production of biohythane from a mixture of swine manure and rice straw at temperatures of 55 °C (hydrogen production) to 37 °C (methane production) and pH of 5.6 and 5.8 respectively, culminated in the production of maximum hydrogen of 73.09 mL/gVS and methane of 235.81 mL/g SV (Liu et al. 2022).

14.3.5 Algae Biomass

Algae biomass is a promising source for biofuels production due to a series of advantages presented, such as a large amount of nutrients such as proteins and minerals, high productivity, contributes to the reduction of greenhouse gases through carbon fixation, does not compete for arable land and has a low amount of lignin in its cell wall, dispensing with delignification steps (Montingelli et al. 2015). Furthermore, algal biomass can be cultivated in wastewater, contributing to the treatment of this effluent and generating bioresources for the production of biofuels and other valuable compounds (Chiu et al. 2015).

Biohythane production from algal biomass was investigated by Mishra et al. (2017) using freshwater algae biomass in two steps under fermentation conditions of 37 °C and pH 7.0, culminating in the production of hydrogen of 116.1 mmol/L and methane of 38 mmol/L. In another study, biohythane was produced from the biomass of *Chlorella* sp. under conditions of 55 °C and pH of 5.5 and 7.0, obtaining the hydrogen production of 190.0 mL H_2/ gVS and methane of 319.8 mL CH_4/ gVS (Jehlee et al. 2017) Algal biomass from microalgae blooms was investigated by

Chen et al. (2020) under fermentation conditions of 35 °C and pH of 6.5, obtaining yields in the production of hydrogen of 39.4 mL/g VS and methane of 276.8 mL/g VS.

14.4 Biohythane-Producing Microorganisms

Most studies on biohythane production use consortia of bacteria from inoculum from various sources (Table 14.1), and comparative studies between the use of pure or mixed cultures are not yet well documented in the literature. The two-step production of biohythane is most commonly described in the literature (Table 14.1), being conducted in two bioreactors, where the first step (acetogenic) corresponds to hydrogen production and the second (methanogenic) to methane production. This production method is justified due to the differences between the predominant microorganisms in each stage, being a result of its physiology, nutritional requirements and resistance changes in the culture medium during the fermentation. Thus, under anaerobic conditions, hydrogen can be produced in the first step by a wide range of microorganisms, which convert the organic compounds from the cultivation substrate into organic acids releasing hydrogen, and subsequently, these organic acids are used by other microorganisms in the second stage for the production of methane.

Strict or facultative anaerobic bacteria belonging to the thermophilic and mesophilic domains are reported to be efficient hydrogen producers (Shanmugam et al. 2021a, b). In the studies reported here (Table 14.1), *Clostridium* sp. and *Thermonaerobacterium sp* were the microorganisms most described as predominant in the biohydrogen production step using mixed cultures (Kongjan et al., 2013; Krishnan et al., 2016; Luo et al., 2011; Mamimin et al., 2015; Krishnan et al. 2016; O-Thong et al. 2016; Mamimin et al. 2019a, b; Seengenyoung et al. 2019).

The different species belonging to the genus *Clostridium* sp. are obligate anaerobic spore-forming bacteria capable of fermenting various carbohydrates as well as cellulosic and lignocellulosic biomass into acetate, butyrate and hydrogen at thermophilic temperatures, obtaining a theoretical hydrogen yield of 2,1–2,2 mol H_2/mol glucose (Kongjan et al. 2011). These microorganisms are carriers of the enzyme Fe-Fe hydrogenase, essential for the production of hydrogen, which is produced during its exponential growth phase, shifting its metabolism to solvent production when the population reaches the stationary phase.

Several species of *Thermoanaerobacterium* spp. have been reported in the literature as being predominant in the hydrogen production process, such as *Thermoanaerobacterium thermosaccharolyticum* (Luo et al. 2011), *Thermoanaerobacterium aciditolerans* (Kongjan et al. 2013), *Thermoanaerobacterium polysaccharolyticum*, *Thermoanaerobacterium zeae*, *Thermoanaerobacterium lactoethylicum* and *Thermoanaerobacterium aotearoense* (O-Thong et al. 2008).

The genus *Thermoanaerobacterium* spp., is extremely diverse, being composed of extreme thermophilic and strictly anaerobic bacteria with optimum growth temperature and pH around 55–70 °C and 5.2–7.8 respectively (O-Thong et al. 2008). They can use a wide range of substrates for hydrogen production such as sucrose, glucose, cellulose, cellobiose, starch and xylan (Krishnan et al. 2016), achieving a hydrogen production of 2.53 mol H_2 / mol glucose.

Although less frequent, other hydrogen-producing microorganisms were also reported to predominate during the first stage of fermentation such as *Mobilitalea sp* (Ali et al. 2022), *Thermoanaerobacter wiegelii, Caldanaerobacter subterraneous Caloramator fervidus* (Kongjan et al. 2011) *e Bacillus* sp.(Mamimin et al. 2019a, b).

Methane is produced in the second fermentation step by several classes of Archaea that produce CH_4 through anaerobic fermentation, mainly of organic acids, bicarbonate, alcohols, volatile fatty acids and CO_2 (Shanmugam et al. 2021a, b). The Archaea *Methanosarcina* sp. and *Methanoculleus* sp. were identified in several studies as predominant in the production of methane in the second stage of fermentation (Luo et al. 2011; Mamimin et al. 2015; Mamimin et al. 2019a, b; O-Thong et al. 2016; Jehlee et al. 2017; (Seengenyoung et al. 2019). The constituent microorganisms of the genus *Methanosarcina* sp., have a coccoid format, and use substrates such as acetate, and compounds containing methyl and H_2 for the production of methane, showing an optimal growth in the temperature range between 50 and 60 °C (Kongjan et al. 2011). *Methanoculleus sp* also has a coccoid cell shape, presenting both mesophilic and thermophilic organisms that are strictly anaerobic. They use substrates such as formate, H_2, CO_2 and alcohol to produce methane, with optimum growth at temperatures in the range of 20–45 °C and a pH that can vary between 6.2 and 8.0. Other methanogenic microorganisms reported by this study were *Methanothermobacter* sp.(Kongjan et al. 2011; Jehlee et al. 2017) and *Clostridium* sp. (Kongjan et al. 2011).

14.5 Process Parameters Affecting Biohythane Production

For the biohythane production process to occur efficiently and with high yields, it is necessary to control some parameters. Among the factors that affect the biohythane production process are: pH, temperature, substrate characteristics, and hydraulic retention time (HRT) (Chen et al. 2022).

The pH is a factor that directly influences microbial reactions (substrate degradation, synthesis of metabolism products and energy use, regulates ion concentration, and growth rate of microorganisms). Biological processes may be negatively affected by a sudden change in pH value (pH shocks) (Ta et al. 2022). In a two-stage process, acidification control in the first reactor can increase process stability, as it avoids the overload and inhibition of methanogenesis in the second reactor (Kongjan et al. 2011). An adequate pH range should be maintained to ensure stability in bioreactor performance and activity of fermentative microorganisms. The hydrogen production rate can be maximized at a pH of 5–6.5, because outside this range it affects hydrogenase activity; for methane, the range of 6.5–8.5 (Yang et al. 2020) is reported as ideal. However, the intermediate acids generated in the hydrogen system and inefficient degradation of organic matter result in a lower pH, which eventually dissociates more acid ions, consequently reducing the pH, until inhibiting methanogenesis. At larger scales, it is indicated that this control is conducted based on hydraulic retention time, higher or lower depending on the fermentation stage. Low HRT can be used to keep the acidic environment in the hydrogen stage, and high retention times to reduce organic load and increase pH to necessary conditions in the methanogenic phase.

Temperature is fundamental in the control of methanogenic archaea and the promotion of hydrogen-producing organisms. Hydrogen-producing organisms operate at room temperature, mesophilic, thermophilic and hyperthermophilic (greater than 80 °C), and methanogenic organisms are sensitive to temperatures above 55 °C. Methanogenic are more stable in mesophilic conditions than in thermophilic conditions. Thus, the reactor in the first stage can be operated under thermophilic and hyperthermophilic conditions to increase hydrogen production, as they perform the suppression of methanogenic and availability of hydrogen-producing organisms at high temperatures. The methanogenic phase can be operated at room temperature or mesophilic temperature (Sasidhar et al. 2022).

The retention time increases the contact time between the substrate and the microbial organisms that aid in the degradation, and controls the production of volatile fatty acids (VFA). Higher values provide enough time for acidogenic bacteria to hydrolyze and solubilize organic compounds, resulting in higher AGV production. A study conducted with the fermentation of food residues showed that AGV concentrations with the increase of hydraulic retention time from 96 to 192 h, but no additional increase was observed when HRT increased to 288 h. Biohydrogen production decreases with the increase of HRT, as it may favor the acidogenic to methanogenic pathway, resulting in lower biohydrogen production. The factors that influence HRT are the type of inoculum and substrate, pH and organic loading rate (Sasidhar et al. 2022). A two-stage process with a hydraulic retention time of 3 days for hydrogen reactor and 12 days for methane reactor, achieved 11% higher energy compared to a single-stage methanogenic process (15 days). The study also revealed that by adjusting the HRT distribution between hydrogen and methane reactors, more energy could be obtained (Luo et al. 2011). Studies indicate that maintaining the constant concentration of raw material and reducing HRT in the first stage of fermentation, prevents inhibition of the substrate, eliminates non-hydrogen-producing bacteria and stabilizes the growth of the hydrogen-producing microbial community (Chu et al. 2020).

14.6 Advantages of Biohythane Production

Hydrogen is considered a sustainable fuel of interest, because of its zero-carbon emission characteristic and high energy content (143 kJ/g). Methane (55 kJ/g) is widely used as a transport fuel in the form of liquefied natural gas, as it has a higher calorific value and lower CO_2 emissions compared to fossil fuels (Mishra et al.

2017). Methane also has the advantage of not producing nitrous oxide and sulfur dioxide (main contributors to air pollution) in its combustion. Biohythane, the mixture of hydrogen (10–25% by volume) and methane acts as the ideal transition fuel for the change to hydrogen, as it provides a significant reduction in nitrous oxide emission compared to natural gas, and is relatively inexpensive and easy to store (Mishra et al. 2017). Biohythane production is an energy-efficient strategy and can be used as an energy source in gas combustion engines (Mamimin et al. 2015).

By combining the advantages of hydrogen and methane, biohythane is considered an ecologically friendly gaseous fuel with high calorific value and is important in the transition from the use of fossil fuels to hydrogen. It is also advantageous for the possibility of integration into the treatment of organic waste (Abdur Rawoof et al. 2021). Thus, comparing the traditional individual production of hydrogen and methane, the production of biohythane from organic waste can recover more energy, cause less environmental damage, and brings more robust and economical processes (Liu et al. 2022). However, the energy yield of hydrogen production at larger scales needs to be higher than the energy provided to perform the process, for a positive energy balance.

14.7 Economic Aspects

14.7.1 Raw Material and/or Substrates

To a large extent, the economic efficiency of biohythane production will depend on the type of substrate and the different treatments that it needs to produce the gaseous mixture. Since biohythane are molecules of low commercial value, the use of pure and easily fermentable compounds such as glucose, fructose, sucrose, maltose, among others, is neither economical nor advantageous for its production (Martinez-Burgos et al. 2021a). Therefore, the strategy is to use low or zero cost raw materials which are sources of proteins, lipids and complex carbohydrates, the latter after physical, chemical or biological treatments can become fermentable sugars. In this sense, the use of agro-industrial and industrial waste (liquid and solid) is relevant, such as: cassava processing wastewater (CPW)(Martinez-Burgos et al. 2019), palm oil mill effluent (POME)(Rosa et al. 2020) food processing effluents in general (Bin et al. 2022), soft drink processing effluents (Do Nascimento Junior et al. 2021), sugarcane vinasse wastewater (Karp et al. 2022), effluent of milk processing, cheese whey, slaughterhouse effluents (Martinez-Burgos et al. 2021b), and different lignocellulosic biomass, among others (Singh et al. 2015).

(Martinez-Burgos et al. 2021a) in a study of biohydrogen production via dark fermentation using effluents (CPW and corn steep liquor-CSL) reported the lowest cost of the medium around US$75/m^3 compared to other mediums that were up to five times more expensive, as they were prepared using pure reagents. In the case of CPW and CSL, for example, the costs are approximately US$90/ton and US$160/

ton, respectively, including freight. In the case of sugarcane vinasse, the effluent obtained from alcohol distillation, the price is quite low, approximately US$0.5/m^3 not including freight. Other effluents such as domestic sewage, and slaughterhouse effluents have no commercial value, however the main inconvenience is the way they are transported.

On the other hand, lignocellulosic biomasses also have low costs when compared to pure reagents, however the tendency is to have a higher cost than effluents, for example sugarcane bagasse has an approximate cost of US$75/ton (Canaonline 2022), in the case of ground rice husk and soybean hulls, it costs approximately US$50/ton and US$260/ton, respectively (MFRural 2022a, b). Other lignocellulosic biomasses such as straw, forestry residues, food processing residues have costs ranging from US$30/ton to US$180/ton (BioBoost 2013). All these substrates have already been used in the production of biomethane and biohydrogen so they can also be used in the production of biohythane.

14.7.2 Waste Pre-treatment

The most promising substrates for the production of biohythane are residues of complex composition, so they must be pre-treated either physically, chemically or enzymatically so that they can be used by microorganisms. However, these treatments are expensive, some because of the high input energy requirements that are necessary or because of the use of highly sophisticated equipment.

In the case of lignocellulosic residues, grinding and sieving processes are required to reduce the size of the biomasses, these processes require large amounts of energy (Fig. 14.1a). According to (Tumuluru and Heikkila 2019) the amount of energy used for milling depends on the final particle diameter and moisture content of the biomass, so the amount of energy for milling a ton of biomass varies between 90 KWh to 205 KWh, when the biomass contains moisture content in the range 9% to 20%. In the case of sieving processes, energy consumption is affected by the particle diameter of the biomass and by the type of biomass. According to (Bhutto et al. 2017) energy expenditure can vary from 5 KWh/ton to 65 KWh/ton. In addition, additional pretreatments are needed because lignocellusic biomasses are recalcitrant due to the lignin content (Vasco-Correa and Shah 2019). Conventional treatments for this type of waste are mainly based on the use of high temperatures (120 °C -200 °C) and pressures up to 20 Mpa and some chemical reagents such as acids and bases which further increases the cost of treatment (Mosier et al. 2005). More recent technologies are also being used for the treatment of lignocellulosic biomasses, such as irradiation systems. According to (Bhutto et al. 2017) radiation doses may vary depending on the type of lignocellulosic biomass which will affect the fermentable sugar content. Thus, irradiation doses can vary from 40 kGy to 2MGy (Khan et al. 1986; Betiku et al. 2009). After the biomass delignification treatments, cellulose and hemicellulose saccharification processes are also necessary, which further adds to the costs of waste pre-treatment.

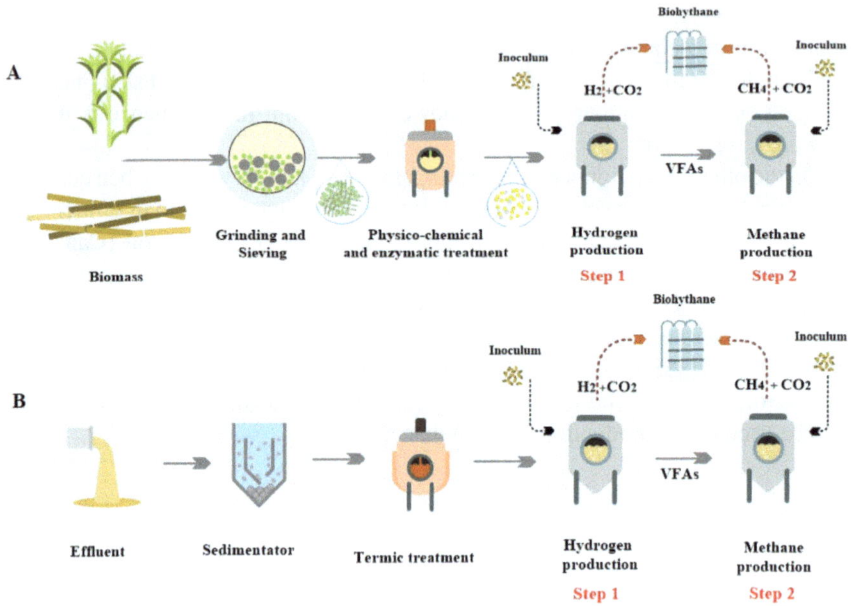

Fig. 14.1 General process for biohythane production using different wastes. (**a**) Lignocellulosic biomass. (**b**) Effluents

According to (Vasco-Correa and Shah 2019) the total cost of pretreatment of lignocellulosic biomass for the production of fermentable sugars is approximately $1600/ton. The fact of having to do so many complex pretreatments on lignocellulosic biomasses is a disadvantage for these residues with respect to effluents, since each pretreatment, in addition to needing a specific infrastructure, also needs energy input. In the case of using effluents as substrates for the production of biohythane, some pre-treatments are also needed, such as solids removal and a thermal pretreatment to eliminate the bacteria that consume the hydrogen that must be produced in the first stage (Fig. 14.1b). In other words, the cost of waste pre-treatment is key to the commercial success of their use in the production of biohythane.

14.7.3 Biohythane Production Plant

In general, biohythane can be produced in one or two stages, as in the first stage hydrogen is produced and in the second stage methane is produced, which is finally mixed with the hydrogen produced in the first stage (Bin et al. 2022). The hydrogen in the first step can be produced by microalgae or bacteria. An advantage of the process is that the sequential production of bio hydrogen and biomethane improves the energy recovery of waste in a range between 10% and 43% as the intermediate metabolites produced in the dark fermentation “volatile fatty acids” are reused in the second step in the generation of methane (Abdur Rawoof et al. 2021). In general, according to (Sgroi et al. 2015) the installation costs and operating costs of a biogas plant depend on its size. Thus, installation costs can range from €0,205—€0,095/Kwh and operating costs can range from €0,065 to €0,030/Kwh. The lowest costs are estimated for plants with capacities up to 999 kW and the highest for plants with capacities up to 100. In the case of biohytane, it is assumed that the costs are a little higher than the production costs of biomethane, since in the first stage (biohydrogen production) the effluents need to be thermally treated to eliminate hydrogen-consuming bacteria. Therefore, additional infrastructure and energy input will be required.

14.8 Conclusions and Perspectives

The production of biohythane could be an interesting alternative to the conventional production of hythane which is nowadays mainly derived from fossil sources. Although the anaerobic digestion of wastewaters and solid wastes are a well-known and widely used as treatment technology, the generation of biohythane as a source of automotive fuel or to produce electricity still needs technological development to be produced and find the productive sector. Given that this process involves multiple microorganisms and substrates, it is crucial to optimize the operating conditions and reactor configurations to achieve the highest biogas production yields.

The main challenge encountered in biohythane production is the low hydrogen productivity. In a conventional fermentation process, only about 7.5–15% of the energy contained in organic waste is converted into hydrogen, the rest of the energy is found in volatile fatty acids. Two-stage fermentation has been an advantageous alternative to single-stage hydrogen and methane production. Efforts to increase production during the two-stage anaerobic digestion (AD) process are mainly related to the definition of the best composition of raw materials, stability of the microbial consortium and configurations of the bioreactors. Among the most available and low cost, biomass and organic waste from agricultural processing, industrial processing waste, organic fractions of urban solid waste, livestock effluents and algae biomass have been commonly used in the production of biohythane in two stages. AD in two stages can make biohythane economically viable. Besides obtaining biogas with higher hydrogen contents, AD is a process of treating effluents and solid organic waste, generating renewable energy and mitigating greenhouse gases.

Acknowledgements This work was supported by the multilateral project BRICS STI FP 2021—BIOGEN (293), financed in Brazil by the National Council for Scientific and Technological Development, CNPq, grant no. 440138/2022-1, also with support of the Coordination of Improvement of Higher Education Personnel—CAPES, project Print-BioADD.

References

Abdur Rawoof SA, Kumar PS, Vo DVN et al (2021) Biohythane as a high potential fuel from anaerobic digestion of organic waste: a review. Renew Sust Energ Rev 152:111700. https://doi.org/10.1016/j.rser.2021.111700

Ahmad S, Pathak VV, Kothari R, Singh RP (2018) Prospects for pretreatment methods of lignocellulosic waste biomass for biogas enhancement: opportunities and challenges. Biofuels 9:575–594. https://doi.org/10.1080/17597269.2017.1378991

Ali MM, Mustafa AM, Zhang X et al (2022) Biohythane production from tofu processing residue via two-stage anaerobic digestion: operational conditions and microbial community dynamics. Biomass Convers Biorefinery 0–5. https://doi.org/10.1007/s13399-022-02736-2

Bauer CG, Forest TW (2001) Effect of hydrogen addition on the performance of methane-fueled vehicles. Part I: effect on S.I. Engine performance. Int J Hydrog Energy 26:55–70. https://doi.org/10.1016/S0360-3199(00)00067-7

Bernet N, Béline F (2009) Challenges and innovations on biological treatment of livestock effluents. Bioresour Technol 100:5431–5436. https://doi.org/10.1016/j.biortech.2009.02.003

Betiku E, Adetunji OA, Ojumu TV, Solomon BO (2009) A comparative study of the hydrolysis of gamma irradiated lignocelluloses. Brazilian J Chem Eng 26:251–255. https://doi.org/10.1590/S0104-66322009000200002

Bhutto AW, Qureshi K, Harijan K et al (2017) Insight into progress in pre-treatment of lignocellulosic biomass. Energy 122:724–745. https://doi.org/10.1016/j.energy.2017.01.005

Bin KS, Khalekuzzaman M, Hossain N et al (2022) Progress in biohythane production from microalgae-wastewater sludge co-digestion: an integrated biorefinery approach. Biotechnol Adv 57:107933. https://doi.org/10.1016/j.biotechadv.2022.107933

BioBoost (2013) Biomass based energy intermediates boosting biofuel production. Retrieved from: https://www.bioboost.eu/uploads/files/bioboost_d1.1-syncom_feedstock_cost-vers_1.0-final.pdf. July 20, 2022

Buitrón G, Kumar G, Martinez-Arce A, Moreno G (2014) Hydrogen and methane production via a two-stage processes (H2-SBR + CH4-UASB) using tequila vinasses. Int J Hydrog Energy 39:19249–19255. https://doi.org/10.1016/j.ijhydene.2014.04.139

Canaonline (2022) Tonelada de bagaco de cana chega a custar 400 reais em 2021. Retrieved from: http://www.canaonline.com.br/conteudo/tonelada-de-bagaco-de-cana-chega-a-custar-400-reais-em-2021.html. July 20, 2022

Castelló E, García y Santos C, Iglesias T et al (2009) Feasibility of biohydrogen production from cheese whey using a UASB reactor: links between microbial community and reactor performance. Int J Hydrog Energy 34:5674–5682. https://doi.org/10.1016/j.ijhydene.2009.05.060

Chen C, Sun C, Xia A et al (2020) Sustainable biohythane production from algal bloom biomass through two-stage fermentation: impacts of the physicochemical characteristics and fermentation performance. Int J Hydrog Energy 45:34461–34472. https://doi.org/10.1016/j.ijhydene.2020.02.027

Chen H, Yang T, Shen Z et al (2022) Can digestate recirculation promote biohythane production from two-stage co-digestion of rice straw and pig manure ? J Environ Manag 319:115655

Chiu SY, Kao CY, Chen TY et al (2015) Cultivation of microalgal chlorella for biomass and lipid production using wastewater as nutrient resource. Bioresour Technol 184:179–189. https://doi.org/10.1016/j.biortech.2014.11.080

Christofoletti CA, Escher JP, Correia JE et al (2013) Sugarcane vinasse: environmental implications of its use. Waste Manag 33:2752–2761. https://doi.org/10.1016/j.wasman.2013.09.005

Chu CY, Vo TP, Chen TH (2020) A novel of biohythane gaseous fuel production from pineapple peel waste juice in two-stage of continuously stirred anaerobic bioreactors. Fuel 279:118526. https://doi.org/10.1016/j.fuel.2020.118526

Cota-Navarro CB, Carrillo-Reyes J, Davila-Vazquez G et al (2011) Continuous hydrogen and methane production in a two-stage cheese whey fermentation system. Water Sci Technol 64:367–374. https://doi.org/10.2166/wst.2011.631

Do Nascimento Junior JR, Torres LAZ, Medeiros ABP et al (2021) Enhancement of biohydrogen production in industrial wastewaters with vinasse pond consortium using lignin-mediated iron nanoparticles. Int J Hydrog Energy 46:27431–27443. https://doi.org/10.1016/j.ijhydene.2021.06.009

Elbeshbishy E, Dhar BR, Nakhla G, Lee HS (2017) A critical review on inhibition of dark biohydrogen fermentation. Renew Sust Energ Rev 79:656–668. https://doi.org/10.1016/j.rser.2017.05.075

Feng Q, Zhu G, Wang K et al (2022) Contribution analysis of different electron transfer pathways to methane production in anaerobic digestion coupled with bioelectrochemical system. Sci Total Environ 849:157745. https://doi.org/10.1016/j.scitotenv.2022.157745

Ghimire A, Luongo V, Frunzo L et al (2022) Biohythane production from food waste in a two-stage process: assessing the energy recovery potential. Environ Technol 43:2190–2196. https://doi.org/10.1080/09593330.2020.1869319

Jehlee A, Khongkliang P, Suksong W et al (2017) Biohythane production from chlorella sp. biomass by two-stage thermophilic solid-state anaerobic digestion. Int J Hydrog Energy 42:27792–27800. https://doi.org/10.1016/j.ijhydene.2017.07.181

Jehlee A, Rodjaroen S, Waewsak J et al (2019) Improvement of biohythane production from chlorella sp. TISTR 8411 biomass by co-digestion with organic wastes in a two-stage fermentation. Int J Hydrog Energy 44:17238–17247. https://doi.org/10.1016/j.ijhydene.2019.03.026

Jönsson LJ, Martín C (2016) Pretreatment of lignocellulose: formation of inhibitory by-products and strategies for minimizing their effects. Bioresour Technol 199:103–112. https://doi.org/10.1016/j.biortech.2015.10.009

Karp SG, Burgos WJM, Vandenberghe LPS et al (2022) Sugarcane : a promising source of green carbon in the circular bioeconomy. Sugar Technol 24:1230–1245. https://doi.org/10.1007/s12355-022-01161-z

Keskin T, Nalakath Abubackar H, Arslan K, Azbar N (2019) Biohydrogen production from solid wastes. Elsevier B.V, Amsterdam

Khan AW, Labrie J-P, McKeown J (1986) Effect of electron-beam irradiation pretreatment on the enzymatic hydrolysis of softwood. Biotechnol Bioeng 28:1449–1453. https://doi.org/10.1002/bit.260280921

Kongjan P, O-Thong S, Angelidaki I (2011) Performance and microbial community analysis of two-stage process with extreme thermophilic hydrogen and thermophilic methane production from hydrolysate in UASB reactors. Bioresour Technol 102:4028–4035. https://doi.org/10.1016/j.biortech.2010.12.009

Kongjan P, O-Thong S, Angelidaki I (2013) Hydrogen and methane production from desugared molasses using a two-stage thermophilic anaerobic process. Eng Life Sci 13:118–125. https://doi.org/10.1002/elsc.201100191

Krishnan S, Singh L, Sakinah M et al (2016) Process enhancement of hydrogen and methane production from palm oil mill effluent using two-stage thermophilic and mesophilic fermentation. Int J Hydrog Energy 41:12888–12898. https://doi.org/10.1016/j.ijhydene.2016.05.037

Kumar R, Kim TH, Basak B et al (2022) Emerging approaches in lignocellulosic biomass pretreatment and anaerobic bioprocesses for sustainable biofuels production. J Clean Prod 333:130180. https://doi.org/10.1016/j.jclepro.2021.130180

Kumari S, Das D (2016) Biologically pretreated sugarcane top as a potential raw material for the enhancement of gaseous energy recovery by two stage biohythane process. Bioresour Technol 218:1090–1097. https://doi.org/10.1016/j.biortech.2016.07.070

Lay CH, Kumar G, Mudhoo A et al (2020) Recent trends and prospects in biohythane research: an overview. Int J Hydrog Energy 45:5864–5873. https://doi.org/10.1016/j.ijhydene.2019.07.209

Li J, He J, Si B et al (2020) A pilot study of biohythane production from cornstalk via two-stage anaerobic fermentation. Int J Hydrog Energy 45:31719–31731. https://doi.org/10.1016/j.ijhydene.2020.08.253

Liu R, Chen X, Zhang K et al (2022) Effect of mixing ratio and total solids content on temperature-phased anaerobic codigestion of rice straw and pig manure: Biohythane production and microbial structure. Bioresour Technol 344:126173. https://doi.org/10.1016/j.biortech.2021.126173
Luo G, Xie L, Zhou Q, Angelidaki I (2011) Enhancement of bioenergy production from organic wastes by two-stage anaerobic hydrogen and methane production process. Bioresour Technol 102:8700–8706. https://doi.org/10.1016/j.biortech.2011.02.012
Mamimin C, Singkhala A, Kongjan P et al (2015) Two-stage thermophilic fermentation and mesophilic methanogen process for biohythane production from palm oil mill effluent. Int J Hydrog Energy 40:6319–6328. https://doi.org/10.1016/j.ijhydene.2015.03.068
Mamimin C, Kongjan P, O-Thong S, Prasertsan P (2019a) Enhancement of biohythane production from solid waste by co-digestion with palm oil mill effluent in two-stage thermophilic fermentation. Int J Hydrog Energy 44:17224–17237. https://doi.org/10.1016/j.ijhydene.2019.03.275
Mamimin C, Probst M, Gómez-Brandón M et al (2019b) Trace metals supplementation enhanced microbiota and biohythane production by two-stage thermophilic fermentation. Int J Hydrog Energy 44:3325–3338. https://doi.org/10.1016/j.ijhydene.2018.09.065
Martinez-Burgos WJ, Sydney EB, Brar SK et al (2019) The effect of hydrolysis and sterilization in biohydrogen production from cassava processing wastewater medium using anaerobic bacterial consortia. Int J Hydrog Energy 44:25551–25564. https://doi.org/10.1016/j.ijhydene.2019.08.085
Martinez-Burgos WJ, Sydney EB, De Paula DR et al (2021a) Hydrogen production by dark fermentation using a new low-cost culture medium composed of corn steep liquor and cassava processing water : process optimization and scale-up. Bioresour Technol 320:124370. https://doi.org/10.1016/j.biortech.2020.124370
Martinez-Burgos WJ, Sydney EB, Medeiros ABP et al (2021b) Agro-industrial wastewater in a circular economy: characteristics, impacts and applications for bioenergy and biochemicals. Bioresour Technol 341:125795. https://doi.org/10.1016/j.biortech.2021.125795
Meena RAA, Rajesh Banu J, Yukesh Kannah R et al (2020) Biohythane production from food processing wastes – challenges and perspectives. Bioresour Technol 298:122449. https://doi.org/10.1016/j.biortech.2019.122449
MFRural (2022a) Casca de arroz moída e palha de arroz moída o ano todo. Retrieved from: https://www.mfrural.com.br/detalhe/46625/casca-de-arroz-moida-e-palha-de-arroz-moida. July 20, 2022. 46625
MFRural (2022b) Farelo de soja, Casca de Soja Moida e Peletizada. Retrieved from: https://www.mfrural.com.br/detalhe/306956/soja-farelo-de-soja-casca-de-soja-moida-e-peletizada. July 20, 2022. 306956
Mishra P, Balachandar G, Das D (2017) Improvement in biohythane production using organic solid waste and distillery effluent. Waste Manag 66:70–78. https://doi.org/10.1016/j.wasman.2017.04.040
Montingelli ME, Tedesco S, Olabi AG (2015) Biogas production from algal biomass: a review. Renew Sust Energ Rev 43:961–972. https://doi.org/10.1016/j.rser.2014.11.052
Mosier N, Wyman C, Dale B et al (2005) Features of promising technologies for pretreatment of lignocellulosic biomass. Bioresour Technol 96:673–686. https://doi.org/10.1016/j.biortech.2004.06.025
Nguyen T, Ta D, Lin C et al (2022) Biohythane production from swine manure and pineapple waste in a single-stage two-chamber digester using gel-entrapped anaerobic microorganisms. Int J Hydrog Energy 47:25245–25255. https://doi.org/10.1016/j.ijhydene.2022.05.259
O-Thong S, Prasertsan P, Intrasungkha N et al (2008) Optimization of simultaneous thermophilic fermentative hydrogen production and COD reduction from palm oil mill effluent by Thermoanaerobacterium-rich sludge. Int J Hydrog Energy 33:1221–1231. https://doi.org/10.1016/j.ijhydene.2007.12.017
O-Thong S, Suksong W, Promnuan K et al (2016) Two-stage thermophilic fermentation and mesophilic methanogenic process for biohythane production from palm oil mill effluent with

methanogenic effluent recirculation for pH control. Int J Hydrog Energy 41:21702–21712. https://doi.org/10.1016/j.ijhydene.2016.07.095

Pasupuleti SB, Venkata Mohan S (2015) Single-stage fermentation process for high-value biohythane production with the treatment of distillery spent-wash. Bioresour Technol 189:177–185. https://doi.org/10.1016/j.biortech.2015.03.128

Ramos LR, Silva EL (2018) Continuous hydrogen production from cofermentation of sugarcane vinasse and cheese whey in a thermophilic anaerobic fluidized bed reactor. Int J Hydrog Energy 43:13081–13089. https://doi.org/10.1016/j.ijhydene.2018.05.070

Ren N, Wang A, Cao G et al (2009) Bioconversion of lignocellulosic biomass to hydrogen: potential and challenges. Biotechnol Adv 27:1051–1060. https://doi.org/10.1016/j.biotechadv.2009.05.007

Rosa D, Medeiros ABP, Martinez-Burgos WJ et al (2020) Biological hydrogen production from palm oil mill effluent (POME) by anaerobic consortia and Clostridium beijerinckii. J Biotechnol 323:17–23. https://doi.org/10.1016/j.jbiotec.2020.06.015

Sasidhar KB, Kumar PS, Xiao L (2022) A critical review on the two-stage biohythane production and its viability as a renewable fuel. Fuel 317:123449. https://doi.org/10.1016/j.fuel.2022.123449

Seengenyoung J, Mamimin C, Prasertsan P, O-Thong S (2019) Pilot-scale of biohythane production from palm oil mill effluent by two-stage thermophilic anaerobic fermentation. Int J Hydrog Energy 44:3347–3355. https://doi.org/10.1016/j.ijhydene.2018.08.021

Sgroi F, Foderà M, Di Trapani AM et al (2015) Economic evaluation of biogas plant size utilizing giant reed. Renew Sust Energ Rev 49:403–409. https://doi.org/10.1016/j.rser.2015.04.142

Shanmugam S, Mathimani T, Rene ER et al (2021a) Biohythane production from organic waste: recent advancements, technical bottlenecks and prospects. Int J Hydrog Energy 46:11201–11216. https://doi.org/10.1016/j.ijhydene.2020.10.132

Shanmugam S, Sekar M, Sivaramakrishnan R et al (2021b) Pretreatment of second and third generation feedstock for enhanced biohythane production: challenges, recent trends and perspectives. Int J Hydrog Energy 46:11252–11268. https://doi.org/10.1016/j.ijhydene.2020.12.083

Si BC, Li JM, Zhu ZB et al (2016) Continuous production of biohythane from hydrothermal liquefied cornstalk biomass via two-stage high-rate anaerobic reactors. Biotechnol Biofuels 9:1–15. https://doi.org/10.1186/s13068-016-0666-z

Sindhu R, Binod P, Pandey A (2016) Biological pretreatment of lignocellulosic biomass - an overview. Bioresour Technol 199:76–82. https://doi.org/10.1016/j.biortech.2015.08.030

Singh A, Sevda S, Abu Reesh IM et al (2015) Biohydrogen production from lignocellulosic biomass: technology and sustainability. Energies 8:13062–13080. https://doi.org/10.3390/en81112357

Sun C, Xia A, Liao Q et al (2019) Life-cycle assessment of biohythane production via two-stage anaerobic fermentation from microalgae and food waste. Renew Sust Energ Rev 112:395–410. https://doi.org/10.1016/j.rser.2019.05.061

Ta DT, Lin CY, Ta TMN, Chu CY (2020) Biohythane production via single-stage anaerobic fermentation using entrapped hydrogenic and methanogenic bacteria. Bioresour Technol 300:122702. https://doi.org/10.1016/j.biortech.2019.122702

Ta DT, Lin CY, Ta TMN, Chu CY (2022) Effect of pH shock on single-stage biohythane production using gel-entrapped anaerobic microorganisms. Int J Hydrog Energy 47:3679–3689. https://doi.org/10.1016/j.ijhydene.2021.11.009

Tumuluru JS, Heikkila DJ (2019) Biomass grinding process optimization using response surface methodology and a hybrid genetic algorithm. Bioengineering 6:2–21. https://doi.org/10.3390/bioengineering6010012

Vasco-Correa J, Shah A (2019) Techno-economic bottlenecks of the fungal pretreatment of lignocellulosic biomass. Fermentation 5:30. https://doi.org/10.3390/fermentation5020030

Vieira S, Barros MV, Sydney ACN et al (2020) Sustainability of sugarcane lignocellulosic biomass pretreatment for the production of bioethanol. Bioresour Technol 299:122635. https://doi.org/10.1016/j.biortech.2019.122635

Yang H, Si B, Huang S et al (2020) Effect of pH control on biohythane production and microbial structure in an innovative multistage anaerobic hythane reactor (MAHR). Int J Hydrog Energy 45:4193–4204. https://doi.org/10.1016/j.ijhydene.2019.12.046
Yeshanew MM, Frunzo L, Pirozzi F et al (2016) Production of biohythane from food waste via an integrated system of continuously stirred tank and anaerobic fixed bed reactors. Bioresour Technol 220:312–322. https://doi.org/10.1016/j.biortech.2016.08.078

Chapter 15
Volatile Fatty Acids Production and Recovery in Biohydrogen Production

Ammini Naduvanthar Anoopkumar, Embalil Mathachan Aneesh, Aravind Madhavan, Raveendran Sindhu, Mukesh Kumar Awasthi, Ashok Pandey, Laya Liz Kuriakose, and Parameswaran Binod

Abstract Replacing the goods procured from fossil matter with other products gained over renewable resources has become a global concern in the current scenario. A major shift to bio-based elements appears vital to evade the predominant negative impacts instigated by various chemical elements on the surrounding. In this regard, with special inference on overcoming the supply limitations, the recent research considers the production of various volatile fatty acids (VFAs) including the Acetic, lactic, butyric and propionic acids with special reference to biohydrogen production. In this regard, here the authors have discussed the methods and effluents of bio-hydrogen production processes in context with the various volatile fatty acids and the key approaches usually employed for VFA recovery. The key inferences

A. N. Anoopkumar · E. M. Aneesh
Centre for Research in Emerging Tropical Diseases (CRET-D), Department of Zoology, University of Calicut, Malappuram, Kerala, India

A. Madhavan
School of Biotechnology, Amrita Vishwa Vidyapeetham, Kollam, Kerala, India

R. Sindhu · L. L. Kuriakose
Department of Food Technology, TKM Institute of Technology, Kollam, Kerala, India

M. K. Awasthi
College of Natural Resources and Environment, Northwest A & F University, Yangling, Shaanxi, China

A. Pandey
Centre for Innovation and Translational Research, CSIR- Indian Institute for Toxicology Research (CSIR-IITR), Lucknow, India

Centre for Energy and Environmental Sustainability, Lucknow, Uttar Pradesh, India

Sustainability Cluster, School of Engineering, University of Petroleum and Energy Studies, Dehradun, India

P. Binod (✉)
Microbial Processes and Technology Division, CSIR-National Institute for Interdisciplinary Science and Technology (CSIR-NIIST), Trivandrum, Kerala, India

C. R. Soccol et al. (eds.), *Biohydrogen - Advances and Processes*, Biofuel and Biorefinery Technologies 13, https://doi.org/10.1007/978-3-031-49818-3_15

obtained in this investigation also validated that the value-added applications of volatile fatty acids (VFAs) in various fields directed the researchers to produce butyric acids, acetic acids, lactic acids and propionic acids from effluents with the aid of the major advanced approaches such as gas stripping with absorption, reverse osmosis, solvent extraction, adsorption, nanofiltration, membrane contractors and electrodialysis.

Keywords Nanofiltration · Electrodialysis · VFA · Acids · Acetic acid

15.1 Introduction

The transition of the current scenario toward a circular economy through the recovery of resources, reuse and recycling may finally reach the focus concerning environmental sustainability. The recovery of resources may primarily reduce the consumption of raw materials thereby simultaneously diminishing the amount of waste generated. The demand of energy recovery puts extensive demands on the growth of next-generation technologies to produce novel elements and materials that have prominent significance in environmental sustainability (Atasoy et al. 2018). Due to the disadvantages of certain conventional strategies, the discussion concerning the new strategies focusing on the paradigm shift unveiled its extensive role in the current scenario.

The Fermentative production of volatile fatty acids and biohydrogen from various resources including advanced feedstocks always offers the way toward a carbon-neutral bioeconomy. The volatile fatty acids are also known to gather prominent attention since they have an extensive application in various fields including the production of fatty acid methyl esters, medium chain fatty acids, electrochemical products and mixed alcohols. From the various pathways and perspectives discussed concerning acetate fermentation, propionate fermentation, and butyrate fermentation, It was evident that the production of hydrogen is critically reliant on the volatile fatty acids composition (Kidanu et al. 2017).

While discussing the energy perspectives of volatile fatty acids and biohydrogen, we can clearly perceive the fact that an anaerobic fermentation process converts about one-third of carbonaceous biomass to hydrogen. That means, both of them are valued energy resources and their co-production of them may usually promote the economic inferences of the platform (Kidanu et al. 2017). Other approaches, like response surface methodology, can also be used to optimize the production of biohydrogen and volatile fatty acids from various raw materials including the food waste. In this regard, it is essential to discuss the methods and various effluents allied with biohydrogen production with special reference on the recovery of volatile fatty acids.

15.2 Methods of Biohydrogen Production

A plethora of following microbial consortia and lower eukaryotes have previously been reported the ability to produce hydrogen; cyanobacteria, anaerobic bacteria, facultative aerobic bacteria and archaea (Kidanu et al. 2017). The following four major strategies are primarily witnessed with the production of hydrogen and are (1) Photofermentation; (2) Biophotolysis (3) microbial electrolysis processing and (4) dark fermentation. However as found for the various reported advanced technologies, the aforesaid technologies have also possessed benefits as well as disadvantages while considering the practicality as well as the energy efficiency. Hence, the selection of precise and accurate inoculum or the biocatalyst for the respective processes could be noted as an integral part that is directly allied with hydrogen production (Chandrasekhar et al. 2015). An investigation conducted by (Mutsvene et al. 2023) analysed the various parameters along with the biohydrogen production rate (Fig. 15.1).

Naturally, either a single microbial species or a mixed group of organisms is involved in hydrogen production and recent years witnessed that mixed cultures are preferable to use since they are known to offer better stability, extensive biochemical functions, followed by operational ease along with the applicability to choose a wide range of substrates (Chandrasekhar et al. 2015).

15.2.1 *Photofermentation*

The method of Photofermentation usually encompasses the formation of biomass along with the production of carbon dioxide (CO2) and hydrogen through light energy conversion. For this process, various nonsulfur photosynthetic bacteria

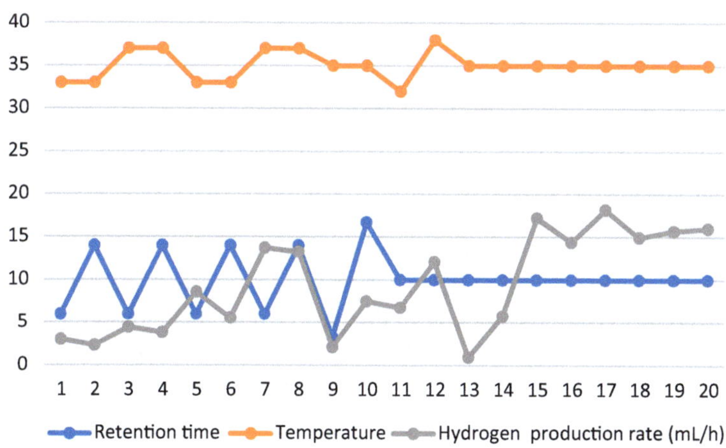

Fig. 15.1 Hydrogen production rate according to the various experimental conditions (Mutsvene et al. 2023)

belonging to the genus *Rhodobacter* would be used and it converts the following organic acids into carbon dioxide (CO2) and hydrogen; butyrate, acetate and lactate. Furthermore, the bacteria belonging to the aforesaid group collect solar energy for transforming the organic acids into various utilizable forms like hydrogen in the absence of ammonium ions (Pandey et al. 2019).

Specifically, the bacteria belonging to the *Rhodobacter* used in this process usually possess nonoxygenic photosynthesis nature. Although in addition to the advantageous effects encompassed by them, certain negative impacts such as diminishing photochemical efficacy followed by reduced catalytic activity have also been linked with the same. While discussing a theoretical perspective, it was evident that this process can entirely convert the organic compounds into hydrogen, however, the practical perspectives have illustrated a little bit different output from the theoretical views (Ghosh et al. 2012).

15.2.2 Biophotolysis

Biophotolysis or Water-Splitting Photosynthesis is generally reported as the most attractive and desirable approach for the production of hydrogen using the following oxygenic photosynthetic microorganisms; cyanobacteria (*Nostoc punctiforme, Synechocystis* sp. and *Anabaena variabilis*), green microalgae (*Chlamydomonas reinhardtii*, *Chlorella* sp., *Scenedesmus* sp., and *Scenedesmus obliquus*). The process is further divided into main categories such as direct and indirect processes. In the direct process, the photosystem II (PS II) and photosystem I (PS I) acts as electron carrier through which the electrons are transferred to ferredoxin (Fd) and subsequently the Fd reduces an enzyme that is responsible for the production of hydrogen (Melis et al. 2000). The indirect way forms the chemical energy from light energy in the form of carbohydrates, which can be reused for hydrogen production.

15.2.3 Microbial Electrolysis Processing

Microbial electrolysis cells (MECs) in recent years have witnessed to represent novel techniques to produce hydrogen from a wide range of substrates as reported by (Cheng and Logan 2007). Microbial electrolysis cells (MECs) are capable of extensive efficiency (more than 90%) in hydrogen production. But the performance of this approach is strictly stuck in to the type of microbial consortia, type of the membrane used, the composition of the substrate, electrode materials, the concentration of the substrate along with the design of the MEC.

15.2.4 Dark Fermentation

Finally, the dark fermentative process is also known to produce hydrogen and recent studies have illustrated the role of the *Escherichia coli, Enterobacter aerogenes, Citrobacter intermedius, Clostridium beijerinckii, Ruminococcus albus, Ruminococcus albus* and *Enterobacter cloacae* in hydrogen production (Hallenbeck and Ghosh 2009). While comparing the output obtained in this technique to the photolysis it was evident that a higher rate of output is obtained in this approach. However, as mentioned for the various other approach introduced, this process has also reported certain disadvantages including the formation of various by-products thereby influencing the yield of hydrogen production in a negative way.

15.2.5 Effluents from the Biohydrogen Production Process

As mentioned earlier there exists a wide range of approaches for hydrogen production in biological ways. Among them, the production of hydrogen by fermentation of organic waste using the microbial consortia is noted to be an environmentally friendly approach with an extensive output rate under realistic conditions (Ma et al. 2017). The Palm oil mill effluent (POME) was found to be an effective substrate for the production of hydrogen. A study by (Mamimin et al. 2017) reported that the following elements were found to be predominant in the hydrogenic effluent; volatile solids, nitrogen, Propionic acid Butyric acid, Lactic acid, Acetic acid, Ethanol, and Propionic acid. The findings concluded by them clearly validated that the propionic acid and butyric acid which are at high concentration the effluent possessed inhibition effect on the production of methane while the acetic, butyric and lactic acid found at lower concentration with positive impacts on the methanogenesis process.

15.2.6 Volatile Fatty Acids in Effluents

The biohydrogen effluent also contains volatile fatty acid (VFA) such as butyric, lactic, acetic, and propionic acids at higher concentrations. In conventional strategies of biohydrogen production, approximately only 5–15% of the energy in the organic waste material is converted to hydrogen and the rest of the energy is included in volatile fatty acids found in the same. Furthermore, about 65% of the energy in organic waste is stored in liquid forms. As mentioned earlier, the acetic, butyric, propionic, and lactic acids are the major liquid forms. In order to save energy from such sources, they were converted into suitable energy carriers or products through methanogenesis under the anaerobic digestion process (Krishnan et al. 2016).

A recent study by (Kumar et al. 2019) reviewed the possibilities of converting the volatile fatty acids by using dark fermentative effluents to polyhydroxyalkanoates during hydrogen production. The presence of acetates and allied products found among the volatile fatty acid (VFA) in the effluents made them an effective candidate for producing various biodegradable biopolymers. The cost-effective production of hydrogen and polyhydroxyalkanoates is recognized to be an additional advantage of the same. Polyhydroxyalkanoates can be gained from a few soluble acid metabolic intermediates, usually called volatile fatty acids. The aforesaid volatile fatty acids are generally in the dark fermentative effluents of bio-hydrogen reactors. The study also reported that about 0.93 g COD/g COD of volatile fatty acids was yielded from the dairy whey effluent while the Pulp/paper mill effluent yielded 0.75 g COD/g COD of volatile fatty acids. Certain advanced technologies like the advanced pressure-driven membrane separation approach reported by (Wainaina et al. 2019) are also used for the recovery of volatile fatty acids from the effluent. Their study also integrated reverse osmosis, nanofiltration and ultrafiltration processes to accomplish the aforesaid target. (Zhou et al. 2013b) recovered acetic acid through reverse osmosis process at a temperature of 40 °C followed by maintaining a pH of 2.9.

15.2.7 Types of VFA in Effluents

The value-added applications of Volatile fatty acids (VFAs) in chemical industries as well as in the bioenergy field have differentiated the same into the following categories; butyric acid, acetic acid, lactic acid and propionic acid The sustainable production of such elements from various renewable resources including biowastes through microbial fermentation has recently attained significant attention for the past few years (Merrylin et al. 2020; Zhang et al. 2022).

15.3 Butyric Acid

The discussions regarding the potential applications of Butyric acids in various fields including the beverage and foodstuffs industries have enhanced day by day. The prominent role of Butyric acids in physiological functions has also been validated by many previous investigations. From this point of a line, (He et al. 2005) investigated the fed-batch production by *Clostridium butyricum* and found that the selected mode could have the ability to upsurge the production rate significantly. One of the central focuses of the current chapter; is precisely to elute the Butyric acids from the waste materials or effluents and simultaneously to eliminate the threat from such effluents towards the environment, (Fu et al. 2012) implemented biotreatment against textile wastewater. According to their final inferences, the addition of starch to the wastewater from the textile industry with a C/N ratio of 30 prominently augmented the percentage as well as the concentration of butyric acid. The rapid drop of toxic effects towards the biological surroundings after the anaerobic digestion and enhanced butyric acid production could hence be considered as a double-edged way in recent research.

15.4 Acetic Acid

The implementation of various advanced approaches, including nanofiltration (NF) as well as microfiltration for the treatment of waste effluents will aid the production of various volatile fatty acids (VFA), especially acetic acid. Such a membranes-based approach performed by (Zacharof and Lovitt 2014) obtained a rapidly enhanced production rate of 21.08 mM for acetic acid from the waste effluents. In addition to this, the application of esterification as well as the thermal treatment for the removal of acetic acid in membrane reactor was previously reported by (Unlu and Hilmioglu 2016). The integration of two specific processes, especially the electrohydrogenesis and fermentation by (Lalaurette et al. 2009) yielded another perspective in which the acetic acid was witnessed to produce the maximum hydrogen level with a value of 1400 ± 170 mL-H2/g-COD. Furthermore, the molar yield electrical energy efficiency and has also been noted to be a peak for acetic acid at a rate of 3.6 mol/mol and 280 ± 40% respectively. The bipolar membrane electrodialysis done by (Yu et al. 2000) analysed that the implemented approach would be is technically feasible and the treatment of wastewater have recovered about 70% (wt%) acetic acid solution.

15.5 Lactic Acid

The various applications of Lactic Acids in chemical, food, textile, cosmetics and medicine industries directed the researchers to find out feasible and sustainable resources. The expanding rate of biodegradable polylactic acid production also upgraded the demand for lactic acid in recent research. But, one of the major things that need to be recalled in this research is that about 10–12 m^3 of wastewater will be generated once the one ton of lactic acid has been produced industrially and it could also be noted to be a bottleneck curbing the healthy growth of lactic acid industry (Zhang et al. 2019). In this regard, many of the studies including (Malmary et al. 2000) have strained to recover lactic acids from aqueous effluents using solvent extraction techniques with special inference on the environmental pollution reduction and the possible uses of various pure solutes in medical and food industry. In addition to the conventional perspectives, recent investigations focussed on the production of Hydrogen from sugarcane in mesophilic environment with special inference on the role of lactic acid production (de Menezes and Silva 2019). Their conclusive remarks by them validated that Lactic acid was the predominant metabolite analysed expanded granular sludge bed reactors (EGSB).

15.6 Propionic Acid

The use of Propionic acid in various industries allied with thermoplastics, perfumes, solvents, medicines, foods, fragrances and seeds has illustrated its significance in recent research. The production of propionic acid at commercial has principally accomplished over petrochemical synthesis. To follow an alternative and environment-friendly approach, the majority of the recent investigations trail the fermentation-based approaches against the effluents. For instance, a study by (Teles et al. 2019), reported the maximum yield of propionic acid production with a rate of 0.79 g of product, when treated with effluents. They observed a strong significant linear correlation between propionic acid production and the consumption of lactic acid, which justifies the fact that the presence of lactic acid primarily favored propionic acid production. However, the influence of lactose, glucose and galactose was not reported as a significant correlation, while acetic acid production turned to be prominently exhibited a significant correlation with the production of propionic acid.

15.6.1 Biochemistry of VFA

The major backbone of the VFAs are primarily directed towards the C2–C6 carboxylic acids that are formed during the anaerobic digestion process (Kim et al. 2018). Precisely, they are the small organic compounds that are previously known to get much more significance than any other because of their potential to use them as the prominent carbon source for specific kinds of microbial consortia, which have a potent role in the production of biodegradable elements and bioactive compounds. The wide range of applications of VFAs in various fields including as a building block for various industries allied with animal feed, food, pharmaceutics, preservation, petrochemicals and flavors enhanced the significance of VFAs-based research in the current scenario. For instance, the oily-colorless carboxylic acid, the Butyric acid is usually reported to be present in farm animal's milk, and the derivative of the same have usually termed as "butanoic acid". The market value of "butanoic acid" will reach at enormous level, precisely, 170 million USD by the year 2026. The production of aforesaid VFA at an industrial scale has usually been accomplished through chemical synthesis, especially through butyraldehyde oxidation. The production of aforesaid VFA at an industrial scale has usually been accomplished through chemical synthesis, especially through butyraldehyde oxidation. Likewise, the colorless oily liquid hexanoic acid usually exhibits a pungent smell and has been produced at the industrial level through the reverse β-oxidation. That means the

Sustainable provision of such elements could be noted as a great defining factor ensuring the environmental as well as the economic stability of the future (Agnihotri et al. 2022).

15.7 Methods of Recovery of VFA

The recovery of volatile fatty acids produced from effluents became more significant in the current scenario. Among the various methods, the Air/gas stripping approach (Atasoy et al. 2018) produced significant results in volatile fatty acids production. This approach principally removes the volatile organic elements from the effluents based on Henry's law. The CaCO3 slurry and N2 gas were the prime elements that make this approach more effective with a recovery efficiency of 0.92–29.5 mmol of VFA. For the effective separation of elements from the aqueous solutions, the solvent extraction approach (Atasoy et al. 2018) can be recognized to be a major approach which obeys the principle concerning the relative solubilities of different immiscible fluids. Nanofiltration as well as Reverse osmosis primarily use semi-permeable membranes to isolate and elute various solutes including organic molecules and salts. Recognizing the element that nanofiltration and reverse osmosis membranes have specific range of molecular weight and pore size cut-offs, it is also suggested to use such approaches for the individual separation of VFA from mixed VFAs. In addition to the discussed approaches, the Fig. 15.2 and Table 15.1

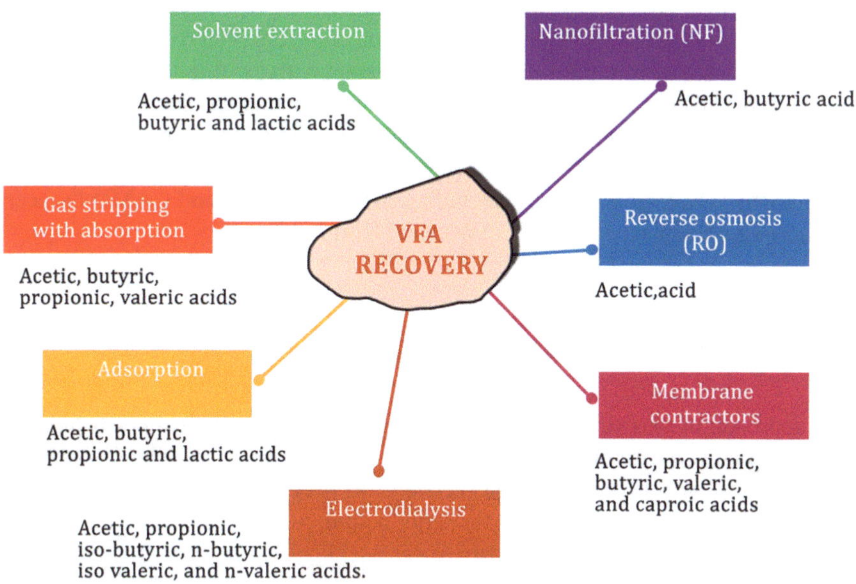

Fig. 15.2 Major approaches employed for VFA recovery

Table 15.1 Various approaches for VFA recovery

Approach	VFA Recovery efficiency and materials used
Gas stripping with absorption	0.92 to 29.5 mmol $CaCO_3$ slurry and N_2 GAS
Reverse osmosis (RO)	Maximum production of acetic acid RO membranes
Solvent extraction	10 wt%. TOA
Adsorption	0.8 mmol Polystyrene-divinylbenzene- resins
Nanofiltration	53.94 mM acetate LF10 NF
Membrane contractors	95% VFA PTFE membranes
Electrodialysis	99% of VFAs Membrane based (ED 64,002)

also illustrate the major common approaches used for the recovery of VFA along with the conditions and type of VFA recovered.

15.8 Solvent Selection

The efficient nature of the solvent selection is greatly reported in many instances, and unlike other strategies, especially distillation, it doesn't require and consume much more energy. In this regard, Booij and Van den Berg (1994) defined the applicability of various extraction approaches in context with the strength and affinity of the various solvents. Proper and efficient solvent selection has been recognized to be a prominent hurdle toward the extensive application of this efficient separation process. A plethora of findings in this regard also executed the possibilities of computer-aided solvent-seeking tools including the COSMO-RS and CAMD in recent studies to get the maximum output (Woo and Kim 2019).

15.9 Gas Stripping with Absorption

The approach of Gas stripping is primarily railed by Henry's law which links the aqueous-phase concentration of a specific element toward its gas-phase concentration. The research focusing on the recovery of VFAs in recent years has implemented the gas-stripping strategies for better output. According to this approach, the elements are stripped with the headspace gas opening over the tube. The gas constituting the VFAs is then transported through the recovery system. The gas stripping technique employed in a recent study verified its potential to purify VFA with efficient results (Huang et al. 2016).

15.10 Adsorption

The adsorption strategy for the VFA recovery has primarily accomplished by passing the VFA-rich element over an ion exchange resin. According to the major process, the carboxylate anion of the specific VFAs that is to be recovered is absorbed into the resin and the resin can be used for many cycles (Edmiston et al. 2018).

15.11 Reverse Osmosis and Nanofiltration

Nanofiltration and reverse osmosis primarily uses semi-permeable membranes for salt and organic molecules separation. A study by (Zhou et al. 2013a) separated the acetic acid from monosaccharides with the aid of aforesaid techniques and attained better separation factors over the glucose and xylose. In addition to this, another investigation unveiled the potential of nanofiltration membranes to produce maximum VFA retention, precisely up to 75%. The performances of the same have been influenced by a lot of the following factors; solute concentration, pH, pressure and temperature. These mentioned factors will affect the specific mechanisms proceeding with the retention of solutes (Zacharof and Lovitt 2014). Because of the effective strategies available in the recent scenario, derived VFAs are extensively used in various fields allied with bioelectricity, hydrogen, biogas production, bioplastics, and biodiesel production.

15.12 Conclusion and Perspectives

The production of volatile fatty acid (VFA) from effluents has gotten much more attention in the current scenario since the wide range usage area followed by the market area of the same has augmented day by day. Photofermentation, biophotolysis, microbial electrolysis processing and dark fermentation are the predominant approaches for biohydrogen production. Furthermore, the value-added applications of volatile fatty acids (VFAs) in various fields directed the researchers to produce butyric acids, acetic acids, lactic acids and propionic acids from effluents with the aid of the major advanced approaches such as solvent extraction, gas stripping with absorption, reverse osmosis, adsorption and nanofiltration. The future directions should be paid towards clearing the various critical challenges faced so far including the mixture complexity, enhanced product cost, and finally the VFAs with low concentration. The advanced approaches and strategies that offers prominent way to combat the critical challenges hence should be considered for practical implementation in future.

References

Agnihotri S, Yin D-M, Mahboubi A, Sapmaz T, Varjani S, Qiao W, Koseoglu-Imer DY, Taherzadeh MJ (2022) A glimpse of the world of volatile fatty acids production and application: a review. Bioengineered 13(1):1249–1275

Atasoy M, Owusu-Agyeman I, Plaza E, Cetecioglu Z (2018) Bio-based volatile fatty acid production and recovery from waste streams: current status and future challenges. Bioresour Technol 268:773–786

Booij K, Van den Berg C (1994) Comparison of techniques for the extraction of lipids and PCBs from benthic invertebrates. Bull Environ Contam Toxicol 53:71–76

Chandrasekhar K, Lee Y-J, Lee D-W (2015) Biohydrogen production: strategies to improve process efficiency through microbial routes. Int J Mol Sci 16(4):8266–8293

Cheng S, Logan BE (2007) Sustainable and efficient biohydrogen production via electrohydrogenesis. Proc Natl Acad Sci 104(47):18871–18873

de Menezes CA, Silva EL (2019) Hydrogen production from sugarcane juice in expanded granular sludge bed reactors under mesophilic conditions: the role of homoacetogenesis and lactic acid production. Ind Crop Prod 138:111586

Edmiston PL, Gilbert AR, Harvey Z, Mellor N (2018) Adsorption of short chain carboxylic acids from aqueous solution by swellable organically modified silica materials. Adsorption 24:53–63

Fu B, Zhang J, Fan J, Wang J, Liu H (2012) Control of C/N ratio for butyric acid production from textile wastewater sludge by anaerobic digestion. Water Sci Technol 65(5):883–889

Ghosh D, Sobro IF, Hallenbeck PC (2012) Optimization of the hydrogen yield from single-stage photofermentation of glucose by Rhodobacter capsulatus JP91 using response surface methodology. Bioresour Technol 123:199–206

Hallenbeck PC, Ghosh D (2009) Advances in fermentative biohydrogen production: the way forward? Trends Biotechnol 27(5):287–297

He G-Q, Kong Q, Chen Q-H, Ruan H (2005) Batch and fed-batch production of butyric acid by clostridium butyricum ZJUCB. J Zhejiang Univ Sci B 6(11):1076–1080

Huang W, Huang W, Yuan T, Zhao Z, Cai W, Zhang Z, Lei Z, Feng C (2016) Volatile fatty acids (VFAs) production from swine manure through short-term dry anaerobic digestion and its separation from nitrogen and phosphorus resources in the digestate. Water Res 90:344–353

Kidanu WG, Trang PT, Yoon HH (2017) Hydrogen and volatile fatty acids production from marine macroalgae by anaerobic fermentation. Biotechnol Bioprocess Eng 22(5):612–619

Kim NJ, Lim SJ, Chang HN (2018) Volatile fatty acid platform: concept and application. Emerging Areas Bioeng 1:173–190

Krishnan S, Singh L, Sakinah M, Thakur S, Wahid ZA, Alkasrawi M (2016) Process enhancement of hydrogen and methane production from palm oil mill effluent using two-stage thermophilic and mesophilic fermentation. Int J Hydrog Energy 41(30):12888–12898

Kumar G, Ponnusamy VK, Bhosale RR, Shobana S, Yoon J-J, Bhatia SK, Banu JR, Kim S-H (2019) A review on the conversion of volatile fatty acids to polyhydroxyalkanoates using dark fermentative effluents from hydrogen production. Bioresour Technol 287:121427

Lalaurette E, Thammannagowda S, Mohagheghi A, Maness P-C, Logan BE (2009) Hydrogen production from cellulose in a two-stage process combining fermentation and electrohydrogenesis. Int J Hydrog Energy 34(15):6201–6210

Ma Z, Li C, Su H (2017) Dark bio-hydrogen fermentation by an immobilized mixed culture of Bacillus cereus and Brevumdimonas naejangsanensis. Renew Energy 105:458–464

Malmary G, Albet J, Putranto A, Hanine H, Molinier J (2000) Recovery of aconitic and lactic acids from simulated aqueous effluents of the sugar-cane industry through liquid–liquid extraction. J Chem Technol Biotechnol Int Res Process Environ Clean Technol 75(12):1169–1173
Mamimin C, Prasertsan P, Kongjan P, Sompong O (2017) Effects of volatile fatty acids in biohydrogen effluent on biohythane production from palm oil mill effluent under thermophilic condition. Electron J Biotechnol 29:78–85
Melis A, Zhang L, Forestier M, Ghirardi ML, Seibert M (2000) Sustained photobiological hydrogen gas production upon reversible inactivation of oxygen evolution in the green alga Chlamydomonas reinhardtii. Plant Physiol 122(1):127–136
Merrylin J, Kannah RY, Banu JR, Yeom IT (2020) Production of organic acids and enzymes/biocatalysts from food waste. In: Food waste to valuable resources. Elsevier, Cambridge, pp 119–141
Mutsvene B, Chetty M, Kumari S, Bux F (2023) Biohydrogen production from brewery wastewater in an anaerobic baffled reactor. A preliminary techno-economic evaluation. South Afr J Chem Eng 43:9–23
Pandey A, Mohan SV, Chang J-S, Hallenbeck PC, Larroche C (2019) Biomass, biofuels, biochemicals: biohydrogen. Elsevier, Cambridge
Teles JC, Stolle EM, Koloda SA, Barana AC (2019) Production of propionic acid by Propionibacterium acidipropionici from agroindustrial effluents. Braz Arch Biol Technol 62:1–9
Unlu D, Hilmioglu ND (2016) Removal of acetic acid from wastewater by esterification in the membrane reactor. Membrane Water Treatment 7(2):143
Wainaina S, Lukitawesa KA, M., Taherzadeh, M.J. (2019) Bioengineering of anaerobic digestion for volatile fatty acids, hydrogen or methane production: a critical review. Bioengineered 10(1):437–458
Woo HC, Kim YH (2019) Eco-efficient recovery of bio-based volatile C2–6 fatty acids. Biotechnol Biofuels 12(1):1–11
Yu L, Guo Q, Hao J, Jiang W (2000) Recovery of acetic acid from dilute wastewater by means of bipolar membrane electrodialysis. Desalination 129(3):283–288
Zacharof M-P, Lovitt R (2014) Recovery of volatile fatty acids (VFA) from complex waste effluents using membranes. Water Sci Technol 69(3):495–503
Zhang C, Yang H-Q, Wu D-J (2019) Study on the reuse of anaerobic digestion effluent in lactic acid production. J Clean Prod 239:118028
Zhang L, Tsui T-H, Loh K-C, Dai Y, Tong YW (2022) Acidogenic fermentation of organic wastes for production of volatile fatty acids. In: Biomass, biofuels, biochemicals. Elsevier, Cambridge, pp 343–366
Zhou F, Wang C, Wei J (2013a) Separation of acetic acid from monosaccharides by NF and RO membranes: performance comparison. J Membr Sci 429:243–251
Zhou F, Wang C, Wei J (2013b) Simultaneous acetic acid separation and monosaccharide concentration by reverse osmosis. Bioresour Technol 131:349–356

Chapter 16
Biohydrogen and Polyhydroxyalkanoates Coupled Production

Ariane Fátima Murawski de Mello, Luciana Porto de Souza Vandenberghe, Zulma Sarmiento Vásquez, Clara Matte Borges Machado, Natália Rodrigues Nitsch, Adriane Bianchi Pedroni Medeiros, Walter José Martinez Burgos, and Carlos Ricardo Soccol

Abstract One of the main environmental concerns of the actuality is the problem of conventional plastics and their lack of biodegradability, being therefore persistent in the environment. Moreover, their degradation generates toxic by-products that threaten natural life. Bioplastics arise as new solutions to replace plastics from fossil origin, and among them, polyhydroxyalkanoates (PHAs) are increasingly gaining attention due to their interesting properties, which are comparable to conventional materials. However, the production costs are still the bottleneck of PHAs' processes establishment at large scale, which have stimulated the scientific community and the productive sector to search for new alternatives. That is why zero-waste-integrated technologies, which will be presented in this chapter, appear as interesting solutions to couple one or more production processes in the same productive unity. One example is the use of agroindustrial wastes in fermentation to produce volatile-fatty acids (VFAs), biohydrogen, methane and carbon dioxide. VFAs are well-known precursors for PHAs synthesis, which is economically and environmentally important to promote a sustainable process. Characteristics of the synthetized PHA depend on the VFA chain length and composition, which can lead to the formation of hydroxybutyrate (HB) and hydroxyvalerate (HV). This last may promote the synthesis of P(HB-*co*-HV), which is interesting from an industrial point of view, as it is more elastic and flexible.

Keywords Bioplastics · Polyhydroxyalkanoates · Biohydrogen · Volatile fatty acids · Agroindustrial residues · Wastewaters · Zero-waste technology

---

A. F. M. de Mello · L. P. de Souza Vandenberghe (✉) · Z. S. Vásquez · C. M. B. Machado · N. R. Nitsch · A. B. P. Medeiros · W. J. M. Burgos · C. R. Soccol
Federal University of Paraná, Department of Bioprocess Engineering and Biotechnology, Centro Politécnico, Curitiba, Paraná, Brazil
e-mail: lvandenberghe@ufpr.br

C. R. Soccol et al. (eds.), *Biohydrogen - Advances and Processes*, Biofuel and Biorefinery Technologies 13, https://doi.org/10.1007/978-3-031-49818-3_16

16.1 Introduction

In 2015, the United Nations adopted the 2030 Agenda for Sustainable Development including 17 Goals—the Sustainable Development Goals (SDGs), which must be followed by countries, enterprises and nongovernmental organizations. Along with hunger, health, education, gender equality and other issues, the environment was the main topic in six of the defined goals for sustainable global progress. One of the targets is to reduce the environmental impact of cities, paying attention to air quality and waste management. The indiscriminate exploration of petroleum to produce fuels and daily used products has led society to a dependence on this fossil source. However, being nonrenewable, petroleum is becoming scarce, making the market unstable. Besides, the use of fuels and products derived from this source also has a negative impact on the environment. The combustion of fossil fuels generates CO_2 and, therefore, impacts on global warming, in the greenhouse effect and in the hole of the ozone layer, beyond other problems (Nicoletti et al. 2015; Damle 2014).

In petroleum refineries, plastics are also produced. Conventional plastics of petrochemical origin are dominant worldwide due to their wide range of uses, from packaging and consumer goods to automotive and electronic parts. Since some applications, such as packaging, are very quickly disposable and due to the great difficulty of degradation of these materials, plastics have aroused great environmental concern (García-Depraect et al. 2021). It is estimated that around 50 Mt of plastic was produced per year during the 1960s and around 460 Mt in 2019, expressing a great increase in manufacture (Law 2017).

An alternative to the production of petroleum derived plastics are biodegradable bioplastics. These have thermal and mechanical properties and performance characteristics similar to those of conventional plastics, but are easily degraded by the action of microorganisms in the environment. Among the bioplastics, polyhydroxyalkanoates (PHAs) are naturally produced by bacteria in the form of intracellular granules for carbon and energy reserve and considered biocompatible and biodegradable (Urtuvia et al. 2014), being potential substitutes for traditional petrochemical plastics. However, the major limitation is the high cost of production. However, the low price of plastic products does not consider the whole lot of costs that are largely invisible, and for which governments, people and the planet are ultimately footing the bill. In fact, the management of plastic waste costs, to collect, sort, dispose and recycle the huge quantities of plastic waste generated every year, are more than US $32 billion (WWF 2021). As an example, in 2015, less than 4% of globally generated plastic waste derived from packaging was recycled (Rosenboom et al. 2022).

Currently, important processes in several segments, such as pharmaceutical, food and beverage, bioethanol and bioenergy industries, are conducted via biotechnological processes with the use of by-products and/or wastes as alternative substrates. Such processes minimize the generation of waste, within the concept of "Zero-Waste Technology," which has brought significant economic and immeasurable environmental advantages. In fact, biomass biorefineries enable the development of a cycle with the valorization of agroindustrial by-products from different chains and food manufacturing processes for the production of biomolecules of medium to high added value of industrial interest. Therefore, the reuse of lignocellulosic residues (sugarcane bagasse, citrus pulp, corn stover, soybean hulls and others) and liquid effluents (residual glycerol, effluents from the food and beverage industry, wastewater) involving the concept of circular bioeconomy is proposed (Lin et al. 2022).

Therefore, alternatives to the production of bioplastics are being discussed and disseminated by the scientific community. This is the case of plant-derived biomasses and other wastes that can be used as a source of carbon and nutrients for the production of bioplastics and, therefore, reducing the production cost and closing the process cycle (Dietrich et al. 2017; Yadav et al. 2020).

In the "Zero-Waste Technology" concept, biohydrogen is recognized as a clean energy source because it is a fuel that does not generate greenhouse gases and is a substitute with less environmental impact than fossil fuels. The production of biohydrogen by dark fermentation has been considered an excellent alternative for the generation of renewable energy, especially when there is reuse of a variety of industrial waste, such as effluents with high organic load, in parallel to the coproduction of compounds with higher added-value (Yang and Wang 2019). The production of biohydrogen and the whole concept of bioenergy can be related to the use of industrial waste and renewable biomass sources (Mishra et al. 2017). Given the issues involved regarding the scarcity of nonrenewable energy sources, the environmental impacts caused by their massive use and the waste disposal burden on the environment, the search for so-called "green" technologies are very promising and have been the focus of research in the development of sustainable biorefineries (Isla et al. 2013). The use of substrates such as industrial waste can reduce the production cost of both PHA and biohydrogen. Furthermore, it is known that volatile organic acids resulting from dark fermentation can be good substrates for PHA production (Colombo et al. 2019).

Therefore, in this context, the integrated production of PHAs and biohydrogen is being studied by different research groups, since the composition of different effluents and by-products is favorable for the synthesis of the biomolecules, following the concept of circular bioeconomy and zero-waste technology (Fig. 16.1).

16.2 Polyhydroxyalkanoates

Plastics are used daily in the most diverse areas of application, ranging from packaging to automotive and medical areas. With the growing population, the demand for plastics is continuously increasing, despite the environmental burden that these materials have, since they are mostly produced from petroleum, a nonrenewable source and not biodegradable. The persistence of these plastics in the environment is raising awareness of governments, enterprises and society in general. It was estimated that 4.8–12.7 million metric tons of plastic waste re-entered the ocean in

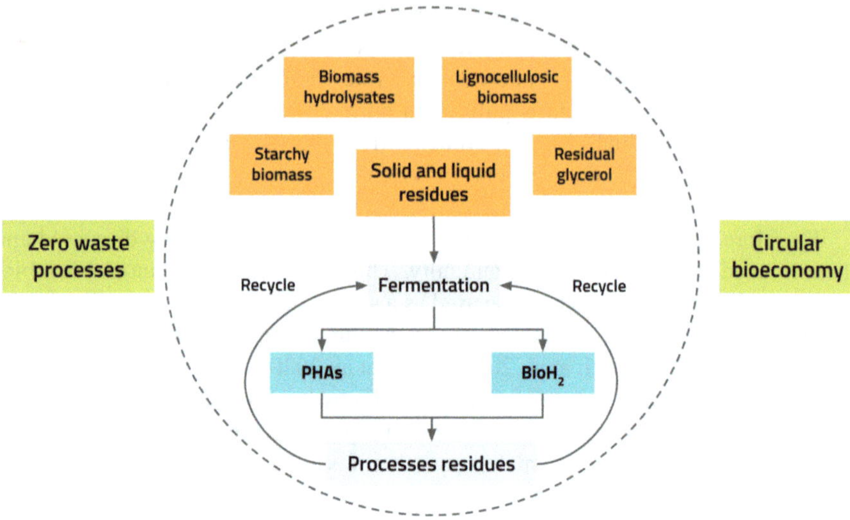

Fig. 16.1 Biohydrogen and PHAs coupled production

2010, and this number can grow by an order of magnitude by 2025 (Jambeck et al. 2015). Recently, microplastics that are generated during plastic degradation have become an issue of increasing concern since they are a risk to ecosystems (Zhang et al. 2021).

Therefore, the production of bioplastics from renewable sources is gaining attention as these materials are biodegradable, present similar properties to conventional plastics and can replace them while reducing the environmental impact. Polyhydroxyalkanoates (PHAs) are an example of biopolymers. They are biological polyesters produced in various organisms (both eukaryotic and prokaryotic) as an intracellular lipid inclusion, serving as an energetic reservoir to the cells. Transgenic plants are one of the eukaryotic sources of PHAs, producing the biopolymer along with the other cell constituents (starch, lignocellulose, pectin, among others). The first report of PHAs derived from plants was in the 1990s, with a polyhydroxybutyrate (PHB) being produced in *Arabidopsis thaliana* with maximum accumulation of 0.01% (Dobrogojski et al. 2018). With technological development in genetic engineering of transgenic plants, there are now reports of organisms that can accumulate up to 40% of the polymers (Tan et al. 2016).

However, the most common mode is the microbial production of PHAs through fermentation. Several species of bacteria, both Gram positive and negative, have been reported as PHAs producers. These bacteria are divided into two groups regarding PHA production: the ones that require nutrient limitation for biopolymer production and do not accumulate PHAs during the growth phase and the ones that do not have these requirements (Raza et al. 2018). *Cupriavidus necator* (formerly known as *Ralstonia eutropha*), *Protomonas extorquens, Pseudomonas oleovorans* and *Pseudomonas putida* are examples of bacteria belonging to the first group (Raza et al. 2018), while *Alcaligenes latus* and recombinant *Escherichia coli* are examples of the second group of bacteria (Muhammadi et al. 2015). These bacteria can use several carbon sources for growth and PHA production, from reducing sugars derived from polysaccharides of agricultural crops and waste, to plant oils and volatile fatty acids (VFAs) and can accumulate up to 90% of PHA (Passanha et al. 2013).

Mixed microbial cultures (MMCs) can also be applied to PHAs' production. In these open and nonsterile systems, external operational conditions are applied for the appropriate selection of PHA-accumulating organisms (Kourmentza et al. 2017). The main advantages of these systems are the possibility of using low-cost substrates such as various wastewaters, no need of sterilization, which reduces operational costs and the possibility of integrating PHA production with wastewater treatment systems in biorefineries (Salehizadeh and Van Loosdrecht 2004). However, yields and productivities of these systems tend to be lower when compared to pure culture production, since nonproducing microorganisms can be present in the system. Besides, the time of operation is usually higher, which can lead to higher costs of maintenance (Tu et al. 2019).

PHAs can be classified into three groups regarding their monomers: short (scl-PHAs, monomers having 1–5 carbon), medium (mcl-PHAs, monomers with 6–16C) and large chain length (lcl-PHAs, monomers with over 17C). Scl and mcl-PHA are the most commonly produced, while lcl-PHAs are rare to be found (Muneer et al. 2020). The change in the number of carbons in the monomers occurs mainly in the R and n sessions of the molecule, giving each type of PHA different properties (Fig. 16.2) (Alcantara et al. 2020). PHB is an example of a scl-PHA and is the most common type of PHA produced by bacteria, possessing properties that are similar to polypropylene (PP) (Parveez et al. 2012; McAdam et al. 2020). Copolymers of PHAs produced by bacteria are also reported, such as polyhydroxybutyrate-*co*-valerate (PHBV). These forms of PHA are largely applied in the biomedical area due to their great biocompatibility and better mechanical properties when compared to PHB (Rivera-Briso and Serrano-Aroca 2018).

The main disadvantage of PHAs, when compared to conventional plastics, is the cost of production and commercialization. The main factors affecting PHA costs are substrate choice and downstream operations. Therefore, the use of waste as carbon

n	R	Compound
1	hydrogen	poly(-3-hydroxypropionate)
	methyl	poly(-3-hydroxybutyrate)
	ethyl	poly(3-horoxyvalerate)
	propyl	poly(-3-hydroxyhexanoate)
	pentyl	poly(-3-hydroxyoctanoate)
	nonyl	poly(3-hydroxydodecanoate)
2	hydrogen	poly(-4-hydroxybutyrate)
3	hydrogen	poly(-5-hydroxyvalerate)

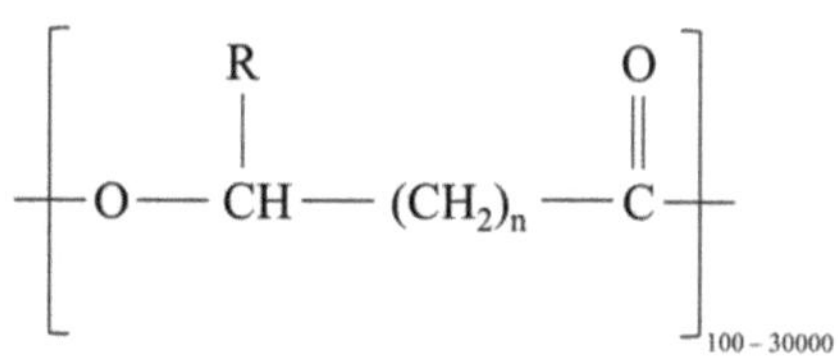

Fig. 16.2 General structure of PHAs and examples of different types of PHA (Adapted from Ojumu et al. (2004))

and nutrient sources for PHA production can help reduce costs and enhance commercialization. Besides, the integration of PHA production with other high-added value products can help to reduce costs of production and recovery of biopolymers (Yadav et al. 2021). Therefore, the integration of the production of PHAs and biohydrogen from various waste sources is being largely researched and will be further discussed.

16.3 Integrated Biorefinery Approach for the Simultaneous Production of Biohydrogen and Polyhydroxyalkanoates

16.3.1 Pathways for the Biosynthesis of Biohydrogen Coupled with Polyhydroxyalkanoates

The use of waste (solid and liquid) for the production of biohydrogen using bacteria, photosynthetic bacteria or microalgae as inoculum is a promising biorefinery process (Karp et al. 2021; Sydney et al. 2019; Vieira et al. 2021), since part of the organic matter and some pollutants are removed (de Nascimento Junior et al. 2021; Martinez-Burgos et al. 2021). Dark fermentation is an anaerobic process and begins with the hydrolysis of the most complex carbohydrates to their monomeric units, that is, monosaccharides. Then glucose is converted into pyruvate, which is a key intermediate in dark fermentation as it is reduced to nicotinamide adenine dinucleotide (NADH). Subsequently, hydrogen is produced by the fragmentation of the pyruvate molecule in CO_2, H_2 and acetyl-CoA in the presence of the enzyme pyruvate ferredoxin oxidoreductase (Fig. 16.3, **step 1**). A characteristic of this route is that part of the electrons is transferred to protons to generate H_2 and the other part to NAD^+, which is transformed into $NADH_2$. Later, $NADH_2$ is used for the production of hydrogen catalyzed by hydrogenases through a second metabolic route (Chen et al. 2006).

In the dark fermentation, additionally to the final product of interest H_2, some intermediate metabolites are also produced, such as alcohols, acetones (Martinez-Burgos et al. 2019; Saady 2013) and volatile fatty acids (VFAs), such as lactic acid, valeric acid, caproic acid, acetic acid, butyric acid and propionic acid (Martinez-Burgos et al. 2021). From the point of view of biohydrogen production, acetic and butyric acids are the most important in dark fermentation, since, stoichiometrically, for each mole of acid generated, two moles of H_2 are produced, while propionic acid is not desired as it consumes H_2 molecules for its generation (Equations step 1).

Thus, de la Rosa et al. (2020) reported concentrations of approximately 6 and 3.5 g/L of butyric acid and acetic acid, respectively, in biohydrogen production, using palm oil mill effluent (POME) as the main substrate. Sydney et al. (2014) described significant concentrations of butyric acid and acetic acid when producing biohydrogen via dark fermentation, 4.5 g/L and 0.25 g/L, respectively. The main problem with the organic acids produced in dark fermentation is that they also

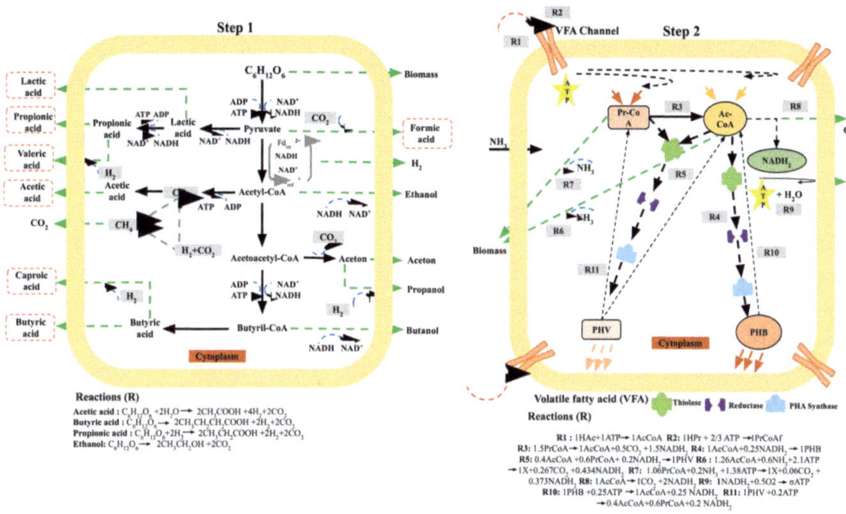

Fig. 16.3 Metabolic pathways for the simultaneous production of biohydrogen and polyhydroxyalkanoates (Adapted from De Donno Novelli et al. (2021) and Saady (2013))

provide a high organic load in the fermented broth, so additional processes or technologies are needed for their use or removal to meet the discharge requirements of these effluents (Choi et al. 2008).

An alternative for the use and removal of VFAs from the fermented broth is to use them for the production of PHAs in an integrated process with hydrogen production. However, it is carried out in more than one step. In the first step, hydrogen and VFAs are produced, and in the second step, PHA is synthesized from the VFAs. According to Pagliano et al. (2017), bacteria such as *Ralstonia eutropha, Bacillus megaterium, Pseudomonas oleovorans, Acinetobacter beijerincki, Rhizobium sp.* and *Nocardia sp.* can use propionic acid, formic acid and acetic acid as a substrate to synthesize PHA, under limited nitrogen and phosphorus conditions. The synthesis of PHA from VFAs starts with their absorption (i.e., acetic acid and propionic acid) via active transport—the energy requirement of this process is one mol of ATP for each mol of acetate or propionate—and then these are converted into acetyl-CoA (ACoA) or propionyl-CoA (PCoA), with energy expenditure.

For the production of PHA, ACoA and PCoA are condensed and reduced by the enzymes, thiolase and reductase, respectively (De Donno Novelli et al. 2021). Three types of compounds can be produced: (i) 3-hydroxybutyrate, (ii) 3-hydroxyvalerate or (iii) 3-hydroxy-2-methylvalerate, the first is formed by 2 ACoA monomers, the second by 1 ACoA and 1 PcoA and the third by 2P-CoA. When the medium is enriched with propionate, a process of selective condensation of ACoA with PCoA is observed, producing PHV, while PHB is produced with ACoA when the PCoA is depleted. According to Dias et al. (2006), PHA accumulation occurs during the anaerobic phase by phosphorus and glycogen accumulating organisms driven by electron acceptor limiting conditions. Finally, microbial biopolymer from ACoA and PCoA can be produced. In fact, cells obtain energy from these compounds mainly when nutrients are scarce in the medium—PHA can be converted to ACoA or PCoA enzymatically (PHA depolymerase) and used by cells as nutrients when necessary (Fig. 16.3, **step 2**).

It is also possible to produce the two bioproducts, biohydrogen and PHA, simultaneously by the same microorganism. However, this is an area that still needs further research. These simultaneous PHA and biohydrogen production processes could be a key to the viability of producing these metabolites at industrial scale, with the strategy of zero-waste processes (Dinesh et al. 2020). In these simultaneous processes, for example, when ACoA is consumed to produce PHA, the rate of degradation of pyruvate increases, which increases the generation of formate and, subsequently, the production of biohydrogen (Thulasidharan et al. 2021; Wang et al. 2012). Among the microorganisms that have been used for the production of PHA and biohydrogen simultaneously are photofermentative bacteria such as *Rhodopseudomonas palustris*, *Rhodobacter sphaeroides* and *Rhodospirillum rubrum* (Melnicki et al. 2009; Vincenzini et al. 1997; Yigit et al. 1999). As an example, Yigit et al. (1999) produced biohydrogen and PHA using the *R. sphaeroides* strain and a medium composed of 30% of wastewater from sugar refinery with L-malic acid and sodium glutamate. The concentrations achieved were 35 mL/L and 0.5 g/L for biohydrogen and PHA, respectively. The process was developed in two stages and the strategy used was nitrogen limitation. Lorrungruang et al. (2006) also produced significant amounts of biohydrogen and PHA using *R. sphaeroides* and a medium based on glucose, fructose and glutamate-malate. A biomass concentration of 35 g/L was reached with a PHA content of 60%.

16.3.2 PHA Production from Biohydrogen VFAs Residue and Oils

Domestic and industrial waste are good alternative carbon sources for PHAs production, as they are rich in organic matter and the environment would benefit from its removal from the effluents. The use of these residues as substrates in PHA production can be a solution to reduce its production costs since the carbon source represents up to 50% of its cost (Kim 2000). Usually, feedstocks rich in organic matter (such as industrial or municipal effluents) go through an acidogenic digestion (AD) process to maximize the VFAs content that will be further incorporated into the PHA production media (Valentino et al. 2017). The AD process generates VFAs as well as $BioH_2$, enabling the integration of both biomolecules production (Wainaina et al. 2019). When using AD-derived VFAs, the use of open mixed cultures is proposed, which are based on the integration of wastewater treatment and the usage of industrial by-products as substrates (Bengtsson et al. 2010). The use of mixed microbial cultures (MMC) also reduces production costs associated with the sterilization process (Pinto-Ibieta et al. 2021). However, using MMC is more time-consuming than using pure cultures, since it is necessary to select the desired culture prior to the fermentation step (Tu et al. 2019). Furthermore, pure cultures provide higher yields than MMC.

In addition to its production cost, PHB homopolymer has a limited application range since it has high crystallinity, brittleness and high melting temperature (Naguib et al. 2011), limiting its processability window. Furthermore, PHB has a glass transition temperature (T_g) around 5 °C (±2 °C) (Wei et al. 2006), that is, the temperature in which a significant loss of stiffness and volume occurs, and the T_g is proportional to material crystallinity level (Greene 2021). The brittleness is related to the formation of rigid amorphous fragments within the PHB (Di Lorenzo et al. 2012). As an alternative, copolymerization between PHB and PHV can improve PHB's mechanical properties (Ahmed et al. 2010), since PHBV has better impact resistance, toughness and flexibility when compared to PHB homopolymer. In fact, increasing the HV content in PHBV up to 50% can lower the melting point of the final copolymer (Wang et al. 2013).

The production of different PHAs monomers can be tailored using different substrates since the carbon source determines the final composition of the biopolymer. For example, in a previous study developed by Wang et al. (2022), it was observed that PHB was produced when butyrate was employed as substrate, whereas PHBV was obtained using valerate. In this sense, the composition of VFAs residues plays a key role in PHA production considering that even-numbered VFAs (like acetate and butyrate) generate HB monomers. However, odd-numbered VFAs (such as propionate and valerate) result in HV units (Montiel-Jarillo et al. 2021). Different oils can be employed (Table 16.1) as a carbon source in PHA production, which also influences the type of monomer unit formed (Flores-Sánchez et al. 2017).

The influence of substrates in PHA monomers synthesis is due to the different metabolic routes employed in the biopolyester formation (Fig. 16.4). Through sugar assimilation, β-ketothiolase catalyzes the successive formation of acetyl-CoA and acetoacetyl-CoA. Then, acetoacetyl-CoA is converted to 3-hydroxybutyryl-CoA by acetoacetyl-CoA reductase, which is then polymerized to PHB by PHA synthase. When fatty acids are used as a carbon source, the β-oxidation pathway is followed (Chen et al. 2015), which comprises successive reactions that convert acyl-CoA into enoyl-CoA, (S)-3-hydroxyacyl and 3-ketoacyl-CoA that can be turned into acyl-CoA (closing the cycle) or acetyl-CoA (Thulasidharan et al. 2021). The latter can also be incorporated into the fatty acid biosynthesis pathway and, in this case, acetyl-CoA will be catalyzed into (R)-3-hydroxyacyl-ACP and then in (R)-3-hydroxyacyl-CoA (Muhammadi et al. 2015). Furthermore, the enoyl-CoA generated in the β-oxidation pathway can be converted into (R)-3-hydroxyacyl-CoA (Thulasidharan et al. 2021). Then, the 3-ketoacyl-CoA synthesized in this pathway can be catalyzed by 3-ketoacyl-ACP reductase, resulting in (R)-3-hydroxyacyl-CoA (Muhammadi et al. 2015). Then, the final molecule obtained, R-hydroxyacyl-CoA, will be polymerized into PHA, also by PHA synthase (Valencia et al. 2021). The difference in the molecule formed by the acetoacetyl-CoA reductase reaction between the metabolic routes occurs possibly because of the variable composition

Table 16.1 PHA production using VFAs and oils residues with different operating conditions

Microorganism	Substrate	Biomass (CDW, g/L)	PHA (% CDW)	PHA (g/L)	Polymer	Operation mode	Working volume	Reference
Marinobacterium nitratireducens	Butyrate	5.06	66.81	3.38	PHB	Batch, shake flasks	30 mL	Wang et al. (2022)
Marinobacterium sediminicola	Valerate	4.50	74.99	3.37	PHB-*co*-HV	Batch, shake flasks	30 mL	Wang et al. (2022)
Cupriavidus necator DSM 428	VFAs from waste paper hydrolysate	1.46	53.50	0.7811	PHB-*co*-HV	Batch, shake flasks	–	Al-Battashi et al. (2019)
Cupriavidus necator H16 (DSM 428)	AD[a] derived VFAs	1.7	44.00	0.748	PHB-*co*-HV	Batch, shake flasks	50 mL	Jawed et al. (2022)
Bacillus megaterium uyuni S29	Acetate	19.05	70.21	13.37	PHB-*co*-HV	Fed-batch, reactor	4 L	Kacanski et al. (2022)
Bacillus tequilensis	Acidogenic fermented food waste	–	36.00	–	PHB-*co*-HV	Batch, shake flasks	100 mL	Venkateswar Reddy et al. (2014)
Methylocystis hirsuta	Biogas and valerate	–	53.80	–	PHB-*co*-HV	Batch, shake flasks	500 mL	López et al. (2018)
MMC[c]	VFAs from cheese whey	5.8[b]	41.50	2.41[b]	PHB-*co*-HV	Fed-batch, reactor	2 L	Portela-Grandío et al. (2021)
	Sludge fermentation liquid	0.46	59.47	0.273	PHA	Batch, reactor	500 mL	Jia et al. (2014)
Cupriavidus necator PHB^{-4}*/pBBREE32b13*[d]	Waste cooking palm oil	22.3	85.00	18.95	P(3HB-*co*-3HHx)	Batch, shake flasks	250 mL	Kamilah et al. (2013)
Cupriavidus necator H16	Fresh cooking palm oil	24.8	70.00	17.36	P(3HB)	Batch, shake flasks	250 mL	Kamilah et al. (2013)
Cupriavidus necator PHB^{-4}*/pBBREE32b13*[d]	Fresh cooking palm oil	23.3	75.00	17.47	P(3HB-*co*-3HHx)	Batch, shake flasks	250 mL	Kamilah et al. (2013)
Cupriavidus necator B-10646	Siberian seed oil	5.8	59.00	3.42	P(3HB-*co*-3HV-*co*-3HHx)	Batch, shake flasks	500 mL	Volova et al. (2020)

Microorganism	Substrate	Biomass (CDW, g/L)	PHA (% CDW)	PHA (g/L)	Polymer	Operation mode	Working volume	Reference
Cupriavidus necator H16	Jatropha oil	19.05	65.20	12.42	P(3HB)	Fed-batch, reactor	7 L	Ng et al. (2010)

[a] Anaerobic digestate
[b] mg/L
[c] Mixed microbial cultures
[d] Recombinant strain

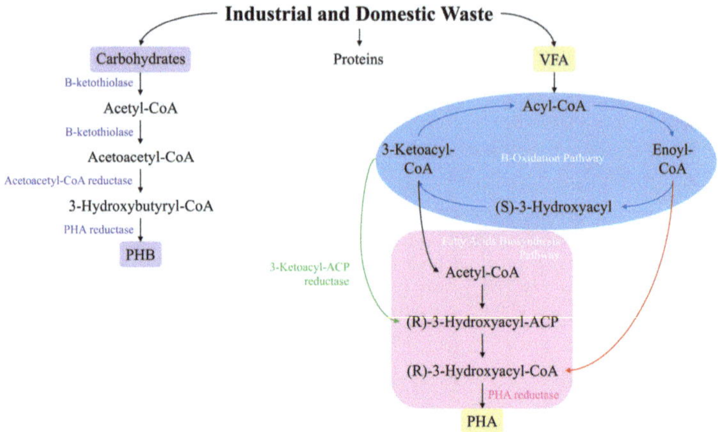

Fig. 16.4 Metabolic routes employed in different PHA monomers synthesis

of the fatty acids, which determines the R-group of R-hydroxyacyl-CoA molecule and, consequently, will determine the PHA final composition.

16.3.3 Lignocellulosic Material and Agroindustrial Wastes as Sources for Coproduction of PHA and Biohydrogen

In Brazil, the agroindustrial chain generates 291 million tons per year of wastes, representing an environmental and social risk (Siqueira et al. 2022). The use of these raw materials as feedstock is pursued in a global low-carbon economy context (Soto et al. 2019). In this sense, lignocellulosic materials can be used to obtain bioproducts (Kumar and Sharma 2017). These materials, such as wheat straw, wheat bran, sugarcane bagasse and rice straw, are composed of cellulose, hemicellulose and lignin (Khatami et al. 2021). To release the sugars trapped within the lignocellulosic matrix, a pretreatment is necessary. This step can be achieved by physical, chemical or physicochemical treatments, and in a second stage, enzymatic hydrolysis is performed (Bittencourt et al. 2021). These treatments facilitate the access of enzymes to polysaccharides and, consequently, the hydrolysis process (Kucharska et al. 2020). The pretreatment followed by hydrolysis can release glucose, xylose, arabinose, organic acids, lignin derivatives (vanillin and phenols), and microbial inhibitors, such as furfural and hydroxymethylfurfural (HMF) (Li and Wilkins 2020). These toxic compounds can inhibit microbial growth; however, it was observed previously that when furfural (0.2 g/L) was added to lignocellulosic rich media, the production of nitrogenase, a key enzyme in photofermentation biohydrogen production, was higher in comparison to media without furfural, resulting in an increase of 19% of BioH_2 production (Yang et al. 2022). Additionally, it was known that higher contents of solids in pretreatment generate higher yields of inhibitory

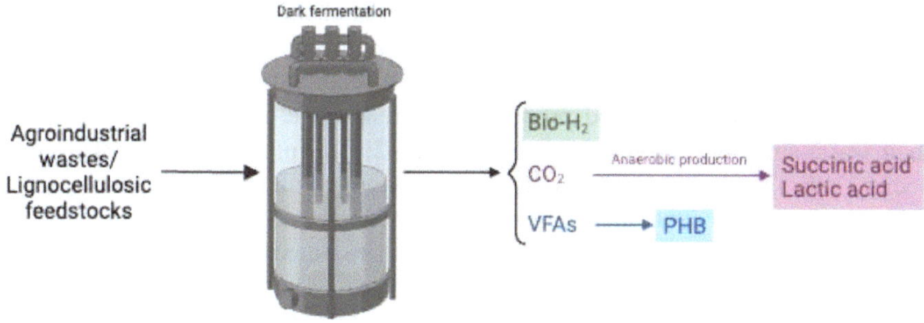

Fig. 16.5 Bio-H_2 and PHA coproduction

compounds (Lee et al. 2017). The inhibition caused by these toxic compounds can be reduced by detoxification processes, which can lead to higher operational costs (Du et al. 2020). Physical pretreatment like milling processes are an alternative, since they produce lower amounts of toxic compounds (Kumar and Sharma 2017).

After one of the treatments mentioned earlier, the biowastes, agroindustrial wastes and lignocellulosic feedstocks can be used to produce biomolecules, such as bioplastics (Bhatia et al. 2021) and biohydrogen (Patel and Shah 2021). Bio-H_2 and PHA coproduction are beneficial for both processes, since the PHB synthesis pathway consumes the ACoA generated by pyruvate, avoiding the accumulation of toxic compounds (Wang et al. 2012). Additionally, the presence of H_2 gas enhances the conversion of pyruvate to ACoA (Thulasidharan et al. 2021) and, consequently, PHB synthesis (Fig. 16.4).

Agroindustrial wastes and lignocellulosic feedstocks can be degraded by anaerobic microorganisms, producing bio-H_2 in a dark fermentation (DF) mode (Bhatia et al. 2021). This fermentation strategy was previously described in the earlier section. DF is the most suitable mode of fermentation in biohydrogen and PHAs production in one-step process, as shown in Fig. 16.5. It is worth mentioning that DF generates low substrate conversion rates (De Gioannis et al. 2013), but it is possible to implement subsequent dark and photo fermentation in a two-step process, generating higher amounts of biomolecules (Dinesh et al. 2020). This strategy is the most used in PHA and bio-H_2 one-step coproduction.

16.4 Research, Patents and Economic Perspectives

The increasing world population is continuously demanding for more energy and daily plastic products. Covid-19 pandemics also played a crucial role in plastic waste generation. Peng et al. (2021) estimated that about 8.4 million ton of plastic waste was generated until August 2021, being 25.9 thousand tons inappropriately disposed in the oceans (Peng et al. 2021). This scenario of petroleum scarcity, negative environmental impact of fossil fuels, rising energy demand and lack of degradability of plastics has led governments, researchers and enterprises to search for alternatives that could mitigate these effects. In this context, the concept of bio-economy has risen. Among various definitions, bioeconomy is usually described as the production of value-added products from renewable bio-based resources and their waste streams (European Commission 2012; Bröring et al. 2020). Therefore, society is facing a transition from the linear economy with take-make-dispose concept to a circular bioeconomy where renewable resources are applied to the production of various products and the waste in the system is minimized through reutilization in a biorefinery manner.

Thus, biofuels and bioplastics pose as great products in these systems. Coupling the manufacturing of these bioproducts is also a great opportunity to reduce dependence on fossil sources, creating multiplatform bio-based refineries. Public policies and fundings are being constantly released in order to stimulate production and commercialization of hydrogen and bioplastics. In March 2022, China released a plan for the development of hydrogen industry in a medium and long-term, in which the production of up to 200 thousand tons of green hydrogen is targeted through the development of four focus areas: collaborative innovation, infrastructure, application of hydrogen in multiple areas and constant improvement of frameworks and policies (HyResource 2022). Also in 2022, the Department of Energy (DOE) of the United States announced a funding of 28 million dollars for the development of projects that would enhance the use of hydrogen, being one focus area the development of clean hydrogen from waste and biomass (US DOE 2022). In the EU, the hydrogen strategy released in 2020 was divided in three phases for technology diffusion until 2050. In the first phase, between the years of 2020 to 2024, the EU targeted the production of up to 1 Mt of renewable hydrogen (Wolf and Zander 2021).

Policies about bioplastics production and commercialization are much scarce, usually appearing as an area of interest in general bioeconomy frameworks. Besides, policies regarding management of conventional plastic waste and disposal also impacts on bioplastics funding and development. Since 2005, China has been promoting a program for the development of biodegradable plastics and, in 2017, banned the importation of plastic waste (Filiciotto and Rothenberg 2021; Moshood et al. 2021). The Biotechnology Strategic Scheme in Japan targeted the replacement of 20% of conventional plastics with plastics derived from renewable sources until 2020 (Philp 2014). Thailand recognized the bioplastics industry as a relevant and strategic sector in 2006 and developed a roadmap in 2008 (Moshood et al. 2021). The EU has several strategies for reducing or dealing with plastic waste, such as the reduction of using cheap plastic bags (released in 2015 in the Circular Economic Action Plan) and the banning of using disposable plastic products that already have a substitute such as straws since 2019. The main focus of these policies is on the disposal of conventional plastics, rather than policies regarding the development of biodegradable plastics (Moshood et al. 2021). However, for the second quarter of 2022, it is expected that there will be a release and adoption of a framework on bio-based, biodegradable and compostable plastics (Lange 2019).

In this context, the market and research about production of biohydrogen and PHAs are growing in recent years. PHAs market was evaluated at 57.8 million dollars in 2021 and is expected to reach U$ 98.5 million until 2026, with a compost annual growth rate (CAGR) of 11.8% (Market Data Forecast 2022). PHA production capacity accounts for 1.8% of global bioplastics global production and is projected to grow to 6.4% until 2026 (European Bioplastics 2022). The market is concentrated mostly in the Europe region, mainly driven by packaging applications (Mordor Intelligence 2022). On the other hand, the global hydrogen market was evaluated at 130 billion dollars in 2020 and is expected to grow up to 201 billion until 2025, representing a CAGR of 9.2%. However, most of this hydrogen is still derived from fossil sources, as it is the most economical method for producing this fuel (Markets and Markets 2021). Biohydrogen companies are concentrated in the European region, but the Asia region is up to show the highest CAGR due to governmental funding on research and development (Transparency Market Research 2021).

The main bottleneck for the growth of both biohydrogen and PHA markets is the cost of production when compared to other technologies. Therefore, the main research efforts are focused on the reduction of these costs either by applying alternative substrates or coupling the production of biomolecules, among several other strategies. In order to reveal the research scenario regarding the coproduction of biohydrogen and PHAs, a bibliometric search was made in the Web of Science database. Keywords (biohydrogen, volatile fatty acids, light and dark fermentation, anaerobic digestion, polyhydroxyalkanoate and specific polymers, copolymers and their abbreviations) were selected based on their relevance and combined with Boolean logic. The search was conducted in the time range of 2000 until June 2022, and 597 documents were retrieved. After reading the title and abstracts of the documents, 558 published works were selected based on their relevance to the theme, and their data about year of publishing and authors' origin countries and regions were analyzed.

In the early 2000s, research about coupled production of biohydrogen and PHAs was still incipient, starting to grow significantly in 2012 (Fig. 16.6). This growth is even more expressive in recent years (2020 and 2021). The decrease in the number of documents in 2022 does not reveal a lack of interest in the technology, but rather reflects the date that the search was made. Therefore, the scenario of research in this theme is positive and with consolidated growth, being mainly boosted by the points previously discussed: scarcity of fossil resources, rising concern about environmental impacts of fuels and plastics, bioeconomy development and public policies that allows a good academic and research environment. As aforementioned, the Europe region is the one that holds the largest market for both biohydrogen and PHAs and is constantly updating their policies concerning these biomolecules, which reflects on the number of European registers found in the database (368). The Asia region is also relevant in this scenario, being the second major contributor with 298 registers. China itself owns 105 registers, being the major single-country contributor as expected, since Chinese scientific production has risen significantly in the last few years and in 2018, China surpassed the United States and became the largest publisher of scholarly works (Tollefson 2018).

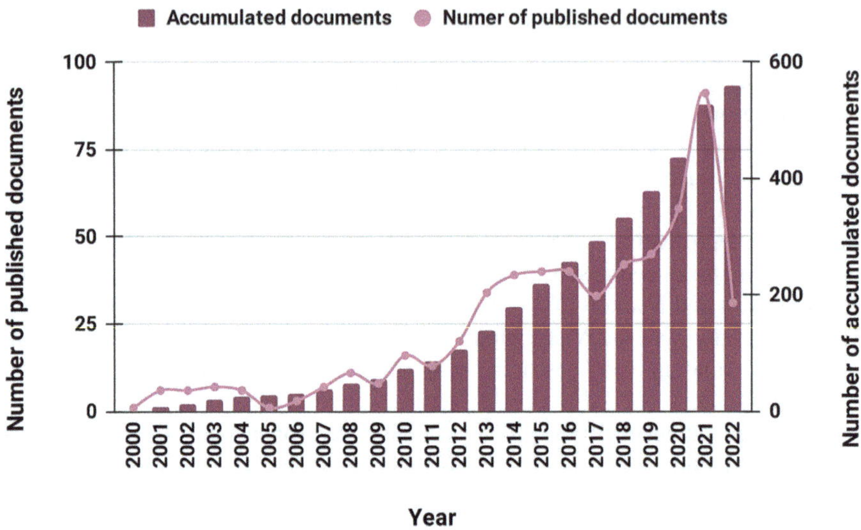

Fig. 16.6 Number of published scholarly works per year regarding the search of coupled biohydrogen and PHAs' production in Web of Science (WoS) database

The search for scholarly works revealed a good scenario for the development of biohydrogen and PHAs coupled production. However, data about this type of document alone are not enough to reveal innovation efforts regarding a certain theme. The bioeconomy is a highly innovative type of economy since new processes and products are constantly being developed from renewable resources in order to substitute conventional technologies. Innovation is a concept that possesses multiple definitions, which share a similar theoretical base: the creation of new knowledge and its dissemination and implementation (Lazzarotti et al. 2020). Innovation in bioeconomy has certain specificities such as a high level of cooperation among different players, being across multiple industries for the generation of multidisciplinary knowledge or across sectors such as private and public. Transferring research efforts and results to a commercial scale in strong partnerships is a crucial step in the innovation for bioeconomy development (Bauer et al. 2018).

One of the ways of measuring innovation is through the patent system, in which the generated knowledge and/or new processes and products are temporarily protected for exclusive exploitation of the technology. Data about patents in a certain theme can help to reveal the scenario and trends of technology development and generate insights about the future direction of the industry. However, there are some relevant points that need to be taken in consideration when analyzing data about patents: not all novelty is patentable, some technologies can be protected by other systems such as industrial secret and the propensity to patent can vary among different sectors (Wydra 2020). Therefore, taking these factors into account, a patent search was performed in an attempt to unveil the scenario of innovation regarding the coupled production of biohydrogen and PHAs.

For this purpose and in order to compare numerical data about scholarly works and patents, the same search conditions described for the WoS database were applied in the Derwent Innovation Index (DII) database. A total of 59 documents were retrieved in the established time range (2000–June 2022) and, after manual revision and sorting through reading titles and abstracts, 54 documents were selected for analysis.

The evolution of published patent documents throughout the years does not follow a determined pattern (Fig. 16.7). In the early 2000s, there was a lack of registers and since 2019 there is a more significant number of documents published per year. The technology of coupling PHAs and biohydrogen production is relatively new and, therefore, can be considered in its initial phase, which can justify this inconsistency in the number of published patent documents in the last few years. Besides, in a recent report, the International Energy Agency (IEA) reveals that there has been a decrease since 2010 in patenting technologies regarding clean energy, among them hydrogen and biofuels (International Energy Agency 2020), which can contribute to this result. The profile of regions of origin of the documents resembles the results obtained in the WoS search. In this case, the Asia region contains the major number of registers (37%), with Europe and North America regions in second place (27% each). These regions, as discussed, are known for relevant policies for the development of both hydrogen and bioplastics and hold the largest markets of these bioproducts. With the development of these policies and funding to research related to the theme, it is expected that technology will start to grow more expressively and consistently in the next few years.

The discrepancy in the number of documents found in each database and the profile of applicants of patent documents reveals the technology behavior in its

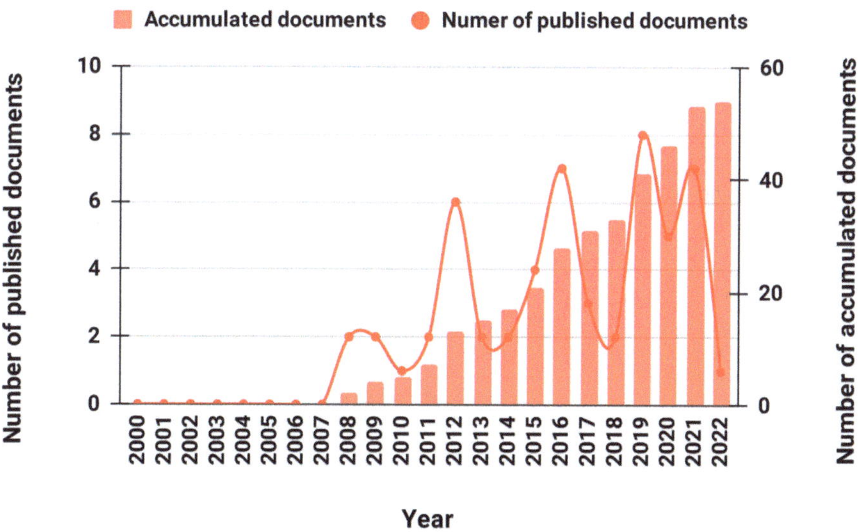

Fig. 16.7 Number of published patent documents per year regarding the search of coupled biohydrogen and PHAs production in Derwent Innovation Index (DII) database

initial phase of development (Fig. 16.8). The effort of cooperation between individuals and institutions is greater than the cooperation among different sectors (academic and private sectors). The number of published patent documents by universities and research institutes, however, is relevant and may imply that some of the scholarly works developed are being scaled-up to pilot scale and being protected. For proper technology development and implementation, the level of cooperation among these sectors needs to increase. As aforementioned, the development of coupled PHAs and biohydrogen production processes is still in an initial phase. Therefore, it can be expected that a larger number of patent documents will be observed in the next few years and also a larger degree of cooperation among actors. Apart from the joint development of new processes among sectors, technology licensing can be an alternative for transferring knowledge to the private sector, in which the original IP owner can transfer the rights of the patented technology to a IP buyer for a certain period of time (Origiin IP Solutions 2022).

The majority of the obtained documents (61%) is protecting production processes and systems in which wastewater can be treated while producing high-added value biomolecules (16%). Paques Biomaterials and Veolia Wastewater pose as great examples of enterprises applying efforts to produce PHAs coupled with biohydrogen from renewable resources. Paques Biomaterials is located in the Netherlands and possesses five pilot projects able to operate with different waste streams with high VFAs content. The polymer produced is PHBV and, in May 2022, the company announced the opening of PHA2USE, a demonstration plant for production of this material (GO!PHA 2022). On the other hand, Veolia is a multinational company with several operating unities around the world, specialized in wastewater treatment. The enterprise has been researching the production of PHAs in wastewater treatment systems for over 10 years and implemented a pilot installation in Brussels, Belgium, for the production of PHA from VFAs rich wastewater applying MMCs (Morgan-Sagastume et al. 2015).

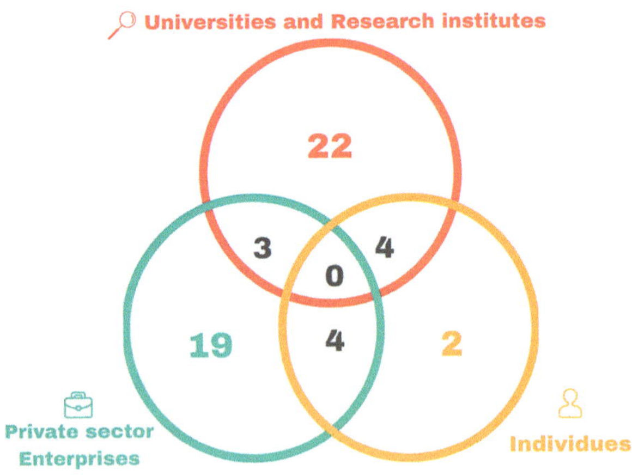

Fig. 16.8 Number of published patent documents per type of applicant regarding the search of coupled biohydrogen and PHAs production in Derwent Innovation Index (DII) database

16.5 Conclusions and Perspectives

The zero-waste technology applied to integrated processes brings new solutions to establish new sustainable biomolecules' production. Different agroindustrial residues can be employed for biohydrogen production. Presented examples of coupled biohydrogen production with sequential PHAs synthesis from previously generated VFAs has gained interest from both scientific community and industry. These materials present important properties that certainly increase their applicability and market value. However, even if these potentialities have already been identified to implement new coupled processes, much effort still needs to be made including the knowledge and technological transfer to the productive sector.

Acknowledgments This work was supported by the multilateral project BRICS STI FP 2021 – BIOGEN (293), financed in Brazil by the National Council for Scientific and Technological Development, CNPq, grant no. 440138/2022-1, also with support of the Coordination of Improvement of Higher Education Personnel – CAPES, project Print-BioADD.

References

Ahmed T, Marçal H, Lawless M, Wanandy NS, Chiu A, Foster JR (2010) Polyhydroxybutyrate and its copolymer with polyhydroxyvalerate as biomaterials: influence on progression of stem cell cycle. Biomacromolecules 11:2707–2715

Al-Battashi H, Annamalai N, Al-Kindi S, Nair AS, Al-Bahry S, Verma JP, Sivakumar N (2019) Production of bioplastic (poly-3-hydroxybutyrate) using waste paper as a feedstock: optimization of enzymatic hydrolysis and fermentation employing Burkholderia sacchari. J Clean Prod 214:236–247. https://doi.org/10.1016/j.jclepro.2018.12.239

Alcantara JMG, Distante F, Storti G, Moscatelli D, Morbidelli M, Sponchioni M (2020) Current trends in the production of biodegradable bioplastics: the case of polyhydroxyalkanoates. Biotechnol Adv 42:107582

Bauer F, Hansen T, Hellsmark H (2018) Innovation in the bioeconomy–dynamics of biorefinery innovation networks. Technol Anal Strateg Manag 30:935–947

Bengtsson S, Piscob AR, Johanssona P, Lemos PC, Reis MAM (2010) Molecular weight and thermal properties of polyhydroxyalkanoates produced from fermented sugar molasses by open mixed cultures. J Biotechnol 147:172–179

Bhatia SK, Jagtap SS, Bedekar AA, Bhatia RK, Rajendran K, Pugazhendhi A, Rao CV, Atabani AE, Kumar G, Yang Y-H (2021) Renewable biohydrogen production from lignocellulosic biomass using fermentation and integration of systems with other energy generation technologies. Sci Total Environ 765:144429

Bittencourt GA, Vandenberghe LPS, Valladares-Diestra K, Herrmann LW, Mello AFM, Sarmiento-Vásquez Z, Karp SG, Soccol CR (2021) Soybean hulls as carbohydrate feedstock for medium to high-value biomolecule production in biorefineries: a review. Bioresour Technol 339:125594

Bröring S, Laibach N, Wustmans M (2020) Innovation types in the bioeconomy. J Clean Prod 266:121939

Chen X, Sun Y, Xiu Z, Li X, Zhang D (2006) Stoichiometric analysis of biological hydrogen production by fermentative bacteria. Int J Hydrogen Energy 31:539–549. https://doi.org/10.1016/j.ijhydene.2005.03.013

Chen GQ, Hajnal I, Wu H, Li L, Ye J (2015) Trends Biotechnol 33(10):565–574

Choi JH, Fukushi K, Yamamoto K (2008) A study on the removal of organic acids from wastewaters using nanofiltration membranes. Sep Purif Technol 59(1):17–25. https://doi.org/10.1016/j.seppur.2007.05.021
Colombo B, Calvo MV, Sciarria TP, Scaglia B, Kizito SS, D'Imporzano G, Adani F (2019) Biohydrogen and polyhydroxyalkanoates (PHA) as products of a two-steps bioprocess from deproteinized dairy wastes. Waste Manag 95:22–31
Damle A (2014) An introduction to the utilization of membrane technology in the production of clean and renewable power. In: Woodhead Publ. Ltd. https://www.semanticscholar.org/paper/An-introduction-to-the-utilization-of-membrane-in-Damle/ffa4bb7511958d6228d04719d313d7121e401a17. Accessed 22 July 2022
De Donno Novelli L, Moreno Sayavedra S, Rene ER (2021) Polyhydroxyalkanoate (PHA) production via resource recovery from industrial waste streams: a review of techniques and perspectives. Bioresour Technol 331(March):124985. https://doi.org/10.1016/j.biortech.2021.124985
De Gioannis G, Muntoni A, Polettini A, Pomi R (2013) A review of dark fermentative hydrogen production from biodegradable municipal waste fractions. Waste Manag 33:1345–1361
de la Rosa O, Múñiz-Marquez DB, Contreras-Esquivel JC, Wong-Paz JE, Rodríguez-Herrera R, Aguilar CN (2020) Improving the fructooligosaccharides production by solid-state fermentation. Biocatal Agric Biotechnol 27:101704. https://doi.org/10.1016/j.bcab.2020.101704
de Nascimento Junior JR, Torres LAZ, Medeiros ABP, Woiciechowski AL, Martinez-Burgos WJ, Soccol CR (2021) Enhancement of biohydrogen production in industrial wastewaters with vinasse pond consortium using lignin-mediated iron nanoparticles. Int J Hydrogen Energy 46(54):27431–27443. https://doi.org/10.1016/j.ijhydene.2021.06.009
Di Lorenzo ML, Gazzano M, Righetti MC (2012) The role of the rigid amorphous fraction on cold crystallization of poly(3-hydroxybutyrate). Macromolecules 45:5684–5691
Dias JML, Lemos PC, Serafim LS, Oliveira C, Eiroa M, Albuquerque MGE, Ramos AM, Oliveira R, Reis MAM (2006) Recent advances in polyhydroxyalkanoate production by mixed aerobic cultures: from the substrate to the final product. Macromol Biosci 6(11):885–906. https://doi.org/10.1002/mabi.200600112
Dietrich K, Dumont M-J, Del Rio LF, Orsat V (2017) Producing PHAs in the bioeconomy—towards a sustainable bioplastic. Sustain Prod Consum 9:58–70
Dinesh GH, Nguyen DD, Ravindran B, Chang SW, Vo D-V, Bach Q-V, Tran HN, Muthuramalingam JB, Mohanrasu K, Murugan RS, Angelin Swetha T, Gurusamy S, Selvaraj A, Arun A (2020) Simultaneous biohydrogen (H2) and bioplastic (poly-β-hydroxybutyrate-PHB) productions under dark, photo, and subsequent dark and photo fermentation utilizing various wastes. Int J Hydrogen Energy 45:5840–5853
Dobrogojski J, Spychalski M, Luciński R, Borek S (2018) Transgenic plants as a source of polyhydroxyalkanoates. Acta Physiol Plant 40:1–17
Du C, Li Y, Zong H, Yuan T, Yuan W, Jiang Y (2020) Production of bioethanol and xylitol from non-detoxified corn cob via a two-stage fermentation strategy. Bioresour Technol 310:123427
European Bioplastics (2022) Forthcoming EU policy framework on bio-based, biodegradable and compostable plastics. In: Eur. Bioplastics. https://www.european-bioplastics.org/forthcoming-eu-policy-framework-on-bio-based-biodegradable-and-compostable-plastics/
European Commission (2012) Directorate General for Research and Innovation. In: Innov. Sustain. growth a bioeconomy Eur. https://www.ecsite.eu/activities-and-services/resources/innovating-sustainable-growth-bioeconomy-europe. Accessed 23 July 2022
Filiciotto L, Rothenberg G (2021) Biodegradable plastics: standards, policies, and impacts. Chem Sustain Energy Mater 14:56–72
Flores-Sánchez A, López-Cuellar MR, Pérez-Guevara F, López UF, Martín-Bufájer JM, Vergara-Porras B (2017) Synthesis of Poly-(R-hydroxyalkanoates) by Cupriavidus necator ATCC 17699 using Mexican Avocado (Persea americana) oil as a carbon source. Int J Polym Sci 6942950
García-Depraect O, Bordel S, Lebrero R, Santos-Beneit F, Borner RA, Borner T, Muñoz R (2021) Inspired by nature: microbial production, degradation and valorization of biodegradable bioplastics for life-cycle-engineered products. Biotechnol Adv 53:107772

GO!PHA (2022) Paques biomaterials opened PHA2USE demonstration plant. https://www.gopha.org/news/paques-biomaterials-opened-pha2use-demonstration-plant. Accessed 23 July 2022
Greene JP (2021) Automotive plastics and composites: materials and processing. Elsevier, Amsterdam
HyResource (2022) China. In: HyResource. https://research.csiro.au/hyresource/policy/international/china/
International Energy Agency (2020) Tracking clean energy innovation: a framework for using indicators to inform policy. https://read.oecd-ilibrary.org/energy/tracking-clean-energy-innovation_cf1ed3eb-en#page7. Accessed 23 July 2022
Isla MA, Comelli RN, Seluy LG (2013) Wastewater from the soft drinks industry as a source for bioethanol production. Bioresour Technol 136:140–147
Jambeck JR, Geyer R, Wilcox C, Siegler TR, Perryman M, Andrady A, Narayan R, Law KL (2015) Plastic waste inputs from land into the ocean. Science 347(80):768–770
Jawed K, Irorere VU, Bommareddy RR, Minton NP, Kovács K (2022) Establishing mixotrophic growth of Cupriavidus necator H16 on CO2 and volatile fatty acids. Fermentation 8:125
Jia Q, Xiong H, Wang H, Shi H, Sheng X, Sun R, Chen G (2014) Production of polyhydroxyalkanoates (PHA) by bacterial consortium from excess sludge fermentation liquid at laboratory and pilot scales. Bioresour Technol 171:159–167
Kacanski M, Pucher L, Peral C, Dietrich T, Neureiter M (2022) Cell retention as a viable strategy for PHA production from diluted VFAs with Bacillus megaterium. Bioengineering 9:122
Kamilah H, Tsuge T, Yang TA, Sudesh K (2013) Waste cooking oil as substrate for biosynthesis of poly(3-hydroxybutyrate) and poly(3-hydroxybutyrate-co-3-hydroxyhexanoate): turning waste into a value-added product. Malays J Microbiol 9:51–59
Karp SG, Medina JDC, Letti LAJ, Woiciechowski AL, Carvalho JC, Schmitt CC, Penha RO, Kumlehn GS, Soccol CR (2021) Bioeconomy and biofuels: the case of sugarcane ethanol in Brazil. Biofuels Bioprod Biorefinig 15(3):899–912. https://doi.org/10.1002/bbb.2195
Khatami K, Perez-Zabaleta M, Owusu-Agyeman I, Cetecioglu Z (2021) Waste to bioplastics: how close are we to sustainable polyhydroxyalkanoates production? Waste Manag 119:374–388
Kim BS (2000) Production of poly-3-hydroxybutyrate from inexpensive substrates. Enzyme Microb Technol 27:774–777
Kourmentza C, Plácido J, Venetsaneas N, Burniol-Figols A, Varrone C, Gavala HN, Reis MAM (2017) Recent advances and challenges towards sustainable polyhydroxyalkanoate (PHA) production. Bioengineering 4:1–43
Kucharska K, Rybarczyk P, Hołowacz I, Konopacka-Łyskawa D, Słupek E, Makoś P, Cieśliński H, Kamiński M (2020) Influence of alkaline and oxidative pre-treatment of waste corn cobs on biohydrogen generation efficiency via dark fermentation. Biomass Bioenergy 141:105691
Kumar AK, Sharma S (2017) Recent updates on different methods of pretreatment of lignocellulosic feedstocks: a review. Bioresour Bioprocess 4(7)
Lange K-B (2019) Bioplastics market data update. In: Eur. Bioplastics. https://www.european--bioplastics.org/market/. Accessed 22 July 2022
Law KL (2017) Plastics in the marine environment. Ann Rev Mar Sci 9:205–229
Lazzarotti F, Dalfovo MS, Hoffmann VE (2020) A bibliometric study of innovation based on Schumpeter. J Technol Manag Innov 6:121–135
Lee J-E, Vadlani PV, Faubion J (2017) Corn bran bioprocessing: development of an integrated process for microbial lipids production. Bioresour Technol 243:196–203
Li M, Wilkins MR (2020) Recent advances in polyhydroxyalkanoate production: feedstocks, strains and process developments. Int J Biol Macromol 156:691–703
Lin Y-L, Zheng N-Y, Wang H-C (2022) Sludge dewatering through H2O2 lysis and ultrasonication and recycle for energy by torrefaction to achieve zero waste: an environmental and economical friendly technology. Renew Sustain Energy Rev 155:111857
López JC, Arnáiz E, Merchán L, Lebrero R, Muñoz R (2018) Biogas-based polyhydroxyalkanoates production by Methylocystis hirsuta: a step further in anaerobic digestion biorefineries. Chem Eng J 333:529–536

Lorrungruang C, Martthong J, Sasaki K, Noparatnaraporn N (2006) Selection of photosynthetic bacterium Rhodobacter sphaeroides 14F for polyhydroxyalkanoate production with two-stage aerobic dark cultivation. J Biosci Bioeng 102(2):128–131. https://doi.org/10.1263/jbb.102.128
Market Data Forecast (2022) Polyhydroxyalkanoate (PHA) market. In: Mark. Data Forecast. https://www.marketdataforecast.com/market-reports/polyhydroxyalkanoate-market
Markets and Markets (2021) Hydrogen generation market. In: Mark. Mark. https://www.marketsandmarkets.com/PressReleases/hydrogen.asp
Martinez-Burgos WJ, Sydney EB, Brar SK, de Andrade Tanobe VO, Medeiros ABP, Carvalho JC, Soccol CR (2019) The effect of hydrolysis and sterilization in biohydrogen production from cassava processing wastewater medium using anaerobic bacterial consortia. Int J Hydrogen Energy 44(47):25551–25564. https://doi.org/10.1016/j.ijhydene.2019.08.085
Martinez-Burgos WJ, Sydney EB, de Paula DR, Medeiros ABP, De Carvalho JC, Molina D, Soccol CR (2021) Hydrogen production by dark fermentation using a new low-cost culture medium composed of corn steep liquor and cassava processing water: process optimization and scale-up. Bioresour Technol 320(September 2020). https://doi.org/10.1016/j.biortech.2020.124370
McAdam B, Fournet MB, McDonald P, Mojicevic M (2020) Production of polyhydroxybutyrate (PHB) and factors impacting its chemical and mechanical characteristics. Polymers 12:1–12
Melnicki MR, Eroglu E, Melis A (2009) Changes in hydrogen production and polymer accumulation upon sulfur-deprivation in purple photosynthetic bacteria. Int J Hydrogen Energy 34(15):6157–6170. https://doi.org/10.1016/j.ijhydene.2009.05.115
Mishra P, Thakur S, Madhab D (2017) Impacts of nano-metal oxides on hydrogen production in anaerobic digestion of palm oil mill effluent: a novel approach. Int J Hydrogen Energy 43(5):2666–2676
Montiel-Jarillo G, Gea T, Artola A, Fuentes J, Carrera J, Suarez-Ojeda ME (2021) Towards PHA production from wastes: the bioconversion potential of different activated sludge and food industry wastes into VFAs through acidogenic fermentation. Waste Biomass Valorization 12:6861–6873
Mordor Intelligence (2022) Polyhydroxyalkanoate (PHA) Market - Growth, Trends, COVID-19 Impact, and Forecasts (2022 - 2027). In: Mordor Intell. https://www.mordorintelligence.com/industry-reports/polyhydroxyalkanoate-market
Morgan-Sagastume F, Hjort M, Cirne D, Gérardin F, Lacroix S, Gaval G, Karabegovic L, Alexandersson T, Johansson P, Karlsson A, Bengtsson S, Arcos-Hernández MV, Magnusson P, Werker A (2015) Integrated production of polyhydroxyalkanoates (PHAs) with municipal wastewater and sludge treatment at pilot scale. Bioresour Technol 181:78–89
Moshood TD, Nawanir G, Mahmud F, Mohamad F, Ahmad MH, Abdul Ghani A (2021) Expanding policy for biodegradable plastic products and market dynamics of bio-based plastics: challenges and opportunities. Sustainability 13(11):6170. https://doi.org/10.3390/su13116170
Muhammadi S, Afzal M, Hameed S (2015) Bacterial polyhydroxyalkanoates-eco-friendly next generation plastic: production, biocompatibility, biodegradation, physical properties and applications. Green Chem Lett Rev 8:56–77
Muneer F, Rasul I, Azeem F, Siddique MH, Zubair M, Nadeem H (2020) Efficient replacement of synthetic polymers. J Polym Environ 28:2301–2323
Naguib HF, Aziz MSA, Sherif SM, Saad GR (2011) Synthesis and thermal characterization of poly(ester-ether urethane)s based on PHB and PCL-PEG-PCL blocks. J Polym Res 18:1217–1227
Ng K-S, Ooi W-Y, Goh L-K, Shenbagarathai R, Sudesh K (2010) Evaluation of jatropha oil to produce poly(3-hydroxybutyrate) by Cupriavidus necator H16. Polym Degrad Stab 95:1365–1369
Nicoletti G, Arcuri N, Nicoletti G, Bruno R (2015) A technical and environmental comparison between hydrogen and some fossil fuels. Energy Convers Manag 89:205–213
Ojumu TV, Yu J, Solomon BO (2004) Production of polyhydroxyalkanoates, a bacterial biodegradable polymer. Afr J Biotechnol 3:18–24
Origiin IP Solutions (2022) Technology licensing: Pros and Cons. In: Origiin IP Solut. https://www.origiin.com/2021/08/27/technology-licensing-pros-and-cons/. Accessed 23 July 2022

Pagliano G, Ventorino V, Panico A, Pepe O (2017) Integrated systems for biopolymers and bioenergy production from organic waste and by-products: a review of microbial processes. Biotechnol Biofuels 10(1):1–24. https://doi.org/10.1186/s13068-017-0802-4
Parveez GKA, Rasid OA, Tarmizi AH, Zamzuri I, Samsul KR, Sambanthamurthi R (2012) Oil palm: production, processing, characterization' and uses. AOCS Press, Urbana
Passanha P, Esteves SR, Kedia G, Dinsdale RM, Guwy AJ (2013) Increasing polyhydroxyalkanoate (PHA) yields from Cupriavidus necator by using filtered digestate liquors. Bioresour Technol 147:345–352
Patel A, Shah AR (2021) Integrated lignocellulosic biorefinery: gateway for production of second generation ethanol and value added products. J Bioresour Bioprod 6(2):108–128. https://doi.org/10.1016/j.jobab.2021.02.001
Peng Y, Wu P, Schartup AT, Zhang Y (2021) Plastic waste release caused by COVID-19 and its fate in the global ocean. Proc Natl Acad Sci 118(47):e2111530118
Philp J (2014) OECD policies for bioplastics in the context of a bioeconomy. Ind Biotechnol 10(1):online
Pinto-Ibieta F, Serrano A, Cea M, Ciudad G, Fermoso FG (2021) Beyond PHA: stimulating intracellular accumulation of added-value compounds in mixed microbial cultures. Bioresour Technol 337:125381
Portela-Grandío A, Lagoa-Costa B, Kennes C, Veiga MC (2021) Polyhydroxyalkanoates production from syngas fermentation effluents: effect of nitrogen availability. J Environ Chem Eng 9:106662
Raza ZA, Abid S, Banat IM (2018) Polyhydroxyalkanoates: characteristics, production, recent developments and applications. Int Biodeterior Biodegrad 126:45–56
Rivera-Briso AL, Serrano-Aroca Á (2018) Poly(3-Hydroxybutyrate-co-3-Hydroxyvalerate): enhancement strategies for advanced applications. Polymers 10:1–28
Rosenboom JG, Langer R, Traverso G (2022) Bioplastics for a circular economy. Nat Rev Mater 7:117–137
Saady NMC (2013) Homoacetogenesis during hydrogen production by mixed cultures dark fermentation: unresolved challenge. Int J Hydrogen Energy 38(30):13172–13191. https://doi.org/10.1016/j.ijhydene.2013.07.122
Salehizadeh H, Van Loosdrecht MCM (2004) Production of polyhydroxyalkanoates by mixed culture: recent trends and biotechnological importance. Biotechnol Adv 22:261–279
Siqueira MU, Contin B, Fernandes PRB, Soares RR, Uhlig PS, Ramos JB (2022) Brazilian agroindustrial wastes as potential textile and other raw materials: a sustainable approach. Mater Circ Econ 4(9)
Soto LR, Byrne E, van Niel EWJ, Sayed M, Villanuevab CC, Hatti-Kaul R (2019) Hydrogen and polyhydroxybutyrate production from wheat straw hydrolysate using Caldicellulosiruptor species and Ralstonia eutropha in a coupled process. Bioresour Technol 272:259–266
Sydney EB, Larroche C, Novak AC, Nouaille R, Sarma SJ, Brar SK, Letti LAJ, Soccol VT, Soccol CR (2014) Economic process to produce biohydrogen and volatile fatty acids by a mixed culture using vinasse from sugarcane ethanol industry as nutrient source. Bioresour Technol 159. https://doi.org/10.1016/j.biortech.2014.02.042
Sydney EB, Neto CJD, de Carvalho JC, de Souza Vandenberghe LP, Sydney ACN, Letti LAJ, Karp SG, Soccol VT, Woiciechowski AL, Medeiros ABP, Soccol CR (2019) Microalgal biorefineries: integrated use of liquid and gaseous effluents from bioethanol industry for efficient biomass production. Bioresour Technol 292:121955. https://doi.org/10.1016/j.biortech.2019.121955
Tan D, Yin J, Chen G-Q (2016) Production of polyhydroxyalkanoates. In: Larroche C, Sanroman MA, Du G, Pandey A (eds) Current developments in biotechnology and bioengineering: bioprocesses, bioreactors and controls, 1st edn. Elsevier, Amsterdam, pp 655–692
Thulasidharan D, Arumugam A, Uppuluri KB (2021) Research and economic perspectives on an integrated biorefinery approach for the simultaneous production of polyhydroxyalkanoates and biohydrogen. Int J Biol Macromol 193:1937–1951
Tollefson J (2018) China declared largest source of research articles. Nature 553:390

Transparency Market Research (2021) Biohydrogen market outlook environment protection and depleting non-renewable energy sources has catalyzing effect on the global biohydrogen market biohydrogen market segmentation biohydrogen market: regional analysis. https://www.transparencymarketresearch.com/biohydrogen-market.html. Accessed 23 July 2022

Tu W, Zhang D, Wang H (2019) Polyhydroxyalkanoates (PHA) production from fermented thermal-hydrolyzed sludge by mixed microbial cultures: the link between phosphorus and PHA yields. Waste Manag 96:149–157

Urtuvia V, Villegas P, González M, Seeger M (2014) Bacterial production of the biodegradable plastics polyhydroxyalkanoates. Int J Biol Macromol 70:208–213

US DOE (2022) Department of Energy announces $28 million to develop clean hydrogen. In: Off. Foss. Energy Carbon Manag. https://www.energy.gov/fecm/articles/us-department-energy-announces-28-million-develop-clean-hydrogen

Valencia AIS, Zamora UR, Rodríguez MM, Ramírez JÁ, Peláez MLS, Ortiz CF (2021) Effect of C/N ratio on the PHA accumulation capability of microbial mixed culture fed with leachates from the organic fraction of municipal solid waste (OFMSW). J Water Process Eng 40

Valentino F, Morgan-Sagastume F, Campanari S, Villano M, Werker A, Majone M (2017) Carbon recovery from wastewater through bioconversion into biodegradable polymers. N Biotechnol 37:9–23

Venkateswar Reddy M, Amulya K, Rohit MV, Sarma PN, Mohan SV (2014) Valorization of fatty acid waste for bioplastics production using Bacillus tequilensis: integration with dark-fermentative hydrogen production process. Int J Hydrogen Energy 39:7616–7626

Vieira S, Schneider J, Martinez-Burgos WJ, Magalhães A, Medeiros ABP, Carvalho JC, de Souza Vandenberghe LP, Soccol CR, Sydney EB (2021) Pretreatments of solid wastes for anaerobic digestion and its importance for the circular economy. In: Handbook of solid waste management. pp 1–23

Vincenzini M, Marchini A, Ena A, De Philippis R (1997) H2 and poly-β-hydroxybutyrate, two alternative chemicals from purple non sulfur bacteria. Biotechnol Lett 19(8):759–762. https://doi.org/10.1023/A:1018336209252

Volova T, Sapozhnikova K, Zhila N (2020) Cupriavidus necator B-10646 growth and polyhydroxyalkanoates production on different plant oils. Int J Biol Macromol 164:121–130

Wainaina S, Lukitawesa, Kumar Awasthi M, Taherzadeh MJ (2019) Bioengineering of anaerobic digestion for volatile fatty acids, hydrogen or methane production: a critical review. Bioengineered 10:437–458

Wang R-Y, Shi Z-Y, Chen J-C, Wu Q, Chen G-Q (2012) Enhanced co-production of hydrogen and poly-(R)-3-hydroxybutyrate by recombinant PHB producing E. coli over-expressing hydrogenase 3 and acetyl-CoA synthetase. Metab Eng 14

Wang Y, Chen R, Cai J, Liu Z, Zheng Y, Wang H, Li Q, He N (2013) Biosynthesis and thermal properties of PHBV produced from levulinic acid by Ralstonia eutropha. PLoS One 8

Wang M-R, Li H-F, Yi J-J, Tao S-Y, Li Z-J (2022) Production of polyhydroxyalkanoates by three novel species of Marinobacterium. Int J Biol Macromol 195:255–263

Wei Z, Liu L, Qi M (2006) Synthesis and characterization of homo- and co-polymers of (R,S)-b-butyrolactone and c-butyrolactone or b-valerolactone initiated with cyclic tin alkoxide. React Funct Polym 66:1411–1419

Wolf A, Zander N (2021) Green hydrogen in Europe: do strategies meet expectations? In: ZBW. https://www.intereconomics.eu/contents/year/2021/number/6/article/green-hydrogen-in-europe-do-strategies-meet-expectations.html. Accessed 29 July 2022

WWF (2021) Plastics: the cost to society, the environment and the economy. Geneva

Wydra S (2020) Measuring innovation in the bioeconomy – conceptual discussion and empirical experiences. Technol Soc 61:101242

Yadav B, Pandey A, Kumar LR, Tyagi RD (2020) Bioconversion of waste (water)/residues to bioplastics- a circular bioeconomy approach. Bioresour Technol 298:122584

Yadav B, Talan A, Tyagi RD, Drogui P (2021) Concomitant production of value-added products with polyhydroxyalkanoate (PHA) synthesis: a review. Bioresour Technol 337:125419

Yang G, Wang J (2019) Changes in microbial community structure during dark fermentative hydrogen production. Int J Hydrogen Energy 44(47):25542–25550. https://doi.org/10.1016/j.ijhydene.2019.08.039

Yang J, Jiang D, Shui X, Lei T, Zhang H, Zhang Z, Zhang X, Zhu S, Zhang Q (2022) Effect of 5-HMF and furfural additives on bio-hydrogen production by photo-fermentation from giant reed. Bioresour Technol 347:126743

Yigit DO, Gunduz U, Turker L, Yucel M, Eroglu I (1999) Identification of by-products in hydrogen producing bacteria; Rhodobacter sphaeroides O.U. 001 grown in the waste water of a sugar refinery. J Biotechnol 70:125–131. https://doi.org/10.7868/s0026365615030118

Zhang K, Hamidian AH, Tubić A, Zhang Y, Fang JKH, Wu C, Lam PKS (2021) Understanding plastic degradation and microplastic formation in the environment: a review. Environ Pollut 274:116554

Chapter 17
Microbial Electrolysis Cells in Biohydrogen Production

Marcela Moreira Albuquerque, Walter José Martinez-Burgos, Gabriela de Bona Sartor, Adriane Bianchi Pedroni Medeiros, Júlio Cesar de Carvalho, and Carlos Ricardo Soccol

Abstract Microbial electrolysis cell (MEC) has been regarded as a promising technology to produce hydrogen using electrochemically active bacteria. The yield of hydrogen can be up to three times higher compared to simple microbial fermentation. A hydrogen yield of 12 mol/mol hexose is theoretically reached compared to 4 mol/mol using dark fermentation. In addition to the large theoretical yield of biohydrogen, MECs demand lower energy for electrohydrogen reactions. This chapter overviews MEC technology basics and critical factors affecting its performance for hydrogen production, such as reactor design, electrode materials, microorganisms, and substrates. Technoeconomic aspects of different methods to produce hydrogen are presented and compared to MECs.

Keywords Bioelectrochemical systems · Biofuel · Agroindustrial residues · Wastewaters

17.1 Introduction

The growing world demand for energy has led to an incessant search for renewable fuels, aiming to replace the current energy matrix, which is still mainly based on fossil fuels (Chen et al. 2022), and thus, circumvent the environmental problems related to its combustion, such as the emission of gases that potentiate the greenhouse effect (Saravanan et al. 2022).

M. M. Albuquerque · W. J. Martinez-Burgos · G. de Bona Sartor · A. B. P. Medeiros (✉) · J. C. de Carvalho · C. R. Soccol
Federal University of Paraná, Department of Bioprocess Engineering and Biotechnology, Centro Politécnico, Curitiba, Paraná, Brazil
e-mail: adrianebpm@ufpr.br

C. R. Soccol et al. (eds.), *Biohydrogen - Advances and Processes*, Biofuel and Biorefinery Technologies 13, https://doi.org/10.1007/978-3-031-49818-3_17

In this scenario, hydrogen has been considered a potential source of energy and is repeatedly cited as the fuel of the future (Gautam et al. 2023) due to its high calorific value, which corresponds to about 122 kJ g^{-1}, about 2.5 times greater than fossil fuels (Argun et al. 2008) and the emission of water vapor as its only combustion product (Gautam et al. 2023; Rani et al. 2022; Sarangi and Nanda 2020).

Currently, hydrogen is mainly produced from fossil fuels and the electrolysis of water (Chi and Yu 2018). The exploration of more sustainable production methods is constant. Biological routes are promising because they are carbon neutral, require a lower amount of energy since they occur at lower temperature and pressure (Ananthi et al. 2022), and allow the use of organic wastes as substrates, concurrently leading to the treatment of these wastes while producing energy (Khan et al. 2017).

The biological production of hydrogen can be classified into water biophotolysis, photofermentation, dark fermentation, and bioelectrochemical production through microbial electrolysis cells (MECs) (Hitam and Jalil 2020). The biophotolysis of water is carried out by photosynthetic microalgae and cyanobacteria that use sunlight to split water molecules into H_2 and O_2 (Eroglu and Melis 2011). Photofermentation is carried out by photosynthetic bacteria that use light energy and organic acids as a carbon source to produce H_2 (Akhlaghi and Najafpour-Darzi 2020). Dark fermentation is carried out by heterotrophic bacteria from the anaerobic metabolism of pyruvate, using residual sources as substrate (Kumar and Das 2001).

Typical biological routes of biohydrogen production have low yields, with the highest theoretical maximum yield of about 4 mol H_2/mol glucose obtained from dark fermentation (Das and Veziroglu 2008). Thus, bioelectrochemical systems called microbial electrolysis cells (MECs) are being studied to obtain higher yields of hydrogen and treatment of effluents (Badia-Fabregat et al. 2019; Hosseinzadeh et al. 2020; Kim and Logan 2019; Son et al. 2021), making it possible, when coupling with dark fermentation, to obtain a maximum theoretical yield of hydrogen of about 12 mol/mol of hexose, three times the theoretical maximum obtained by the dark fermentation alone (Hawkes et al. 2007).

In general, MECs can be double or single chamber (Rani et al. 2022) and consist of an anode a cathode, and may or may not have a membrane separating the electrodes (Fig. 17.1) (Samsudeen et al. 2020). These components can be produced from a series of materials, which must have good electrical conductivity as one of their main characteristics, directly affecting the biohydrogen production rate (Park et al. 2022).

Electrochemically active anaerobic bacteria form biofilms at the anode that oxidize organic matter into volatile fatty acids, CO_2, H^+, and free electrons (Saravanan et al. 2020). The protons (H^+) migrate to the cathode through the membrane, and the electrons are transferred to the cathode through an external circuit, and under the application of a small external voltage, are reduced to hydrogen gas at the cathode in the presence of a catalyst (Kadier et al. 2022). For biohydrogen production to

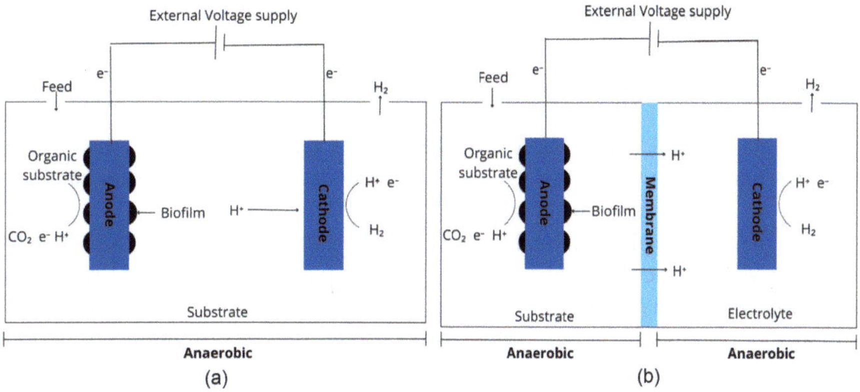

Fig. 17.1 Schematic of a (**a**) single- and (**b**) dual-chambered MEC for H_2 generation

occur, the cathode chamber must be maintained in anaerobic condition (Saravanan et al. 2020).

The ability to use organic waste in MECs to produce biohydrogen, also resulting in its treatment simultaneously, is one of the main factors that have attracted the attention of researchers to the use of MECs (Rani et al. 2022). Some of the categories of residues most used as substrate in MECs are the residues from industrial processing (Jayabalan et al. 2019), dark fermentation effluents (Khongkliang et al. 2017), urban wastewater (Badia-Fabregat et al. 2019) and lignocellulosic biomass (Wang et al. 2021).

In addition to the construction materials of the MECs, their configurations, and the substrates, a series of other factors such as the microorganisms used (Lee et al. 2017), deactivation of methanogens (Fujinawa et al. 2019), temperature (Dange et al. 2021), pH (Khan et al. 2017), hydraulic retention time (Ye et al. 2020) and the applied voltage (Dange et al. 2021) also affect biohydrogen production in MECs.

In the following sections, this chapter overviews the different materials used in the construction of MECs, their configurations, mode of operation, and main factors that affect the production of hydrogen, also carrying out a technoeconomic analysis of the production of biohydrogen by MECs compared to other production methods such as dark fermentation, and presents future perspectives.

17.2 Design of MECs Systems

System performance and, consequently, hydrogen production are factors directly influenced by the design of the MECs. Single-chamber and dual-chamber models are the main configurations and work similarly in various formats such as H-shaped, cube, tubular, or cassette type (Kadier et al. 2016).

17.2.1 *Configurations*

17.2.1.1 Double-Chamber MECs

Since 2005, several studies have explored MECs technology mainly using two-chamber reactors, in which the electrodes are contained in two compartments separated by an ion exchange membrane (Bora et al. 2022).

This membrane plays a fundamental role in the performance of the double-chamber MECs, since it enables greater hydrogen purity by preventing the passage of biological material and contaminating gases to the cathode, in addition to mitigating short circuits in the electrolytic cell (Kadier et al. 2016). However, besides the cost, the pH gradient generated across the membrane is one of the main drawbacks present in this system and can lead to decreases in hydrogen yield (Hu et al. 2008).

In this sense, Jiang et al. (2016) proposed an alternative for pH control in hydrogen production in dual-chamber MECs using hybrid biotic/abiotic electrodes. In the study, the operating mode used was the periodic polarity inversion (PPR), in which the same electrode sometimes works as an anode and sometimes as a cathode. Different PPR intervals (0.5, 1, 2, 4, and 6 h) and voltages (0.7/−0.7 V, 0.8/−0.8 V, and 1.0/−1, 0 V) were tested at 25 °C. As a result, a controlled range of pH (6.4–8.6) was obtained. In addition, utilizing the PPR mode interval, the H_2 production rate increased with increasing voltage, resulting in a value of 7.4 ± 0.5 mL after 40 h.

17.2.1.2 Single-Chamber MECs

In single-chamber MECs, both electrodes are immersed in the same electrolyte (Bora et al. 2022). Due to the absence of a membrane separating the anode and cathode, the single-chamber MECs avoid problems arising from increased internal resistance. Consequently, there are lower potential losses (Hu et al. 2008). Furthermore, the construction and maintenance of this type of system are simplified, which also translates into lower costs (Hu et al. 2008).

However, despite the advantages described, without the membrane, the gas obtained may contain contaminants such as CO_2, CH_4, and H_2S (Rozendal et al. 2007). In particular, the presence of acetoclastic methanogens that convert acetate into methane and hydrogenotrophic methanogens, which use part of the hydrogen produced at the cathode to produce methane, is largely responsible for the loss of energy produced in the MECs (Montpart et al. 2015).

Hence, due to the risk of reduced H_2 yield in long-term operations, Cui et al. (2019) evaluated the inhibition of methanogenesis in single-chamber MECs, under alkaline conditions. For this, 50 mM of bicarbonate buffer and 1 g/L of acetate were used to conduct experiments at different pH values (8.5, 9.5, 10.5, and 11.2). In the end, the pH 11.2 obtained the best hydrogen yield (2.64 ± 0.16–3.36 ± 0.10 mol H_2

per mol of acetate in 50 days of operation) with a current density of 83.7 ± 1.5 A m^{-3}. In addition, the percentage of CH_4 was less than 15%.

Lu et al. (2016) also obtained good yields when using a gas-permeable but hydrophobic membrane to extract the hydrogen produced from a single-chamber MEC. The cell was fed with 1 g/L of sodium acetate in a 50 mM phosphate buffer solution, and operating conditions were 25 ± 1 °C under a voltage of 0.6 V. As a result, a higher H_2 yield was obtained (2.62–3.39 mol H_2/mol acetate) than under conditions without using the hydrophobic membrane (0.79 mol H_2/mol acetate), and methane production was interrupted.

Exposure to UV light, reported in a study by Hou et al. (2014), was another proposed alternative to mitigate the action of methanogens. In this study, a comparison was made between single-chamber MEC operating with and without the incidence of UV light on the cathode under a voltage of 0.8 V. The water substrate is a 50:50 mixture of primary clarifier run-off (from wastewater treatment) and sodium acetate medium. In the end, it was observed that with UV radiation, H_2 concentrations greater than 91% were obtained, and yields ranged from 2.87 ± 0.03 to 3.70 ± 0.11 mol H_2/mol acetate. In contrast, methane concentrations increased up to 94% without UV radiation, and consequently, hydrogen yield was lower (3.78 ± 0.12 to 0.03 ± 0.004 mol H_2/mol acetate).

17.2.2 Components

The correct choice of electrodes and membranes is also essential to determine the success of MECs. Regarding electrodes, these materials must have specific fundamental characteristics such as high conductivity, biological, chemical, and physical stability, be economically viable, and have a large surface area (Park et al. 2022).

17.2.2.1 Anode

The colonization and subsequent biofilm formation on the anode surface by exoelectrogenic bacteria promotes the release of electrons to the anode after the degradation of organic matter. However, not all electrons are captured by the anode, which leads to changes in the current density, directly influencing the hydrogen production rate (Kumar et al. 2017).

The efficiency in transferring electrons to the anode depends mainly on the interaction between microorganisms and electrodes (Guo et al. 2015). Notably, a better microbial adhesion in the anode cavities increases its electronically active surface area and, consequently, its performance. In addition, the anode must have a high electrical conductivity to allow the passage of electrons to the external circuit, as well as being preferably hydrophobic since contact with the aqueous medium can deteriorate the material (Kumar et al. 2013).

Carbon-based anodes are more commonly used because they present, besides the characteristics already mentioned, low cost, high stability, and low overpotential (Cheng et al. 2022). Materials that stand out comprise paper, fabrics, felt, carbon fiber brush or mesh, graphite granules and felt, etc. (Kadier et al. 2016). Zikmund et al. (2018) realized a comparative study of dual-chamber MEC performance when using graphite fiber brush carbon flat felt anodes. The MEC was operated with a voltage of 0.9 V. The highest hydrogen production rate was obtained in the brush anode (0.38 ± 0.02 m^3 H_2/m^3 d), but with the agitation of the electrolyte present in the felt anode, this presented a higher rate of 0.41 ± 0.04 m^3 H_2/m^3 d.

Regardless of the advantages of using carbon-based anodes, the hydrophobic nature of this material can prevent interactions with water and gas exchange, consequently compromising microbial fixation and thus reducing the electrically active surface area (Park et al. 2023). Thus, the study by Park et al. (2023) sought to minimize the problems related to the hydrophobicity of carbon-based anodes. In the mentioned study, a carbon felt anode was coated with a highly conductive and hydrophilic poly (3,4-ethylenedioxythiophene) bifunctional material: polystyrene sulfonate (PEDOT:PSS). The MEC operation was under agitation (350 rpm), 30 °C, and with an applied voltage of 0.5 V. The H_2 production rate constant using the modified anode was 1.57 against 0.68 for the conventional carbon felt anode, thus demonstrating the potential of the PEDOT:PSS anode to increase the performance of the MEC.

17.2.2.2 Cathode

It is at the cathode that the hydrogen evolution reaction (HER) takes place, so the cathode must be suitable support for H^+ reduction. Most of the materials used in the conductive base of the cathode are of low catalytic activity, such as stainless steel, carbon, and titanium. Therefore, a catalyst material, usually platinum, is used to coat these materials to increase the rate of HER (Hydrogen Evolution Reaction) (Park et al. 2022).

Despite presenting good results, the use of platinum makes the process more expensive due to its low overpotential. It is also a finite resource and can be partially or wholly deactivated by CO and sulfur, causing damage to the environment (Jafary et al. 2015). Therefore, new catalysts and electrodes are being studied and developed (Park et al. 2022).

Indeed, alloys, polymers, oxides, sulfides, metallic nanoparticles, and carbides, among others, developed from transition metals (such as nickel, iron, cobalt, and molybdenum) are a viable alternative for catalysts (Neto et al. 2019). Other materials, such as graphene and carbon nanotubes, also promise high performance (Jayabalan et al. 2019). In fact, the work of Son et al. (2021) compared the effect of five different stainless-steel mesh (SSM) cathodes in a single-chamber MEC. Cathodes were tested using SSM, flame oxidized SSM, activated carbon with polyvinylidene fluoride binder, nickel with activated carbon binder and polyvinylidene fluoride, and finally, activated carbon, nickel, and poly (1,1,2,2-tetrafluoroethylene). The experiments were carried out at 30 °C under a voltage of 0.9 V. Among the tested cathodes, the combination of nickel, activated carbon, and poly (1,1,2,2-tetrafluoroethylene) binder obtained the best hydrogen production rate in terms of HPR (1.88 L/L d), with a purity of 97.5%, which corroborates the idea that the production of H_2 can be efficient when modifying the cathode material.

Another study (Jayabalan et al. 2020), tested the feasibility of using two graphene hybrid metal oxide nanocomposites (NiOrGO and Co_3O_4.rGO) as catalysts in reactions in a dual-chamber MEC where the anode was fed with wastewater of the sugar industry and the cathode with phosphate buffer (50 mM). All experiments were carried out at a temperature of 32 °C under a voltage of 1.0 V. As a result, the catalyst NiOrGO presented an HPR of 4.38 ± 0.11 mmol/L/d, being a rate 2.68 times larger than the uncoated nickel cathode.

It is also possible to use a biocathode, in which microorganisms containing the enzyme hydrogenase coat the cathode and are used as reaction catalysts for the formation of hydrogen (Cheng et al. 2022). A biocathode can be obtained through a half-biological start-up by reversing the polarity of the bioanode or through a complete biological start-up with microorganisms on both electrodes (Jafary et al. 2015). For example, Dai et al. (2019) developed a biocathode on a carbon paper support, starting from inoculum using the full-start technique in a single-chamber MEC. This initialization process took 163 h and, in the end, produced a current density of 14.75 A/m^2. The MEC was operated under a voltage of 0.7 V at 20 °C and fed with aerobic sludge from the municipal wastewater treatment aeration tank and a nutrient solution. As a result, when using the produced biocathode, the hydrogen production rate was 0.428 ± 0.054 m^3 H_2/m^3d, which is higher than the rate obtained using Pt/C cathode (0.377 ± 0.052 m^3 H_2/m^3d), showing, therefore, that the use of biocathodes is promising in MEC systems.

17.2.2.3 Membranes

As previously seen, the membranes allow better operating conditions at the cathode and allow hydrogen production free from interfering gases. Several types of membranes are used in MECs, such as proton exchange membrane (PEM), anion exchange membrane (AEM), cation exchange membrane (CEM), and bipolar membrane, among others (Kadier et al. 2016). Membranes based on perfluorosulfonic acid (Nafion) are the most used PEMs due to their chemical stability and high conductivity (Gong et al. 2016). However, its high cost and risk of deterioration at high temperatures have led to the investigation of other materials (Gong et al. 2016). Chacón-Carrera et al. (2019) compared the efficiency of two dual-chamber MEC systems, which differed only in the type of membrane used. In one of the configurations, Nafion® 117 was used, and in the other, a single-bonded FBM bipolar membrane FUMASEP®. For both configurations, the MEC was fed with wastewater from the wastewater treatment (enriched with 20 mM of sodium acetate) and the tests were performed at 25 °C. The result showed that the MEC-Nafion system presented a COD removal of 87.4%, higher than that obtained by the MEC-BPM (44.8%). Concerning hydrogen production, only the Nafion membrane produced hydrogen (10.3 mg H_2/g COD). PEM membranes prepared from styrene, polybenzimidazole impregnated with phosphoric acid (PA), polyphosphazenes, and sulfonated aromatic compounds are potential alternatives (Walkowiak-Kulikowska et al. 2017). Hybrid composite membranes are also emerging, formed from organic and inorganic materials. The resulting materials can reduce the loss of conductivity while providing thermal and mechanical stability (De Bonis et al. 2016).

Zhao et al. (2021) obtained good hydrogen yield results when using nonwoven fabric (NWC) with a porosity of 69% and polyvinylidene difluoride (PVDF) with a porosity of 59% as an alternative to PEM in a double-sided chamber MEC. The MEC setup for both tests used graphite felt and titanium mesh as anode and cathode, respectively. The tests were carried out with a voltage of 0.8 V and temperature of 30±2 °C. As a result, hydrogen production in the NWC-MEC configuration was 2.5 to 2.1 m^3/m^3d, and in the PVDF-MEC was 2.2 to 2.0 m^3/m^3d. Both membranes were more efficient in hydrogen production when compared to PEM-MECs type systems (2.2 to 1.6 m^3/m^3d) during 30 days of operation.

A nonhighly selective poly(vinyl) alcohol/chitosan membrane was used in a study by González-Pabón et al. (2021) as an alternative to the Nafion-type membrane. In the study, a double-chamber MEC was used. The anode material consisted of graphite felt and nickel foam was used as the cathode. The experiments were conducted at 30 °C and agitation (175 rpm), and cyclic voltammetry was performed between 300 and 700 mV at a sweep rate of 1 mV/s. MEC with PVA/CS showed a hydrogen production rate and hydrogen yield of 1277 ± 46 mL H_2/L_{cat} d and 974 ± 116 mL $H_2/g_{acetate}$, respectively—similar to Nafion membranes. However, biofouling is a challenge related to the use of membranes in MEC. As the rate of hydrogen production directly depends on the rate of transfer of protons and electrons through the membrane, three-dimensional metallic nanoparticles (3D) are being used to avoid the formation of encrustations and thus guarantee a better efficiency of the MEC (Abd-Elrahman et al. 2022).

The objective of the study by Park et al. (2021) was precisely to investigate the effect of antibiofouling on membranes. For that purpose, different combinations of surface coatings of PEM membranes were tested: silver nanoparticles (AgNP), polydopamine (PDA), or a combination of both. To carry out the evaluations, a double-chamber MEC with a carbon felt electrode (anode) and titanium plate with platinum (cathode) with a voltage of 0.5 V. The combined use of both coatings (PDA_Ag) showed a better performance, with a biofouling reduction of 80.74% compared to the original PEM. Furthermore, after 6 months of operation using this coating combination, the hydrogen recovery was 68.12%, while without the coating, the rate was only 3.29%.

17.3 Main Parameters That Affect Hydrogen Production in Microbial Electrolysis Cells

17.3.1 Substrates

The current consumption model and population growth led to the overexploitation of resources, unbridled industrial production, and the exponential generation of waste in the industry as a whole (Nanda and Berruti 2021). These contaminating residues, rich in nutrients (Gautam et al. 2023), are often discarded in the environment without prior treatment, affecting air quality, soil, water resources, and, consequently, the biota (Porwal et al. 2015).

Thus, using these residues in energy production appears as an alternative to circumvent the problem associated with their production, simultaneously resulting in their treatment and energy recovery (Jayabalan et al. 2019). Thus, the generation of biohydrogen in MECs has been studied using various residues and environmental contaminants as substrates, and their composition and concentration directly affect the yield of biohydrogen in MECs (Koroglu et al. 2019).

As seen in Table 17.1, some of the residues that have been most commonly used in the production of biohydrogen in MECs are wastewater from industrial processing (Badia-Fabregat et al. 2019; Iskandar et al. 2018; Jayabalan et al. 2019; Kadier et al. 2022; Samsudeen et al. 2020; Rani et al. 2022), dark fermentation effluents (Khongkliang et al. 2017, 2019; Li et al. 2014; Marone et al. 2017) urban effluents (Koroglu et al. 2019) and lignocellulosic biomass (Wang et al. 2021; Zhang et al. 2019).

17.3.1.1 Industrial Processing Effluents

Industrial processing effluents are the most exploited substrates to produce biohydrogen in MECs. Effluents from the sugar, dairy, palm oil, distilleries, and biodiesel processing industries are quite abundant and often pose serious environmental problems due to their composition and high chemical oxygen demand (COD) that, when incorrectly disposed of, may pose serious risks to the environment (Lee et al. 2019).

Jayabalan et al. (2019) used effluent from the sugarcane and raw sugar processing industry to produce biohydrogen in a double-chamber MEC. The maximum hydrogen production of 1.59 mmol/L/d and removal of 49.5% of the chemical oxygen demand was obtained under optimal operating conditions, such as an applied voltage of 1 V, anolyte pH of 6, 5, and nickel foam as cathode material, indicating the potential of this effluent to produce biohydrogen.

Dairy industry effluents are classified as high-resistance effluents due to the significant concentration of proteins, carbohydrates, fats, and recalcitrant compounds, translated into]high levels of chemical oxygen demand (COD) (Stasinakis et al.

Table 17.1 Substrates and microorganisms used in biohydrogen production in microbial electrolysis cells

Substrate	Inoculum	Predominant microorganisms	Single or double chamber	H_2 production	Reference
Sugar industry effluents	[a]	[a]	Double	1.59 mmol/L/d	Jayabalan et al. (2019)
Dairy industry wastewater	Anaerobic biogas digester	[a]	Single	32 ± 5 mL/L/d	Rani et al. (2022)
Palm oil mill effluent	POME anaerobic sludge sample	*Geobacter, Pseudomonas* sp, *Klebsiella.*	Single	1.16 m^3 H_2/m^3d	Kadier et al. (2022)
Distillery wastewater	[a]	[a]	Double	40.05 ± 0.5 mL	Samsudeen et al. (2020)
Crude glycerol	Sludge from a running alkaline air cathode MFC	*Enterococcus* sp., *Acetobacterium* sp., and *Geoalkalibacter* sp	[a]	0.46 LH_2/L/d	Badia-Fabregat et al. (2019)
Urban wastewater	Sludge from wastewater treatment plant	[a]	Double	0.18 ± 0.03 m^3/m^3/d	Chen et al. (2019)
Urban wastewater	Anaerobic sludge	[a]	Double	5.1 ± 0.2 L/d	Guerrero-Sodric et al. (2023)
Corn stalk	Cow dung	[a]	Single	129.8 mL H_2/g-corn stalk	Li et al. (2014)
Cassava starch	Peatland soil	*Thermoanaerobacterium* sp., *Geobacillus* sp., *Thermacetogenium* sp. E *Pseudomonas* sp.	Single	205 mL H_2 $gCOD^{-1}$	Khongkliang et al. (2017)
Palm oil mill effluent	Anaerobic microbial sludge from a POME based biogas plant	*Geobacter* sp., *Desulfovibrio* sp., and *Thermoanaerobacterium sp.*	Single	134 mL-H_2 $gCOD^{-1}$	Khongkliang et al. (2019)
Rice straw	Municipal sewage sludge	[a]	Double	2.56 mmol	Zhang et al. (2019)
Lignocellulosic hydrolysate	Sludge from a local wastewater treatment plant	*Enterococcus* spp.	Single	10.9 mol H_2/mol of glucose	Wang et al. (2021)

[a] Not specified in the study

2022). Without proper treatment, its release into the environment can lead to serious environmental problems (Abdulgader et al. 2019). Rani et al. (2022) employed dairy industry effluent as substrate in a single-chamber MEC, aiming at treating this effluent and producing biohydrogen, obtaining a maximum production of 32 ± 5 mL/L/d of hydrogen and removal of 95 ± 2% under conditions applied voltage (0.8 V) and hydraulic retention time of 2 days.

Palm oil, the most consumed edible oil in the world, has enormous socioeconomic importance for countries such as Malaysia and Indonesia, its largest producers (Iskandar et al. 2018). However, its production entails a high consumption of water (from 5 to 7.5 tons of water per ton of palm oil), with about 50% of this water becoming effluent from palm oil processing (POME) (Liew et al. 2015), having large percentages of oil and grease and high chemical oxygen demand (COD). In this context, Kadier et al. (2022) demonstrated the viability of using a MEC for the treatment of POME together with the generation of biohydrogen, obtaining a maximum hydrogen production of 1.1747 m^3 H_2/m^3 under optimized conditions of incubation temperature of 30 °C, initial pH of 6.63 and effluent dilution of 50.71% COD.

As Samsudeen et al. (2020) demonstrated, distillery effluents have also been used to produce biohydrogen in MECs. In this work, the authors used this effluent in a double chamber MEC, obtaining a cumulative hydrogen production of 40.05 ± 0.5 mL with the application of a current density of 811.7 ± 20 mA/m^2 and COD removal of 75.6% ± 1.5%.

Crude glycerol is an important byproduct of the biodiesel industry, and its production is increasing due to the growing demand for this biofuel, to the point that in the near future, the production of glycerol might be more significant than its demand (Badia-Fabregat et al. 2019), requiring technologies for its valorization. Badia-Fabregat et al. (2019) obtained a biohydrogen production of 0.46 L H_2/L/d using crude glycerol at a concentration of 3.1 g/L as the sole carbon source in a MEC at alkaline pH with an application of a voltage of 0.8 V at an ambient temperature of 20 °C, thus demonstrating the possibility of using excess crude glycerol to produce biohydrogen through microbial electrolysis cells.

17.3.1.2 Dark Fermentation Effluents

Dark fermentation is one of the main biological routes used to produce biohydrogen, and it is possible to use different substrates (Li et al. 2014). However, dark fermentation provides low hydrogen yields due to the formation of volatile fatty acids such as acetic, propionic, butyric, isobutyric, valeric, and isovaleric acids (Magrini et al. 2021), which contains much more energy than H_2, resulting in reduced fermentation of the raw material (Jung et al. 2022).

Thus, biohydrogen production in two stages through the coupling of MECs to dark fermentation systems has been extensively studied, proving to be an excellent way to increase biohydrogen yield and consume volatile fatty acids from dark fermentation. In this process, a wide range of fermented substrates from dark fermentation can be used, such as corn stalks (Li et al. 2014), industrial wastewater from cheese production, juices, paper, sugar, and distillates (Marone et al. 2017), cassava starch (Khongkliang et al. 2017), palm oil industry effluent (Khongkliang et al. 2019), algal biomass (Nguyen et al. 2020) and others.

Li et al. (2014) in order to achieve higher yields in corn stalk fermentation, integrated dark fermentation and MECs, reaching a maximum yield of 129.8 mL of H_2/g of corn stalk at pH 7.0 and temperature of 36 °C during the first stage. The second step was performed in a single-chamber microbial electrolysis cell; the dark fermentation effluent was diluted, culminating in a yield of 257.3 mL of H_2/g of corn stalk under an applied voltage of 0.8 V, with the removal of 44% of COD in the MEC.

Using the cassava starch processing effluent in a dark fermentation system coupled to a single-chamber MEC, Khongkliang et al. (2017) obtained in the first stage a maximum yield of 260 mL H_2 gCOD and COD removal efficiency of 38% with pH of 5.5, temperature of 55 °C and 24 h HRT. The second step took place in a MEC with a pH of 6.0, applied voltage of 0.6 V, at 55 °C, and with a 48 h HRT obtaining a hydrogen production of 205 mL H_2/gCOD and efficiency in the 32% COD removal. Khongkliang et al. (2019) produced biohydrogen in two stages from palm oil effluent. Under optimal conditions of COD concentration of 66 g/L and pH of 6.5, the dark fermentation obtained a hydrogen production yield of 56 mL H_2/gCOD. Hydrogen production in the second stage yielded 134 mL H_2/gCOD with an applied voltage of 0.7 V under the same conditions of COD load, temperature, and pH.

17.3.1.3 Urban Effluents

Urban wastewater is widely available and mainly composed of carbon, nitrogen, and phosphorus. Conventional treatment processes are usually satisfactory for removing these nutrients, allowing their safe release into water bodies. However, they have the main disadvantage of being energy-intensive (Acién et al. 2016).

The scientific community has not exploited urban wastewater to produce biohydrogen in MECs as frequently as other residues, such as industrial effluents and effluents from dark fermentation. However, this technology can be used to carry out a more efficient treatment of wastewater in urban areas, reducing treatment costs and performing energy recovery (Sales and Lee 2015).

Chen et al. (2019) used a tubular microbial electrolysis cell to obtain hydrogen from domestic wastewater, achieving a hydrogen production rate of about 0.18 ± 0.03 m^3/m^3d under a current of 0.74 ± 0.19 A/m^2 in a 3-h HRT.

Guerrero-Sodric et al. (2023) produced biohydrogen on a pilot scale (135 L) using urban wastewater as substrate. The pilot plant was set up using a 220 L stainless steel tank containing nine double chamber MECs, which were operated by varying conditions such as temperature (18–30 °C), hydraulic residence time (1.9–3.9 d), applied potential (0.8–1.1 V), achieving an average hydrogen production of 5.1 ± 0.2 L/d with a hydraulic retention time of 1 day. The optimal applied voltage was around 9 V and a temperature of 30 °C.

17.3.1.4 Lignocellulosic Biomass

Lignocellulosic biomass has significant energy potential and growing availability (Wang et al. 2021), and can be reused to produce biofuels (Ashokkumar et al. 2022). However, its use in microbial electrolysis cells for biohydrogen production is still little explored, having been recently studied by Zhang et al. (2019) and Wang et al. (2021).

Zhang et al. (2019) used rice straw pretreated with 1% sodium hydroxide, plant residues, and a mixture of these two substrates for biohydrogen production in a single-chamber MEC through simultaneous saccharification and fermentation. The MEC was operated at a temperature of 35 °C and pH 7 for 3 days, yielding 2.56 ± 0.07 mmol H_2/L d of hydrogen from the mixed substrate.

Wang et al. (2021) used enzymatically pretreated lignocellulosic biomass in a 10 L single-chamber MEC under a temperature of 35 °C and application of 1.36 V. They obtained a maximum hydrogen yield of 10.9 mol H_2/mol of glucose.

17.3.2 Hydrogen-Producing Microorganisms in Microbial Electrolysis Cells

The production of biohydrogen in MECs is mainly mediated by microorganisms capable of hydrolyzing organic matter and thus providing monomers, and electrochemically active bacteria that oxidize these monomers into electrons that are transferred to the anode and into H^+ protons, which in the absence of oxygen recombine to form hydrogen gas at the cathode (Gautam et al. 2023; Saravanan et al. 2020).

These bacteria form biofilms on the anode, being able to transfer electrons directly through its membrane or indirectly through chemical mediators and microbial nanowires (Venkata et al. 2014). Inoculation of microbial electrolysis cells can be performed using pure cultures or consortia, with the use of consortia being the most recommended since the greater diversity of microorganisms leads to better removal of organic compounds, resulting in better efficiency and stability of the process (Hasany et al. 2016). In contrast, pure cultures are more often used when the objective is to specifically investigate the performance of a particular species in the production of biohydrogen in MECs and when the objective is to understand its molecular mechanisms of electron transfer (Gorby et al. 2006).

Bacteria capable of producing biohydrogen in MECs are widely available in wastewater (Hasany et al. 2016), it can be seen in Table 17.1 that most of the works cited used inoculum from anaerobic bioreactors for effluent treatment (Chen et al. 2019; Guerrero-Sodric et al. 2023; Kadier et al. 2022; Khongkliang et al. 2017, 2019; Rani et al. 2022; Zhang et al. 2019), as well as soil samples (Khongkliang et al. 2017) animal dung (Liu et al. 2014) and microbial fuel cell activated sludge (MFC) (Badia-Fabregat et al. 2019).

Hydrogen-producing consortia harbor a symbiosis of microorganisms consisting of exoelectrogenic and strictly anaerobic bacteria such as *Shewanella spp.*, *Geobacter* spp., *Pseudomonas* spp., *Rhodoferax spp.*, *Rhodopseudomonas spp.*, etc. (Gautam et al. 2023). However, these consortia can also be composed of methanogenic bacteria, which inhibit biohydrogen production (Saravanan et al. 2020). For this reason, some techniques can be used to inhibit methanogens in these cultures, such as thermal shock, pH reduction, the addition of chemical inhibitors such as 2-bromomethane sulfonate (2-BES) lumazine (Gautam et al. 2023; Hasany et al. 2016) and among other techniques.

Kadier et al. (2022), using palm oil mill effluent as substrate and anaerobic sludge from a biogas plant (also based on palm oil effluent), observed the predominance of microorganisms of the species *Geobacter* sp (30.8%), that act as electron carriers between the bacterial cell and the electrodes, *Pseudomonas* sp (11.1%) and *Klebsiella* (9.4%) that carry out oxidation of the substrate by increasing the rate of electron transfer.

Using the same substrate and inoculum, Khongkliang et al. (2019) observed the dominance of bacteria of the species *Geobacter* sp., *Desulfovibrio* sp., and *Thermoanaerobacterium thermosaccharolyticum.*

Using crude glycerol as substrate and an inoculum from a microbial fuel cell (MFC) Badia-Fabregat et al. (2019) verified the presence of microorganisms of the species *Enterococcus* sp (32.4%), *Acetobacterium* sp. (16.2%), and *Geoalkalibacter* sp (9.9%). *Enterococcus sp.* can act both in the oxidation of substrates and in the transfer of electrons to the anode, while bacteria of the genus *Acetobacterium* sp. are able to use glycerol producing acetic acid that can later be oxidized by other microorganisms such as *Geoalkalibacter* sp which plays an important role in transferring electrons to the anode.

Khongkliang et al. (2017) using wastewater from cassava processing as a substrate and soil samples as inoculum, verified the formation of a biofilm consisting of bacteria of the species *Thermoanaerobacterium* sp., *Geobacillus* sp., *Thermacetogenium* sp. and *Pseudomonas* sp.

17.3.3 Temperature

Temperature directly impacts biohydrogen production in MECs, as it influences the enzymatic activity of microorganisms and biofilm formation, and consequently substrate degradation, electron transfer, and biohydrogen generation (Dange et al. 2021). According to the literature, mesophilic temperatures around 28 °C are ideal for the metabolism of exoelectrogenic microorganisms (Saravanan et al. 2022), having been demonstrated by the work of Kadier et al. (2022) a maximum production of hydrogen under the optimized temperature of 30.23 °C.

17.3.4 pH

Similarly to temperature, pH also affects biofilm formation on the anode, in addition to the kinetics and thermodynamics of the reaction, and, consequently, directly affects biohydrogen production (Gautam et al. 2023). A more alkaline pH in a range of 7.0–9.0 results in better MECs performance since the anodic potential is higher at higher pHs, also favoring the growth of exoelectrogenic microorganisms and thus transferring a higher amount of electrons to the anode (Khan et al. 2017).

17.3.5 Hydraulic Retention Time and Applied Electrical Potential

In biohydrogen production by MECs, the hydraulic retention time (HRT) is also a factor that requires optimization to improve performance. Very low hydraulic retention times can lead to inefficient oxidation of substrates by microorganisms. In contrast, very high hydraulic retention times can lead to methane production, with the reduction of HRT being a strategy to circumvent the unwanted methane production in MECs (Gautam et al. 2023).

The production of hydrogen in MECs is a nonspontaneous process. Therefore, it is necessary to apply a small electrical potential of about 0.5–1 V so that the thermodynamic barrier to biohydrogen production occurs at the cathode of the MECs (Dange et al. 2021; Hassan et al. 2018). In this case, the electric potential is relatively smaller than the potential applied in water electrolysis.

17.4 Technoeconomic Analysis

The use of wastewater for biohydrogen production is one of the promising alternatives for effluent disposal because it contains large organic loads (De Carvalho et al. 2022), and, consequently, large amounts of energy. The energy content of important agroindustrial wastes can be significant, e.g., 5 kWh/m^3 for slaughterhouse effluents, 9 kWh/m^3 for milk processing effluent, 155 kWh/m^3 for cheese whey, 281 kWh/m^3 for palm oil mill effluent, 195 kWh/m^3 for cassava processing wastewater, and 132 kWh/m^3 for vinasse wastewater (Martinez-Burgos et al. 2021a). Different processes have been used to produce biohydrogen with effluents, such as photo fermentation, biophotolysis, and dark fermentation (Martinez-Burgos et al. 2021b). However, in these processes, the effluents need to be supplemented with macro- and micronutrients such as sources of carbon, nitrogen, potassium, phosphorus, and sometimes amino acids and vitamins (Martinez-Burgos et al. 2022; Sharma and Melkania 2018). Dos Reis et al. (2015) and Krishnan et al. (2016) supplemented media to produce hydrogen via dark fermentation with reducing agents such as Cysteine-HCl. Vitamins and minerals are also used for supplementing culture media because they can alter metabolic pathways due to their being enzymatic cofactors of hydrogenases (Lu et al. 2013). Combining effluents to reach an ideal substrate composition is a strategy developed by Martinez-Burgos et al. (2021c) to efficiently produce biohydrogen, using cassava processing wastewater and corn steep liquor. The first effluent contains large amounts of carbon sources, while the second contains significant amounts of minerals, nitrogen, and amino acids. Supplementing the effluents can be considered a disadvantage of these processes because it increases the production costs of biohydrogen.

Biohydrogen production via dark or photofermentation technologies integrated to microbial electrolysis cells (MECs) can be attractive. According to Karthikeyan et al. (2017), single-chamber MECs are compact and economical. These systems are emerging technologies for the production of biohydrogen and are considered promising both for the production of the energy vector and for the treatment of effluents. Among the advantages of these systems is the fact that they can produce hydrogen with purities above 95% (Kadier et al. 2022). Call and Logan (2008) and Lu et al. (2009) obtained biogas with hydrogen concentrations of 96% in single-chamber MECs systems. Furthermore, MECs systems have conversion efficiencies of approximately 80% (Ajayi et al. 2009). Unlike dark fermentation, these systems can use fermentable and nonfermentable substrates (Liu et al. 2010). Other advantages of MECs are that they can remove around 80% of organic matter when wastewater is used (Kadier et al. 2022). Thus, Yu et al. (2017) reported the removal of 96% of organic matter from soybean oil refinery wastewater using a MECs system.

In MECs systems, the potential difference needed to produce hydrogen is approximately 0.12 V, while in conventional electrolytic technologies, it is approximately ten times higher (1.23 V), and these systems have low energy consumption (Aiken et al. 2019). Since materials used for the cathodes and anodes are critical components of MECs (Karthikeyan et al. 2017), they have a significant economic impact. Common materials used for cathodes are carbon with some aliases to metals such as platinum and nickel (Call and Logan 2008; Karthikeyan et al. 2017; Lu et al. 2009) and in the case of the anode, carbon, ammonia treated, graphite fiber is generally used (Call and Logan 2008; Karthikeyan et al. 2017; Lu et al. 2009). The use of metals in cathodes increases the cost of biohydrogen production with these systems; in fact, according to Rozendal et al. (2008), approximately 47% of the cost of MECs is due to the use of noble materials in the electrodes. Another critical component of MECs is the cation or proton exchange membranes. These prevent the crossing of substrates between the two half cells, thus minimizing the loss of hydrogen. However, membranes generally limit pH division due to the concentration of some ions such as Na^+, K^+, and Ca^{2+} (Rozendal et al. 2006). As a result, the anolyte and catholyte can easily become acidified and alkaline, respectively, suppressing microbial activity (Karthikeyan et al. 2017). These membranes are also expensive, and it is estimated that each m^2 of ion exchange membrane has a value of 400 €, which represents around 38% of the cost of a MECs system (Karthikeyan et al. 2017; Rozendal et al. 2008).

Production costs are a major barrier to hydrogen production using biological methods on a large scale. Currently, producing hydrogen from fossil fuels remains the cheapest option, even though it is the most polluting. Dark fermentation has a limitation in terms of yields because stoichiometrically, the maximum yield reached can be 4 mol H_2/mol glucose when the intermediate product is acetic acid (Rosa et al. 2020; Sydney et al. 2020), despite being a quick process that can be developed in between 20 h and 40 h (Martinez-Burgos et al. 2019). In the case of photofermentation and biophotolysis, in addition to obtaining lower yields, the process takes longer, which makes industrialization difficult. Thus, the cost of producing hydrogen from natural gas ranges from US$ 0.51/kg H_2 to US$ 1.7/kg H_2 (IEA 2021). Regarding hydrogen produced from renewable sources, the production cost ranges from US$ 4.30/kg H_2 to US$ 5.80/kg H_2 (Fan et al. 2022). For hydrogen produced via electrolysis but using renewable energy, the cost ranges from US$ 3/kg H_2 to US$ 8/kg (IEA 2021). Finally, in MEC systems, technology costs are the main inconvenience for large-scale use (Aiken et al. 2019). According to Rozendal et al. (2008), MECs systems are approximately 10 times more expensive than anaerobic digestion systems, directly impacting the energy vector production cost. According to Aiken et al. (2019), the investment costs for MECs systems are still high, and these systems to be economically viable, the cost of the anode and current collector must decrease by 90%, and the organic load rates must remain between 800 and 1400 g-COD/m^3/d.

17.5 Conclusion and Perspectives

Microbial electrolysis cells (MEC) are a promising technology to produce hydrogen using electrochemically active bacteria. The yield of hydrogen can be up to three times higher compared to simple microbial fermentation. However, although many advances have been made to improve MECs, this technology is still difficult to apply on a larger scale. The main challenges are in the cell materials and design since electrodes and membranes require expensive materials. To overcome these difficulties, studies using MECs have focused on the following aspects: selection of adequate microbial consortia and substrates, improvement of electrode materials, and reactor designs. Wastewater from industrial processing and dark fermentation effluents are common substrates used in biohydrogen production in MECs. It is also observed that integrated technologies have been attractive options for the energy sector. By integrating with conventional dark fermentation or photofermentation technologies, electrobiochemical biohydrogen production can become viable due to the lower energy demand and higher biohydrogen yield.

While MECs offer the potential for improved biohydrogen yields, technical and economic analyses are essential to assess their viability compared to other production methods, such as lower yield but inexpensive dark fermentation. In conclusion, microbial electrolysis cells hold promise as a sustainable and efficient approach for biohydrogen production, offering the potential for clean energy generation and waste treatment. Continued research and development in this field will contribute to making hydrogen a viable future fuel, reducing our dependence on fossil fuels and mitigating environmental impacts.

Acknowledgments This work was supported by the multilateral project BRICS STI FP 2021 – BIOGEN (293), financed in Brazil by the National Council for Scientific and Technological Development, CNPq, grant no. 440138/2022-1, also with support of the Coordination of Improvement of Higher Education Personnel – CAPES, project Print-BioADD.

References

Abd-Elrahman NK, Al-Harbi N, Al-Hadeethi Y, Alruqi AB, Mohammed H, Umar A, Akbar S (2022) Influence of nanomaterials and other factors on biohydrogen production rates in microbial electrolysis cells—a review. Molecules 27(23). MDPI. https://doi.org/10.3390/molecules27238594

Abdulgader M, Yu J, Zinatizadeh AA, Williams P, Rahimi Z (2019) Process analysis and optimization of single stage flexible fibre biofilm reactor treating milk processing industrial wastewater using response surface methodology (RSM). Chem Eng Res Des 149:169–181. https://doi.org/10.1016/J.CHERD.2019.07.011

Acién FG, Gómez-Serrano C, Morales-Amaral MM, Fernández-Sevilla JM, Molina-Grima E (2016) Wastewater treatment using microalgae: how realistic a contribution might it be to significant urban wastewater treatment? Appl Microbiol Biotechnol 100(21):9013–9022. Springer Verlag. https://doi.org/10.1007/s00253-016-7835-7

Aiken DC, Curtis TP, Heidrich ES (2019) Avenues to the financial viability of microbial electrolysis cells [MEC] for domestic wastewater treatment and hydrogen production. Int J Hydrogen Energy 44(5):2426–2434. https://doi.org/10.1016/j.ijhydene.2018.12.029

Ajayi FF, Kim KY, Chae KJ, Choi MJ, Kim SY, Chang IS, Kim IS (2009) Study of hydrogen production in light assisted microbial electrolysis cell operated with dye sensitized solar cell. Int J Hydrogen Energy 34(23):9297–9304. https://doi.org/10.1016/j.ijhydene.2009.09.096

Akhlaghi N, Najafpour-Darzi G (2020) A comprehensive review on biological hydrogen production. Int J Hydrogen Energy 45(43):22492–22512. https://doi.org/10.1016/J.IJHYDENE.2020.06.182

Ananthi V, Ramesh U, Balaji P, Kumar P, Govarthanan M, Arun A (2022) A review on the impact of various factors on biohydrogen production. Int J Hydrogen Energy. https://doi.org/10.1016/j.ijhydene.2022.08.046

Argun H, Kargi F, Kapdan IK, Oztekin R (2008) Biohydrogen production by dark fermentation of wheat powder solution: effects of C/N and C/P ratio on hydrogen yield and formation rate. Int J Hydrogen Energy 33(7):1813–1819. https://doi.org/10.1016/j.ijhydene.2008.01.038

Ashokkumar V, Venkatkarthick R, Jayashree S, Chuetor S, Dharmaraj S, Kumar G, Chen WH, Ngamcharussrivichai C (2022) Recent advances in lignocellulosic biomass for biofuels and value-added bioproducts - a critical review. Bioresource Technol 344. Elsevier Ltd. https://doi.org/10.1016/j.biortech.2021.126195

Badia-Fabregat M, Rago L, Baeza JA, Guisasola A (2019) Hydrogen production from crude glycerol in an alkaline microbial electrolysis cell. Int J Hydrogen Energy 44(32):17204–17213. https://doi.org/10.1016/J.IJHYDENE.2019.03.193

Bora A, Mohanrasu K, Angelin Swetha T, Ananthi V, Sindhu R, Chi NTL, Pugazhendhi A, Arun A, Mathimani T (2022) Microbial electrolysis cell (MEC): reactor configurations, recent advances and strategies in biohydrogen production. Fuel 328. https://doi.org/10.1016/j.fuel.2022.125269

Call D, Logan BE (2008) Hydrogen production in a single chamber microbial electrolysis cell lacking a membrane. Environ Sci Technol 42(9):3401–3406. https://doi.org/10.1021/es8001822
Chacón-Carrera RA, López-Ortiz A, Collins-Martínez V, Meléndez-Zaragoza MJ, Salinas-Gutiérrez J, Espinoza-Hicks JC, Ramos-Sánchez VH (2019) Assessment of two ionic exchange membranes in a bioelectrochemical system for wastewater treatment and hydrogen production. Int J Hydrogen Energy 44(24):12339–12345. https://doi.org/10.1016/j.ijhydene.2018.10.153
Chen J, Xu W, Wu X, Jiaqiang E, Lu N, Wang T, Zuo H (2019) System development and environmental performance analysis of a pilot scale microbial electrolysis cell for hydrogen production using urban wastewater. Energy Convers Manag 193:52–63. https://doi.org/10.1016/j.enconman.2019.04.060
Chen H, Wu J, Huang R, Zhang W, He W, Deng Z, Han Y, Xiao B, Luo H, Qu W (2022) Effects of temperature and total solid content on biohydrogen production from dark fermentation of rice straw: performance and microbial community characteristics. Chemosphere 286. https://doi.org/10.1016/j.chemosphere.2021.131655
Cheng D, Ngo HH, Guo W, Chang SW, Nguyen DD, Zhang S, Deng S, An D, Hoang NB (2022) Impact factors and novel strategies for improving biohydrogen production in microbial electrolysis cells. Bioresource Technol 346. Elsevier Ltd. https://doi.org/10.1016/j.biortech.2021.126588
Chi J, Yu H (2018) Water electrolysis based on renewable energy for hydrogen production. Cuihua Xuebao/Chin J Catal 39(3):390–394. Science Press. https://doi.org/10.1016/S1872-2067(17)62949-8
Cui W, Liu G, Zeng C, Lu Y, Luo H, Zhang R (2019) Improved hydrogen production in the single-chamber microbial electrolysis cell with inhibition of methanogenesis under alkaline conditions. RSC Adv 9(52):30207–30215. https://doi.org/10.1039/c9ra05483a
Dai HY, Yang HM, Liu X, Song XL, Liang ZH (2019) Hydrogen production using "direct-starting" biocathode microbial electrolysis cell and the analysis of microbial communities. Acta Metallurgica Sinica (English Letters) 32(3):297–304. https://doi.org/10.1007/s40195-018-0785-6
Dange P, Pandit S, Jadhav D, Shanmugam P, Gupta PK, Kumar S, Kumar M, Yang YH, Bhatia SK (2021) Recent developments in microbial electrolysis cell-based biohydrogen production utilizing wastewater as a feedstock. Sustainability (Switzerland) 13(16). MDPI. https://doi.org/10.3390/su13168796
Das D, Veziroglu TN (2008) Advances in biological hydrogen production processes. Int J Hydrogen Energy 33(21):6046–6057. https://doi.org/10.1016/j.ijhydene.2008.07.098
De Bonis C, Simari C, Kosma V, Mecheri B, D'Epifanio A, Allodi V, Mariotto G, Brutti S, Suarez S, Pilar K, Greenbaum S, Licoccia S, Nicotera I (2016) Enhancement of proton mobility and mitigation of methanol crossover in sPEEK fuel cells by an organically modified titania nanofiller. J Solid State Electrochem 20(6):1585–1598. https://doi.org/10.1007/s10008-016-3167-x
De Carvalho JC, Molina-Aulestia DT, Martinez-Burgos WJ, Karp SG, Manzoki MC, Medeiros ABP, Rodrigues C, Scapini T, de Souza Vandenberghe LP, Vieira S, Woiciechowski AL, Soccol VT, Soccol CR (2022) Agro-industrial wastewaters for algal biomass production, bio-based products, and biofuels in a circular bioeconomy. Fermentation 8(12). https://doi.org/10.3390/fermentation8120728
Dos Reis CM, Carosia MF, Sakamoto IK, Amâncio Varesche MB, Silva EL (2015) Evaluation of hydrogen and methane production from sugarcane vinasse in an anaerobic fluidized bed reactor. Int J Hydrogen Energy 40(27):8498–8509. https://doi.org/10.1016/j.ijhydene.2015.04.136
Eroglu E, Melis A (2011) Photobiological hydrogen production: recent advances and state of the art. Bioresource Technol 102(18):8403–8413. https://doi.org/10.1016/j.biortech.2011.03.026
Fan JL, Yu P, Li K, Xu M, Zhang X (2022) A levelized cost of hydrogen (LCOH) comparison of coal-to-hydrogen with CCS and water electrolysis powered by renewable energy in China. Energy 242:123003. https://doi.org/10.1016/j.energy.2021.123003

Fujinawa K, Nagoya M, Kouzuma A, Watanabe K (2019) Conductive carbon nanoparticles inhibit methanogens and stabilize hydrogen production in microbial electrolysis cells. Appl Microbiol Biotechnol 103(15):6385–6392. https://doi.org/10.1007/s00253-019-09946-1

Gautam R, Nayak JK, Ress NV, Steinberger-Wilckens R, Ghosh UK (2023) Bio-hydrogen production through microbial electrolysis cell: structural components and influencing factors. Chem Eng J 455. Elsevier B.V. https://doi.org/10.1016/j.cej.2022.140535

Gong C, Zheng X, Liu H, Wang G, Cheng F, Zheng G, Wen S, Law W-C, Tsui C-P, Tang C-Y (2016) A new strategy for designing high-performance sulfonated poly(ether ether ketone) polymer electrolyte membranes using inorganic proton conductor-functionalized carbon nanotubes

González-Pabón MJ, Cardeña R, Cortón E, Buitrón G (2021) Hydrogen production in two-chamber MEC using a low-cost and biodegradable poly(vinyl) alcohol/chitosan membrane. Bioresource Technol 319. https://doi.org/10.1016/j.biortech.2020.124168

Gorby YA, Yanina S, McLean JS, Rosso KM, Moyles D, Dohnalkova A, Beveridge TJ, Seop Chang I, Hong Kim B, Shik Kim K, Culley DE, Reed SB, Romine MF, Saffarini DA, Hill EA, Shi L, Elias DA, Kennedy DW, Pinchuk G et al (2006) Electrically conductive bacterial nanowires produced by Shewanella oneidensis strain MR-1 and other microorganisms. www.pnas.orgcgidoi10.1073pnas.0604517103

Guerrero-Sodric O, Baeza JA, Guisasola A (2023) Exploring key operational factors for improving hydrogen production in a pilot-scale microbial electrolysis cell treating urban wastewater. Chem Eng J 469:144001. https://doi.org/10.1016/j.cej.2023.144001

Guo K, Prévoteau A, Patil SA, Rabaey K (2015) Engineering electrodes for microbial electrocatalysis. Curr Opin Biotechnol 33:149–156. Elsevier Ltd. https://doi.org/10.1016/j.copbio.2015.02.014

Hasany M, Mardanpour MM, Yaghmaei S (2016) Biocatalysts in microbial electrolysis cells: a review. Int J Hydrogen Energy 41(3):1477–1493. Elsevier Ltd. https://doi.org/10.1016/j.ijhydene.2015.10.097

Hassan M, Fernandez AS, San Martin I, Xie B, Moran A (2018) Hydrogen evolution in microbial electrolysis cells treating landfill leachate: dynamics of anodic biofilm. Int J Hydrogen Energy 43(29):13051–13063. https://doi.org/10.1016/j.ijhydene.2018.05.055

Hawkes FR, Hussy I, Kyazze G, Dinsdale R, Hawkes DL (2007) Continuous dark fermentative hydrogen production by mesophilic microflora: principles and progress. Int J Hydrogen Energy 32(2):172–184. https://doi.org/10.1016/j.ijhydene.2006.08.014

Hitam CNC, Jalil AA (2020) A review on biohydrogen production through photo-fermentation of lignocellulosic biomass. Biomass Convers Biorefin. Springer Science and Business Media Deutschland GmbH. https://doi.org/10.1007/s13399-020-01140-y

Hosseinzadeh A, Zhou JL, Altaee A, Baziar M, Li D (2020) Effective modelling of hydrogen and energy recovery in microbial electrolysis cell by artificial neural network and adaptive network-based fuzzy inference system. Bioresource Technol 316:123967. https://doi.org/10.1016/J.BIORTECH.2020.123967

Hou Y, Luo H, Liu G, Zhang R, Li J, Fu S (2014) Improved hydrogen production in the microbial electrolysis cell by inhibiting methanogenesis using ultraviolet irradiation. Environ Sci Technol 48(17):10482–10488. https://doi.org/10.1021/es501202e

Hu H, Fan Y, Liu H (2008) Hydrogen production using single-chamber membrane-free microbial electrolysis cells. Water Res 42(15):4172–4178. https://doi.org/10.1016/j.watres.2008.06.015

IEA (2021) Global Hydrogen Review 2021, executive summary. Retrieved from: <https://www.iea.org/reports/global-hydrogen-review-2021/executive-summary> (Jun 7, 2023). 4(1):88–100

Iskandar MJ, Baharum A, Anuar FH, Othaman R (2018) Palm oil industry in South East Asia and the effluent treatment technology—a review. Environ Technol Innov 9:169–185. Elsevier B.V. https://doi.org/10.1016/j.eti.2017.11.003

Jafary T, Daud WRW, Ghasemi M, Kim BH, Md Jahim J, Ismail M, Lim SS (2015) Biocathode in microbial electrolysis cell; present status and future prospects. Renew Sustain Energy Rev 47:23–33. Elsevier Ltd. https://doi.org/10.1016/j.rser.2015.03.003

Jayabalan T, Matheswaran M, Naina Mohammed S (2019) Biohydrogen production from sugar industry effluents using nickel based electrode materials in microbial electrolysis cell. Int J Hydrogen Energy 44(32)

Jayabalan T, Matheswaran M, Preethi V, Naina Mohamed S (2020) Enhancing biohydrogen production from sugar industry wastewater using metal oxide/graphene nanocomposite catalysts in microbial electrolysis cell. Int J Hydrogen Energy 45(13):7647–7655. https://doi.org/10.1016/j.ijhydene.2019.09.068

Jiang Y, Liang P, Zhang C, Bian Y, Sun X, Zhang H, Yang X, Zhao F, Huang X (2016) Periodic polarity reversal for stabilizing the pH in two-chamber microbial electrolysis cells. Appl Energy 165:670–675. https://doi.org/10.1016/j.apenergy.2016.01.001

Jung JH, Sim YB, Ko J, Park SY, Kim GB, Kim SH (2022) Biohydrogen and biomethane production from food waste using a two-stage dynamic membrane bioreactor (DMBR) system. Bioresource Technol 352. https://doi.org/10.1016/j.biortech.2022.127094

Kadier A, Kalil MS, Abdeshahian P, Chandrasekhar K, Mohamed A, Azman NF, Logroño W, Simayi Y, Hamid AA (2016) Recent advances and emerging challenges in microbial electrolysis cells (MECs) for microbial production of hydrogen and value-added chemicals. Renew Sustain Energy Rev 61:501–525. Elsevier Ltd. https://doi.org/10.1016/j.rser.2016.04.017

Kadier A, Singh R, Song D, Ghanbari F, Zaidi NS, Prihartini Aryanti PT, Jadhav DA, Islam MA, Kalil MS, Nabgan W, Hamid AA, Hasan HA, Ma PC (2022) A novel pico-hydro power (PHP)-Microbial electrolysis cell (MEC) coupled system for sustainable hydrogen production during palm oil mill effluent (POME) wastewater treatment. Int J Hydrogen Energy. https://doi.org/10.1016/j.ijhydene.2022.09.023

Karthikeyan R, Cheng KY, Selvam A, Bose A, Wong JWC (2017) Bioelectrohydrogenesis and inhibition of methanogenic activity in microbial electrolysis cells - a review. Biotechnol Adv 35(6):758–771. https://doi.org/10.1016/j.biotechadv.2017.07.004

Khan MZ, Nizami AS, Rehan M, Ouda OKM, Sultana S, Ismail IM, Shahzad K (2017) Microbial electrolysis cells for hydrogen production and urban wastewater treatment: a case study of Saudi Arabia. Appl Energy 185:410–420. https://doi.org/10.1016/J.APENERGY.2016.11.005

Khongkliang P, Kongjan P, Utarapichat B, Reungsang A, O-Thong S (2017) Continuous hydrogen production from cassava starch processing wastewater by two-stage thermophilic dark fermentation and microbial electrolysis. Int J Hydrogen Energy 42(45):27584–27592. https://doi.org/10.1016/j.ijhydene.2017.06.145

Khongkliang P, Jehlee A, Kongjan P, Reungsang A, O-Thong S (2019) High efficient biohydrogen production from palm oil mill effluent by two-stage dark fermentation and microbial electrolysis under thermophilic condition. Int J Hydrogen Energy 44(60):31841–31852. https://doi.org/10.1016/j.ijhydene.2019.10.022

Kim KY, Logan BE (2019) Nickel powder blended activated carbon cathodes for hydrogen production in microbial electrolysis cells. Int J Hydrogen Energy 44(26):13169–13174. https://doi.org/10.1016/J.IJHYDENE.2019.04.041

Koroglu EO, Ozdemir OK, Ozkaya B, Demir A (2019) An integrated system development including PEM fuel cell/biogas purification during acidogenic biohydrogen production from dairy wastewater. Int J Hydrogen Energy 44(32):17297–17303. https://doi.org/10.1016/j.ijhydene.2019.01.291

Krishnan S, Singh L, Sakinah M, Thakur S, Wahid ZA, Alkasrawi M (2016) Process enhancement of hydrogen and methane production from palm oil mill effluent using two-stage thermophilic and mesophilic fermentation. Int J Hydrogen Energy 41(30):12888–12898. https://doi.org/10.1016/j.ijhydene.2016.05.037

Kumar N, Das D (2001) Continuous hydrogen production by immobilized Enterobacter cloacae IIT-BT 08 using lignocellulosic materials as solid matrices. www.elsevier.com/locate/enzmictec

Kumar GG, Sarathi VGS, Nahm KS (2013) Recent advances and challenges in the anode architecture and their modifications for the applications of microbial fuel cells. Biosens Bioelectron 43(1):461–475. https://doi.org/10.1016/j.bios.2012.12.048

Kumar G, Bakonyi P, Zhen G, Sivagurunathan P, Koók L, Kim SH, Tóth G, Nemestóthy N, Bélafi-Bakó K (2017) Microbial electrochemical systems for sustainable biohydrogen production: surveying the experiences from a start-up viewpoint. Renew Sustain Energy Reviews 70:589–597. Elsevier Ltd. https://doi.org/10.1016/j.rser.2016.11.107

Lee B, Park JG, Shin WB, Tian DJ, Jun HB (2017) Microbial communities change in an anaerobic digestion after application of microbial electrolysis cells. Bioresource Technol 234:273–280. https://doi.org/10.1016/j.biortech.2017.02.022

Lee ZS, Chin SY, Lim JW, Witoon T, Cheng CK (2019) Treatment technologies of palm oil mill effluent (POME)and olive mill wastewater (OMW): a brief review. Environ Technol Innov 15. Elsevier B.V. https://doi.org/10.1016/j.eti.2019.100377

Li XH, Liang DW, Bai YX, Fan YT, Hou HW (2014) Enhanced H_2 production from corn stalk by integrating dark fermentation and single chamber microbial electrolysis cells with double anode arrangement. Int J Hydrogen Energy 39(17):8977–8982. https://doi.org/10.1016/j.ijhydene.2014.03.065

Liew WL, Kassim MA, Muda K, Loh SK, Affam AC (2015) Conventional methods and emerging wastewater polishing technologies for palm oil mill effluent treatment: a review. J Environ Manag 149:222–235. Academic Press. https://doi.org/10.1016/j.jenvman.2014.10.016

Liu H, Hu H, Chignell J, Fan Y (2010) Microbial electrolysis: novel technology for hydrogen production from biomass. Biofuels 1(1):129–142. https://doi.org/10.4155/bfs.09.9

Liu Z, Li Q, Zhang C, Wang L, Han B, Li B, Zhang Y, Chen H, Xing XH (2014) Effects of operating parameters on hydrogen production from raw wet steam-exploded cornstalk and two-stage fermentation potential for biohythane production. Biochem Eng J 90:234–238. https://doi.org/10.1016/j.bej.2014.06.013

Lu L, Ren N, Xing D, Logan BE (2009) Hydrogen production with effluent from an ethanol-H2-coproducing fermentation reactor using a single-chamber microbial electrolysis cell. Biosens Bioelectron 24(10):3055–3060. https://doi.org/10.1016/j.bios.2009.03.024

Lu ZH, Li J, Zhu A, Yao Q, Huang W, Zhou R, Zhou R, Chen X (2013) Catalytic hydrolysis of ammonia borane via magnetically recyclable copper iron nanoparticles for chemical hydrogen storage. Int J Hydrogen Energy 38(13):5330–5337. https://doi.org/10.1016/j.ijhydene.2013.02.076

Lu L, Hou D, Wang X, Jassby D, Ren ZJ (2016) Active H2 harvesting prevents methanogenesis in microbial electrolysis cells. Environ Sci Technol Lett 3(8):286–290. https://doi.org/10.1021/acs.estlett.6b00209

Magrini FE, de Almeida GM, da Maia Soares D, Fuentes L, Ecthebehere C, Beal LL, da Silveira MM, Paesi S (2021) Effect of different heat treatments of inoculum on the production of hydrogen and volatile fatty acids by dark fermentation of sugarcane vinasse. Biomass Convers Biorefin 11(6):2443–2456. https://doi.org/10.1007/s13399-020-00687-0

Marone A, Ayala-Campos OR, Trably E, Carmona-Martínez AA, Moscoviz R, Latrille E, Steyer JP, Alcaraz-Gonzalez V, Bernet N (2017) Coupling dark fermentation and microbial electrolysis to enhance bio-hydrogen production from agro-industrial wastewaters and by-products in a bio-refinery framework. Int J Hydrogen Energy 42(3):1609–1621. https://doi.org/10.1016/j.ijhydene.2016.09.166

Martinez-Burgos WJ, Sydney EB, Brar SK, de Andrade Tanobe VO, Pedroni Medeiros AB, De Carvalho J, Soccol CR (2019) The effect of hydrolysis and sterilization in biohydrogen production from cassava processing wastewater medium using anaerobic bacterial consortia. Int J Hydrogen Energy 44(47):25551–25564. https://doi.org/10.1016/j.ijhydene.2019.08.085

Martinez-Burgos WJ, Bittencourt Sydney E, Bianchi Pedroni Medeiros A, Magalhães AI, de Carvalho JC, Karp SG, de Souza Vandenberghe LP, Junior Letti LA, Thomaz Soccol V, de Melo Pereira GV, Rodrigues C, Lorenci Woiciechowski A, Soccol CR (2021a) Agro-industrial wastewater in a circular economy: characteristics, impacts and applications for bioenergy and biochemicals. Bioresource Technol 341:125795. https://doi.org/10.1016/j.biortech.2021.125795

Martinez-Burgos WJ, De Souza Candeo E, Pedroni Medeiros AB, Cesar de Carvalho J, de Andrade Tanobe VO, Soccol CR, Sydney EB (2021b) Hydrogen: current advances and patented technologies of its renewable production. J Clean Prod 286:124970. https://doi.org/10.1016/j.jclepro.2020.124970

Martinez-Burgos WJ, Sydney EB, de Paula DR, Medeiros ABP, de Carvalho JC, Molina D, Soccol CR (2021c) Hydrogen production by dark fermentation using a new low-cost culture medium composed of corn steep liquor and cassava processing water: process optimization and scale-up. Bioresource Technol 320(September 2020). https://doi.org/10.1016/j.biortech.2020.124370

Martinez-Burgos WJ, do Nascimento Junior JR, Medeiros ABP, Herrmann LW, Sydney EB, Soccol CR (2022) Biohydrogen production from agro-industrial wastes using clostridium beijerinckii and isolated bacteria as inoculum. Bioenergy Res 15(2):987–997. https://doi.org/10.1007/s12155-021-10358-1

Montpart N, Rago L, Baeza JA, Guisasola A (2015) Hydrogen production in single chamber microbial electrolysis cells with different complex substrates. Water Res 68:601–615. https://doi.org/10.1016/j.watres.2014.10.026

Nanda S, Berruti F (2021) Municipal solid waste management and landfilling technologies: a review. Environ Chem Lett 19(2):1433–1456. Springer Science and Business Media Deutschland GmbH. https://doi.org/10.1007/s10311-020-01100-y

Neto SA, Moreira TFM, Olivi P (2019) Preparation and characterization of active and cost-effective nickel/platinum electrocatalysts for hydrogen evolution electrocatalysis. Int J Hydrogen Energy 44(16):8079–8088. https://doi.org/10.1016/j.ijhydene.2019.02.083

Nguyen PKT, Das G, Kim J, Yoon HH (2020) Hydrogen production from macroalgae by simultaneous dark fermentation and microbial electrolysis cell. Bioresource Technol 315. https://doi.org/10.1016/j.biortech.2020.123795

Park SG, Rajesh PP, Hwang MH, Chu KH, Cho S, Chae KJ (2021) Long-term effects of anti-biofouling proton exchange membrane using silver nanoparticles and polydopamine on the performance of microbial electrolysis cells. Int J Hydrogen Energy 46(20):11345–11356. https://doi.org/10.1016/j.ijhydene.2020.04.059

Park SG, Rajesh PP, Sim YU, Jadhav DA, Noori MT, Kim DH, Al-Qaradawi SY, Yang E, Jang JK, Chae KJ (2022) Addressing scale-up challenges and enhancement in performance of hydrogen-producing microbial electrolysis cell through electrode modifications. Energy Rep 8:2726–2746. https://doi.org/10.1016/J.EGYR.2022.01.198

Park SG, Rhee C, Jadhav DA, Eisa T, Al-Mayyahi RB, Shin SG, Abdelkareem MA, Chae KJ (2023) Tailoring a highly conductive and super-hydrophilic electrode for biocatalytic performance of microbial electrolysis cells. Sci Total Environ 856. https://doi.org/10.1016/j.scitotenv.2022.159105

Porwal HJ, Mane AV, Velhal SG (2015) Biodegradation of dairy effluent by using microbial isolates obtained from activated sludge. Water Resour Ind 9:1–15. https://doi.org/10.1016/j.wri.2014.11.002

Rani G, Banu JR, Kumar G, Yogalakshmi KN (2022) Statistical optimization of operating parameters of microbial electrolysis cell treating dairy industry wastewater using quadratic model to enhance energy generation. Int J Hydrogen Energy 47(88):37401–37414. https://doi.org/10.1016/j.ijhydene.2022.03.120

Rosa D, Medeiros ABP, Martinez-Burgos WJ, Do Nascimento JR, De Carvalho JC, Sydney EB, Soccol CR (2020) Biological hydrogen production from palm oil mill effluent (POME) by anaerobic consortia and Clostridium beijerinckii. J Biotechnol 323(April):17–23. https://doi.org/10.1016/j.jbiotec.2020.06.015

Rozendal RA, Hamelers HVM, Buisman CJN (2006) Effects of membrane cation transport on pH and microbial fuel cell performance. Environ Sci Technol 40(17):5206–5211. https://doi.org/10.1021/es060387r

Rozendal RA, Hamelers HVM, Molenkamp RJ, Buisman CJN (2007) Performance of single chamber biocatalyzed electrolysis with different types of ion exchange membranes. Water Res 41(9):1984–1994. https://doi.org/10.1016/j.watres.2007.01.019

Rozendal RA, Hamelers HVM, Rabaey K, Keller J, Buisman CJN (2008) Towards practical implementation of bioelectrochemical wastewater treatment. Trends Biotechnol 26(8):450–459. https://doi.org/10.1016/j.tibtech.2008.04.008

Sales CM, Lee PKH (2015) Resource recovery from wastewater: application of meta-omics to phosphorus and carbon management. Curr Opin Biotechnol 33:260–267. Elsevier Ltd. https://doi.org/10.1016/j.copbio.2015.03.003

Samsudeen N, Spurgeon J, Matheswaran M, Satyavolu J (2020) Simultaneous biohydrogen production with distillery wastewater treatment using modified microbial electrolysis cell. Int J Hydrogen Energy 45(36):18266–18274. https://doi.org/10.1016/j.ijhydene.2019.06.134

Sarangi PK, Nanda S (2020) Biohydrogen production through dark fermentation. Chem Eng Technol 43(4):601–612. Wiley-VCH Verlag. https://doi.org/10.1002/ceat.201900452

Saravanan A, Karishma S, Kumar PS, Yaashikaa PR, Jeevanantham S, Gayathri B (2020) Microbial electrolysis cells and microbial fuel cells for biohydrogen production: current advances and emerging challenges. Biomass Convers Biorefin. Springer. https://doi.org/10.1007/s13399-020-00973-x

Saravanan A, Senthil Kumar P, Jeevanantham S, Karishma S, Vo DVN (2022) Recent advances and sustainable development of biofuels production from lignocellulosic biomass. Bioresource Technol 344:126203. https://doi.org/10.1016/J.BIORTECH.2021.126203

Sharma P, Melkania U (2018) Enhancement effect of amino acids on hydrogen production from organic fraction of municipal solid waste using co-culture of Escherichia coli and Enterobacter aerogenes. Energy Convers Manag 163(February):260–267. https://doi.org/10.1016/j.enconman.2018.02.072

Son S, Koo B, Chai H, Tran HVH, Pandit S, Jung SP (2021) Comparison of hydrogen production and system performance in a microbial electrolysis cell containing cathodes made of non-platinum catalysts and binders. J Water Process Eng 40:101844. https://doi.org/10.1016/J.JWPE.2020.101844

Stasinakis AS, Charalambous P, Vyrides I (2022) Dairy wastewater management in EU: produced amounts, existing legislation, applied treatment processes and future challenges. J Environ Manag 303. Academic Press. https://doi.org/10.1016/j.jenvman.2021.114152

Sydney EB, Duarte ER, Martinez Burgos WJ, De Carvalho JC, Larroche C, Soccol CR (2020) Development of short chain fatty acid-based artificial neuron network tools applied to biohydrogen production. Int J Hydrogen Energy 45(8):5175–5181. https://doi.org/10.1016/j.ijhydene.2019.10.128

Venkata S, Velvizhi G, Vamshi Krishna K, Lenin Babu M (2014) Microbial catalyzed electrochemical systems: a bio-factory with multi-facet applications. Bioresource Technol 165(C):355–364. Elsevier Ltd. https://doi.org/10.1016/j.biortech.2014.03.048

Walkowiak-Kulikowska J, Wolska J, Koroniak H (2017) Polymers application in proton exchange membranes for fuel cells (PEMFCs). Phys Sci Rev 2(8). De Gruyter. https://doi.org/10.1515/psr-2017-0018

Wang L, Long F, Liang D, Xiao X, Liu H (2021) Hydrogen production from lignocellulosic hydrolysate in an up-scaled microbial electrolysis cell with stacked bio-electrodes. Bioresource Technol 320. https://doi.org/10.1016/j.biortech.2020.124314

Ye Y, Ngo HH, Guo W, Chang SW, Nguyen DD, Zhang X, Zhang S, Luo G, Liu Y (2020) Impacts of hydraulic retention time on a continuous flow mode dual-chamber microbial fuel cell for recovering nutrients from municipal wastewater. Sci Total Environ 734. https://doi.org/10.1016/j.scitotenv.2020.139220

Yu N, Xing D, Li W, Yang Y, Li Z, Li Y, Ren N (2017) Electricity and methane production from soybean edible oil refinery wastewater using microbial electrochemical systems. Int J Hydrogen Energy 42(1):96–102. https://doi.org/10.1016/j.ijhydene.2016.11.116

Zhang L, Wang YZ, Zhao T, Xu T (2019) Hydrogen production from simultaneous saccharification and fermentation of lignocellulosic materials in a dual-chamber microbial electrolysis cell. Int J Hydrogen Energy 44(57):30024–30030. https://doi.org/10.1016/j.ijhydene.2019.09.191

Zhao N, Liang D, Li X, Meng S, Liu H (2021) Hydrophilic porous materials provide efficient gas-liquid separation to advance hydrogen production in microbial electrolysis cells. Bioresource Technol 337. https://doi.org/10.1016/j.biortech.2021.125352

Zikmund E, Kim K-Y, Logan BE (2018) Hydrogen production rates with closely-spaced felt anodes and cathodes 4 compared to brush anodes in two-chamber microbial electrolysis cells. http://www.elsevier.com/open-access/userlicense/1.0/

Chapter 18
Gas Mass Transfer and System Pressure in Biohydrogen Production

Wenjian Dong, Youli Yang, Chao Liu, Chao Li, Bilal Abdullahi Shuaibu, Lin Luo, and Binghua Yan

Abstract The process of hydrogen production interacts highly with gas–liquid mass transfer and bubble behaviors as well as the gaseous pressures. In this chapter, basic mathematical models for analysis of liquid–gas mass transfer kinetics, the relationship between gas solubility and mass transfer, and liquid product composition will be discussed. Then, basic information of system pressure and effects of headspace composition and gas partial pressures on biohydrogen production will be summarized. The routes to enhance hydrogen production metabolic pathway through regulating the gas composition and partial pressure in the system will also be proposed. The information of gas mass transfer and system pressure will certainly pose insight for possible enhancement of biohydrogen production.

Keywords Gas–liquid mass transfer · Solubility · Hydrogen partial pressure · Regulation

18.1 Introduction

With the development of global economy, energy, as an important driving force, has been rapidly consumed (Posso et al. 2022; Nemestothy et al. 2020), which results the energy crisis (Dincer and Acar 2015) and environment problem (Danish et al. 2019). In the current energy consumption structure, fossil energy is still in an important position (Höök and Tang 2013), and the annual consumption of fossil energy is increasing (Lucas et al. 2015). The overuse of fossil fuels and their nonrenewability

W. Dong · Y. Yang · C. Liu · B. A. Shuaibu · L. Luo · B. Yan (✉)
College of Environment and Ecology, Hunan Agricultural University, Changsha, Hunan, China

C. Li
College of Mechanical and Electrical Engineering, Hunan Agricultural University, Changsha, Hunan, China

C. R. Soccol et al. (eds.), *Biohydrogen - Advances and Processes*, Biofuel and Biorefinery Technologies 13, https://doi.org/10.1007/978-3-031-49818-3_18

have led to an increasingly serious energy crisis. Furthermore, pollutants, e.g., CO_2, SO_2, NOx, resulted by the combustion of fossil energy, are the main culprits of global warming, photochemical smog and haze (Hallenbeck et al. 2009; Chen et al. 2008; Rupprecht 2009). In response to global climate change, dozens of countries and regions have put forward climate targets of "zero carbon" or "carbon neutrality" and introduced corresponding policy plans in recent years, the global energy transition is accelerating (Yang et al. 2017; Liu et al. 2018). Among the various renewable energy sources, hydrogen is considered one of the highest priority fuels for future use due to its high energy efficiency and cleanliness.

Hydrogen energy is an easily renewable "energy-containing energy." The energy density of hydrogen is 122 kJ/g, higher than usual hydrocarbon fuel about 2.75 times (Arimi et al. 2015). Early 1971, Jones suggested that hydrogen could eventually replace fossil fuels (Jones 1971). As a multipurpose energy carrier and chemical raw material, hydrogen can realize the effective connection of multiple energy types (renewable energy, nuclear energy and fossil energy). In the end of the twenty-first century, hydrogen energy predicted by experts will become the mainstream energy. There are different methods of biohydrogen production such as consumption market, which will supply more than 75% of the global energy demand (Armor 1999). According to global environmental protection measures, there is a growing demand for cleaner fuels and more effective energy conversion technologies. Hydrogen plays a key role in such a system, both as a fuel and as an energy carrier (Baykara 2005). In conclusion, the use of hydrogen as a fuel source and energy carrier is likely to become an essential aspect of future energy development, as it can effectively connect multiple energy types and meet the increasing demand for cleaner fuels and more efficient energy conversion technologies in line with global environmental protection measures.

Traditional hydrogen production technology includes hydrogen production from coal gasification (Wang et al. 2019), natural gas (Steinberg 1989; Ersoz 2008; Muradov 2001; Holladay et al. 2009), pyrolysis of hydrogen (Dincer and Acar 2015; Lee et al. 2022a; Baykara 2004; Castro Oliveira et al. 2022), solar hydrolysis of hydrogen (Bard and Fox 2002; Fujishima and Honda 1972), electrolysis water hydrogen (Zhou et al. 2018a; Chen et al. 2021) and other physical and chemical methods. At present, hydrogen production from fossil energy has a mature technological route and relatively low cost, which is the main source of hydrogen (Voldsund et al. 2016). However, its high energy consumption will also bring a series of environmental problems (Das 2001). Therefore, with increasing environmental pressures and a sharp increase in demand for hydrogen, the development of efficient and clean hydrogen production technologies has become particularly urgent (Das 2009).

Biohydrogen production is a technology that utilizes the metabolic properties of microorganisms to produce hydrogen. By manipulating the biological systems of these microorganisms, hydrogen gas can be generated as a product of their metabolic activities. There are different methods of biohydrogen production such as photobiological hydrogen production (e.g., green algae and cyanobacteria hydrogen production) (Pathy et al. 2022), dark fermentation (DF) hydrogen production (Das et al. 2020; Kongjan et al. 2010; Ghimire et al. 2015) and photofermentation hydrogen production (Ghosh et al. 2017; Hallenbeck and Liu 2016). Before the 1990s, the research of biohydrogen production mainly focused on photofermentation. However, photosynthetic microorganisms require high quality of light source and continuous light input, which leads to no breakthrough in the research for decades. Indeed, in practical projects, it is difficult to obtain natural light continuously, and photosynthetic microorganisms have low conversion efficiency of light energy. Compared with other biological hydrogen production technologies, DF has the characteristics of wide source of raw materials, simple operating conditions and large potential for hydrogen production, which is easier to realize industrial production (Das 2009). DF technology can use waste biomass as raw materials, such as industrial waste, municipal waste, factory sewage and kitchen waste for biohydrogen production, which can effectively combine waste resources and new energy development (Das 2009; Guo et al. 2010). Therefore, DF might be the most promising method for hydrogen production from biomass.

18.2 Biological Metabolism of Hydrogen Production Via Dark Fermentation

18.2.1 The Metabolic Pathway of Hydrogen Production

Hydrogen production by DF is a complex biological process, the essence of which is to use microbial metabolic activity to convert organic substrates into hydrogen. As shown in Fig. 18.1, after entering the DF system, organic matter was degraded by three steps: hydrolysis, fermentation and acetogenesis.

In this process, protons (H^+) can act as electron acceptors, neutralizing electrons produced by microbial oxidation of substrates to produce hydrogen (Eq. (18.1)) (Yang and Wang 2018). Generally, the DF starts with the breakdown of carbohydrates. Carbohydrates are converted to glucose by hydrolysis reaction. Figure 18.2 shows the metabolic pathway of hydrogen production from glucose as the typical substrate in biohydrogen system.

Initially, anaerobic microorganisms utilize glucose molecules in the action of nicotinamide adenine dinucleotide ions (NAD^+) to degrade to pyruvate, nicotinamide adenine dinucleotide (NADH) and H^+, a process known as glycolysis (Eq. (18.2)) (Li and Fang 2007). Then, there are three possible ways to produce hydrogen. The first is the pyruvate-formate lyase (PFL) pathway, where pyruvate produces acetyl coenzyme A (acetyl-CoA) and formic acid through PFL. Under acidic conditions, the formate hydrogen lyase (FHL) is activated and formic acid is further converted to H_2 and CO_2 (Eqs. (18.3 and 18.4)) (Das 2001). This fermentation pathway is commonly found in facultative anaerobic bacteria. Theoretically, the hydrogen yield of the obtained substrate is 2 mol H_2/mol-glucose via this pathway.

The second pathway is the pyruvate-ferredoxin oxidoreductase (PFOR) pathway, through which pyruvate is converted to acetyl-CoA and CO_2 by CoA and ferredoxin

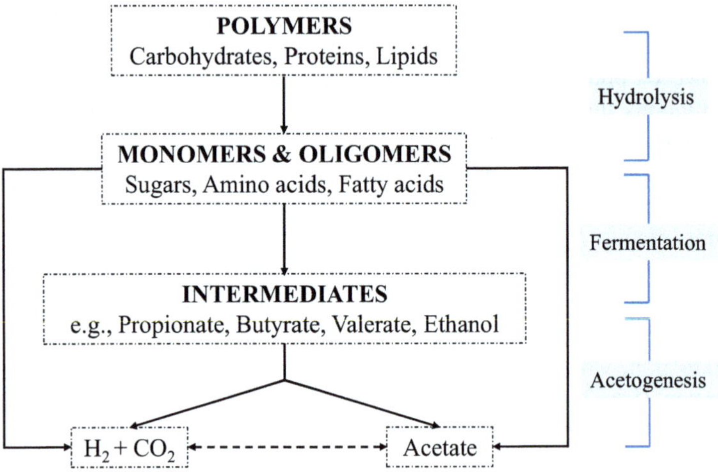

Fig. 18.1 Flow chart of hydrogen production by dark fermentation (adapted from Angenent et al. (2004))

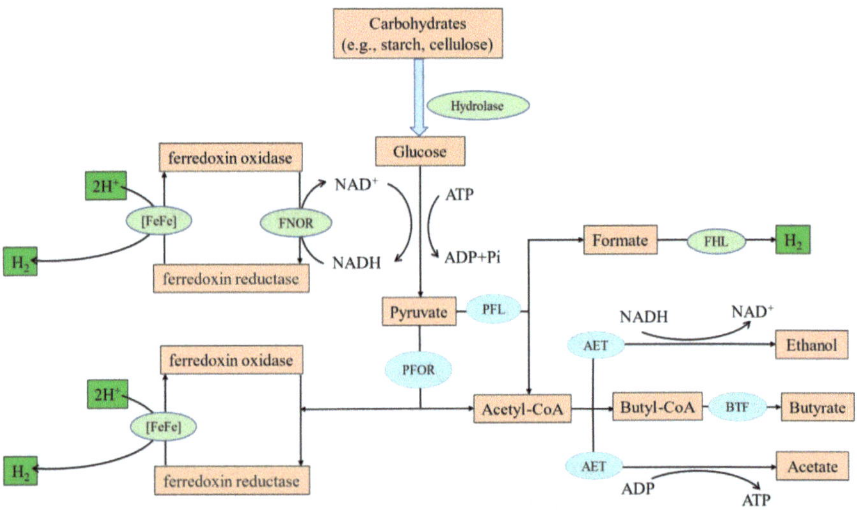

Fig. 18.2 Metabolic pathways of hydrogen production by dark fermentation (adapted from Yang and Wang (2018) and Zhou et al. (2018b)). The abbreviations used in this figure are described as follows. *[FeFe]* [FeFe]–Hydrogenases, *PFL* Pyruvate-Formate Lyase, *FNOR* Ferredoxin Dependent NADP⁺ Oxidoreductase, *PFOR* Pyruvate-Ferredoxin Oxidoreductase, *AET* Acetate-Ethanol Type Fermentation, *BTF* Butyrate Type Fermentation, *FHL* Formate Hydrogen Lyase, *NAD⁺* Nicotinamide Adenine Dinucleotide Ions, *NADH* Nicotinamide Adenine Dinucleotide, *ADP* Adenosine Diphosphate, *ATP* Adenosine Triphosphate

oxidase (Fd_{ox}) to produce ferredoxin reductase (Fd_{red}). The Fd_{red} transfers electrons to hydrogenase, which eventually produces hydrogen gas (Eqs. 18.5 and 18.6). For the third pathway, NADH produced in glycolytic pathway is converted into energy for cell growth and activity or is used to convert into NAD^+ during metabolism. The remaining NADH can be converted back to NAD^+ and release hydrogen under the action of hydrogenases (Eq. (18.7)) (Das 2001; Tanisho 1998). The latter two fermentation pathways are commonly found in strictly anaerobic bacteria. Meanwhile, in the PFL and PFOR pathways, acetyl-CoA produced by pyruvate decomposition is usually eventually converted into some soluble metabolite products, such as acetate, butyrate and ethanol (Eqs. 18.8 and 18.9) (Hawkes et al. 2007).

$$2H^+ + 2e^- \leftrightarrow H_2 \tag{18.1}$$

Glycolysis

$$C_6H_{12}O_6 + 2NAD^+ \rightarrow 2CH_3COCOOH + 2NADH + 2H^+ \tag{18.2}$$

Pyruvate-formate lyase (PFL pathway)

$$2CH_3COCOOH + 2H\text{-}CoA \rightarrow 2Acetyl\text{-}CoA + 2HCOOH \tag{18.3}$$

$$2HCOOH \rightarrow 2H_2 + 2CO_2 \tag{18.4}$$

Pyruvate-ferredoxin oxidoreductase (PFOR pathway)

$$CH_3COCOOH + CoA + 2Fd_{ox} \rightarrow Acetyl\text{-}CoA + 2Fd_{red} + CO_2 \tag{18.5}$$

$$Fd_{red} + 2H^+ \rightarrow Fd_{ox} + H_2 \tag{18.6}$$

Additional H2 formation

$$NADH + H^+ \rightarrow H_2 + NAD^+ \tag{18.7}$$

Reactions of acetyl-CoA

$$Acetyl\text{-}CoA + H_2O \rightarrow CH_3COOH + H\text{-}CoA \tag{18.8}$$

$$Acetyl\text{-}CoA + 2NADH + 2H^+ \rightarrow CH_3CH_2OH + H\text{-}CoA + 2NAD^+ \tag{18.9}$$

18.2.2 Metabolites and Their Effects in Biohydrogen Production from Dark Fermentation

The biohydrogen production in DF can be divided into butyric acid fermentation, propionic acid fermentation and ethanol fermentation according to the different components of the end metabolic products (Lee et al. 2008; Taheri et al. 2018; Ren et al. 2007). According to literature reports, different dominant flora can be formed by the system under different environmental conditions, thus showing different fermentation types (Lee et al. 2008).

Butyrate fermentation: The bacteria that can carry out butyric acid fermentation are mainly some strictly anaerobic bacteria or facultative anaerobic bacteria, and the dominant population is *Clostridium*. The main terminal metabolites of butyric acid fermentation are butyric acid, acetic acid, hydrogen and carbon dioxide.

Propionate fermentation: Compared with the metabolic pathway of butyric acid production, the metabolic pathway of propionic acid production has strong reducibility, which is conducive to NADH + H^+ oxidation. In anaerobic biological treatment, acid degradation of organic matter often takes place in propionic acid fermentation. And propionate metabolism is mainly carried out under the action of *Propionibacterium*. The main liquid metabolites of propionic acid fermentation route are propionic acid and acetic acid. However, propionic acid bacilli generally do not contain hydrogenase and have very few gas products.

Ethanol fermentation: The genera of ethanol-based fermentative bacteria mainly included *Bacteroides*, *Pseudomonas fermentans*, *Fusobacteria*, *Bacteroides* and *Clostridium* were absolutely dominant. The end metabolites of ethanol-type fermentation are mainly ethanol and acetic acid, and there is a lot of H_2 in fermentation gas. Compared with the propionic acid production pathway, bacterial ethanol fermentation has the same oxidation capacity for NADH + H^+. For every 1 mol glucose oxidized, 4 mol NAD^+ can be regenerated, while the butyric acid production pathway can only regenerate 2 mol NAD^+. The coupling reaction of ethanol and acetic acid has a strong ability to adjust the balance of NADH + H^+/NAD^+. Ethanol fermentation has a stronger stability and higher hydrogen production capacity than butyric acid fermentation.

While low concentrations of these end products may be beneficial for biohydrogen production, high concentrations acidify the system and reduce H_2 production rates and yields. During acidification, organic acids, as by-products of the DF, will lead to a decrease in pH of external cells. These acids exist in dissociated or undissociated forms, but both inhibit the production of biological hydrogen (Chong et al. 2009a; Chong et al. 2009b). Dissociation of the acid increases the ionic strength of the medium due to cell lysis of HPB (Ciranna et al. 2014), which affects H_2 production and leads to a transition from acidophilic to solvent (Chong et al. 2009a). As for undissociated acids, they are able to penetrate microbial cell membranes and release intracellular protons at higher intracellular pH (Chong et al. 2009b). The influx of protons into the cytoplasm can cause a pH imbalance, resulting in a decrease in intracellular pH that may affect metabolic activities, while also causing cell death and inhibiting biohydrogen production (Khanal 2004; Grupe and Gottschalk 1992). In addition, maintenance of intracellular pH by proton pumps may be accompanied by depletion of intracellular ATP (Grupe and Gottschalk 1992), potentially reducing the level of energy required for cell growth and inhibiting bacterial cell growth.

In general, the physical solubility of hydrogen in aqueous solutions is very limited (0.0016 g H_2/kg-H_2O at 1 atm and 293 K) (Nemestothy et al. 2020). If hydrogen supersaturation occurs in the fermentation broth of hydrogen production reactor, it may be attributed to insufficient mass transfer between liquid and gas phases in the reactor (Beckers et al. 2015). However, as the amount of hydrogen in the liquid increases, the biological response to hydrogen production will be suppressed (Pauss et al. 1990; Sivagurunathan et al. 2016). At the same time, the metabolic flux of hydrogen microorganisms may also be redirected and diverted to the production of other end products, such as lactate, ethanol, acetone and butanol (Noblecourt et al. 2017; Saady 2013). Therefore, hydrogen accumulation in the liquid phase may prevent microbial growth and reduce biohydrogen productivity.

18.3 Gas–Liquid Mass Transfer in Biohydrogen Production

18.3.1 System Pressure in Biohydrogen Production

In the biohydrogen production system, CO_2 and H_2 will exist in the form of bubbles in the fermentation broth or in the reactor headspace. Currently, the actual production of biohydrogen is much lower than the theoretical value (4 mol/mol-glucose). One study pointed out that there is no direct relationship between hydrogen production and the composition of the headspace (Beckers et al. 2015). Headspace pressure in the fermentation system is a critical factor affecting hydrogen production (Levin 2004; Sonnleitner et al. 2012). Reducing the headspace partial pressure of H_2 could be considered as an approach toward improvement of H_2 productivity (Levin 2004).

$$C_6H_{12}O_6 + 2H_2O \rightarrow 2CH_3COOH + 2CO_2 + 4H_2 \quad \Delta G = -196\text{KJ} \quad (18.10)$$

As Eq. (18.10) shows, according to Le Chatelier's principle, when the reaction products are removed in time, the balance of the reaction will shift to the right. By reducing the total pressure of the system to reduce the H_2 partial pressure, the reaction balance can shift to the right, thus enhancing the H_2 production.

Hydrogen producing bacteria are sensitive to hydrogen partial pressure. According to Henry's law Eq. (18.11), the partial pressure of hydrogen in the gas phase will affect the amount of hydrogen dissolved in the fermentation broth. To increase the hydrogen production, it is necessary to reduce the amount of hydrogen dissolved in the liquid phase.

$$K_H^i = \frac{P_G^i}{C_F^i} \quad (18.11)$$

K_H^i is the coefficient of Henry for gas *i*, P_G^i is the partial pressure of *i* in the gas phase and C_F^i is the (dissolved) *i* concentration in the liquid phase.

The DF mainly produces hydrogen by ferric oxide protein or reducing coenzyme. From the perspective of chemical thermodynamics, the partial pressure of hydrogen at high concentration is unfavorable to the reduction of protons, which will lead to the oxidation of hydrogen more easily, thus obtaining a lower hydrogen yield. Indeed, when the hydrogen partial pressure increases, it is not conducive to the further degradation of some compounds, such as propionic acid and some amino acids. Moreover, the high hydrogen partial pressure can affect the ratio of electron carriers ($NADH/NAD^+$) that regulates the metabolic pathways in microbial cells, which will affect microbial growth and hydrogen production ability (Valdez-Vazquez et al. 2006; Tiwari et al. 2006). When the H_2 partial pressure reaches 0.54 atm in the continuous DF system, hydrogen production is inhibited (Valdez-Vazquez et al. 2006). In another study, as the headspace H_2 partial pressure increased to 2.5 atm, the fermentation system gradually shifted toward ethanol production, resulting in a decrease in hydrogen yield (Lamed et al. 1988). Therefore, it is crucially important to keep the proper hydrogen partial pressure for hydrogen production.

18.3.2 Mass Transfer Characteristic in Biohydrogen Production

Since headspace pressure can be more easily monitored in fermentation system such as pressure sensor and gas chromatography analyzing component gas partial pressure, researchers prefer to describe the effect of headspace hydrogen partial pressure on hydrogen production system. As previously mentioned, hydrogen partial pressures can affect biological metabolic pathway, resulting in unstable hydrogen yield. However, it also was found that the dissolved hydrogen in the fermentation broth would directly affect the hydrogen production efficiency. With the increase of hydrogen concentration in liquid phase, the hydrogen production is inhibited and the effect of biological hydrogen production is reduced (Levin 2004; Valdez-Vazquez and Poggi-Varaldo 2009; Oh et al. 2009). According to Henry's law Eq. (18.11), the partial pressure of hydrogen and the concentration of dissolved hydrogen should be in equilibrium. However, because of mass transfer limitations, the gas partial pressure in the headspace and the dissolved gas in the liquid phase cannot reach the equilibrium state, so the Henry's law cannot apply to biohydrogen system. The relationship between hydrogen production performance and dissolved hydrogen concentration was verified in a study under three different total pressure conditions. The results show that the headspace hydrogen partial pressure should not be used as the research parameter, but the dissolved hydrogen concentration should be studied (Laurent et al. 2012). Dissolved hydrogen is often mistakenly thought to be in equilibrium (Balachandar et al. 2013). However, in general, when the hydrogen production rate exceeds the interfacial mass transfer rate, a high concentration of dissolved hydrogen may appear in fermentation broth (Pauss et al. 1990). This phenomenon is called supersaturation and the supersaturation concentration has been shown to be several times the equilibrium concentration suggested by Henry's law (Sivagurunathan et al. 2016). The high concentration of dissolved hydrogen further limits hydrogen production thermodynamically (Bundhoo and Mohee 2016; Verhaart et al. 2010), which prevents hydrogen production until the rate of hydrogen production and the concentration of dissolved hydrogen in the system reach an equilibrium (Balachandar et al. 2013). The high concentration of dissolved hydrogen can also lead to unwanted changes in microbial communities, resulting in lower hydrogen production (Bakonyi et al. 2017). More fundamentally, near stoichiometric yield can only be achieved under near equilibrium conditions, which means very slow production rate and/or very low partial pressure of H_2. Therefore, the hydrogen partial pressure in liquid phase should become the main criterion for inhibition of hydrogen production (Hawkes 2002; Mizuno et al. 2000). Presently, hydrogen sensors are mainly used to detect the hydrogen partial pressure in the liquid phase. Such sensors need high sensitivity and selectivity, and also need to withstand the harsh environment in the anaerobic fermentation system (such as corrosivity, temperature, pH). Due to the immature technology of this hydrogen sensor currently, the hydrogen sensor has not been widely used to monitor the hydrogen partial pressure in the liquid phase.

Theories of liquid to gas or gas to liquid mass transfer in biological processes have been developed in many engineering textbooks. Biological hydrogen production system is a typical multiphase flow system with coexistence of gas, liquid and solid three phases. In biological hydrogen production, the transfer of hydrogen from liquid phase to gas phase is very important. Gaseous substances are formed in the liquid phase and tend to escape into the gas phase. The mass transfer restriction mainly exists at the gas–liquid interface, which is a narrow liquid layer characterized by concentration gradient. The mass transfer restriction will lead to the failure of the normal transfer of dissolved gases from liquid to gas phase (Beckers et al. 2015). The high partial pressure of hydrogen in liquid phase is the key factor affecting the continuous hydrogen production by fermentation (Levin 2004; Sonnleitner et al. 2012). In addition, after the generated hydrogen is transferred from the liquid phase to the gas phase, the hydrogen in the gas phase also should be removed in time to ensure the stability of the headspace pressure of the system. Releasing hydrogen from the headspace is simple and improving the mass transfer of hydrogen between gas and liquid phase is the key. Figure 18.3 shows the mass transfer of gas in a biohydrogen production system.

For mass transfer processes, the mass transfer coefficient can also reflect the enhancement degree of this specific mass transfer process (under unit area, unit concentration or pressure difference, the quantity of substances transferred from one phase to another in unit time). The mass transfer calculation formula is shown in Eq. (18.12).

$$N_A = k\left(C_{Ai} - C_A\right) \tag{18.12}$$

Where k is called the mass transfer coefficient, or simply the mass transfer coefficient. N_A is the mass transfer flux, and (C_{Ai}–C_A) is the concentration difference between the two phases. The expression form of mass transfer coefficient also changes with different systems and different expression ways of driving force.

In addition to the mass transfer coefficient, the mass transfer efficiency is sometimes used to stand for the characteristics of the mass transfer process, especially in polymer devolatilization, the devolatilization efficiency is often used to stand for the

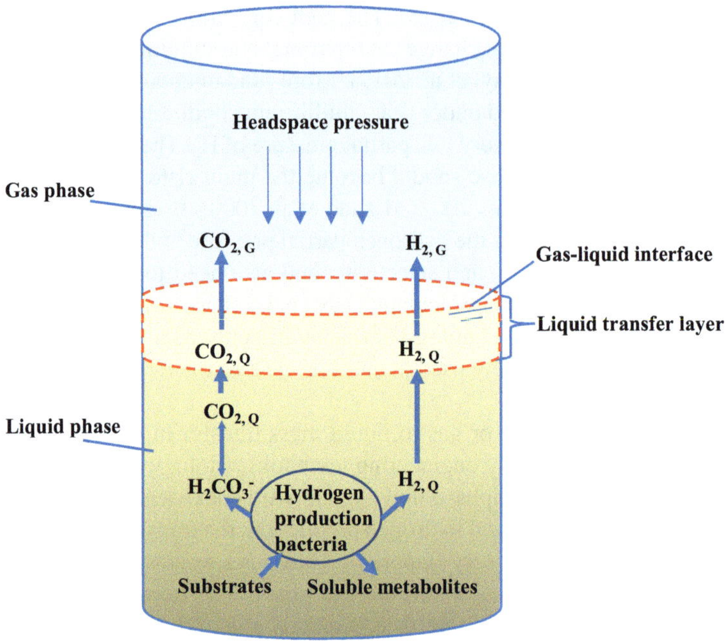

Fig. 18.3 Mass transfer of gases in biohydrogen production systems (adapted from Beckers et al. (2015))

advantages and disadvantages of the mass transfer performance of the process. Ignoring the changes in density and volume caused by mass transfer, the definition is as follows Eq. (18.13).

$$E_f = \frac{C_0 - C_{av}}{C_0 - C_e} \tag{18.13}$$

Where, C_0 represents the initial concentration of substances, C_{av} represents the concentration of substances at the gas–liquid interface, C_e represents equilibrium concentration of substances in the gas–liquid phase and E_f represents the mass transfer efficiency. Eq. (18.13) characterizes the proportion of the transferred quantity to the largest amount that can be transferred thermodynamically. When the mass transfer time is short, as shown in Eq. (18.14).

$$\frac{E_f}{t} = K_L a \tag{18.14}$$

Where E_f is the mass transfer efficiency, t is the mass transfer time, K_L is the liquid-side mass transfer coefficient (m s^{-1}), and a is the interface area (m^2). Because it is difficult to measure the foam gas–liquid specific surface area (a) in practical application, K_La is often viewed as a monolithic factor in the description of mass transfer.

According to the classical definition of K_La, namely, the global volumetric mass transfer coefficient (Pauss et al. 1990). The mass transfer efficiency of gas–liquid interface is usually expressed by K_La coefficient. As shown in Eq. (18.15),

$$Q = K_L a \cdot He^{-1} \cdot \left(P_G - P_G^{\circ}\right) \tag{18.15}$$

The mass flow rate Q indicates transmission through a specific gas–liquid interface and depends on the global mass transfer coefficient K_La (which is related to the mixing conditions of the medium and the gas–liquid interface area) and the potential transfer of $\left(P_G - P_G^{o}\right)$ (which is related to the partial pressure of the gas and the liquid concentration of the dissolved gas at equilibrium). Henry's constant (*He*) depends only on the properties considered as gases (Laurent et al. 2012). Therefore, it is reasonable to use K_La to describe the mass transfer efficiency in biohydrogen production. A study has shown that K_La is a function of the gas nature, the liquid physiochemical and hydrodynamic properties, which is not related to glucose concentration (Pauss et al. 1990). However, it is difficult to directly measure K_La currently. The studies of mass transfer in biohydrogen production are still not thorough enough currently. Indeed, the mass transfer problem is common in bioreactors. Mass transfer involves many factors and there is still no satisfactory solution.

18.3.3 The Influencing Factors of Mass Transfer

Main factors affecting mass transfer include operation conditions (temperature, pressure, pH value, etc.), physicochemical properties of fermentation broth (viscosity and composition of fermentation broth, flow state, product inhibition, etc.) and bioreactor configuration (type, design parameter, stirring conditions, etc.) (Zhang et al. 2017). The influencing factors of gas–liquid mass transfer in biohydrogen production are shown in Fig. 18.4.

Temperature and pressure affect the physicochemical properties of the fermentation broth and microbial activity of hydrogen production, which affect mass transfer ability (Guo et al. 2010; Lee et al. 2008). pH is considered as one of most important and sensitive operating parameters that affect the performance of dark fermentation, which also can affect microbial activity (Ginkel et al. 2001; Kumar et al. 2016). Therefore, the suitable operation conditions are the basic guarantee for biohydrogen production.

In general, factors such as the internal hydrodynamics of a bioreactor can determine the efficiency with which the gas is transported across the gas–liquid interface. In bioreactors, parameters such as liquid viscosity play a significant role. Although hydrogen gas is characterized by its very limited physical solubility in aqueous solution (Nemestothy et al. 2020). Supersaturation may occur in fermentation broth, attributing to insufficient mass transfer between the liquid and gas phases of the reactor (Beckers et al. 2015; Kraemer and Bagley 2006; Zhang et al. 2012). In this case, when the interphase gas exchange (desorption) is not fast enough (Frigon and

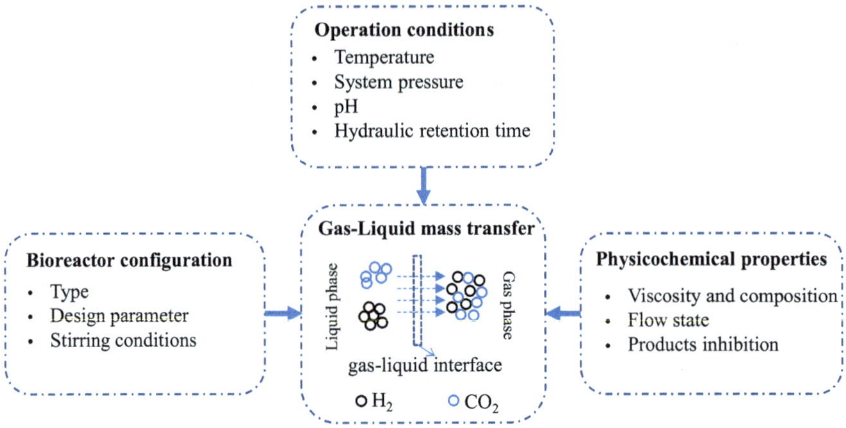

Fig. 18.4 The influencing factors of mass transfer

Guiot 1995; Maluta et al. 2019), under the given equilibrium conditions calculated according to Henry's law (Eq. (18.11)), the amount of H_2 present in the dissolved state is more (and proportionally less present in the gaseous state) than expected. The mass transfer efficiency varies with the flow state of the fermentation liquor in the reactor. Generally, the more stable the fluid is, the more difficult it is to transfer the dissolved hydrogen in the liquid phase to the gas phase (Zhang et al. 2012). The rheological properties of heterogeneous flow are closely related to the structure and size of the reactor. A new type of flat plate bioreactor enhances the mass transfer effect through swinging movement, the maximum hydrogen production rate is increased to 492 mL, the maximum hydrogen production rate is 11 mL L^{-1} h^{-1} and the substrate conversion efficiency is 44.4% (Gilbert et al. 2011). The mass transfer coefficient is affected by the viscosity, density, surface tension of the reaction fluid, diffusion coefficient of the solute and the nature of the biochemical reaction product (Zhang et al. 2017). Generally, stirring increases the mass transfer efficiency because it can break up the gas bubbles generated during the reaction to allow the fermentation broth to mix well. Microorganisms remain in suspension, which increases their contact with the fermentation broth. However, rapid stirring can also generate a large amount of heat while promoting reaction rates, which can alter the optimal temperature conditions of the system and be detrimental to biochemical reaction processes. Therefore, excess heat should be released to maintain a better reaction temperature.

The reactor configuration affects the internal mixing conditions, thereby affecting the mass transfer performance of the fermentation broth. It also affects the growth and metabolism of hydrogen production bacteria, as well as the substrate conversion efficiency in the process of biohydrogen production (Kongjan and Angelidaki 2010), which is one of the key factors for hydrogen production by fermentation (Laurent et al. 2012; Lalit Babu et al. 2009; Castelló et al. 2020). In the stirred reactor, the hydrogen production rate was significantly increased from 87.5

to 121.4 mL g VS^{-1} (Clark et al. 2012). The mixing condition can increase the contact between bacteria and substrate, which also is conducive to the mass transfer of hydrogen from the fermentation broth. However, much of the current research on biological hydrogen production has been conducted in a static state under biological control, such as feedstock screening, feedstock pretreatment, high-efficiency bacterial screening, addition of trace elements and metal ions (Ghimire et al. 2015; Song et al. 2012). For different hydraulic characteristics, the mode of operation and the flow regime are less studied.

18.4 Regulation of Biological Hydrogen Production

18.4.1 Regulation of Hydrogen Concentration in the Hydrogen Production System

The generated H_2 exist in the fermentation broth and the headspace. The concentration of dissolved hydrogen in the fermentation broth directly affects the metabolic, biochemical and enzymatic reactions of the organism, i.e., the "hydrogen partial pressure in liquid phase" affects the mass transfer efficiency. At present, the technology of online hydrogen detection is not mature enough (Kraemer and Bagley 2006), so few studies directly measure the concentration of hydrogen in fermentation broth. However, in this regard, various technologies were used to counteract the inhibitory effect of dissolved hydrogen (Sivagurunathan et al. 2016; Bundhoo and Mohee 2016), including stripping with N_2 or CO_2, reducing total pressure, increasing stirring, increasing surface area and recycling H_2 enriched biogas (H_2, CO_2) (Bakonyi et al. 2017; Dreschke et al. 2019). Through the regulation of the concentration of dissolved hydrogen, the yield of biological hydrogen production is significantly increased, which is also an important research direction of biohydrogen production. As shown in Table 18.1, the advantages and disadvantages of regulatory techniques are compared.

Stripping with inert gases can alleviate inhibition of dissolved hydrogen in biohydrogen production, the result showed that stripping with N_2 could increase the hydrogen production rate by almost 70% (Mizuno et al. 2000; Hussy et al. 2005). However, stripping with inert gases also causes dilution of the generated biogas (H_2) (Sonnleitner et al. 2012). And N_2 as the stripping gas is not a satisfactory solution in terms of process upgrade because separating N_2 and H_2 is technically difficult. Biogas (H_2, CO_2) recirculation and inert gas injection have similar effects, which can accelerate gas–liquid mass transfer. Also, biogas recirculation not caused dilution to the biogas (H_2) and CO_2 is relatively easy to separate from hydrogen, which is a promising technology (Bakonyi et al. 2017; Frigon and Guiot 1995). However, the use of CO_2 will reduce the system pH, inhibit microbial activity and reduce hydrogen production rate, so alkaline agents need to be added to maintain the pH value (Willquist et al. 2009). Although stirring treatment has been proved to be

Table 18.1 Regulation techniques of hydrogen concentration enhancing biohydrogen production

Regulation techniques	Advantages	Disadvantages	Ref.
Gas stripping	1. Accelerating system mass transfer 2. Relieving the inhibition of hydrogen partial pressure	1. Dilution of the generated biogas 2. The difficult of separating N_2 and H_2 3. Stripping with CO_2 reduces system pH	Sonnleitner et al. (2012), Bakonyi et al. (2017), Mizuno et al. (2000), Frigon and Guiot (1995), Hussy et al. (2005), Willquist et al. (2009)
Vigorous stir	1. Accelerating system mass transfer 2. Increasing the contact area between microorganisms and substrates	1. Increasing operating costs 2. Increasing system temperature 3. Breaking down the biological flocs	Lamed et al. (1988), Dreschke et al. (2019)
Bubble inducer	1. Inducing bubble formation 2. Increasing the attached area of microorganisms	1. Combining with other technologies, e.g., depressurization measures	Fritsch et al. (2008)

beneficial to mass transfer in experimental studies, thereby improving the yield of biological hydrogen production. However, the vigorous stirring is extremely challenging in industrial applications and lead to an inescapable increase in the operating cost of reactors (Dreschke et al. 2019). In addition, the vigorous stirring was not enough to eliminate hydrogen supersaturation (Zhang et al. 2012). Moreover, high shear forces from vigorous stirring can disrupt the structure of microbial flocs (Kaparaju et al. 2008) and may also damage the structure of microbial communities (Stroot et al. 2001). The structure of microbial communities, in turn, can disrupt symbiotic relationships among organisms and have a detrimental impact on reactor performance (McMahon et al. 2001). A material called bubble inducer can also enhance the formation of dissolved hydrogen bubbles. Some experiments show that by inserting porous solid particles to force the generation of bubbles and timely reduce the total pressure of the system, the negative effect of hydrogen partial pressure in the biological hydrogen production fermentation system can be reduced. The addition of bubble inducing material is beneficial to degassing to some extent, but it cannot completely replace degassing. These findings can explain the better performance of the fermentation system using carrier materials, not only because of increased biomass retention (Zhang et al. 2007; Peintner et al. 2010; Han et al. 2012; Lee et al. 2004), but also enhanced bubble formation and better hydrogen removal. The rationale for this technique is that energy is necessary for the formation of bubbles in supersaturated liquids at the surface (heterogeneous nucleation) than in homogeneous solutions (homogenous nucleation) and can therefore occur at lower levels of supersaturation (Sear 2007). Therefore, foam formation results from providing a high surface area in the reactor, which increases gas–liquid mass transfer without changing other operating conditions (Fritsch et al. 2008). Therefore, the application of bubble inducer in biological hydrogen production fermentation system shows a high potential.

The above-mentioned methods are useful for relieving the inhibition of dissolved hydrogen, but they also suffer from unavoidable drawbacks. The hydrogen accumulation concentration in the fermentation system is a very important factor affecting the rate of biological hydrogen production. Reducing the hydrogen concentration through an external driving force can fundamentally solve the problem of hydrogen inhibition.

18.4.2 Selection of Reactor Configurations in Dark Fermentation Hydrogen Production

The design and configuration of the reactor plays a crucial role in hydrogen production as it affects operating conditions and microbial growth (Banu et al. 2021). Therefore, the regulation of process parameters will significantly improve the biological hydrogen fermentation. Table 18.2 compares the advantages and disadvantages of different reactor configurations in biohydrogen production.

Continuously stirred-tank reactor (CSTR) is the most used reactor for continuous hydrogen production due to simple structure and convenient operation. Moreover, continuous stirring is beneficial to the uniformity of the medium and provides good contact between the microorganism and the substrate (Balachandar et al. 2013). At the same time, continuous stirring also improved the removal efficiency of hydrogen in the fermentation broth, which reduced the pressure in the system and increased the hydrogen production rate. Nonetheless, during its operation, the concentration of biomass in the fermentation broth is limited by hydraulic retention time (HRT), which must be an indisputable influencing factor on the rate of hydrogen production (Li and Fang 2007). Notably, the activity of bacteria is limited at very low concentrations of biomass (Hallenbeck 2012). Thereby, the HRT must be less than the biological growth cycle in biohydrogen production (Balachandar et al. 2013). Therefore, some researchers have proposed that various biological or physical fixation techniques can be used to promote the retention of microorganisms, which improve the hydrogen production capacity of CSTR. Sivagurunathan et al. (Sivagurunathan et al. 2015) studied the effect of immobilization on biologic hydrogen production under different HRT. They used immobilized hydrogen-producing bacteria, *Clostridium* species, to treat beer industry wastewater in CSTR. At HRT of 1.5 h (organic loading rate of 320 g L^{-1} d^{-1}-hexose equivalent), the maximum hydrogen production rate was 55 L L^{-1} d^{-1}. In addition, the design of stirrer and mixing speed is also important to regulate the biological hydrogen production in CSTR. A well-designed agitator ensures adequate agitation and good mass transfer while reducing the formation of shear tension on the biomass (Difuntorum and Danao 2017). For example, Ding et al. (Difuntorum and Danao 2017) studied the influence of impeller and its stirring speed on biological hydrogen production

Table 18.2 Reactor configuration enhancing biohydrogen production

Configuration	Advantages	Disadvantages	Ref.
CSTR	1. High mass transfer efficiency 2. Simple structure and convenient operation 3. Promoting the contact between microbial and substrate	1. High operating cost 2. Maintaining difficultly high biomass in the system 3. Foam reduces mass transfer efficiency	Li and Fang (2007), Balachandar et al. (2013)
MBR	1. Maintaining a high biomass concentration 2. Promoting mass transfer between microbial and substrate	1. Membrane fouling 2. High operating cost 3. High mass transfer resistance at the bottom of the bed	Li and Fang (2007)
UASB	1. Strong hydrogen production capacity. 2. Strong system buffer capacity. 3. Microbes are easy to attach substrates improving mass transfer efficiency	1. High solid retention time 2. Long startup time 3. Substrate accumulation easily leads to excessive local pressure	Qyyum et al. (2022)
AFBR	1. High mass transfer efficiency 2. Strong system buffer capacity 3. Gas is easy to maintain stability and uniformity	1. Maintaining the reactor in fluidized state need high energy 2. Substrate accumulation easily leads to excessive local pressure	Zhang et al. (2007); Barros et al. (2010)

during CSTR operation, and the results showed that impeller speed had a great influence on biological hydrogen production.

Membrane bioreactor reactor (MBR) is a typical reactor for fermentation process, which combines the characteristics of activated sludge treatment and membrane filtration. The use of MBR eliminates the main disadvantage of CSTR reactors, namely, the elution of biomass. This membrane separation process can retain microorganisms in the reactor and increase the biomass in the reactor (Li and Fang 2007). High solid retention times (SRT) generally improve substrate conversion. However, SRT might have a negative effect on the rate of hydrogen production (Oh et al. 2004). In addition, according to diffusion or convection phenomena, solute migration across the membrane is stimulated by a transmembrane chemical potential gradient (Eibl et al. 2009). It involves mass transfer and hydrodynamic properties (Pinelo et al. 2009). Recently, liquid or gas membrane bioreactors (L/G MBR) have been developed to enhance the DF. Among them, hollow fiber membrane assemblies are used to retain biohydrogen production, *in situ* liquid–gas separation and hydrogen-producing bacteria in a single unit. Marie et al. (Renaudie et al. 2021) obtained the optimal hydrogen rate (1.22 mol/mol-glucose) with a feed concentration of 13.1 g-glucose/L through L/G MBR reactor. MBR can also use two-phase (gas–liquid) cross-flow. Because of the high liquid velocity that occurs in the falling film formed between the bubble and the membrane surface, as well as the eddy currents created as the air bubble rises (Bérubé and Lei 2006). This allows membrane components to be configured to achieve high and uniform membrane scour in two-phase flow to control scaling. However, Zhang et al. (Zhang et al. 2009) found that bubble size and frequency have a great influence on hydrodynamic characteristics. Electrodialysis (ED) is another potential membrane-based technology proposed in recent years, which can be combined with DF to further increase hydrogen production through continuous removal of volatile fatty acids (VFA) (Hassan et al. 2021). However, membrane contamination and high operating cost further limit the application of MBR bioreactor for hydrogen production.

There are three common forms of microorganisms in the packed-bed bioreactors reactor (PBR): particles, biofilm or biomass particles attached to gels (Show et al. 2012). The most typical is Up-flow anaerobic blanket reactor (UASB). The UASB reactor has a better hydrogen production capacity because of the self-agglomerating granular bacterial community with high degradation capacity (Qyyum et al. 2022). They retain more in the reactor and are more resistant to toxic conditions. The compactness of microbes in the particles also makes the conversion of the substrate to hydrogen more likely. Interestingly, Afridi et al. (Afridi et al. 2018) found that the mass transfer rate of UASB was directly related to particle size. Specifically, the mass transfer rate of the bottom region is the fastest compared with the middle and top regions of the reactor. In addition, the upward movement of bubbles formed during fermentation will cause turbulence of the medium to achieve mixing (Łukajtis et al. 2018). However, the main limitations of UASB reactors are a large amount of biomass present in the effluent, high SRT and long start-up time. It is worth noting that the mass transfer rate is positively correlated with the organic loading rate. Both convection and molecular mass transfer exist in the UASB reactor, but are inversely proportional to the height of the UASB reactor. The higher fluid collection efficiency is attributed to the faster mass transfer rate in the reactor (Afridi et al. 2018). At present, some researchers have added carriers to UASB to increase biomass and improve the hydrogen production capacity of the system. Lo et al. (Lo et al. 2009) studied the effect of adding water distributor on UASB hydrogen production reactor. The results showed that the water distributor improved the mass transfer efficiency of the reactor and accelerated the formation of hydrogen producing granular sludge. At 0.5 h HRT, the maximum specific hydrogen production rate and hydrogen conversion rate were 7.89 L H_2/(h·L) and 3.42 mol H_2/mol-sucrose, respectively.

Anaerobic fluidized bed reactor (AFBR) combines the characteristics of CSTR and fixed-bed reactors. Microorganisms are deposited on a solid surfaces on granules or a biofilm (Łukajtis et al. 2018). Compared with CSTR, these reactors have good mixing ability and less shear force. Therefore, these reactors have good mass transfer efficiency. Due to microbial deposition on solid surfaces, biomass also has a lower probability of being washed away compared and substrate conversion is higher in the UASB reactor. However, the main parameter affecting the higher performance efficiency of AFBR is the choice of supporting material. When the reactor is fed or discharged, AFBR enables fermentation bacteria to adhere to the support material, thus allowing higher biomass in the reactor (Barros et al. 2010). A literature survey shows that the supporting materials tested in AFBR for biohydrogen production include activated carbon, diatomaceous earth and expanded clay. Zhang et al. (Zhang et al. 2007) used particulate activated carbon as AFBR support for hydrogen production by fermentation and obtained a specific hydrogen production rate of 2.36 L H_2/(h·L). The results show that when the substrate concentration is 10–30 g/L, HRT is 0.5–4 h, the pH of the reaction system is 4.0, and the biomass in the system can reach 21.5 g-VSS/L. On the other hand, Encina et al. (Encina and Hidalgo 2005) found that biomass adhesion would affect the size and density of biological particles, thus causing the reduction of bed hydrodynamic behavior. The biomass concentration and biofilm density at the bottom of the bed were higher than those at the top, resulting in increased mass transfer resistance in this area.

18.4.3 Coupling with Microbial Electrolysis Cell in Biohydrogen Production

The microbial electrolysis cell (MEC) is a technology in which electrogenic microorganisms utilize substrates to produce hydrogen (Kadier et al. 2016). Single-chamber and two-chamber reactors are two common types of MEC configurations. The MEC has two electrodes as anode and cathode mounted in the associated pole chamber and connected with an external circuit. In MEC, organic matter is consumed by electrogenic bacteria and produces CO_2, electrons and protons. The bacteria first transfer the electrons to the anode, and the protons are released into the solution. The electrons released in the anode then migrate from the anode to the cathode via an external circuit at an external potential and combine with free protons in solution to form H_2 (Shanthi Sravan et al. 2021). Figure 18.5 shows the coupling of MEC and DF to enhance biological hydrogen production.

MECs are not independent biohydrogen production technologies, because anode-respiring bacteria cannot utilize complex forms of organic wastewater and waste (Mukherjee and Mohan 2022). However, MECs can efficiently utilize pure strains, cocultured strains and strains mixed with several substrates (such as glucose, cellulose, volatile carboxylate (C2–C4) and acetate lactate) while producing 70% to 90% H_2 (Lee et al. 2022b). Therefore, when MEC is combined with DF, their advantages become significant. MEC can use DF effluent to further produce biohydrogen. Chookaew et al. (2014) investigated the potential of using DF effluent as a substrate in MEC. The results show that when DF effluent is used as the substrate of MEC (applied voltage: 1.0 V), the highest yield at 106 mL H_2/g-COD and COD removal efficiency is increased from 20% to 41% of that of DF alone. Marone et al. (2017) used DF and MEC coupling to assess the potential for biohydrogen production from six different effluents. It is also reported that compared with single DF, the total hydrogen production of the coupled process increases 13 times. The maximum total hydrogen production is 1608.6 + 266.2 mL H_2/g COD, and the maximum COD removal rate is 78.5% + 5.7%. These results indicate that the

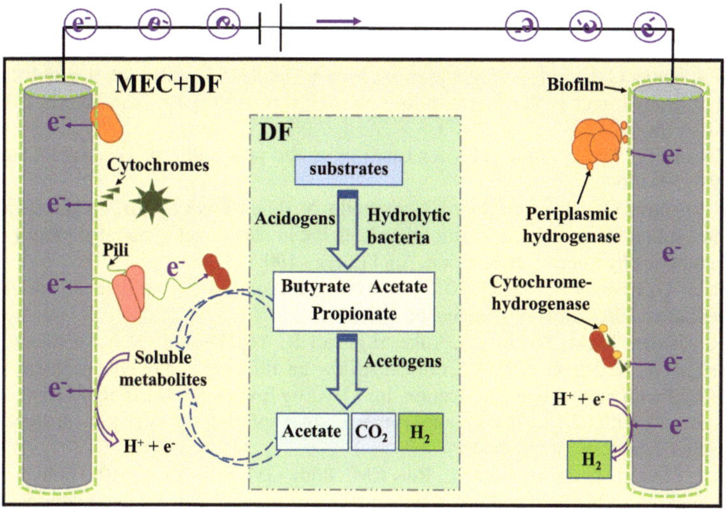

Fig. 18.5 Dark fermentation coupled with microbial electrolytic cell for biohydrogen production

coupling process of DF and MEC is a very promising option for enhancing biohydrogen production.

18.5 Conclusion and Perspectives

Hydrogen is certainly accepted as an energy carrier and a substitute for fossil fuels. There are various technologies for hydrogen production today, but none of them is fully mature. In the present situation, the potential of biohydrogen production by dark fermentation is unquestionable. However, biohydrogen production in dark fermentation is not satisfactory in terms of yield and purity currently. Gas mass transfer is the one of prominent problems in dark fermentation. But in the complex environment of fermentation broth, the mass transfer involves many factors. Indeed, hydrogen production can be improved by some measures, e.g., gas stripping and stirring, but there are inevitable drawbacks to those measures. There is no satisfactory solution yet. Dark fermentation coupled with other technologies such as microbial electrolytic cells mentioned in this chapter might be a good way out for biological hydrogen production.

References

Afridi ZUR, Wu J, Li ZH, Akand R, Cao ZP, Poncin S, Li HZ (2018) Novel insight of spatial mass transfer conditions of upflow anaerobic reactor. J Clean Prod 204:390–398

Angenent LT, Karim K, Al-Dahhan MH, Wrenn BA, Domiguez-Espinosa R (2004) Production of bioenergy and biochemicals from industrial and agricultural wastewater. Trends Biotechnol 22(9):477–485

Arimi MM, Knodel J, Kiprop A, Namango SS, Zhang Y, Geißen S-U (2015) Strategies for improvement of biohydrogen production from organic-rich wastewater: a review. Biomass Bioenergy 75:101–118

Armor JN (1999) The multiple roles for catalysis in the production of H2. Appl Catal A Gen 176(2):159–176

Bakonyi P, Buitrón G, Valdez-Vazquez I, Nemestóthy N, Bélafi-Bakó K (2017) A novel gas separation integrated membrane bioreactor to evaluate the impact of self-generated biogas recycling on continuous hydrogen fermentation. Appl Energy 190:813–823

Balachandar G, Khanna N, Das D (2013) Biohydrogen production from organic wastes by dark fermentation. Elsevier, Biohydrogen, pp 103–144

Banu JR, Usman TMM, Kavitha S, YukeshKannah R, Yogalakshmi KN, Sivashanmugam P, Bhatnagar A, Kumar G (2021) A critical review on limitations and enhancement strategies associated with biohydrogen production. Int J Hydrog Energy 46(31):16565–16590

Bard AJ, Fox MA (2002) Artificial photosynthesis: solar splitting of water to hydrogen and oxygen. Acc Chem Res 28(3):141–145

Barros AR, Cavalcante de Amorim EL, Reis CM, Shida GM, Silva EL (2010) Biohydrogen production in anaerobic fluidized bed reactors: effect of support material and hydraulic retention time. Int J Hydrog Energy 35(8):3379–3388

Baykara S (2004) Experimental solar water thermolysis. Int J Hydrog Energy 29(14):1459–1469

Baykara S (2005) Hydrogen as fuel: a critical technology? Int J Hydrog Energy 30(5):545–553

Beckers L, Masset J, Hamilton C, Delvigne F, Toye D, Crine M, Thonart P, Hiligsmann S (2015) Investigation of the links between mass transfer conditions, dissolved hydrogen concentration and biohydrogen production by the pure strain clostridium butyricum CWBI1009. Biochem Eng J 98:18–28

Bérubé P, Lei E (2006) The effect of hydrodynamic conditions and system configurations on the permeate flux in a submerged hollow fiber membrane system. J Membr Sci 271(1–2):29–37

Bundhoo MAZ, Mohee R (2016) Inhibition of dark fermentative bio-hydrogen production: a review. Int J Hydrog Energy 41(16):6713–6733

Castelló E, Nunes Ferraz-Junior AD, Andreani C, Anzola-Rojas MP, Borzacconi L, Buitrón G, Carrillo-Reyes J, Gomes SD, Maintinguer SI, Moreno-Andrade I, Palomo-Briones R, Razo-Flores E, Schiappacasse-Dasati M, Tapia-Venegas E, Valdez-Vázquez I, Vesga-Baron A, Zaiat M, Etchebehere C (2020) Stability problems in the hydrogen production by dark fermentation: possible causes and solutions. Renew Sust Energ Rev:119

Castro Oliveira M, Iten M, Matos HA (2022) Review of thermochemical technologies for water and energy integration systems: energy storage and recovery. Sustainability 14(12):7506

Chen C-Y, Yang M-H, Yeh K-L, Liu C-H, Chang J-S (2008) Biohydrogen production using sequential two-stage dark and photo fermentation processes. Int J Hydrog Energy 33(18):4755–4762

Chen Z, Wei W, Ni B-J (2021) Cost-effective catalysts for renewable hydrogen production via electrochemical water splitting: recent advances. Curr Opin Green Sustain Chem 27:100398

Chong M-L, Sabaratnam V, Shirai Y, Hassan MA (2009a) Biohydrogen production from biomass and industrial wastes by dark fermentation. Int J Hydrog Energy 34(8):3277–3287

Chong M-L, Abdul Rahman NA, Yee PL, Aziz SA, Rahim RA, Shirai Y, Hassan MA (2009b) Effects of pH, glucose and iron sulfate concentration on the yield of biohydrogen by clostridium butyricum EB6. Int J Hydrog Energy 34(21):8859–8865

Chookaew T, Prasertsan P, Ren ZJ (2014) Two-stage conversion of crude glycerol to energy using dark fermentation linked with microbial fuel cell or microbial electrolysis cell. New Biotechnol 31(2):179–184

Ciranna A, Ferrari R, Santala V, Karp M (2014) Inhibitory effects of substrate and soluble end products on biohydrogen production of the alkalithermophile Caloramator celer: kinetic, metabolic and transcription analyses. Int J Hydrog Energy 39(12):6391–6401

Clark IC, Zhang RH, Upadhyaya SK (2012) The effect of low pressure and mixing on biological hydrogen production via anaerobic fermentation. Int J Hydrog Energy 37(15):11504–11513

Danish HST, Baloch MA, Mahmood N, Zhang J (2019) Linking economic growth and ecological footprint through human capital and biocapacity. Sustain Cities Soc:47

Das D (2001) Hydrogen production by biological processes: a survey of literature. Int J Hydrog Energy 26(1):13–28

Das D (2009) Advances in biohydrogen production processes: an approach towards commercialization. Int J Hydrog Energy 34(17):7349–7357

Das S, Mishra A, Ghangrekar MM (2020) Production of hydrogen peroxide using various metal-based catalysts in electrochemical and bioelectrochemical systems: mini review. J Hazard Toxic Radioact Waste 24(3):06020001

Difuntorum JK, Danao LAM (2017) Parametric Study of Vertical Axis Wind Turbine Rotor Configurations Using CFD. ASME 2017 11th International Conference on Energy Sustainability collocated with the ASME 2017 Power Conference Joint With ICOPE-17, the ASME 2017 15th International Conference on Fuel Cell Science, Engineering and Technology, and the ASME 2017 Nuclear Forum2017

Dincer I, Acar C (2015) Review and evaluation of hydrogen production methods for better sustainability. Int J Hydrog Energy 40(34):11094–11111

Dreschke G, Papirio S, Lens PNL, Esposito G (2019) Influence of liquid-phase hydrogen on dark fermentation by Thermotoga neapolitana. Renew Energy 140:354–360

Eibl R, Eibl D, Pörtner R, Catapano G, Czermak P, Catapano G, Czermak P, Eibl R, Eibl D, Pörtner R (2009) Bioreactor design and scale-up. Cell and tissue reaction engineering: with a contribution by Martin fussenegger and wilfried weber. Springer, Berlin, pp 173–259

Encina PG, Hidalgo M (2005) Influence of substrate feed patterns on biofilm development in anaerobic fluidized bed reactors (AFBR). Process Biochem 40(7):2509–2516

Ersoz A (2008) Investigation of hydrocarbon reforming processes for micro-cogeneration systems. Int J Hydrog Energy 33(23):7084–7094

Frigon J-C, Guiot SR (1995) Impact of liquid-to-gas hydrogen mass transfer on substrate conversion efficiency of an upflow anaerobic sludge bed and filter reactor. Enzym Microb Technol 17(12):1080–1086

Fritsch M, Hartmeier W, Chang J (2008) Enhancing hydrogen production of clostridium butyricum using a column reactor with square-structured ceramic fittings. Int J Hydrog Energy 33(22):6549–6557

Fujishima A, Honda K (1972) Electrochemical photolysis of water at a semiconductor electrode. Nature 238(5358):37–38

Ghimire A, Frunzo L, Pirozzi F, Trably E, Escudie R, Lens PNL, Esposito G (2015) A review on dark fermentative biohydrogen production from organic biomass: process parameters and use of by-products. Appl Energy 144:73–95

Ghosh S, Dairkee UK, Chowdhury R, Bhattacharya P (2017) Hydrogen from food processing wastes via photofermentation using purple non-sulfur bacteria (PNSB) – A review. Energy Convers Manag 141:299–314

Gilbert JJ, Ray S, Das D (2011) Hydrogen production using Rhodobacter sphaeroides (O.U. 001) in a flat panel rocking photobioreactor. Int J Hydrog Energy 36(5):3434–3441

Ginkel SV, Sung S, Lay JJ (2001) Biohydrogen production as a function of pH and substrate concentration. Environ Sci Technol 35(24):4726–4730

Grupe H, Gottschalk G (1992) Physiological events in clostridium acetobutylicum during the shift from acidogenesis to solventogenesis in continuous culture and presentation of a model for shift induction. Appl Environ Microbiol 58(12):3896–3902

Guo XM, Trably E, Latrille E, Carrère H, Steyer J-P (2010) Hydrogen production from agricultural waste by dark fermentation: a review. Int J Hydrog Energy 35(19):10660–10673

Hallenbeck PC (2012) Microbial technologies in advanced biofuels production. Springer

Hallenbeck PC, Liu Y (2016) Recent advances in hydrogen production by photosynthetic bacteria. Int J Hydrog Energy 41(7):4446–4454

Hallenbeck PC, Ghosh D, Skonieczny MT, Yargeau V (2009) Microbiological and engineering aspects of biohydrogen production. Indian J Microbiol 49(1):48–59
Han W, Wang B, Zhou Y, Wang DX, Wang Y, Yue LR, Li YF, Ren NQ (2012) Fermentative hydrogen production from molasses wastewater in a continuous mixed immobilized sludge reactor. Bioresour Technol 110:219–223
Hassan GK, Jones RJ, Massanet-Nicolau J, Dinsdale R, Abo-Aly MM, El-Gohary FA, Guwy A (2021) Increasing 2 -Bio- (H2 and CH4) production from food waste by combining two-stage anaerobic digestion and electrodialysis for continuous volatile fatty acids removal. Waste Manag 129:20–25
Hawkes F (2002) Sustainable fermentative hydrogen production: challenges for process optimisation. Int J Hydrog Energy 27(11–12):1339–1347
Hawkes F, Hussy I, Kyazze G, Dinsdale R, Hawkes D (2007) Continuous dark fermentative hydrogen production by mesophilic microflora: principles and progress. Int J Hydrog Energy 32(2):172–184
Holladay JD, Hu J, King DL, Wang Y (2009) An overview of hydrogen production technologies. Catal Today 139(4):244–260
Höök M, Tang X (2013) Depletion of fossil fuels and anthropogenic climate change—A review. Energy Policy 52:797–809
Hussy I, Hawkes F, Dinsdale R, Hawkes D (2005) Continuous fermentative hydrogen production from sucrose and sugarbeet. Int J Hydrog Energy 30(5):471–483
Jones LW (1971) Liquid hydrogen as a fuel for the future. Science 174(4007):367–370
Kadier A, Simayi Y, Abdeshahian P, Azman NF, Chandrasekhar K, Kalil MS (2016) A comprehensive review of microbial electrolysis cells (MEC) reactor designs and configurations for sustainable hydrogen gas production. Alex Eng J 55(1):427–443
Kaparaju P, Buendia I, Ellegaard L, Angelidakia I (2008) Effects of mixing on methane production during thermophilic anaerobic digestion of manure: lab-scale and pilot-scale studies. Bioresour Technol 99(11):4919–4928
Khanal SK, Chen WH, Li L, Sung S. (2004). Biological hydrogen production: effects of pH and intermediate products. Int J Hydrog Energy 29(11):1123–1131
Kongjan P, Angelidaki I (2010) Extreme thermophilic biohydrogen production from wheat straw hydrolysate using mixed culture fermentation: effect of reactor configuration. Bioresour Technol 101(20):7789–7796
Kongjan P, OT S, Kotay M, Min B, Angelidaki I (2010) Biohydrogen production from wheat straw hydrolysate by dark fermentation using extreme thermophilic mixed culture. Biotechnol Bioeng 105(5):899–908
Kraemer JT, Bagley DM (2006) Supersaturation of dissolved H(2) and CO (2) during fermentative hydrogen production with N(2) sparging. Biotechnol Lett 28(18):1485–1491
Kumar G, Zhen G, Kobayashi T, Sivagurunathan P, Kim SH, Xu KQ (2016) Impact of pH control and heat pre-treatment of seed inoculum in dark H 2 fermentation: a feasibility report using mixed microalgae biomass as feedstock. Int J Hydrog Energy 41(7):4382–4392
Lalit Babu V, Venkata Mohan S, Sarma PN (2009) Influence of reactor configuration on fermentative hydrogen production during wastewater treatment. Int J Hydrog Energy 34(8):3305–3312
Lamed RJ, Lobos JH, Su TM (1988) Effects of stirring and hydrogen on fermentation products of clostridium thermocellum. Appl Environ Microbiol 54(5):1216–1221
Laurent B, Serge H, Julien M, Christopher H, Philippe T (2012) Effects of hydrogen partial pressure on fermentative biohydrogen production by a chemotropic clostridium bacterium in a new horizontal rotating cylinder reactor. Energy Procedia 29:34–41
Lee HS, Salerno MB, Rittmann BE (2008) Thermodynamic evaluation on H2 production in glucose fermentation. Environ Sci Technol 42(7):2401–2407
Lee HS, Xin W, Katakojwala R, Venkata Mohan S, Md TN (2022b) Microbial electrolysis cells for the production of biohydrogen in dark fermentation- A review. Bioresour Technol 127934
Lee JE, Shafiq I, Hussain M, Lam SS, Rhee GH, Park Y-K (2022a) A review on integrated thermochemical hydrogen production from water. Int J Hydrog Energy 47(7):4346–4356

Lee KS, Lo YS, Lo YC, Lin PJ, Chang JS (2004) Operation strategies for biohydrogen production with a high-rate anaerobic granular sludge bed bioreactor. Enzym Microb Technol 35(6–7):605–612

Levin D (2004) Biohydrogen production: prospects and limitations to practical application. Int J Hydrog Energy 29(2):173–185

Li C, Fang HHP (2007) Fermentative hydrogen production from wastewater and solid wastes by mixed cultures. Crit Rev Environ Sci Technol 37(1):1–39

Liu G, Li M, Zhou B, Chen Y, Liao S (2018) General indicator for techno-economic assessment of renewable energy resources. Energy Convers Manag 156:416–426

Lo YC, Lee KS, Lin PJ, Chang JS (2009) Bioreactors configured with distributors and carriers enhance the performance of continuous dark hydrogen fermentation. Bioresour Technol 100(19):4381–4387

Lucas SDM, Peixoto G, Mockaitis G, Zaiat M, Gomes SD (2015) Energy recovery from agro-industrial wastewaters through biohydrogen production: kinetic evaluation and technological feasibility. Renew Energy 75:496–504

Łukajtis R, Hołowacz I, Kucharska K, Glinka M, Rybarczyk P, Przyjazny A, Kamiński M (2018) Hydrogen production from biomass using dark fermentation. Renew Sust Energ Rev 91:665–694

Maluta F, Paglianti A, Montante G (2019) Modelling of biohydrogen production in stirred fermenters by computational fluid dynamics. Process Saf Environ Prot 125:342–357

Marone A, Ayala-Campos OR, Trably E, Carmona-Martínez AA, Moscoviz R, Latrille E, Steyer J-P, Alcaraz-Gonzalez V, Bernet N (2017) Coupling dark fermentation and microbial electrolysis to enhance bio-hydrogen production from agro-industrial wastewaters and by-products in a bio-refinery framework. Int J Hydrog Energy 42(3):1609–1621

McMahon KD, Stroot PG, Mackie RI, Raskin L (2001) Anaerobic codigestion of municipal solid waste and biosolids under various mixing conditions–II: Microbial population dynamics. Water Res 35(7):1817–1827

Mizuno O, Dinsdale R, Hawkes FR, Hawkes DL, Noike T (2000) Enhancement of hydrogen production from glucose by nitrogen gas sparging. Bioresour Technol 73(1):59–65

Mukherjee T, Mohan SV (2022) Bio-waste to hydrogen production technologies. Elsevier, Advanced Biofuel Technologies, pp 389–407

Muradov N (2001) Hydrogen via methane decomposition: an application for decarbonization of fossil fuels. Int J Hydrog Energy 26(11):1165–1175

Nemestothy N, Belafi-Bako K, Bakonyi P (2020) Enhancement of dark fermentative H_2 production by gas separation membranes: A review. Bioresour Technol 302

Noblecourt A, Christophe G, Larroche C, Santa-Catalina G, Trably E, Fontanille P (2017) High hydrogen production rate in a submerged membrane anaerobic bioreactor. Int J Hydrog Energy 42(39):24656–24666

Oh SE, Iyer P, Bruns MA, Logan BE (2004) Biological hydrogen production using a membrane bioreactor. Biotechnol Bioeng 87(1):119–127

Oh S-E, Zuo Y, Zhang H, Guiltinan MJ, Logan BE, Regan JM (2009) Hydrogen production by clostridium acetobutylicum ATCC 824 and megaplasmid-deficient mutant M5 evaluated using a large headspace volume technique. Int J Hydrog Energy 34(23):9347–9353

Pathy A, Nageshwari K, Ramaraj R, Pragas Maniam G, Govindan N, Balasubramanian P (2022) Biohydrogen production using algae: potentiality, economics and challenges. Bioresour Technol 360:127514

Pauss A, Andre G, Perrier M, Guiot SR (1990) Liquid-to-gas mass transfer in anaerobic processes: inevitable transfer limitations of methane and hydrogen in the biomethanation process. Appl Environ Microbiol 56(6):1636–1644

Peintner C, Zeidan AA, Schnitzhofer W (2010) Bioreactor systems for thermophilic fermentative hydrogen production: evaluation and comparison of appropriate systems. J Clean Prod 18:S15–S22

Pinelo M, Jonsson G, Meyer AS (2009) Membrane technology for purification of enzymatically produced oligosaccharides: molecular and operational features affecting performance. Sep Purif Technol 70(1):1–11

Posso F, Galeano M, Baranda C, Franco D, Rincón A, Zambrano J, Cavaliero C, Lópes D (2022) Towards the hydrogen economy in Paraguay: green hydrogen production potential and end-uses. Int J Hydrog Energy 47(70):30027–30049

Qyyum MA, Ihsanullah I, Ahmad R, Ismail S, Khan A, Nizami A-S, Tawfik A (2022) Biohydrogen production from real industrial wastewater: potential bioreactors, challenges in commercialization and future directions. Int. J, Hydrogen Energy

Ren N, Xing D, Rittmann BE, Zhao L, Xie T, Zhao X (2007) Microbial community structure of ethanol type fermentation in bio-hydrogen production. Environ Microbiol 9(5):1112–1125

Renaudie M, Clion V, Dumas C, Vuilleumier S, Ernst B (2021) Intensification and optimization of continuous hydrogen production by dark fermentation in a new design liquid/gas hollow fiber membrane bioreactor. Chem Eng J 416:129068

Rupprecht J (2009) From systems biology to fuel--Chlamydomonas reinhardtii as a model for a systems biology approach to improve biohydrogen production. J Biotechnol 142(1):10–20

Saady NMC (2013) Homoacetogenesis during hydrogen production by mixed cultures dark fermentation: unresolved challenge. Int J Hydrog Energy 38(30):13172–13191

Sear RP (2007) Nucleation: theory and applications to protein solutions and colloidal suspensions. J Phys Condens Matter 19(3)

Shanthi Sravan J, Tharak A, Annie Modestra J, Seop Chang I, Venkata MS (2021) Emerging trends in microbial fuel cell diversification-critical analysis. Bioresour Technol 326:124676

Show KY, Lee DJ, Tay JH, Lin CY, Chang JS (2012) Biohydrogen production: current perspectives and the way forward. Int J Hydrog Energy 37(20):15616–15631

Sivagurunathan P, Sen B, Lin C-Y (2015) High-rate fermentative hydrogen production from beverage wastewater. Appl Energy 147:1–9

Sivagurunathan P, Kumar G, Bakonyi P, Kim S-H, Kobayashi T, Xu KQ, Lakner G, Tóth G, Nemestóthy N, Bélafi-Bakó K (2016) A critical review on issues and overcoming strategies for the enhancement of dark fermentative hydrogen production in continuous systems. Int J Hydrog Energy 41(6):3820–3836

Song Z-X, Dai Y, Fan Q-L, Li X-H, Fan Y-T, Hou H-W (2012) Effects of pretreatment method of natural bacteria source on microbial community and bio-hydrogen production by dark fermentation. Int J Hydrog Energy 37(7):5631–5636

Sonnleitner A, Peintner C, Wukovits W, Friedl A, Schnitzhofer W (2012) Process investigations of extreme thermophilic fermentations for hydrogen production: effect of bubble induction and reduced pressure. Bioresour Technol 118:170–176

Steinberg M (1989) Modern and prospective technologies for hydrogen production from fossil fuels. Int J Hydrog Energy 14(11):797–820

Stroot PG, McMahon KD, Mackie RI, Raskin L (2001) Anaerobic codigestion of municipal solid waste and biosolids under various mixing conditions–I. Digester performance. Water Res 35(7):1804–1816

Taheri E, Amin MM, Pourzamani H, Fatehizadeh A, Ghasemian M, Bina B (2018) Comparison of acetate-butyrate and acetate-ethanol metabolic pathway in biohydrogen production. J Med Sign Sens 8(2):101–107

Tanisho S (1998) Effect of CO2 removal on hydrogen production by fermentation. Int J Hydrog Energy 23(7):559–563

Tiwari MK, Guha S, Harendranath CS, Tripathi S (2006) Influence of extrinsic factors on granulation in UASB reactor. Appl Microbiol Biotechnol 71(2):145–154

Valdez-Vazquez I, Poggi-Varaldo HM (2009) Hydrogen production by fermentative consortia. Renew Sust Energ Rev 13(5):1000–1013

Valdez-Vazquez I, Rios-Leal E, Carmona-Martinez A, Munoz-Paez KM, Poggi-Varaldo HM (2006) Improvement of biohydrogen production from solid wastes by intermittent venting and gas flushing of batch reactors headspace. Environ Sci Technol 40(10):3409–3415

Verhaart MR, Bielen AA, van der Oost J, Stams AJ, Kengen SW (2010) Hydrogen production by hyperthermophilic and extremely thermophilic bacteria and archaea: mechanisms for reductant disposal. Environ Technol 31(8–9):993–1003

Voldsund M, Jordal K, Anantharaman R (2016) Hydrogen production with CO2 capture. Int J Hydrog Energy 41(9):4969–4992

Wang Y, Li G, Liu Z, Cui P, Zhu Z, Yang S (2019) Techno-economic analysis of biomass-to-hydrogen process in comparison with coal-to-hydrogen process. Energy 185:1063–1075

Willquist K, Claassen PAM, van Niel EWJ (2009) Evaluation of the influence of CO2 on hydrogen production by Caldicellulosiruptor saccharolyticus. Int J Hydrog Energy 34(11):4718–4726

Yang G, Wang J (2018) Various additives for improving dark fermentative hydrogen production: a review. Renew Sust Energ Rev 95:130–146

Yang S, Liu Z, Tang Z, Wang Y, Chen Q, Sun Y (2017) Performance analysis of solar energy integrated with natural-gas-to-methanol process. Energy Convers Manag 150:375–381

Zhang F, Zhang Y, Chen M, Zeng RJ (2012) Hydrogen supersaturation in thermophilic mixed culture fermentation. Int J Hydrog Energy 37(23):17809–17816

Zhang K, Cui Z, Field RW (2009) Effect of bubble size and frequency on mass transfer in flat sheet MBR. J Membr Sci 332(1–2):30–37

Zhang Z, Tay J, Show K, Yan R, Teeliang D, Lee D, Jiang W (2007) Biohydrogen production in a granular activated carbon anaerobic fluidized bed reactor. Int J Hydrog Energy 32(2):185–191

Zhang Z, Zhou X, Hu J, Zhang T, Zhu S, Zhang Q (2017) Photo-bioreactor structure and light-heat-mass transfer properties in photo-fermentative bio-hydrogen production system: A mini review. Int J Hydrog Energy 42(17):12143–12152

Zhou H, Yu F, Zhu Q, Sun J, Qin F, Yu L, Bao J, Yu Y, Chen S, Ren Z (2018a) Water splitting by electrolysis at high current densities under 1.6 volts. Energy Environ Sci 11(10):2858–2864

Zhou M, Yan B, Wong JWC, Zhang Y (2018b) Enhanced volatile fatty acids production from anaerobic fermentation of food waste: a mini-review focusing on acidogenic metabolic pathways. Bioresour Technol 248(Pt A):68–78

Chapter 19
Computational Modeling and Optimization Strategies for Biohydrogen Production

N. Dlangamandla, A. K. Puri, and K. Permaul

Abstract Biohydrogen has gained recognition as a promising and clean alternative energy source. However, this promising technology faces challenges in terms of microbial metabolism, optimization, and scaling-up the processes. This chapter focuses on the computational models and optimization strategies that are crucial for biohydrogen production. Simplified microbial processes from previous studies have been reviewed in this chapter. Dark and photofermentations have been demonstrated as suitable methods to produce biohydrogen from different types of biomasses, depending on the operational conditions. Studies have shown that response surface methodology (RSM) is a key computational tool to simplify and optimize this microbial process, including the influence of kinetic models for scaling-up the process. Both theoretically and experimentally, sequential dark and photofermentations have been demonstrated to exhibit high yields and productivity rates for the production of biohydrogen. It is crucial to conduct further research into key areas, including optimizing slow metabolism, selecting appropriate feedstocks, and advancing computational models. These strategies can enhance the sustainability and economic viability of biohydrogen production. Biohydrogen production models offer the potential for renewable and sustainable hydrogen production. However, challenges remain in terms of optimizing conversion efficiency, scaling up production, reducing costs, and developing the necessary infrastructure for storage and distribution. Ongoing research and technological advancements are aimed at addressing these challenges and making biohydrogen a viable option for clean energy production. It is important to also note that the effectiveness of ANNs for biohydrogen production depends on the specific application, available data, and the expertise of

N. Dlangamandla · A. K. Puri · K. Permaul (✉)
Department of Biotechnology and Food Science, Durban University of Technology, Durban, South Africa
e-mail: kugen@dut.ac.za

C. R. Soccol et al. (eds.), *Biohydrogen - Advances and Processes*, Biofuel and Biorefinery Technologies 13, https://doi.org/10.1007/978-3-031-49818-3_19

the modelers. Additionally, other modeling approaches such as mechanistic models or hybrid models, combining ANNs with other techniques, may complement the strengths and weaknesses of ANNs to provide more comprehensive insights into biohydrogen production processes.

Keywords Artificial neural networks (ANNs) · Biohydrogen · Dark fermentation · Photofermentation · Metabolism · Kinetic models

19.1 Introduction

Hydrogen is regarded as one of the primary alternative sources of energy, especially relevant to the global depletion of fossil fuels. The advantage of hydrogen use is that it can be generated as a renewable source of energy. Biohydrogen production is considered to be sustainable and friendly to the environment (Boshagh and Rostami 2021; Patel et al. 2021; Rao and Basak 2021). Biohydrogen can be produced from various feedstocks such as agricultural biomass, spent algal biomass, and food waste (Boshagh et al. 2022). Municipal solid waste contributes to an increase in greenhouse gases due to the accumulation of biogas in landfills (Tian et al. 2013; Zarea et al. 2019). Generally, biogas produced from biowaste (Fig. 19.1) contains approximately 80% methane (CH_4) and 16% carbon dioxide (CO_2) and contributes to global emissions (Giwa et al. 2018; Hochman et al. 2015). Although methane has a greater global warming potential than carbon dioxide, it can be converted to hydrogen and other by-products through bioprocessing (Patel et al. 2021; Patel et al. 2018). The metabolic pathways involved in biohydrogen production are crucial in these bioprocessing procedures (Hallenbeck and Ghosh 2012; Kumar et al. 2013; Patel et al. 2021). For biohydrogen production, metabolism of fermentable sugars is observed when pyruvate is degraded into formate, with the release of acetyl-CoA, while formate is transformed to hydrogen (H_2) or to polyhydroxyalkanoates (PHAs) through aceto-acetyl CoA (Hallenbeck and Ghosh 2012).

Recent studies have shown different processes of biohydrogen production, including dark fermentation (DF) and photofermentation (PF) (Chezeau and Vial 2019; Patel et al. 2021). These processes require specific bioreactors to optimize and enhance the production (Boshagh et al. 2022). Therefore, to scale up the process from laboratory scale to pilot scale and eventually to a commercial scale, certain modeling and optimization techniques such as response surface methodology (RSM) and microbial kinetics can be employed. These computational models and simulations are conducted to further evaluate and validate the feasibility of the process before final product completion. Both fermentation processes depend on the major metabolic intermediates produced. However, the challenges are the time, cost, and number of experiments needed for these investigations. Therefore, suitable and reliable modeling methodologies and simulation systems are required for simplified modeling and optimization to scale up the production in these bioreactors.

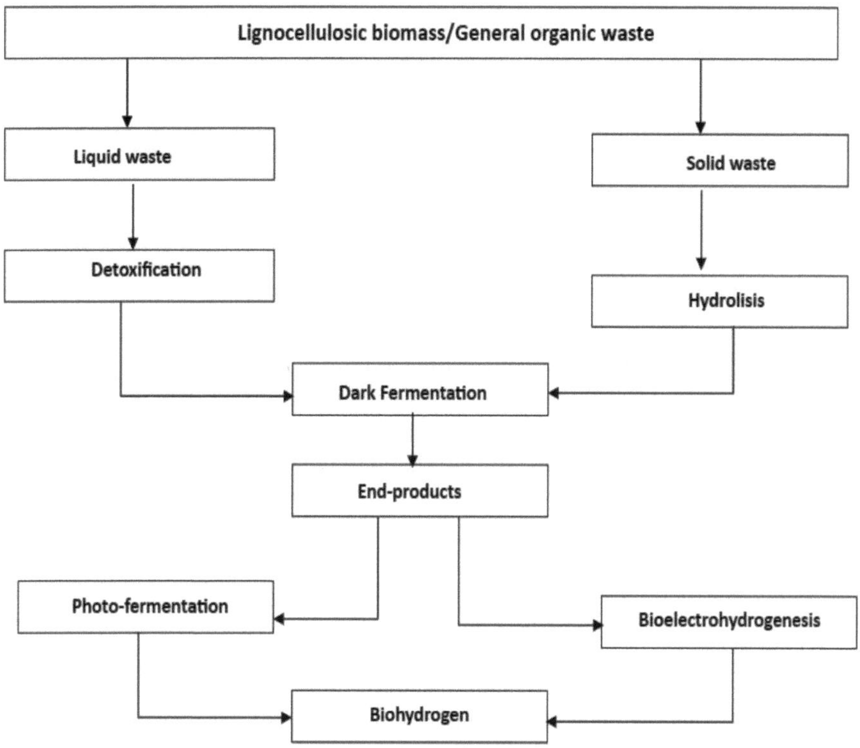

Fig. 19.1 General scheme for processing of biomass into biohydrogen

The primary objective of this chapter is to review the strategies for biohydrogen production using simplified modeling and simulation systems to reduce the development time, lower the costs of the processes, and reduce the number of experiments for investigations. The biohydrogen yield, productivity, and sustainability and finally the advantages and disadvantages of biohydrogen are also discussed.

19.2 Conventional Biohydrogen Production

Biohydrogen can be produced through different methods such as fermentation (DF and PF), biophotolysis, bioelectrochemical system, and gasification (Osman et al. 2020). However, DF and PF processes are considered to be the most effective and sustainable methods for biohydrogen production, adding value to the hydrogen economy, which is currently seen a viable future alternative energy source. Recent studies have shown the potential of biohydrogen production from bio-methanol to reduce carbon emissions, as demonstrated through modeling and simulation processes (Rajesh Banu et al. 2020).

DF and PF are the key fermentation processes, which are simpler to conduct and have less complicated control parameters, with higher yield and productivity rates as compared with other processes. The difference between DF and PF lies in their dependence on light. A summary of selected studies listing hydrogen yield from dark and/or photofermentation using varied substrates in batch culture is presented in Table 19.1.

19.2.1 Dark Fermentation

DF is a well-known process for the production of biohydrogen under anaerobic conditions, using selected microorganisms (Pandey et al. 2022). This process can be conducted in batch, continuous, and fed-batch reactors, with the production of biohydrogen and carbon dioxide as well as other by-products that can impact the yield of biohydrogen. The batch operating process is the most researched and frequently used strategy for biohydrogen production. In contrast, continuous systems are sustainable processes with improved product quality, productivity rates, and biohydrogen yields (Rao and Basak 2021). The fed-batch process has advantages over batch and continuous operation because toxic compound inhibitors, which influence substrate utilization, biomass growth rate, and product formation, including the production of biohydrogen, are not as inhibitory at lower substrate concentrations (Rao and Basak 2021). The low yield of biohydrogen in the DF process is mainly attributed to the formation of acetic acid, propionic acid, and butyric acid (Rao and Basak

Table 19.1 Biohydrogen production using dark and/or photofermentation

Substrate	Organism	Reactor type	Temperature	Hydrogen yield	Reference
Corn straw and duckweed	Mixed culture	Batch	30.94 °C	103.33 mL H_2/g	Zhang et al. (2023)
Garden wastes	*Escherichia coli*	Batch	33 °C	97 mL H_2/g	Ramprakash and Incharoensakdi (2022)
Rice mill wastewater	*Enterobacter aerogenes* RM 08	Batch	33 °C	45 mL H_2/g	Ramprakash and Muthukumar (2015)
Shrub landscaping waste	*Buxus megistophylla*	Batch	–	73.82 mL H_2/g	Yue et al. (2021)
Rice straw	*Bacillus cereus*	Batch	–	1.82 mol H_2/mol	Dinesh et al. (2020)
Food	*Rhodobacter sphaeroides* AV1b	Batch	–	35 mL H_2/g	Ghimire et al. (2016)
Cheese whey permeate	Mixed culture	Batch	–	55.1 g PHA/g VSS	Colombo et al. (2019)

2021). Equations 19.1 to 19.4 describe the key chemical reactions during the dark fermentation process that produces biohydrogen using glucose as a substrate.

$$C_6H_{12}O_6 + 2H_2O \rightarrow 6CO_2 + 4H_2 \quad (19.1)$$

$$C_6H_{12}O_6 + 2H_2O \rightarrow 2CH_3COOH(\text{Acetic acid}) + 2CO_2 + 4H_2 \quad (19.2)$$

$$C_6H_{12}O_6 \rightarrow CH_3COOH + CH_3CH_2COOH(\text{Propionic acid}) + CO_2 + H_2 \quad (19.3)$$

$$C_6H_{12}O_6 + 6H_2O \rightarrow 2CH_3CH_2CH_2COOH(\text{Butyric acid}) + 2CO_2 + 4H_2 \quad (19.4)$$

A variety of organic wastes and biomasses are generally used as substrates to produce CO_2 and H_2 during dark fermentation. The most commonly used cultures that utilize these substrates for biohydrogen production include *Clostridium* sp., facultative *Enterobacter* sp., *Bacillus* sp., some thermophilic bacteria, and anaerobic acidogenic sludge microorganisms. Acetic acid, propionic acid, and butyric acid are formed as by-products, which can affect the yield and productivity of biohydrogen production during dark fermentation (Rao and Basak 2021). Other factors that may affect the biohydrogen production during DF include operating conditions such as pH and temperature. Biohydrogen production occurs during the acidogenic stage of anaerobic metabolism (Rao and Basak 2021; Wang and Yin 2018). Studies have reported an optimal temperature between 25–80 °C and pH range of 5.5–6.5 (Argun and Kargi 2011). Therefore, the computational modeling and optimization strategies should consider these conditions, including the bacterial species. The optimizations and simulation of the feedstocks can be crucial in developing the process of biohydrogen production. Depending on their constituents, some substrates may affect the biohydrogen yield and productivity. Substrates with high sugar or carbohydrates tend to produce more H_2 than substrates with a high amount of fat or protein. Pre-treatment processes are commonly used to enhance biohydrogen production. Numerous studies and reviews on pre-treatment processes have been conducted to enhance the efficiency and maximize the yield of H_2 production.

19.2.2 Photofermentation

PF is a well-known method for producing biohydrogen, utilizing the abilities of purple non-sulfur (PNS) bacteria to convert volatile fatty acids (VFAs) to H_2 and CO_2 (as illustrated in Eqs. 19.5–19.9) under anoxygenic conditions (Argun and Kargi 2011). Additionally, these bacteria can also use carbon sources such as glucose, sucrose, succinate rather than VFAs for H_2 production. *Rhodobacter sphaeroides* O.U.001, *R. capsulatus*, *R. sphaeroides* RV, *R. sulfidophilus*, *Rhodopseudomonas palustris*, and *Rhodospirillum rubrum* are well-known to produce biohydrogen using PF. The efficiency of the PF process is influenced by several factors such as light intensity, inoculum age, types of substrates, and temperature (Rao and Basak 2021).

$$C_2H_4O_2\,(\text{Acetic acid}) + 2H_2O \rightarrow 4H_2 + 2CO_2 \tag{19.5}$$

$$C_3H_6O_3\,(\text{Lactic acid}) + 3H_2O \rightarrow 6H_2 + 3CO_2 \tag{19.6}$$

$$C_3H_6O_2\,(\text{Propionic acid}) + 4H_2O \rightarrow 7H_2 + 3CO_2 \tag{19.7}$$

$$C_4H_8O_2\,(\text{Butyric acid}) + 6H_2O \rightarrow 10H_2 + 4CO_2 \tag{19.8}$$

$$C_4H_6O_5\,(\text{Malic acid}) + 3H_2O \rightarrow 6H_2 + 4CO_2 \tag{19.9}$$

The biohydrogen produced through PF is evaluated based on its yield and light efficiency, as the process requires an external source of light energy to drive the reaction through a positive free energy change (Argun and Kargi 2011; Chen et al. 2010). Therefore, proper distribution of light around the system for optimal use of radiant energy ($h\nu$) is crucial (Argun and Kargi 2011). PF can be performed in batch, continuous, and fed-batch modes using suspended and immobilized-cell systems. However, batch systems have been widely used due to their simple design and operating conditions. Process parameters such as pH, temperature, wavelength, and intensity of light have been reported in several publications to optimize PF (Basak and Das 2007; Koku et al. 2002).

Zhang et al. (2020a) compared PF and DF using corn stover as the substrate and *E. aerogenes* as the bacterial strain. For PF, operational conditions were maintained at an initial pH of 6.5, at 30 °C, and light intensity of 3500 lux. The maximum biohydrogen yield was 141.42 mL/g TS, while the energy conversion efficiency for photofermentation was 10.12%. Li et al. (2020) evaluated biohydrogen yield using concentrated *Chlorella* solution as the raw material and photosynthetic bacterium (PSB) as the strain. The results obtained in this study indicated that the maximum biohydrogen yield was obtained with optimal substrate concentration of 25 g/L, while the experimental data fitted with the Gompertz model. Zhang et al. (2020b) investigated lighting and mixing in a photofermentation to obtain optimal conditions for generating high biohydrogen yield and productivity. This study reported a biohydrogen yield and light conversion efficiency of 84.7 mL/g TS and 36.32%, respectively. The highest light intensity was 7000 lux with a mixing speed of 150 rpm.

Moreover, continuous photofermentation has also been reported to yield positive and feasible results in generating biohydrogen. Many recent studies have conducted continuous photofermentation (Argun and Kargi 2011; Hu et al. 2020; Li et al. 2020) based on the combination of photofermentation and dark fermentation. Hu et al. (2020) investigated biohydrogen yield and productivity rate using continuous photofermentation, where starch agricultural leftover was used as substrate and photosynthetic bacteria as the biocatalyst. A maximum biohydrogen yield of 642 mL and highest production rate of 77.78 mL/(L·h) was achieved with a continuous system at 30 °C and a light intensity of 3000 lux.

Fed-batch photofermentation system has also been reported for biohydrogen production with specific advantages compared to other process. However, only few studies have been reported on the fed-batch single system. Ren et al. (2009) reported biohydrogen production using repeated fed-batch, with *Rhodopseudomonas faecalis* strain RLD-53 as the culture and acetate as the sole source of carbon. The maximum biohydrogen yield was 3.17 mol H_2/mol acetate, while the biohydrogen productivity rate was 37.2 mL H_2/(L·h). The conclusion based on the results of this investigation indicated that the higher efficiency for biohydrogen production can be achieved through substrate feeding and pH control of the system.

19.2.3 Combination of Dark and Photofermentation

Sequential dark and photofermentation has been reported as a promising process to achieve maximum biohydrogen production (Table 19.2), in comparison with individual processes that have low outputs (Nasr et al. 2015). Theoretically, the two-stage process can result in the maximum biohydrogen yield of 12 mol H_2/mol glucose, considering acetic acid as the only product of dark fermentation (Manish and Banerjee 2008). The sequential dark and photofermentation reaction of glucose can be expressed as:

$$C_6H_{12}O_6 + 6H_2O = 12H_2 + 6CO_2 \Delta G^\circ = 3.2\,\text{kJ} \tag{19.10}$$

The above reaction can be made thermodynamically feasible by combining the PF and DF processes. A flow diagram of the combined DF/PF process is shown in Fig. 19.2 where feedstock is initially hydrolyzed under acidic conditions, with or without enzyme addition, followed by dark fermentation. This is followed by photofermentation where light is required, with the final process being the recovery of biohydrogen.

Nasr et al. (2015) reported biohydrogen production using starch wastewater as the substrate in a sequential dark photofermentation process, with *Clostridium and Rhodopseudomonas palustris* as biocatalysts. The results showed a maximum biohydrogen yield of 166.83 ± 27.79 mL H_2/g COD using the combined process. More recently, Policastro et al. (2022) investigated combined dark and photofermentation processes for enhancement of biohydrogen yield using winery wastewater as the substrate. The authors achieved a maximum biohydrogen yield of 310 mL H_2/g during the final stage of this process. However, these studies have identified a few challenges related to the utilization of sequential dark and photofermentation technology on actual waste as the substrates.

Table 19.2 Combined/sequential dark and photofermentation (DPF) for biohydrogen production

Substrate	Organism	Reactor type	DF hydrogen yield	Organism	PF hydrogen yield	DPF hydrogen yield	Reference
Palm oil mill effluent	*Clostridium butyricum* LS2	Batch	0.784 mL H_2/mL	*Rhodopseudomonas palustris*	2.86 mL H_2/mL	3.064 mL H_2/mL	Mishra et al. (2016)
Molasses	*Caldicellulosiruptor saccharolyticus*	Batch	2.10 mol H_2/mol	*Rhodobacter capsulatus* (DSM1710)	3.71 mol H_2/mol	5.81 mol H_2/mol	Özgür et al. (2010)
Potato juice	*Clostridium butyricum*	Batch	–	*Rhodopseudomonas palustris*	–	8.3 mmol H_2/g COD	Hitit et al. (2017)
Cassava	*Clostridium* sp. (cattle dung)	Batch	1.60 mol H_2/mol	*Rhodobacter sphaeroides* ZX-5	4.91 mol H_2/mol	6.51 mol H_2/mol	Zong et al. (2009)
Cassava	Anaerobic activated sludge		10.73 mmol H_2/g	*Rhodopseudomonas palustris*	5.89 mmol H_2/g	16.62 mmol H_2/g	Su et al. (2009)
Corn	*Clostridium acetobutylicum*	Fed-batch	–	*Rhodobacter sphaeroides*	–	2.62 mol H_2/mol	Zagrodnik and Łaniecki (2017)
Wheat	*Clostridium beijerinckii* DSMZ-791	Continuous system		*Rhodobacter sphaeroides* RV		90 mL/g starch	Argun et al. (2009)

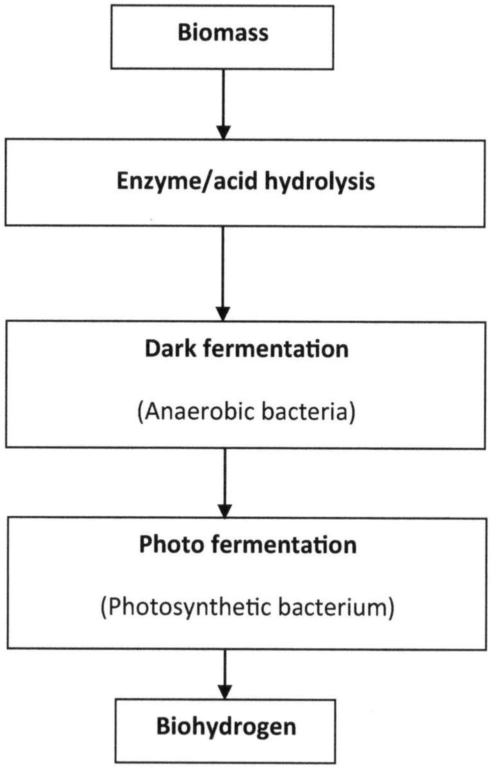

Fig. 19.2 Schematic process showing sequential dark and photofermentation for biohydrogen production

19.3 Microbial Kinetics

The modeling and optimization of biohydrogen production is used to enhance and improve the quality of the final product. These kinetic models are classified into structured and unstructured models. The structured models incorporate metabolic pathways and are more complicated than the unstructured models. Kinetic models such as the modified Gompertz model and the modified logistic model are based on mathematical equations that can be simulated to fit experimental data by linear or nonlinear regression, specifically for biohydrogen production using dark fermentation or photofermentation. The modified Gompertz model is a function to describe microbial growth, especially for batch production of biohydrogen without depending on the substrate type and concentration. Even though the modified logistic model has a function that fits a sigmoid curve and is comparable to the Gompertz model, the modified model also incorporates a lag time (λ) to account for a latency phase (Boshagh and Rostami 2021). The modified Gompertz model and the logistic model for bacterial growth can be expressed in Eqs. 19.11 and 19.12, respectively:

$$\mathrm{H}(t) = H_{\max} \exp\left\{-\exp\left[\frac{R_{\max}\exp(1)}{H_{\max}}(\lambda - t) + 1\right]\right\} \tag{19.11}$$

$$H(t) = \frac{H_{\max}}{1 + \exp\left[4R_{\max} \cdot \frac{(\lambda - t)}{H_{\max} + 2}\right]} \tag{19.12}$$

where $H(t)$ is the cumulative biohydrogen production at time (t) (usually expressed in mL), $H_{\max}$ is the biohydrogen production potential (mL), $R_{\max}$ is the maximum production rate (mL/h), and λ is the lag time (h). However, $H_{\max}$ for the logistic model can be achieved in the stationary phase.

There are also simpler models (such as Monod expression) used as basic models to describe the microbial kinetics. Monod expression (Chezeau and Vial 2019) can be written as:

$$\mu = \mu_{\max} \frac{S}{K_s + S} \tag{19.13}$$

where S is the substrate concentration (mol/L) used during the experiment, while μ is the specific growth rate (per hour), $\mu_{\max}$ is the maximum specific microbial growth rate, and K_s is the substrate constant (mol/L). However, during fermentation experiments, the factors affecting the experiments such inhibitors, which are required to be added in the computational model, which already include substrate consumption rate and the metabolites formation rate. Therefore, these revised models (Wang et al. 2021b) can be written as:

$$\frac{\mathrm{d}S}{\mathrm{d}t} = -\mu Y_s \mathrm{XI}_{\mathrm{pH}}\mathrm{CF} + \mathrm{XC}_{\mathrm{s}} \tag{19.14}$$

$$\mu = \mu_{\max} \frac{S}{K_s + S} \exp\left(-\frac{S}{K_{\mathrm{I}}}\right) \tag{19.15}$$

$$\frac{\mathrm{d}X}{\mathrm{d}t} = \mu \mathrm{XI}_{\mathrm{pH}} - K_d \mathrm{XCF} \tag{19.16}$$

$$\mathrm{I}_{\mathrm{pH}} = \exp\left[-3\left(\frac{\mathrm{pH} - \mathrm{pH}_{\mathrm{L}}}{\mathrm{pH}_{\mathrm{H}} - \mathrm{pH}_{\mathrm{L}}}\right)^2\right] \tag{19.17}$$

$$\frac{\mathrm{d}P_L}{\mathrm{d}t} = -F_{P_L/S}\frac{\mathrm{d}S}{\mathrm{d}t} + \mathrm{XP} \tag{19.18}$$

$$\frac{\mathrm{d}P_L^*}{\mathrm{d}t} = \mathrm{AA} - F_{P_L/S}\frac{\mathrm{d}S}{\mathrm{d}t} + \mathrm{XP} \tag{19.19}$$

$$\frac{\mathrm{d}P_g}{\mathrm{d}t} = -F_{P_g/S}\frac{\mathrm{d}S}{\mathrm{d}t} \quad (19.20)$$

$$\mathrm{CF} = 1 - \exp\left(-\frac{S}{K_\mathrm{I}}\right) \quad (19.21)$$

where Y_s is the substrate yield (biohydrogen yield), C_s is the substrate constant (per hour), X is the biomass concentration (mol/L), while I_{pH} is the pH factor, K_I is the inhibition factor (i.e., when bacterial growth rate less than to 50% of max), $\mathrm{d}P_L^* / \mathrm{d}t$ refers to the production rate of liquid products including acetic acid (AA) in the cultivation media (mol/L·h), $\mathrm{d}P_g/\mathrm{d}t$ refers to gaseous product (mol/L·h), K_d is the death rate (per h), $\mathrm{pH_H}$ and $\mathrm{pH_L}$ are higher than 7.4 and lower than 4.8 pH values at the end of the dark fermentation experiment, while $F_{\mathrm{P_L}/S}$ is the product yield coefficient in liquid phase, while $F_{\mathrm{P_g}/\mathrm{S}}$ is the yield coefficient in gas phase and CF is the correction factor. The summarized models above (Eqs. 19.13–19.21) further reinforce the Monod equation advantage as being a more simplified model. There are different models that have been developed for the Monod equation, which can be used to describe the substrate consumption and non-competitive or competitive inhibition (Wang et al. 2021b). Moreover, the temperature and pH have an impact on the biohydrogen-producing bacteria. Further expression of pH inhibition on the growth rate is derived from the Haldane-Andrews model (Chezeau and Vial 2019). This model was used (Jun et al. 2008) to predict biohydrogen production process in a batch system using sucrose as the substrate:

$$\mathrm{I}_{\mathrm{pH}} = \frac{\left[\mathrm{H}^+\right]}{K_a + \left[\mathrm{H}^+\right] + \frac{\left[\mathrm{H}^+\right]^2}{K_b}} \quad (19.22)$$

where K_a is the H^+ saturation constant and K_b is the H^+ inhibition constant. The alternative equation is indicated in Eq. 19.17.

19.3.1 Product Formation Models

The general specific product formation rate can be expressed using the Luedeking-Piret model (Luedeking and Piret, 1959), which was initially developed to describe lactic acid formation. The Luedeking-Piret model has been used to verify the correlation between the growth rate of biohydrogen-producing bacteria, substrate/biomass hydrolysis rate, and metabolite production rates in the process (Chezeau and Vial 2019). In Eq. 19.23, the first term ($Y_{s/x}\,\mu$) is associated with growth rate, while the second term (μ_{PX}) is associated with specific product formation rate. The Luedeking-Piret model can be expressed as:

$$q_p \frac{dp}{dt} = Y_{p/x} \frac{dX}{dt} + \beta X \tag{19.23}$$

where P is the product, $Y_{P/x}$ is the growth associated yield coefficient, and β is the non-growth associated product yield coefficient (Akbari and Vaziri 2017). The Luedeking-Piret model and the Gompertz model were used in combination to predict the hydrogen gas evolution and *Rhodobacter capsulatus* growth, respectively, in continuous stirred tank photo-bioreactors (Deseure et al. 2021). A good correlation was observed between the predicted values and experimental data. Bacterial growth was modeled using a simpler equation from Gompertz function (Eq. 19.24):

$$\ln\left(\frac{X}{X_0}\right) = ae^{-e^{(b-ct)}} \tag{19.24}$$

where X is the bacterial density and X_0 is the initial bacterial density (Gibson et al. 1988). These models can be further modified to account for inhibition and pH that can affect the bioprocess of biohydrogen production (Chezeau and Vial 2019; Van Niel et al. 2003).

19.4 Statistical Models for Experimental Design

19.4.1 Response Surface Methodology (RSM)

Statistical models are increasingly being used to plan experiments to ensure that optimal conditions and a reduction in the number of experiments are achieved. The design of experiments methodology (i.e., factorial designs) and response surface methodology (RSM) are the main statistical models used for biohydrogen production. Additionally, RSM has been widely utilized to enhance the statistical significance of the operational parameters of the experimental findings (Jabbari et al. 2021; Pandey et al. 2021; Wang et al. 2021c). An application of RSM is to scale up laboratory-scale experiments using statistical tools. RSM employs straightforward polynomial formulations, allowing for the quantitative correlation of the researched operational parameters with the results (Chezeau and Vial 2019). To quantify the relationship between the fermentation parameters and response, the Plackett-Burman design technique, Box-Behnken design (BBD), or central composite design (CCD) is used to analyze the fermentation parameters (Chezeau and Vial 2019).

RSM can be expressed as a quadratic polynomial:

$$Y_i = b_0 + \sum_{i=1}^{n} b_i X_i + \sum_{i=1}^{n} b_{ii} X_i^2 + \sum_{i<j=1}^{n} b_{ij} X_i X_j \tag{19.25}$$

where Y_i is the output, while b_0, b_i, and b_{ii} (b_{ij}) are the coefficients and X_i represents the experimental inputs that influence the output Y after manipulation of the n inputs (Chezeau and Vial 2019; Wang et al. 2021b). These coefficients can be determined using RSM at different parameters to obtain optimum conditions for fermentation. After optimization of biohydrogen production using RSM, the results are further validated in reaction vessels of varied volumes. The optimal laboratory-scale data can be further upscaled and verified in large bioreactors (Pandey et al. 2021).

RSM have been extensively used in recent studies to investigate the optimum parameters for improved yield of biohydrogen and productivity rate using simple sugars or organic waste materials in shake flasks (i.e., at laboratory scale). Al-Mohammedawi et al. (2018) reported higher biohydrogen production of 41.74 mL/L·h from DL-malic acid via a mesophilic fermentative process using *R. sphaeroides* 158 DSM. Specifically, they used RSM and a three-factor, three-level BBD matrix to successfully optimize the pH, carbon-to-nitrogen ratio, and light intensity during photofermentation. Pandey et al. (2021) investigated a sequential dark and photofermentation using wet biomass hydrolysate of *Spirulina platensis*, where 1510 ± 35 mL/L of biohydrogen was produced in 48 h using a co-culture of *Rhodobacter sphaeroides* and *Bacillus firmus* (Pandey et al. 2021). After the optimal parameters were obtained, central composite design (CCD) was used to further optimize the screener variables to enhance biohydrogen production (Pandey et al. 2021). Wang et al. (2021b) optimized conditions for dark fermentation for enhanced biohydrogen production. In this study, the optimal operation conditions indicated that the process could produce biohydrogen with a yield of 106.2 cm^3/g together with by-products such as propionic acid (2.8 kg/m^3), acetic acid (1.9 kg/m^3), and lower amounts of butyric acid, valeric acid, and bioethanol. However, the study was done under anaerobic conditions using *Clostridium propionicum* (Wang et al. 2021c). Das and Basak (2022) have also reported similar results, where the biohydrogen production was 1580 mL/L with a yield of 6.31 mol of biohydrogen per mol of carbohydrates. However, in this study potato waste was used as substrate as compared to microalgal biomass (Das and Basak 2022; Wang et al. 2021c). Taherdanak et al. (2015) reported the effect of iron and nickel nanoparticles on biohydrogen production using RSM and CCD. The authors reported 200% higher biohydrogen production (149.8 mL/g) using starch as substrate with no Fe and Ni (Taherdanak et al. 2015). Lazaro et al. (2017) developed a new strategy for dark microaerobic fermentation, whereby the optimal operational conditions were obtained using RSM. In this study, a highest biohydrogen production of 173.5 μmol was achieved. Cheng et al. (2011) reported biohydrogen production using cassava as the substrate. The authors reported a maximum biohydrogen yield of 489 mL/g starch with photofermentation, while dark fermentation yielded 351 mL H_2/g starch.

19.4.2 Artificial Neural Networks (ANN)

During the last few years, ANN has received the great attention as a powerful artificial intelligence (AI) tool to solve complex nonlinear problems. ANN models are inspired by the structure and function of the human brain, which comprises interconnected nerve cells. The model simulates the human brain by learning and making predictions based on patterns and data inputs. In the context of extremely complex biohydrogen production process, the relationship between dependent and independent variables exhibits a strong nonlinear nature. Therefore, ANN models have been employed to optimize process parameters, predict and control process variables affecting production rates, and identify key factors influencing hydrogen production.

General in developing, the ANN model variables with a substantial impact are normal included as input variables, and the result can be used as an output variable. Due to their distinct measurement ranges, all the variables (both input and output ones) were scaled in the rank [0.1, 0.9] to prevent saturation issues with the sigmoidal function in the network. The following steps are used to obtain the new scaled variables:

$$X^* = 0.8\frac{X - X_{\min}}{X_{ma} - X_{\min}} + 0.1 \tag{19.26}$$

$$X = \frac{\left(X_{\max} - X_{\min}\right)\left(X^* - 0.1\right)}{0.8} + X_{\min} \tag{19.27}$$

where X is the variable in a group to be scaled, $X_{\min}$ is the minimum variable in a group to be scaled, $X_{\max}$ is the maximum variable in a group to be scaled, and X^* is the scaled variable.

The ANN model used by Yadav et al. (2021) suggested temperature (37.89 °C), headspace (65.25%), and butyric acid supplementation (9.21 g/L) as optimum parameters to predict 9.39 L/L of biohydrogen production, which was very close to the experimental response of 9.25 L/L (Yadav et al. 2021). Safarian et al. (2021) used ANN model to show that gasifier temperature, solid-to-biomass ratio (SBR), moisture content, and hydrogen have the highest impacts on biohydrogen production. All ANN models designed by Sydney et al. (2020) predicted biohydrogen production, hydrogen production rate, and H_2 yield with high accuracy ($R^2 > 0.987$). In fact, several studies have showed the superior prediction capability and accuracy of ANN models over RSM (Wang et al. 2021c). Whiteman and Gueguim Kana (2014) obtained R^2 values of 0.75 and 0.91 for RSM and ANN models, respectively, while studying biohydrogen production using sugar cane molasses. A combined hybrid ANN-RSM model suggested that starch 6.2 kg/m^3, pH 6.7, Fe 11.7 g/m^3, and sludge 24.6 g/m^3 could lead to biohydrogen yield of 106.2 cm^3/g (Wang et al. 2021c).

19.5 Advantages and Disadvantages of Biohydrogen Production Models

Biohydrogen production models involve using microorganisms to produce hydrogen gas through biological processes (Kapdan and Kargi 2006; Wang et al. 2021a). There are several approaches to biohydrogen production, each with its own advantages and disadvantages. In the dark fermentation process, anaerobic microorganisms break down organic substrates in the absence of light to produce hydrogen gas (Kumar et al. 2019). It has the advantage of being able to use a wide range of feedstocks, including agricultural waste and organic matter. However, dark fermentation suffers from low hydrogen yields and the production of unwanted byproducts such as volatile fatty acids (Bina et al. 2019). Photofermentation utilizes photosynthetic bacteria to convert organic compounds into hydrogen gas in the presence of light (Li et al. 2022). This process has the advantage of higher hydrogen yields compared with dark fermentation. However, it requires specific light conditions, and the efficiency of conversion is influenced by factors like light intensity and nutrient availability. In the algae-based systems, certain species of microalgae can produce hydrogen through a process called photobiological water splitting (Khetkorn et al. 2017). Algae capture solar energy and use it to generate hydrogen by splitting water molecules. Algae-based systems have the advantage of high growth rates and the ability to utilize wastewater or CO_2 as feedstocks. However, scaling up the production and harvesting of algae can be challenging, and maintaining consistent hydrogen yields can be difficult (Frowijn and Van Sark 2021; Khetkorn et al. 2017; Pragya et al. 2013). Syngas fermentation, a mixture of carbon monoxide and hydrogen, can be generated from various carbon-rich feedstocks, such as biomass or industrial waste gases. Syngas fermentation involves using specific microorganisms to convert syngas into hydrogen gas (Okolie et al. 2020). It offers the advantage of utilizing waste gases and has the potential for high hydrogen yields. However, syngas production can be energy-intensive and may require additional processing steps. Bioelectrochemical systems employ microbial electrolysis cells or microbial fuel cells to produce hydrogen gas (Almatouq et al. 2020). Microorganisms oxidize organic matter or hydrogen at the anode, generating electrons that are transferred to the cathode, where they reduce protons to produce hydrogen gas. Bioelectrochemical systems have the advantage of producing high-purity hydrogen and the ability to treat wastewater. However, they require additional infrastructure and are sensitive to electrode fouling and system optimization. The advantages of ANNs for biohydrogen is the pattern recognition, nonlinearity relationship, and adaptability and leaning of the models (Mert 2021). ANNs are excellent at recognizing complex patterns and relationships within large datasets. They can analyze large amounts of data and identify hidden patterns that may not be apparent using traditional statistical methods. In the context of biohydrogen production, ANNs can help identify key factors and variables that affect hydrogen production, leading to optimized process conditions. ANNs are capable of modeling nonlinear relationships between input and output variables (Mert 2021; Petrone et al. 2013). In biohydrogen production, where multiple interacting factors influence the process, ANNs can capture the complex nonlinear relationships between various input parameters (such as pH, temperature, substrate concentration) and hydrogen production. This flexibility allows ANNs to model and predict hydrogen production accurately. ANNs have the ability to adapt and learn from new data. They can continuously update their internal parameters and adjust their predictions based on new information. In the case of biohydrogen, ANNs can adapt to changes in environmental conditions, substrate composition, or process parameters, allowing for more robust and flexible control and optimization of hydrogen production (Petrone et al. 2013). However, the disadvantage of ANNs is the requirement of a significant amount of training data to learn effectively. In some cases, acquiring large and reliable datasets for biohydrogen production may be challenging, especially for emerging or less-studied processes. Insufficient or low-quality data can result in inaccurate predictions or unreliable models (Yadav et al. 2021). ANNs are prone to overfitting, which occurs when the model becomes too complex and starts to memorize the training data instead of generalizing from it. Overfitting can lead to poor performance when the model is applied to new, unseen data. Regularization techniques and careful validation are necessary to mitigate overfitting and ensure the model's reliability. ANNs are often considered "black-box" models because they provide little insight into the underlying mechanisms or relationships between input and output variables. While ANNs can make accurate predictions, understanding the exact cause-and-effect relationships within the system may be challenging. In biohydrogen production, where process optimization and control are crucial, interpretability can be essential for implementing practical improvements.

19.6 Strategies for Biohydrogen Production

Different strategies have been developed to improve biohydrogen production. Computer simulation tools have been discussed in the sections above aimed at enhancing and optimizing biohydrogen production, including cost reduction. Sequential dark and photofermentation have been shown as a promising microbial system to enhance biohydrogen production (Pandey et al. 2021; Zhang et al. 2018). In most of these processes, varying of light intensity has improved the efficiency of the process. Therefore, bioreactor design must consider the light efficiency, as it affects the generated biohydrogen (Argun and Kargi 2011). Several studies have reported that photosynthetic bacterium have an ability to produce biohydrogen from low to higher light intensity (Niño-Navarro et al. 2020; Sağır and Hallenbeck 2019). Another factor that plays a crucial role in improving biohydrogen production is the metabolic pathways to produce the final product. Patel et al. discussed different pathways that can enhance the biohydrogen yield using different substrates. However, processes such as biophotolysis have challenges that are related to the sensitivity of enzymes to oxygen, slow development, and low output which are drawbacks of the process. Therefore, the application of combined dark and photofermentation with optimal conditions can improve the feasibility of the process for biohydrogen production (Srivastava et al. 2020).

19.7 Conclusion and Perspectives

Biohydrogen as a renewable source of energy has the ability to add value to the current and future markets. Hydrogen yield is around 4 mol H_2/mol hexose for dark fermentation, while the combined system can produce 12 mol H_2/mol hexose. Computational modeling and optimization strategies are key to upscale the processes with optimized conditions and significant cost reduction. Computational modeling approaches, coupled with optimization techniques, have proven to be powerful tools in understanding and improving biohydrogen production processes. Kinetic models of the processes can also result in improving the biohydrogen production. Several studies have shown the feasibility of the sequential dark and photofermentation for biohydrogen production with higher biohydrogen yield and productivity rate using biomass as the carbon source. Researchers can gain valuable insights into the complex interplay of multiple factors affecting biohydrogen production, leading to enhanced process efficiency and productivity. Therefore, more optimization studies are required to further improve the process for commercialization. Additionally, the creation of multiobjective optimization frameworks can aid in determining the best conditions for a process while taking into account a variety of performance metrics, including hydrogen production, substrate utilization efficiency, and economic viability. By leveraging these computational modeling and optimization strategies and embracing future developments, researchers can drive future innovations in biohydrogen production, ultimately leading to more sustainable and efficient processes. Strategies to improve biohydrogen production's sustainability and economic viability include optimizing conversion efficiency, scaling production, reducing costs, and developing infrastructure. Research and technological advancements have aimed to make biohydrogen a viable option for clean energy production. ANNs' effectiveness depends on application, data, and modeler expertise. Mechanistic and hybrid models can complement ANNs for more comprehensive insights.

Acknowledgments This study is part of a multilateral BRICS Joint Science and Technology Research Collaboration project supported by the National Research Foundation (NRF) of South Africa, under grant numbers (UID 150503 and 116019), also with support of the Coordination of Improvement of Higher Education Personnel – CAPES of Brazil, project Print-BioADD.

References

Akbari L, Vaziri BM (2017) Comprehensive modeling of photo-fermentation process for prediction of hydrogen production. Int J Hydrog Energy 42:14449–14457

Al-Mohammedawi HH, Znad H, Eroglu E (2018) Synergistic effects and optimization of photo-fermentative hydrogen production of Rhodobacter sphaeroides DSM 158. Int J Hydrog Energy 43:15823–15834

Almatouq A, Babatunde AO, Khajah M, Webster G, Alfodari M (2020) Microbial community structure of anode electrodes in microbial fuel cells and microbial electrolysis cells. J Water Process Eng 34:101140

Argun H, Kargi F (2011) Bio-hydrogen production by different operational modes of dark and photo-fermentation: an overview. Int J Hydrog Energy 36:7443–7459

Argun H, Kargi F, Kapdan IK (2009) Effects of the substrate and cell concentration on bio-hydrogen production from ground wheat by combined dark and photo-fermentation. Int J Hydrog Energy 34:6181–6188

Basak N, Das D (2007) The prospect of purple non-sulfur (PNS) photosynthetic bacteria for hydrogen production: the present state of the art. World J Microbiol Biotechnol 23:31–42

Bina B, Amin MM, Pourzamani H, Fatehizadeh A, Ghasemian M, Mahdavi M, Taheri E (2019) Biohydrogen production from alkaline wastewater: the stoichiometric reactions, modeling, and electron equivalent. MethodsX 6:1496–1505

Boshagh F, Rostami K (2021) Kinetic models of biological hydrogen production by Enterobacter aerogenes. Biotechnol Lett 43:435–443

Boshagh F, Rostami K, Van Niel EWJ (2022) Application of kinetic models in dark fermentative hydrogen production–a critical review. Int J Hydrog Energy 47:21952–21968

Chen C-Y, Yeh K-L, Lo Y-C, Wang H-M, Chang J-S (2010) Engineering strategies for the enhanced photo-H2 production using effluents of dark fermentation processes as substrate. Int J Hydrog Energy 35:13356–13364

Cheng J, Su H, Zhou J, Song W, Cen K (2011) Hydrogen production by mixed bacteria through dark and photo fermentation. Int J Hydrog Energy 36:450–457

Chezeau B, Vial C (2019) Modeling and simulation of the biohydrogen production processes. *Biohydrogen*. Elsevier

Colombo B, Calvo MV, Sciarria TP, Scaglia B, Kizito SS, D'imporzano G, Adani F (2019) Biohydrogen and polyhydroxyalkanoates (PHA) as products of a two-steps bioprocess from deproteinized dairy wastes. Waste Manag 95:22–31

Das SR, Basak N (2022) Optimization of process parameters for enhanced biohydrogen production using potato waste as substrate by combined dark and photo fermentation. Biomass Convers Bioref:1–21

Deseure J, Obeid J, Willison JC, Magnin J-P (2021) Reliable determination of the growth and hydrogen production parameters of the photosynthetic bacterium Rhodobacter capsulatus in fed batch culture using a combination of the Gompertz function and the Luedeking-Piret model. Heliyon 7:e07394

Dinesh GH, Nguyen DD, Ravindran B, Chang SW, Vo D-VN, Bach Q-V, Tran HN, Basu MJ, Mohanrasu K, Murugan RS (2020) Simultaneous biohydrogen (H2) and bioplastic (poly-β-hydroxybutyrate-PHB) productions under dark, photo, and subsequent dark and photo fermentation utilizing various wastes. Int J Hydrog Energy 45:5840–5853

Frowijn LS, Van Sark WG (2021) Analysis of photon-driven solar-to-hydrogen production methods in The Netherlands. Sustain Energy Technol Assessm 48:101631

Ghimire A, Valentino S, Frunzo L, Pirozzi F, Lens PNL, Esposito G (2016) Concomitant bio-hydrogen and poly-hydroxybutyrate production from dark fermentation effluents by adapted Rhodobacter sphaeroides and mixed photofermentative cultures. Bioresour Technol 217:157–164

Gibson AM, Bratchell N, Roberts TA (1988) Predicting microbial growth: growth responses of salmonellae in a laboratory medium as affected by pH, sodium chloride and storage temperature. Int J Food Microbiol 6:155–178

Giwa A, Adeyemi I, Dindi A, Lopez CG-B, Lopresto CG, Curcio S, Chakraborty S (2018) Techno-economic assessment of the sustainability of an integrated biorefinery from microalgae and jatropha: a review and case study. Renew Sust Energ Rev 88:239–257

Hallenbeck PC, Ghosh D (2012) Improvements in fermentative biological hydrogen production through metabolic engineering. J Environ Manag 95:S360–S364

Hitit ZY, Lazaro CZ, Hallenbeck PC (2017) Increased hydrogen yield and COD removal from starch/glucose based medium by sequential dark and photo-fermentation using Clostridium butyricum and Rhodopseudomonas palustris. Int J Hydrog Energy 42:18832–18843

Hochman G, Wang S, Li Q, Gottlieb PD, Xu F, Li Y (2015) Cost of organic waste technologies: a case study for New Jersey. AIMS Energy 3:450–462

Hu B, Li Y, Zhu S, Zhang H, Jing Y, Jiang D, He C, Zhang Z (2020) Evaluation of biohydrogen yield potential and electron balance in the photo-fermentation process with different initial pH from starch agricultural leftover. Bioresour Technol 305:122900

Jabbari B, Jalilnejad E, Ghasemzadeh K, Iulianelli A (2021) Modeling and optimization of a membrane gas separation based bioreactor plant for biohydrogen production by CFD–RSM combined method. J Water Process Eng 43:102288

Jun Y-S, Yu S-H, Ryu K-G, Lee T-J (2008) Kinetic study of pH effects on biological hydrogen production by a mixed culture. J Microbiol Biotechnol 18:1130–1135

Kapdan IK, Kargi F (2006) Bio-hydrogen production from waste materials. Enzym Microb Technol 38:569–582

Khetkorn W, Rastogi RP, Incharoensakdi A, Lindblad P, Madamwar D, Pandey A, Larroche C (2017) Microalgal hydrogen production–a review. Bioresour Technol 243:1194–1206

Koku H, Eroğlu I, Gündüz U, Yücel M, Türker L (2002) Aspects of the metabolism of hydrogen production by Rhodobacter sphaeroides. Int J Hydrog Energy 27:1315–1329

Kumar G, Mathimani T, Rene ER, Pugazhendhi A (2019) Application of nanotechnology in dark fermentation for enhanced biohydrogen production using inorganic nanoparticles. Int J Hydrog Energy 44:13106–13113

Kumar P, Patel SK, Lee J-K, Kalia VC (2013) Extending the limits of bacillus for novel biotechnological applications. Biotechnol Adv 31:1543–1561

Lazaro CZ, Hitit ZY, Hallenbeck PC (2017) Optimization of the yield of dark microaerobic production of hydrogen from lactate by Rhodopseudomonas palustris. Bioresour Technol 245:123–131

Li S, Tabatabaei M, Li F, Ho S-H (2022) A review of green biohydrogen production using anoxygenic photosynthetic bacteria for hydrogen economy: challenges and opportunities. Int J Hydrog Energy

Li Y, Zhang Z, Xia C, Jing Y, Zhang Q, Li S, Zhu S, Jin P (2020) Photo-fermentation biohydrogen production and electrons distribution from dark fermentation effluents under batch, semi-continuous and continuous modes. Bioresour Technol 311:123549

Luedeking R, Piret EL (1959) A kinetic study of the lactic acid fermentation. Batch process at controlled pH. J Biochem Microbiol Technol Eng 4:393–412

Manish S, Banerjee R (2008) Comparison of biohydrogen production processes. Int J Hydrog Energy 33:279–286

Mert I (2021) Agnostic deep neural network approach to the estimation of hydrogen production for solar-powered systems. Int J Hydrog Energy 46:6272–6285

Mishra P, Thakur S, Singh L, Ab Wahid Z, Sakinah M (2016) Enhanced hydrogen production from palm oil mill effluent using two stage sequential dark and photo fermentation. Int J Hydrog Energy 41:18431–18440

Nasr M, Tawfik A, Ookawara S, Suzuki M, Kumari S, Bux F (2015) Continuous biohydrogen production from starch wastewater via sequential dark-photo fermentation with emphasize on maghemite nanoparticles. J Ind Eng Chem 21:500–506

Niño-Navarro C, Chairez I, Christen P, Canul-Chan M, García-Peña E (2020) Enhanced hydrogen production by a sequential dark and photo fermentation process: effects of initial feedstock composition, dilution and microbial population. Renew Energy 147:924–936

Okolie JA, Nanda S, Dalai AK, Berruti F, Kozinski JA (2020) A review on subcritical and supercritical water gasification of biogenic, polymeric and petroleum wastes to hydrogen-rich synthesis gas. Renew Sust Energ Rev 119:109546

Osman AI, Deka TJ, Baruah DC, Rooney DW (2020) Critical challenges in biohydrogen production processes from the organic feedstocks. Biomass Convers Bioref:1–19

Özgür E, Afsar N, De Vrije T, Yücel M, Gündüz U, Claassen PA, Eroglu I (2010) Potential use of thermophilic dark fermentation effluents in photofermentative hydrogen production by Rhodobacter capsulatus. J Clean Prod 18:S23–S28

Pandey A, Sinha P, Pandey A (2021) Hydrogen production by sequential dark and photofermentation using wet biomass hydrolysate of Spirulina platensis: response surface methodological approach. Int J Hydrog Energy 46:7137–7146

Pandey AK, Pilli S, Bhunia P, Tyagi R, Surampalli RY, Zhang TC, Kim S-H, Pandey A (2022) Dark fermentation: production and utilization of volatile fatty acid from different wastes-a review. Chemosphere 288:132444

Patel SK, Kondaveeti S, Otari SV, Pagolu RT, Jeong SH, Kim SC, Cho B-K, Kang YC, Lee J-K (2018) Repeated batch methanol production from a simulated biogas mixture using immobilized Methylocystis bryophila. Energy 145:477–485

Patel SK, Das D, Kim SC, Cho B-K, Kalia VC, Lee J-K (2021) Integrating strategies for sustainable conversion of waste biomass into dark-fermentative hydrogen and value-added products. Renew Sust Energ Rev 150:111491

Petrone R, Zheng Z, Hissel D, Péra M-C, Pianese C, Sorrentino M, Becherif M, Yousfi-Steiner N (2013) A review on model-based diagnosis methodologies for PEMFCs. Int J Hydrog Energy 38:7077–7091

Policastro G, Giugliano M, Luongo V, Napolitano R, Fabbricino M (2022) Enhancing photo fermentative hydrogen production using ethanol rich dark fermentation effluents. Int J Hydrog Energy 47:117–126

Pragya N, Pandey KK, Sahoo P (2013) A review on harvesting, oil extraction and biofuels production technologies from microalgae. Renew Sust Energ Rev 24:159–171

Rajesh Banu J, Kavitha S, Yukesh Kannah R, Bhosale RR, Kumar G (2020) Industrial wastewater to biohydrogen: possibilities towards successful biorefinery route. Bioresour Technol 298:122378

Ramprakash B, Incharoensakdi A (2022) Dark fermentative hydrogen production from pretreated garden wastes by Escherichia coli. Fuel 310:122217

Ramprakash B, Muthukumar K (2015) Comparative study on the performance of various pretreatment and hydrolysis methods for the production of biohydrogen using Enterobacter aerogenes RM 08 from rice mill wastewater. Int J Hydrog Energy 40:9106–9112

Rao R, Basak N (2021) Fermentative molecular biohydrogen production from cheese whey: present prospects and future strategy. Appl Biochem Biotechnol 193:2297–2330

Ren N-Q, Liu B-F, Zheng G-X, Xing D-F, Zhao X, Guo W-Q, Ding J (2009) Strategy for enhancing photo-hydrogen production yield by repeated fed-batch cultures. Int J Hydrog Energy 34:7579–7584

Safarian S, Ebrahimi Saryazdi SM, Unnthorsson R, Richter C (2021) Modeling of hydrogen production by applying biomass gasification: artificial neural network modeling approach. Fermentation 7:71

Sağır E, Hallenbeck PC (2019) Photofermentative hydrogen production. *Biohydrogen*. Elsevier

Srivastava N, Srivastava M, Mishra P, Kausar MA, Saeed M, Gupta VK, Singh R, Ramteke P (2020) Advances in nanomaterials induced biohydrogen production using waste biomass. Bioresour Technol 307:123094

Su H, Cheng J, Zhou J, Song W, Cen K (2009) Improving hydrogen production from cassava starch by combination of dark and photo fermentation. Int J Hydrog Energy 34:1780–1786

Sydney EB, Duarte ER, Burgos WJM, De Carvalho JC, Larroche C, Soccol CR (2020) Development of short chain fatty acid-based artificial neuron network tools applied to biohydrogen production. Int J Hydrog Energy 45:5175–5181

Taherdanak M, Zilouei H, Karimi K (2015) Investigating the effects of iron and nickel nanoparticles on dark hydrogen fermentation from starch using central composite design. Int J Hydrog Energy 40:12956–12963

Tian H, Gao J, Hao J, Lu L, Zhu C, Qiu P (2013) Atmospheric pollution problems and control proposals associated with solid waste management in China: a review. J Hazard Mater 252:142–154

Van Niel EW, Claassen PA, Stams AJ (2003) Substrate and product inhibition of hydrogen production by the extreme thermophile, Caldicellulosiruptor saccharolyticus. Biotechnol Bioeng 81:255–262

Wang J, Yin Y (2018) Fermentative hydrogen production using various biomass-based materials as feedstock. Renew Sust Energ Rev 92:284–306

Wang Y, Jing Y, Lu C, Kongjan P, Wang J, Awasthi MK, Tahir N, Zhang Q (2021a) A syntrophic co-fermentation model for bio-hydrogen production. J Clean Prod 317:128288

Wang Y, Tang M, Ling J, Wang Y, Liu Y, Jin H, He J, Sun Y (2021b) Modeling biohydrogen production using different data driven approaches. Int J Hydrog Energy 46:29822–29833

Wang Y, Yang G, Sage V, Xu J, Sun G, He J, Sun Y (2021c) Optimization of dark fermentation for biohydrogen production using a hybrid artificial neural network (ANN) and response surface methodology (RSM) approach. Environ Prog Sustain Energy 40:e13485

Whiteman J, Gueguim Kana E (2014) Comparative assessment of the artificial neural network and response surface modelling efficiencies for biohydrogen production on sugar cane molasses. Bioenergy Res 7:295–305

Yadav S, Singh V, Mahata C, Das D (2021) Optimization for simultaneous enhancement of biobutanol and biohydrogen production. Int J Hydrog Energy 46:3726–3741

Yue T, Jiang D, Zhang Z, Zhang Y, Li Y, Zhang T, Zhang Q (2021) Recycling of shrub landscaping waste: exploration of bio-hydrogen production potential and optimization of photo-fermentation bio-hydrogen production process. Bioresour Technol 331:125048

Zagrodnik R, Łaniecki M (2017) Hydrogen production from starch by co-culture of clostridium acetobutylicum and Rhodobacter sphaeroides in one step hybrid dark-and photofermentation in repeated fed-batch reactor. Bioresour Technol 224:298–306

Zarea MA, Moazed H, Ahmadmoazzam M, Malekghasemi S, Jaafarzadeh N (2019) Life cycle assessment for municipal solid waste management: a case study from Ahvaz, Iran. Environ Monitor Assessm 191:1–13

Zhang Q, Zhang Z, Wang Y, Lee D-J, Li G, Zhou X, Jiang D, Xu B, Lu C, Li Y (2018) Sequential dark and photo fermentation hydrogen production from hydrolyzed corn Stover: a pilot test using 11 m^3 reactor. Bioresour Technol 253:382–386

Zhang T, Jiang D, Zhang H, Jing Y, Tahir N, Zhang Y, Zhang Q (2020a) Comparative study on biohydrogen production from corn Stover: photo-fermentation, dark-fermentation and dark-photo co-fermentation. Int J Hydrog Energy 45(6):3807–3814

Zhang Z, Zhang H, Li Y, Lu C, Zhu S, He C, Ai F, Zhang Q (2020b) Investigation of the interaction between lighting and mixing applied during the photo-fermentation biohydrogen production process from agricultural waste. Bioresour Technol 312:123570

Zhang X, Zhang Q, Li Y, Zhang H (2023) Modeling and optimization of photo-fermentation biohydrogen production from cosubstrates basing on response surface methodology and artificial neural network integrated genetic algorithm. Bioresour Technol 374:128789

Zong W, Yu R, Zhang P, Fan M, Zhou Z (2009) Efficient hydrogen gas production from cassava and food waste by a two-step process of dark fermentation and photo-fermentation. Biomass Bioenergy 33:1458–1463

Chapter 20
Various Routes for Hydrogen Production and Its Utilization for Sustainable Economy

Avinash Anand, Karan Kumar, and Vijayanand S. Moholkar

Abstract Hydrogen as a fuel source has the potential to give a long-term solution to the challenge of ensuring a constant supply of cleaner, environmentally acceptable fuel. Steam methane reforming (SMR) produces 80–85% of the hydrogen with a 74–85% energy efficiency. Steam methane reforming (SMR) and other fossil fuel-based technologies are neither green nor environmentally sustainable. If hydrogen (H_2) is produced from renewable sources, then it could only be considered a clean and renewable fuel. Hydrogen can be produced by a number of processes, including thermocatalytic reformation of hydrogen-rich organic compounds, electrolysis of water, and biological processes. It is essential to develop renewable hydrogen production techniques, such as biohydrogen, in order for hydrogen to become a promising energy carrier. This chapter presents a critical analysis and comparison of various routes for hydrogen production that includes conventional and biological processes. Further, the present infrastructure for storage, transportation, and utilization of hydrogen gas has been discussed in detail. Lastly, we have also provided the future scope of hydrogen production technology and its utilization for achieving sustainable development goals.

A. Anand
Department of Chemical Engineering, Indian Institute of Technology Guwahati, Guwahati, Assam, India

K. Kumar
School of Energy Science and Engineering, Indian Institute of Technology Guwahati, Guwahati, Assam, India

V. S. Moholkar (✉)
Department of Chemical Engineering, Indian Institute of Technology Guwahati, Guwahati, Assam, India

School of Energy Science and Engineering, Indian Institute of Technology Guwahati, Guwahati, Assam, India
e-mail: vmoholkar@iitg.ac.in

C. R. Soccol et al. (eds.), *Biohydrogen - Advances and Processes*, Biofuel and Biorefinery Technologies 13, https://doi.org/10.1007/978-3-031-49818-3_20

Keywords Hydrogen · Biophotolysis · Steam reforming · Electrolysis · Dark fermentation · Microbial production · H_2 storage · H_2 utilization · SMR

20.1 Introduction

The continuous combustion of fossil fuels in equipment like furnaces, boilers, and IC engines releases very harmful greenhouse gases (GHG) into the atmosphere, posing a significant threat to the global atmosphere and resulting in climate change (Malani et al. 2019; Kumar et al. 2023a, b). Increasing concerns regarding climate change and the depletion of petroleum reserves are driving renewed exploration for alternative energy sources, particularly renewable fuels (Kumar et al. 2021; Singh et al. 2022; Kumar and Moholkar 2023). H_2 is one probable candidate, which is also regarded as the cleanest fuel with only water as the by-product. It has a high heat of combustion of 142 MJ/kg (approximately three times greater than that of hydrocarbon fuel) (Appel and Schulz 1998). Table 20.1 lists the heating values of the various common fuels employed as energy source. The main issue with using hydrogen as an alternate fuel is its scarcity in nature and the necessity for low-cost manufacturing technologies. Present methods for hydrogen production employed can be distinguished into two categories: conventional and renewable technologies. The process of pyrolysis and hydrocarbon reforming are included in the first category, which is based on fossil fuels. SMR, POX, and ATR are the chemical processes used in the hydrocarbon reforming process (Nikolaidis and Poullikkas 2017). Even though hydrogen produced through this route has demonstrated its capacity to be used as a clean energy resource, only a small portion of the same is presently being used for this purpose. Most of the hydrogen has been used in the culinary, metallurgical, petrochemical, and electronics processing sectors as feedstock (Muradov 1993). According to published statistics (Balat and Kırtay 2010), roughly 49% of hydrogen generated through hydrocarbon reforming is utilized in (NH_3) ammonia manufacturing, followed by 8% in methyl alcohol (CH_3OH) production, 37% in petroleum refining, and 6% in other miscellaneous applications. H_2 storage has become the primary impediment in its use as a vehicle fuel due to the low density of hydrogen. Steam methane reforming (SMR) is the most cost-effective and widely used method for generating hydrogen from natural gas. A typical SMR hydrogen plant with a capacity of one million m^3 of hydrogen per day is said to emit roughly 0.3–0.4 million SCO (standard cubic meters) of CO_2 into the atmosphere per day (Muradov and

Table 20.1 Lower heating (LHV) and higher heating (HHV) in MJ/kg values for various fuels

Fuel	At ambient pressure and temperature	LHV (MJ/kg)	HHV (MJ/kg)
Methane	Gas	50	55.5
Hydrogen	Gas	120	142
Methanol	Liquid	18	20
Gasoline	Liquid	45	48
Diesel	Liquid	43	45
Ethane	Gas	48	52

Vezirolu 2005). It has been stipulated that CO_2 capture imposed approximately 25–30% additional cost to the SMR hydrogen production (Muradov 1993). The technologies that manufacture hydrogen from renewable resources, such as biomass or water, fall in the second group. These technologies, which use biomass as a feedstock, may be split into two categories: thermochemical and biological processes (Kothari 2017). Nowadays, mostly H_2 is produced from steam reforming of natural gas (Hosseini and Wahid 2016). Although prototype hydrogen vehicles have been developed, the lack of widespread infrastructure for H_2 as transportation fuel and ongoing challenge of in-vehicle storage capacity pose significant hurdles. Moreover, H_2 fuel cells are expensive to manufacture, are prone to fragility, and have limited service life. However, biohydrogen production from carbohydrate-rich organic wastes and industrial wastewater has emerged as a highly promising and sustainable alternatives to fossil fuels, one that can address the greenhouse effect. Biological methods such as direct and indirect biophotolysis, dark fermentation, and photo fermentation can all produce hydrogen (Gadhe et al. 2013). Among these, dark fermentation is an important method due to its higher rate of biohydrogen generation as well as the flexibility of the organic substrates used. However, the main constraint of this method is the low yield of hydrogen per mole of organic substrate consumed. Fermentative processes generally yield higher rates (Kumar et al. 2022a, b 2022) than photosynthetic hydrogen production, do not require light, use a variety of renewable sources, use less energy, and are technically much simpler and more stable, which seems to have more potential for practical applications (Das 2001). Hydrogen production from various sources is shown in Fig. 20.1. Hydrogen generation from organic waste and residues has enormous potential to meet future energy needs. Till this date, majority of biohydrogen production research has been done on a lab scale. Only a few pilot-scale trials have been reported, and no industrial scale biohydrogen production has been reported to our knowledge (Balachandar et al. 2020). The sustainability of the hydrogen production process has to be accompanied by the potential for efficient scale-up at commercial levels. Despite encouraging results on lab-scale research, significant R&D effort is required for hydrogen

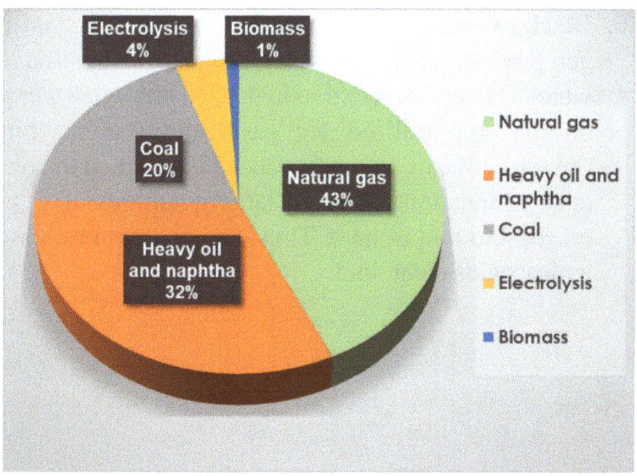

Fig. 20.1 Hydrogen production from different sources

production through biological routes on commercial scale. In addition, material and energy analyses are necessary to assess the process performance.

Hydrogen is regarded as the cleanest fuel. However, it is not the case, as the technique or method of bulk production of hydrogen may release significant quantities of greenhouse gases. There are three potential technologies for bulk production of hydrogen to be used as transportation fuel, viz., (1) coal gasification followed by water gas shift reaction, (2) steam reforming of methane followed by water gas shift reaction, and (3) electrolysis of water. The key factor in all these technologies is the effective sequestration and storage of CO_2 without release into the atmosphere. In order to distinguish between the above methodologies, hydrogen has been given a color code. This color code is explained as follows:

1. **Green hydrogen:** This is produced through electrolysis of water, where electricity produced through renewable sources is utilized. This process is essentially "green" as it causes no emissions of carbon dioxide to atmosphere. Hence, the hydrogen produced through this process is also designated green color.
2. **Blue hydrogen:** This is produced using conventional or fossil hydrocarbons as substrates. The carbon dioxide produced in this process is sequestrated and stored underground.
3. **Gray hydrogen:** This is produced from the combined process of steam methane reforming (SMR) and water gas shift reaction using natural gas as the feedstock. However, the CO_2 produced in the process is directly released to the atmosphere.
4. **White hydrogen:** It is a naturally available hydrogen in the atmosphere.
5. **Red hydrogen:** This is produced by high-temperature catalytic cracking of water, in which the required energy is produced using nuclear power heating.
6. **Pink hydrogen:** Pink hydrogen is essentially a product of electrolysis of water, where the required electricity is provided by a nuclear power station.
7. **Purple hydrogen:** This is produced through water-splitting by chemo-thermal electrolysis using energy supplied by nuclear power station.
8. **Turquoise hydrogen:** This is a relatively novel technology based on thermal cracking or pyrolysis of methane. A peculiarity of this process is that it removes carbon in the form of a solid rather than CO_2.
9. **Black or brown hydrogen:** This is the most conventional technique for hydrogen production. This is based on coal gasification to produce carbon monoxide, which is later converted to hydrogen through water gas shift reaction. Depending on the coal utilized for gasification—either bituminous (black) or lignite (brown)—the hydrogen produced is assigned the black or brown color. However, coal gasification produced large quantities of CO_2 (along with CO), which is released to atmosphere. Thus, this process has least prospects for production of hydrogen as clean fuel.

20.2 Conventional Routes of Hydrogen Production from Fossil Fuels

Conventional routes for hydrogen production from fossil fuels are shown in Fig. 20.2. Pyrolysis and hydrocarbon reforming are two common methods for producing hydrogen from fossil fuels. These technologies have been extensively studied and investigated and also widely implemented. At present, fossil fuels continue to play a major role in global hydrogen supply since fuel costs, which are at acceptable levels and are closely linked to production costs.

20.2.1 Hydrocarbon Reforming Methods

Hydrocarbon reforming (CHs) is the method through which hydrocarbons are transformed into hydrogen gas by some reforming techniques. Steam and oxygen, in addition to hydrocarbons, can be used as reactants in the reforming process (Chen et al. 2008). Natural gas reforming is an advanced technology for the hydrogen production process. Nowadays, 95% of the hydrogen produced in the USA is through the route of natural gas reforming in large central plants. Reforming plants consist of hydrodesulfurization unit (HDS), catalytic reforming unit (CRU), and the auxiliary units such as pumps, compressors, clean-up sections expanders, heat exchangers, cooling towers and combustors, etc. (Ersoz 2008).

20.2.1.1 Steam Methane Reforming Method (SMR)

In this process, methane obtained from natural gas is heated with steam in the presence of catalyst to produce hydrogen and carbon monoxide. SMR is an endothermic process, and energy must be supplied to the process for the reaction to proceed (Shen 2002; Shinnar 2003).

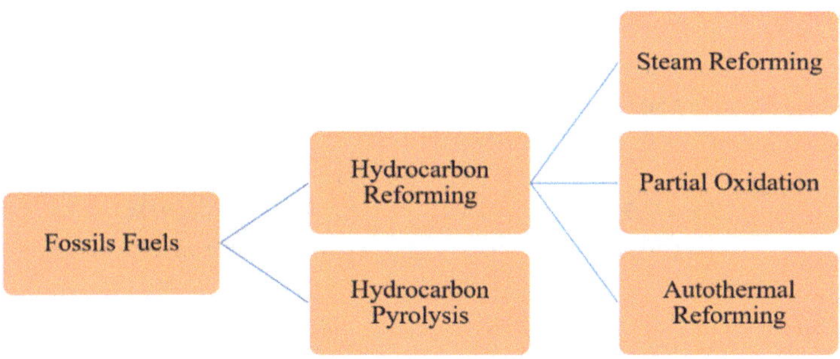

Fig. 20.2 Hydrogen production methods from fossil fuels

In a water-gas shift reaction, the carbon monoxide and steam are reacted using a catalyst to produce carbon dioxide and more hydrogen. Carbon dioxide and other impurities in the product stream are removed using pressure swing adsorption. SR can also be used to produce hydrogen from other fuels such as gasoline, propane, and ethyl alcohol (Yao et al. 2006).

Water gas shift reaction

$$CO + H_2O \rightarrow CO_2 + H_2 \quad \Delta H = -41\,kJ/mol$$

SMR reaction

$$CH_4 + H_2O \rightarrow CO + 3H_2 \, \Delta H = +206\,kJ/mol$$

The generation of hydrogen from coal gasification is cost-effective. The major disadvantage is that the process generates a lot of CO_2 (Adamson and Pearson 2000). Several catalytic systems are being investigated to assess their viability for dry reforming, focusing on operating parameters, performance characteristics, and coke deposition (Galadima and Muraza 2015). Methane has high caloric value which leads to an interest in the development of large-scale steam reformers. Steam reforming of methane is an extensively used method of generating syngas from natural gas. Utilization of waste CO_2 in the production of syngas from natural gas is achieved through dry reforming of methane (DRM). Although DRM has yet to be implemented commercially as a standalone process, the industrial sector has been combining SMR and DMR for several years. This combined method has a lower carbon footprint (Gangadharan et al. 2012). Nickel (Ni)-based catalysts are predominantly utilized in SMR due to their superior reactivity compared to natural catalysts. While dolomite and olivine are noteworthy natural catalysts with the potential for hydrogen-rich gas production, and in situ CO_2 capture, the efficiency of H_2 production is higher with Ni-based catalysts when compared to natural catalysts. The different catalysts used for hydrogen production, their operating conditions, and the typical yields are summarized in Table 20.2.

Table 20.2 Effect of catalysts on hydrogen production during SMR

Catalyst	Working conditions	H_2 production	References
Ni/MgO	$T = 650$ °C	57%	Adhikari et al. (2008)
Ni/YSZ	$T = 900$ °C	75%	Laosiripojana and Assabumrungrat (2007)
Ni/CeO_2	$T = 873$ K	44%	Adhikari et al. (2008)
Ni-based catalyst	$T = 650$ °C	92%	Martavaltzi et al. (2010)
Pt/Al_2O3	$T = 498$ K	70%	Adhikari et al. (2008)
Ni/TiO_2	$T = 550$ °C	31%	Adhikari et al. (2008)
Ni/Al_2O	$T = 325$ °C	$H_2/CO = 0.8$	Hou and Hughes (2001)

20.2.1.2 Partial Oxidation (POX) Method

Partial oxidation is a type of chemical reaction. This reaction takes place when a sub-stoichiometric fuel-air mixture is partly burned in a reformer, generating hydrogen-rich syngas, which can then be put to further use. In the POX process, the feedstock is partially burned in a simple pre-combustion chamber in the presence of a small amount of air and converted into carbon monoxide and hydrogen. POX is an exothermic reaction in which some amount of heat of combustion is released. The main benefit of the POX process is the ability of the system to acclimate almost any carbonaceous feedstock, which can compose of very heavy molecular weight organic constituents such as petroleum residue and coke (Speight 2015). POX method involves the transformation of oxygen, steam, and hydrocarbons to carbon oxides and hydrogen. There are two types of partial oxidation methods, thermal partial oxidation method (TPOX) and catalytic partial oxidation method (CPOX). TPOX reaction temperatures are dependent on the oxygen-fuel ratio (Steinberg 1989). During catalytic process, the energy is provided under control and its thermal efficiency is 60–75% (Holladay et al. 2009). The equations of catalytic and non-catalytic reforming processes are presented below (Nikolaidis and Poullikkas 2017).

$$\text{Reformer}: \quad C_nH_m + \frac{1}{2}O_2 \rightarrow nCO + \frac{1}{2}mH_2 \quad (\text{catalytic})$$

$$C_nH_m + nH_2O \rightarrow nCO + \frac{1}{2}mH_2 \quad (\text{non}-\text{catalytic})$$

Simultaneous emissions of CO and CO_2 is the major limitation of POX method.

20.2.1.3 Autothermal Reforming Method

It is the combination of steam reforming method (SMR) and partial oxidation method (POX).

ATR is made up of exothermic POX reactions and endothermic steam reforming processes (Chen et al. 2008). In order to generate syngas, ATR employs both oxygen and steam in reaction with hydrocarbon. The reaction occurs in a particular chamber using a partially oxidized hydrocarbon. Because of the oxidation, the process is exothermic. The energy produced by the partial oxidation reaction (POX) is utilized to make syngas in the SMR process. It is ideal to balance the reactions such that they are effectively thermoneutral, requiring no net heat input. The overall reaction of ATR can be described by the following set of equations (Lamb and Pollet 2020).

$$CH_4 + \frac{1}{4}O_2 + \frac{1}{2}H_2O \rightarrow CO + \frac{5}{2}H_2$$

$$CH_4 + \frac{1}{2}O_2 + \frac{1}{2}CO_2 \rightarrow \frac{3}{2}H_2 + \frac{3}{2}CO + \frac{1}{2}H_2O$$

$$C_nH_m + \frac{m}{4}O_2 + 2\left(n - \frac{m}{4}\right)H_2O \rightarrow nCO + 2nH_2$$

$$C_nH_m + \left(\frac{n-1}{2}\right)O_2 + CO_2 \rightarrow (n+1)CO + \frac{m}{2}H_2$$

Thermoneutrality of ATR depends on values of n and m. In an ideal configuration, the first half of the reactor is filled with a POX catalyst, followed by SR catalyst in the second half of the reactor.

20.2.2 Hydrocarbon Pyrolysis (CHs)

In pyrolysis of hydrocarbon, the only supply of hydrogen is the hydrocarbons itself, which decomposes thermally by the general reaction. It is a type of endothermic reaction.

$$C_nH_m \rightarrow nC + \frac{1}{2}mH_2$$

Methane decomposition is also known as methane cracking or methane pyrolysis. It is a chemical process of splitting methane into its elemental components of hydrogen and solid carbon. In this decomposition process, methane enters the process and hydrogen leaves in gaseous state. Pyrolysis is carried out in absence of oxygen. This process is less complex than classical steam methane reforming (Amin et al. 2011; Lee 2015).

20.3 Hydrogen Production from Biological Processes

During biohydrogen production, some anaerobes are used as matrix. These microbes degrade organic waste to produce biohydrogen. Anaerobic photosynthesis and anaerobic fermentation are two types of biological hydrogen production methods. H_2-producing microbes are classified as follows: (i) photosynthetic organisms, including photosynthetic bacteria, cyanobacteria, or green algae and (ii) compatible and specific anaerobic H_2-producing bacteria, such as *Clostridium* sp., *Escherichia coli (E. coli)*, and *Enterobacter aerogenes*. Research on lab-scale hydrogen production from biomass routes has given promising results, which demonstrate potential of green hydrogen economy. Recent literature reports a number of studies on the conversion of waste to energy. Fermentative biohydrogen from various cheap feedstock such as forest or agricultural residue (Pattra et al. 2008), wastewater (Kapdan and Kargi 2006; Aydin et al. 2021), municipal solid waste (Dzulkarnain et al. 2022), and food waste (Han et al. 2016) has been reported (Wu et al. 2012). The amount of biohydrogen production from sucrose is influenced by various metabolites and metabolic pathways (Ren et al. 2008). Acetic acid and butyric acid are important metabolites, which are formed during biohydrogen production. Typically, 1 mole of glucose acetate fermentation yields 4 moles of hydrogen, while 1 mole butyrate fermentation yields 2 moles of hydrogen (O-Thong et al. 2008). Transition metals like, zinc (Zn), cadmium (Cd), chromium (Cr), copper (Cu), nickel (Ni), and lead (Pb) have been overwhelmingly reported to inhibit the anaerobic fermentation for H_2 production (Mudhoo and Kumar 2013).

20.3.1 Biophotolysis

Biophotolysis is a bio-mechanism-driven process that uses same pathway as in algal and plant photosynthesis but adapts them for the generation of biohydrogen gas. Hydrogenase enzymes are absent in green plants for H_2 production, and only deduction of CO_2 occurs. On the other hand, algae contain hydrogen-producing enzymes and can produce biohydrogen under particular conditions. Blue green algae and green algae are able to break molecules of water into hydrogen ion and oxygen via indirect and direct biophotolysis, respectively. Cyanobacteria or blue green algae are gram-positive bacteria that carry out similar photosynthesis as higher plants. They may be found in both the maritime environment and other soils/ecosystems. (Subramanian 1998). These bacteria are nitrogen-fixing bacteria, and they can break water into hydrogen and oxygen through biophotolysis. The hydrogenase and nitrogenase enzymes are involved (Smith et al. 1992). Hydrogen production by diazotrophic cyanobacteria, either by nitrogenase or hydrogenase, has been considered a suitable sunlight-assisted biological system (Fay 1970). ATP-dependent nitrogenase activity allows cyanobacteria to fix atmospheric nitrogen. The hydrogenase enzyme is protected from oxygen by nitrogenase, which is found in heterocysts. The nitrogen fixed in the heterocysts is transferred to the surrounding vegetative cells as glutamine under normal physical circumstances. Glutamate synthetase (GS), which is likewise found in heterocysts, catalyzes conversion of ammonium to glutamine. Algae are probably better at producing hydrogen than cyanobacteria, which requires more energy-intensive enzymes like nitrogenase.

Some typical examples of different hydrogen-producing microorganisms included in different categories are listed below:

Green algae: *Chlamydomonas reinhardtii, C. moewusii, Scenedesmus obliquus*
Cyanobacteria heterocystous: *Anabaena CA, A. variabilis, Nostoc muscorum, N. spongiaeforme, Westiellopsis prolifica, Anabaena azollae, A. cylindrica*

Nonheterocystous: *Oscillatoria* Miami BG7, *Mastigocladus laminosus*, *Phormidium valderianum*, *Plectonema boryanum*, *Aphanothece halophytica*

Photosynthetic bacteria: *R. capsulatus*, *R. sulidophilus*, *Rhodopseudomonas sphaeroides*, *R. palustris*, *R. capsulatus*, *Chlorobium limicola*, *Chloroflexus aurantiacus*, *Thiocapsa roseopersicina*, *Rhodobacter sphaeroides*, *Rhodospirillum rubrum*, *Chromatium* sp. Miami PSB 1071

Fermentative bacteria: *E. cloacae*, *Desulfovibrio vulgaris*, *Megasphaera elsdenii*, *Escherichia coli*, *Enterobacter aerogenes*, *Clostridium butyricum*, *Clostridium pasteurianum*

20.3.2 Direct Biophotolysis

Direct biophotolysis is a process employing microalgae photosynthetic systems for the conversion of solar energy into chemical energy in the form of hydrogen (refer to Fig. 20.3). Green algae use sunlight to split water molecules into hydrogen ions and oxygen in the direct biophotolysis process. The hydrogen ions are subsequently transformed to H_2 gas by the hydrogenase enzyme (Kapdan and Kargi 2006). Oxygen-sensitive enzyme should keep the oxygen content of the media (containing substrate and other additives) below 0.1%. A possible solution to keep oxygen content below 0.1% is to use of green algae *Chlamydomonas reinhardtii* that can deplete oxygen during oxidative respiration (Melis et al. 2000). Due to large substrate consumption during process, the overall efficiency of this process is low. The mutants of *Chlamydomonas reinhardtii* have recently been reported to exhibit strong O_2 tolerance and enhance hydrogen production.

Fig. 20.3 Direct biophotolysis

20.3.3 *Indirect Biophotolysis*

The schematic depiction of indirect photolysis is shown in Fig. 20.4. The following are general reactions that cyanobacteria or blue green algae use to generate hydrogen from water via indirect biophotolysis:

$$12H_2O + 6CO_2 + \text{light energy} \rightarrow C_6H_{12}O_6 + 6O_2$$

$$C_6H_{12}O_6 + 12H_2O + \text{light energy} \rightarrow 12H_2 + 6CO_2$$

Hydrogen is produced by both nitrogenase and hydrogenase enzymes, and the rate of hydrogen production is same as hydrogenase-based hydrogen production in green algae. Algal H_2 generation might be regarded a cost-effective and long-term solution in terms of water utilization as renewable resource and CO_2 consumption as one of the air pollutants. The primary disadvantages of this route for biohydrogen generation are low H_2 production capability, the demand for a large surface area to absorb enough light, and the lack of waste utilization as substrate.

20.3.4 *Photofermentation*

Photofermentation is a biochemical process in which degradation of substrate takes place in the presence of light (shown in Fig. 20.5). Solar energy and organic acids are used in this process in nitrogen-deficient circumstances. Some photosynthetic bacteria can convert organic acids (succinic, butyric, acetic, and lactic acids) into hydrogen and carbon dioxide due to the presence of nitrogenase. For example:

$$CH_3COOH + 2H_2O + \text{light energy} \rightarrow 4H_2 + 2CO_2$$

Increased light intensity enhances H_2 production rate but at the same time decreases overall conversion efficiency in the process. If industrial effluents are

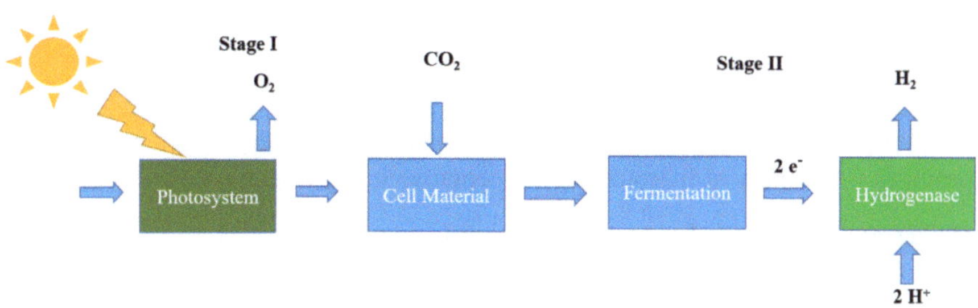

Fig. 20.4 Indirect biophotolysis

Fig. 20.5 Photofermentation

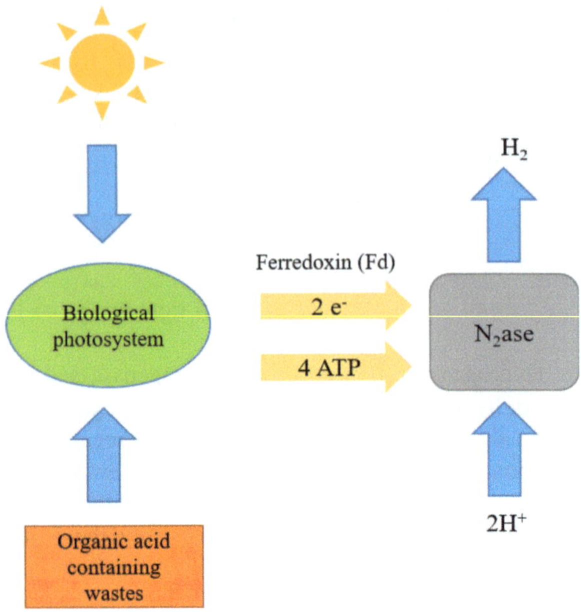

utilized to generate H_2, a severe problem may arises due to the colors of wastewater, which may impede penetration of light and the presence of harmful substances such as heavy or transition metals, etc. Proper pretreatment of industrial wastes are needed for further utilization as substrate for biohydrogen production from this process.

20.3.5 Dark Fermentation

The glycolysis pathway converts glucose to pyruvic acid (pyruvate) during anaerobic fermentation. Pyruvate is then transformed to acetyl-CoA, which may serve as feed for anaerobic bacteria, such as *C. butyricum* to produce H_2 and CO_2. Alternatively, several microbes may convert acetyl-CoA to alcohols (e.g., ethyl alcohol) and acids (e.g., acetate and butyrate) under distinct circumstances. Butyric acid and acetic acid are made from NADH. Excess NADH is converted to NADP, which releases H_2. Hydrogen fermentation may be classified into three forms based on the content of the final product: (i) H_2 fermentation with butyric acid, (ii) H_2 fermentation with propionic acid, and (iii) ethanol type H_2 fermentation. Table 20.3 provides a summary of literature on fermentative biohydrogen production.

H_2 Fermentation with Butyric Acid Some *Clostridium* species e.g., *C. pasteurianum*, *C. butyricum*, and *C. acetobutylicum*, carry out this form of H_2 fermentation under anaerobic conditions (Kumar et al. 2022a, 2023b, c). During H_2 fermentation with butyrate formation, simple carbohydrates (e.g., hexose, sucrose, starch) are

Table 20.3 Various substrates used for fermentative biohydrogen production process

Substrate	Inoculum	Substrate concentration		Yield (mol-H_2/mol-substrate	References
		Range	Optimal		
Xylose	Municipal sewage sludge	10–100 g/L	20 g/L	2.25	Lin and Cheng (2006)
	Clostridium butyricum CGS5	6–35 g/L	20 g/L	–	Lo et al. (2008)
Glucose	Seed sludge	0–295 g/L	–	2.2	Fang and Liu (2002)
	Digested sludge	1.1–300 g/L	2.1 g/L	3.2	Fang et al. (2006)
	Enterobacter cloacae IIT-BT 08	–	–	2	Kumar and Das (2001)
	Ruminococcus albus	–	–	2.5	Ntaikou et al. (2009)
Sucrose	*C. butyricum CGS5*	5–35 g/L	20 g/L	2.8	Lo et al. (2008)
	Thermoanaero-bacter Thermosaccharo-lyticum PSU-2	5.6–60 g/L	20 g/L	2.5	O-Thong et al. (2008)
	E. cloacae IIT-BT 08	0–1 g/L	–	6	Kumar and Das (2000)
Starch	*Municipal digester sludge*	8-35 g/L	32 g/L	1.8	Lee et al. (2008)
	Thermococcus kodakarensis KODI	2–12 g/L	–	1.9	Hussy et al. (2005)
	Anaerobic sludge	5–55 g/L	20 g/L	2.2	Lin et al. (2008)
Non-fat dry milk	*Anaerobic digester sludge*	10–100 g-COD/L	4 g-COD/L	0.005	Lin and Cheng (2006)
Wheat straw	*Mixed culture from cow dung*	0–15 g/L	–	0.003	Fan et al. (2006b)
Food waste	*Anaerobic sludge*	3.2–10 g-COD/L	6.4 g-COD/L	1.8	Shin (2004)
Beer less biomass	*Mixed culture*	0–50 g/L	20 g/L	0.0003	Fan et al. (2006a)
Corn stalk wastes	*Mixed culture*	5–20 g/L	15 g/L	0.007	Zhang et al. (2007a)
	Clostridium sp. *X9*	0–20 g/L	–	0.006	Zhang et al. (2007b)

being metabolized to form CO_2, H_2, butyrate, acetate, and very less amount of propionate as end-products. The main reactions are:

$$C_6H_{12}O_6 + 2H_2O \rightarrow 2CH_3COOH + 2CO_2 + 4H_2$$

$$C_6H_{12}O_6 \rightarrow C_4H_9COOH + CO_2 + 2H_2$$

Propionic Acid-Type Fermentation Fermentation of this type takes place during the anaerobic treatment of sewage which contains more nitrogenous compounds (e.g., gelatin, beef extract, and yeast extract). Complex carbohydrates that are difficult to breakdown (e.g., cellulose) frequently undergo H_2 fermentation with propionic acid production in absence of oxygen. The most commonly used bacteria for this sort of fermentations are *Propionibacterium*. Another essential feature of these bacteria is that they do not have the hydrogenase enzyme. As a result, in propionic acid-type fermentation, no hydrogen is produced. This type of fermentation produces propionate and acetate as end-products.

Ethanol-Type H_2 Fermentation In this fermentation method, carbohydrates are first converted to pyruvic acid by using *Saccharomyces cerevisiae*. The pyruvic acid is transformed to acetaldehyde and eventually ethanol or ethyl alcohol via the EMP or Entner-Doudoroff route (Kumar et al. 2023c). However, the major end-products of ethyl alcohol-type H_2 fermentation are ethanol (C_2H_5OH), CH_3COOH (acetic acid), H_2, CO_2, and a little amount of butyric acid and succinic acid.

20.3.6 Two Stage Process (Integration of Dark and Photo Fermentation)

Two-stage fermentation is a hydrogen generation method that involves dark and light fermentation stages (refer to Fig. 20.6). This is the latest technique of H_2 production, which can effectively enhance the H_2 production. The fermentation broth is rich in volatile fatty acids (VFAs) that may be utilized for light fermentation, reducing the inhibitory effect of VFAs on hydrogen generation by dark fermentation. Furthermore, photosynthetic bacteria can lower the (COD) chemical oxygen demand value of wastewater by using organic acids in light fermentation. The microorganisms have different role in their respective environments. In the first stage, dark fermentative bacteria produce H_2 and soluble organic metabolites, and in second stage light fermentative bacteria (which depends on light) metabolize these small organic acid molecules to produce H_2 (Argun et al. 2009). The integrated method produces 12 moles H_2 per mole of glucose, which is maximum theoretical yield. However, if the dark fermentation broth is utilized directly for light fermentation, it leads to some problems such as production of excessive ammonia and high volatile acid concentrations during the process. Due to these reasons, the efficiency

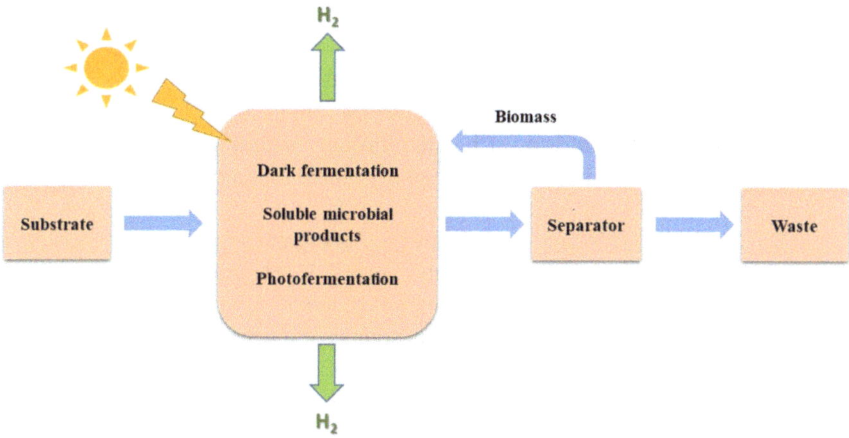

Fig. 20.6 Scheme for combined dark and photofermentation process for H_2 production from biomass

of H_2 production decreases. Before using dark fermentation broth for light fermentation to produce H_2, reasonable precautions should be taken to keep the concentrations of these chemicals within a particular range (Jeong and Kim 1999).

20.3.7 Biocatalyzed Electrolysis

Biocatalyzed electrolysis is a type of electrolytic process. In this process, the organic material (which is substrate) gets oxidized at a biological anode which is electrically connected via proton reduction at the cathode. This results in the formation of hydrogen gas. The conversion of organic material to hydrogen can be described by two half reactions, viz., (i) firstly, the transformation of organic substrates into electrons, protons, and bicarbonate and (ii) secondly, the assimilation of protons and electrons to produce H_2 gas. For example, in the case of biocatalyzed electrolysis of acetate (as substrate) is given below:

$$\text{Anode}: CH_3COO^- + 4H_2O \rightarrow 2HCO_3^- + 9H^+ + 8e^-$$

$$\text{Cathode}: 8H^+ + 8e^- \rightarrow 4H_2$$

$$\text{Overall}: CH_3COO^- + 4H_2O \rightarrow 2HCO_3^- + H^+ + 4H_2.$$

Biocatalyzed electrolysis is an alternate method for biohydrogen production through wastewater treatment. The desired salient features of this method are as follows:

(a) It must able to convert dissolved organic compounds, which is byproducts of dark fermentation.

(b) It must cope with the endothermic character of many dissolved organic compounds for H_2 production.
(c) Hypothetically, biocatalyzed electrolysis should involve the mixed cultures that can grow on moderate temperature. The one thing is to keep in mind while designing them is that it must happen through natural selection and must be capable of transforming a broad range of soluble organic compounds to inculcate the possibility of hydrogen production from wastewaters (Roskoski et al. 1975).Typically, a biocatalyzed electrolysis of organic material has a theoretical electromotive force (EMF) of roughly −0.12 V. The negative EMF suggests that drive this process's electrical energy is required. A power supply is included in the electrical circuit to provide this source of electrical energy. Essentially, throughout the process, electrolysis of dissolved organic material is resulted into hydrogen gas and bicarbonate, with electrochemically active microorganisms acting as the catalyst, hence the name biocatalyzed electrolysis. Overall, photoheterotrophic fermentations and biocatalyzed electrolysis both involve conversion of organic material through endothermic processes. However, biocatalyzed electrolysis depends on electrical power instead of sunlight to provide energy input.

Biocatalyzed electrolysis, the method discussed in this book chapter, has all of these features. It calculates the needed energy input using electrical energy rather than sunlight. The use of mixed consortia of electrochemically active microorganisms is critical to the operation of biocatalyzed electrolysis systems (Kumar et al. 2023b, c). Extracellular transfer of electron of the cell is possible in electrochemically active microorganisms. This mixotrophic microbial culture has capability to utilize electrode as the acceptor of electron while growing on surface of electrode for the oxidation of dissolved organic molecules in wastewater. During this process, the electrochemically active microbes release high energy electrons that result in a low electrode potential at the electrode. The generation of low potential on electrodes helps in production of hydrogen in the biocatalyzed electrolysis. Therefore it becomes indispensable to have detailed understanding of exocellular electron transfer by electrochemically active microorganisms in order to understand the biocatalyzed electrolysis (Stellwagen and Babul 1975).

A typical biocatalyzed electrolysis device shares structural similarities with microbial fuel cells (MFCs), consisting of two chambers separated by a cation exchange membrane. However, as compared to MFCs, two significant changes are necessary to initiate biocatalyzed electrolytic reactions: (i) oxygen must be kept away from the cathode in order to make the process entirely anaerobic and (ii) the electrical circuit must have a power source in order to deliver requisite electrical energy. This low energy need is a key benefit of biocatalyzed electrolysis over water electrolysis, as the cost of power is a large component of the cost of hydrogen generated by water electrolysis. Biocatalyzed electrolysis offers several significant advantages over other biological hydrogen production processes. Biological anodes may readily be operated under non-sterile circumstances because electrochemically active consortia can be spontaneously chosen from a wide variety of inoculum

(Rozendal et al. 2006). In biocatalyzed electrolysis, H_2 can be produced as a pure gas within the cathode chamber, contrasting with conventional H_2-producing bioreactors where a mixture of H_2 and CO_2 is typically generated. Biocatalyzed electrolysis might possibly replace dark fermentation in many circumstances, in addition to functioning as an effective second stage process to dark fermentation (e.g., for complex wastewaters).

20.4 Overall Comparisons of Hydrogen Production Methods and Their Critical Analysis

The progress of the global civilization depends on the availability of energy. Fossil energy has made significant contributions to the world's fast industrialization, but it has also resulted in many issues, including pollution, resource waste, ecological degradation, and a public health crisis. Global climate change and environmental degradation have grown increasingly significant as the number of resource-based businesses has increased, the social economy has developed rapidly, and the need for energy has increased. These eventualities have fuelled research in renewable energy. Hydrogen is definitely a potential renewable and clean alternate fuel, but there are technological obstacles in the generation, storage, transportation, and application of hydrogen as petroleum substitute on commercial scale. Table 20.4 compares the merits and demerits of various conventional methods for hydrogen production. As noted earlier, hydrogen is mostly produced in industry by steam reforming of methane (SMR) and reforming of electrolyzed water. However, these technologies not only extensively utilize primary energy but also pollute the environment. Thus, there is a necessity to develop a more capable, clean, and cheap hydrogen production process. Biological hydrogen production, which includes direct bio-decomposition, indirect bio-decomposition, photofermentation, and dark fermentation, has become a potential H_2 production process in recent times. Table 20.5 provides a summary of advantages and disadvantages of different biological methods for hydrogen production. Biological hydrogen production can generate clean hydrogen energy from industrial waste and forest or agricultural waste as well as home sewage in cities, rendering it the greatest solution to reduce pollution and produce clean fuel (Luongo et al. 2017). The advantages of biological hydrogen generation include various substrates, moderate H_2 production conditions, and a rapid hydrogen production rate. Meanwhile, the issues of high hydrogen production costs, low energy conversion efficiency, and less substrate conversion rate remain unsolved (Nikolaidis and Poullikkas 2017).

Table 20.4 Comparison on hydrogen production from different conventional methods

Process	Advantages	Disadvantages
Hydrocarbon pyrolysis	Free from harmful emission of gases	Carbon by-product, dependence on non-conventional fuels
Steam methane reforming(SMR)	Most advanced technology, infrastructure developed and available	Emission of greenhouse gases, dependence on non-conventional fuels
Partial oxidation(POX)	Developed technology and infrastructure	Emission of CO_2, dependence on non-renewable sources
Autothermal reforming (ATR)	Developed technology and infrastructure	Emission of harmful gases like CO_2 and CO, dependence on non-renewable resources

20.5 Hydrogen Storage, Transportation, and Utilization

Storage of hydrogen is a critical challenge in H_2-producing systems, particularly, when using hydrogen energy on a large scale. Hydrogen storage applications may be split into two categories: mobile and stationary applications. Stationary storage systems are mostly used for on-site storage at the place of manufacture as well as static energy generation (shown in Fig. 20.7). Mobile applications are designed to carry hydrogen from a storage location to a vehicle or to utilize hydrogen in automobiles. Because hydrogen has a lower heating value, it requires a very large storage tank. To overcome these issues, at least one of the three qualities listed below must be present in order to store enough hydrogen: high storage pressure, low storage temperature, or the use of a substance that attracts hydrogen molecules. Physical and material-based hydrogen storage systems are the most common types (refer to flow chart in Fig. 20.8). Physical hydrogen storage comprises compressed gas storage, cold/cyro-compressed hydrogen storage, and liquid hydrogen storage. Two subgroups of material-based hydrogen storage technologies are chemisorption and physisorption. Hydrogen distribution has an impact on the cost, energy consumption, and emissions related with hydrogen channels. There are two key phases for end users in the event of centralized hydrogen generation: transmission of hydrogen from factories to main city gates and transportation of fuel from main city gates to fueling stations end consumers, etc. The three primary channels for distribution are gaseous-based hydrogen delivery, liquid-based hydrogen delivery, and material-based hydrogen carriers, all of which are dependent on different sorts of storage systems. Nowadays, hydrogen can be used in each and every sector industries such as chemical industry, fertilizer industry, metal industries, and power industries (Moradi and Groth 2019). The economic benefits of application of hydrogen in diverse sectors are summarized in Fig. 20.9.

Table 20.5 Comparative study on different biological hydrogen production technologies

Methods	Advantages	Disadvantages	Outlook
Dark fermentation	Waste resources can be utilized in a number of ways. The reactor is simple to use and does not require sterilizing. Immobilized mixed culture produces a higher gas yield	Strict anaerobic conditions	Improvement of metabolic inherent limitations through metabolic engineering. The two-step hydrogen method generates greater energy while lowering chemical oxygen requirement
Photofermentation	Different waste resources can be used. H_2 may be recovered from dark fermentation waste liquor	Need of photo biochemical reactor. Nitrogenase is inefficient in producing hydrogen. Low efficiency of photosynthetic conversion. The surface area is quite large	Metabolic engineering enhanced strain performance. Nitrogenase was replaced by hydrogenase. Mutants are currently being researched. Advancement in materials science and technology
Two-stage process (dark photofermentation)	Waste substrate can be utilized in a variety of ways. At a later stage, the dark fermentative inhibitor VFAs (volatile fatty acids) were transformed to hydrogen. A continuous process might result in higher hydrogen generation	Bacteria must be controlled or sterilized. Reaction conditions are controlled by temperature and biomass. Control complexity is converted to dark light conditions	The characteristics of transition substrates are characterized and quantified. Reaction conditions must be well controlled. Process control and reactor design on a continuous basis
Thermochemical	Applicable to all types of solid waste, lower harmful, and greenhouse gas emissions	High-pressure and high-temperature extremes. A large quantity of catalysts is required. Separating products is a challenging task	Optimization of reaction conditions and reactor size. Catalyst research and development. Protect nature and control pollution caused by by-products

20.6 Conclusions and Future Perspectives

There are several hydrogen generation processes available today, including conventional techniques of hydrogen production from non-renewable resources or fossil fuels, classical water electrolysis, gasification, pyrolysis of biomass, and the use of novel photocatalysts to break water molecules and make H_2 fuel. These technologies, however, have issues of high power consumption and limited efficiency.

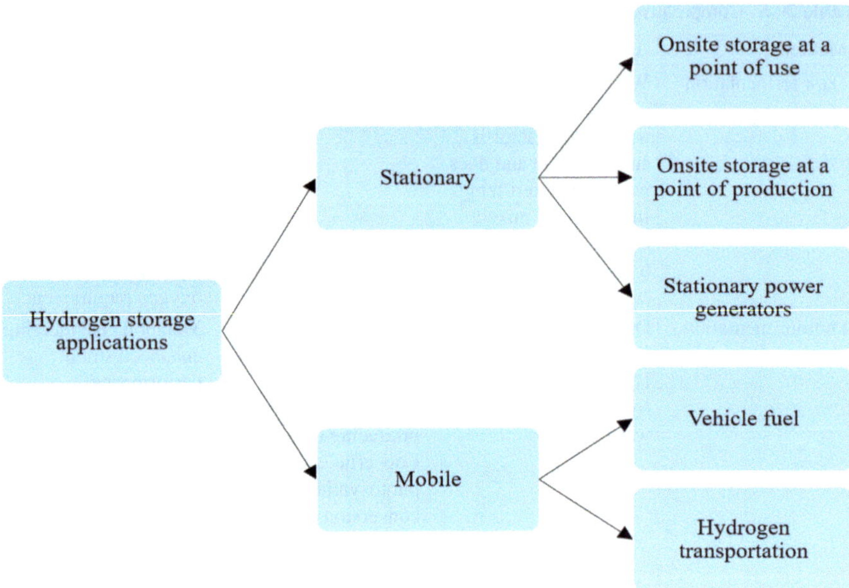

Fig. 20.7 Different types of hydrogen storage applications

Alternative processes must be developed in order to minimize reliance on non-renewable resources so that merchandises can meet rising H_2 demand, at least in the transportation sector in the coming years. Thermal conversion of biomass offers several advantages over fermentative hydrogen generation technology, including cleanliness, energy savings, and the absence of mineral-based resources. Biomass plays an increasingly prominent position in the development of the energy sector as a type of renewable energy. Although substantial progress has been achieved in the field of biohydrogen generation, the conditions for realistic large-scale commercial production remain unmet. The main issues are as follows: (1) genetic and metabolic features of present fermentative hydrogen-producing microorganisms, which

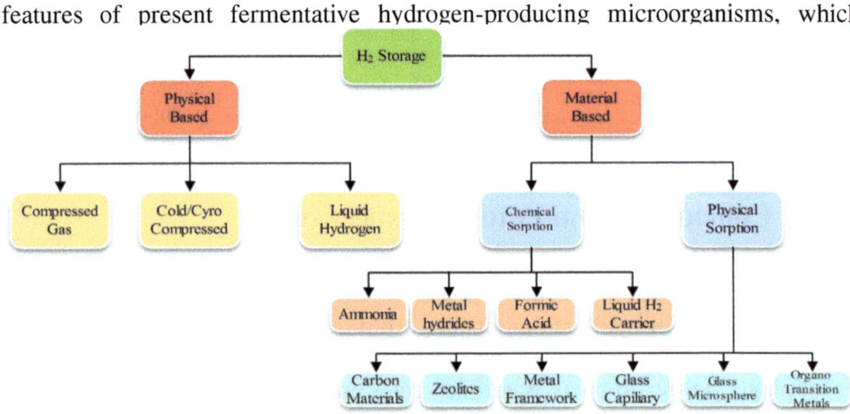

Fig. 20.8 Hydrogen (H_2) storage methods

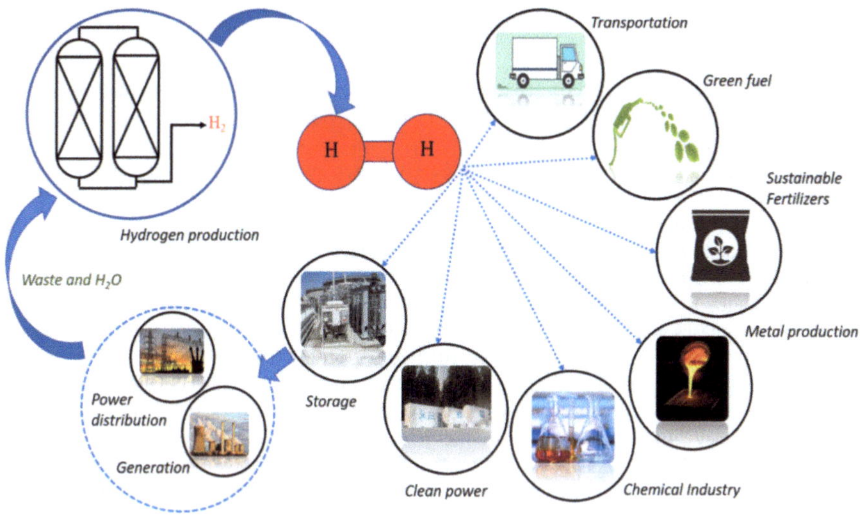

Fig. 20.9 Benefits of hydrogen (H_2) economy

determine the hydrogen production efficiency, are insufficient to meet hydrogen energy need. (2) Anaerobic fermentation produces volatile fatty acids (VFAs) or organic acids during the H_2 generation process, which lowers the pH of the reaction system and inhibits the fermentation process. The normal growth and metabolism of natural gas limits hydrogen generation capacity; (3) presence of oxygen inhibits hydrogenase enzymes in the H_2 production process through biophotolysis of water.

The following aspects should be prioritized in the future research and development in biological hydrogen production: (1) novel strain development using genetic engineering and screening of efficient natural hydrogen-producing strains; (2) novel immobilization methods for microbial strains; (3) the extraction of hydrogen production enzymes and genes from highly efficient H_2-producing bacteria using metabolic engineering; and (4) research on the enzymes involved in the metabolic pathway of hydrogen generation. To conclude: hydrogen holds high potential as a clean alternative to fossil fuels. The advent of commercial production of hydrogen through renewable and sustainable routes will mark new era of hydrogen energy that will provide simultaneous solutions to the issues of energy security and environmental pollution leading to climate change.

Acknowledgments One of the authors, Mr. Karan Kumar, is grateful to the Prime Minister's Research Fellowship provided by the Ministry of Education, Government of India.

Compliance with Ethical StandardsWe confirm that we have no conflict of interest to declare, and this work is original and has not been published elsewhere, or is it currently under consideration for publication elsewhere.

References

Adamson K-A, Pearson P (2000) Hydrogen and methanol: a comparison of safety, economics, efficiencies and emissions. J Power Sources 86:548–555. https://doi.org/10.1016/S0378-7753(99)00404-8

Adhikari S, Fernando SD, Haryanto A (2008) Hydrogen production from glycerin by steam reforming over nickel catalysts. Renew Energy 33:1097–1100. https://doi.org/10.1016/j.renene.2007.09.005

Amin AM, Croiset E, Epling W (2011) Review of methane catalytic cracking for hydrogen production. Int J Hydrog Energy 36:2904–2935. https://doi.org/10.1016/j.ijhydene.2010.11.035

Appel J, Schulz R (1998) Hydrogen metabolism in organisms with oxygenic photosynthesis: hydrogenases as important regulatory devices for a proper redox poising? J Photochem Photobiol B Biol 47:1–11. https://doi.org/10.1016/S1011-1344(98)00179-1

Argun H, Kargi F, Kapdan IK (2009) Effects of the substrate and cell concentration on bio-hydrogen production from ground wheat by combined dark and photo-fermentation. Int J Hydrog Energy 34:6181–6188. https://doi.org/10.1016/j.ijhydene.2009.05.130

Aydin MI, Karaca AE, Qureshy AMMI, Dincer I (2021) A comparative review on clean hydrogen production from wastewaters. J Environ Manag 279:111793. https://doi.org/10.1016/j.jenvman.2020.111793

Balachandar G, Varanasi JL, Singh V, Singh H, Das D (2020) Biological hydrogen production via dark fermentation: a holistic approach from lab-scale to pilot-scale. Int J Hydrog Energy 45:5202–5215. https://doi.org/10.1016/j.ijhydene.2019.09.006

Balat H, Kırtay E (2010) Hydrogen from biomass – present scenario and future prospects. Int J Hydrog Energy 35:7416–7426. https://doi.org/10.1016/j.ijhydene.2010.04.137

Chen H, Lee H, Chen S, Chao Y, Chang M (2008) Review of plasma catalysis on hydrocarbon reforming for hydrogen production—interaction, integration, and prospects. Appl Catal B Environ 85:1–9. https://doi.org/10.1016/j.apcatb.2008.06.021

Das D (2001) Hydrogen production by biological processes: a survey of literature. Int J Hydrog Energy 26:13–28. https://doi.org/10.1016/S0360-3199(00)00058-6

Dzulkarnain ELN, Audu JO, Wan Dagang WRZ, Abdul-Wahab MF (2022) Microbiomes of biohydrogen production from dark fermentation of industrial wastes: current trends, advanced tools and future outlook. Bioresour Bioproc 9:16. https://doi.org/10.1186/s40643-022-00504-8

Ersoz A (2008) Investigation of hydrocarbon reforming processes for micro-cogeneration systems. Int J Hydrog Energy 33:7084–7094. https://doi.org/10.1016/j.ijhydene.2008.07.062

Fan Y-T, Zhang G-S, Guo X-Y, Xing Y, Fan M-H (2006a) Biohydrogen-production from beer lees biomass by cow dung compost. Biomass Bioenergy 30:493–496. https://doi.org/10.1016/j.biombioe.2005.10.009

Fan Y-T, Zhang Y-H, Zhang S-F, Hou H-W, Ren B-Z (2006b) Efficient conversion of wheat straw wastes into biohydrogen gas by cow dung compost. Bioresour Technol 97:500–505. https://doi.org/10.1016/j.biortech.2005.02.049

Fang H, Li C, Zhang T (2006) Acidophilic biohydrogen production from rice slurry. Int J Hydrog Energy 31:683–692. https://doi.org/10.1016/j.ijhydene.2005.07.005

Fang HHP, Liu H (2002) Effect of pH on hydrogen production from glucose by a mixed culture. Bioresour Technol 82:87–93. https://doi.org/10.1016/S0960-8524(01)00110-9

Fay P (1970) Photostimulation of nitrogen fixation in *Anabaena cylindrica*. Biochimica et Biophysica Acta (BBA). Bioenergetics 216:353–356. https://doi.org/10.1016/0005-2728(70)90226-4

Gadhe A, Sonawane SS, Varma MN (2013) Optimization of conditions for hydrogen production from complex dairy wastewater by anaerobic sludge using desirability function approach. Int J Hydrog Energy 38:6607–6617. https://doi.org/10.1016/j.ijhydene.2013.03.078

Galadima A, Muraza O (2015) From synthesis gas production to methanol synthesis and potential upgrade to gasoline range hydrocarbons: a review. J Nat Gas Sci Eng 25:303–316. https://doi.org/10.1016/j.jngse.2015.05.012

Gangadharan P, Kanchi KC, Lou HH (2012) Evaluation of the economic and environmental impact of combining dry reforming with steam reforming of methane. Chem Eng Res Des 90:1956–1968. https://doi.org/10.1016/j.cherd.2012.04.008

Han W, Yan Y, Shi Y, Gu J, Tang J, Zhao H (2016) Biohydrogen production from enzymatic hydrolysis of food waste in batch and continuous systems. Sci Rep 6:38395. https://doi.org/10.1038/srep38395

Holladay JD, Hu J, King DL, Wang Y (2009) An overview of hydrogen production technologies. Catal Today 139:244–260. https://doi.org/10.1016/j.cattod.2008.08.039

Hosseini SE, Wahid MA (2016) Hydrogen production from renewable and sustainable energy resources: promising green energy carrier for clean development. Renew Sust Energ Rev 57:850–866. https://doi.org/10.1016/j.rser.2015.12.112

Hou K, Hughes R (2001) The kinetics of methane steam reforming over a Ni/α-Al2O catalyst. Chem Eng J 82:311–328. https://doi.org/10.1016/S1385-8947(00)00367-3

Hussy I, Hawkes F, Dinsdale R, Hawkes D (2005) Continuous fermentative hydrogen production from sucrose and sugarbeet. Int J Hydrog Energy 30:471–483. https://doi.org/10.1016/j.ijhydene.2004.04.003

Jeong G-S, Kim B-W (1999) The influence of light/dark cycle at low light frequency on the desulfurization by a photosynthetic microorganism. J Biosci Bioeng 87:481–488. https://doi.org/10.1016/S1389-1723(99)80097-8

Kapdan IK, Kargi F (2006) Bio-hydrogen production from waste materials. Enzym Microb Technol 38:569–582. https://doi.org/10.1016/j.enzmictec.2005.09.015

Kothari R (2017) A critical review on factors influencing fermentative hydrogen production. Front Biosci 22:1195–1220. https://doi.org/10.2741/4542

Kumar K, Moholkar VS (2023) Mechanistic aspects of enhanced kinetics in Sonoenzymatic processes using three simultaneous approaches. In: Moholkar VS, Mohanty K, Goud VV (eds) Sustainable energy generation and storage. Springer Nature Singapore, Singapore, pp 41–57

Kumar K, Roy K, Moholkar VS (2021) Mechanistic investigations in sonoenzymatic synthesis of n-butyl levulinate. Process Biochem 111:147–158. https://doi.org/10.1016/j.procbio.2021.09.005

Kumar K, Shah H, Moholkar VS (2022a) Genetic algorithm for optimization of fermentation processes of various enzyme productions. In: Optimization of sustainable enzymes production. Chapman and Hall/CRC

Kumar K, Mohammad AS, Pragati et al (2022b) Computational investigations in inhibition of alcohol/aldehyde dehydrogenase in lignocellulosic hydrolysates. bioRxiv 2022.11.19.517192; https://doi.org/10.1101/2022.11.19.517192

Kumar K, Patro P, Raut U, Yadav V, Barbora L, Moholkar VS (2023a) Elucidating the molecular mechanism of ultrasound-enhanced lipase-catalyzed biodiesel synthesis: a computational study. Biomass Conv Bioref. https://doi.org/10.1007/s13399-023-04742-4

Kumar K, Jadhav SM, Moholkar VS (2023b) Acetone-Butanol-Ethanol (ABE) fermentation with Clostridial Co-cultures for Enhanced Biobutanol Production. bioRxiv 2023.12.08.570763. https://doi.org/10.1101/2023.12.08.570763

Kumar K, Barbora L, Moholkar VS (2023c) Genomic insights into clostridia in bioenergy production: Comparison of metabolic capabilities and evolutionary relationships. Biotechnology and Bioengineering, 1–16. https://doi.org/10.1002/bit.28610

Kumar N, Das D (2000) Enhancement of hydrogen production by *Enterobacter cloacae* IIT-BT 08. Process Biochem 35:589–593. https://doi.org/10.1016/S0032-9592(99)00109-0

Kumar N, Das D (2001) Continuous hydrogen production by immobilized *Enterobacter cloacae* IIT-BT 08 using lignocellulosic materials as solid matrices. Enzym Microb Technol 29:280–287. https://doi.org/10.1016/S0141-0229(01)00394-5

Lamb JJ, Pollet BG (2020) Future prospects of selected hydrogen and biomass energy technologies. In: Hydrogen, biomass and bioenergy. Elsevier, pp 155–162

Laosiripojana N, Assabumrungrat S (2007) Catalytic steam reforming of methane, methanol, and ethanol over Ni/YSZ: the possible use of these fuels in internal reforming SOFC. J Power Sources 163:943–951. https://doi.org/10.1016/j.jpowsour.2006.10.006
Lee DH (2015) Hydrogen production via the Kværner process and plasma reforming. In: Compendium of hydrogen energy. Elsevier, pp 349–391
Lee K, Hsu Y, Lo Y, Lin P, Lin C, Chang J (2008) Exploring optimal environmental factors for fermentative hydrogen production from starch using mixed anaerobic microflora. Int J Hydrog Energy 33:1565–1572. https://doi.org/10.1016/j.ijhydene.2007.10.019
Lin C, Cheng C (2006) Fermentative hydrogen production from xylose using anaerobic mixed microflora. Int J Hydrog Energy 31:832–840. https://doi.org/10.1016/j.ijhydene.2005.08.010
Lin C, Chang C, Hung C (2008) Fermentative hydrogen production from starch using natural mixed cultures. Int J Hydrog Energy 33:2445–2453. https://doi.org/10.1016/j.ijhydene.2008.02.069
Lo Y-C, Chen W-M, Hung C-H, Chen S-D, Chang J-S (2008) Dark H2 fermentation from sucrose and xylose using H2-producing indigenous bacteria: feasibility and kinetic studies. Water Res 42:827–842. https://doi.org/10.1016/j.watres.2007.08.023
Luongo V, Ghimire A, Frunzo L, Fabbricino M, d'Antonio G, Pirozzi F, Esposito G (2017) Photofermentative production of hydrogen and poly-β-hydroxybutyrate from dark fermentation products. Bioresour Technol 228:171–175. https://doi.org/10.1016/j.biortech.2016.12.079
Malani RS, Umriwad SB, Kumar K, Goyal A, Moholkar VS (2019) Ultrasound–assisted enzymatic biodiesel production using blended feedstock of non–edible oils: kinetic analysis. Energy Convers Manag 188:142–150. https://doi.org/10.1016/j.enconman.2019.03.052
Martavaltzi CS, Pampaka EP, Korkakaki ES, Lemonidou AA (2010) Hydrogen production via steam reforming of methane with simultaneous CO_2 capture over CaO–$Ca_{12}Al_{14}O_{33}$. Energy Fuel 24:2589–2595. https://doi.org/10.1021/ef9014058
Melis A, Zhang L, Forestier M, Ghirardi ML, Seibert M (2000) Sustained Photobiological hydrogen gas production upon reversible inactivation of oxygen evolution in the green alga *Chlamydomonas reinhardtii*. Plant Physiol 122:127–136. https://doi.org/10.1104/pp.122.1.127
Moradi R, Groth KM (2019) Hydrogen storage and delivery: review of the state of the art technologies and risk and reliability analysis. Int J Hydrog Energy 44:12254–12269. https://doi.org/10.1016/j.ijhydene.2019.03.041
Mudhoo A, Kumar S (2013) Effects of heavy metals as stress factors on anaerobic digestion processes and biogas production from biomass. Int J Environ Sci Technol 10:1383–1398. https://doi.org/10.1007/s13762-012-0167-y
Muradov N (1993) How to produce hydrogen from fossil fuels without CO2 emission. Int J Hydrog Energy 18:211–215. https://doi.org/10.1016/0360-3199(93)90021-2
Muradov N, Vezirolu T (2005) From hydrocarbon to hydrogen? Carbon to hydrogen economy. Int J Hydrog Energy 30:225–237. https://doi.org/10.1016/j.ijhydene.2004.03.033
Nikolaidis P, Poullikkas A (2017) A comparative overview of hydrogen production processes. Renew Sust Energ Rev 67:597–611. https://doi.org/10.1016/j.rser.2016.09.044
Ntaikou I, Gavala HN, Lyberatos G (2009) Modeling of fermentative hydrogen production from the bacterium *Ruminococcus albus*: definition of metabolism and kinetics during growth on glucose. Int J Hydrog Energy 34:3697–3709. https://doi.org/10.1016/j.ijhydene.2009.02.057
O-Thong S, Prasertsan P, Karakashev D, Angelidaki I (2008) Thermophilic fermentative hydrogen production by the newly isolated Thermoanaerobacterium thermosaccharolyticum PSU-2. Int J Hydrog Energy 33:1204–1214. https://doi.org/10.1016/j.ijhydene.2007.12.015
Pattra S, Sangyoka S, Boonmee M, Reungsang A (2008) Bio-hydrogen production from the fermentation of sugarcane bagasse hydrolysate by Clostridium butyricum. Int J Hydrog Energy 33:5256–5265. https://doi.org/10.1016/j.ijhydene.2008.05.008
Ren N, Wang A, Gao L, Xin L, Lee D-J, Su A (2008) Bioaugmented hydrogen production from carboxymethyl cellulose and partially delignified corn stalks using isolated cultures. Int J Hydrog Energy 33:5250–5255. https://doi.org/10.1016/j.ijhydene.2008.05.020
Roskoski R, Lim CT, Roskoski LM (1975) Human brain and placental choline acetyltransferase: purification and properties. Biochemistry 14:5105–5110. https://doi.org/10.1021/bi00694a013

Rozendal R, Hamelers H, Euverink G, Metz S, Buisman C (2006) Principle and perspectives of hydrogen production through biocatalyzed electrolysis. Int J Hydrog Energy 31:1632–1640. https://doi.org/10.1016/j.ijhydene.2005.12.006

Shen J (2002) Influence of preparation method on performance of cu/Zn-based catalysts for low-temperature steam reforming and oxidative steam reforming of methanol for H2 production for fuel cells. Catal Today 77:89–98. https://doi.org/10.1016/S0920-5861(02)00235-3

Shin H (2004) Hydrogen production from food waste in anaerobic mesophilic and thermophilic acidogenesis. Int J Hydrog Energy 29:1355–1363. https://doi.org/10.1016/j.ijhydene.2003.09.011

Shinnar R (2003) The hydrogen economy, fuel cells, and electric cars. Technol Soc 25:455–476. https://doi.org/10.1016/j.techsoc.2003.09.024

Singh N, Kumar K, Goyal A, Moholkar VS (2022) Ultrasound-assisted biodiesel synthesis by in–situ transesterification of microalgal biomass: optimization and kinetic analysis. Algal Res 61:102582. https://doi.org/10.1016/j.algal.2021.102582

Smith G, Ewart G, Tucker W (1992) Hydrogen production by cyanobacteria. Int J Hydrog Energy 17:695–698. https://doi.org/10.1016/0360-3199(92)90089-F

Speight JG (2015) Gasification processes for syngas and hydrogen production. In: Gasification for synthetic fuel production. Elsevier, pp 119–146

Steinberg M (1989) Modern and prospective technologies for hydrogen production from fossil fuels. Int J Hydrog Energy 14:797–820. https://doi.org/10.1016/0360-3199(89)90018-9

Stellwagen E, Babul J (1975) Stabilization of the globular structure of ferricytochrome c by chloride in acidic solvents. Biochemistry 14:5135–5140. https://doi.org/10.1021/bi00694a018

Subramanian G (ed) (1998) Cyanobacterial biotechnology. Science Publications, Enfield, NH

Wu TY, Hay JXW, Kong LB, Juan JC, Jahim JM (2012) Recent advances in reuse of waste material as substrate to produce biohydrogen by purple non-sulfur (PNS) bacteria. Renew Sust Energ Rev 16:3117–3122. https://doi.org/10.1016/j.rser.2012.02.002

Yao C, Wang L, Liu Y, Wu G, Cao Y, Dai W, He H, Fan K (2006) Effect of preparation method on the hydrogen production from methanol steam reforming over binary Cu/ZrO catalysts. Appl Catal A Gen 297:151–158. https://doi.org/10.1016/j.apcata.2005.09.002

Zhang M-L, Fan Y-T, Xing Y, Pan C-M, Zhang G-S, Lay J-J (2007a) Enhanced biohydrogen production from cornstalk wastes with acidification pretreatment by mixed anaerobic cultures. Biomass Bioenergy 31:250–254. https://doi.org/10.1016/j.biombioe.2006.08.004

Zhang Z, Tay J, Show K, Yan R, Teeliang D, Lee D, Jiang W (2007b) Biohydrogen production in a granular activated carbon anaerobic fluidized bed reactor. Int J Hydrog Energy 32:185–191. https://doi.org/10.1016/j.ijhydene.2006.08.017

Chapter 21
Biohydrogen Downstream Processing and Storage Systems

Surjith Ramasamy and Kannan Pakshirajan

Abstract Biohydrogen is a clean and renewable source of energy, which has the potential to replace the conventional fossil fuels. However, purification and storage of biohydrogen are some important factors which need to be addressed for its application in energy sector. Biohydrogen produced using any biobased method consists of other gaseous byproducts, particularly carbon dioxide. Absorption, adsorption, permeation through membrane, and other methods are, therefore, widely used to obtain biohydrogen in pure form. Reliable storage systems for storing biohydrogen in the form of compressed hydrogen gas, liquid hydrogen, and solid or material-based systems are essential for energy supply, transportation, and other applications. Different and advanced purification and storage technologies with respect to biohydrogen are discussed in this chapter. Recent developments, including integrating biohydrogen production and purification, are also presented.

Keywords Biohydrogen · Hydrogen · Renewable energy · Purification · Storage · Downstream processing

21.1 Introduction

Biohydrogen is a sustainable and clean fuel due to its many advantages, such as biological synthesis route (direct bio-photolysis, indirect bio-photolysis, dark fermentation, microbial electrolysis, and light-fermentation), carbon neutral process, high calorific value (142–120 MJ/kg), direct generation of electricity from biohydrogen by fuel cells, and non-toxic product (H_2O) upon combustion (Nagarajan et al. 2017; Hallenbeck and Ghosh 2012). Despite all these advantages, purification

S. Ramasamy · K. Pakshirajan (✉)
Department of Biosciences and Bioengineering, Indian Institute of Technology Guwahati, Guwahati, India
e-mail: pakshi@iitg.ac.in

C. R. Soccol et al. (eds.), *Biohydrogen - Advances and Processes*, Biofuel and Biorefinery Technologies 13, https://doi.org/10.1007/978-3-031-49818-3_21

and storage of hydrogen often limits its large-scale applications (Bakonyi et al. 2013; Sivagurunathan et al. 2016). Purification is one of the most important step in biohydrogen production due to the complex gaseous mixture containing H_2 (35–65 vol. %), CO_2, H_2S, N_2, H_2O, etc., as well as high capital investment costs (50–80%). Moreover, continuous recovery of the produced gaseous mixture is required to avoid biomass growth inhibition and formation of other by-products such as ethanol, acetate, and butanol in the fermentation system. Due to the presence of high concentrations of impurities, biohydrogen produced from the fermentation process cannot be directly utilized in industrial combustion engine or fuel cells which require 75–80% pure H_2; therefore, new technologies are needed for further upgradation (White et al. 2006; Winter 2009). Hence, different downstream strategies have been developed to purify or upgrade biohydrogen from its mixture, which include pressure swing adsorption, chemical absorption, cryogenic distillation, membrane separation, etc. These technologies can be applied based on various factors such as gas mixture's composition, bioreactor productivity, purity of the biohydrogen, and the type and scale of operation (Rahman et al. 2016).

The design and configuration of a suitable downstream processing system along with the fermenter are required for continuous purification of biohydrogen and its storage (Bakonyi et al. 2014). Composition of the gas mixture plays a vital role in the design and configuration of these systems for biohydrogen purification (Lin et al. 2006). Table 21.1 summarizes the various production processes along with gas composition used in biohydrogen production. Among these biological routes, dark fermentation is advantageous for biohydrogen production due to its simple process control requirements, high volumetric productivity, high stability, etc. However, this technology also requires further hydrogen purification steps. The gas composition can change with the type of biological synthesis routes such as dark fermentation and photo-fermentation as well as depending upon the process parameters pH, seed microorganisms, temperature, micronutrients, substrate concentration, and macronutrients (Bakonyi et al. 2013; Lin et al. 2006; Vi et al. 2017). However, as the gas composition in the fermenter can change even at optimum process conditions, it affects the overall separation efficiency (Bakonyi et al. 2013). The selectivity of adsorbent and membrane, as well as their durability and operating conditions, are the main factors for considering biohydrogen purification in pressure swing adsorption, absorption, and membrane technology (Rahman et al. 2016; Liu et al. 2010; Ye et al. 2019). The development of downstream technologies that neither depend on gas composition nor require high operating costs for successful biohydrogen purification is of recent interest in the field of hydrogen production.

In addition to biohydrogen purification, effective storage of hydrogen is critical for its energy application. Hydrogen can be stored in different forms such as compressed gas, liquefied gas, cryogenic gas, or solid hydrogen (Stephenson 1947). Materials used and operating conditions in the storage system play an important role (González-Coloma et al. 2004). For proper choice of materials and operating conditions, several aspects need to be considered including storage, transportation, periodic inspection, gas quality management, and maintenance. This chapter presents an overview of various available downstream processes, problems associated

Table 21.1 Various production processes along with gas compositions used in biohydrogen production

Production technology	Mechanism	Gas composition (%)	Merits	Limitations
Dark fermentation	In dark fermentation, hydrogen is produced in the absence of solar light, oxygen, and water $\mathbf{C_6H_{12}O_6 + 6H_2O \rightarrow 6CO_2 + 12H_2}$	60–63% H_2 and 9–85.3% CO_2	No requirement of light energy and continuous use of carbohydrate	Need for separating CO_2 during the H_2 production process
Light fermentation	Fermentative conversion of organic substrate into H_2 by photosynthetic bacteria $\mathbf{CH_3COOH + 2H_2O \rightarrow 2CO_2 + 4H_2}$	4–57% H_2 and 45–72% CO_2	Removal of environmental pollutants	Light fermentation requires partial nitrogen condition
Direct bio-photolysis	Bio-hydrogen is produced directly from water by utilizing the micro-algae photosynthesis system $\mathbf{2H_2O + solar\ energy \rightarrow 2H_2 + O_2}$	Lower than 50% (H_2) and higher than 50% (CO_2)	Water is the only substrate to obtain H_2 and is not related to carbon metabolic pathways	Hydrogenase activity is highly sensitive to O_2, and therefore, it inhibits the hydrogen production
Indirect bio-photolysis	Cyanobacteria and microalgae generate hydrogen from stored starch and glycogen $\mathbf{C_6H_{12}O_6 + 6H_2O \rightarrow 6CO_2 + 12H_2}$	Lower than 50% (H_2) and higher than 50% (CO_2)	The fixed carbon source through the duration of cell growth	Difficulty in improving photosynthetic efficiency
Microbial electrolysis	Microbial electrolysis is a process for the production of green hydrogen from organic matter Anodic reaction $\mathbf{C_6H_{12}O_6 + 6H_2O \rightarrow 6CO_2 + 12H_2 + e^-}$ Cathodic reaction $\mathbf{2H^+ + 2e^- \rightarrow H_2}$	63–82% H_2 and 18–37% CO_2	Biocathodes do not require a mediator for hydrogen production	High cost of catalyst

with the operating conditions, and process performance. An overview of the hydrogen storage systems and their drawbacks with conventional storage system are also presented.

21.2 Biohydrogen Purification

Biohydrogen can be separated by pressure swing adsorption, membrane technology, and absorption. As mentioned earlier, separation efficiency and hydrogen purification are limited by adsorbent and membrane properties. Therefore, in the absence of any economic concern, the main design considerations of purification technology are as follows: efficient adsorbent utilization, high adsorption capacity, high diffusivity, and selectivity over conventional separation technologies.

21.2.1 Pressure Swing Adsorption

Pressure swing adsorption (PSA) is a widely used technology in large-scale applications to produce high purity biohydrogen (95–99.999%) from its mixture for successful application in fuel cells (Vasiliev et al. 2007). In this process, the various types of layers of adsorbents are generally used in columns for the purification of biohydrogen from the gaseous mixture; absorbents such as alumina or silica (water), activated carbon (CO_2), and zeolite (CO and N_2) are used (Basu et al. 2010). PSA system involves four steps: adsorption, depressurization, blowdown, and pressurization. Delgado et al. (2017) examined the use of HKUST-1 and ZIF-8 with polyvinyl alcohol adsorbent for the separation of biohydrogen from CO_2/H_2 mixture and obtained high purification efficiency. Similarly, Kuroda et al. (2018) utilized hydroxyl aluminum silicate clay-based adsorbent for biohydrogen purification from the multicomponent gas mixture (H_2, 33.0%; CH_4, 4.8%; CO_2, 12.0%; CO, 10.2%; H_2S, 200 ppm). Table 21.2 summarizes the various materials used in PSA along with their purification efficiency.

Several recent strategies have been studied to further enhance the process performance and decrease the cost of the system, including operation in series, low-cost adsorbent materials, formulation of structured adsorbents, PSA cycle, adsorbent selectivity, and reduction in column size (Ye et al. 2019; Delgado et al. 2017). Compact design and low-pressure drop are the key factors in PSA to produce H_2 in a continuous mode of operation. The development of novel and low-cost adsorbents for enhanced H_2 purification by PSA deals with equilibrium and kinetics of adsorption. Various research have been undertaken to develop novel adsorbents based on metal-organic frameworks (mesoporous materials) with a good adsorption capacity (Delgado et al. 2017). The understanding of equilibrium adsorption and kinetics of the biohydrogen production can be useful for the design of PSA for hydrogen purification. The main limitation of this technology is that pressure drop decreases the

Table 21.2 Various adsorbent materials used in pressure swing adsorption for biohydrogen purification

Column type	Material used	Gas composition	Operating conditions	Purification efficiency	Advantages
Single column	HKUST-1 and ZIF-8 with polyvinyl alcohol	40% H_2 and 60% CO_2	Pressures up to 30–50 bar and temperatures 298–338 K	99.993%	High selectivity toward CO_2
Single column	Activated carbon	CO 4000 ppm and CO_2 5 mol%	Pressure 5 atmosphere and temperature 302.15 K	99.999%	High selectivity toward CO and CO_2
Single column	Hydroxyl aluminum silicate clay	H_2, 33.0%; CH_4, 4.8%; CO_2, 12.0%; CO, 10.2%; H_2S, 200 ppm	Pressure 5.3–9.3 kPa and temperature 150 °C	98.4% CO_2	Suitable for CO_2 separation from gas mixtures
Double column	CuBTC	CO_2, 0.18; H_2, 0.78; CH_4, 0.04 (molar ratio)	Pressure 2 bar and temperature 303 K	99.99%	The adsorbent has high adsorption capacity for CO_2, CH_4, CO and N_2
Single column	Activated carbon	CO_2 30% and CO, 63 ppm	Pressure range 0.1–5 bar and temperature 98–423 K	99.981%	High adsorption capacity for CO_2 and CO

working capacity of the adsorption column, and it is applicable only for large-scale purification (Moran and Talu 2018). The scale-down of such a process is difficult for biohydrogen production in small-scale bioreactor level. Therefore, the research on novel adsorbent materials and their application in real-time biohydrogen systems should be continued for the development of an efficient and low-cost PSA system for the purification of biohydrogen from multicomponent mixtures.

21.2.2 Metal Hydride Absorption

Metal hydrides are selectively used to separate biohydrogen from its mixture. In this process, the gaseous mixture is supplied into the column, where biohydrogen is absorbed by the alloy bed with the production of metal hydride and heat transferred to a cooling liquid. Highly pure hydrogen is obtained through desorption from the column (Fig. 21.1) (Dunikov et al. 2016; Miura et al. 2013). Potential of metal hydride use for hydrogen separation from gas mixtures containing CO_2 were successfully utilized by Miura et al. (2013). Dunikov et al. (2016) successfully used metal hydride for the purification of hydrogen with a recovery efficiency of 94%

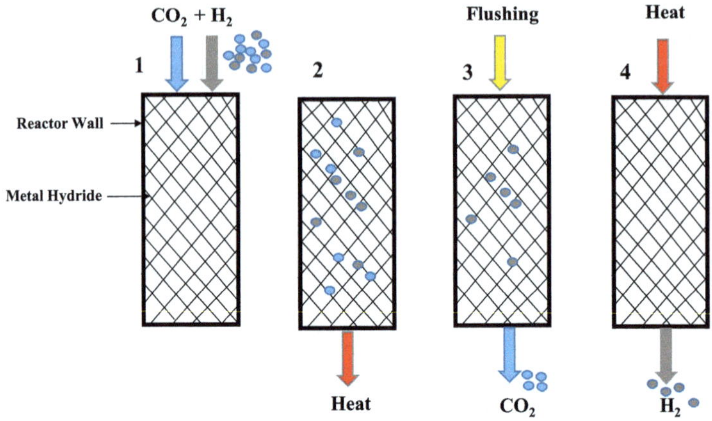

Fig. 21.1 Schematic showing metal hydride absorption for biohydrogen purification

from 59 vol % of hydrogen mixture. Similarly, Miura et al. (2012) reported the application of metal hydride for the purification of hydrogen with good separation efficiency, and the purified hydrogen was successfully utilized in a fuel cell. This finding demonstrated the successful application of metal hydride for biohydrogen purification in a small-scale bioreactor system and subsequent use of the pure biohydrogen in fuel cell for electricity generation. However, this technology is rarely applied in biohydrogen purification due to possible damage of metal hydride by H_2S poisoning, inert gas blanketing, retardation, and corrosion (water, vapor, CO_2) in large-scale systems.

21.2.3 Cryogenic Separation

Cryogenic separation is another widely used technology at large-scale to produce a highly pure biohydrogen from its mixture for application in fuel cells (Dunikov et al. 2016). In this method, feed is separated using heat exchanger and phase separator to increase the concentration of hydrogen. It is used to produce 90–98 vol % purity of biohydrogen from the multicomponent gas mixture. Aasadnia et al. (2021) utilized the cryogenic method for the purification of biohydrogen. It was observed that the power requirement (68.6%) drastically increased with an increase in the hydrogen concentration in the feed mixture (60.0%) as well as a decrease in the hydrogen recovery rate. In addition, it does not require any chemical additives (Miura et al. 2013). The main problem with this process is the phase change of CO_2 gas to liquid form at a pressure of 1 atm and H_2S separation (due to H_2S solidification under cryogenic conditions). Various strategies were followed to reduce its limitations at a large-scale level. An integrated system involving membrane and cryogenic separation is used to reduce the cost of operation for removing H_2S and

CO_2 from the multicomponent gas mixtures (Agarwal and Balasubramaniam 1996). Aasadnia et al. (2021) also utilized an integrated system of heat exchanger and phase separator followed by the cryogenic system to produce a hydrogen-rich gas steam from its mixture.

21.2.4 Membrane Technology

Membrane technology is attractive due to its selectivity, relative ease of scalability, portability, and diffusivity. The membrane technology is mainly based on three parameters: raw material selection for membrane fabrication, design, preparation of membrane modules, and membrane operating conditions. Various types of membranes can be used to separate biohydrogen from its mixture such as inorganic and organic membranes. The membranes used are further classified into two categories, i.e., H_2-selective/CO_2-rejective and CO_2-selective/H_2-rejective (Du et al. 2021; Hallenbeck and Ghosh 2009). These membranes were successfully used for the separation of individual gases (CO_2 and H_2) with high selectivity and purification efficiency. The main mechanism involved in membrane separation is solubility and diffusivity differences of penetrating gas (Bakonyi et al. 2013). However, the separation of biohydrogen from the multicomponent gas mixture is the main factor for the application of membrane technology for biohydrogen purification. For example, the presence of one gas molecule affects the permeability and concentration polarization of other gas molecules in the membrane separation (Budd et al. 2005). Several CO_2-selective membranes have been reported for the purification of hydrogen from mixed gases by membrane separation. Car et al. (2008) reported the use of PEBAX/PEG organic membranes for the purification of both pure (H_2, CO_2, N_2) and multicomponent gas mixture (H_2/CO_2). It was observed that purification efficiency drastically decreased with the multicomponent gas mixture than with pure gas. Similarly, Yave et al. (2009) reported that gas composition affects the purification efficiency of the membrane poly amide-b-ethylene oxide/polyethylene glycol. In contrast, Chen et al. (2010) obtained enhanced separation of CO_2/H_2 mixture in comparison with the pure gas by using PEO-based membranes. Table 21.3 shows the use of various types of membranes for purification of biohydrogen. The selectivity of polysulfone is high compared to other polymeric membranes reported in the literature. The major limitations of this technology are sensitivity to temperature and chemicals, and high membrane cost. Therefore, further research on the use of low-cost materials with low sensitivity to chemicals and temperature is needed for the development of an efficient and low-cost membrane system for the purification of biohydrogen from multicomponent gas mixtures.

Inorganic membranes prepared using silica, ceramic, and platinum/palladium alloys are used for separating hydrogen from the mixture of gases at higher temperature. Inorganic membranes offer 99.999% purity along with high mechanical and thermal strength. Cost of the membrane increases due to addition of high-value metals. Membranes coated with high density of palladium or its alloys are capable

Table 21.3 Various types of membranes used for the purification of biohydrogen

Type of membrane	Material used	Gas composition (%)	Operating conditions	Selectivity (H_2/CO_2)	References
Organic membrane	Polyimide	10/90 vol.% H_2/CO_2	Feed pressure, 4–6 bar; temperature, 30 °C; mixture composition	2.93	David et al. (2011)
Organic membrane	Polyethersulfone	CO_2/H_2: 50%/50% molar fraction	Temperature, 35 °C; total pressure, 20 atm	2.65	Li et al. (2008)
Organic membrane	UBE polyimide module	H_2/CO_2: 65 vol%/35 vol%	Feed pressure, 2.2 bar; temperature, 55 °C	1.62	Bakonyi et al. (2013)
Organic membrane	Polysulfone	39% H_2 and 49% CO_2	2 to 10 bar and room temperature	4.4	Hamid et al. (2019)

of attaining high purity hydrogen. Metallic membranes are subtle to gases such as carbon monoxide and hydrogen sulfide. Among inorganic membranes, separation flux of microporous membrane is directly proportional to the pressure, whereas thin film palladium membrane flux is proportional to the square root of pressure. Membranes with palladium are used in membrane reactor for hydrogen purification (Adhikari and Fernando 2006). Kajama et al. (2015) prepared inorganic membrane by impregnating platinum in silica membrane to separate hydrogen from mixture of gases. Silica membrane with platinum showed highest selectivity value of 2.93 for H_2/N_2 at 0.9 barg and 25 °C, whereas the selectivity value for H_2/He was 1.96 at 1.6 barg and 300 °C (Kajama et al. 2016).

21.2.5 *Integrated Strategies for Biohydrogen Production and Purification*

Reactors such as gas separation membrane bioreactors are proposed for hydrogen production and purification. Figure 21.2 shows the integrated strategy for hydrogen production, purification, and application. In this reactor, in situ or in line hydrogen separation unit is connected to the head space of the reactor to enrich the biohydrogen. Biochemicals and microbes from the production media may block the membrane and reduce its efficiency; water traps are essential for an uninterrupted process (Bakonyi et al. 2013; Bakonyi et al. 2017; Ramírez-Morales et al. 2013; Ramírez-Morales et al. 2015; Nor Azira and Umi Aisah 2019). High partial pressure of the hydrogen is inhibitory to the activity of microbes and enzymes, and, therefore, monitoring the hydrogen concentration and its removal from the head space is essential. Ramírez-Morales et al. (2013) showed that the continuous removal of

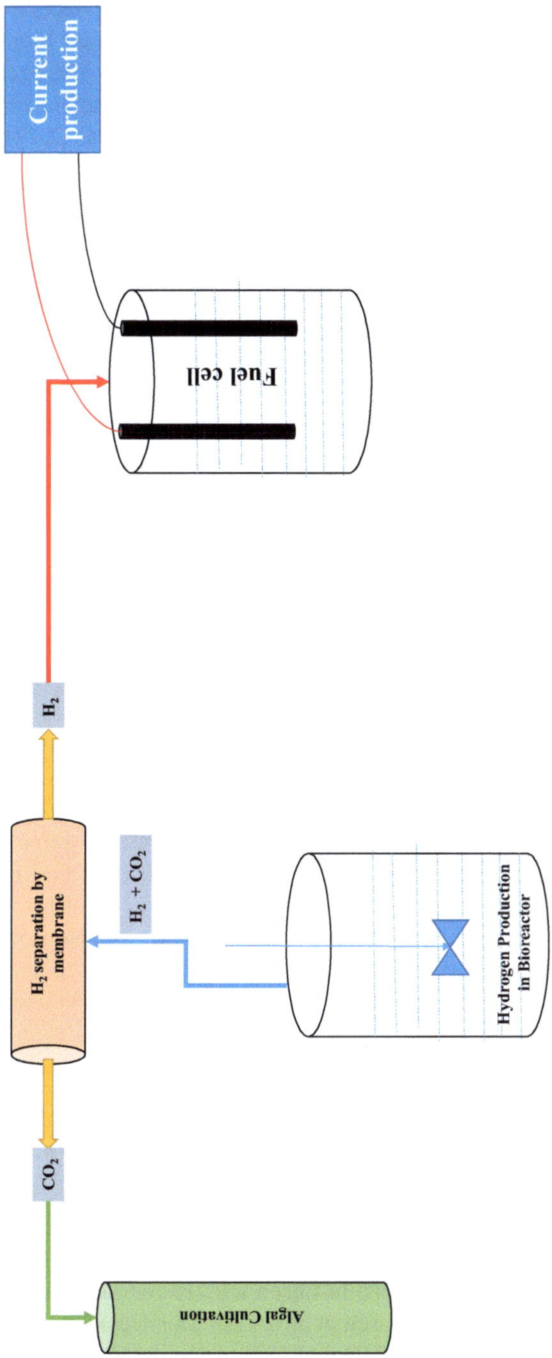

Fig. 21.2 Schematic showing integrated system for biohydrogen production and purification (Usman 2022)

hydrogen and recycling the CO_2 into the reactor reduces the inhibitory effect of hydrogen concentration (Ramírez-Morales et al. 2013). The separated waste CO_2 was used as a feed for algae in the photobioreactor, and the algal biomass produced was used as the feed in the dark fermentation step for biohydrogen production. Polydimethylsiloxane (PDMS) membrane integrated with the bioreactor was used to separate the biohydrogen from the bioreactor (Bakonyi et al. 2017).

Double membrane bioreactor is also used for continuous hydrogen production and purification. In continuous operation mode, the effect of different HRT on the performance of the microfilter in gas rejection, hydrogen productivity, and yield was studied. Decrease in HRT reduced the yield, whereas the hydrogen production rate increased. The hydrogen purity was 67.3 vol % in the off gas from the fermenter, treated using the in situ membrane, and the value was 30% higher than the untreated gas (Bakonyi et al. 2015). Bélafi-Bakó et al. (2006) used an integrated system for biohydrogen production and purification through a two-step gas separation process. Heterotrophic archaebacterium was used for producing the hydrogen in the fermenter. Porous and non-porous separators were used for removing hydrogen-nitrogen and hydrogen-carbon dioxide, respectively, from the feed. Through this method, gas mixture with 73% pure hydrogen was obtained (Bélafi-Bakó et al. 2006).

21.3 Biohydrogen Storage

Biohydrogen storage is important for the application of biohydrogen in various energy sectors. Biohydrogen is stored in different types of storage systems such as physical storage system (compressed, liquefied, cryogenic, and solid hydrogen storage) and material-based storage system (metal hydrides, adsorptive materials, and chemical storage materials). Each storage system has its own merits and demerits. Material-based storage system is of recent interest, due to advances in the development of porous materials and metal organic frameworks. However, no perfect hydrogen storage system is reported to be the best for industrial applications. The main characteristics of each storage technologies and their limitations are discussed in the following sections.

21.3.1 Liquid Hydrogen Storage

Liquefied hydrogen is usually produced at very low temperature (−253 °C). This storage system is widely used for medium- to large-scale storage (Moradi and Groth 2019). In addition, liquid hydrogen has been utilized for gas delivery using trucks with a capacity of 60,000 L (Hassen et al. 2021). This technique has a high gravimetric and volumetric density (70.9 kg-H_2/m^3) as compared to other storage system such as compressed (40 kg-H_2/m^3) and cryo-compressed storage (39.1 kg-H_2/m^3).

Moreover, this technique consumes high energy (2.343 kWh/L) compared to other storage system such as compressed (0.792 kWh/L) and cryo-compressed storage (1.34 kWh/L). The main limitation of these types of storage system is maintenance of low temperature. Therefore, this technology cannot be used for space and aviation applications. Total energy consumption for liquid hydrogen technologies is around 35% of the total energy content of the stored hydrogen due to the requirement of special types of vessel and effective insulation setup or layer to decrease the evaporation loss from the vessel during prolonged period of storage (Kiyobayashi et al. 2002; Yanxing et al. 2019).

21.3.2 Compressed Hydrogen Storage

Hydrogen has a low density at room temperature. Compressed hydrogen storage is widely used than liquefied hydrogen storage to store the hydrogen with a low density. The most common materials used are metallic (aluminum and steel), polymeric parts (polyamide or polyethylene-based polymers), and composite materials (carbon fiber, glass, aramid). However, the cost of the system (manufacturing and development) is the main limiting factor for its application. High pressure (700 to 1000 bar) is required for successful application of this system in vehicle market (AbuAisha et al. 2021). Total energy consumption for compressed hydrogen system is around 10% of the total energy content of the stored hydrogen.

Pressure vessels of compressed hydrogen storage are classified into five types such as fully metallic, metallic enclosure, metallic liner, fully composite and linerless pressure vessel. The common operating pressure varies from 25 MPa to 30 MPa for the application of compressed natural gas and oxygen-air for breathing apparatus. The design and configuration of a proper hydrogen storage system along with other factors, such as test pressure, lifetime, and safety, need to be considered for its successful application (Barthelemy et al. 2017). Composition of the material plays a vital role in the design and configuration of storage system. The selection of the hydrogen storage system is based on its final application. For industrial-scale operation, the operating pressure varies from 20 to 30 MPa in metallic cylinders, which have low storage (1 wt%) efficiency. In case of European standard, the mass storage efficiency should be 4.8 wt%, which can be achieved by using carbon fiber composite along with high operating pressure (70 MPa) (González-Coloma et al. 2004). Table 21.4 shows the comparison of various types of hydrogen storage systems.

21.3.3 Cryo-compressed Storage

Cryo-compressed storage system is developed to reduce boil off loss by combining both the cryogenic and compressed storage systems. Biohydrogen is stored at cryogenic temperatures (20 K) and high pressures (30 MPa) using an insulation tank.

Table 21.4 Comparison of various types of hydrogen storage system

Types of storage system	Storage system	Conditions	Storage density (kg-H_2/m^3)	Energy consumption (kWh/L)	References
Physical storage system	Liquefied hydrogen	20 K and atmospheric pressure or higher	70.9	2.343	Hassen et al. (2021); Faye et al. (2022)
Physical storage system	Compressed hydrogen	20 MPa to 700 MPa	40[a]	1.32[a]	Moradi and Groth (2019); Barthelemy et al. (2017)
Physical storage system	Cryo-compressed hydrogen	20 to 40 K and 0.4 MPa	39.1	0.792	Yanxing et al. (2019); Faye et al. (2022)

Hydrogen density at normal conditions-0.089 kg/m^3
[a]Hydrogen density at 700 Mpa

The Bayerische Motoren Werke (BMW) Group, Germany, successfully reported the use of cryo-compressed hydrogen storage for hydrogen-based vehicles with high mass storage efficiency (80 g/L) (Verhelst et al. 2006). The composite pressure vessel with metallic inner can be used for storing the cryo-compressed hydrogen system. Various studies have been carried out to check the effect of combined pressure and temperature on composite material tank. The major advantages of this storage system are flexibility and compatibility with liquid and gases.

21.3.4 Material-Based Storage System

Hydrogen storage in solid or material system offers various merits compared to the other storage systems due to its high volumetric density. There are various types of materials used in this storage system, including alkali metal borohydrides, metallic hydride, and complex hydride. Figure 21.3 shows the various types of commercially available hydrogen storage systems. The most important factor for selecting solid material for hydrogen storage is thermodynamic properties of the material (equilibrium conditions and its kinetics). The mechanism involved in this method is dissociation of hydrogen from the surface of the material followed by diffusion and binding with the metal based on ionic, covalent, or metallic interaction (Faye et al. 2022; Black and Black 2005).

Metals and alloys bind with the hydrogen to form hydrides; $LaNi_5H_6$, $ZrV_2H_{5.5}$, $CeNi_3H_4$, and $TiFeH_2$ are some of the hydrides used in this type of storage. The metallic hydrides include two parts: Part I is made up of rare earth metal or alkali earth metal to form stable hydride, and part II is made up of transition metals to form unstable hydrides to assist in hydrogen dissociation. The absorption and desorption of hydrides are done at normal temperature and pressure. Volumetric

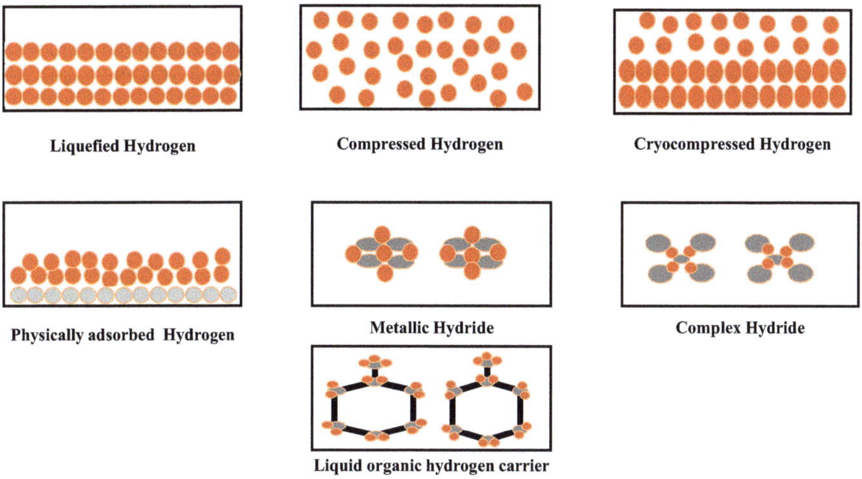

Fig. 21.3 Schematic showing various types of hydrogen storage systems (Hamid et al. 2019)

density of the hydride is very high; for example, $LaNi_5H_6$ has a capacity of 115 kg/m^3. The gravimetric density of hydrogen for metallic hydrides is less than 3 wt%, as this hydride works on the ambient temperature and pressure. Hydride formation is an exothermic reaction, and, therefore, the release of hydrogen from the metallic hydride requires heat (Zhou 2005).

In complex hydrides, elements belonging to groups I, II, and III are used and include magnesium, boron, and aluminum. The maximum volumetric density of the complex hydrides such as Mg_2FeH_6 and $Al(BH_4)_3$ is 150 kg/m^3. Complexes such as $NaAlH_4$ are capable of both absorption and desorption at normal temperature. The major drawback of this method is slow release of hydrogen from the complex as it follows cascade decomposition and special conditions for each and every step. Further research on adsorption and desorption cycles are essential for successful application of this storage system (Zhou 2005).

21.4 Safety and Regulations for Hydrogen Fuel

Physical, physiological, and chemical changes are the dangers associated with hydrogen, and safe handling and prevention of danger are essential for commercialization of hydrogen. International Standardization Organization (ISO) has formulated guidelines for safe handling of hydrogen as a fuel in the form of liquid and gas, and also the safety aspects of hydrogen. ISO 13984:1999 deals with design and installation of fueling and dispensing of hydrogen, and ISO 14687:2019 and ISO/CD 14687 deal with the quality and characteristics of the hydrogen fuel. ISO 16111:2018 deals with requirements for the material, design, construction, and testing of transportable hydrogen gas storage systems using metal hydride assemblies.

ISO/TR 15916:2015 is assigned for safety by identifying basic safety concerns, hazards, risks, and properties of hydrogen (Du et al. 2021). The European Industrial Gases Association (EIGA) has recommended separation distance for the hydrogen tank, which specifies the safe distance between the human premises and the storage tank to reduce any injury or accident. Similarly, EIGA recommends not to install liquid hydrogen inside the buildings, whereas installation under the ground requires special requirements. EIGA's guidelines involve avoiding copper, tin, alloys, and zinc for transporting hydrogen, as they are prone to corrosion in the presence of chlorine and ammonia (Aziz 2021).

21.5 Conclusion and Perspectives

This chapter dealt with different methods used for purification and storage of biohydrogen along with their advantages and disadvantages. Material-based storage systems provide an edge over other storage systems due to its temperature independent operation. The vast amount of research on biohydrogen production, downstream processing, storage, and applications are carried out individually, whereas integrated operations are in need for reducing the energy consumption and production cost. In addition to these, safety aspects for hydrogen production, storage, and development of sensors to monitor the hydrogen leakage are certain bottlenecks that need to be addressed for the commercialization of hydrogen.

References

AbuAisha M, Rouabhi A, Billiotte J, Hadj-Hassen F (2021) Non–isothermal two–phase hydrogen transport in rock salt during cycling in underground caverns. Int J Hydrog Energy 46:6632–6647. https://doi.org/10.1016/j.ijhydene.2020.11.152

Adhikari S, Fernando S (2006) Hydrogen membrane separation techniques. Ind Eng Chem Res 45:875. https://doi.org/10.1021/ie050644l

Agarwal A, Balasubramaniam R (1996) Role of surface passive films in the hydrogen embrittlement of iron aluminides. Bull Mater Sci 19:91–102. https://doi.org/10.1007/BF02744791

Aasadnia M, Mehrpooya M, Ghorbani B (2021) A novel integrated structure for hydrogen purification using the cryogenic method. J Clean Prod 278:123872

Aziz M (2021) Liquid hydrogen: a review on liquefaction, storage, transportation, and safety. Energies 14. https://doi.org/10.3390/en14185917

Bakonyi P, Nemestóthy N, Bélafi-Bakó K (2013) Biohydrogen purification by membranes: an overview on the operational conditions affecting the performance of non-porous, polymeric and ionic liquid based gas separation membranes. Int J Hydrog Energy 38:9673–9687. https://doi.org/10.1016/j.ijhydene.2013.05.158

Bakonyi P, Nemestóthy N, Simon V, Bélafi-Bakó K (2014) Fermentative hydrogen production in anaerobic membrane bioreactors: a review. Bioresour Technol 156:357–363. https://doi.org/10.1016/j.biortech.2014.01.079

Bakonyi P, Nemestóthy N, Lankó J, Rivera I, Buitrón G, Bélafi-Bakó K (2015) Simultaneous biohydrogen production and purification in a double-membrane bioreactor system. Int J Hydrog Energy 40:1690–1697. https://doi.org/10.1016/j.ijhydene.2014.12.002

Bakonyi P, Buitrón G, Valdez-Vazquez I, Nemestóthy N, Bélafi- Bakó K (2017) A novel gas separation integrated membrane bioreactor to evaluate the impact of self-generated biogas recycling on continuous hydrogen fermentation. Appl Energy 190:813–823. https://doi.org/10.1016/j.apenergy.2016.12.151

Barthelemy H, Weber M, Barbier F (2017) Hydrogen storage: recent improvements and industrial perspectives. Int J Hydrog Energy 42:7254–7262. https://doi.org/10.1016/j.ijhydene.2016.03.178

Basu S, Khan AL, Cano-Odena A, Liu C, Vankelecom IFJ (2010) Membrane-based technologies for biogas separations. Chem Soc Rev 39:750–768. https://doi.org/10.1039/b817050a

Bélafi-Bakó K, Búcsú D, Pientka Z, Bálint B, Herbel Z, Kovács KL, Wessling M (2006) Integration of biohydrogen fermentation and gas separation processes to recover and enrich hydrogen. Int J Hydrog Energy 31:1490–1495. https://doi.org/10.1016/j.ijhydene.2006.06.022

Black J, Black J (2005) Biological performance of materials. In: Fabbrizzi L, Poggi A (eds) Biological performance of materials. Springer, Berlin, pp 161–183. https://doi.org/10.1201/9781420057843

Budd PM, Msayib KJ, Tattershall CE, Ghanem BS, Reynolds KJ, McKeown NB, Fritsch D (2005) Gas separation membranes from polymers of intrinsic microporosity. J Memb Sci 251:263–269. https://doi.org/10.1016/j.memsci.2005.01.009

Car A, Stropnik C, Yave W, Peinemann KV (2008) Pebax®/polyethylene glycol blend thin film composite membranes for CO2 separation: Performance with mixed gases. Sep Purif Technol 62(1):110–117

Chen H, Xiao Y, Chung TS (2010) Synthesis and characterization of poly (ethylene oxide) containing copolyimides for hydrogen purification. Polymer 51(18):4077–4086

David OC, Gorri D, Urtiaga A, Ortiz I (2011) Mixed gas separation study for the hydrogen recovery from H2/CO/N2/CO2 post combustion mixtures using a Matrimid membrane. J Memb Sci 378:359–368. https://doi.org/10.1016/j.memsci.2011.05.029

Delgado D, Minakshi M, Kim DJ, Kyeong CW (2017) Influence of the oxide content in the catalytic power of Raney nickel in hydrogen generation. Anal Lett 50:2386–2401. https://doi.org/10.1080/00032719.2017.1300806

Du Z, Liu C, Zhai J, Guo X, Xiong Y, Su W, He G (2021) A review of hydrogen purification technologies for fuel cell vehicles. Catalysts 11. https://doi.org/10.3390/catal11030393

Dunikov D, Borzenko V, Blinov D, Kazakov A, Lin CY, Wu SY, Chu CY (2016) Biohydrogen purification using metal hydride technologies. Int J Hydrog Energy 41:21787–21794. https://doi.org/10.1016/j.ijhydene.2016.08.190

Faye O, Szpunar J, Eduok U (2022) A critical review on the current technologies for the generation, storage, and transportation of hydrogen. Int J Hydrog Energy 47:13771. https://doi.org/10.1016/j.ijhydene.2022.02.112

González-Coloma A, Reina M, Medinaveitia A, Guadaño A, Santana O, Martínez-Díaz R, Ruiz-Mesía L, Alva A, Grandez M, Díaz R, Gavín JA, De La Fuente G (2004) Structural diversity and defensive properties of norditerpenoid alkaloids. J Chem Ecol 30:1393–1408. https://doi.org/10.1023/B:JOEC.0000037747.74665.0a

Hallenbeck PC, Ghosh D (2009) Advances in fermentative biohydrogen production: the way forward? Trends Biotechnol 27:287–297. https://doi.org/10.1016/j.tibtech.2009.02.004

Hallenbeck PC, Ghosh D (2012) Improvements in fermentative biological hydrogen production through metabolic engineering. J Environ Manag 95:S360. https://doi.org/10.1016/j.jenvman.2010.07.021

Hamid MAA, Chung YT, Rohani R, Junaidi MUM (2019) Miscible-blend polysulfone/polyimide membrane for hydrogen purification from palm oil mill effluent fermentation. Sep Purif Technol 209:598–607. https://doi.org/10.1016/j.seppur.2018.07.067

Hassen W, Selimefendigil F, Ben Khedher N, Kolsi L, Borjini MN, Alresheedi F (2021) Control of magnetohydrodynamic mixed convection and entropy generation in a porous cavity by using double rotating cylinders and curved partition, ACS. Omega 6:35607–35618. https://doi.org/10.1021/acsomega.1c05334
Kajama MN, Nwogu NC, Okon EO, Shehu H, Orakwe I, Gobina E (2015) Hydrogen separation using Pt-alumina impregnated membrane. Energy Power Eng 7(9)
Kajama MN, Nwogu NC, Gobina E (2016) Preparation and characterization of inorganic membranes for hydrogen separation. Int J Hydrog Energy 41:8221. https://doi.org/10.1016/j.ijhydene.2015.11.179
Kiyobayashi T, Takeshita HT, Tanaka H, Takeichi N, Züttel A, Schlapbach L, Kuriyama N (2002) Hydrogen adsorption in carbonaceous materials - how to determine the storage capacity accurately. J Alloys Compd 330–332:666–669. https://doi.org/10.1016/S0925-8388(01)01436-0
Kuroda S, Nagaishi T, Kameyama M, Koido K, Seo Y, Dowaki K (2018) Hydroxyl aluminium silicate clay for biohydrogen purification by pressure swing adsorption: Physical properties, adsorption isotherm, multicomponent breakthrough curve modelling, and cycle simulation. Int J Hydrog Energy 43(34):16573–16588
Li X, De Feyter S, Chen D, Aldea S, Vandezande P, Du Prez F, Vankelecom IFJ (2008) Solvent-resistant nanofiltration membranes based on multilayered polyelectrolyte complexes. Chem Mater 20:3876–3883. https://doi.org/10.1021/cm703072k
Lin H, Van Wagner E, Freeman BD, Toy LG, Gupta RP (2006) Plasticization-enhanced hydrogen purification using polymeric membranes. Science 311:639–642. https://doi.org/10.1126/science.1118079
Liu Y, Ren L, He Y, Cheng HP (2010) Titanium-decorated graphene for high-capacity hydrogen storage studied by density functional simulations. J Phys Condens Matter 22:5. https://doi.org/10.1088/0953-8984/22/44/445301
Miura S, Fujisawa A, Ishida M (2012) A hydrogen purification and storage system using metal hydride. Int J Hydrog Energy 37(3):2794–2799
Miura S, Fujisawa A, Tomekawa S, Taniguchi Y, Hanada N, Ishida M (2013) A hydrogen purification and storage system using CO adsorbent and metal hydride. J Alloys Compd 580:S414. https://doi.org/10.1016/j.jallcom.2013.03.154
Moradi R, Groth KM (2019) Hydrogen storage and delivery: review of the state of the art technologies and risk and reliability analysis. Int J Hydrog Energy 44:12254–12269. https://doi.org/10.1016/j.ijhydene.2019.03.041
Moran A, Talu O (2018) Limitations of portable pressure swing adsorption processes for air separation. Ind Eng Chem Res 57:11981–11987. https://doi.org/10.1021/acs.iecr.8b02237
Nagarajan D, Lee DJ, Kondo A, Chang JS (2017) Recent insights into biohydrogen production by microalgae – from biophotolysis to dark fermentation. Bioresour Technol 227:373–387. https://doi.org/10.1016/j.biortech.2016.12.104
Nor Azira AM, Umi Aisah A (2019) Purification of biohydrogen from fermentation gas mixture using two-stage chemical absorption. E3S Web Conf 90. https://doi.org/10.1051/e3sconf/20199001012
Rahman KM, Kawahara N, Matsunaga D, Tomita E, Takagi Y, Mihara Y (2016) Local fuel concentration measurement through spark-induced breakdown spectroscopy in a direct-injection hydrogen spark-ignition engine. Int J Hydrog Energy 41:14283–14292. https://doi.org/10.1016/j.ijhydene.2016.05.280
Ramírez-Morales JE, Tapia-Venegas E, Nemestóthy N, Bakonyi P, Bélafi-Bakó K, Ruiz-Filippi G (2013) Evaluation of two gas membrane modules for fermentative hydrogen separation. Int J Hydrog Energy 38:14042–14052. https://doi.org/10.1016/j.ijhydene.2013.08.092
Ramírez-Morales JE, Tapia-Venegas E, Toledo-Alarcón J, Ruiz-Filippi G (2015) Simultaneous production and separation of biohydrogen in mixed culture systems by continuous dark fermentation. Water Sci Technol 71:1271–1285. https://doi.org/10.2166/wst.2015.104
Sivagurunathan P, Kumar G, Bakonyi P, Kim SH, Kobayashi T, Xu KQ, Lakner G, Tóth G, Nemestóthy N, Bélafi-Bakó K (2016) A critical review on issues and overcoming strategies

for the enhancement of dark fermentative hydrogen production in continuous systems. Int J Hydrog Energy 41:3820–3836. https://doi.org/10.1016/j.ijhydene.2015.12.081
Stephenson M (1947) Some aspects of hydrogen transfer. Antonie Van Leeuwenhoek 12:33–48. https://doi.org/10.1007/BF02272650
Usman MR (2022) Hydrogen storage methods: review and current status. Renew Sust Energ Rev 167:112743. https://doi.org/10.1016/j.rser.2022.112743
Vasiliev LL, Kanonchik LE, Kulakov AG, Babenko VA (2007) Hydrogen storage system based on novel carbon materials and heat pipe heat exchanger. Int J Therm Sci 46:914–925. https://doi.org/10.1016/j.ijthermalsci.2006.11.020
Verhelst S, Sierens R, Verstraeten S (2006) A critical review of experimental research on hydrogen fueled SI engines. SAE. Tech. Pap.,, p 13. https://doi.org/10.4271/2006-01-0430
Vi LVT, Salakkam A, Reungsang A (2017) Optimization of key factors affecting bio-hydrogen production from sweet potato starch. Energy Procedia 138:973–978. https://doi.org/10.1016/j.egypro.2017.10.092
White CM, Steeper RR, Lutz AE (2006) The hydrogen-fueled internal combustion engine: a technical review. Int J Hydrog Energy 31:1292–1305. https://doi.org/10.1016/j.ijhydene.2005.12.001
Winter CJ (2009) Hydrogen energy - abundant, efficient, clean: A debate over the energy-system-of-change. Int J Hydrog Energy 34:S1. https://doi.org/10.1016/j.ijhydene.2009.05.063
Yanxing Z, Maoqiong G, Yuan Z, Xueqiang D, Jun S (2019) Thermodynamics analysis of hydrogen storage based on compressed gaseous hydrogen, liquid hydrogen and cryo-compressed hydrogen. Int J Hydrog Energy 44:16833–16840. https://doi.org/10.1016/j.ijhydene.2019.04.207
Yave W, Car A, Peinemann KV, Shaikh MQ, Rätzke K, Faupel F (2009) Gas permeability and free volume in poly (amide-b-ethylene oxide)/polyethylene glycol blend membranes. J Membr Sci 339(1-2):177–183
Ye S, Luo F, Zhang Q, Zhang P, Xu T, Wang Q, He D, Guo L, Zhang Y, He C, Ouyang X, Gu M, Liu J, Sun X (2019) Highly stable single Pt atomic sites anchored on aniline-stacked graphene for hydrogen evolution reaction. Energy Environ Sci 12:1000–1007. https://doi.org/10.1039/c8ee02888e
Zhou L (2005) Progress and problems in hydrogen storage methods. Renew Sust Energ Rev 9:395–408. https://doi.org/10.1016/j.rser.2004.05.005

Chapter 22
A Life Cycle Assessment and Economic Perspective of Biohydrogen Production

Bikash R. Tiwari and Satinder Kaur Brar

Abstract The energy and transportation sectors are majorly dependent on fossil fuels to meet its ever-growing demand. Hydrogen, a non-carbon-based fuel is a suitable alternative as it eliminates the carbon emissions linked to fuel combustion. However, currently hydrogen is primarily produced from natural gas, coal, electrolysis, heavy oil, and naphtha. A shift to biological processes for production of hydrogen may be a more environment-friendly alternative. In order to establish the environmental performance of biohydrogen production over current commercial technologies, life cycle assessment (LCA) studies are conducted. LCA can evaluate the potential environmental impacts associated with various biohydrogen production processes by considering the material, energy, fuel, and emissions involved. In addition to environmental performance, the economics of biohydrogen production should be competitive enough with commercial hydrogen production processes to make biohydrogen the primary choice for market use. Hence, economic analysis is undertaken to evaluate the cost of production based on commercial scale production of biohydrogen. This chapter presents the LCA and economic assessment studies with regard to biohydrogen production to enable stakeholders and policy-makers in decision-making.

Keywords Life cycle assessment · Biohydrogen · Fermentation · Microbial electrolysis cell

B. R. Tiwari
Institut National de la Recherche Scientifique - Centre Eau Terre Environnement, Université du Québec, Quebec City, QC, Canada

S. K. Brar (✉)
Department of Civil Engineering, Lassonde School of Engineering, York University, North York, TO, Canada
e-mail: Satinder.Brar@lassonde.yorku.ca

C. R. Soccol et al. (eds.), *Biohydrogen - Advances and Processes*, Biofuel and Biorefinery Technologies 13, https://doi.org/10.1007/978-3-031-49818-3_22

22.1 Introduction

Global energy consumption is likely to increase at an exponential rate with the rise in economic and social framework driven by rapid industrialization. It is reported that since 1970 the global energy consumption increased from around 9000 million tonnes oil equivalent (Mtoe) to approximately 14,000 Mtoe in 2018, which is further expected to rise to 20,000 Mtoe in 2050 (Mead 2017; Enerdata 2014). About one-third of the current energy consumption is attributed to transportation (Pathy et al. 2022) which is heavily dependent on fossil fuels. This dependency on fossil fuels creates some critical concerns. It is known that the reserves of fossil fuels are limited. More importantly, the combustion of fossil-derived transportation fuels is associated with environmental pollution, greenhouse gas (GHG) emissions being the major contributor. For example, 2.9 kg of GHG are emitted per litre of diesel consumed (Ağbulut and Sarıdemir 2021). Moreover, the excessive dependency on fossil fuels has far-reaching consequences including, but not limited, to climate change which can severely damage the ecosystem. As a cautionary measure, the Intergovernmental Panel on Climate Change (IPCC) has set a target of 1.5 °C to limit the temperature rise as compared to the preindustrial levels (Pörtner et al. 2019). Further, IPCC expects to achieve net zero carbon dioxide emissions by 2050 (Pörtner et al. 2019). However, the commonly used transportation fuels are carbon based, and hence it is inevitable that their combustion will generate carbon emission. On the contrary, a non-carbon-based fuel can easily avoid this emission and reduce the adverse impacts on the environment credited to its combustion. In this context, hydrogen is projected as a cleaner substitute to the conventional transportation fuels since its combustion is expected to generate water. It is speculated that in the near future the success of hydrogen-based energy can translate to a carbon neutral economy by drastically reducing the carbon emissions.

According to International Energy Agency report (2021), around 95% of the commercial hydrogen is derived from natural gas, coal, electrolysis, heavy oil, and naphtha (Ibrahim et al. 2021). This suggests that although the combustion-associated emissions are avoided, the commercial hydrogen production is still associated with GHG emission. Hence, a greener alternative would be to use biological processes for hydrogen production which can be carried out under ambient temperature and pressure.

Biohydrogen is commonly produced by fermentation of organic substrates (Osman et al. 2020), biophotolysis with microalgae (Nagarajan et al. 2017), and microbial electrochemical cell (MEC) (Heidrich et al. 2013). Dark fermentation is found to be more practically feasible at a large scale owing to its higher rate of hydrogen production as compared with biophotolysis (Yun et al. 2018). However, fermentation is an energy-intensive technology. Microalgae-based biohydrogen production is attractive due to low lignin content in biomass, eliminating energy and cost-intensive pretreatment methods (Behera et al. 2020). Nevertheless, the cultivation of microalgae for large-scale biohydrogen production is a challenge considering the high water demand and significant initial investment (Anwar et al. 2019). In order to be sustainable at a commercial scale, not only the feasibility of the technology is important but also the environmental performance along with the cost economics should be optimized. Although biohydrogen is considered a green fuel, the complete production process should also be taken into consideration to evaluate the environmental performance. Environmental assessment is suggested for a process/technology taking into account the feedstock used, energy and chemical input, and waste generated to estimate the associated environmental impacts (Panigrahi et al. 2022; Tiwari et al. 2023). Specific approaches such as life cycle assessment (LCA) can be employed to estimate the environmental impacts of biohydrogen production. LCA can be used for biohydrogen production to achieve three major goals: (i) to evaluate the impacts of biohydrogen production on biotic and abiotic components of the environment, (ii) to identify specific processes in the biohydrogen production which contribute to highest environmental impacts, and (iii) to assemble critical data on biohydrogen production for transition to a global hydrogen economy. Similarly, methodologies like life cycle cost analysis and techno-economic analysis (TEA) are used to assess the economics of the process. Both economic and environmental assessment can identify the corresponding lacuna and hotspot of the complete process. Eventually they can assist technology developers, policy-makers, and entrepreneurs in decision-making. This chapter will provide an insight into the environmental and economic perspectives of the various biohydrogen production processes.

22.2 Life Cycle Assessment

LCA is a globally accepted assessment method to quantify the direct and indirect environmental impacts associated with a product or process owing to the energy, material, and waste consumption or generation. A standard LCA protocol is based on ISO14040 and ISO14044, which consist of four stages: (1) goal and scope definition; (2) life cycle inventory (LCI) analysis; (3) life cycle impact assessment (LCIA); and (4) life cycle interpretation (ISO 2006).

22.2.1 Goal and Scope Definition

The goal of the study deals with the purpose for conducting the LCA; the scope identifies the potential audience and defines the system under consideration (Tiwari and Brar 2022). For example, the goal of an LCA for a process/technology producing hydrogen may be to quantify the environmental impacts associated with one stage or all stages of the production considering the material and energy inputs and waste generated.

The functional unit provides a qualitative and quantitative aspect to the output or service provided by a process (Corominas et al. 2020). In addition, it provides a reference for normalizing the inventory data and the associated environmental impacts (Corominas et al. 2020). Any product is easily identified by the amount produced in a process and hence, the mass of hydrogen produced is one of the most commonly used functional units (Ochs et al. 2010; Djomo and Blumberga 2011). Since hydrogen is mostly projected as a potential energy source, the functional unit of '1 MJ of hydrogen produced' can also be used (Ferreira et al. 2013a). A modern approach presents a functional unit based on final application of hydrogen produced. For example, a biohydrogen plant built with an aim to utilize biohydrogen for electricity generation can have functional unit in 'kWh of electricity generated per unit hydrogen' (Suwanmanee et al. 2018). Similarly, if hydrogen will be utilized as transportation fuel, the functional unit can be expressed in terms of 'km of vehicle traveled per unit hydrogen fuel' (Hwang and Chang 2010). Since all these functional units are based on the output of the process/plant, they are identified as output-based functional units (Quek and Balasubramanian 2014). In a few instances, the functional unit used are input based, i.e. the normalized quantity of the substrate utilized to produce hydrogen such as '1 g of microalgae' (Pacheco et al. 2015) and '1 kg volatile solids of food waste'.

The system boundary includes all the upstream (material and energy inputs for infrastructure, operation, and maintenance) and downstream (extraction and purification, by-product recovery, waste emission) processes pertinent to the goal and scope of the LCA (Tiwari and Brar 2022). The most comprehensive system boundary is the cradle to grave which includes all the upstream and downstream process from raw material to final product (Fig. 22.1). A stakeholder interested only in the impacts limited till the production step will consider a cradle to gate system boundary. Similarly, if the point of interest is only limited to the utilization of the hydrogen produced, a gate to grave boundary is ideal. However, if the goal is to evaluate the emission from a particular sub-process such as only the hydrogen production

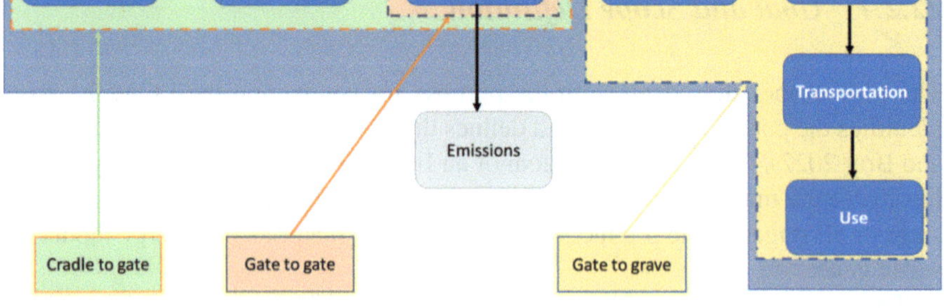

Fig. 22.1 System boundary consideration for LCA of biohydrogen refinery

step, the system boundary will include resources limited to that specific sub-process. This system boundary is denoted as gate to gate (Fig. 22.1). Another classification of system boundary is also used in LCA studies where a fuel is under consideration. These are well to tank (includes production, transportation, and distribution of the fuel), tank to wheel (includes operation of the vehicle), and the comprehensive well to wheel (includes all stages of life cycle, production to use).

22.2.2 Life Cycle Inventory

Life cycle inventory is the collection of environmentally relevant input and output data specific to the system boundary as defined in the previous stage. In case of hydrogen production, the inventory includes data regarding the substrate utilized for hydrogen production, chemicals required for various processes (such as cultivation of microbes, maintaining pH, extraction), energy consumed (operation of fermenters, centrifuge, drying, heating, etc.), infrastructure, and waste emission to land, water, and air. The primary source of foreground data is based on real-time values obtained during a plant operation or lab-/pilot-scale experiments and sometimes based on previously published literature. Although energy data should be obtained from metred/measured readings, it is sometimes possible to estimate data with the help of simple equations such as energy required for heating, pumps, stirring, etc. (Bhar et al. 2022). Even for some operations like filtration and centrifugation, pre-established thumb rule or energy values specified by manufacturers can be used to express power consumption in kWh/tonne dry material or kWh/m^3 of the sample (Piccinno et al. 2016). In some cases, where some parameters in lab-scale cannot be linearly extrapolated for corresponding large-scale application, data can be generated based on mathematical modelling. Standard chemical equations can also be used for chemical reactions to derive data regarding some emissions or by-products such as methane, carbon dioxide, etc. Since biohydrogen production is mostly restricted to lab-scale and pilot-scale studies, it can be challenging to find industrial data for the same. Hence, inventory data for full-scale plant operation can be obtained using process simulation software such as Aspen Plus and SuperPro which can model, evaluate, and optimize integrated processes for a broad spectrum of industries (Tiwari et al. 2023). Background data can be mostly derived from available LCI databases like Eco invent and GREET (Mehmeti et al. 2018).

22.2.3 Life Cycle Impact Assessment

The inventory data is further processed and interpreted in terms of environmental impacts grouped under different impact categories which represent critical threats to the environment. Different methods such as ReCiPe, CML, Impact 2002+, Eco-indicator 99, and IPCC 2007 are employed for impact assessment (Pradhan et al.

2019). Each method has a different combination of impact categories based on the interpretation of inventory data. Impact categories can also be expressed as mid-point and endpoint indicators. The midpoint indicators represent impacts in the early stage of the cause-effect chain directed on each individual problem, while endpoint indicators which are generated by aggregating different midpoint indicators represent the end of cause-effect chain directed at damage levels (Tiwari and Brar 2022; Tiwari et al. 2023). The most widely relevant impact category is the global warming potential (GWP) due to its social and political perspective (Tiwari and Brar 2022). GWP is majorly attributed to the direct and indirect emission of carbon dioxide, methane, and nitrous oxide from natural or anthropogenic activities (Panigrahi et al. 2022). While direct emissions during biohydrogen production arise from various unit processes like fermentation, anaerobic digestion, etc., indirect emissions are ascribed to production of electricity and chemicals and transportation. Apart from GWP, eutrophication potential is also considered a highly relevant category which is associated with the emission of phosphorus, nitrogen, and organic matter (Panigrahi et al. 2021). Moreover, impacts are categorized according to the specific abiotic component it affects, such as freshwater and marine eutrophication and terrestrial, marine, and freshwater ecotoxicity (Tiwari and Brar 2022). Furthermore, few categories are dedicated to resources that are utilized in the system boundary such as water consumed, land use, mineral resource utilization, fossil depletion (Tiwari et al. 2023). Life cycle assessment is usually conducted using some commercial and open-source software such as SimaPro, GaBi, and OpenLCA (Tiwari et al. 2021). These software make the arduous process of impact assessment convenient as they are equipped with substantial database of life cycle inventory which can translate the material, energy, and processes to their corresponding environmental impacts.

22.2.4 Life Cycle Interpretation

Once the life cycle inventory and processes included in the system boundary are translated to their corresponding life cycle impacts, the results are discussed to divulge reasons for adverse and/or favourable environmental impacts. In addition, comparisons are drawn in relation to other LCA studies in the same field of work. This allows to draw conclusions and recommendations to propose means to reduce adverse environmental impacts and enhance sustainability of the process. Moreover, alternatives for a material, energy source, or sub-process may be suggested to reduce the overall adverse impacts. In addition, sensitivity and scenario analysis are conducted to understand the extent of variation in results corresponding to a change in input parameters over a certain range (Bhar et al. 2022; Tiwari et al. 2023). Furthermore, it is recommended to include uncertainty analysis such as Monte Carlo simulation to account for assumptions in inventory or process. These assumptions are usually undertaken to simplify the LCA model which can eventually translate to less accurate data in terms of emissions or impacts (Michiels and Geeraerd 2020).

22.3 Environmental Assessment of Biohydrogen Production: Case Studies

One of the earliest studies on environmental assessment of biohydrogen production from sugarcane compared dark fermentation, photofermentation, two-stage dark and photofermentation, and biocatalysed electrolysis to steam methane reforming (Manish and Banerjee 2008). A commonly followed schematic of the dark fermentation process for LCA consideration is provided in Fig. 22.2. It was estimated that while commercially preferred steam reforming process contributed around 12.8 kg CO_2 eq/kg H_2, dark fermentation (−87 kg CO_2 eq/kg H_2), photofermentation (−21.9 kg CO_2 eq/kg H_2), two-stage process (−19.5 kg CO_2 eq/kg H_2), and biocatalysed electrolysis (−17.5 kg CO_2 eq/kg H_2) were carbon negative (Manish and Banerjee 2008). This GHG reduction was attributed to the avoided fossil energy contributed by the generation of by-products such as methane and bagasse. However, the system boundary did not include the waste collection as well as hydrogen compression and storage.

A wide range of substrates ranging from synthetic carbon sources, organic waste, and wastewater have been used to generate biohydrogen through different technologies. In order to understand the effect of change in feedstock on the environmental performance of biohydrogen production, an LCA was performed for two-stage integrated thermophilic and photofermentation of three different substrates, viz. wheat straw (WS), sweet sorghum stalk (SSS), and steam potato peels (SPP) (Djomo and

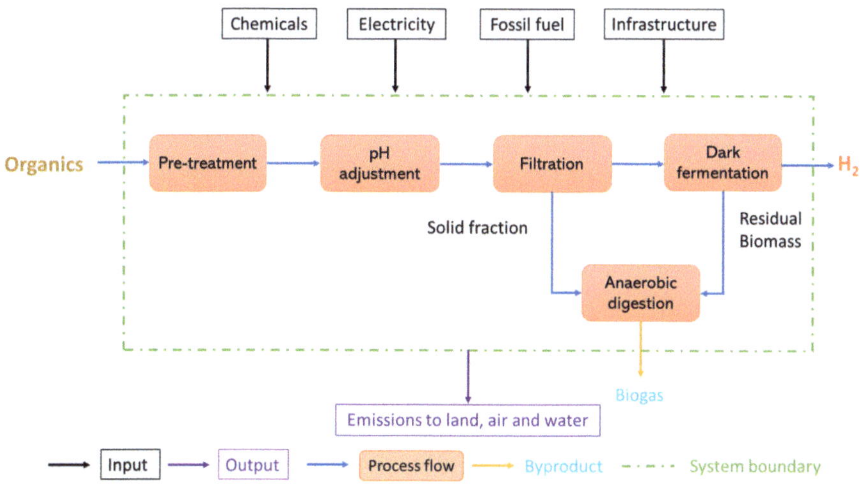

Fig. 22.2 Typical system boundary for LCA of dark fermentation for hydrogen production

Blumberga 2011). Midpoint indicators such as resource depletion and climate change were observed to be the largest contributors to the environmental impacts for all the feedstocks. The resource depletion was highest for SSS (87.30 MJ/kg H_2) followed by WS (79.09 MJ/kg H_2) and SPP (45.70 MJ/kg H_2) (Djomo and Blumberga 2011). The same trend was true for GHG emissions as well at 4.05, 3.48, and 1.28 kg CO_2 eq/kg H_2 for SSS, WS, and SPP. Among the unit processes involved, fermentation was the largest contributor to environmental impacts, contributing around 41–43% to the resource depletion category and 42–44% to the climate change category. This high impact from fermentation was attributed to two major inputs, i.e. electricity and steam. Apart from fermentation, the compression and storage stage cumulatively accounted for around 25–26% in climate change and 26–28% in resource depletion. Electricity produced from burning of natural gas was the highest contributor to both the resource depletion (64%) and climate change category (63%) (Djomo and Blumberga 2011).

In a similar study, it was estimated that the GHG emission with and without energy recovery was around 9.8 and 19 kg CO_2 eq/kg H_2, respectively, with electricity input identified as a major contributor for biohydrogen production by dark fermentation (Pi et al. 2016). The environmental performance of a dark fermentation system utilizing rice husk as a substrate as well as source of heat and electricity in a combined heat and power plant (CHP) was evaluated (Table 22.1). It was found that the associated GHG emission of 10.92 kg CO_2 eq was almost half of that reported for a gasification process for hydrogen production (20.10 kg CO_2 eq) (Reaño 2020). However, dark fermentation has significantly higher eutrophication potential owing to the organic content of the wastewater generated as a result of fermentation. Three parameters were identified which can reduce the GHG emission, i.e. increase in hydrogen yield and efficiency of the CHP plant and reduction in the travel distance from the rice husk source to the plant site (Reaño 2020).

Microalgae have been a preferred substrate for biohydrogen research considering its lack of lignin as opposed to other lignocellulosic substrates. Microalgal hydrogen production has been subjected to many LCA studies to evaluate the inherent environmental impacts (Fig. 22.3). When biohydrogen was produced by photoautotrophic cyanobacteria *Anabaena* sp., GHG emission was estimated to be around 115 kg CO_2 eq/MJ H_2 (Ferreira et al. 2012). The residual *Anabaena* sp. biomass was utilized again for hydrogen production through dark fermentation which led to a 12.1% increase in CO_2 production. A variation to the previous study was presented when the hydrolysate of a microalga *Scenedesmus obliquus* grown in a photobioreactor was used as a substrate for dark fermentation by *Clostridium butyricum.* This amassed a GWP of 670 kg CO_2/MJ H_2 (Ferreira et al. 2013a). The largest amount of CO_2 emission was ascribed to the microalgal culture using a photobioreactor, responsible for almost 99% of the overall energy consumed. However, a change in the microalgal culture site from a photobioreactor to an outdoor raceway pond reduced emission by an order of 10 to 5.1–6.2 kg CO_2 eq/MJ H_2 (Ferreira et al. 2013b). This reduction is due to the utilization of sunlight and atmospheric CO_2 as compared to artificial sources. Even for this study, the culture of microalgae was the stage with the highest energy requirement emitting around 3.6 kg CO_2 eq/MJ H_2.

Table 22.1 Life cycle assessment data for biohydrogen production process

Process	Functional unit	System boundary	LCIA[a] method	References
Fermentation of sugarcane juice	1 kg of H_2 production			Manish and Banerjee (2008)
Fermentation of potato steam peels	1 kg of H_2 produced	Cradle to grave	Eco-indicator 99	Ochs et al. (2010)
Fermentation sweet sorghum stalk, wheat straw, potato steam peels	460 tonne of H_2 produced	Well to tank	IMPACT 2002+	Djomo and Blumberga (2011)
Fermentation of microalgae	1 MJ of H_2 produced	Cradle to gate	Inventory analysis	Ferreira et al. (2013a)
Fermentation of microalgae	1 MJ of H_2 produced	Cradle to gate	Inventory analysis	Ferreira et al. (2013b)
Fermentation of microalgae	1 g of microalgae, 1 MJ of H_2 produced	Cradle to gate	Ecoinvent, experiment	Pacheco et al. (2015)
MEC[b] with urban wastewater	1 kg of H_2 produced	Cradle to gate	IMPACT 2002+	Chen et al. (2019)
Fermentation followed by MEC of corn stover	1 kg of H_2 produced	Cradle to gate	ReCiPe 2016	Mehmeti et al. (2018)
Hydrogen production from biowaste	1 L of H_2 g^{-1} organic waste		Only GHG	Hosseinzadeh et al. (2022)
Hydrogen production from rice husk	1 kg of H_2 production	Gate to gate	CML 2001	Reaño (2020)
Hydrogen production from food waste	1 km of transportation distance	Cradle to grave	Eco-indicator 99	Patterson et al. (2013)
Digestible materials, woody materials, straw, and glycerol	1 MJ of H_2	Cradle to grave		Wulf and Kaltschmitt (2013)

[a]Life cycle impact assessment
[b]Microbial electrolysis cell

However, the storage, distribution, and transportation were not included in the system boundary.

Another variation to the microalgae-based hydrogen production was presented when it was combined with the scope for pigment extraction (Pacheco et al. 2015). Centrifugation is the conventional choice for microalgae harvesting, which is often associated with high operational energy requirement. Hence, electrocoagulation was evaluated as an alternative to centrifugation in this study. Moreover, a comparison was made between electrical oven drying and sun drying from microalgal biomass. Both these comparisons for biomass drying and harvesting were conducted to evaluate if the dependency on fossils for the corresponding operation will affect the environmental impact associated with the microalgal biohydrogen process. Hence, LCA was conducted for four different pathways, i.e. Pathway A, harvesting by centrifugation and oven dewatering (0.237 MJ/g microalgae, 24.9 g CO_2/g microalgae);

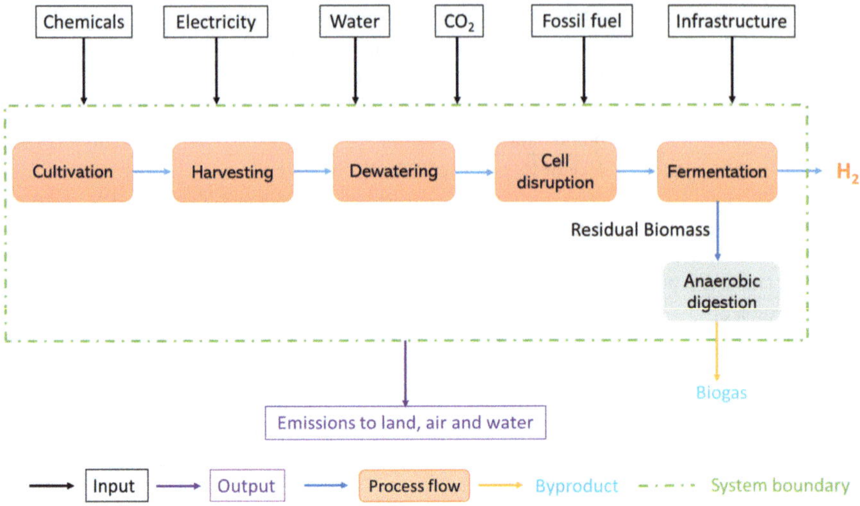

Fig. 22.3 Typical system boundary for microalgal biohydrogen production

Pathway B, harvesting by electrocoagulation, centrifugation, and oven dewatering (0.028 MJ/g microalgae, 2.99 g CO_2/g microalgae); Pathway C, harvesting by centrifugation and solar dewatering; (0.235 MJ/g microalgae, 24.6 g CO_2/g microalgae); and Pathway D, harvesting by electrocoagulation, centrifugation, and solar dewatering (0.026 MJ/g microalgae, 2.71 g CO_2/g microalgae). The study concluded that electrocoagulation-assisted harvesting prior to centrifugation could reduce energy requirements by 90%, whereas solar drying-based dewatering could reduce the electricity consumption by 86%. This is eventually reflected in the drastically reduced GHG emission for Pathways B and D (Pacheco et al. 2015).

MEC is an upcoming technology for biohydrogen production which is still in lab-scale (Fig. 22.4). One of the major advantages of the MEC over conventional electrolysis for hydrogen production is the low electricity requirement in the range of 0.6–1 kWh/m^3 H_2 (Rozendal et al. 2008; Steinbusch et al. 2009). Moreover, there is no need for an additional purification stage for the hydrogen produced. In comparison to dark fermentation where the yield is limited to around 4 mol H_2/mol-glucose, MEC can manage a higher biohydrogen recovery up to 8.55 mol H_2/mol-glucose at an applied potential of 0.6 V (Cheng and Logan 2011). A very detailed LCA was performed for a pilot-scale MEC system (Chen et al. 2019) which consisted of three units, (i) pretreatment system based on a continuous flow pilot-scale 1000 L MEC system (Cusick et al. 2011); (ii) MEC reactor, configuration, and operation are based on 130 L MEC reactor developed in Barcelona (Baeza et al. 2017); and (iii) post treatment system, designed as per a bioelectrochemical system operated for a flow of 2200 m^{3}d^{-1} (Foley et al. 2010). The complete biohydrogen production system boundary was divided into five different stages for impact assessment. The preconstruction phase, preoperation phase, MEC construction phase, MEC operation phase, post construction phase, and post operation phase received

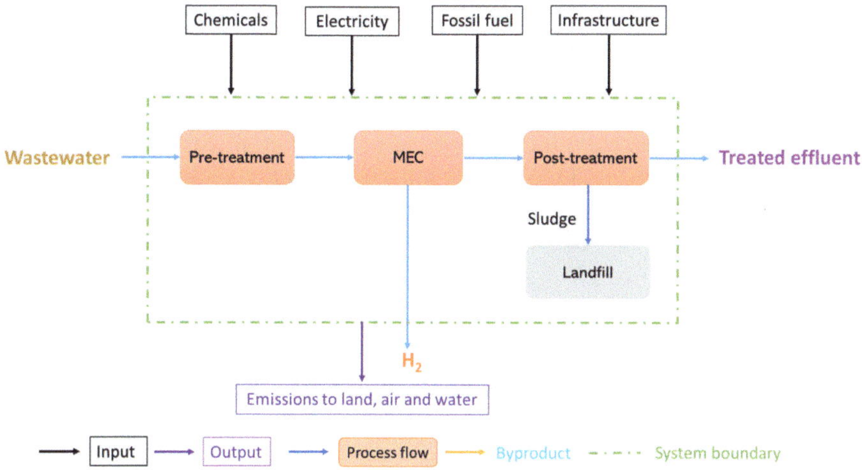

Fig. 22.4 Typical system boundary for biohydrogen production in MEC

impact scores of 76.0, 5.0, 253.5, 17.6, 25.6, and 1.5 mPt, respectively. The highest adverse impact for human health was reported for MEC construction phase (148 mPt). The CO_2 emission was found to be highest for the MEC construction phase (260.4 kg CO_2 eq/kg H_2) followed by preconstruction and MEC operation, at 63.5, and 52.8 kg CO_2 eq/kg H_2, respectively. It is one of the rare cases where infrastructure in construction was considered for LCA of biohydrogen production, and it was found that construction has a considerable contribution to impacts associated with MECs. The resources category also followed the same trend for the MEC construction, pre-construction, and MEC operation phase achieving impact scores of 983.7, 3953.2, and 722.1 MJ/kg H_2, respectively. It was observed that with the increase in electricity transformation efficiency and cathodic gas recovery, the emissions reduced. For example, an increase in cathodic gas recovery from 82% to 95% led to a decrease in GWP from 52.8 to 45.6 kg CO_2 eq/kg H_2. Similarly, CO_2 emissions reduced from 52.8 to 39.2 kg CO_2 eq/kg H_2, with the increase in electricity transformation efficiency from 66.7% to 90%. However, the trend was opposite for the applied potential.

Apart from the above discussed technologies, anaerobic digestion, such as the two-stage biohydrogen/biomethane production, is also an interesting avenue for biohydrogen production from organic waste. An LCA was conducted for a functional unit of 1 km distance travelled by a passenger vehicle fuelled by biomethane and biohydrogen/ biomethane (Patterson et al. 2013). It was observed that both biomethane and biohydrogen/biomethane did not have adverse impacts in the categories of carcinogens (-5.94 E-5 Pt to -7.70 E-5 Pt) and ecotoxicity (-2.76E-5 Pt to -3.33E-5 Pt). Even for the climate change category (-2.48E-6 Pt to -3.99E-6 Pt) and fossil fuel use (8.96E-6 Pt to 2.25E-5 Pt), although impacts are adverse, they are still significantly lower as compared with diesel production. It was found that biomethane and biohydrogen/biomethane have fossil fuel advantages over diesel

(3.67E-5 Pt, 2.87E-5 Pt, and 5.89E-5 Pt, respectively). However, for all other impact categories, diesel utilization had lower burdens than either biomethane or biomethane/biohydrogen use. A similar environmental assessment study was conducted for a pilot-scale up-flow anaerobic sludge blanket fixed-film (UASFF) reactor producing biohydrogen from palm oil mill effluent (Akhbari et al. 2021). The estimated GWP was around 818 kg CO_2 eq/ kg H_2 with electricity accounting for around 98% of the total CO_2 emission. The majority of the GWP is attributed to the fossil-derived electricity utilized for the operation of pumps and maintaining optimum temperature.

22.4 Economic Assessment of Biohydrogen Production: Case Studies

On broader terms, the capital and operational costs are determining parameters for the economic feasibility of biohydrogen production, which varies with the process/technology, raw material, equipment, as well as transportation distance. The more detailed cost distribution may include the cost of land (installation of fermenter and photobioreactor setup with downstream processes, open raceway algal pond), equipment cost (fermenters, photobioreactor, gas separation, extraction, storage, pumps, condensers, cooler, etc.), material cost (substrate, chemicals, water, enzymes), power cost (operation of pumps, centrifuge, stirring, temperature control, mixing, etc.), maintenance cost, and labour cost. Since fermentation-based hydrogen production is an energy-driven process, the corresponding expenditure is directly associated with the operating cost which are relatively higher than the capital cost (Pathy et al. 2022). A major contributor to the fermentation-based hydrogen production is levied on pure substrates, such as glucose which can significantly increase the final cost of production. Considering that fermentation can utilize substrates ranging from organic biomass, lignocellulosic feed, wastewater, etc. and the cost of hydrogen storage is comparable to conventional processes, a profit of 0.81 $/$m^3$ can be expected (Kannah et al. 2019). It was estimated that with food waste as substrate for a pilot-scale dark fermentation-based hydrogen production, the return on investment (ROI) of 26% could be achieved with a payback period of 5 years (Han et al. 2016a). The ROI has a direct relation with the reactor volume, where an increase from 10 to 50 m^3 could result in an increase in ROI from 37% to 47% (Han et al. 2016b). Subsequently researchers arrived at a hydrogen production cost in the range of 10–30 $/kg H_2 (Han et al. 2016a, b). Moreover, with increasing the hydrogen yield to 2.26 mol H_2/mol hexose at an organic loading rate of 100 kg chemical oxygen demand (COD)/m^3/d, the hydrogen production cost could be further reduced to 3.2 $/kg H_2 (Yun et al. 2018) (Table 22.2). Still, the current selling price of hydrogen is around 0.5–3.2 $/kg (Bartels et al. 2010); hence, further efforts are required to gain competitive advantage using dark fermentation for hydrogen production.

In the case of biophotolysis with cyanobacteria driven by light from solar energy and water, the cost levied on operation and cultivation is low as compared with the

Table 22.2 Cost of biohydrogen production

Technology	Substrate	Cost, $/kg H_2	Reference
Dark fermentation	Biowaste	2.3	Hosseinzadeh et al. (2022)
Dark fermentation	Food waste	3.2	Yun et al. (2018)
Dark fermentation	Food waste	1.02 $/m³	Han et al. (2016a)
Dark fermentation	Food waste	1.34 $/m³	Han et al. (2016b)
Dark fermentation	Wastewater	2.7 $/m³	Li et al. (2012)
Dark fermentation	Algae	18.7	Sathyaprakasan and Kannan (2015)
Photofermentation	Algae	3.7	
Direct photolysis	Algae	1342	
Indirect photolysis	Algae	2	
Direct photolysis	Algae	2.13	Nikolaidis and Poullikkas (2017)

capital cost for installation of photobioreactors considering a large-scale deployment (Ramprakash et al. 2022). The cost for direct and indirect photolysis was estimated to be around 2.13 $/kg H_2 and 1.42 $/kg H_2 (Nikolaidis and Poullikkas 2017). However, the cultivation is very time-consuming, and infrastructure is required for gas separation, storage, and transportation (Ramprakash et al. 2022). It is clear that cost of biohydrogen production could be significantly reduced if the primary substrate is a waste biomass or wastewater. In addition, a focused technological advancement regarding the separation of hydrogen from the mixture of gases is expected to further reduce the cost of production. A study observed that genetically engineering the microalgae *Chlamydomonas reinhardtii* could increase the hydrogen production fivefold as compared to the actual strain by biophotolysis (Kruse and Hankamer 2010). This can significantly reduce the cost of hydrogen production in comparison with fermentation. Another study reported that a 10% photon transformation efficiency for the microalgae-driven biophotolysis could present an economically favourable scenario (Nagarajan et al. 2013). Even the photon conversion efficiency of 5% can be suitable considering that steam generated is recycled co-product generation (Nagarajan et al. 2013).

A cost comparison for hydrogen production by direct/indirect photolysis, photofermentation, and dark fermentation based on a theoretical estimate was presented (Sathyaprakasan and Kannan 2015). The cost for hydrogen production was significantly higher, almost 500-fold for direct photolysis (4930 United Arab Emirates Dirham/kg) in comparison to the other three processes considered in the study. This can be attributed to the large land footprint and high operating cost which includes the cost levied on labour, culture production, and general supplies. Fermentation requires less land area as compared to biophotolysis. However, as mentioned previously, a significant part of the fermentation-based hydrogen production cost was due to the use of glucose as substrate. The cost of dark fermenters was assumed to be higher than the photofermenters as well as the pond for biophotolysis. Eventually, the cost for indirect biophotolysis was found to be the lowest among the four technologies followed by photofermentation and dark fermentation.

With the rapid growth and interest in biohydrogen production routes, it is expected that the operational cost for dark fermentation can be reduced to make it more economically suitable. The economic assessment studies usually consider only significant cost-implicating factors and optimistic assumptions which may present a slight variation to the real world economics (Pathy et al. 2022). Some factors are liable to vary such as the fuel cost, land cost, and labour cost. Hence, the temporal and spatial consideration should also be kept in mind while comparing two studies carried out at different places and time. A change in the economic scenario is also expected for integrated processes for biohydrogen production as well as cost reduction from the utilization of waste as substrate and cost recovery from by-products generated during hydrogen production.

22.5 Future Perspectives and Conclusion

Hydrogen could potentially replace a part of the fossil fuel demand in the current industrialization era. Each biohydrogen production technology has its own advantages and disadvantages. Simple reactor configuration and the acceptability of wide range of substrates are the merits of dark fermentation (Tian et al. 2019). On the other hand, the hydrogen yield is low in comparison to photofermentation and MEC. Dark fermentation and photolysis have more chances of being scaled up than MEC due to the complexity in architecture and associated material cost of bioelectrochemical systems (Noori et al. 2020). Use of low-cost catalyst (Tiwari et al. 2017), membranes (Tiwari et al. 2016), and electrodes (Ghangrekar and Tiwari 2022) can make hydrogen production economically more favourable in a commercial scale. The combination of two different technologies may eliminate individual drawbacks and increase the hydrogen yield which can subsequently improve the economic and environmental performance of biohydrogen production. For example, wastewater generated from fermentation is still laden with organic matter which can be utilized as a substrate in MECs to increase the hydrogen yield (Pathy et al. 2022). In addition to optimizing operating conditions of bioreactors, recent techniques such as implementation of genomics and proteomics may help achieve higher substrate consumption and conversion to hydrogen (Ramprakash et al. 2022). Selection of the best strain of microbes such as bacterial strain with high conversion efficiency of organics to biohydrogen or a fast-growing strain of algae can also yield better results. The substrate is one of the cost-determining factors of biohydrogen production, i.e. use of commercial substrates such as glucose may not be economically favourable in a large scale. The use of algae can be a cheaper alternative, and considering their growth rate, the biomass is higher than terrestrial biomass (Pathy et al. 2022). A more sustainable alternative can be wastewater which can not only improve the environmental performance but also contribute to the reduction in overall cost of biohydrogen production. Even wastewater can be used as the medium to grow algal biomass. Efforts directed towards reducing the lag time in hydrogen production can also reduce the operating cost as well as the energy demand of the process, which can eventually translate into economic and environmental benefits (Pathy et al. 2022). The simultaneous production of value-added chemicals along with hydrogen such as pigments, biomethane, etc., can further enhance the profitability, such as astaxanthin (~5112.78 €/kg) and β-carotene (~1095.41 €/kg) from *Spirogyra* microalgal species (Pacheco et al. 2015). A rather overlooked subject during comparison with fossil fuel is the cost associated with carbon capture and storage. It is estimated that by 2040, the global emission of CO_2 could reach a maximum of 44 billion tonnes. Moreover, considering a carbon capture and storage cost of 16$/tonne CO_2, the corresponding expenditure is around 704 billion dollars (Arranz 2016; Theo et al. 2016). However, biohydrogen production from renewable sources could reduce this cost aspect drastically. It is expected that the current advances in research and innovation could provide an economically favourable biohydrogen production model while maintaining sustainability also.

References

Ağbulut Ü, Sarıdemir S (2021) A general view to converting fossil fuels to cleaner energy source by adding nanoparticles. Int J Ambient Energy 42:1569–1574

Akhbari A, Onn CC, Ibrahim S (2021) Analysis of biohydrogen production from palm oil mill effluent using a pilot-scale up-flow anaerobic sludge blanket fixed-film reactor in life cycle perspective. Int J Hydrogen Energy 46:34059–34072

Anwar M, Lou S, Chen L et al (2019) Recent advancement and strategy on bio-hydrogen production from photosynthetic microalgae. Bioresour Technol 292:121972

Arranz AM (2016) Hype among low-carbon technologies: Carbon capture and storage in comparison. Glob Environ Chang 41:124–141

Baeza JA, Martínez-Miró À, Guerrero J et al (2017) Bioelectrochemical hydrogen production from urban wastewater on a pilot scale. J Power Sources 356:500–509

Bartels JR, Pate MB, Olson NK (2010) An economic survey of hydrogen production from conventional and alternative energy sources. Int J Hydrogen Energy 35:8371–8384

Behera B, Nageshwari K, Darshini M, Balasubramanian P (2020) Evaluating the harvesting efficiency of inorganic coagulants on native microalgal consortium enriched with human urine. Water Sci Technol 82:1217–1226

Bhar R, Tiwari BR, Sarmah AK et al (2022) A comparative life cycle assessment of different pyrolysis-pretreatment pathways of wood biomass for Levoglucosan production. Bioresour Technol 127:305

Chen J, Xu W, Wu X et al (2019) System development and environmental performance analysis of a pilot scale microbial electrolysis cell for hydrogen production using urban wastewater. Energy Convers Manag 193:52–63

Cheng S, Logan BE (2011) High hydrogen production rate of microbial electrolysis cell (MEC) with reduced electrode spacing. Bioresour Technol 102:3571–3574

Corominas L, Byrne DM, Guest JS et al (2020) The application of life cycle assessment (LCA) to wastewater treatment: a best practice guide and critical review. Water Res 184. https://doi.org/10.1016/j.watres.2020.116058

Cusick RD, Bryan B, Parker DS et al (2011) Performance of a pilot-scale continuous flow microbial electrolysis cell fed winery wastewater. Appl Microbiol Biotechnol 89:2053–2063

Djomo SN, Blumberga D (2011) Comparative life cycle assessment of three biohydrogen pathways. Bioresour Technol 102:2684–2694

Enerdata (2014) Total Energy Consumption. https://yearbook.enerdata.net/#energy-consumption-data.html
Ferreira AF, Marques AC, Batista AP et al (2012) Biological hydrogen production by Anabaena sp.–yield, energy and CO2 analysis including fermentative biomass recovery. Int J Hydrogen Energy 37:179–190
Ferreira AF, Ortigueira J, Alves L et al (2013a) Energy requirement and CO2 emissions of bioH2 production from microalgal biomass. Biomass Bioenergy 49:249–259
Ferreira AF, Ortigueira J, Alves L et al (2013b) Biohydrogen production from microalgal biomass: energy requirement, CO2 emissions and scale-up scenarios. Bioresour Technol 144:156–164
Foley JM, Rozendal RA, Hertle CK et al (2010) Life cycle assessment of high-rate anaerobic treatment, microbial fuel cells, and microbial electrolysis cells. Environ Sci Technol 44:3629–3637
Ghangrekar MM, Tiwari BR (2022) Development of low-cost microbial fuel cell for converting waste to electricity and abating pollution. In: Environmental management in India: waste to wealth. Springer, pp 167–198
Han W, Liu Z, Fang J et al (2016a) Techno-economic analysis of dark fermentative hydrogen production from molasses in a continuous mixed immobilized sludge reactor. J Clean Prod 127:567–572
Han W, Yan Y, Gu J et al (2016b) Techno-economic analysis of a novel bioprocess combining solid state fermentation and dark fermentation for H2 production from food waste. Int J Hydrogen Energy 41(22619–22):625
Heidrich ES, Dolfing J, Scott K et al (2013) Production of hydrogen from domestic wastewater in a pilot-scale microbial electrolysis cell. Appl Microbiol Biotechnol 97:6979–6989
Hosseinzadeh A, Zhou JL, Li X et al (2022) Techno-economic and environmental impact assessment of hydrogen production processes using bio-waste as renewable energy resource. Renew Sustain Energy Rev. 156:111991
Hwang JJ, Chang WR (2010) Life-cycle analysis of greenhouse gas emission and energy efficiency of hydrogen fuel cell scooters. Int J Hydrogen Energy 35(11947–11):956
Ibrahim IA, Ötvös T, Gilmanova A et al (2021) International energy agency. Kluwer Law International BV
ISO ISO (2006) International Standard ISO 14044, Environmental Management—Life Cycle Assessment—Requirements and guidelines. Int J Life Cycle Assess:652–668
Kannah RY, Kavitha S, Sivashanmugham P et al (2019) Biohydrogen production from rice straw: effect of combinative pretreatment, modelling assessment and energy balance consideration. Int J Hydrogen Energy 44:2203–2215
Kruse O, Hankamer B (2010) Microalgal hydrogen production. Curr Opin Biotechnol 21:238–243
Li Y-C, Liu Y-F, Chu C-Y et al (2012) Techno-economic evaluation of biohydrogen production from wastewater and agricultural waste. Int J Hydrogen Energy 37(15704–15):710
Manish S, Banerjee R (2008) Comparison of biohydrogen production processes. Int J Hydrogen Energy 33:279–286
Mead I (2017) International energy outlook 2017. US Energy InformationAdministration
Mehmeti A, Angelis-Dimakis A, Arampatzis G et al (2018) Life cycle assessment and water footprint of hydrogen production methods: from conventional to emerging technologies. Environments 5:24
Michiels F, Geeraerd A (2020) How to decide and visualize whether uncertainty or variability is dominating in life cycle assessment results: a systematic review. Environ Model Softw 133:104841
Nagarajan S, Chou SK, Cao S et al (2013) An updated comprehensive techno-economic analysis of algae biodiesel. Bioresour Technol 145:150–156
Nagarajan D, Lee D-J, Kondo A, Chang J-S (2017) Recent insights into biohydrogen production by microalgae–From biophotolysis to dark fermentation. Bioresour Technol 227:373–387
Nikolaidis P, Poullikkas A (2017) A comparative overview of hydrogen production processes. Renew Sustain energy Rev 67:597–611

Noori MT, Ganta A, Tiwari BR (2020) Recent advances in the design and architecture of bioelectrochemical systems to treat wastewater and to produce choice-based byproducts. J Hazardous Toxic Radioact Waste 24:4020023
Ochs D, Wukovits W, Ahrer W (2010) Life cycle inventory analysis of biological hydrogen production by thermophilic and photo fermentation of potato steam peels (PSP). J Clean Prod 18:S88–S94
Osman AI, Deka TJ, Baruah DC, Rooney DW (2020) Critical challenges in biohydrogen production processes from the organic feedstocks. Biomass Convers Biorefinery:1–19
Pacheco R, Ferreira AF, Pinto T et al (2015) The production of pigments & hydrogen through a Spirogyra sp. biorefinery. Energy Convers Manag 89:789–797
Panigrahi S, Sharma HB, Tiwari BR et al (2021) Insight into understanding the performance of electrochemical pretreatment on improving anaerobic biodegradability of yard waste. Renew Energy. https://doi.org/10.1016/j.renene.2021.08.123
Panigrahi S, Tiwari BR, Brar SK, Dubey BK (2022) Thermo-chemo-sonic pretreatment of lignocellulosic waste: Evaluating anaerobic biodegradability and environmental impacts. Bioresour Technol 361:127675
Pathy A, Nageshwari K, Ramaraj R et al (2022) Biohydrogen production using algae: potentiality, economics and challenges. Bioresour Technol 127:514
Patterson T, Esteves S, Dinsdale R et al (2013) Life cycle assessment of biohydrogen and biomethane production and utilisation as a vehicle fuel. Bioresour Technol 131:235–245
Pi AE, Dai Q, Han J, Wang M (2016) Life cycle analysis of emerging hydrogen production technologies. Argonne National Laboratory, Lemont, pp 1–5
Piccinno F, Hischier R, Seeger S, Som C (2016) From laboratory to industrial scale: a scale-up framework for chemical processes in life cycle assessment studies. J Clean Prod 135:1085–1097
Pörtner H, Roberts D, Masson-Delmotte V et al (2019) IPCC intergovernmental panel on climate change. Switz, Geneva
Pradhan S, Tiwari BR, Kumar S, Barai SV (2019) Comparative LCA of recycled and natural aggregate concrete using particle packing method and conventional method of design mix. J Clean Prod 228:679–691. https://doi.org/10.1016/j.jclepro.2019.04.328
Quek A, Balasubramanian R (2014) Life cycle assessment of energy and energy carriers from waste matter–a review. J Clean Prod 79:18–31
Ramprakash B, Lindblad P, Eaton-Rye JJ, Incharoensakdi A (2022) Current strategies and future perspectives in biological hydrogen production: A review. Renew Sust Energ Rev 168:112773
Reaño RL (2020) Assessment of environmental impact and energy performance of rice husk utilization in various biohydrogen production pathways. Bioresour Technol 299:122590
Rozendal RA, Hamelers HVM, Rabaey K et al (2008) Towards practical implementation of bioelectrochemical wastewater treatment. Trends Biotechnol 26:450–459
Sathyaprakasan P, Kannan G (2015) Economics of bio-hydrogen production. Int J Environ Sci Dev 6:352
Steinbusch KJJ, Arvaniti E, Hamelers HVM, Buisman CJN (2009) Selective inhibition of methanogenesis to enhance ethanol and n-butyrate production through acetate reduction in mixed culture fermentation. Bioresour Technol 100:3261–3267
Suwanmanee U, Saebea D, Hacker V et al (2018) Conceptual design and life cycle assessment of decentralized power generation by HT-PEMFC system with sorption enhanced water gas shift loop. Energy Convers Manag 171:20–30
Theo WL, Lim JS, Hashim H et al (2016) Review of pre-combustion capture and ionic liquid in carbon capture and storage. Appl Energy 183:1633–1663
Tian H, Li J, Yan M et al (2019) Organic waste to biohydrogen: a critical review from technological development and environmental impact analysis perspective. Appl Energy 256:113961
Tiwari BR, Bhar R, Dubey BK et al (2023) Life Cycle Assessment of Microbial 2,3-Butanediol Production from Brewer's Spent Grain Modeled on Pinch Technology. ACS Sustain Chem Eng. https://doi.org/10.1021/acssuschemeng.3c00616

Tiwari BR, Brar SK (2022) A life cycle assessment perspective to conventional and modular wastewater treatment. Modul Treat Approach Drink Water Wastewater:187–205

Tiwari BR, Noori MT, Ghangrekar MM (2017) Carbon supported nickel-phthalocyanine/MnOx as novel cathode catalyst for microbial fuel cell application. Int J Hydrogen Energy. https://doi.org/10.1016/j.ijhydene.2017.07.201

Tiwari BR, Noori MT, Ghangrekar MM (2016) A novel low cost polyvinyl alcohol-Nafion-borosilicate membrane separator for microbial fuel cell. Mater Chem Phys 182:86–93. https://doi.org/10.1016/j.matchemphys.2016.07.008

Tiwari BR, Rouissi T, Brar SK, Surampalli RY (2021) Critical insights into psychrophilic anaerobic digestion: Novel strategies for improving biogas production. Waste Manag 131:513–526

Wulf C, Kaltschmitt M (2013) Life cycle assessment of biohydrogen production as a transportation fuel in Germany. Bioresour Technol 150:466–475. https://doi.org/10.1016/j.biortech.2013.08.127

Yun Y-M, Lee M-K, Im S-W et al (2018) Biohydrogen production from food waste: current status, limitations, and future perspectives. Bioresour Technol 248:79–87

Chapter 23
Biohydrogen and the Circular Economy: Insertion of Biohydrogen Production in Biorefineries Aiming at Zero-Waste Processes

Arindam Sinharoy and Kannan Pakshirajan

Abstract Renewable energy derived from alternative waste resources has the potential to gradually replace fossil fuels or at least supplement the ever-growing energy demand. Hydrogen is one such clean renewable energy source which can be produced from a wide range of substrates by both biological and physico-chemical processes. However, biological methods employing different classes of microorganisms seem to be advantageous over other methods from the economical and sustainability aspects. In this regard, producing biohydrogen from organic waste from different industrial, agricultural, and domestic sources is an interesting option. Hydrogen is an attractive tool due to its diverse application potential, choice of diverse substrate for its production, and capacity to reduce carbon emissions. Along with hydrogen production, this biological route can also be utilized for the production of valuable chemicals, food, feed, and other materials under the biorefinery concept. Moreover, integration of biohydrogen production processes with other systems such as methane production, microbial fuel cells, and algal bioprocess can boost the economic prospect of the entire system. This will not only provide cheap substrates for energy generation but will also fulfil the goals of circular economy, i.e. reduce, reuse, and recycle. The principle of circular economy is derived from the Paris climate conference and is built on the concept of carbon balance or net zero release to manage the energy and carbon flow.

A. Sinharoy
Department of Microbiology, University of Galway, Galway, Ireland

K. Pakshirajan (✉)
Department of Biosciences and Bioengineering, Indian Institute of Technology Guwahati, Guwahati, Assam, India
e-mail: pakshi@iitg.ac.in

C. R. Soccol et al. (eds.), *Biohydrogen - Advances and Processes*, Biofuel and Biorefinery Technologies 13, https://doi.org/10.1007/978-3-031-49818-3_23

Graphical Abstract

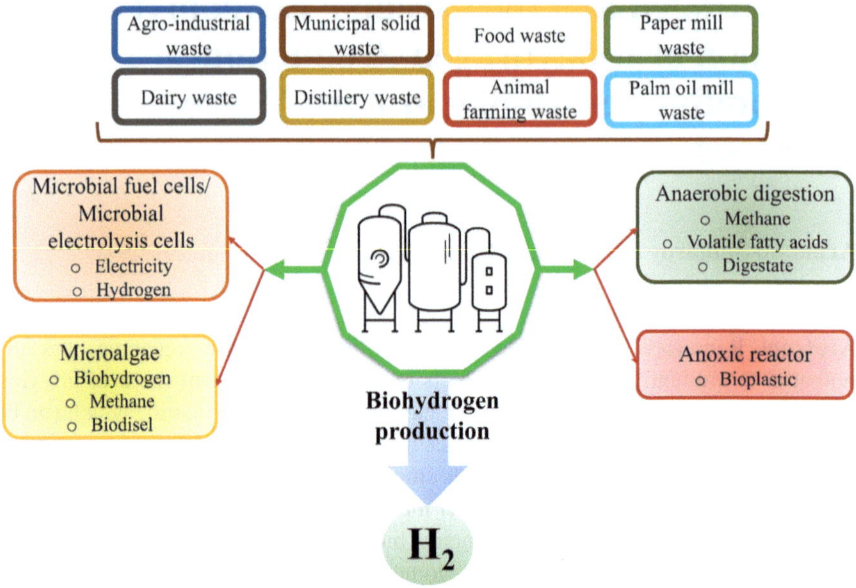

Keywords Biohydrogen · Bioeconomy · Waste to energy · Circular economy · Waste management

23.1 Introduction

Most of our energy demand for industry, transportation, and domestic usage is fulfilled by fossil fuel sources. The negative impact of this fossil fuel-derived energy is evident and well-known (Ellabban et al. 2014; Mohtasham 2015). Fossil fuel burning causes production of carbon dioxide, sulfur dioxide (SO_2), nitrogen oxides (NO_x), and particulate matters, which are the leading reasons for global warming (Nicoletti et al. 2015). In order to control global warming and greenhouse gas emission, limiting the utilization of fossil fuel reserves for commercial energy production is essential; this is, however, not feasible because by the middle of this century the world's energy demand is expected to increase by twofold (Haghi et al. 2018). In addition, political instability and war (or war-like situation) in many of the oil- and natural gas-producing countries hinder the smooth supply of these fossil fuels, causing energy insecurity in oil-importing countries (San-Akca et al. 2020). As only a few countries have considerable oil reserves to export, this monopoly over oil supply causes drastic price fluctuations worldwide, which not only impact the energy sector but also affects the economy in the majority of countries.

Renewable energy sources have the potential to solve many of these problems, including ensuring energy security, as these energy sources are derived from nature and replenish themselves over time without debilitating the Earth's resources (Stankiewicz 2006; Mathews and Tan 2014). Renewable energy sources include solar, tidal, wave, geothermal, wind, and biomass-based energy. Among the different renewable energy sources, biofuels derived from biomass have gained a lot of traction due to ease of availability, carbon neutrality, and reduction of waste, in addition to helping to reduce reliance on fossil fuel-based energy. Bio-based fuels include gaseous products such as methane, hydrogen, and liquid biofuel such as ethanol, methanol, butanol, biodiesel, etc. (Roy et al. 2016; Sinharoy et al. 2020a). Among these bioenergy sources, hydrogen seems to be promising due to its highest energy content. Furthermore, it is considered as clean fuel as it does not produce any of the pollutants that fossil fuels emit upon combustion. Not only as energy source, hydrogen is also important as reductive agent in various chemical processes, including wastewater treatment such as reduction of sulfate (Sinharoy et al. 2020b) and selenium oxyanions (Sinharoy and Lens 2020).

Biohydrogen can be produced by many methods using different classes of microorganisms (Table 23.1). These biohydrogen production methodologies can be broadly categorized into two groups: fermentative and photolytic methods, which could be further sub-categorized into photo fermentation, dark fermentation, and

Table 23.1 Biohydrogen production methods along with their biochemical reaction

Method	Biochemical reaction	Theoretical yield of H_2
Dark fermentation	$C_6H_{12}O_6 + 6H_2O \rightarrow 6CO_2 + 12H_2$ $C_6H_{12}O_6 + 2\ H_2O \rightarrow 2CH_3COOH + 2CO_2 + 4H_2$ $C_6H_{12}O_6 \rightarrow CH_3CH_2CH_2COOH + 2CO_2 + 2H_2$ $C_6H_{12}O_6 \rightarrow CH_3COOH + CH_3CH_2COOH + CO_2 + H_2$	12 mol of H_2 per mol of glucose 4 mol of H_2 per mol of glucose (for acetic acid production) 2 mol of H_2 per mol of glucose 1 mol of H_2 per mol of glucose
Photo fermentation	$C_6H_{12}O_6 + 6H_2O \rightarrow 6CO_2 + 12H_2$ $2CH_3COOH + 4H_2O \rightarrow 8H_2 + 4CO_2$	12 mol of H_2 per mol of glucose 8 mol of H_2 per mol of glucose
Direct photolysis	$2H_2O$ + light energy $\rightarrow O_{2\,+}\ 4H^+ + 4e^-$	2 mol of H_2 per mol of water lysed
Indirect photolysis	$6CO_2 + 12H_2O$ + light energy $\rightarrow C_6H_{12}O_6 + 6O_2$ $C_6H_{12}O_6 + 2H_2O \rightarrow 4H_2 + 2CH_3COOH + 2CO_2$ $2CH_3COOH + 4H_2O$ + light energy $\rightarrow 8H_2 + 4CO_2$	4 mol of H_2 per mol of glucose/ acetic acid
Bioelectrochemical system	Anode: $CH_3COOH + 2H_2O \rightarrow 2CO_2 + 8e^- + 8H^+$ Cathode: $8e^- + 8H^+ \rightarrow 4H_2$ Overall: $CH_3COOH + 2H_2O \rightarrow 2CO_2 + 4H_2$	4 mol of H_2 per mol of acetic acid
Biological water gas shift reaction	$CO + H_2O \rightarrow CO_2 + H_2$	1 mol of H_2 per mol of CO

direct and indirect biophotolysis, respectively (Manish and Banerjee 2008; Das 2009; Show et al. 2011). Dark fermentation is performed by homoacetogenic microorganisms, whereas photo fermentation is carried out by purple sulfur and nonsulfur bacteria. The biophotolysis (both direct and indirect ones) is performed by photosynthetic organisms such as algae and cyanobacteria. In addition to these methods, biological water gas shift (WGS) reaction can be used for production of hydrogen using carbon monoxide as the substrate (Sinharoy et al. 2022). The microorganisms capable of performing this WGS reaction contains hydrogenase and carbon monoxide dehydrogenase enzyme complex, and involve Wood–Ljungdahl pathway (also known as the reductive acetyl–CoA pathway) for the production of hydrogen.

Although biohydrogen has become one of the major renewable fuels with production reaching more than 50 million tonnes every year and its market share is increasing every year by 10% (Nagarajan et al. 2021; Ahmed et al. 2021), its commercialization is very challenging. Moreover, the majority (>99%) of the hydrogen produced today is from non-renewable sources such as natural gas, coal, and petroleum (Osman et al. 2020). The difficulties faced by hydrogen industry are mainly related to the techno-economic and environmental issues related to its production. Moreover, transportation and storage issues negatively impact its successful commercialization (Ahmed et al. 2021). In this regard, use of waste resources such as organic-rich wastewater, solid organic waste, and waste biomass for biohydrogen production could serve as better alternatives from economic and environmental point of view. Furthermore, integration of biohydrogen production process with production of other value-added products such as alcohol, volatile fatty acids, long chain fatty acids, bioplastics, etc. under biorefinery approach can improve the process economics (Banu et al. 2020).

23.2 Biohydrogen Production Routes

Biohydrogen production from organic waste can be achieved through various routes as mentioned earlier in the introduction section (Table 23.1). Among these methods, fermentation, more particularly dark fermentation, is the most commonly reported method for converting organic waste to biohydrogen (Sinharoy et al. 2020a). Along with hydrogen, other by-products such as acetic acid, butyric acid, and propionic acid are produced which reduce the yield of hydrogen on biomass. The maximum hydrogen yield can be 4 mol of hydrogen per mol of glucose utilized in case of only acetic acid as the by-product which is quite low in comparison with theoretical hydrogen yield of 12 mol in the absence of acid production. The value can, however, further reduce to 2 mol or 1 mol depending upon butyric and propionic acid formation, respectively, during the process. The key enzyme involved for hydrogen production is hydrogenase with either a Ni–Fe core or Fe–Fe core (Manish and Banerjee 2008).

Photo fermentation process uses photosynthetic bacteria such as sulfur and non-sulfur purple bacteria for the production of biohydrogen from organic wastes in the presence of light (Das 2009; Show et al. 2011). In this process both sugar and organic acids can be utilized for biohydrogen production by photo fermentative microorganisms. However, the H_2 yield is higher (8 mol H_2 per mol acetic acid) in case of organic acid compared with other substrates or the dark fermentation process. Hence, this process is often used as the second stage in a two-stage biohydrogen production process to utilize organic acid by-products produced in the first stage (dark fermentation). The role of light in this process is to help generate adenosine triphosphate (ATP) required for microbial growth. For this purpose, hydrogen production using both solar and artificial lights has been reported in the literature.

Photolysis is a process in which photosynthetic aerobic microorganisms such as algae and cyanobacteria perform direct water splitting reaction to produce hydrogen (Tamburic et al. 2013). Hydrogenase and nitrogenase enzymes play an important role in this process. In case of direct photolysis, biohydrogen can be produced directly from water and light. The microorganisms perform the water splitting reaction by taking up solar energy at 680 nm wavelength and the electrons generated are transferred through photosystem I and photosystem II. The indirect photolysis is called so because it involves a two-step reaction: In the first step, the solar energy is converted to chemical energy (as carbohydrate) which is then converted to CO_2 and H_2 in the second step. N_2 limiting condition favours a high carbohydrate accumulation in the first step of indirect photolysis. Although both these steps occur under light condition, the first step is aerobic, and the second one is conducted under anaerobic condition (Ghirardi et al. 2014).

Biohydrogen production using bioelectrochemical system is another new technique with conversion of organic wastes (Rosenbaum et al. 2010). This process utilizes microbial electrolysis cells (MEC) consisting of two electrodes, a cathode and an anode, placed either in a single chamber or in two different chambers separated by ion-selective membrane. The organics containing wastewater for treatment is supplied in the anode chamber, whereas the cathode chamber is filled with solutions of various compositions (Arun et al. 2020). The working principle involves the oxidation of the organic matter in the anode compartment to produce electrons which are then transported to the cathode compartment where they combine with protons to form H_2. This type of system is operated under anaerobic condition and is sensitive to oxygen.

Another less explored method for biohydrogen production is water gas shift reaction which produces equimolar concentrations of CO_2 and H_2 from CO and water (Sinharoy et al. 2019). The conventional thermo-catalytic reaction for production of H_2 from CO is performed under high temperature and pressure conditions using costly, perishable catalysts. However, the same reaction can be performed by a class of microorganism containing hydrogenase and carbon monoxide dehydrogenase enzyme systems (Sinharoy and Pakshirajan 2021). This type of carboxydotrophic hydrogenogenic microbes are rare in nature and only been isolated from extreme environments such as hot springs, thermophilic hot vent, deep ocean trench, etc. (Omae et al. 2019). The substrate, i.e. CO, can be produced from any type of waste through biomass gasification which produces syngas containing CO, CO_2, and H_2 (Alfano and Cavazza 2018). The H_2 and CO_2 can be separated for use in many other applications, whereas the CO can be used for hydrogen production by water gas shift reaction. In a different strategy, the proportion of H_2 in the syngas can be enhanced by passing it through a bioreactor containing carboxydotrophic hydrogenogens (Sinharoy et al. 2022). The advantage of this process is that whole biomass can be converted to gaseous substrate with very little loss in biomass in the form of biochar or solid particles. This is in contrast to biological biomass conversion methods which utilize only the organic fraction and requires other undesired recalcitrant compounds (such as lignin) present in the biomass to be removed. Hence, overall conversion of biomass is higher in this gasification–fermentation route to biohydrogen production.

23.3 Biohydrogen Production from Different Waste

Due to the exponential population growth, generation of both wastewater and solid waste from anthropogenic activities are increasing significantly (Bräutigam et al. 2014; Sinharoy et al. 2020c; Baskaran et al. 2020). Safe disposal of such wastes is one of the major challenges faced by civic authorities throughout the world. Hence, conversion of these wastes combined with bioenergy generation such as for the production of biohydrogen help in transition towards a zero-waste circular economy (Banu et al. 2020; Paul et al. 2022). Utilization of different organic wastes for biohydrogen production is well-reported in the literature (Table 23.2). Most of these studies pertain to fermentative biohydrogen production using acidogenic microorganisms. The complex nature of the waste due to the exponential population growth seldom poses difficulty for the microorganisms to efficiently utilize the substrate for biohydrogen production (Dahiya et al. 2021). To solve such problem, often pretreatment of the wastes and other strategies such as optimization of process parameters are taken up by researchers. In the following sections, biohydrogen production from different wastes is discussed in detail.

23.3.1 Agro-industrial Waste

Agricultural industry produces large quantities of waste which calls for sustainable waste management practices. For example, the total amount of waste generated worldwide by top five crop residues, namely, corncob, oil palm residue, rice husk, rice straw, sugarcane bagasse, and wheat straw, was 2.09 billion tonnes in the year 2017–2018 (Millati et al. 2019). However, the annual generation of agricultural waste amounts to 350 million tonnes in India (ICAR 2020). This data reveals regional and seasonal variations in the amount of agro-industrial waste generated.

Table 23.2 Selected studies on biohydrogen production from different type of waste

Type of waste	Bioreactor	Inoculum	pH	Temp. (°C)	HRT (h)	Substrate concentration	Hydrogen production/ yield	Treatment efficiency	References
Switchgrass (steam explosion treated)	UASB	Granular sludge from brewery wastewater treatment plant	5.0	37	10	5 g COD/L/day	Yield = 99.86 mL/g VS	NM	Veeravalli et al. (2014)
Rice bran hydrolysate	Batch Duran bottles	*Clostridium saccharoperbutylacetonicum*	6.0	30	NA	~30 g/L	Yield = 3.37 mol/mol sugar Productivity = 7.58 mmol/(L h)	Maximum 85%	Dada et al. (2013)
Oat straw hydrolysate	Anaerobic sequencing batch reactors	Anaerobic granular sludge from full-scale UASB	5.5	35	24 h (cycle time of 12 h)	5 g COD/L	Yield = 2 mol/mol sugar Rate = 278 mL/L/h	NM	Arreola-Vargas et al. (2013)
Mixed agro-municipal waste	Batch modified Erlenmeyer flasks	Anaerobic sludge pretreated at 100 °C for 30 min	7.9	35	72	40.45 g COD/L	Yield = 57.73 mL/g TVS	NM	Sekoai and Kana (2013)
Rice slurry	Batch serum bottles	Anaerobic digester sludge pretreated at 100 °C for 30 min	4.5	37	NA	5.5 g carbohydrate/L	Yield = 346 mL/g carbohydrate Rate of 2.1 L/(g VSS day)	NM	Fang et al. (2006)
Sugar beet pulp (enzymatically hydrolysed)	Semi-continuous reactor	Anaerobic sludge pretreated at 80 °C for 1 h	5.5	35	NM	OLR = 9.95 g/dm^3/day	Yield = 52.11 dm^3/kg VS	NM	Cieciura-Włoch et al. (2020)
Beverage wastewater	High rate stirred tank reactor	Enriched anaerobic mixed culture	6.3	37	1.5	OLR = 320 g/L day	Yield = 1.50 mol/mol substrate	86.3–98.3%	Sivagurunathan et al. (2015)
Palm oil mill effluent	UASB	*Clostridium* sp. LS2	5.5	37	16	OLR = 30 g/L day	Yield = 0.38 mol/mol substrate Rate = 8.76 L/L day	56–62%	Singh et al. (2013)

(continued)

Table 23.2 (continued)

Type of waste	Bioreactor	Inoculum	pH	Temp. (°C)	HRT (h)	Substrate concentration	Hydrogen production/ yield	Treatment efficiency	References
Palm oil mill effluent	Stirred tank reactor	Anaerobic digester sludge	5.5	55	48	12.7	Yield = 1.58 mol/mol substrate Rate = 2.1 L/L day	81.2%	Ismail et al. (2011)
Cheese whey	Batch serum bottles	Anaerobic sludge	6.8	55	NA	26 g/L as sugar	Yield = 1.56 mol/mol glucose Rate = 3.45 mL/h	NM	Kargi et al. (2012)
Cheese whey	UASB	Sludge from an acidogenic lab-scale reactor	5.0	30	24 h	OLR = 20 g COD/L/day	Rate = 122 mL/L/day	NM	Castelló et al. (2009)
Dairy wastewater	Batch serum bottles	Anaerobic sludge from full-scale anaerobic digester	5.5	37	NA	15.3 g COD/L	Yield = 13.54 mmol/g COD Rate = 29.91 mmol/g VSS/day	NM	Gadhe et al. (2013)
Distillery wastewater	Anaerobic sequencing batch biofilm reactor	Enriched anaerobic mixed consortia	7.0	28	24 h	OLR = 9.6 kg COD/m^3/day	Rate = 6.98 mol/kg COD day	NM	Mohan et al. (2008b)
Alcohol production wastewater	Anaerobic sequencing batch reactor	Sludge treated at 100 °C for 15 min	5.5	55	NM	OLR = 68 kg/m^3/day	Yield = 130 mL/g COD removed Rate = 2100 mL/L/day	NM	Intanoo et al. (2012)
Food waste (ultrasonic treated)	Batch serum bottles	Anaerobic sludge from full-scale anaerobic digester	5.5	37	NA	NM	Yield = 149 mL/g VS added Rate = 5.23 mL/h	NA	Gadhe et al. (2014)
Food waste	Leaching bed reactor	Heat shock treated anaerobic sludge	5.5–7.0	37	25 days	OLR = 11.9 kg VS/m^3/day	Yield = 310 mL/g VS Rate = 151.3 mL/L/h	72.5% VS	Han and Shin (2004)
Vegetable kitchen waste	Intermittent-continuous stirred tank reactor	Kitchen waste compost	6.0	55	4 days	OLR = 19 and 28 g COD/L/day	Yield = 1.7 mmol/g COD Rate = 1 L/L/day	62% carbohydrate and 37% cellulose	Lee et al. (2010)

Thick juice dark fermenter effluent	Pilot scale solar tubular photo-bioreactor	*Rhodobacter capsulatus*	6.4	30–35	NM	3700 mg/L COD	Yield = 0.4 mol/mol acetic acid Rate = 0.15 mol/(m^3 h)	40% COD	Boran et al. (2012)
Sewage sludge	Thermophilic flow-through reactor	Sludge treated at 100 °C for 30 min	6.1	55	3 days	42.6 g/L	Yield = 3.07 mmol/g TS	22.2–37.9% VS	Woo and Song (2010)
Oil palm trunk hydrolysate with Slaughter-house wastewater	Batch serum bottles	*Thermoanaerobacterium thermosaccharolyticum*	6.5	60	NA	14.12 g/L sugar and 22.76 g/L COD	Yield = 114 mL/g COD	NM	Khamtib and Reungsang (2014)
Cornstalk waste (acid pretreated)	Batch serum bottles	Cow dung compost	7.0	36	NA	NM	Yield = 149.7 mL/g VS Rate = 7.6 mL/h	NM	Zhang et al. (2007)
Organic fraction of municipal solid waste	Stirred tank reactor	Activated sludge	5.5 and 6.5	38	NM	NM	Yield = 68.5 to 88.5 L/kg VS	NM	Baldi et al. (2019)
Organic fraction of municipal solid waste	Batch reactor	Anaerobic inoculum	7.6	37	NM	NM	Yield = 17.2 mL/g VS	NM	Paillet et al. (2021)

NA: not applicable; NM: not mentioned

There is also a large variation in the amount of waste generated from the different crops. For example, the amount of rice straw and rice husk is the largest among the different crop residues produced with China, India, and Indonesia being the top three countries producing the waste (188, 144, and 49 million tonnes, respectively) (USDA F 2020). In case of wheat straw EU, China, and the USA produce 128, 110, and 40 million tonnes of the agri-waste, respectively (USDA N 2016).

Agro-industrial wastes can be categorized into two types: the first one being agricultural residues which include field residue (e.g. crop stalk, stem, husk, leaves, bagasse, etc.) and the second category being industrial residues (e.g. fruit and vegetable peels, oil cakes, etc.). The first category of the agro-industrial wastes, i.e. field residues, are often discarded or used as animal feed. Sometimes they are burnt in field which causes severe environmental pollution. The industrial residues are also underutilized, and due to its high content of COD, BOD, and other nutrients, proper treatment is essential for safe disposal of the waste. Hence, considering their easy availability at no cost, such agro-industry waste can serve as alternate substrate for biohydrogen production, thereby ensuring both safe disposal and energy recovery as hydrogen.

Many previous studies have been reported on biohydrogen production using different agricultural wastes as substrate. Veeravalli et al. (2014) studied biohydrogen production from switchgrass liquor (pretreated by steam explosion) and using an upflow anaerobic sludge blanket reactor. The maximum hydrogen yield of 99.86 (±5.6) ml/g VS was obtained under optimum experimental conditions, i.e. pH 5.0, 10 h HRT and 1.75 g/L linoleic acid concentration. The results showed that lowering the HRT and pH inhibited the activity of methanogens in the reactor and addition of linoleic acid enhanced hydrogenogenic organisms such as *Ruminococcaceae* and *Clostridiaceae* sp. in the reactor. In another study on biohydrogen production from rice slurry, a very high value of specific hydrogen production rate (2.1 L/g VSS day) and a high hydrogen yield of 346 ml/g carbohydrate were reported at optimum experimental conditions of pH 4.5 and 37 °C temperature (Fang et al. 2006). The stoichiometric yield on substrate was 62.6% which is significantly higher than other reported values. The inoculum used in this study was anaerobic sludge pretreated at 100 °C for 30 min. Microbial community analysis of the inoculum revealed that it was enriched with the *Clostridium* sp. bacterium.

Tawfik and Salem (2014) examined the effect of HRT (4–20 h) on biohydrogen production from pretreated rice straw using a pilot-scale upflow anaerobic staged reactor (80 L volume) under fixed organic lading rate (30 g COD/L day) and sludge retention time (1.9 days). The H_2 production increased from 0.4 to 3.6 L H_2/day when the HRT was increased from 4 to 20 h, corresponding to an increase in H_2 yield from 0.03 mol H_2/g COD (at 4 h HRT) to 2.1 mol H_2/g COD (at 20 h HRT). However, the COD removal efficiency decreased from 77.7% to 12.2% with a decrease in HRT from 20 h to 4 h due to low residence time of wastewater inside the reactor. In another approach, Anam et al. (2012) explored photo fermentation method for hydrogen production from bagasse and soy sauce wastewater using a marine photosynthetic bacterial consortium predominated by *Rhodobium marinum*. The maximum cumulative hydrogen production obtained in this study was 200

(±67) ml H_2 with soy sauce wastewater and 41 (±16) ml H_2 from the hydrolysed bagasse as substrate.

23.3.2 *Municipal Solid Wastes*

Municipal solid waste (MSW) includes waste from homes, commercial institutions and official establishments, construction and demolition waste, medical waste, and municipal waste (World bank 2022). According to World Bank data, the amount of MSW generation throughout the world is nearly 2.01 billion tonnes per year which is expected to rise to 3.40 billion tonnes by the middle of this century (World bank 2022). The average value of solid waste generated per person per day is 0.74 kg but varies largely (in the range 0.11–4.54 kg) depending upon the source country (World bank 2022). Although unsafe disposal of MSW is more of an issue in developing countries, the solid waste generation is much larger in developed or economically well to do countries as they produce 34% of all the MSW amounting to 683 million tonnes (World bank 2022). According to a report by the Environmental Protection Agency (EPA), the USA generated 292.4 million tonnes of municipal solid waste, out of which 62.94% was organic waste constituting of food waste, yard trimming, wood, paper and paper board, etc., whereas 35.5% was inorganic (or non-compostable forms) wastes such as metals, glass, plastic, rubber, leather, etc. (EPA 2018).

The safe disposal of MSW is a challenge owing to the exponential rate of MSW generated and the ever-growing human population. In general, the civic municipal authorities play a primary role in collection and disposal of MSW throughout the world. Conventional methods for MSW disposal involve land filling and open-air dumping. However, 33% of the total MSW generated worldwide is disposed of in an unsafe manner and can potentially harm the environment (World bank 2022). If these wastes are segregated into recyclable, compostable (organic), inorganic, and hazardous wastes, they could be disposed of more scientifically under the ambit of hierarchical approach of solid waste management, i.e. reduce, reuse, recycle, recover, and dispose.

From the point of view of biohydrogen generation, the organic fraction of the MSW can be utilized by fermentative microorganisms for hydrogen production. There are quite a few previous studies on this approach that are listed in Table 23.2. Process parameters such as temperature, pH, organic loading rate, total solids, hydraulic retention time, etc. have significant influence on the biohydrogen production. In a study conducted by Escamilla-Alvarado et al. (2013), effect of temperature, solids content, and solid retention time on biohydrogen production from organic fraction of municipal solid waste was evaluated under batch and semi-continuous modes of operation. During the semi-continuous mode, the effect due to total solids (TS) content was found to be the most significant on the biohydrogen production compared with the other two parameters. Maximum biohydrogen productivity of 123 ml H_2/kg/day was obtained in the reactor fed with 20.9% TS containing feedstock. However, no significant lag phase was observed during the batch experiment compared with the semi-continuous operation mode due to low pH condition and accumulation of lactic acid which inhibited the hydrogen production. Due to this reason, the batch process showed three times higher specific energetic potential in comparison to the semi-continuous mode of operation.

In another study, a two-stage approach was followed for simultaneous production of hydrogen and methane from organic fraction of municipal solid waste (Kvesitadze et al. 2012). The first stage, i.e. the hydrogen-producing reactor, was operated under thermophilic condition (55 °C) to inhibit methanogenic activity. The maximum H_2 production of 104 L/kg volatile solids was achieved at pH 9.0 with the pretreated substrate, which was significantly higher than the value obtained (82.5 L/kg VS) with untreated substrate. The purity of hydrogen produced in the system was also high with only <0.1% of methane present in the gas stream. In the second stage, i.e. methane-producing reactor, the maximum cumulative methane production was 520 L/kg VS with methane constituting 78.6% of the biogas produced.

23.3.3 *Food Waste*

Food waste is one of the major constituents of municipal solid waste. It consists of both raw and cooked food, and it contains high amount of carbohydrate and protein. According to the Food and Agriculture Organization (FAO) of the United Nations, more than 30% of the food produced for human consumption end up as waste which amounts to almost 1.3 billion tonnes per year (Bräutigam et al. 2014). Food waste occurs throughout the supply chain starting from production-harvesting to final consumption. In developed countries, the food waste occurs in the downstream stages of the supply chain and at the consumer level, and hence food waste generation is mainly related to the supplier and customer behaviour. However, in case of developing and underdeveloped countries, most of the wastage is in the upstream stage of the supply chain, including during harvesting, processing, and storing of food products. The reasons for food wastage in developing countries can be various: poor technology, lack of infrastructure and storing facility, difficult environmental conditions, packaging and transporting issues, etc. (Kummu et al. 2012; Thi et al. 2015).

The characteristics of food waste are favourable to biological degradation as most of it is organic in nature with a very high carbon content. Ramos et al. (2012) examined cafeteria food waste for biohydrogen production under mesophilic temperature (36 °C) in a batch system. The maximum H_2 production of 2.90 mmol/day was obtained at 90 g/L of TS and 5.5 pH. However, the total solids and COD removal were minimum (10% and 14%, respectively) at this experimental condition. In another study, effect of process parameters such as temperature (35–55 °C) and pH (4–8) on biohydrogen production from food processing wastewater was investigated (Lin et al. 2011). These parameters were found to have great influence on the hydrogen production: under thermophilic condition (55 °C), the amount of specific hydrogen production was more than twice the amount produced under mesophilic condition (35 °C). The maximum hydrogen production rate of 26.7 ml H_2/g VSS/h was observed at an optimum pH of 6.

The effect of inoculum to substrate ratio on biohydrogen production was examined in a study with food waste as the substrate and activated sludge from wastewater treatment plant as the inoculum (Cappai et al. 2018). The optimum ratio was found to be 0.14 g of substrate to inoculum with maximum biohydrogen yield of 88.8 L H_2/kg VS and hydrogen production rate of 10.8 L H_2/kg VS/h.

Simultaneous production of hydrogen and methane was also attempted by codigestion strategy with sewage sludge and food waste as substrates (Siddiqui et al. 2011). The experimental setup consisted of a two-stage reactor with the first stage for hydrogen production and the second stage for methane production from the digestate obtained from the first stage. The two-stage system reported a 0.13 m^3 hydrogen/kg VS and 0.67 m^3 methane/kg VS production, which was higher than the value obtained using a single-stage system (0.48 m^3 methane/kg VS). The volatile solids removal was 69–89% higher in the two-stage system than in the single-stage system.

23.3.4 Paper Mill Wastewater

Paper and pulp industry is one of the major industries processing large quantities of lignocellulosic materials. The discharge from this industry contains huge amounts of suspended materials such as plant parts, cellulosic fibres, organic debris, etc. (Ashrafi et al. 2015). Wastewater generated from paper–pulp industry is characterized by high COD content, presence of lignin and phenolic derivatives of lignin, dark colour, and high toxicity. Both the wastewater and paper mill sludge have been utilized previously for the production of bioenergy compounds such as methane, ethanol, hydrogen, etc. However, due to the presence of highly recalcitrant compounds, pretreatment of the wastewater is necessary for its complete utilization. Dessì et al. (2018) examined paper and pulp wastewater for biohydrogen production over a wide temperature range (37–80 °C), and observed that even a small increase in the temperature affected the hydrogen yield and productivity. The optimum temperature was observed to be 70 °C with a maximum H_2 yield of 3.6 mmol/g COD at that temperature. Microbial community analysis showed that *Thermoanaerobacterium* sp. is predominant over other bacteria at such high temperature. The authors suggested that since the temperature of the paper and pulp wastewater at discharge is already high (50–80 °C), hydrogen production could be successfully achieved using this strategy without the need for any external heating.

Enzymatic pretreatment of paper and pulp industry wastewater for hydrogen production by *Enterobacter aerogenes* using crude cellulase enzyme obtained from *Trichoderma reesei* was examined in a study by Lakshmidevi and Muthukumar (2010). The enzymatic hydrolysate (22 g/L of total sugars) showed a maximum hydrogen yield of 2.03 mol H_2/mol sugar and specific hydrogen production rate of 225 mmol of H_2/g cell/h.

Codigestion of paper mill sludge with food waste was carried out for simultaneous production of hydrogen and methane in a two-stage process (Lin et al. 2013). The first stage was operated under mesophilic condition for hydrogen production and the second stage under thermophilic condition for methane production with the remaining substrate from the first stage. The best hydrogen yield of 64.5 ml/g VS and methane yield of 432.3 ml/g VS were obtained when both the substrates were at 1:1 ratio (VS basis). The soluble COD removal in the system ranged in between 71 and 87%. The bioreactor system performed well in terms of maintaining a stable pH of the wastewater in the reactors as there was no acid accumulation in the system.

23.3.5 Dairy Industry Wastewater

Dairy industry is one of the major producers of wastewater worldwide (Kolev Slavov 2017). For example, dairy industry in Europe processes 170 billion L of milk per year with 2.5 L of wastewater generated per L of milk processed. Dairy wastewater is another good substrate for biohydrogen production (Gadhe et al. 2014). Composition of the wastewater is complex, and it contains lipids along with carbohydrate and proteins which are difficult to treat by anaerobic methods. Dairy industry produces wastewater with varying contents of organic matter depending upon the location and process being followed in a particular plant. For example, wastewater obtained from cheese-making process is complex with a very high content of COD (28–65 g/L), fat (14–24.5 g/L), and acidity (3.9–6.1 g/L) (Erdirencelebi 2011).

Gadhe et al. (2014) studied biohydrogen production using dairy wastewater as the substrate and with pretreated (treatment at 90 °C for 20 min) anaerobic sludge as the inoculum. The study revealed good potential of dairy wastewater for biohydrogen production with a maximum specific growth rate of 0.64 h^{-1}, half saturation concentration of 15.89 g COD/L, and maximum predicted biomass concentration of 7.26 g VSS/L using Gompertz model. Moreover, the maximum hydrogen production predicted by the Gompertz equation was 2283.6 mM/L.

Cheese whey industry is a major dairy processing industry, but it generates significant amount of wastewater during different stages of production. On an average, 9–10 L of wastewater is generated for every kg of cheese whey produced. Owing to its high content of nutrients, the wastewater can cause serious pollution of water bodies due to eutrophication, if released without treatment. Dessì et al. (2020) used cheese whey for biohydrogen production in two UASB reactors operated at different temperatures (20 and 35 °C) and pH (5 and 4). The maximum H_2 yield (0.9 mol/mol glucose equivalent) was obtained at 35 °C and pH 5.0 during batch operation. The H_2 production further increased to 2.0 mol/mol glucose equivalent under continuous mode of operation at the same experimental conditions.

23.3.6 Animal Farming Waste

Increase in population has created high demand for animal products such as milk, egg, and meat for consumption and has, therefore, necessitated increase in animal farming practices (Abdeshahian et al. 2016). As a result of this, treatment and safe disposal of livestock manure from animal farms rearing cattle, poultry, swine, etc. is much needed. One of the common disposal mechanisms is through production of biogas and composting (Purdy et al. 2018). However, biohydrogen production could be of more beneficial due to its high energy potential and clean combustion property which reduces carbon emission from the treatment process (Table 23.2). The following are some studies reported on biohydrogen production from various livestock manure.

An acidified (with 0.2% HCl) and thermally pretreated dairy manure produced a maximum of 31.5 ml/g total volatile solids at optimum conditions of 70 g/L of substrate and pH 5.0 (Xing et al. 2010). The liquid effluent from the system contained mostly different VFAs with acetic acid and butyric acid being dominant (78.2–81.4%) among them. A thermophilic (55 °C) stirred tank type digester treating cattle manure together with waste from agricultural slaughter industry at different ratio yielded a high amount of hydrogen (10.3–30.1 ml/g of VS) (Gilroyed et al. 2010). The maximum H_2 production was observed in case of a mixture of cattle manure and agricultural slaughter industry waste of ratio 90:10; above 10% w/w of the agricultural slaughter industry waste, the hydrogen production was low, and the production of unwanted ammonia and fatty acids were high.

Perera and Nirmalakhandan (2010) demonstrated increase in biohydrogen production (by 10%) from sucrose by the addition of dairy manure as a co-substrate. Dairy manure addition improved buffering capacity of the system, provided additional nutrients, and enhanced the growth of hydrogenogenic organisms in the system, which helped to increase the hydrogen production and hydrogen yield on the substrate.

A lesser used animal waste, i.e. elephant dung, was explored for biohydrogen production under thermophilic condition (55 °C) (Saripan and Reungsang 2014). The elephant dung is a complex substrate containing high amount of cellulose, hemicellulose, and lignin which are highly recalcitrant to biodegradation. However, in this study anaerobic consortium dominated by *Thermoanaerobacterium thermosaccharolyticum* and *Clostridium* species was able to produce hydrogen from this substrate. The maximum hydrogen yield and hydrogen production rate obtained in this study were 7.22 (±0.62) mmol/g substrate and 73.4 (±3.8) ml/L h, respectively, at pH 7.0 and substrate concentration of 0.25 g/L.

23.3.7 Distillery Industry Wastewater

Distillery industry wastewater is generated in large amounts (8–10 L per l of alcohol produced). In addition to high amount of sugars, dextrin and hemicelluloses, the wastewater contains significant amount of recalcitrant phenolic derivatives of lignin which makes it difficult for biodegradation (Mohana et al. 2009; Dhote et al. 2021). However, there are a few studies which used specific microorganisms capable of utilizing distillery industry effluent as the substrate for biohydrogen production. One such study reported *Enterobacter cloacae* for the production of H_2 from distillery industry effluent, and the effect of Fe^{2+}, Cu^{2+}, and Mg^{2+} addition on the process efficiency (Mishra and Das 2014). The experimental results showed a maximum H_2 yield of 7.38 (±0.24) mol/kg COD and hydrogen production rate of 80 mL/L h with supplementation of nutrients such as yeast extract (2.50 g/L), malt extract (1.50 mg/L), iron sulfate (0.25 g/L), magnesium sulfate (2.50 g/L), and copper chloride (3.50 g/L).

Biohydrogen production from alcohol industry wastewater was examined using an anaerobic sequencing batch reactor operated under a fixed temperature of 55 °C and pH 5.5 (Intanoo et al. 2012). The bioreactor system reported a high hydrogen yield and specific hydrogen production rate of 130 ml/g COD and 2.1 L/L day, respectively, at a COD loading rate of 68 kg/m^3 day. Optimum nutrient condition in the reactor for hydrogen production was found to be 100:6:0.5 for COD:N:P.

Pretreatment of the wastewater can further improve its biodegradability for enhancing the biohydrogen production. In a previous study, the effect of ultrasonication of distillery wastewater at a constant frequency of 20 kHz on dark fermentative hydrogen production was investigated (Gadhe et al. 2014). The pretreated wastewater showed enhanced hydrogen yield and hydrogen production rate by 101 and 103%, respectively. The maximum values for both these parameters were 10.95 mmol/g COD and 6.67 mmol/L/h under the optimum conditions of 56 g/L COD, 0.12 W/ml ultrasonication intensity, and 17 min ultrasonication time.

23.3.8 Palm Oil Mill Wastewater

Palm oil mill wastewater is a high strength wastewater generated during palm oil production. However, this industry is located only in few countries producing palm oil, namely, Indonesia, Malaysia, Thailand and Nigeria, etc. (Lee et al. 2019). As per an estimate, 1.5 m^3 of water is required to process a tonne of fresh palm fruit for the oil production, thereby generating large quantities of wastewater. The wastewater contains both water-soluble and suspended fraction from palm fruits, including carbohydrates (simple sugars, cellulose), lipids, oil residues, short palm fibre, cell walls, organelles, amino acids, proteins, etc. (Ahmad et al. 2016). Hence, a lot of efforts are focused towards treating the wastewater to avoid environmental pollution.

Palm oil mill wastewater has been evaluated for biohydrogen production. For example, Chong et al. (2009) studied biohydrogen production by *Clostridium butyricum* EB6, isolated from anaerobically digested palm oil mill sludge, with raw palm oil mill effluent as the sole substrate. The maximum H_2 yield of 31.9 ml/g COD and volumetric H_2 production rate of 1034.7 ml/l h was achieved at 37 °C temperature and 5.5 pH. Acid-hydrolysed palm fruit empty branch was used for biohydrogen production in a study by Chong et al. (2013). The results revealed maximum production of 690 ml/L along with H_2 yield of 1.98 mol/mol of xylose at pH 5.5 and initial total sugar concentration of 5 g/L.

An upflow anaerobic sludge blanket-fixed film bioreactor was used for hydrogen production from pre-settled palm oil mill effluent using heat pretreated anaerobic sludge as the inoculum (Mohammadi et al. 2014). The reactor was operated in closed-loop fed-batch mode for 5 days at 24 h HRT, 38 °C temperature, and 5.5 pH. The specific hydrogen production rate in the reactor was 0.514 L/g VSS/day.

Biohydrogen production from acid-hydrolysed palm oil waste has been reported using photo fermentation using a novel photosynthetic bacterium, *Rhodobacter sphaeroides* S10 (Pattanamanee et al. 2012). The effect of different nutrient supplement on biohydrogen production was examined and the results showed significant effect due to yeast extract followed by, respectively, molybdenum, magnesium, EDTA, and iron. The volumetric production rate reached 22.4 ml/L along with 7.0 ml/g substrate of H_2 yield under optimum condition.

23.3.9 Biodiesel Production Waste

Biodiesel has become one of the major biofuels in recent times. The biodiesel production process involves transesterification reaction with animal fats or vegetable oils with alcohol in the presence of a catalyst (acid, base, or enzyme catalyst) (Roy et al. 2016). The products of this reaction are ethyl esters of fatty acids (biodiesel) and glycerol. The increase in biodiesel production has also led to increase in production of biodiesel by-product or crude glycerol, which can be further utilized for producing an array of products including biohydrogen (Haron et al. 2018). The following are some studies on biohydrogen production using glycerol as a substrate.

Two hyperthermophilic bacteria, *Thermotoga maritima* and *Thermotoga neapolitana*, were examined for biohydrogen production using glycerol at 80 °C (Maru et al. 2012). The biohydrogen yield values were 2.75 and 2.65 mol/mol glycerol consumed by *T. maritima* and *T. neapolitana*, respectively, with 2.5 g/L of initial glycerol concentration. In a similar study, 2.73 (±0.14) mol hydrogen/mol of glycerol was reported using *Thermotoga neapolitana* strain at an elevated temperature of 75 °C (Ngo et al. 2011).

In a different study, enhanced biohydrogen production from crude glycerol was achieved following a two-stage dark fermentation–photo fermentation strategy

(Chookaew et al. 2015). *Klebsiella* sp. TR17 and *Rhodopseudomonas palustris* TN1 were used in dark fermentation and photo fermentation, respectively. The maximum hydrogen yield values were 5.74 and 0.68 mmol H_2/g COD, respectively, during dark and photo fermentation.

From the afore-mentioned examples, it is clear that the prospect of hydrogen production from various agricultural and industrial wastes is promising from the economic and environmental point of view. Both liquid waste and solid residues are used for hydrogen production; however, yield and conversion efficiency are dependent on several factors including carbon content of the feedstock, biodegradability, presence of recalcitrant compound in the feedstock, microorganisms used, and operating conditions (pH, temperature, etc.). Hence, in order to improve the process efficiency, these factors need to be optimized or at least considered before suitable feedstock selection. Moreover, some more specific challenges may exist for biohydrogen production from a specific type of waste-based substrate which may need to be addressed if only that specific substrate is selected as feedstock. For example, ammonia presence in corn stalk hinders its use in biohydrogen production. Sometimes even the inoculum needs pretreatment, particularly when using mixed microbial consortium as the inoculum as presence of methanogenic organisms in the inoculum may cause methane production along with hydrogen reducing the desired product yield. In addition to inoculum pretreatment, optimizing operating conditions such as pH, temperature, substrate loading rate, and hydraulic retention time can also help eliminate methanogenic activity in hydrogen-producing reactors. Another issue with this type of waste-fed biohydrogen production is due to the low yield; it is only feasible to operate such plants at smaller scale. Hence, comprehensive techno-economic analysis needs to be undertaken for developing an economically and commercially viable biohydrogen production process. Apart from economic considerations, carbon emission and environmental impact assessment of each process should be carried out as these parameters have become critical indicators of a sustainable process in current scenario.

23.4 Hydrogen-Based Waste-Fed Biorefinery Concept

Biorefinery mimics the chemical refinery in terms of production of chemicals, food, feed, and other materials along with energy production (Katakojwala and Mohan 2022). However, the process utilizes biomass-based resources rather than petrochemicals (Cherubini 2010). It also integrates a sequence of biomass conversion and processing steps under a unitary setup to produce a spectrum of products. Although the steps in any biorefinery could vary depending on the desired products and processes involved, they can be broadly classified under few common steps such as biomass collection–handling–storage, biomass pretreatment, biomass conversion to main products and by-products, separation and purification of products, and product upgrading. Since there are no industrial-scale biorefinery currently operating anywhere, the biorefinery concept is mostly based on lab findings and limited to their feasibility studies.

The biohydrogen production process through dark fermentation in itself can be considered as a biorefinery as it produces other by-products such as ethanol, acetic acid, butyric acid, propionic acid, etc. (Reungsang et al. 2013). In case of biohydrogen production from waste, the biorefinery could be very well integrated with the bioconversion process as a large portion of the biomass remains unutilized at the end of the biohydrogen production process. This residual biomass could be easily used for production of other energy compounds such as methane, ethanol, etc. or other value-added products such as bioplastics, long chain fatty acids, etc. (Ebrahimian et al. 2022). Moreover, the by-products of fermentative hydrogen production, i.e. volatile fatty acids and alcohols, could be used for producing the same products. Direct use of effluent from biohydrogen production process without any pretreatment is also feasible for producing other products. Table 23.3 shows combined production of multiple products along with biohydrogen through biorefinery approach.

The integration of different processes with dark fermentative biohydrogen production is depicted in Fig. 23.1. The integration of anaerobic methanogenic reactor as subsequent step in a two-stage biohydrogen production process utilizing different wastes has already been mentioned (Banu et al. 2020). This process uses two different types of inoculum: one hydrogen producing and the other methanogenic. Another strategy is to use different experimental conditions (pH and temperature) in the two reactors to provide ideal condition for the production of H_2 and CH_4. For example, a two-stage fermentation process combining thermophilic hydrogen production and mesophilic methane production from organic fraction of municipal solid wastes has been demonstrated (Chu et al. 2008). In this study, the first stage reactor was operated at 55 °C temperature, 5.5 pH, and a short HRT of 1.3 day, whereas the methanogenic reactor was operated at 35 °C, 7.2 pH and an HRT of 5 days. Thus, hydrogen and methane production were achieved sequentially using the two-stage reactor system. Similarly, in another study, hydrogen and methane production could be achieved in two sequential reactors operated at different pH (pH 6.0 and 7.0 for H_2 and CH_4 production, respectively) (Mohan et al. 2008a).

The two-stage reactor system has also been used to increase the yield of hydrogen on substrate by photo fermentation in the second stage (Rai and Singh 2016). Similar to the study, the addition of photo fermentation stage following dark fermentation increased the hydrogen yield by 134% and substrate utilization by 67% (Cai et al. 2019). In another study, biohydrogen production from sugarcane bagasse by integrating dark fermentation using *Enterobacter aerogenes* MTCC 2822 along with photo fermentation using *Rhodopseudomonas* BHU 01 has been demonstrated (Rai et al. 2014).

Many studies have reported production of bioplastics using alternative substrates. In this regard, effluent from dark fermentation process can be used as substrate for polyhydroxyalkanoate (PHA) production. Yan et al. (2010) reported biohydrogen production from Taihu blue algae and used the effluent from hydrogen production process for PHA production. In another study, biohydrogen and PHAs were

Table 23.3 Integration of biohydrogen production with biorefinery approach for multiple product formation

Type of waste used	Bioreactor system	Hydrogen production/yield	Other products (yield/concentration)	References
Combined food waste and sewage sludge	Two-stage system (first stage for H_2, second stage for CH_4 production)	Yield = 0.13 m^3/kg VS	Methane (0.67 m^3/kg VS)	Siddiqui et al. (2011)
Winery wastewater	Batch serum bottle	Specific H_2 production = 528 mL/L	Methane (207 mL/g COD)	Carrillo-Reyes et al. (2019)
Raw cheese whey	Two-stage process with CSTR (for H_2 production) followed by periodic anaerobic baffled reactor (for CH_4 production)	Yield = 0.041 m^3/kg COD Rate = 7.53 L/day	Methane (75.6 L/day)	Antonopoulou et al. (2008)
Synthetic wastewater containing sucrose and fructose	Packed-bed bioreactors with immobilized anaerobic sludge	Yield = 1.04 mol/mol hexose Rate = 16 mmol/L/h	Ethanol (6 mmol/L/h)	Wu et al. (2007)
Cheese processing wastewater	Stirred tank reactor	Yield = 1.7–10.1 mM/g COD	Acetic acid (365–1344 mg/L), butyric acid (729–1057 mg/L)	Yang et al. (2007)
Starchy and sugary wastes	Batch reactor	Yields = 199–291 L/kg dry matter	Volatile fatty acids (9.7–14.5 g/L)	Weide et al. (2019)
Platanus orientalis leaves	Biohydrogen production by photo fermentation	Cumulative H_2 production = 368.8 mL	Acetic acid (2966 mg/L), butyric acid (624 mg/L), propionic acids (422 mg/L) and ethanol (1624 mg/L)	Zhang et al. (2017)
Platanus orientalis leaves	Biohydrogen production by dark-fermentation	Cumulative H_2 production = 312 mL	Acetic acid (1365 mg/L), butyric acid (558 mg/L), propionic acids (866 mg/L) and ethanol (1352 mg/L)	Zhang et al. (2017)
Wheat straw hydrolysate	Two-stage process with CSTR (for H_2 production) followed by UASB (for CH_4 production)	Rate = 5.2 L/L/day	Methane (2.6 L/L/day)	Pawar et al. (2013)

Cheese whey wastewater	Two-stage H_2 production process with dark fermentation followed by photo fermentation	Yield = 2.04 mol/mol lactose and rate = 1.09 mmol/L/h (with free cells) Yield = 3.50 mol/mol lactose and rate = 1.91 mmol/L/h for immobilized cells	Acetic acid (1900 mg/L) and butyric acid (537.2 mg/L) with free cells in first stage; H_2 (1.63 mol/mol acetic acid) in second stage Acetic acid (2100 mg/L), butyric acid (718 mg/L) with immobilized cells in first stage; H_2 (2.69 mol/mol acetic acid) in second stage	Rai et al. (2012)
Sucrose	Two-stage process combining dark-fermentation and photo fermentation	Yield = 3.80 mol/mol sucrose	H_2 (yield of 6.22 mol/mol of sucrose in photo fermentation)	Chen et al. (2008)
Cheese whey waste	Two-step process (first step dark-fermentation and second step polyhydroxyalkanoates (PHA) production from dark fermentation effluent)	Yield = 1.37 and 1.93 mol/mol sugars	Organic acid in first stage (14.6 and 12.6 g/L/day) PHA in second stage (yield of 0.74 mg/mg COD)	Colombo et al. (2019)
Spent wash effluent	Two-step process with anaerobic reactor in first step (for dark fermentation) and CSTR in second step (for PHA production from dark fermentation effluent)	Yield = 142 L/kg COD	VFA (12 g/L) in first stage PHA (40% dry cell weight) in second stage	Amulya et al. (2014)
Different organic substrates (acetate, propionate, malate, and lactate)	Single-stage batch system (using *Rhodopseudomonas palustris*)	Maximum H_2 production 131.3 mL on lactate, 105.5 mL on acetate, 104.3 mL on malate and 79.2 mL on propionate	PHB (17.1–11.6% and 11.8–2.5% cell dry weight for acetate and propionate, respectively)	Wu et al. (2012)

(continued)

Table 23.3 (continued)

Type of waste used	Bioreactor system	Hydrogen production/yield	Other products (yield/concentration)	References
Synthetic wastewater	Two-stage process with first stage extraction of lipids and pigment from microalgae *Nannochloropsis* sp. followed by H_2 production by dark fermentation with remaining biomass	Yield = 60.6 mL/g dry biomass	Lipids (33 g lipids/100 g dry biomass) Pigment carotenoids (70% of total weight)	Nobre et al. (2013)
Lipid extracted microalga biomass	Two-stage process, in first stage H_2 production with lipid extracted biomass and CH_4 production with effluent of dark fermentation	Yield = 97.6 mL/g VS	Methane (cumulative methane production of 1060 mL/L)	Singh et al. (2022)
Sugarcane bagasse	Two-step process, acid treatment to depolymerize followed by dark fermentation for H_2 production	Yield = 0.15 L/g COD	Nanocrystalline cellulose (0.15 g/g biomass) and cellulose (0.34 g/g biomass) during first stage Acetic acid (2.2 g/L), butyric acid (0.65 g/L) and propionic acid (0.35 g/L) during dark fermentation	Katakojwala and Mohan (2022)
Organic fraction of municipal solid waste	Bacteria fungal co-culture	96.3 L/kg substrate of cumulative hydrogen production	Butanol (80.2 g/kg substrate), acetone (30.2 g/kg substrate) and ethanol (108.1 g/kg substrate)	Ebrahimian et al. (2022)

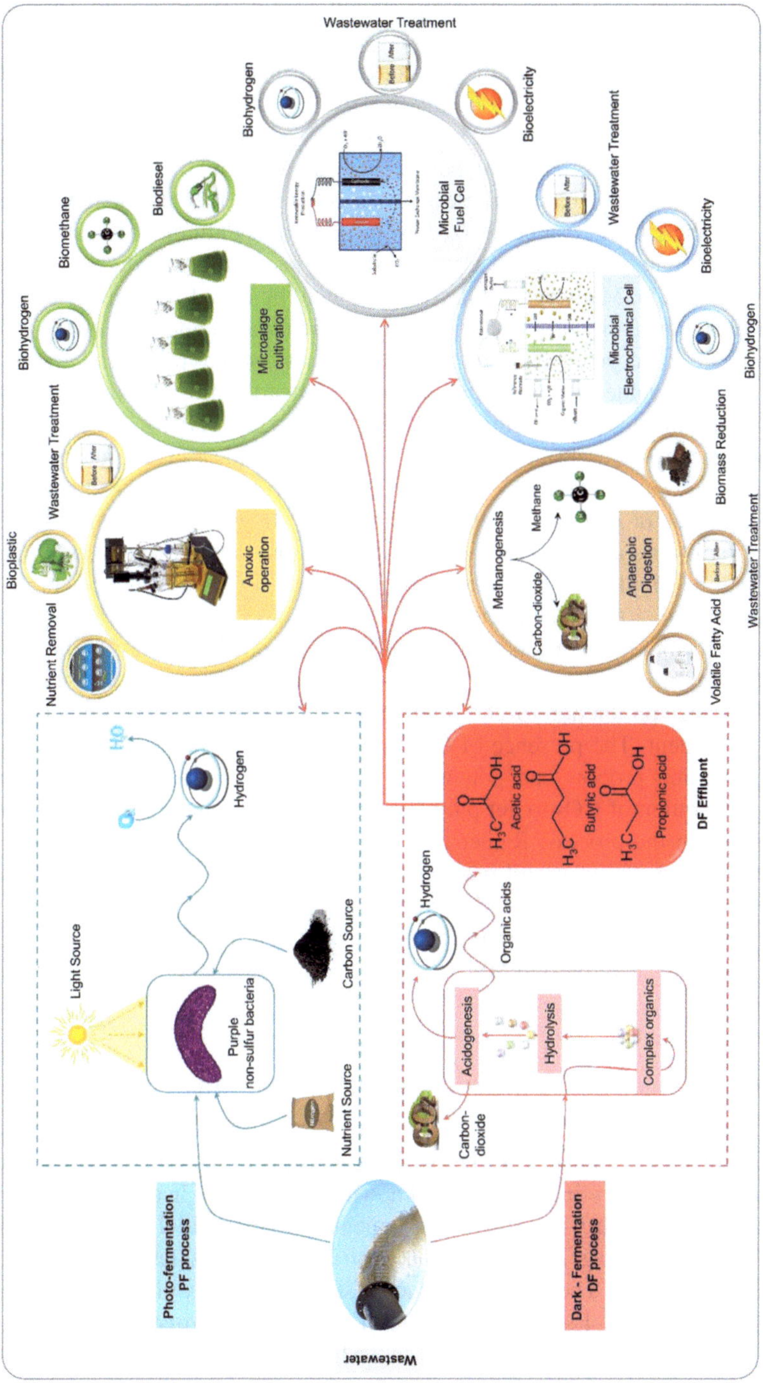

Fig. 23.1 Integration of different processes and products with biohydrogen production from wastewater towards a waste-fed biorefinery (reused with permission from Banu et al. 2020), Copyright Bioresource Technology)

produced simultaneously using spent wash effluent as substrate (Amulya et al. 2014). In this study, a two-step approach was followed wherein the first stage was operated with the spent wash effluent as the substrate for biohydrogen production and the acidogenic effluent from the first stage was subsequently used as the substrate for PHA production by *Bacillus tequilensis*.

In addition to the aforementioned processes and products, effluent obtained from biohydrogen production process can be fed either to microbial fuel cell for production of electricity or to microbial electrolysis cell for further production of H_2. Furthermore, the effluent can be utilized as the growth medium for cultivating oleaginous microorganisms such as algae, yeast, or bacteria to produce biodiesel (Dahiya et al. 2021). Many studies have highlighted the use of wastewater as substrate for oleaginous microorganisms. The spent microbial biomass following lipid extraction can be recycled back to the anaerobic hydrogen-/methane-producing reactor or can be subjected to thermal treatment for production of other products such as bio-oil, biochar, etc. (Osman et al. 2020; Nobre et al. 2013). The biohydrogen or methane production step also produces carbon dioxide as a co-product as the biogas produced can contain 40–70% CO_2. This CO_2 could be separated and used for growing algae or other CO_2-utilizing organisms. Other studies on algae-based biogas upgrading are also reported, which suggested that biogas containing H_2/CH_4 and CO_2 can be supplied to algae for selective uptake of CO_2 to produce pure H_2/CH_4 (Dahiya et al. 2021). The algal biomass grown using CO_2 from biogas can be further used for biodiesel production and other value-added products. Figure 23.2 depicts such an integrated biorefinery system where algal process can be combined with biohydrogen production for the valorization of dark fermentative biohydrogen production effluent. The presence of CO_2 along with H_2 in the product stream of dark fermentation process can potentially limit its application in certain field, requiring separation of CO_2 to get pure form of H_2 which imposes additional cost and difficulty in the process. Rather than separating CO_2, both the gases can be fermented together to get useful products such as alcohols and acids. In a recent study, biorefinery approach was followed to produce succinic acid from H_2/CO_2 by *Citrobacter amalonaticus* strain and bioplastic (PHA) from VFA using enriched mixed consortium (Amulya and Mohan 2022).

23.5 Challenges and Future Research Directions

There are many challenges in the way of successful commercialization of biohydrogen-based biorefinery. The well-known challenges for hydrogen as energy are cost associated with its production at large scale, large initial investment, and problems with its storage, transportation, distribution, and safety-related issues (Eljack and Kazi 2021). The cost of hydrogen is still significantly higher than energy obtained from fossil fuel sources. The cost of producing hydrogen from biomass (including waste sources) are also more than the hydrogen produced from fossil fuel which still remains the simplest raw material for commercial hydrogen production

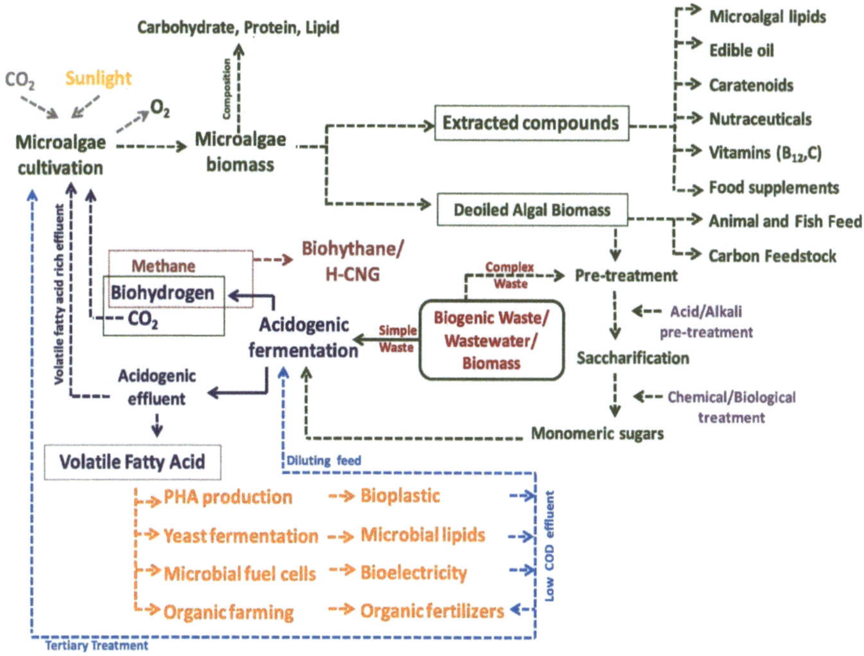

Fig. 23.2 Closed-loop biorefinery using algal processes along with biohydrogen production (reused with permission from Dahiya et al. (Dahiya et al. 2021), Copyright Bioresource Technology)

(>95%). Hence, making the biohydrogen production process commercially viable is one of the main objectives for future research work. Other ongoing research work in the area of hydrogen storage and transportation are towards improving efficiency, durability, refuelling time, and cost involved (Abdalla et al. 2018). Another aspect that needs to be taken care of is the sustainability of the biohydrogen production process. Although hydrogen production from renewable resources is considered as sustainable, often many of these processes are found to score poorly on sustainability scale. In this regard, life cycle assessment (LCA) can be used to analyse sustainability of the process.

The use of waste to produce biohydrogen itself can pose many challenges for its commercialization. For example, all types of waste cannot be used in a single plant as most of them will be suited to utilize only a specific type of waste for hydrogen production. Ensuring continuous supply of waste (or biomass) for hydrogen production could be another challenge due to seasonal variation in availability of such substrates. Hence, it would be meaningful to collaborate with specific industry generating the waste that is intended to be treated using hydrogenogenic process. Many of the waste substrates contain various recalcitrant or toxic compounds which are detrimental to their utilization for biohydrogen production. In most cases, pretreatment (physical, chemical, or biological method) needs to be applied to make the substrate biodegradable. For example, lignocellulosic wastes contain lignin which needs to be removed for better utilization of cellulose or hemicellulose present for the production of hydrogen.

From the process point of view, hydrogen is not produced in pure form, and hence it is necessary to remove other gaseous products for producing high-quality hydrogen for a specific application. Hydrogen purification by removal of other gases is not easy and may incur additional cost. Similarly, separation of VFA from the effluent of dark fermentation process is a cost-intensive process and may require additional technology such as membrane separation. Many recent studies have reported novel selective silicone membrane-based process for simultaneous extraction of VFAs from the dark fermentation stage itself (Dessì et al. 2020). In some cases, even a single VFA of interest could be extracted using selective membranes (Ravishankar et al. 2021). This is helpful in two ways: first, ease of recovery and second, avoids acidification and maintains optimum pH inside the reactor for hydrogen production.

Integration of biohydrogen production process (dark fermentation) with other bioconversion techniques (e.g. photo fermentation, methane production, microbial fuel cell, etc.) can pose some challenges (Banu et al. 2020). For example, the acidogenic effluent from the dark fermentation process is not suitable for photo fermentative microorganisms or methanogens. Methanogenic microorganisms have optimum pH near neutral value (7.0) which may require adjusting the pH of the acidogenic effluent from the hydrogen-producing stage when two stages need to be integrated in the process. Microbial electrochemical cells can avoid this problem, but the cost (particularly installation cost) associated with MFCs is significantly higher than other bioprocesses. Hence, future research should be directed towards these aspects to make the process more economical.

Other areas for future research need to be focused on optimization and intensification of the biohydrogen production process. Many approaches, including media optimization using statistical design of experiments, have already been reported. Optimization of key operational parameters such as temperature, pH, oxidation–reduction potential (ORP), etc. have led to improved biohydrogen production. But there is still scope for improving the yield of hydrogen on substrate as optimum conditions for biohydrogen production vary for different microorganisms. Another strategy for enhancing biohydrogen production is addition of micronutrients and metals to the production medium (Sarma et al. 2013; de Oliveira and Ferreira-Leitão 2016). These micronutrients are often related to the different important enzymes responsible for biohydrogen production. Recently, addition of nanoparticles to enhance biohydrogen production has gained renewed interest among the researchers (Engliman et al. 2017; Zhao et al. 2021). The effect of nanoparticles is mostly similar to the addition of inorganic nutrients to improve certain enzyme activity, but nanoparticles are highly effective even at low concentration. Future research towards understanding their mechanism of action, reducing their cost through sustainable production strategy, and their recycling/reuse are of interest in this area.

23.6 Conclusions and Perspectives

The demand for energy is increasing worldwide at an alarming level along with the increase in population. A lot of research is being undertaken towards fulfilling this ever-growing energy demand. Bioenergy, particularly biohydrogen produced from alternative resources such as different industrial wastes, could potentially provide a solution. Hydrogen as a fuel is nonmetallic and non-toxic and can produce higher energy equivalent than fossil fuel-based energy. Moreover, using waste resources for hydrogen production can serve to treat industrial waste and clean our environment. This waste to energy concept will help in achieving a circular bioeconomy and will fulfil sustainable development goals. Among the different processes, dark fermentation is mostly applied for hydrogen production from industrial effluent. Furthermore, integration of waste-based biohydrogen production with other bioprocesses to produce an array of energy compounds, chemicals, and food-feed products will make the process more commercially viable. Future research aimed towards improving the yield and hydrogen production rate through different strategies, such as selection of efficient microorganisms, optimization of media component and operational parameters, use of additives (micronutrients or nanoparticles), etc. are still required. Moreover, cost–benefit analysis and life cycle analysis of the hydrogen production process need to be undertaken for understanding the commercial viability and sustainability of the process.

References

Abdalla AM, Hossain S, Nisfindy OB, Azad AT, Dawood M, Azad AK (2018) Hydrogen production, storage, transportation and key challenges with applications: a review. Energy Convers Manag 165:602–627. https://doi.org/10.1016/j.enconman.2018.03.088

Abdeshahian P, Lim JS, Ho WS, Hashim H, Lee CT (2016) Potential of biogas production from farm animal waste in Malaysia. Renew Sust Energ Rev 60:714–723. https://doi.org/10.1016/j.rser.2016.01.117

Ahmad A, Buang A, Bhat AH (2016) Renewable and sustainable bioenergy production from microalgal co-cultivation with palm oil mill effluent (POME): a review. Renew Sust Energ Rev 65:214–234. https://doi.org/10.1016/j.rser.2016.06.084

Ahmed SF, Rafa N, Mofijur M, Badruddin IA, Inayat A, Ali MS, Farrok O, Khan Y (2021) Biohydrogen production from biomass sources: metabolic pathways and economic analysis. Front Energy Res 9:753878. https://doi.org/10.3389/fenrg.2021.753878

Alfano M, Cavazza C (2018) The biologically mediated water–gas shift reaction: structure, function and biosynthesis of monofunctional [NiFe]-carbon monoxide dehydrogenases. Sustain Energy Fuels 2(8):1653–1670. https://doi.org/10.1039/C8SE00085A

Amulya K, Mohan SV (2022) Green hydrogen based succinic acid and biopolymer production in a biorefinery: adding value to CO_2 from acidogenic fermentation. Chem Eng J 429:132163. https://doi.org/10.1016/j.cej.2021.132163

Amulya K, Reddy MV, Mohan SV (2014) Acidogenic spent wash valorization through polyhydroxyalkanoate (PHA) synthesis coupled with fermentative biohydrogen production. Bioresour Technol 158:336–342. https://doi.org/10.1016/j.biortech.2014.02.026

Anam K, Habibi MS, Harwati TU, Susilaningsih D (2012) Photofermentative hydrogen production using *Rhodobium marinum* from bagasse and soy sauce wastewater. Int J Hydrog Energy 37(20):15436–15442. https://doi.org/10.1016/j.ijhydene.2012.06.076

Antonopoulou G, Stamatelatou K, Venetsaneas N, Kornaros M, Lyberatos G (2008) Biohydrogen and methane production from cheese whey in a two-stage anaerobic process. Ind Eng Chem Res 47(15):5227–5233. https://doi.org/10.1021/ie071622x

Arreola-Vargas J, Celis LB, Buitrón G, Razo-Flores E, Alatriste-Mondragón F (2013) Hydrogen production from acid and enzymatic oat straw hydrolysates in an anaerobic sequencing batch reactor: performance and microbial population analysis. Int J Hydrog Energy 38(32):13884–13894. https://doi.org/10.1016/j.ijhydene.2013.08.065

Arun S, Sinharoy A, Pakshirajan K, Lens PNL (2020) Algae based microbial fuel cells for wastewater treatment and recovery of value-added products. Renew Sust Energ Rev 132:110041. https://doi.org/10.1016/j.rser.2020.110041

Ashrafi O, Yerushalmi L, Haghighat F (2015) Wastewater treatment in the pulp-and-paper industry: a review of treatment processes and the associated greenhouse gas emission. J Environ Manag 158:146–157. https://doi.org/10.1016/j.jenvman.2015.05.010

Baldi F, Iannelli R, Pecorini I, Polettini A, Pomi R, Rossi A (2019) Influence of the pH control strategy and reactor volume on batch fermentative hydrogen production from the organic fraction of municipal solid waste. Waste Manage Res 37(5):478–485. https://doi.org/10.1177/0734242X19826371

Banu JR, Kavitha S, Kannah RY, Bhosale RR, Kumar G (2020) Industrial wastewater to biohydrogen: possibilities towards successful biorefinery route. Bioresour Technol 298:122378. https://doi.org/10.1016/j.biortech.2019.122378

Baskaran D, Sinharoy A, Pakshirajan K, Rajamanickam R (2020) Gas-phase trichloroethylene removal by Rhodococcus opacus using an airlift bioreactor and its modeling by artificial neural network. Chemosphere 247:125806. https://doi.org/10.1016/j.chemosphere.2019.125806

Boran E, Özgür E, Yücel M, Gündüz U, Eroglu I (2012) Biohydrogen production by Rhodobacter capsulatus in solar tubular photobioreactor on thick juice dark fermenter effluent. J Clean Prod 31:150–157. https://doi.org/10.1016/j.jclepro.2012.03.020

Bräutigam KR, Jörissen J, Priefer C (2014) The extent of food waste generation across EU-27: different calculation methods and the reliability of their results. Waste Manage Res 32(8):683–694. https://doi.org/10.1177/0734242X14545374

Cai J, Zhao Y, Fan J, Li F, Feng C, Guan Y, Wang R, Tang N (2019) Photosynthetic bacteria improved hydrogen yield of combined dark-and photo-fermentation. J Biotechnol 302:18–25. https://doi.org/10.1016/j.jbiotec.2019.06.298

Cappai G, De Gioannis G, Muntoni A, Spiga D, Boni MR, Polettini A, Pomi R, Rossi A (2018) Biohydrogen production from food waste: influence of the inoculum-to-substrate ratio. Sustainability 10(12):4506. https://doi.org/10.3390/su10124506

Carrillo-Reyes J, Albarrán-Contreras BA, Buitrón G (2019) Influence of added nutrients and substrate concentration in biohydrogen production from winery wastewaters coupled to methane production. Appl Biochem Biotechnol 187(1):140–151. https://doi.org/10.1007/s12010-018-2812-5

Castelló E, Santos CG, Iglesias T, Paolino G, Wenzel J, Borzacconi L, Etchebehere C (2009) Feasibility of biohydrogen production from cheese whey using a UASB reactor: links between microbial community and reactor performance. Int J Hydrog Energy 34(14):5674–5682. https://doi.org/10.1016/j.ijhydene.2009.05.060

Chen CY, Yang MH, Yeh KL, Liu CH, Chang JS (2008) Biohydrogen production using sequential two-stage dark and photo fermentation processes. Int J Hydrog Energy 33(18):4755–4762. https://doi.org/10.1016/j.ijhydene.2008.06.055

Cherubini F (2010) The biorefinery concept: using biomass instead of oil for producing energy and chemicals. Energy Convers Manag 51(7):1412–1421

Chong ML, Rahim RA, Shirai Y, Hassan MA (2009) Biohydrogen production by *Clostridium butyricum* EB6 from palm oil mill effluent. Int J Hydrog Energy 34(2):764–771. https://doi.org/10.1016/j.ijhydene.2008.10.095

Chong PS, Jahim JM, Harun S, Lim SS, Mutalib SA, Hassan O, Nor MTM (2013) Enhancement of batch biohydrogen production from prehydrolysate of acid treated oil palm empty fruit bunch. Int J Hydrog Energy 38(22):9592–9599. https://doi.org/10.1016/j.ijhydene.2013.01.154

Chookaew T, Sompong O, Prasertsan P (2015) Biohydrogen production from crude glycerol by two stage of dark and photo fermentation. Int J Hydrog Energy 40(24):7433–7438. https://doi.org/10.1016/j.ijhydene.2015.02.133

Chu CF, Li YY, Xu KQ, Ebie Y, Inamori Y, Kong HN (2008) A pH-and temperature-phased two-stage process for hydrogen and methane production from food waste. Int J Hydrog Energy 33(18):4739–4746. https://doi.org/10.1016/j.ijhydene.2008.06.060

Cieciura-Włoch W, Borowski S, Domański J (2020) Dark fermentative hydrogen production from hydrolyzed sugar beet pulp improved by iron addition. Bioresour Technol 314:123713. https://doi.org/10.1016/j.biortech.2020.123713

Colombo B, Calvo MV, Sciarria TP, Scaglia B, Kizito SS, D'Imporzano G, Adani F (2019) Biohydrogen and polyhydroxyalkanoates (PHA) as products of a two-steps bioprocess from deproteinized dairy wastes. Waste Manag 95:22–31. https://doi.org/10.1016/j.wasman.2019.05.052

Dada O, Yusoff WMW, Kalil MS (2013) Biohydrogen production from ricebran using *Clostridium saccharoperbutylacetonicum* N1-4. Int J Hydrog Energy 38(35):15063–15073. https://doi.org/10.1016/j.ijhydene.2013.07.048

Dahiya S, Chatterjee S, Sarkar O, Mohan SV (2021) Renewable hydrogen production by dark-fermentation: current status, challenges and perspectives. Bioresour Technol 321:124354. https://doi.org/10.1016/j.biortech.2020.124354

Das D (2009) Advances in biohydrogen production processes: an approach towards commercialization. Int J Hydrog Energy 34(17):7349–7357. https://doi.org/10.1016/j.biortech.2016.07.141

de Oliveira FM, Ferreira-Leitão VS (2016) Optimization of biohydrogen yield produced by bacterial consortia using residual glycerin from biodiesel production. Bioresour Technol 219:365–370. https://doi.org/10.1016/j.biortech.2016.07.141

Dessì P, Porca E, Lakaniemi AM, Collins G, Lens PNL (2018) Temperature control as key factor for optimal biohydrogen production from thermomechanical pulping wastewater. Biochem Eng J 137:214–221. https://doi.org/10.1016/j.bej.2018.05.027

Dessì P, Asunis F, Ravishankar H, Cocco FG, De Gioannis G, Muntoni A, Lens PNL (2020) Fermentative hydrogen production from cheese whey with in-line, concentration gradient-driven butyric acid extraction. Int J Hydrog Energy 45(46):24453–24466. https://doi.org/10.1016/j.ijhydene.2020.06.081

Dhote L, Kumar S, Singh L, Kumar R (2021) A systematic review on options for sustainable treatment and resource recovery of distillery sludge. Chemosphere 263:128225. https://doi.org/10.1016/j.chemosphere.2020.128225

Ebrahimian F, Denayer JF, Karimi K (2022) Efficient coproduction of butanol, ethanol, and biohydrogen from municipal solid waste through a cocultivated biorefinery. Energy Convers Manag 255:115303. https://doi.org/10.1016/j.enconman.2022.115303

Eljack F, Kazi MK (2021) Prospects and challenges of green hydrogen economy via multi-sector global symbiosis in Qatar. Front Sustain 1:612762. https://doi.org/10.3389/frsus.2020.612762

Ellabban O, Abu-Rub H, Blaabjerg F (2014) Renewable energy resources: current status, future prospects and their enabling technology. Renew Sust Energ Rev 39:748–764. https://doi.org/10.1016/j.rser.2014.07.113

Engliman NS, Abdul PM, Wu SY, Jahim JM (2017) Influence of iron (II) oxide nanoparticle on biohydrogen production in thermophilic mixed fermentation. Int J Hydrog Energy 42(45):27482–27493

EPA (2018). https://www.epa.gov/facts-and-figures-about-materials-waste-and-recycling/national-overview-facts-and-figures-materials

Erdirencelebi D (2011) Treatment of high-fat-containing dairy wastewater in a sequential UASBR system: influence of recycle. J Chem Technol Biotechnol 86(4):525–533. https://doi.org/10.1002/jctb.2546

Escamilla-Alvarado C, Ponce-Noyola T, Ríos-Leal E, Poggi-Varaldo HM (2013) A multivariable evaluation of biohydrogen production by solid substrate fermentation of organic municipal wastes in semi-continuous and batch operation. Int J Hydrog Energy 38(28):12527–12538. https://doi.org/10.1016/j.ijhydene.2013.02.124

Fang HH, Li C, Zhang T (2006) Acidophilic biohydrogen production from rice slurry. Int J Hydrog Energy 31(6):683–692. https://doi.org/10.1016/j.ijhydene.2005.07.005

Gadhe A, Sonawane SS, Varma MN (2013) Optimization of conditions for hydrogen production from complex dairy wastewater by anaerobic sludge using desirability function approach. Int J Hydrog Energy 38(16):6607–6617. https://doi.org/10.1016/j.ijhydene.2013.03.078

Gadhe A, Sonawane SS, Varma MN (2014) Ultrasonic pretreatment for an enhancement of biohydrogen production from complex food waste. Int J Hydrog Energy 39(15):7721–7729. https://doi.org/10.1016/j.ijhydene.2014.03.105

Ghirardi ML, King PW, Mulder DW, Eckert C, Dubini A, Maness PC, Yu J (2014) Hydrogen production by water biophotolysis. In: Microbial BioEnergy: hydrogen production. Springer, Dordrecht, pp 101–135. https://doi.org/10.1007/978-94-017-8554-9_5

Gilroyed BH, Li C, Hao X, Chu A, McAllister TA (2010) Biohydrogen production from specified risk materials co-digested with cattle manure. Int J Hydrog Energy 35(3):1099–1105. https://doi.org/10.1016/j.ijhydene.2009.11.072

Haghi E, Raahemifar K, Fowler M (2018) Investigating the effect of renewable energy incentives and hydrogen storage on advantages of stakeholders in a microgrid. Energy Policy 113:206–222. https://doi.org/10.1016/j.enpol.2017.10.045

Han SK, Shin HS (2004) Performance of an innovative two-stage process converting food waste to hydrogen and methane. J Air Waste Manag Assoc 54(2):242–249. https://doi.org/10.1080/10473289.2004.10470895

Haron R, Mat R, Abdullah TAT, Rahman RA (2018) Overview on utilization of biodiesel by-product for biohydrogen production. J Clean Prod 172:314–324. https://doi.org/10.1016/j.jclepro.2017.10.160

ICAR (2020). https://icar.org.in/sites/default/files/Creating-Wealth-From-Agricultural-Waste.pdf

Intanoo P, Rangsunvigit P, Namprohm W, Thamprajamchit B, Chavadej J, Chavadej S (2012) Hydrogen production from alcohol wastewater by an anaerobic sequencing batch reactor under thermophilic operation: Nitrogen and phosphorous uptakes and transformation. Int J Hydrog Energy 37(15):11104–11112. https://doi.org/10.1016/j.ijhydene.2012.04.129

Ismail I, Hassan MA, Rahman AA, Soon CS (2011) Effect of retention time on biohydrogen production by microbial consortia immobilised in polydimethylsiloxane. Afr J Biotechnol 10(4):601–609. https://doi.org/10.5897/AJB10.825

Kargi F, Eren NS, Ozmihci S (2012) Effect of initial bacteria concentration on hydrogen gas production from cheese whey powder solution by thermophilic dark fermentation. Biotechnol Prog 28(4):931–936. https://doi.org/10.1002/btpr.1558

Katakojwala R, Mohan SV (2022) Multi-product biorefinery with sugarcane bagasse: process development for Nanocellulose, lignin and biohydrogen production and lifecycle analysis. Chem Eng J 446:137233. https://doi.org/10.1016/j.cej.2022.137233

Khamtib S, Reungsang A (2014) Co-digestion of oil palm trunk hydrolysate with slaughterhouse wastewater for thermophilic bio-hydrogen production by *Thermoanaerobacterium thermosaccharolyticm* KKU19. Int J Hydrog Energy 39(13):6872–6880. https://doi.org/10.1016/j.ijhydene.2014.02.073

Kolev Slavov A (2017) General characteristics and treatment possibilities of dairy wastewater–a review. Food Technol Biotechnol 55(1):14–28. https://doi.org/10.17113/ftb.55.01.17.4520

Kummu M, De Moel H, Porkka M, Siebert S, Varis O, Ward PJ (2012) Lost food, wasted resources: global food supply chain losses and their impacts on freshwater, cropland, and fertiliser use. Sci Total Environ 438:477–489. https://doi.org/10.1016/j.scitotenv.2012.08.092

Kvesitadze G, Sadunishvili T, Dudauri T, Zakariashvili N, Partskhaladze G, Ugrekhelidze V, Tsiklauri G, Metreveli B, Jobava M (2012) Two-stage anaerobic process for bio-hydrogen and bio-methane combined production from biodegradable solid wastes. Energy 37(1):94–102. https://doi.org/10.1016/j.energy.2011.08.039
Lakshmidevi R, Muthukumar K (2010) Enzymatic saccharification and fermentation of paper and pulp industry effluent for biohydrogen production. Int J Hydrog Energy 35(8):3389–3400. https://doi.org/10.1016/j.ijhydene.2009.12.165
Lee ZK, Li SL, Kuo PC, Chen IC, Tien YM, Huang YJ, Chuang CP, Wong SC, Cheng SS (2010) Thermophilic bio-energy process study on hydrogen fermentation with vegetable kitchen waste. Int J Hydrog Energy 35(24):13458–13466. https://doi.org/10.1016/j.ijhydene.2009.11.126
Lee ZS, Chin SY, Lim JW, Witoon T, Cheng CK (2019) Treatment technologies of palm oil mill effluent (POME) and olive mill wastewater (OMW): a brief review. Environ Technol Innov 15:100377. https://doi.org/10.1016/j.eti.2019.100377
Lin YH, Juan ML, Hsien HJ (2011) Effects of temperature and initial pH on biohydrogen production from food-processing wastewater using anaerobic mixed cultures. Biodegradation 22(3):551–563. https://doi.org/10.1007/s10532-010-9427-z
Lin Y, Wu S, Wang D (2013) Hydrogen-methane production from pulp & paper sludge and food waste by mesophilic–thermophilic anaerobic co-digestion. Int J Hydrog Energy 38(35):15055–15062. https://doi.org/10.1016/j.ijhydene.2012.01.051
Manish S, Banerjee R (2008) Comparison of biohydrogen production processes. Int J Hydrog Energy 33(1):279–286. https://doi.org/10.1016/j.ijhydene.2007.07.026
Maru BT, Bielen AAM, Kengen SWM, Constantí M, Medina F (2012) Biohydrogen production from glycerol using *Thermotoga* spp. Energy Procedia 29:300–307. https://doi.org/10.1016/j.egypro.2012.09.036
Mathews JA, Tan H (2014) Economics: manufacture renewables to build energy security. Nature 513(7517):166–168. https://doi.org/10.1038/513166a
Millati R, Cahyono RB, Ariyanto T, Azzahrani IN, Putri RU, Taherzadeh MJ (2019) Agricultural, industrial, municipal, and forest wastes: an overview. In: Taherzadeh MJ, Bolton K, Wong J, Pandey A (eds) Sustainable resource recovery and zero waste approaches. Elsevier, pp 1–22. https://doi.org/10.1016/B978-0-444-64200-4.00001-3
Mishra P, Das D (2014) Biohydrogen production from *Enterobacter cloacae* IIT-BT 08 using distillery effluent. Int J Hydrog Energy 39(14):7496–7507. https://doi.org/10.1016/j.ijhydene.2013.08.100
Mohammadi P, Ibrahim S, Annuar MSM (2014) High-rate fermentative hydrogen production from palm oil mill effluent in an up-flow anaerobic sludge blanket-fixed film reactor. Chem Eng Res Des 92(10):1811–1817. https://doi.org/10.1016/j.cherd.2014.04.023
Mohan SV, Mohanakrishna G, Sarma PN (2008a) Integration of acidogenic and methanogenic processes for simultaneous production of biohydrogen and methane from wastewater treatment. Int J Hydrog Energy 33(9):2156–2166. https://doi.org/10.1016/j.ijhydene.2008.01.055
Mohan SV, Mohanakrishna G, Ramanaiah SV, Sarma PN (2008b) Simultaneous biohydrogen production and wastewater treatment in biofilm configured anaerobic periodic discontinuous batch reactor using distillery wastewater. Int J Hydrog Energy 33(2):550–558. https://doi.org/10.1016/j.ijhydene.2007.10.013
Mohana S, Acharya BK, Madamwar D (2009) Distillery spent wash: treatment technologies and potential applications. J Hazard Mater 163(1):12–25. https://doi.org/10.1016/j.jhazmat.2008.06.079
Mohtasham J (2015) Renewable energies. Energy Procedia 74:1289–1297. https://doi.org/10.1016/j.egypro.2015.07.774
Nagarajan D, Dong CD, Chen CY, Lee DJ, Chang JS (2021) Biohydrogen production from microalgae-major bottlenecks and future research perspectives. Biotechnol J 16:2000124. https://doi.org/10.1002/biot.202000124

Ngo TA, Kim MS, Sim SJ (2011) High-yield biohydrogen production from biodiesel manufacturing waste by *Thermotoga neapolitana*. Int J Hydrog Energy 36(10):5836–5842. https://doi.org/10.1016/j.ijhydene.2010.11.057

Nicoletti G, Arcuri N, Nicoletti G, Bruno R (2015) A technical and environmental comparison between hydrogen and some fossil fuels. Energy Convers Manag 89:205–213. https://doi.org/10.1016/j.enconman.2014.09.057

Nobre BP, Villalobos F, Barragan BE, Oliveira AC, Batista AP, Marques PASS, Mendes RL, Sovová H, Palavra AF, Gouveia L (2013) A biorefinery from *Nannochloropsis* sp. microalga–extraction of oils and pigments. Production of biohydrogen from the leftover biomass. Bioresour Technol 135:128–136. https://doi.org/10.1016/j.biortech.2012.11.084

Omae K, Fukuyama Y, Yasuda H, Mise K, Yoshida T, Sako Y (2019) Diversity and distribution of thermophilic hydrogenogenic carboxydotrophs revealed by microbial community analysis in sediments from multiple hydrothermal environments in Japan. Arch Microbiol 201(7):969–982. https://doi.org/10.1007/s00203-019-01661-9

Osman AI, Deka TJ, Baruah DC, Rooney DW (2020) Critical challenges in biohydrogen production processes from the organic feedstocks. Biomass Conv Bioref 1–19. https://doi.org/10.1007/s13399-020-00965-x

Paillet F, Barrau C, Escudié R, Bernet N, Trably E (2021) Robust operation through effluent recycling for hydrogen production from the organic fraction of municipal solid waste. Bioresour Technol 319:124196. https://doi.org/10.1016/j.biortech.2020.124196

Pattanamanee W, Choorit W, Kantachote D, Chisti Y (2012) Repeated-batch production of hydrogen using *Rhodobacter sphaeroides* S10. Int J Hydrog Energy 37(21):15855–15866. https://doi.org/10.1016/j.ijhydene.2012.08.041

Paul T, Sinharoy A, Baskaran D, Pakshirajan K, Pugazhenthi G, Lens PNL (2022) Bio-oil production from oleaginous microorganisms using hydrothermal liquefaction: a biorefinery approach. Crit Rev Environ Sci Technol 52(3):356–394. https://doi.org/10.1080/10643389.2020.1820803

Pawar SS, Nkemka VN, Zeidan AA, Murto M, van Niel EW (2013) Biohydrogen production from wheat straw hydrolysate using *Caldicellulosiruptor saccharolyticus* followed by biogas production in a two-step uncoupled process. Int J Hydrog Energy 38(22):9121–9130. https://doi.org/10.1016/j.ijhydene.2013.05.075

Perera KRJ, Nirmalakhandan N (2010) Enhancing fermentative hydrogen production from sucrose. Bioresour Technol 101(23):9137–9143. https://doi.org/10.1016/j.biortech.2010.06.145

Purdy A, Pathare PB, Wang Y, Roskilly AP, Huang Y (2018) Towards sustainable farming: feasibility study into energy recovery from bio-waste on a small-scale dairy farm. J Clean Prod 174:899–904. https://doi.org/10.1016/j.jclepro.2017.11.018

Rai PK, Singh SP (2016) Integrated dark-and photo-fermentation: recent advances and provisions for improvement. Int J Hydrog Energy 41(44):19957–19971. https://doi.org/10.1016/j.ijhydene.2016.08.084

Rai PK, Singh SP, Asthana RK (2012) Biohydrogen production from cheese whey wastewater in a two-step anaerobic process. Appl Biochem Biotechnol 167(6):1540–1549. https://doi.org/10.1007/s12010-011-9488-4

Rai PK, Singh SP, Asthana RK, Singh S (2014) Biohydrogen production from sugarcane bagasse by integrating dark-and photo-fermentation. Bioresour Technol 152:140–146. https://doi.org/10.1016/j.biortech.2013.10.117

Ramos C, Buitrón G, Moreno-Andrade I, Chamy R (2012) Effect of the initial total solids concentration and initial pH on the bio-hydrogen production from cafeteria food waste. Int J Hydrog Energy 37(18):13288–13295. https://doi.org/10.1016/j.ijhydene.2012.06.051

Ravishankar H, Dessì P, Trudu S, Asunis F, Lens PNL (2021) Silicone membrane contactor for selective volatile fatty acid and alcohol separation. Process Saf Environ Prot 148:125–136. https://doi.org/10.1016/j.psep.2020.09.052

Reungsang A, Sittijunda S, Angelidaki I (2013) Simultaneous production of hydrogen and ethanol from waste glycerol by *Enterobacter aerogenes* KKU-S1. Int J Hydrog Energy 38(4):1813–1825. https://doi.org/10.1016/j.ijhydene.2012.11.062

Rosenbaum M, Cotta MA, Angenent LT (2010) Aerated *Shewanella oneidensis* in continuously fed bioelectrochemical systems for power and hydrogen production. Biotechnol Bioeng 105(5):880–888. https://doi.org/10.1002/bit.22621
Roy AS, Chingkheihunba A, Pakshirajan K (2016) An overview of production, properties, and uses of biodiesel from vegetable oil. In: Soccol C, Brar S, Faulds C, Ramos L (eds) Green fuels technology. Green energy and technology. Springer, Cham, pp 83–105. https://doi.org/10.1007/978-3-319-30205-8_4
San-Akca B, Sever SD, Yilmaz S (2020) Does natural gas fuel civil war? Rethinking energy security, international relations, and fossil-fuel conflict. Energy Res Soc Sci 70:101690. https://doi.org/10.1016/j.erss.2020.101690
Saripan AF, Reungsang A (2014) Simultaneous saccharification and fermentation of cellulose for bio-hydrogen production by anaerobic mixed cultures in elephant dung. Int J Hydrog Energy 39(17):9028–9035. https://doi.org/10.1016/j.ijhydene.2014.04.066
Sarma SJ, Brar SK, Le Bihan Y, Buelna G, Rabeb L, Soccol CR, Naceur M, Rachid B (2013) Evaluation of different supplementary nutrients for enhanced biohydrogen production by *Enterobacter aerogenes* NRRL B 407 using waste derived crude glycerol. Int J Hydrog Energy 38(5):2191–2198. https://doi.org/10.1016/j.ijhydene.2012.11.110
Sekoai PT, Kana EG (2013) A two-stage modelling and optimization of biohydrogen production from a mixture of agro-municipal waste. Int J Hydrog Energy 38(21):8657–8663. https://doi.org/10.1016/j.ijhydene.2013.04.130
Show KY, Lee DJ, Chang JS (2011) Bioreactor and process design for biohydrogen production. Bioresour Technol 102(18):8524–8533. https://doi.org/10.1016/j.biortech.2011.04.055
Siddiqui Z, Horan NJ, Salter M (2011) Energy optimisation from co-digested waste using a two-phase process to generate hydrogen and methane. Int J Hydrog Energy 36(8):4792–4799. https://doi.org/10.1016/j.ijhydene.2010.12.118
Singh L, Wahid ZA, Siddiqui MF, Ahmad A, Rahim MHA, Sakinah M (2013) Application of immobilized upflow anaerobic sludge blanket reactor using *Clostridium* LS2 for enhanced biohydrogen production and treatment efficiency of palm oil mill effluent. Int J Hydrog Energy 38(5):2221–2229. https://doi.org/10.1016/j.ijhydene.2012.12.004
Singh H, Rout S, Das D (2022) Dark fermentative biohydrogen production using pretreated Scenedesmus obliquus biomass under an integrated paradigm of biorefinery. Int J Hydrog Energy 47(1):102–116. https://doi.org/10.1016/j.ijhydene.2010.12.118
Sinharoy A, Lens PNL (2020) Biological removal of selenate and selenite from wastewater: options for selenium recovery as nanoparticles. Curr Pollution Rep 6(3):230–249. https://doi.org/10.1007/s40726-020-00146-4
Sinharoy A, Pakshirajan K (2021) Methane free biohydrogen production from carbon monoxide using a continuously operated moving bed biofilm reactor. Int J Hydrog Energy 46(1):306–313. https://doi.org/10.1016/j.ijhydene.2020.09.250
Sinharoy A, Baskaran D, Pakshirajan K (2019) Sustainable biohydrogen production by dark fermentation using carbon monoxide as the sole carbon and energy source. Int J Hydrog Energy 44(26):13114–13125. https://doi.org/10.1016/j.ijhydene.2019.03.130
Sinharoy A, Kumar M, Pakshirajan K (2020a) An overview of bioreactor configurations and operational strategies for dark fermentative biohydrogen production. In: Singh L, Yousuf A, Mahapatra DM (eds) . Elsevier, Bioreactors - sustainable design and industrial applications in mitigation of GHG emissions, pp 249–288. https://doi.org/10.1016/B978-0-12-821264-6.00014-0
Sinharoy A, Pakshirajan K, Lens PNL (2020b) Biological sulfate reduction using gaseous substrates to treat acid mine drainage. Curr Pollution Rep 6(4):328–344. https://doi.org/10.1007/s40726-020-00160-6
Sinharoy A, Baskaran D, Pakshirajan K (2020c) Process integration and artificial neural network modeling of biological sulfate reduction using a carbon monoxide fed gas lift bioreactor. Chem Eng J 391:123518. https://doi.org/10.1016/j.cej.2019.123518

Sinharoy A, Pakshirajan K, Lens PNL (2022) Syngas fermentation for bioenergy production: advances in bioreactor systems. In: Sinharoy A, Lens PNL (eds) Renewable energy technologies for energy efficient sustainable development. Applied environmental science and engineering for a sustainable future. Springer, Cham, pp 325–358. https://doi.org/10.1007/978-3-030-87633-3_13

Sivagurunathan P, Sen B, Lin CY (2015) High-rate fermentative hydrogen production from beverage wastewater. Appl Energy 147:1–9. https://doi.org/10.1016/j.apenergy.2015.01.136

Stankiewicz A (2006) Energy matters: alternative sources and forms of energy for intensification of chemical and biochemical processes. Chem Eng Res Des 84(7):511–521. https://doi.org/10.1205/cherd.05214

Tamburic B, Dechatiwongse P, Zemichael FW, Maitland GC, Hellgardt K (2013) Process and reactor design for biophotolytic hydrogen production. Phys Chem Chem Phys 15(26):10783–10794. https://doi.org/10.1039/C3CP51866C

Tawfik A, Salem AH (2014) Optimization of hydrogen production from pretreated rice straw waste in a mesophilic up-flow anaerobic staged reactor. Int J Energy Res 38(9):1155–1161. https://doi.org/10.1002/er.3128

Thi NBD, Kumar G, Lin CY (2015) An overview of food waste management in developing countries: current status and future perspective. J Environ Manag 157:220–229. https://doi.org/10.1016/j.jenvman.2015.04.022

USDA F (2020) Grain: world markets and trade. United States Department of Agriculture Foreign Agricultural Service

USDA N (2016) World agricultural supply and demand estimates. Internet site: http://usda.mannlib.cornell.edu/reports/waobr/wasde-bb

Veeravalli SS, Chaganti SR, Lalman JA, Heath DD (2014) Optimizing hydrogen production from a switchgrass steam exploded liquor using a mixed anaerobic culture in an upflow anaerobic sludge blanket reactor. Int J Hydrog Energy 39(7):3160–3175. https://doi.org/10.1016/j.ijhydene.2013.12.057

Weide T, Brügging E, Wetter C, Ierardi A, Wichern M (2019) Use of organic waste for biohydrogen production and volatile fatty acids via dark fermentation and further processing to methane. Int J Hydrog Energy 44(44):24110–24125. https://doi.org/10.1016/j.ijhydene.2019.07.140

Woo JH, Song YC (2010) Biohydrogen production from sewage sludge using a continuous hydrogen fermentation system with a heat treatment vessel. KSCE J Civ Eng 14(5):673–679. https://doi.org/10.1007/s12205-010-0686-3

World bank (2022). https://datatopics.worldbank.org/what-a-waste/trends_in_solid_waste_management.html#:~:text=The%20world%20generates%202.01%20billion,from%200.11%20to%204.54%20kilograms

Wu KJ, Chang CF, Chang JS (2007) Simultaneous production of biohydrogen and bioethanol with fluidized-bed and packed-bed bioreactors containing immobilized anaerobic sludge. Process Biochem 42(7):1165–1171. https://doi.org/10.1016/j.procbio.2007.05.012

Wu SC, Liou SZ, Lee CM (2012) Correlation between bio-hydrogen production and polyhydroxybutyrate (PHB) synthesis by *Rhodopseudomonas palustris* WP3-5. Bioresour Technol 113:44–50. https://doi.org/10.1016/j.biortech.2012.01.090

Xing Y, Li Z, Fan Y, Hou H (2010) Biohydrogen production from dairy manures with acidification pretreatment by anaerobic fermentation. Environ Sci Pollut Res 17(2):392–399. https://doi.org/10.1007/s11356-009-0187-4

Yan Q, Zhao M, Miao H, Ruan W, Song R (2010) Coupling of the hydrogen and polyhydroxyalkanoates (PHA) production through anaerobic digestion from Taihu blue algae. Bioresour Technol 101(12):4508–4512. https://doi.org/10.1016/j.biortech.2010.01.073

Yang P, Zhang R, McGarvey JA, Benemann JR (2007) Biohydrogen production from cheese processing wastewater by anaerobic fermentation using mixed microbial communities. Int J Hydrog Energy 32(18):4761–4771. https://doi.org/10.1016/j.ijhydene.2007.07.038

Zhang ML, Fan YT, Xing Y, Pan CM, Zhang GS, Lay JJ (2007) Enhanced biohydrogen production from cornstalk wastes with acidification pretreatment by mixed anaerobic cultures. Biomass Bioenergy 31(4):250–254. https://doi.org/10.1016/j.biombioe.2006.08.004
Zhang Z, Li Y, Zhang H, He C, Zhang Q (2017) Potential use and the energy conversion efficiency analysis of fermentation effluents from photo and dark fermentative bio-hydrogen production. Bioresour Technol 245:884–889. https://doi.org/10.1016/j.biortech.2017.09.037
Zhao C, Sharma A, Ma Q, Zhu Y, Li D, Liu Z, Yang Y (2021) A developed hybrid fixed-bed bioreactor with Fe-modified zeolite to enhance and sustain biohydrogen production. Sci Total Environ 758:143658. https://doi.org/10.1016/j.scitotenv.2020.143658

Chapter 24
Patents in Biohydrogen Production

Susan Grace Karp, Luiz Alberto Junior Letti, Leonardo Wedderhoff Herrmann, Rafaela de Oliveira Penha, Walter José Martinez Burgos, Jéssica Aparecida Viesser, Júlio César de Carvalho, and Carlos Ricardo Soccol

Abstract The global demand for hydrogen shall surpass 200 Mt by 2030, as the US International Energy Agency reported. Hydrogen obtained from renewable sources, or biohydrogen, is one of the most promising energy sources that can avoid fossil CO_2 emissions to the atmosphere. However, there are still important technological challenges to be overcome before biohydrogen reaches a considerable scale in the world energy matrix. This chapter presents an overview of biohydrogen technologies and a technological landscape of biohydrogen production and purification outlined from patent documents. A patent search allowed the identification of the major players, technology segments, and technological trends related to pure microorganisms and microbial consortia, fermentation processes, and bioreactor models. This information provides the panorama and perspectives for this emerging green alternative in the developing bioeconomy.

Keywords Technological development · Patent landscape · Biofuels · Bioenergy

24.1 Introduction

The global energy matrix is highly dependent (about 80% of demand) on fossil and non-renewable energy sources. The use of such fuels promotes pollution of urban air and the generation of particles with potentially toxic effects and, in addition, is responsible for the emission of greenhouse gases (GHG) contributing to global warming. Projections also indicate that fossil fuel sources will be depleted in a few decades (about 4 decades for oil, 6 for natural gas, and 16 for coal) (Abe et al. 2019).

S. G. Karp · L. A. J. Letti · L. W. Herrmann · R. de Oliveira Penha · W. J. M. Burgos · J. A. Viesser · J. C. de Carvalho · C. R. Soccol (✉)
Department of Bioprocess Engineering and Biotechnology, Federal University of Paraná, Curitiba, Paraná, Brazil
e-mail: soccol@ufpr.br

C. R. Soccol et al. (eds.), *Biohydrogen - Advances and Processes*, Biofuel and Biorefinery Technologies 13, https://doi.org/10.1007/978-3-031-49818-3_24

In this context, the need for a paradigm shift toward a cleaner, renewable, and sustainable energy matrix is evident.

The establishment of a sustainable energy matrix requires some aspects to be met, such as (1) security of energy supply; (2) existence of local renewable sources; (3) reduction of CO_2 emissions; (4) significant improvement in urban air quality; and (5) creation of a new industrial and technological energy matrix. In this context, H_2 presents itself as one of the most promising alternatives, and it also has other advantages, such as: (1) high energy density when compared to other fuels (more than double the value for many classic fuels); (2) is non-toxic; and (3) generates only water as a combustion product (Abe et al. 2019). In fact, in 2020, 120 Mt. of H_2 was demanded for energy generation, and projections indicate that in 2050 it will be 550 Mt. (Hermesmann and Müller 2022).

Hydrogen can be obtained from both non-renewable and renewable sources. The so-called gray hydrogen is obtained by burning fossil sources (coal or natural gas)—which negatively impacts the environment, enhancing global warming. The blue hydrogen is also provided by the burning of fossil fuels, but it also comprises the capture and storage of the generated carbon. Turquoise hydrogen comes from the pyrolysis of methane, but unlike blue hydrogen, it does not generate carbon (so it does not require capture and storage steps). Finally, green hydrogen results from the electrolysis of water. It is the most promising bet by enthusiasts of an energy matrix based on H_2, because, in addition to using renewable sources, it is theoretically the only "near-zero" emission technology. Still, it is the only one which does not consume O_2 (net value for the whole process) and does not generate water (e.g., blue and turquoise H_2 consume O_2, which negatively impacts global warming). On the other hand, it is the most expensive technology due to the high cost of renewable sources. There are also yellow hydrogen (obtained from nuclear energy) and hydrogen from fermentation or pyrolysis of biomass; the latter can also be an interesting alternative, as it has the potential for a negative carbon balance (Newborough and Cooley 2021; Agaton et al. 2022).

However, several factors are still barriers to be overcome in order to promote a real paradigm shift in the energy matrix based on H_2, such as the following: (1) low participation of developing countries in research and implementation of advances related to H_2 technology. Electricity costs in these countries, even when there are abundant renewable sources, are very significant; (2) requirement of specific infrastructure and logistics for a network capable of sustaining an economy based on H_2 energy for various sectors (transport, steel, chemical, among others); (3) lack of public policies with a gradual reduction of tax incentives for the use of fossil fuels and a gradual increase in incentives for the use of renewable fuels; and (4) apprehension of various sectors of society regarding the use of H_2 (e.g., fear of leaks, accidents, and explosions) (Agaton et al. 2022).

In addition to economic and social factors, some technological factors represent bottlenecks toward the use of H_2 on global scales, namely, (1) unsatisfactory performance of some processes for obtaining of H_2; (2) storage and transport difficulties (H_2 needs to be pressurized, liquefied, and maintained at extremely low temperatures); (3) storage requires reservoirs built with materials capable of resisting the chemical attack promoted by H_2; and (4) requirement of high gas purity (99.99%) for storage in fuel cells (Agaton et al. 2022). Due to these and other obstacles, many technologies have been developed and patented over the last few years. The objective of this chapter, therefore, is to show and discuss the most recent technological advances that seek to enable the use of H_2 as an energy vector, toward the establishment of a clean, technically feasible, environmentally sustainable, economically attractive, and socially safe energy matrix. The first section of the chapter is dedicated to the description of biohydrogen production routes and applications; the second section is the patent landscape of biohydrogen technologies; and the third section highlights the conclusions and future perspectives based on scientific literature and technological information.

24.2 Overview of Biohydrogen Production and Use

24.2.1 Renewable Hydrogen Production Routes

Biohydrogen can be produced from different renewable sources such as through biomethane reform, from biological fermentations using bacteria or microalgae and/or subjecting lignocellulosic biomass to thermochemical processes (Martinez-Burgos et al. 2021a). The hydrogen obtained from the dissociation of the water molecule is not biological; however, it is renewable. Figure 24.1 shows the possible routes for the production of renewable hydrogen. Although different routes are currently known, around 95% of the hydrogen consumed in the world comes from conventional sources, such as coal and oil through hydrocarbon reforming and hydrocarbon pyrolysis processes; only 5% of the hydrogen comes from renewable sources, specifically water, through electrolysis processes, which is the only large-scale renewable hydrogen production process.

The production of biohydrogen from reforming processes, partial oxidation, or autothermal reforming of biomethane are gaining significant importance

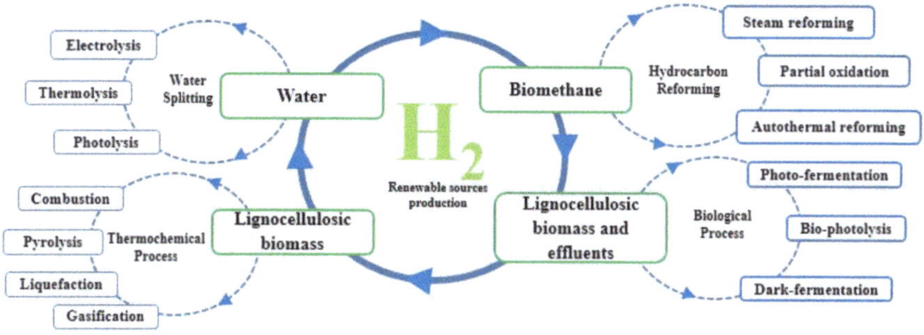

Fig. 24.1 Hydrogen production routes from renewable sources. Adapted from (Martinez-Burgos et al. 2021a)

worldwide, since biomethane production and reforming processes are mature and widely used technologies (Badoga and Dalai 2017; Vieira et al. 2021).

Thermochemical processes are also being used to produce biohydrogen from different types of biomass. In these processes, pyrolysis and gasification are especially important, they generate CO_2 and CH_4 and the latter can be processed via steam reforming to produce hydrogen. Other thermochemical techniques such as liquefaction and combustion are less used, since in addition to generating some polluting by-products, extreme pressure conditions of up to 20 MPa are required in the case of liquefaction (Akhtar et al. 2018; Chen et al. 2021).

Biological processes for hydrogen production are also gaining attention due to the increase in sustainable development policies in the world. Furthermore, these hydrogen production processes are developed at ambient temperatures and pressures, so energy consumption is lower, and the required infrastructure is less complex when compared to processes developed at high pressures and temperatures. Among the biological processes are bio-photolysis, photo-fermentation, and dark fermentation. However, the latter is the most promising, since it is a relatively fast reaction where generally the substrates used are residues and the yields are better when compared to bio-photolysis and photo-fermentation (Anwar et al. 2019). The main limitation of biological hydrogen production is yield, since the maximum theoretical yield is 4 mol of hydrogen for each mol of glucose (Sydney et al. 2020).

Different types of effluents have been tested in dark fermentation experiments, such as sugarcane vinasse (Bittencourt et al. 2014), cassava processing wastewater (CPW) (Martinez-Burgos et al. 2019), palm oil mill effluent (POME) (Rosa et al. 2020), soft drink wastewater (Junior et al. 2021), corn steep liquor (CSL) (Martinez-Burgos et al. 2021c), whey (Martinez-Burgos et al. 2021b), domestic sewage, and other agro-industrial and industrial effluents. When effluents are used as substrates in the production of hydrogen via dark fermentation, this serves as a first treatment of the effluent as it manages to remove up to 80% of the chemical oxygen demand (COD), representing circular economy processes (Martinez-Burgos et al. 2021d; Sydney et al. 2021; Karp et al. 2022). Other substrates such as pretreated lignocellulosic biomasses (wheat straw, corn stover, barley straw, carrot pulp, miscanthus) and some pure carbohydrates (glucose, xylose, fructose, sucrose, cellobiose, maltose, starch, ribose, xylose, trehalose, raffinose, arabinose, galactose, lactose, mannose, rhamnose, etc.) (Elsharnouby et al. 2013) have also been used in dark fermentation.

24.2.2 Microorganisms and Metabolic Pathways

Biological production of hydrogen is carried out by several types of microorganisms, including photosynthetic bacteria, cyanobacteria, algae, or fermentative bacteria (Mahidhara et al. 2019). These microorganisms have been employed in the conversion of biowastes to biohydrogen using some biological approaches such as biophotolysis (direct and indirect processes), microbial electrolysis cells, and fermentation (dark and photo-fermentation) (Brar et al. 2022; Cheng et al. 2022). Among them, dark and photo-fermentation are the most widely used processes for biohydrogen production (Cheng et al. 2022).

The production of hydrogen by microbial metabolism is dependent on the hydrogen-producing enzymes that catalyze a redox reaction, as shown in Eq. (24.1), inside the individual microorganism (Mathews and Wang 2009).

$$2H^{+} + 2e^{-} \leftrightarrow H_2 \tag{24.1}$$

To perform this redox reaction, most biohydrogen-involving enzymes have complex metalloclusters as active sites and require special maturation proteins (Ogata et al. 2016). The four main hydrogen-involving enzymes used by microbial systems are nitrogenases, [NiFe]-hydrogenases, [FeFe]-hydrogenases, and [Fe]-hydrogenases (Mathews and Wang 2009; Brar et al. 2022). According to Madden et al. (2012), [FeFe]-hydrogenases are the most efficient hydrogen-producing enzymes due to their 10–100 times higher activity than [NiFe]-hydrogenases.

Microalgae and cyanobacteria are commonly used to produce hydrogen through light- and water-dependent processes known as direct or indirect biophotolysis (Brar et al. 2022). Two classes of enzymes are related to hydrogen microbial metabolism: hydrogenases and nitrogenases. Hydrogenase enzymes are described in both cyanobacteria and microalgae, whereas nitrogenases are found in cyanobacteria only (Carrieri et al. 2011; Osman et al. 2020). Green microalgae can generate hydrogen by splitting water through the photosystem II (PSII), under anoxic and light conditions, by the photochemical oxidation of water and transfer of electrons to [Fe]-hydrogenase during direct biophotolysis processes (Show et al. 2018; El-Dalatony et al. 2020). In addition, hydrogenase enzymes are also involved in biohydrogen production through fermentation or respiration under anoxic conditions (indirect biophotolysis) in microalgae metabolism (El-Dalatony et al. 2020; Brar et al. 2022). For instance, *Chlamydomonas reinhardtii* has been studied for biohydrogen production due to the decarboxylation of pyruvate to acetyl-CoA by pyruvate ferredoxin oxidoreductase (PFR1) via the reduction of ferredoxin (Fd) during anaerobic fermentation (Mus et al. 2007). Then, the reduced ferredoxin (Fd) is used for the reduction of hydrogenase enzyme NADP+ to NADPH, which is responsible for the production of hydrogen as described in Eq. (24.2) (Osman et al. 2020).

$$2H^{+} + 2Fd(red) \leftrightarrow H_2 + 2Fd(ox) \tag{24.2}$$

Other reported hydrogen-producing microalgae are *Chlorella fusca*, *Chlorococcum littorale*, *Platymonas subcordiformis*, and *Scenedesmus obliquus* (Nagarajan et al. 2017).

In cyanobacteria, both nitrogenase and hydrogenase enzymes participate in hydrogen metabolism. Nitrogenases generate hydrogen as a by-product of the energy process of fixing atmospheric nitrogen and reducing it to ammonium ions (Tsygankov 2007; Carrieri et al. 2011). Cyanobacteria of the genera *Anabaena*,

Nostoc, *Calothrix* and *Oscillatoria* are nitrogen fixing according to the reaction in Eq. (24.3) (Nagarajan et al. 2017).

$$N_2 + 8e^- + 8H^+ + 16ATP \rightarrow 2NH_3 + H_2 \uparrow + 16ADP + 16Pi \quad (24.3)$$

There are three types of nitrogenases synthesized by cyanobacteria classified based on their central metallic cofactor called [Mo]-nitrogenase, [V]-nitrogenase, and [Fe]-nitrogenase (Tsygankov 2007; Nagarajan et al. 2017). Furthermore, another enzyme involved in hydrogen catalysis in cyanobacteria metabolism is the hox-encoded Ni–Fe hydrogenase (Hox hydrogenase). This enzyme catalyzes a reversible reaction that inter-converts protons and electrons with hydrogen gas, interacting with the redox partner NADH and/or NADPH as the electron donor, as detailed in Eq. (24.4) (Carrieri et al. 2011).

$$2H^+ + NAD(P)H \leftrightarrow H_2 + NAD(P)^+ \quad (24.4)$$

Non-nitrogen-fixing cyanobacteria (*Gloeobacter*, *Synechococcus*, and *Synechocystis*) are also capable of hydrogen generation through the action of nitrogenase as an ATP supplier to hydrogenase by the following Eq. (24.5) (Kim and Kim 2011).

$$8H^+ + 8e^- + 4ATP \rightarrow 4H_2 \uparrow + 4ADP \quad (24.5)$$

Biohydrogen production using microbial electrolysis cells (MECs) is a new technology gaining popularity in recent years. In this process, electrons are generated by the oxidation of organic matter (e.g., glucose, acetate, butyrate, glycol, renewable biomass, and wastewaters) which are transported to the anode. Then, they are transported to the cathode where, upon combining with protons, hydrogen is generated (Tian et al. 2019; El-Dalatony et al. 2020). Electrochemically active bacteria such as *Acetobacterium woodii*, *Geobacter sulfurreducens*, *Ochrobactrum anthropi*, *Shewanella oneidensis*, *Sphingomonas* sp., *Rhodopseudomonas palustris*, and *Rhodoferax ferrireducens* generate carbon dioxide (CO_2), electrons (transferred to anode), and H^+ (released in the solution) by oxidizing organic matter. The global equation to produce hydrogen using MEC for acetate is shown in Eq. (24.6) (Osman et al. 2020).

$$CH_3COOH + 2H_2O \rightarrow 2CO_2 + 4H_2 \quad (24.6)$$

A good method to produce hydrogen along with other organic alcohols/acids as by-products is through fermentation processes of biowaste materials (e.g., agricultural residues, food waste, sewage sludge, and wastewaters) using several microbial species under anaerobic conditions (Osman et al. 2020). This anaerobic fermentation is divided into (1) dark fermentation, (2) photo-fermentation, and (3) integrated dark photo-fermentation depending on the microorganisms' need for light (Cheng et al. 2022). Comparatively, in the dark process, biohydrogen is produced through the acidogenesis of biowaste hydrolysates with fatty acids, which are degraded to hydrogen and acetate by hydrogenase enzymes. In contrast, for the biohydrogen production in the photo-fermentation process, photosynthetic bacteria convert simple sugars like glucose and fatty acids to biohydrogen by using sunlight as energy and the nitrogenase enzymes as catalysts (Mishra et al. 2019).

Enterobacter spp., *Clostridium* spp., and *Escherichia coli*, as well as mixed cultures originated from anaerobic sludge, bovine manure, or organic composts, are commonly used as inoculum in dark fermentation processes to obtain high hydrogen yield (Cheng et al. 2022; Ren et al. 2022). Biohydrogen production during dark fermentation is primarily driven by anaerobic metabolism of pyruvate obtained from the glycolytic breakdown of carbohydrate-derived sugars (Hallenbeck 2005; Ren et al. 2022). Then, enteric bacteria (e.g., *Enterobacter* and *E. coli*) produce hydrogen from one formate molecule, whereas strict anaerobes (e.g., *Clostridium*) generate hydrogen from the reduced form of ferredoxin (Fd [red]) (Hallenbeck 2005). The decomposition of formate into hydrogen in the metabolic pathway of *Enterobacter* spp. is catalyzed by the formate hydrogenlyase complex 1 (FHL-1), which is composed by formate dehydrogenase (FDH-H) and hydrogenase 3 (Hyd-3) microbial enzymes (Petrosyan et al. 2020). Similarly, *E. coli* can consume different substrates, such as glucose, galactose, sucrose, xylose, arabinose, fructose, glycerol, formate, and acetate for hydrogen production. However, the metabolic pathway of glycerol and xylose are relatively different in producing pyruvate (Trchounian et al. 2017). Differently, *Clostridium* spp. have a strictly anaerobic metabolism and degrade two pyruvate molecules to acetyl-CoA, CO_2, and Fd (red) by the action of pyruvate ferredoxin oxidoreductase (*pfor*) enzyme. Next, the Fd (red) directly feeds electrons to [Fe-Fe]-hydrogenases generating two moles of hydrogen (Ren et al. 2022).

In the photo-fermentation process, hydrogen productivity depends on the intensity of light, the carbon source used, and the type of microorganisms that are employed (Brar et al. 2022). Purple non-sulfur bacteria (PNSB), including *Rhodospirillum rubrum*, *Rhodopseudomonas palustris*, *Rhodobacter sphaeroides*, and *Rhodobacter capsulatus*, are photosynthetic bacteria and can convert organic compounds to biohydrogen in the presence of light energy and anaerobic conditions (Cheng et al. 2022). Their hydrogen metabolism is based on the organic acid oxidation through the tricarboxylic acid cycle (TCA) and the generated electrons are transferred to nitrogenase via the respiratory chain of cytochromes and ferredoxins (Fd), resulting in the formation of hydrogen, as shown in Eq. (24.7) (Trchounian et al. 2017; Brar et al. 2022). For instance, the hydrogen production by *R. sphaeroides* is coupled to [Mo-Fe]-nitrogenase, which participates in nitrogen fixation and also catalyzes the reduction of H^+ to hydrogen (Trchounian et al. 2017).

$$CH_3COOH + 2H_2O + \text{light} \rightarrow 4H_2 + 2CO_2 \tag{24.7}$$

Several metabolic pathway engineering strategies have been employed for enhanced production of biohydrogen in microorganisms using molecular biology, modern analytical techniques, and genomic approaches. Smith et al. (2012) developed a new enzymatic pathway for the delivery of electrons to the [FeFe] hydrogenase from

Clostridium pasteurianum for the production of hydrogen from NADPH with significant improvement in its productivity. The integration of fast hydrogen-producing enzymes into non-native hosts could provide a way to increase hydrogen yields to a commercially competitive level (Mathews and Wang 2009).

24.2.3 Recovery, Storage, and Uses of Hydrogen

Hydrogen Recovery and Purification

Pressure swing adsorption (PSA) is one of the most efficient gas purification systems that can produce very high purity hydrogen at approximately 99.9999% (Luberti and Ahn 2022). This system is also used in air drying, separation of linear and branched paraffins, and carbon capture. Especially for biohydrogen produced by microbial bioprocesses, previous cleaning steps are usually required to remove H_2S and other reactive volatiles.

PSA systems are based on the selective absorption of impurities (Vieira et al. 2021). The system is composed of a bed with absorbent materials such as silica gel, activated carbons, and zeolites. The principle of the system is based on the phenomenon that under high pressure, gases tend to get stuck on solid surfaces, that is, to be "adsorbed." The higher the pressure, the more gas is adsorbed and when the pressure drops, the gas is released or desorbed (Zhu et al. 2018). After the gas is purified, it is stored, and storage research is being carried out to develop safe, compact, and economical materials that can be used in the safe storage and transport of the hydrogen vector.

Hydrogen Storage Methods

According to Niaz et al. (2015), hydrogen can be stored as compressed, liquid, adsorbed, or absorbed hydrogen—that is, this gas can also be captured and released in some materials, which involve molecular adsorption, diffusion, attraction, and dissociation processes. Figure 24.2 shows the different hydrogen storage methods and mechanisms. The main forms of storing hydrogen are compressed hydrogen and liquid hydrogen. The first is based on compressing the gas in high pressure tanks (approximately 69 MPa) and is the most recommended method when gas is going to be used as a fuel, as the gas can be stored in small spaces, in addition to maintaining its energy efficiency (Dillon and Heben 2001; Niaz et al. 2015). The second method is the storage of hydrogen at 20 K, i.e., in liquid form. Compared to compressed gas tanks, liquid hydrogen tanks can store a higher concentration of hydrogen (0.07 kg/L vs. 0.03 kg/L). The main bottlenecks of this method are the great energy loss attributed to the high liquefaction rate, hydrogen boil-off, and the cost of the tank, as it needs to be well insulated to maintain sub-zero temperatures (Niaz et al. 2015).

Hydrogen Storage as Chemical Compounds

Hydrogen can also be stored in the form of compounds which, after chemical reactions, can release the gas. The compounds used for storage are ammonia (NH_3),

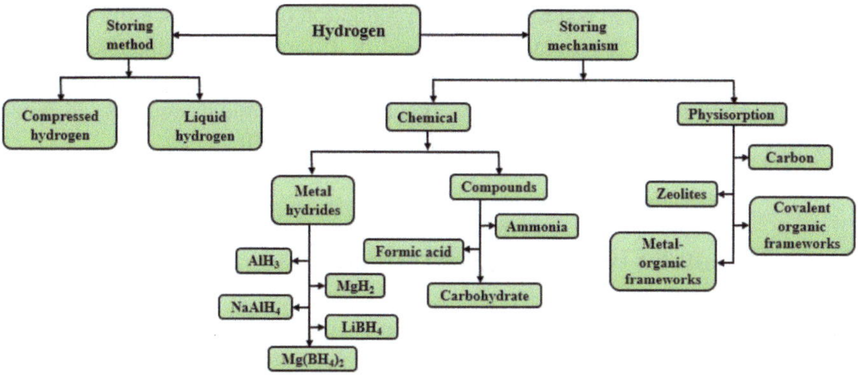

Fig. 24.2 Hydrogen storage methods and mechanisms. Note: Original artwork based on information provided by Niaz et al. (2015)

metal hydrides, formic acid, carbohydrates, and/or synthetic hydrocarbons, among others. NH_3 is the second most commonly produced chemical in the world (Niaz et al. 2015) and can be mixed with water to facilitate its storage in liquid form at atmospheric pressures and temperatures. In addition, NH_3 can be reformed to produce hydrogen without harmful residues (Kim and Kwon 2011). Formic acid (CH_2O_2) is also a compound of interest for hydrogen storage, as the H_2 produced from the reaction of CH_2O_2 and catalyzed by ruthenium is free of carbon monoxide (Grasemann and Laurenczy 2012). Carbohydrates are also hydrogen carriers (~14.8 wt%) and can have complete conversion under modest reaction conditions; however, large amounts of CO_2 can be generated (Niaz et al. 2015).

Hydrogen Storage by Physisorption

Another known mechanism for hydrogen storage is "physisorption," a process characterized by H_2 molecules being weakly adsorbed on the surface of a material (Thommes et al. 2015). Covalent organic frameworks, carbon, zeolites, and metal-organic frameworks, among others, are possible materials. In the case of carbon, the hydrogen molecules are absorbed through van der Waals bonding. Carbon structures are characterized by having a high surface area, among which are fullerenes, nanotubes, graphene, etc. In the case of zeolites, hydrogen is forced to move into the molecular cavities of the material at high pressures and temperatures, and when the material is cooled, the H_2 molecule is trapped in the porous material and its release also requires high pressures and high temperatures. For metal-organic structures, the geometry and size of the pore must be known because the absorption behavior is a function of these variables. Finally, covalent organic structures are building units held together by covalent bonds; their main feature is their crystalline structure which provides high surface area absorption.

Main Uses of Hydrogen

Hydrogen is a raw material used in different industrial processes (Fig. 24.3). Currently, approximately 44% of the hydrogen produced worldwide is used in petroleum refining, 38% is destined to the production of fertilizers, 8.0% to chemical products, and the remaining 10% is used in different applications such as the metal industry, food industry, and as a fuel. This last sector is gaining more importance since hydrogen has a high energy content (122 kJ/g) and its combustion generates only water and no polluting gases (Sharma et al. 2020).

24.3 Patents in Biohydrogen Production

24.3.1 Patent Landscape

In order to investigate the current innovations related to the biohydrogen production field, a patent search in the Derwent Innovations Index database was performed. The search was run with the terms [(TI = (hydrogen) OR TI = (biohydrogen)) AND IP = (C12P*)] to search for relevant patents published between the dates January 1, 2018, and July 18, 2022 (filed from 2016 on). The term "biohydrogen" alone is not a common word used for characterizing hydrogen produced through biological processes in the Derwent platform, requiring a combination of the term "hydrogen" and the International Patent Classification code (IPC) C12P to reach a more significant and specific number of documents. The IPC codes are similar to tags referring to specific technologies within areas and subareas. The "C" refers to the chemistry and metallurgy great area of knowledge; the "C12" relates to fermentation, biochemistry, and microbiology areas; and the "C12P" is a code for fermentation processes destined to produce a chemical compound. Combining hydrogen and biohydrogen with the C12P code resulted in 478 total patents, which were manually filtered and classified according to year of deposit, country of deposit, type of assignee, and type of technology associated with biohydrogen. After filtering, 186 documents were considered relevant, which will be discussed further.

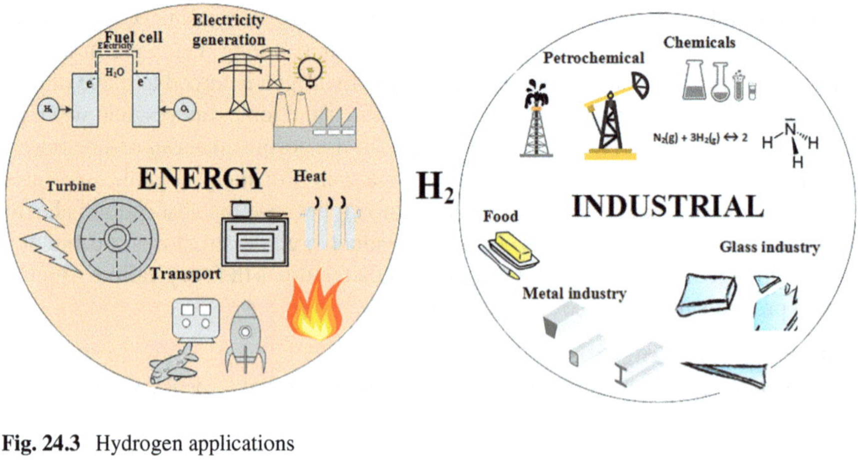

Fig. 24.3 Hydrogen applications

Figure 24.4a shows the distribution of patents related to biohydrogen between the years 2016 and 2022. The number of patents in 2022 and 2021 is probably underestimated because there is a secrecy period of usually 18 months between deposit and publication. Evaluating the years when the disclosure already expired, 2019 presented a significantly higher number of documents (63), more than twofold higher as compared with the years 2018 and 2020. This could indicate a period when the investments and interest in biohydrogen process development were increasing, just before the COVID-19 pandemics altered the priorities of research.

Regarding the countries of deposit, as presented in Fig. 24.4b, China is the one with the highest number of patents, reaching 113 documents, followed by South Korea with 24 and Japan with 13. China alone is responsible for 61% of the patents, and combining the Asian countries (China, South Korea, Japan, and India), they reach more than 82% of the filed inventions, demonstrating a huge interest in developing and protecting biohydrogen technologies.

Analyzing the type of assignee, which is the entity that holds the patent property, also provides an overview of the segments that are interested in the technology. Figure 24.5a shows the classification of the assignees into universities or institutes, companies, and individuals, or a combination of them, according to the submission data. There were 128 documents filed only by universities or institutes, representing 69% of the total, and adding the collaboration between universities and companies, the percentage reaches 72%, indicating that recent biohydrogen technologies mostly come from the academic sector, without participation of other partners. This information implies that possibly most of the processes are developed at laboratory or pilot scale and present a low technology readiness level. Only 21% of patent documents were filed by companies or industries, while individuals represented 6% of the assignees. Hydrogen fermentation technologies are not easily accessible, and this can be explained by the difficulty to provide an anaerobic environment for photo or dark fermentation, the use of residues as substrates, and the difficulty to collect and classify the produced biogas (Hallenbeck et al. 2012; Dolly et al. 2015; Ghimire et al. 2015; Lin et al. 2018).

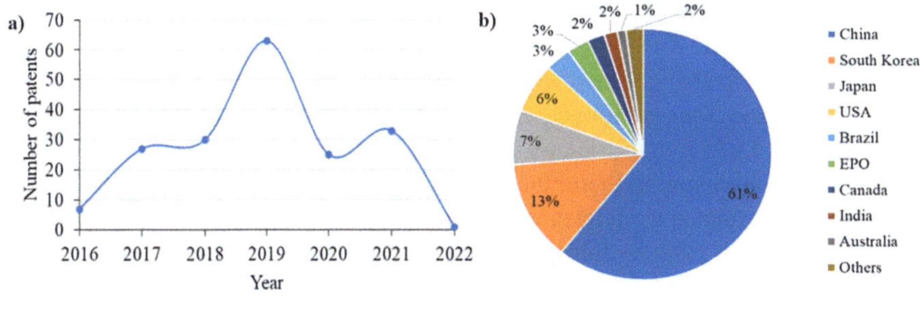

Fig. 24.4 Classification of biohydrogen associated patents according to (**a**) year of deposit and (**b**) country of deposit

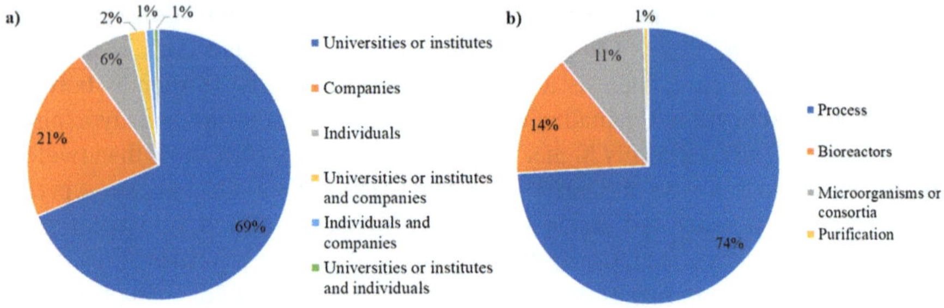

Fig. 24.5 Classification of biohydrogen associated patents according to (**a**) type of assignee and (**b**) type of technology

As for the type of technology, the documents were separated into process, which involves culture media, fermentation steps, and strategies applied to hydrogen production; bioreactor, which relates exclusively to the device, recipient, or fermenter adapted to the production; microorganisms or consortia, which protects the genetically modified strain or the modification that induces biohydrogen formation; and purification, related to the unit operation applied to purify the biogas to hydrogen. The majority of the evaluated patents were about processes, resulting in 138 documents or 74% of the total as presented in Fig. 24.5b, followed by bioreactors with 14%, microorganisms or consortia with 11%, and purification with one single patent. In fact, much of the focus when dealing with new fermentation processes is the development and optimization of culture media, operating methods, and physicochemical parameters (Aly et al. 2018; Yang and Wang 2018). The protection of bioreactors and microorganisms also faces the difficulties of finding and adapting the vessels and cells to an anaerobic environment, different substrates, and conditions (Dolly et al. 2015; Ghimire et al. 2015; Łukajtis et al. 2018), and the purification represents a final step that, to date, is not the main technology focus in the retrieved patent documents.

24.3.2 Microorganisms

Focusing on the patents related to microorganisms and consortia, seven protected modified bacteria, four mentioned archaea species, three were related to consortia, one was related to a fungus, one was related to a microalga, and one was related to a cyanobacterium. Among them, the species most frequently mentioned within the four documents related to archaea was *Thermococcus onnurineus*, a sulfur-reducing hyperthermophilic carboxydotrophic strain (Kang et al. 2018a, b, 2020, 2022) which appeared in other six patents from this research. The most mentioned bacteria were from the *Clostridium* genus (*Clostridium beijerinckii* and *Clostridium butyricum*), within three of the documents related to microorganisms (Liu et al. 2018; Liu and Li 2019; Ding and Hu 2022) and other ten documents from this research.

Clostridium bacteria are often used for hydrogen production and have been tested in several conditions and substrates with great productivity (Chen et al. 2005; Seppälä et al. 2011; Aly et al. 2018). The remaining species identified and protected were *Rhodobacter sphaeroides*, *Thermoanaerobacterium thermosaccharolyticum*, *Escherichia coli*, *Bacillus* sp., and *Piromyces* sp., and some documents did not report the genus or species protected. Three of the selected patents protected consortia, a group of two or more species, sometimes not identified, that act and grow in symbiosis to produce the desired product, in this case biohydrogen (Huang et al. 2018; Murayama et al. 2018; Romero et al. 2018). Additionally, nine process-related documents reported the production of hydrogen by the HAU-M1 photosynthetic bacteria flora group, which appears to be a well-known hydrogen-producing microbial consortium in the literature (Shui et al. 2022). Lastly, there were reports of three recombinant vectors for stimulating biohydrogen production from different cells, all of them using the *E. coli* plasmid pET32a coupled with enzymatic genes, classified together with the microorganisms and consortia patents due to their application in modifying cells for producing desired chemicals (Wang et al. 2018, 2019, 2020).

24.3.3 Production Process

The production process of biohydrogen can be performed using biophotolysis, dark and photo-fermentation, and microbial electrolysis cells (MECs). As explained previously, the biophotolysis consists in the use of microorganisms, mainly green algae and cyanobacteria, that are able to use light energy to split water molecules and produce H_2 (Ferraren-De Cagalitan and Abundo 2021). Photo-fermentation is performed by purple non-sulfur bacteria that are able to produce hydrogen from organic acids and CO_2 under anoxygenic conditions and using light as energy source. Dark fermentation for hydrogen production can be achieved by different fermentative microorganisms using a broad range of substrates (Mishra et al. 2019). Finally, a MEC is composed of an anode and a cathode and uses exoelectrogenic bacteria to convert organic matter to CO_2, electrons (e^-), and protons (H^+). The e^- are transferred to the anode and the H^+ are released to the electrolyte solution. Then, the e^- goes from the anode to the cathode and combines with free H^+, producing H_2 (Ferraren-De Cagalitan and Abundo 2021).

Several patents related to the different processes of biohydrogen production can be found, where the conditions, microorganisms, substrates, types of process, and equipment can vary. Table 24.1 shows a few examples of patents with different types of processes. Since the production of hydrogen can be done using different types of organic matter, it is possible to often observe patents using a wide range of wastes/residues as substrate, such as lignocellulosic material (Lu et al. 2021; Wei et al. 2021), algal biomass (Zhang et al. 2019b), kitchen waste (Zhao et al. 2018), glycerol (Pachapur et al. 2019), and others. The use of wastes is frequently observed in dark and photo-fermentation processes.

Table 24.1 Patents in different types of processes for biohydrogen production

Type of process	Microorganism	Substrate	Process condition	Patent
Dark fermentation	*Klebsiella* sp.	Lignocellulosic material	30–37 °C; 72–120 h	CN113881710-A (Wei et al. 2021)
Dark fermentation	Consortium	Açaí waste	35 °C; pH 7	BR102017017381-A2 (Amorim et al. 2017)
Photo-fermentation	Photosynthetic bacteria HAU-M1	Corn stalk	28–32 ° C; pH 5–9; 2800–3200 Lux.	US20210324421-A1 (Lu et al. 2021)
Photo-fermentation	Photosynthetic bacteria HAU-M1	Digested *Chlorella* biomass	30–37 °C; pH 7; 3000 Lux	CN110129399-A (Zhang et al. 2019b)
Dark fermentation	Consortium	Kitchen waste	37 °C; 60 rpm	CN108611376-A (Zhao et al. 2018)
Dark fermentation	Consortium	Sugar-containing wastewater	50–60 °C; pH 5–7	CN111763694-A (Liu et al. 2020)
Dark fermentation	*Enterobacter aerogenes*, *Clostridium butyricum*, *Chlamydomonas reinhardtii*, or combinations thereof	Glycerol	–	US20190218576-A1 (Pachapur et al. 2019)
Biophotolysis	*Chlamydomonas reinhardtii* or *Chlorella* sp.	Carbon dioxide	25 °C	CN107663529-A (Ge et al. 2018)
Photo-fermentation	Photosynthetic bacteria HAU-M1	Banana waste	28–32 °C; 2500 Lux	CN113604534-A (Yang et al. 2021)
Dark fermentation	Consortium	Cassava liquid waste and corn steep liquor	15–45 °C; pH 6; 6–96 h	BR1020190210559-B1 (Soccol et al. 2019)

A vast number of technologies could be found in fermentative processes, while the number of patents containing processes of biophotolysis was not as expressive. The patent CN107663529-A (Ge et al. 2018) brings a biophotolysis process of hydrogen production using photosynthetic microalgae. Interestingly, the inventors co-culture the microalgae with sulfur-oxidizing bacteria so that the oxygen produced by the algae is consumed by the bacteria, and the carbon dioxide generated by the bacteria is consumed by the algae. Patents CN113881710-A (Wei et al. 2021), BR102017017381-A2 (Amorim et al. 2017), US20210324421-A1 (Lu et al. 2021), CN110129399-A (Zhang et al. 2019b), and CN108611376-A (Zhao et al. 2018) are a few examples of patents containing the production of hydrogen through fermentation processes (Table 24.1).

24.3.4 Bioreactors

The reactor configuration is one of the most important factors in the efficiency of biohydrogen production (Banu et al. 2021). Photo-bioreactors, dark fermentation bioreactors, continuous stirred tank reactors (CSTR), anaerobic sequencing batch reactors, membrane bioreactors, fixed-bed bioreactors, fluidized-bed bioreactors, microbial electrolysis cells, hybrid bioreactors, and multi-stage bioreactors are some of the options that can be used for biohydrogen production (Show et al. 2011).

The patent IN201811028439-A (Mukherjee and Singh 2020) presented a fixed-bed bioreactor, useful for producing biohydrogen using thermophilic microorganisms. For that purpose, the reactor comprises a surface and base sparger for high aeration, a water jacket and a heat transfer structure for the high-temperature maintenance, a porous support for biofilm development, and a pressure pump for continuously recirculating the reaction mixture in a closed loop. According to the authors, the bioreactor affords maximum biohydrogen production with porous polyethylene support carriers, high aeration capacity, and ideal temperature control.

The document WO2018207964-A1 (Yoo et al. 2018) provides an invention containing a reactor capable of precisely controlling temperature using a steam inlet/outlet for supplying/discharging steam inside the reactor that does not kill microorganisms while maximizing the reactivity of the microorganisms. As a solution, the inventors developed an inner Y-shaped baffle, making it easy to control the temperature and allow a smooth gas supply. The patent KR2050061-B1 (Lee et al. 2019) also brings an injector in the bioreactor. In this case, the authors used an injector with a differential pressure on the inlet and outlet side for injecting carbon monoxide (carbon source) to the medium as needed. This injecting system allowed the inventors to produce hydrogen gas continuously with improved operation and monitoring of the process.

The inventors of the patent PL434798-A1 (Sokołowski and Cenian 2020) used a series connection of reactors to stabilize the production of hydrogen. In the first reactor, the substrate (cabbage substrate solution) and the inoculum were added. In the second reactor, a lignocellulosic substrate and the digestate (with adjusted pH from the reactor 1) were added to finally perform a dark fermentation at oxidation-reduction potential (ORP) of 180–240 mV, maintaining a constant hydrogen content of 20–60% in the outlet gas. A series of reactors were also used in the patent CN109762849-A (Pan et al. 2019). The inventors produced biohydrogen using kasugamycin (an antibiotic) residue in a system formed by a series connection between a CSTR, an upflow anaerobic sludge blanket (UASB) reactor, and a reactor for photosynthetic bacteria (PSB) light fermentation. The fermentation liquid that leaves the PSB reactor is recirculated in the UASB reactor for continuously producing hydrogen. In the proposed system, the CSTR reactor carries out the hydrogen production through anaerobic dark fermentation, the gas is collected in the UASB reactor, and the photosynthetic bacteria produce low molecular acid substances using the anaerobic fermentation hydrogen production.

The inventors of the patent CN109321444-A (Xia et al. 2019) presented a downflow two-step anaerobic fermentation biofilm reactor. The reactor is able to produce hydrogen in its top part and methane in its bottom part. For hydrogen production, the top part of the reactor is filled with spherical ceramic particles and the reactor is fed with sewage water.

One of the focuses of the patent CN110144293-A (Zhang et al. 2019a) was a device able to provide light for a photosynthetic hydrogen production. For that purpose, the authors used a drum bioreactor, and the light supply was done with a light source inside the reactor cylinder and with the use of triangular prisms in the surface of the cylinder for better distribution of light. According to the authors, the device has a simple structure, convenient operation, high processing efficiency, small occupied area, short time consumption, strong applicability, energy saving and environmental protection, and improved liquid hydrogen production amount.

24.4 Conclusion and Future Perspectives

Technologies related to biohydrogen production received increased attention until 2019, as represented by the number of filed patent documents. This growing interest was probably affected by the COVID-19 pandemics when the priorities changed. However, an increase of technological development for the following years is expected if the tendency of the pre-pandemics period is maintained. Asian countries (China, South Korea, Japan, and India) were the most representative developers of biohydrogen technologies. Regarding the sectors of society where the technologies were developed, represented by the type of assignee, universities and research institutes were responsible for around 70% of patent filings, which indicates that most technologies were not necessarily transferred to the industrial sector and consequently to the market. Most patent documents related to biohydrogen (74%) were about the production process, which includes culture media, fermentation steps, and strategies applied to hydrogen production. Bioreactor models and microorganisms or consortia represented 14% and 11% of the patent documents, respectively, and purification technologies were the focus of only one patent document.

The fact that biohydrogen technologies are still in a maturing stage is confirmed by market data that indicate that only 5% of the hydrogen used worldwide comes from renewable sources, and this refers specifically to water submitted to electrolysis processes, which is the only large-scale production process of renewable hydrogen.

The growing number of patent documents related to production processes conducted at optimized conditions, novel microorganisms and microbial consortia, and diverse models and configurations of bioreactors to provide improved hydrogen yields, however, points out to an increased participation of biohydrogen in the future energy matrix, once these technologies in fact evolve to higher readiness levels to be implemented at commercial scale.

References

Abe JO, Popoola API, Ajenifuja E, Popoola OM (2019) Hydrogen energy, economy and storage: review and recommendation. Int J Hydrog Energy 44:15072–15086. https://doi.org/10.1016/J.IJHYDENE.2019.04.068

Agaton CB, Batac KIT, Reyes EM (2022) Prospects and challenges for green hydrogen production and utilization in The Philippines. Int J Hydrog Energy 47:17859–17870. https://doi.org/10.1016/J.IJHYDENE.2022.04.101

Akhtar A, Krepl V, Ivanova T (2018) A combined overview of combustion, pyrolysis, and gasification of biomass. Energy Fuels 32:7294–7318. https://doi.org/10.1021/acs.energyfuels.8b01678

Aly SS, Imai T, Hassouna MS et al (2018) Identification of factors that accelerate hydrogen production by *Clostridium butyricum* RAK25832 using casamino acids as a nitrogen source. Int J Hydrog Energy 43:5300–5313. https://doi.org/10.1016/j.ijhydene.2017.08.171

Amorim ELC, Andrade SLP, Santos ACS (2017) Biohydrogen production from açaí waste (seeds and rachils) in anaerobic reactors. Patent document BR102017017381-A2

Anwar M, Lou S, Chen L et al (2019) Recent advancement and strategy on bio-hydrogen production from photosynthetic microalgae. Bioresour Technol 292:121972. https://doi.org/10.1016/j.biortech.2019.121972

Badoga S, Dalai A (2017) Liquid fuels from oil sands. In: Sustainable utilization of natural resources. CRC Press, Boca Raton, pp 121–143

Banu RJ, Usman MTM, Kavitha S et al (2021) A critical review on limitations and enhancement strategies associated with biohydrogen production. Int J Hydrog Energy 46:16565–16590. https://doi.org/10.1016/J.IJHYDENE.2021.01.075

Bittencourt E, Larroche C, Cristine N et al (2014) Economic process to produce biohydrogen and volatile fatty acids by a mixed culture using vinasse from sugarcane ethanol industry as nutrient source. Bioresour Technol 159:380–386. https://doi.org/10.1016/j.biortech.2014.02.042

Brar KK, Cortez AA, Pellegrini VOA et al (2022) An overview on progress, advances, and future outlook for biohydrogen production technology. Int J Hydrog Energy 47:37264. https://doi.org/10.1016/j.ijhydene.2022.01.156

Carrieri D, Wawrousek K, Eckert C et al (2011) The role of the bidirectional hydrogenase in cyanobacteria. Bioresour Technol 102:8368–8377. https://doi.org/10.1016/j.biortech.2011.03.103

Chen WM, Tseng ZJ, Lee KS, Chang JS (2005) Fermentative hydrogen production with *Clostridium butyricum* CGS5 isolated from anaerobic sewage sludge. Int J Hydrog Energy 30:1063–1070. https://doi.org/10.1016/j.ijhydene.2004.09.008

Chen L, Wang P, Shen Y, Guo M (2021) Spent lithium-ion battery materials recycling for catalytic pyrolysis or gasification of biomass. Bioresour Technol 323:124584. https://doi.org/10.1016/j.biortech.2020.124584

Cheng D, Ngo HH, Guo W et al (2022) Advanced strategies for enhancing dark fermentative biohydrogen production from biowaste towards sustainable environment. Bioresour Technol 351:127045. https://doi.org/10.1016/j.biortech.2022.127045

Dillon AC, Heben MJ (2001) Hydrogen storage using carbon adsorbents: past, present and future. Appl Phys A Mater Sci Process 142:133–142. https://doi.org/10.1007/s003390100788

Ding D, Hu P (2022) Hydrogen-producing anaerobic bacteria comprising basophilic *Clostridium* isolated and screened from the bottom sludge, where the basophilic *Clostridium* species is rod-shaped. Patent document CN114317351-A

Dolly S, Pandey A, Pandey BK, Gopal R (2015) Process parameter optimization and enhancement of photo-biohydrogen production by mixed culture of *Rhodobacter sphaeroides* NMBL-02 and *Escherichia coli* NMBL-04 using Fe-nanoparticle. Int J Hydrog Energy 40:16010–16020. https://doi.org/10.1016/j.ijhydene.2015.09.089

El-Dalatony MM, Zheng Y, Ji MK et al (2020) Metabolic pathways for microalgal biohydrogen production: current progress and future prospectives. Bioresour Technol 318:124253. https://doi.org/10.1016/j.biortech.2020.124253

Elsharnouby O, Hafez H, Nakhla G, El Naggar MH (2013) A critical literature review on biohydrogen production by pure cultures. Int J Hydrog Energy 38:4945–4966. https://doi.org/10.1016/j.ijhydene.2013.02.032

Ferraren-De Cagalitan DDT, Abundo MLS (2021) A review of biohydrogen production technology for application towards hydrogen fuel cells. Renew Sust Energ Rev 151:111413. https://doi.org/10.1016/J.RSER.2021.111413

Ge B, Xi L, He J, Huang F (2018) A kind of phycomycete co-cultures the method for improving photosynthetic microalgae hydrogen output. Patent document CN107663529-A

Ghimire A, Frunzo L, Pontoni L et al (2015) Dark fermentation of complex waste biomass for biohydrogen production by pretreated thermophilic anaerobic digestate. J Environ Manag 152:43–48. https://doi.org/10.1016/j.jenvman.2014.12.049

Grasemann M, Laurenczy G (2012) Formic acid as a hydrogen source - recent developments and future trends. Energy Environ Sci 5:8171–8181. https://doi.org/10.1039/c2ee21928j

Hallenbeck PC (2005) Fundamentals of the fermentative production of hydrogen. Water Sci Technol 52:21–29. https://doi.org/10.2166/wst.2005.0494

Hallenbeck PC, Abo-Hashesh M, Ghosh D (2012) Strategies for improving biological hydrogen production. Bioresour Technol 110:1–9. https://doi.org/10.1016/j.biortech.2012.01.103

Hermesmann M, Müller TE (2022) Green, turquoise, blue, or grey? Environmentally friendly hydrogen production in transforming energy systems. Prog Energy Combust Sci 90:100996. https://doi.org/10.1016/J.PECS.2022.100996

Huang Y, Hong D, Hu S (2018) Hydrogen-producing biological compound comprises protein nucleic *Chlorella*, *Enterococcus faecium* and *Rhodopseudomonas palustris*. Patent document CN108179113-A

Junior JR do N, Torres LAZ, Medeiros ABP, et al (2021) Enhancement of biohydrogen production in industrial wastewaters with vinasse pond consortium using lignin-mediated iron nanoparticles. Int J Hydrog Energy 6:27431–27443. https://doi.org/10.1016/j.ijhydene.2021.06.009

Kang SG, Kim TW, Kwon KK, et al (2018a) New *Thermococcus onnurineus* strain having enhanced ability for converting carbon monoxide and producing hydrogen. Patent document WO2018008786-A1

Kang SG, Lee J, Lee HS, et al (2018b) *Thermococcus onnurineus* strain used for hydrogen production from formate, has mutation in formate transporter (TON1573) and shows greater hydrogen production ability form formate than wild-type strain efficiently with lower cost. Patent document WO2018105790-A1

Kang SK, Hyun LJ, Lee HS, et al (2020) New strain *Thermococcus onnurineus* 2437T strain 156T-LD2437 useful for producing biohydrogen under continuous culture conditions. Patent document KR2020118304-A

Kang SK, Hyun LJ, Lee HS, et al (2022) New *Thermococcus onnurineus* strain WTF-350T (KCTC 14212BP) useful for producing hydrogentle. Patent document KR2022061417-A

Karp SG, Burgos WJM, Vandenberghe LPS (2022) Sugarcane: a promising source of green carbon in the circular bioeconomy. Sugar Tech 24:1230. https://doi.org/10.1007/s12355-022-01161-z

Kim DH, Kim MS (2011) Hydrogenases for biological hydrogen production. Bioresour Technol 102:8423–8431. https://doi.org/10.1016/j.biortech.2011.02.113

Kim JH, Kwon OC (2011) A micro reforming system integrated with a heat-recirculating micro-combustor to produce hydrogen from ammonia. Int J Hydrog Energy 36:1974–1983. https://doi.org/10.1016/j.ijhydene.2010.11.043

Lee HH, Kang JH, Yoo YD, Kang SK (2019) Monitoring and operation guidance device for hydrogen gas continuous production plant. Patent document KR2050061-B1

Lin CY, Nguyen TML, Chu CY et al (2018) Fermentative biohydrogen production and its byproducts: a mini review of current technology developments. Renew Sust Energ Rev 82:4215–4220. https://doi.org/10.1016/j.rser.2017.11.001

Liu H, Li K (2019) New *Clostridium butyricum* LQ25 useful for producing hydrogen and reducing dissimilatory iron(III). Patent document CN109825459-A

Liu H, Wang H, Li K (2018) New hydrogen-producing bacteria *Clostridium beijerinckii* WANY51 used for producing hydrogen for butanol production. Patent document CN107653208-A
Liu B, Yin T, Du J, et al (2020) Method for producing energy by coupling high-temperature hydrogen production and microalgae oil production. Patent document CN111763694-A
Lu C, Zhang Q, Zhang Z, et al (2021) Method for biohydrogen production. Patent document US20210324421-A1
Luberti M, Ahn H (2022) Review of Polybed pressure swing adsorption for hydrogen purification. Int J Hydrog Energy 47:10911–10933. https://doi.org/10.1016/j.ijhydene.2022.01.147
Łukajtis R, Hołowacz I, Kucharska K et al (2018) Hydrogen production from biomass using dark fermentation. Renew Sust Energ Rev 91:665–694. https://doi.org/10.1016/j.rser.2018.04.043
Madden C, Vaughn MD, Díez-Pérez I et al (2012) Catalytic turnover of [FeFe]-hydrogenase based on single-molecule imaging. J Am Chem Soc 134:1577–1582. https://doi.org/10.1021/ja207461t
Mahidhara G, Burrow H, Sasikala C, Ramana CV (2019) Biological hydrogen production: molecular and electrolytic perspectives. World J Microbiol Biotechnol 35:1–13. https://doi.org/10.1007/s11274-019-2692-z
Martinez-Burgos WJ, Sydney EB, Brar SK et al (2019) The effect of hydrolysis and sterilization in biohydrogen production from cassava processing wastewater medium using anaerobic bacterial consortia. Int J Hydrog Energy 44:25551–25564. https://doi.org/10.1016/j.ijhydene.2019.08.085
Martinez-Burgos WJ, de Souza CE, Pedroni Medeiros AB et al (2021a) Hydrogen: current advances and patented technologies of its renewable production. J Clean Prod 286:124970. https://doi.org/10.1016/j.jclepro.2020.124970
Martinez-Burgos WJ, Do Nascimento Junior JR, Medeiros ABP et al (2021b) Biohydrogen production from agro-industrial wastes using *Clostridium beijerinckii* and isolated bacteria as inoculum. Bioenergy Res 15:987–997. https://doi.org/10.1007/s12155-021-10358-1
Martinez-Burgos WJ, Sydney EB, de Paula DR et al (2021c) Hydrogen production by dark fermentation using a new low-cost culture medium composed of corn steep liquor and cassava processing water: process optimization and scale-up. Bioresour Technol 320:124370. https://doi.org/10.1016/j.biortech.2020.124370
Martinez-Burgos WJ, Sydney EB, Medeiros AP et al (2021d) Agro-industrial wastewater in a circular economy: characteristics, impacts and applications for bioenergy and biochemicals. Bioresour Technol 341:125795. https://doi.org/10.1016/j.biortech.2021.125795
Mathews J, Wang G (2009) Metabolic pathway engineering for enhanced biohydrogen production. Int J Hydrog Energy 34:7404–7416. https://doi.org/10.1016/j.ijhydene.2009.05.078
Mishra P, Krishnan S, Rana S et al (2019) Outlook of fermentative hydrogen production techniques: an overview of dark, photo and integrated dark-photo fermentative approach to biomass. Energy Strateg Rev 24:27–37. https://doi.org/10.1016/j.esr.2019.01.001
Mukherjee G, Singh AK (2020) High temperature fixed bed bioreactor for biohydrogen production. Patent document IN201811028439-A
Murayama T, Sasaki K, Kodaka H, et al (2018) Dominating hydrogen-producing complex microorganism group useful for producing hydrogen, by subjecting aqueous solution containing digestive liquid obtained by methane fermentation of organic waste to pressurized hot-water treatment. Patent document JP2018148818-A
Mus F, Dubini A, Seibert M et al (2007) Anaerobic acclimation in *Chlamydomonas reinhardtii*: anoxic gene expression, hydrogenase induction, and metabolic pathways. J Biol Chem 282:25475–25486. https://doi.org/10.1074/jbc.M701415200
Nagarajan D, Lee DJ, Kondo A, Chang JS (2017) Recent insights into biohydrogen production by microalgae – from biophotolysis to dark fermentation. Bioresour Technol 227:373–387. https://doi.org/10.1016/j.biortech.2016.12.104
Newborough M, Cooley G (2021) Green hydrogen: the only oxygen and water balanced fuel. Fuel Cells Bull 2021:16–19. https://doi.org/10.1016/S1464-2859(21)00169-3

Niaz S, Manzoor T, Pandith AH (2015) Hydrogen storage: materials , methods and perspectives. Renew Sust Energ Rev 50:457–469. https://doi.org/10.1016/j.rser.2015.05.011

Ogata H, Lubitz W, Higuchi Y (2016) Structure and function of [NiFe] hydrogenases. J Biochem 160:251–258. https://doi.org/10.1093/jb/mvw048

Osman AI, Deka TJ, Baruah DC, Rooney DW (2020) Critical challenges in biohydrogen production processes from the organic feedstocks. Biomass Convers Biorefinery 13:8383. https://doi.org/10.1007/s13399-020-00965-x

Pachapur V, Sarma SJ, Maiti S, et al (2019) Process for hydrogen production from glycerol. Patent document US20190218576-A1

Pan Z, Zhang H, Tian L, et al (2019) A kind of kasugamycin residue fermentation and hydrogen production device and its produce hydrogen methods. Patent document CN109762849-A

Petrosyan H, Vanyan L, Trchounian A, Trchounian K (2020) Defining the roles of the hydrogenase 3 and 4 subunits in hydrogen production during glucose fermentation: a new model of a H2-producing hydrogenase complex. Int J Hydrog Energy 45:5192–5201. https://doi.org/10.1016/j.ijhydene.2019.09.204

Ren Y, Si B, Liu Z et al (2022) Promoting dark fermentation for biohydrogen production: potential roles of iron-based additives. Int J Hydrog Energy 47:1499–1515. https://doi.org/10.1016/j.ijhydene.2021.10.137

Romero JG, Lopez AL, Depraect OG, et al (2018) Obtaining a stable and robust microbial consortium for producing hydrogen and/or hydrolysate by selection, adaptation, conservation of microbial consortium, inoculum reactivation, and production of hydrogen and hydrolysates. Patent document WO2018088884-A1

Rosa D, Medeiros ABP, Martinez-burgos WJ et al (2020) Biological hydrogen production from palm oil mill effluent (POME) by anaerobic consortia and *Clostridium beijerinckii*. J Biotechnol 323:17–23. https://doi.org/10.1016/j.jbiotec.2020.06.015

Seppälä JJ, Puhakka JA, Yli-Harja O et al (2011) Fermentative hydrogen production by *Clostridium butyricum* and *Escherichia coli* in pure and cocultures. Int J Hydrog Energy 36:10701–10708. https://doi.org/10.1016/j.ijhydene.2011.05.189

Sharma S, Basu S, Shetti N, Aminabhavi T (2020) Waste-to-energy nexus for circular economy and environmental protection: recent trends in hydrogen energy. Sci Total Environ 713:136633. https://doi.org/10.1016/j.scitotenv.2020.136633

Show KY, Lee DJ, Chang JS (2011) Bioreactor and process design for biohydrogen production. Bioresour Technol 102:8524–8533. https://doi.org/10.1016/J.BIORTECH.2011.04.055

Show KY, Yan Y, Ling M et al (2018) Hydrogen production from algal biomass – advances, challenges and prospects. Bioresour Technol 257:290–300. https://doi.org/10.1016/j.biortech.2018.02.105

Shui X, Jiang D, Zhang H et al (2022) Effect of light perturbation on the photo-fermentative hydrogen production of biomass from giant reed. J Clean Prod 351:131481. https://doi.org/10.1016/J.JCLEPRO.2022.131481

Smith PR, Bingham AS, Swartz JR (2012) Generation of hydrogen from NADPH using an [FeFe] hydrogenase. Int J Hydrog Energy 37:2977–2983. https://doi.org/10.1016/j.ijhydene.2011.03.172

Soccol CR, Burgos WJM, Medeiros ABP, et al (2019) Process for the production of bio-hydrogen from agro-industrial effluents and a microbial consortium for the production of bio-hydrogen. Patent document BR1020190210559-B1

Sokołowski G, Cenian A (2020) Method of stabilizing hydrogen production by adjusting the pH with digestate of fermented plants in the process of dark fermentation. Patent document PL434798-A1

Sydney EB, Duarte ER, Martinez Burgos WJ et al (2020) Development of short chain fatty acid-based artificial neuron network tools applied to biohydrogen production. Int J Hydrog Energy 45:5175–5181. https://doi.org/10.1016/j.ijhydene.2019.10.128

Sydney EB, de Carvalho JC, Letti LAJ et al (2021) Current developments and challenges of green technologies for the valorization of liquid, solid, and gaseous wastes from sugarcane ethanol production. J Hazard Mater 404:124059. https://doi.org/10.1016/j.jhazmat.2020.124059
Thommes M, Kaneko K, Neimark AV et al (2015) Physisorption of gases, with special reference to the evaluation of surface area and pore size distribution (IUPAC technical report). Pure Appl Chem 87:1051–1069. https://doi.org/10.1515/pac-2014-1117
Tian H, Li J, Yan M et al (2019) Organic waste to biohydrogen: a critical review from technological development and environmental impact analysis perspective. Appl Energy 256:113961. https://doi.org/10.1016/j.apenergy.2019.113961
Trchounian K, Sawers RG, Trchounian A (2017) Improving biohydrogen productivity by microbial dark- and photo-fermentations: novel data and future approaches. Renew Sust Energ Rev 80:1201–1216. https://doi.org/10.1016/j.rser.2017.05.149
Tsygankov AA (2007) Nitrogen-fixing cyanobacteria: a review. Appl Biochem Microbiol 43:250–259. https://doi.org/10.1134/S0003683807030040
Vieira S, Schneider J, Martinez-Burgos WJ, et al (2021) Pretreatments of solid wastes for anaerobic digestion and its importance for the circular economy. In: Handbook of solid waste management, pp 1–23
Wang H, Liu W, Zhao Y, et al (2018) New hydrogen production function gene vector pET32a-fdhF-1 useful for producing hydrogen, which is pET32a plasmid of *Escherichia coli* formate dehydrogenase gene fdhF. Patent document CN108220319-A
Wang H, Liu W, Cheng S (2019) New functional gene vector pET32a-fdhD for assisting hydrogen production, comprising fdhD gene of *Bacillus cereus* assisted formate dehydrogenase. Patent document CN110093362-A
Wang H, Liu W, Cheng S (2020) Recombinant vector for hydrogen-producing recombinant strain for assisting hydrogen production, comprises initial vector and fhlA gene, where initial vector is pET32a vector. Patent document CN111471700-A
Wei S, Zhang Q, Zhang C, et al (2021) Method for co-producing biological hydrogen and microalgae grease by utilizing lignocellulose fermentation. Patent document CN113881710-A
Xia A, Wei P, Liao Q, et al (2019) The method and biofilm reactor of two step anaerobic fermentation preparing hydrogen alkane gas of streaming drop. Patent document CN109321444-A
Yang G, Wang J (2018) Various additives for improving dark fermentative hydrogen production: a review. Renew Sust Energ Rev 95:130–146. https://doi.org/10.1016/j.rser.2018.07.029
Yang X, Liu L, Zhang N, et al (2021) Method for producing hydrogen by using waste bananas through photosynthetic organisms. Patent document CN113604534-A
Yoo Y, Kim S, Jung W, et al (2018) Apparatus for producing hydrogen by microorganisms and hydrogen production method using same. Patent document WO2018207964-A1
Zhang Q, Jing Y, Zhang H, et al (2019a) A kind of photosynthetic hydrogen preparation device and the technique for preparing hydrogen. Patent document CN110144293-A
Zhang Q, Liu H, Jiang D, et al (2019b) A method of promoting HAU-M1 photosynthetic bacteria group highly effective hydrogen yield using bead algal biomass. Patent document CN110129399-A
Zhao M, Gao C, Xu J, et al (2018) A method of it improving kitchen garbage anaerobic dry fermentation and generates hydrogen. Patent document CN108611376-A
Zhu X, Shi Y, Li S, Cai N (2018) Two-train elevated-temperature pressure swing adsorption for high-purity hydrogen production. Appl Energy 229:1061–1071. https://doi.org/10.1016/j.apenergy.2018.08.093